Handbook of Formulas for STRESS AND STRAIN

HANDBOOK OF FORMULAS FOR STRESS AND STRAIN

WILLIAM GRIFFEL
Mechanical Engineering Scientist
Department of the Army
Picatinny Arsenal

FREDERICK UNGAR PUBLISHING CO.
NEW YORK

Second Printing, 1976

Printed in the United States of America

Library of Congress Card Catalog No. 66-17539
ISBN 0-8044-4332-7

For my granddaughter
SANDRA FRANCES

FOREWORD

Calculations of stresses and strains in structural members are associated with much arduous labor, if precise values are required. Often, in fact, in the great majority of cases such precision is not warranted. When one contemplates the unknowns in applied loads, in metal characteristics, and even in the dimensions of assembled structures, it becomes obvious that even the most rigorous calculation methods may result in values of questionable accuracy.

By means of charts, diagrams, curves and simplified formulas much of the drudgery of such computations can be shortcircuited without the loss of necessary accuracy, thus saving many hours of costly effort.

With the development of the modern computer the preparation of such charts and curves, with increased accuracy, has become a practical possibility. Precisely, this is what the author has achieved in this volume. Mr. Griffel has amassed a wealth of material which is applicable to a wide variety of problems in modern engineering. While knowledge of stress theory on the part of the reader is assumed, no extensive mathematical facility is required to make full use of the book. However, to satisfy those who want to study the more intimate details of the technology, sufficient treatment of the theory is given to meet their needs.

Much of the material presented here has appeared in published papers and articles by the author, and in consequence has enjoyed the tests of practical experience. This is the first presentation of the complete material in book form, making the work easily accessible and widening its value as an engineering tool.

The author has presented his material in a manner found in no other book. The *Handbook of Formulas for Stress and Strain* will be of tremendous value to the engineer having to make design calculations of a difficult order. He will find with the use of this book that most of his problems are greatly simplified.

KENNETH LOFGREN
Professor Emeritus of Mechanical Engineering
The Cooper Union

PREFACE

With present-day engineering literature overflowing with reference books, covering the field of the strength of materials, one may wonder, why another book?

After surveying practically all the reference books on the market it was found that, despite the development of high-speed digital computers, the scope of these books, as well as the presentation of basic equations, has not changed from what it was 25 years ago. Basic formulas for *flat plates, beams with combined loading, out of plane loading*, etc. are still quite cumbersome and tedious.

For example, the textbook equation for the deflection of a circular plate with outer edge fixed and supported, inner edge fixed, with a uniform load over entire actual surface is:

$$y = \frac{3w(m^2 - 1)}{16m^2Et^3}\left[a^4 + 3b^4 - 4a^2b^2 - 4a^2b^2 \log\frac{a}{b} + \frac{16a^2b^4}{a^2 - b^2}\left(\log\frac{a}{b}\right)^2\right]$$

To simplify this and other equations pertaining to *flat plate* design, it was conceived to assume a constant Poisson's ratio of 0.3, a value generally used for steel, aluminum, magnesium (See Appendix A). Thus, the above equation reduces to $y = Wa^2/Et^3\,K$, where K is a dimensionless factor and a function of plate dimensions a, b, only. Numerical values of K, for $a/b = 1, 1.25, 1.5, 2, 3, 4, 5$, were calculated and tabulated for fast reference.

An engineer who is interested in beams under combined axial and transverse loading is just as frustrated as the flat plate designer. For example, the equation for the deflection of a cantilever beam under axial tension and uniform transverse load is:

$$y = \frac{wj}{P}[j(1 - \tfrac{1}{2}U^2 - \operatorname{sech} U) - L(\tanh U - U)]$$

where $j = \sqrt{EI/P}$, $U = L/j$, and sech U, tanh U are hyperbolic functions that require tables. A simplified form of this basic equation, used in this book, is: $y = WL^3K/CEI$ where C and K are dimensionless constants that were tabulated for easy reference for 14 different loading and support conditions.

A very comprehensive chapter on beams on elastic foundation is another good example of what this book can do. Here too, the basic equations for deflection, moment, and slope assume a very complicated form of hyperbolic and trigonometric functions. Here, too, the equations were simplified and expressed as a function of a dimensionless factor K. The values of K, for various load and support conditions, are given as ordinates to curves which were plotted for various lengths of beam. Thus, these curves in conjunction with a simple equation provide an easy method for finding the deflection. Similar curves are given for the moment and slope of the beam.

Another major contribution of the book is the chapter on concentration stresses around holes. Formulas with graphs for stress distribution and stress concentration factors for plates with elliptical, circular, square, and triangular holes are given. Various types of loading is assumed.

Because of the complexity of subject matter, *out of plane loading of circular beams* has a very meager treatment in the engineering literature. With the exception of one ASME paper, this author does not know a single book where this subject is discussed in sufficient detail to give the engineer a thorough understanding of the subject. It suffices to quote one equation in order to show why a simplification is necessary. The equation for the angle of slope of a circular cantilever beam with a concentrated load acting at any point along the center line of the beam and perpendicular to its plane of curvature is:

$$\theta = \frac{PR^3}{EI}\{[(1-c)/2][\cos j \sin^2\alpha - \sin j(\alpha - j + \sin\alpha\cos\alpha)] + c[\cos j - \cos\alpha - (\alpha - j)\sin j]\}$$

A simplification of this equation took the following form: $\theta = (PR^3/EI)K$. Values of K for several most commonly used loading conditions were tabulated. Similar tabulations of K factor are given for deflection and moment.

In the case of straight beams with transverse loadings, this book contains a complete tabulation of equations for shear, moment, deflection and stresses in terms of deflection for 31 different loading and support conditions. These are basic textbook equations with one important exception, namely, in each case, diagrams for moment, sheer, and elastic line are presented for better interpretation of equations involved.

The equations for deflection of straight beams with transverse loadings are also presented graphically, thus allowing the designer a quick and accurate estimate of the deformation at any section of the beam. Thirty-one graphs cover the whole subject.

The chapter on *pressure vessels* presents a very comprehensive analysis of vessels of various end closures. It goes beyond the textbook presentation of basic equations. A detailed analysis of pressure vessels with spherical end closures, for instance, is given in a very simplified form. The designer using this method does not have to dwell upon the theory of shells in order to calculate the discontinuity stresses at the juncture of the head and cylindrical portion of the vessel.

Also, there is a chapter on pressure vessels with flat closure. Here the simplification went one step further; namely, the solutions to pressure vessel problems are given in the form of graphs.

Vibration problems exist as a permanent and inseparable element in all the stress problems. Therefore, a chapter such as "Vibration Frequency Charts" is of great importance to an engineer interested in *formulas on stress and strain.* Vibration at or near resonance on such common structures as plates and beams creates high stress and hastens the time to eventual failure. Standard practice is to design these elements so that their natural frequencies will not excite resonance.

Tables of frequency constants are given to allow solutions for frequencies and modes of vibration for beams and plates with various boundary conditions. The chapter covers: (1) Rectangular plate; (2) Square plate; (3) Cantilever rectangular plate; (4) Cantilever skew plate; (5) Circular plate; (6) Ring; (7) Uniform beam; (8) Continuous beam; (9) Variable section beam; (10) Combined beam.

The critical speed of a shaft is a function of the static deflection. Therefore, in the case of a multiple-loaded shaft, the computation is rather involved. To help the engineer to design a shaft with an operating speed either above or below the critical speed, the basic Dunkerley's formula has been solved by the use of critical speed constants K which are functions of the static deflection of the shaft. The result has been fitted into 2 nomograms which determine critical speed of steel, aluminum, and magnesium shafts within the accuracy of a slide rule.

The chapter on *rigid frames* contains a compendium of frame formulas. A simple method for solving certain types, and a roundup of moment equations, with diagrams, for 31 cases of the more common frame configurations is included. This chapter also contains an article on elastic-center stress analysis in an unsymmetrical frame.

The last chapter should be of particular interest to a design engineer whose work requires the use of equations of mass and area properties for various geometrical shapes. A separate section is devoted to properties of ogival shapes, which are commonly used in the design of missile nose cones. Also, the solid and thin-shelled tangent ogives are included. This section was reprinted, by permission, from Navy Report No. 7827.

All the tables and formulas presented in this book were published and copyrighted from time to time in: (1) *Machine Design*, (2) *Product Engineering*, (3) *Design News*, (4) *Materials Design and Engineering* (England), (5) *Product Design and Values Engineering* (Canada), (6) *Design Engineering* (Canada), (7) *Journal of Applied Mechanics*. Thus, opportunity afforded for criticism has been of considerable advantage.

To the publishers of the above journals and others who have generously permitted the use of material, the author wishes to express his thanks.

William Griffel

TABLE OF CONTENTS

Chapter I

RELATION BETWEEN STRESS AND STRAIN

Chapter II

FLEXURE OF STRAIGHT AND CURVED BEAMS

Chapter IV

THE THEORY OF ELASTIC ENERGY

Chapter V

FLAT PLATES

Chapter VI

COLUMN DESIGN

Chapter VII

PRESSURE VESSELS

LIST OF TABLES

Chapter II

FLEXURE OF STRAIGHT AND CURVED BEAMS

Chapter III

TORSION

Chapter IV

THE THEORY OF ELASTIC ENERGY

Chapter V

FLAT PLATES

Chapter VI

COLUMN DESIGN

Chapter VII

PRESSURE VESSELS

Chapter I

RELATION BETWEEN STRESS AND STRAIN

(*Ref. 1*)

1. DEFINITION OF STRESS

Let F be a force across a small area A of a body in equilibrium. This force F can be resolved into two components, N perpendicular to the area, and S_s in the plane of the area. We will assume, for simplicity, that the force F shown in Fig. 1 is the action of the matter to the right of the area on the matter to the left. By Newton's Third Law the action of the matter to the left of the area on the matter to the right is the force F reversed. The normal component force N is called *tension* if its direction is away from the matter on which it acts, as shown in Fig. 1. If the force N is toward the matter on which it acts it is called compression which, actually, may be regarded as a negative tension.

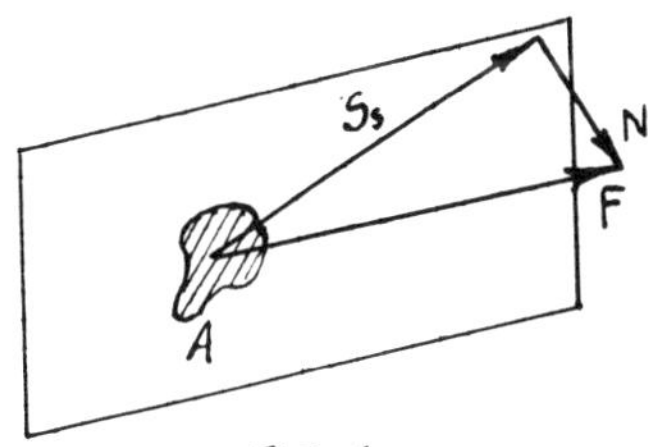

FIG. 1

The component S_s in the plane of the area is called *shear force* and it can have any direction in the area depending on the direction of F in space.

The mean normal force per unit area of A is N/A, called *mean normal stress.* If N is a tension this normal stress is called *tensile stress*, whereas if N is a thrust the stress is a compressive stress.

The mean shear force per unit area A is S/A, and this is called the *mean shear stress.*

If the area A is infinitely small and concentrated around a point B the mean normal stress and the mean shear stress are then called the normal stress and the shear stress at the point B.

Imagine an infinitely small cube enclosing point A of the body being isolated. There are three normal stresses acting upon the six faces of the cube. If the normal stresses alone act upon the faces of the cube, they are called the principal stresses. If all three principal stresses are different from zero, the state of the stress is called triaxial. If one of the principal stresses is zero, the state of the stress is called biaxial. If two principal stresses are zero, the state of stress is called uniaxial.

If the state of stress (i.e., the value of all three principal stresses) is known at any point of a structural part, the strength of this part can be estimated.

2. ELASTICITY

The ability of a structural element to resist deformation is called stiffness. Although we often speak of rigid bodies there are no absolutely rigid bodies.

A body whose shape or size is altered by stress is said to be strained. To every kind of *stress* there is corresponding *strain.* If the stresses are not too great, the strained body will recover its original shape and size when the stress is removed. This property of bodies of returning, after unloading, to their initial form is called *elasticity.* A body is *perfectly elastic* if it recovers its original shape completely after unloading. Some bodies, such as steel, recover completely after very large stresses, while others, such as cast iron and lead, do not completely recover from comparatively small stress. The strain that remains when the stress is removed is called permanent set.

Throughout this book it will be assumed that the bodies undergoing the action of external forces are perfectly elastic.

3. HOMOGENEOUS AND ISOTROPIC BODIES

It will be assumed that the matter of an elastic body is homogeneous and isotropic. A *homogeneous* material is one having the same chemical composition throughout, whereas an *isotropic* material has the same internal structure throughout (the same elastic properties in every direction). Thus, a material may be homogeneous but not isotropic.

Metals possess a high degree of homogeneity, whereas timber, concrete, and stone are less homogeneous.

4. HOOKE'S LAW

Hooke's law may be stated briefly in the following form: Within the elastic limit the stress producing any strain is proportional to that strain. In more general terms this law can be expressed as follows: Each force acting on an elastic body produces its own strains independently of the other forces and the total strains due to several forces is the sum of the strains due to each force separately.

5. MODULUS OF ELASTICITY IN TENSION

The block of dimensions $a\,b\,c$, in Fig. 2, is under the action of a uniform tensional stress S. As no other stresses are acting on the six faces of the block it is known by experiment that the length a increases and b and c decrease. If the new lengths of edges of the block will be designated as $a + u$, $b - v$, $c - w$, then the three ratios u/a, v/b, w/c are called the strains of the block in the three directions, the first being an extensional strain and the two latter compressive strains. It is always convenient to treat a compressive strain as a negative extensional strain. Then, the strains are: $\alpha = u/a$, $\beta = -v/b$, $\gamma = -w/c$. From Hooke's law we obtain: $S = E\alpha$ or $\alpha = S/E$ where E is a constant called Young Modulus of elasticity. It is an experimental parameter that varies in magnitude depending upon the material. Since S is measured generally in pounds per square inch and α in inches per inch, the modulus E is measured in pounds per square inch.

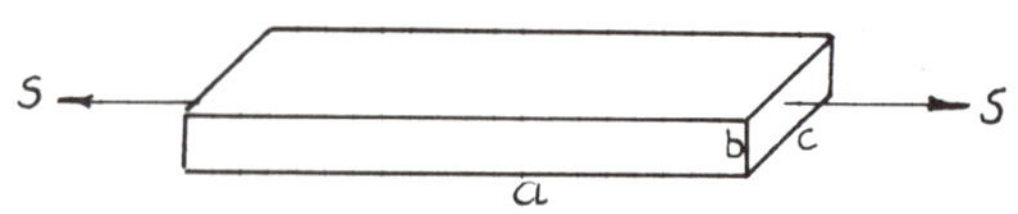

FIG. 2

6. POISSON'S RATIO

It was shown in paragraph 5 that the stress S produces an extension α in the direction of this stress and lateral contractions β, γ. It is known by experiment that

$$\beta = \gamma = -\alpha = -\frac{S}{E}\nu$$

where ν is a constant called *Poisson's Ratio* for the material.

Most metals have values of ν varying from 0.25 to 0.35. A value of 0.30 is commonly used for steel. Concrete has a small lateral contraction as indicated by a magnitude of ν, which varies from 1/8 to 1/12. Rubber on the other hand, has a value of ν equal to about 0.40.

Values of ν are: Glass 0.244, brass 0.333, copper 0.333, cast iron 0.270, wrought iron 0.278, steel 0.303, lead 0.430.

7. THE BULK MODULUS

Assume a uniform tensile stress S, and that no shear stress acts on every face of a rectangular block. In such a case the three normal stresses acting upon the six faces of the rectangle are the same and so is the extensional strain.

Under such system of stresses there is no change of shape, but necessarily an increase of volume. Let V be the original volume of the block, and v the increase in volume under stresses. Then Hooke's law for this case is expressed by the equation:

$$S = kv/V$$

where k is a constant called the bulk modulus of the material.

8. STRAIN DUE TO SIMULTANEOUS STRESSES

When several stresses exist simultaneously in any element of a body, each stress produces its own strain just as if no other stress were present. This means that the total stress at any point is merely the sum of the stresses due to each force acting separately.

9. RELATION BETWEEN THE ELASTIC CONSTANTS FOR AN ISOTROPIC BODY

Bulk Modulus:

$$k = E/3(1 - 2\nu)$$

Modulus of rigidity, or shear modulus:

$$G = E/2(1 + \nu)$$

Poisson's Ratio:

$$\nu = (E - 2G)/2G$$

Modulus of elasticity in tension or compression:

$$E = 2G(1 + \nu)$$

With reference to the equation for the bulk modulus it is important to note that since equal tensile stresses produce an increase in volume, it is necessary for $(1 - 2\nu)$ to be positive. Theoretically, therefore, the value of the Poisson Ratio, ν, cannot exceed 0.5.

10. THE YIELD POINT AND ULTIMATE STRENGTH OF MATERIAL

In finding the tensile strength of any given material, small bars usually of circular cross section are gripped at the ends, and put in tension till they break. The graph showing the relation between stress and strain is shown in Fig. 3. It should be noted that in Fig. 3 the magnitude of normal stress at a cross section of the tensile test specimen $S = P/A_o$, where A_o is the original cross-sectional area, is laid off along the ordinate axis, and the unit elongations of the specimen $\varepsilon = \Delta L/L_o$, where L_o is its original length of the specimen, are laid off along the abscissa axis. The diagram in Fig. 3 is called the conventional tensile test diagram because stresses and unit elongations are calculated in relation to the original cross-sectional area and the original length of the specimen.

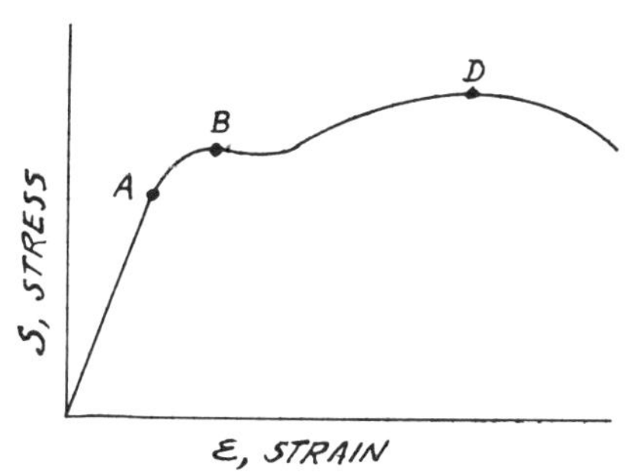

FIG. 3

The portion OA is a straight line, but AB is slightly curved. Point A corresponds to proportional limit, whereas point B, at which the strain begins to increase greatly for very little increase of stress, is called the yield point. Point B determines the elastic limit. At point D the stress reaches its maximum value. This highest stress related to the original cross-sectional area of the specimen is called the ultimate (tensile) strength.

An important characteristic of the properties of the material, especially under impact load, is the area under the tensile test diagram which represents the work of external forces required to produce rupture of the specimen.

Tensile test diagrams for many grades of steel as well as alloys of nonferrous metals do not have a pronounced yield point. The characteristic shape of the tensile test diagram for such materials is shown in Fig. 4.

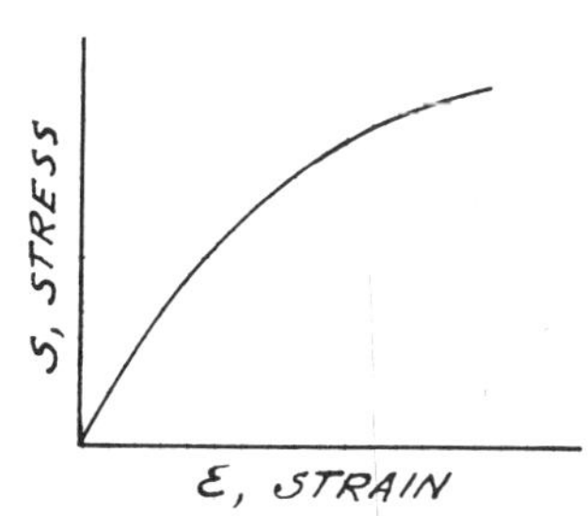

FIG. 4

11. FACTOR OF SAFETY

At times, both the actual loads acting on a part and the properties of materials which it is made of differ from those assumed in design. These factors reducing the strength of a part are of random character in most cases and cannot be estimated beforehand. Since, however, parts and whole structures must operate safely under these unfavorable conditions as well, it is necessary to take certain precautionary measures. For this reason working stresses are assigned which are lower than the ultimate stresses at which a part cannot continue to give normal service. Thus

$$n = \frac{\text{Ultimate stress}}{\text{Allowable stress}}$$

is called the factor of safety.

12. ELASTIC FATIGUE

As a rule the elastic properties of a body and its strength do not alter when the stresses are within the elastic limit. However, there are circumstances in which this is not true. Experiments show that when a body is subjected for some time to rapidly alternating forces, the properties of the material may be very much changed. Such a body may fracture under a stress far below that at which it would have broken before the application of the alternating force. The mere repeated application and removal of a load, if the change takes place rapidly, is known to weaken the body. However, in most cases serious weakening of a member is dependent on the total number of such stress changes, which usually is a rather large number, frequently in the millions.

When the load is such that the stress changes alternately from tensile to compressive, the condition is termed *stress reversals*. Such cases are far more serious than simple applications and removal of loads.

13. SYMBOLS AND UNITS

In a handbook of this scope it was found that the use of a master tabulation of symbols and units was not practical. The preferred method, used throughout this book, is to indicate, in each discussion and at the head of each table of formulas, the notation and units there employed. Furthermore, to facilitate comparison, it was sometimes considered advantageous to adopt a notation identical with that used in articles to which reference was made, even though such notation might differ from that used elsewhere in this book. The symbols listed below, however, have been used in all cases:

E = modulus of elasticity (psi)
G = modulus of rigidity (psi)
I = moment of inertia of an area (in.^{4})
J = polar moment of inertia of an area (in.^{4})
M = bending moment (in.-lb.)
S = unit stress, with subscript indicating kind or direction (psi)
T = twisting moment (in.-lb.)
U = strain energy (in.-lb.)
ν = Poisson's Ratio (nondimensional)
g = acceleration of gravity (ft. per sec. per sec.)

In all formulas, unless other units are specified, the unit of distance is the inch and the unit of force is the pound. Therefore, all areas are in square inches, all moments of inertia are inches fourth, all distributed loads are in pounds per linear inch or pounds per square inch, all moments are in inch-pounds, and all stresses are in pounds per square inch.

REFERENCE

Section	*Ref. No.*	
	1	PRESCOTT, J.: *Applied Elasticity*, Dover Publications, Inc., N.Y. 1961

Chapter II

FLEXURE OF STRAIGHT AND CURVED BEAMS

14. BASIC CONCEPT IN BENDING

In the following discussion it is assumed that the beam is horizontal and the loads and reactions are vertical.

Bending implies deformation of a bar produced by loads perpendicular to its axis as well as couples acting in a plane passing through the axis of the bar. Fibers on the convex side lengthen, fibers on the concave side shorten. The neutral surface is normal to the plane of the loads and contains the centroids of all sections. Fibers which neither lengthen nor shorten are said to lie in the neutral surface.

If the plane of action of external forces passes through one of the principal centroidal axes of inertia of the cross section of the bar, bending is described as plane. If bending loads act in a plane that does not pass through any of the principal centroidal axes of inertia of the cross section of the bar, oblique bending takes place. Thus, in the case of a plane bending, the axis of the beam continues to remain in the plane of action of external forces after deformation. In oblique bending, also called unsymmetrical bending, the plane of deformation does not coincide with the plane of loading.

When a beam bends under the action of a load, the cross sections of the beam displace perpendicular to the initially straight axis and at the same time rotate, (Fig. 1). Longitudinal displacements of points on on the neutral surface are usually negligible.

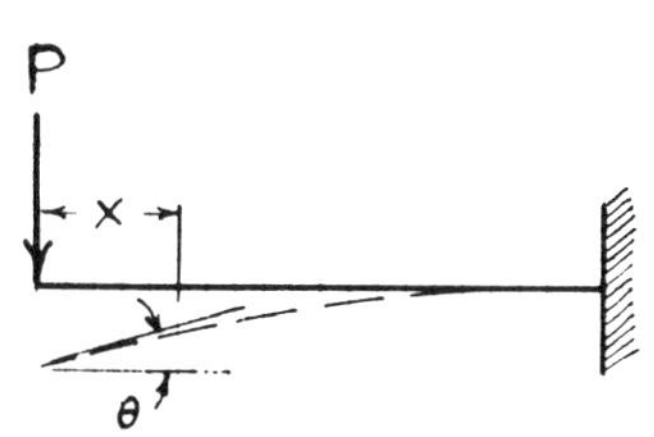

Fig. 1

The vertical deflection, y, is largely due to bending, that due to shear being usually negligible. The angle θ through which the cross section rotates with respect to its original position is called the angle of rotation of the section or the change in slope of the elastic curve, and is expressed in radians. Bearing in mind that the cross section after rotation is perpendicular to the deflected axis of the beam we conclude that instead of determining the angle of rotation of the section it is possible to determine the angle of equal magnitude between the tangent to the deflected axis at a given point and the original axis of the beam.

The general differential equation of the elastic curve is

$$EI\frac{d^2y}{dx^2} = M_b \tag{1}$$

where M_b represents the bending moment equation in terms of the distance from the left end of the beam to any section (considering the coordinate system as shown in Fig. 1).

By integrating equation (1) once, we obtain the equation for angles of rotation:

$$EIy' = \int M_b\,dx + C \tag{2}$$

where C is the constant of integration. By integrating a second time we obtain the equation of deflection:

$$EIy = \int dx \int M_b\,dx + Cx + D \tag{3}$$

where D is the second constant of integration.

The constants C and D are determined from the boundary conditions (supporting conditions of the beam). For example, for the beam shown in Fig. 1 both the deflection and the angle of rotation at the built-in end must be zero. For a beam with supported ends the deflection must be zero at both ends of the beam.

Having determined the constants of integration the deflection and the angle of rotation of any section can be found from equations (1) and (2).

There are also other methods for calculating the beam deflection, such as the elastic energy method. In that case:

$$y = \int \frac{M_b m}{EI} dx$$

where M_b has the same meaning as in equation (1) and m is the equation of the bending moment due to a load of 1 lb. acting vertically at the section where y is to be found. The integration indicated must be performed over each portion of the beam for which either M_b or m is expressed by a different equation.

The use of the general differential equation for calculating the deflection of a beam will be shown in the following example.

Example: Determine y_{max} and θ_{max} for a cantilever beam shown in Fig. 1.

Solution: With the coordinate axis as indicated, the bending moment at a section having abscissa x is:

$$M_x = -Px$$

and

$$EIy'' = -Px.$$

The first integration gives:

$$EIy' = \frac{-Px^2}{2} + C.$$

The second integration gives:

$$EIy = \frac{-Px^3}{6} + Cx + D.$$

For determining C and D we have the following boundary conditions:

$$(1) \text{ at } x = L, y = 0$$

$$(2) \text{ at } x = L, y' = 0.$$

From the second condition we obtain $C = \frac{PL^2}{2}$.

From the first condition we obtain:

$$O = \frac{-PL^3}{6} + \frac{PL^3}{2} + D$$

or

$$D = \frac{-PL^3}{3}.$$

It is evident that y_{max} and θ_{max} will occur at $x = 0$, thus:

$$\begin{aligned} y_{max} &= -PL^3/3EI \\ \theta_{max} &= PL^2/2EI \end{aligned} \qquad (4)$$

The negative value of y indicates that the centroid of section displaces downward, i.e., in the direction of negative values of ordinates Y. The positive value of θ indicates that the section rotates counterclockwise.

It is interesting to note that

$$C = EI\theta_o \qquad (5)$$

$$D = EIy_o \qquad (6)$$

where y_o is the deflection at the origin and θ_o is the angle of rotation at the origin. Relations (5) and (6) will be always valid provided bending moments are calculated by considering forces lying between the section and the origin.

15. FLEXURE OF STRAIGHT BARS

ASSUMPTIONS

The equations of this article are based on the following assumptions:

1. The material of the beam is homogeneous and isotropic. (A constant density and a constant modulus of elasticity in tension and compression are assumed.)
2. The beam is of uniform cross section and straight.
3. All loads and reactions are perpendicular to the axis of the beam and lie in the longitudinal plane of symmetry.
4. The beam is long in proportion to its depth.
5. The beam is not stressed beyond the proportional limit of the material.

Throughout this handbook, except in vibration problems, loads are treated as static loads acting upon structures assumed to be at rest. The dynamic factors are then automatically compensated for by using appropriate working stresses, factors of safety, etc., which are determined experimentally.

SIGN CONVENTION

The determination and interpretation of algebraic signs, in problems of stress analysis, frequently constitute a major difficulty. The magnitude determination of an unknown amounts to one half of the answer. The correct interpretation of the algebraic sign is the other half.

The meaning of signs is essentially relative. One of the important rules of mathematics is that an equation may be multiplied by -1, changing the sign of all the terms in the equation, without in any way changing its meaning. A plus or minus sign by itself has no meaning. It assumes significance only in its relationship to other signs.

In this book we assume the coordinate system of a deflected beam so that X axis is along the undeflected axis of the beam. Thus, the Y coordinate of any point of the deflected beam axis is numerically equal to the deflection of the point. It should be emphasized again that the positive and negative directions of the X and Y axes have been selected arbitrarily. However, once the choice of axes is made, the mathematical treatment of the problem must be developed consistent with that convention.

16. MAXIMUM DEFLECTION AND DEFORMATION OF BEAMS

(Ref. 1)

External loads acting on a body produce changes in the shape and size of the body called deformation or strain. The amount of deformation under a given load is governed by the mechanical properties of the material and the geometrical configuration of the body.

The deformation and deflection of load-resisting members of machines and structures are important in themselves and also in determining the forces and moments in statically indeterminate members and structures.

At times, deformations become important to the designer and may govern the design itself. Such would be the case when two materials of different mechanical properties act together in a composite member, or when, for special reasons, the deformation must remain within specified limits.

The deformation of a beam is most easily expressed in terms of the *deflection* of the beam from its original unloaded position and *slope.*

The slope actually represents the tangent of the angle of inclination of the deflection curve. For very small deformations, such as being encountered in practice, the value of the angle expressed in units of radians is very nearly equal to the tangent of the angle. Hence, the slope is expressed in radians, which is a dimensionless unit of angular measure.

The general formulas that apply here are:

$$(1)\quad y = \frac{WL^3}{KEI}$$

$$(2)\quad y = \frac{M_oL^2}{KEI}$$

$$(3)\quad \theta = \frac{WL^2}{K_1EI}$$

$$(4)\quad \theta = \frac{M_oL}{K_1EI}$$

In the derivation of the deflection formula it was assumed that the material of the beam follows Hooke's law. Consequently, in every problem involving beam deflection it is to be emphasized that it is necessary to ascertain that the maximum bending stress in the beam is below the proportional limit of the material. This is easily done by the use of the flexure formula: $S = Mc/I$

In the above equations:

y = maximum deflection (in.)
θ = slope (radians)
W = total applied load (lb.)
w = unit applied load (lb./in.)
M = maximum bending moment (in.-lb.)
M_o = applied moment (in.-lb.)
K = deflection factor } nondimensional
K_1 = slope factor } *See* Table 1
L = span of beam (in.)
I = moment of inertia (in.^{4})
E = modulus of elasticity (psi)
S = bending stress (psi)
c = the distance from the neutral axis to the outer fiber of the beam (in.)

Numerical values for constants K, K_1, are tabulated in Table 1. Also, the expressions for bending moments M will be found in Table 1.

The cases listed in Table 1 have been so selected that, by superposition, almost any combination of loading and support encountered in practice may be solved. The use of the tabulated constants is illustrated in the following example.

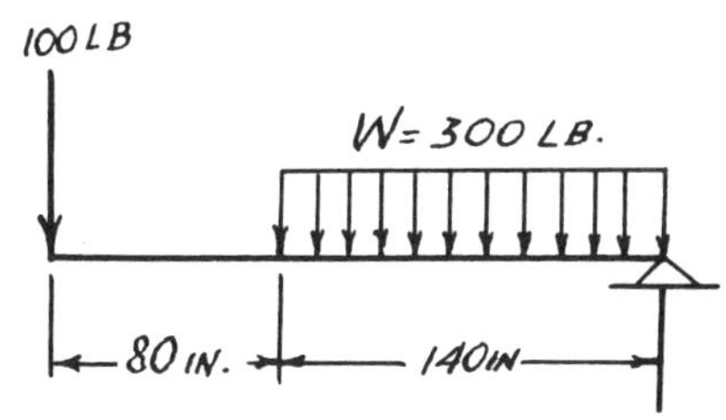

Example: The beam shown above has a rectangular section 2 in. wide and 4 in. deep, and is of aluminum, for which $E = 10 \times 10^6$ psi. It is required to determine the deflection, y of the left end.

Solution: This problem will be solved by superposition, using the constants of Table 1.

Let the deflection y_W be the one produced by the distributed load and y_P produced by the concentrated load. Thus the required deflection $y = y_W + y_P$

Compute y_W

$y_W = -80\theta$

From Table 1, Case 6, $K_1 = -24$.

Therefore

$$y_W = -80\left(\frac{-1}{24}\frac{(300)(140^2)}{EI}\right) = \frac{19{,}000{,}600}{EI}.$$

Compute y_P

The 100-lb. load produces (1) a deflection at the free end assuming the beam fixed at the left support, and (2) the deflection due to the fact that it actually assumes a slope there. The first part of the deflection is given by the formula for max y, Case 1, $K = -3$; the second part is found by multiplying the overhang (80 in.) by the slope produced at the left end of the 140-in. span by a couple equal to $100 \times 80 = 8{,}000$ applied at that point.

By using the formula for θ and $K_1 = -3$ we obtain:

$$y_P = -\frac{1}{3}\frac{(100)(80^3)}{EI} + (-80)\left[-\frac{1}{3}\frac{(-8000)(140)}{EI}\right] = \frac{-47{,}000{,}000}{EI}.$$

$$\text{Finally, } y = \frac{-27{,}400{,}000}{EI} = -0.200 \text{ in.}$$

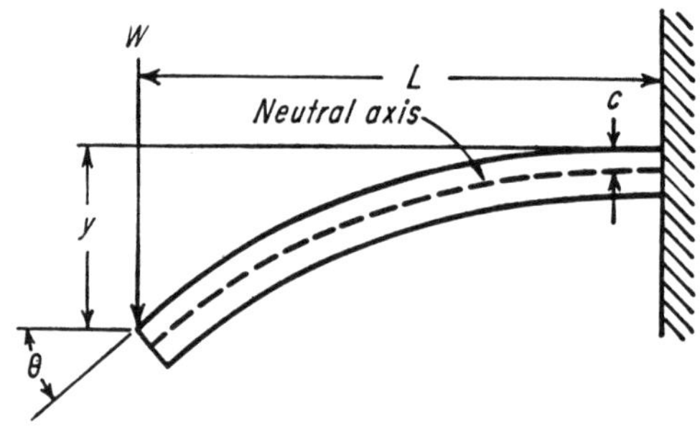

TABLE 1

	Beam Type	K for deflection	K_1 for slope	Max M	K_2
1	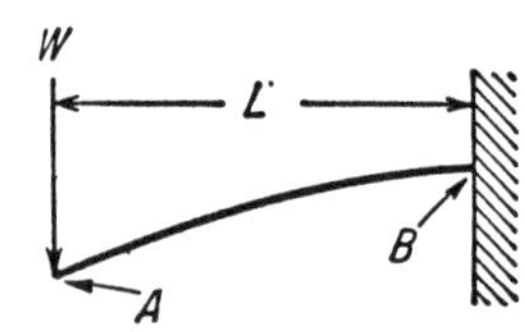	-3 at A	2 at A	-WL at B	-3
2	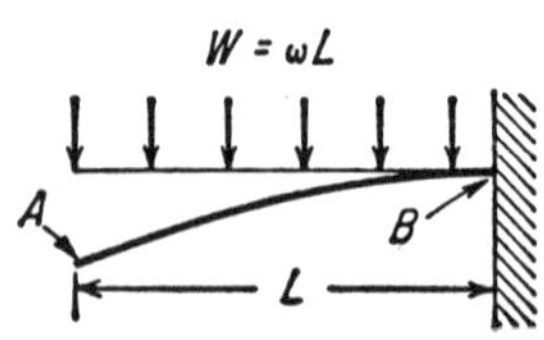	-8 at A	6 at A	-WL/2 at B	-4
3	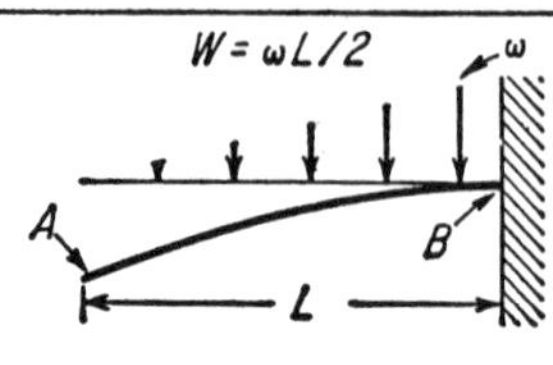	-15 at A	12 at A	-WL/3 at B	-5

	Beam Type	K for deflection	K_1 for slope	Max M	K_2
4	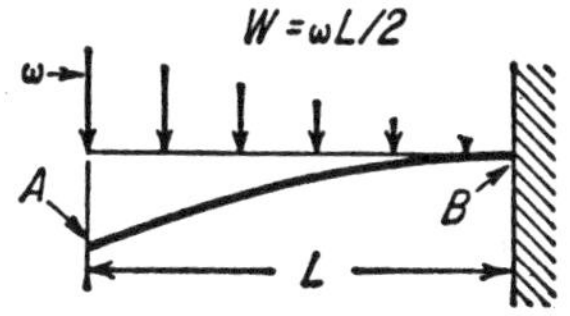	− 5.45 at A	4 at A	$-\frac{2}{3}WL$ at B	−3.6
5	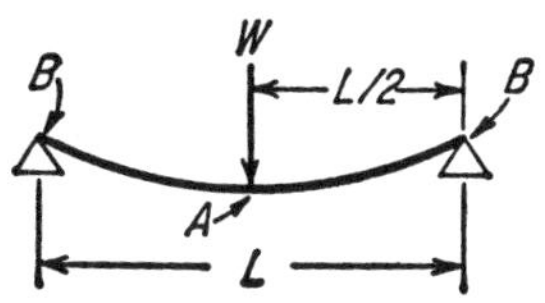	−48 at A	−16 at B +16 at A	WL/4 at C	−12
6	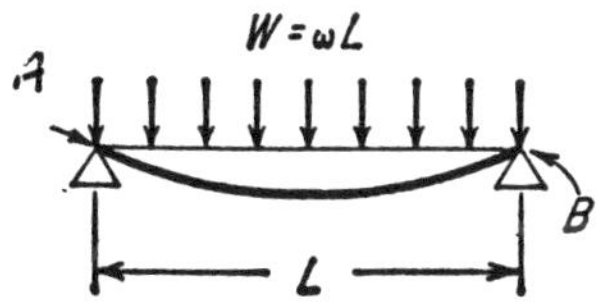	−76.8 at L/2	+24 at B −24 at A	WL/8 at L/2	−9.6
7	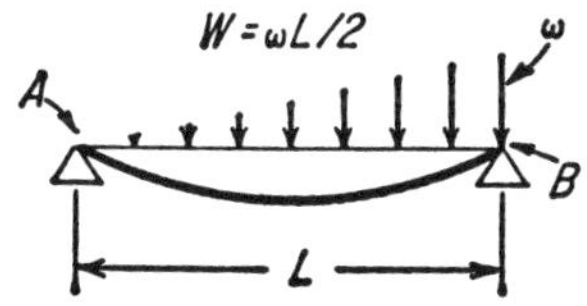	−76.7	−25.7 at A +25.7 at B	0.128 WL	−10.2
8	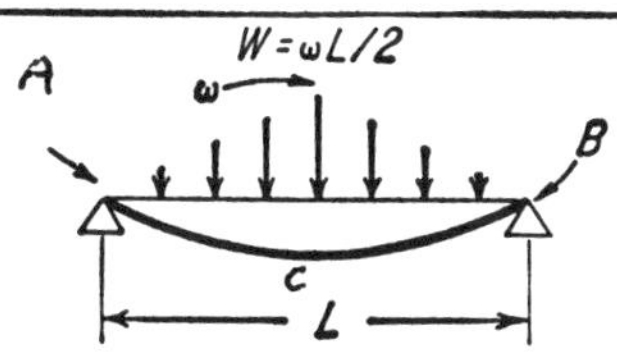	− 60 at A	−19.2 at A +19.2 at B	WL/6 at C	−10
9	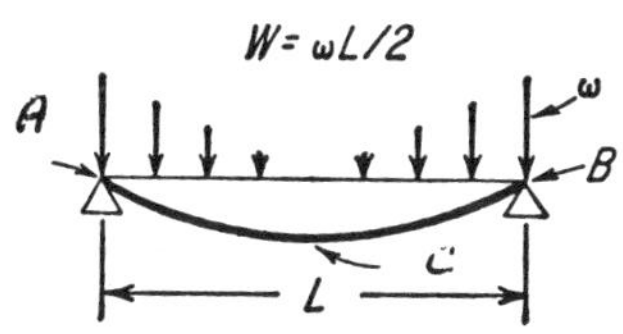	− 106.7 at A	+32 at B −32 at A	WL/2 at C	−8.9
10	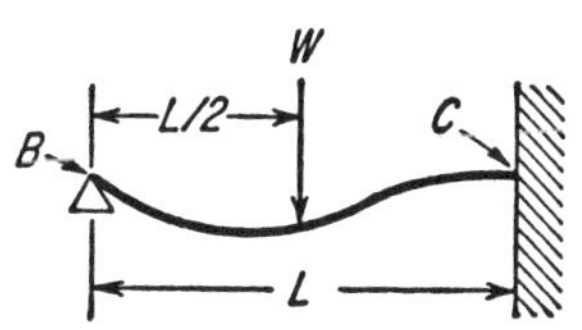	−107.3	−32 at B	− 3WL/16 at C	−20.5

	Beam Type	K for deflection	K_1 for slope	Max M	K_2
11	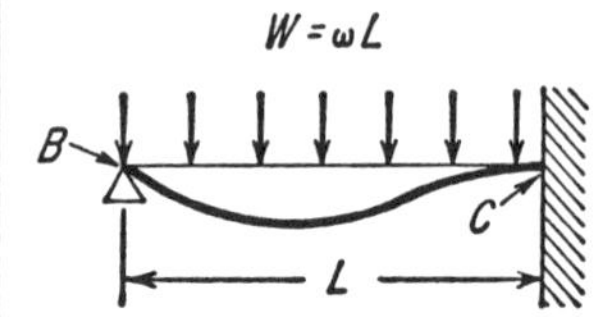	−185	−48 at B	−WL/8 at C	−23
12	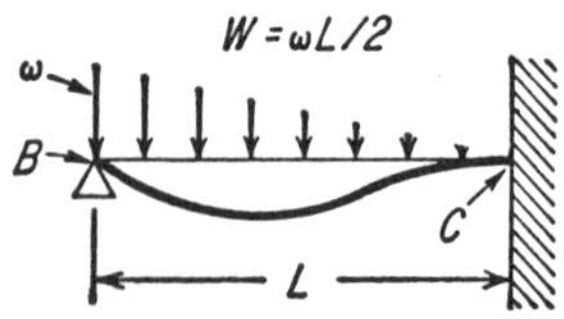	−164	−40 at B	−7WL/60 at C	−19.1
13	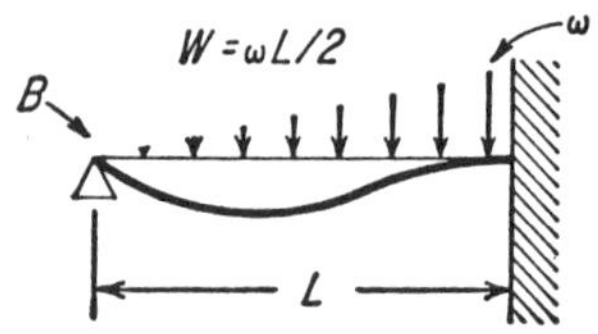	−210	−60 at B	WL/16.7	−12.6
14	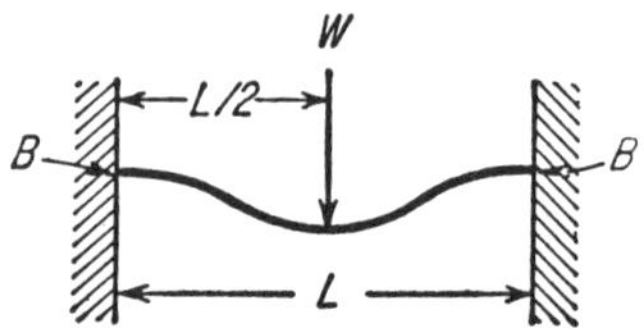	−192 at L/2	—	WL/8 at A,B	−24
15	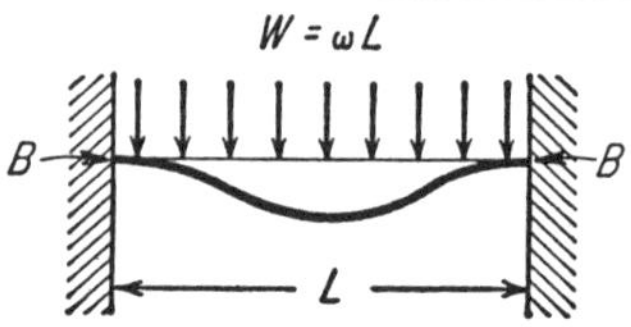	−384 at L/2	—	WL/12 at B	−32
16	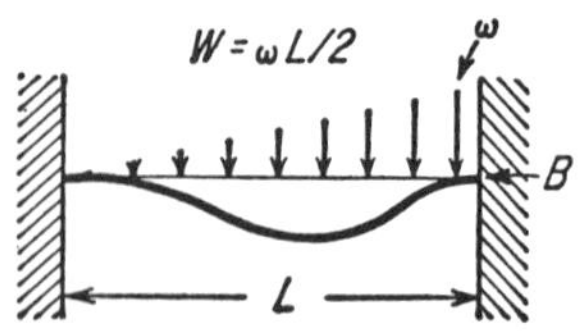	−382	—	WL/10 at B	−38.2

17. DEFLECTION IN TERMS OF FIBER STRESS

(*Ref. 2*)

In the solution of problems in certain statically indeterminate structures, it is convenient to have an expression for beam deflection in terms of the extreme fiber stress on the section due to bending. To illustrate, consider a beam of span L carrying a load W. The maximum deflection for a given condition is

$$y = \frac{WL^3}{K_1 EI}$$

where K_1 is a factor that designates a specific loading-support condition.

To express this deflection equation in terms of fiber stress we may proceed as follows: From the general equation for fiber stress, we have $S = Mc/I$. Solving for M and letting $d = 2c$ (depth of section), we have $M = 2SI/d$. For a given loading condition $M = WL/K_3$. Equating these values of M, and solving for W, we find:

$$W = \frac{K_3 SI}{dL}.$$

Finally by substituting this value of W in the above equation we obtain:

$$y = \frac{K_3}{K_1}\frac{2SL^2}{dE}.$$

Let $$\frac{K_1}{K_3} = K_2; \therefore y = \frac{2SL}{K_2 dE}.$$

Table 1 contains a tabulation of numerical values for K_2, for 18 different most commonly used loading-support conditions.

18. GENERAL FORMULAS FOR DEFLECTION AND SLOPE OF BEAMS WITH UNIFORM CROSS SECTION

When a beam is subjected to several loads such as moment M, concentrated load P, and uniform load q, the deflection caused by the combined load is usually obtained by superposition of equations assuming that the beam is subjected (1) only to M, (2) only to P, (3) only to q. This procedure is quite tedious and time-consuming. A simplification of the double integration method for calculating beam deflection was obtained, resulting in one general equation from which the deflection of the beam, with any supports subjected to any type of loading and any number of applied loads, may be obtained. However, it should be pointed out that this simplified procedure works only when the reactions are known. In the general case of a statically indeterminate beam the procedure outlined is not applicable.

A very useful application of this general equation will be shown in the calculation of the deflection and slope of beams, the cross sections of which vary by steps. (*See* Stepped Beams.)

BEAMS OF UNIFORM CROSS SECTION

The usual method of determining the elastic line of a beam is to start with the functional equation:

$$EI\, d^2y/dx^2 = M \tag{7}$$

and integrate it twice, whereby we obtain the equation of the deflection curve $y = f(x)$. Most of the current methods, such as the moment area or conjugate beam, are based on carrying out in some way this double integration. The constants of integration, C, D, are easily determined from the boundary conditions. As a rule:

$$C = EI\theta_o$$

$$D = EIy_o$$

where Y_o and θ_o are the deflection and angle of rotation at the origin of coordinate system.

For example, if we assume a cantilever beam with a uniform load q and the origin of the coordinates at the free end, we obtain the following relationships:

$$EI\frac{d^2y}{dx^2} = -q\frac{x^2}{2} \tag{8}$$

$$EI\frac{dy}{dx} = -q\frac{x^3}{6} + C \tag{9}$$

$$EIy = -q\frac{x^4}{24} + Cx + D. \tag{10}$$

For $x = 0$

$$EIy_o = D \tag{11}$$

$$EI\theta_o = C. \tag{12}$$

It is interesting to note that the relationship (11) and (12) will be valid on the condition that the same origin be used for the x coordinate no matter what portion of the beam is considered.

The sign convention throughout this paper is based on the assumption that the upward deflection is positive and clockwise movement to the left of a section is considered positive.

Assume a cantilever beam loaded as shown in Fig. 7. Consider four segments of the beam (A–B, A–C, A–D, A–E) as defined by their coordinates, x_1, x_2, x_3, x_4. The bending moments in each segment must be expressed by different equations, which in turn requires a derivation of a differential equation of the elastic line for each segment. The resulting eight constants of integration may be determined from appropriate boundary conditions. As this requires an involved procedure, we apply a special approach whereby the number of integration constants can be reduced to two, regardless of the number of segments. This is achieved by following the following three steps:

1. Obtain the equation for the bending moment at a given section by considering the external forces lying between that section and the origin.

2. Integrate the equation without opening the brackets.

3. For convenience, introduce a factor $(x - a)^0 = 1$.

1. Segment A–B; (x_1y_1)

$$EI\frac{d^2y_1}{dx_1^2} = 0$$

$$EI\frac{dy_1}{dx_1} = C_1$$

$$EIy_1 = C_1x_1 + D_1.$$

2. Segment A–C; (x_2y_2)

$$EI\frac{d^2y_2}{dx_2^2} = M(x_2 - a)^0$$

$$EI\frac{dy_2}{dx_2} = M(x_2 - a) + C_2$$

$$EIy_2 = M\frac{(x_2 - a)^2}{2} + C_2x_2 + D_2.$$

3. Segment A–D; (x_3,y_3)

$$EI\frac{d^2y_3}{dx_3^2} = M(x_3 - a)^0 + P(x_3 - b)$$

$$EI\frac{dy_3}{dx_3} = M(x_3 - a) + P\frac{(x_3 - b)^2}{2} + C_3$$

$$EIy_3 = M\frac{(x_3 - a)^2}{2} + P\frac{(x_3 - b)^3}{6} + C_3x_3 + D_3.$$

4. Segment A–E; (x_4,y_4)

$$EI\frac{d^2y_4}{dx_4^2} = M(x_4 - a)^0 + P(x_4 - b) + q\frac{(x_4 - c)^2}{2}$$

$$EI\frac{dy_4}{dx_4} = M\frac{(x_4 - a)}{1} + P\frac{(x_4 - b)^2}{2} + q\frac{(x_4 - c)^3}{6} + C_4$$

$$EIy_4 = M\frac{(x_4 - a)^2}{2} + P\frac{(x_4 - b)^3}{6} + q\frac{(x_4 - c)^4}{24} + C_4x_4 + D_4.$$

where a, b, c are the distances from the origin to the points of application of the moment M, the concentrated load P, and the point where the uniform load begins.

The eight constants of integration will be determined from the following boundary conditions:

1. $x_1 = x_2 = a \quad y_1 = y_2; \quad \theta_1 = \theta_2$
2. $x_2 = x_3 = b \quad y_2 = y_3; \quad \theta_2 = \theta_3$
3. $x_3 = x_4 = c \quad y_3 = y_4; \quad \theta_3 = \theta_4$
4. $x_4 = L \quad y_4 = 0\,; \quad \theta_4 = 0.$

These conditions yield:

$$C_1 = C_2 = C_3 = C_4 = C$$

$$D_1 = D_2 = D_3 = D_4 = D.$$

The two constants, C, D, that remain, may be determined from the fourth condition. This was already calculated previously. (*See* equation [11] [12].)

$$C = EI\theta_o;\ D = EIy_o$$

Thus, considering a case when several moments and forces are applied we obtain a general formula:*

$$EIy = EIy_o + EI\theta_ox + \sum\frac{M(x - a)^2}{2} + \sum\frac{P(x - b)^3}{6} + \sum\frac{q(x - c)^4}{24} \qquad (13)$$

By differentiating equation (13) we obtain:

$$EI\theta = EI\theta_o + \sum M(x - a) + \sum\frac{P(x - b)^2}{2} + \sum\frac{q(x - c)^3}{6}. \qquad (14)$$

Since the reactions of the supports are included in the external forces and moments, M, P, q, lying between the

*The method presented in this article is better known in Great Britain than in U.S.A. Credit for its development is given to R. Macaulay.

section in question and the origin, equations (13) and (14) are applicable to a beam with any type of supports. It is important to note that these equations are based on the positive directions of forces and moments as indicated in Fig. 1. Any deviation from these directions will introduce a negative sign in the appropriate member of the equation.

In equations (13) and (14) the term that contains the distributed load q is valid only if the distributed load continues at least to the section where Y and θ are to be determined. If the load breaks off before the section, it should be continued to the section and an equivalent load of opposite direction should be added.

Thus, one equation is all we need for calculating the deflection of a beam with any supports, any type of loading, and any number of applied loads.

To demonstrate the versatility of equation (13) and (14), a few cases of conventional loading will be discussed and equations for y and θ derived.

Example 1: Determine the maximum deflection for a beam with both ends fixed subjected to a center load (Fig. 3).

Solution: The origin of the coordinates for this particular case may be placed at the left or right end of the beam; in both cases $y_o = 0$ and $\theta_o = 0$. Consequently, we can, by using the general equation (13), calculate y_{max} at $x = L/2$.

As was mentioned before, equation (13) and (14) require to take all forces and moments lying between the origin and the section under consideration which is $x = L/2$ in the case of this example. This implies that the reactive force $R = P/2$ and reactive moment $M = PL/8$ at the built-in end should be used too. Since the direction of the reactive moment is counterclockwise, it will enter in the general equation with a minus sign. Thus, y_{max}, for $x = L/2$, from equation (13) is:

$$EIy_{x=L/2} = -\frac{PL}{8}\frac{(L/2-0)^2}{2} + \frac{P}{2}\frac{(L/2-0)^3}{6} = -\frac{PL^3}{192}$$

or

$$y_{max} = -\frac{PL^3}{192EI}$$

From our sign convention, the minus sign designates a downward deflection ($a = b = c = 0$).

Example 2: Find the maximum deflection of the beam in Fig. 4.

Solution: Here too, the location of the origin is arbitrary; y_{max} is at $x = L/2$ and $a = b = c = 0$, thus from equation (13):

$$EIy_{x=L/2} = -\frac{qL^2}{12}\frac{(L/2-0)^2}{2} + \frac{qL}{2}\frac{(L/2-0)^3}{6} - \frac{q(L/2-0)^4}{24} = -\frac{qL^4}{384}$$

or

$$y_{max} = -\frac{qL^4}{384\,EI}$$

Note that the downward load q is entered with a minus sign.

Example 3: Find the maximum deflection and maximum angle of rotation for a cantilever beam loaded as shown in Fig. 5.

Solution: In this case it is preferable to place the origin at the built-in end, then; $y_o = 0$, $\theta_o = 0$ and the y_{max}, θ_{max} are at $x = L$. With $a = 0$, $b = 0$, $c = 0$ we obtain from equation (13):

$$EIy_{x=L} = -\frac{qL^2}{2}\frac{(L-0)^2}{2} + \frac{qL(L-0)^3}{6} - \frac{q(L-0)^4}{24} = -\frac{qL^4}{8}$$

From equation [8]:

$$EI\theta_{x=L} = -\frac{qL^2}{2}(L-0) + \frac{qL(L-0)^2}{2} - \frac{q(L-0)^3}{6} = -\frac{qL^3}{6}.$$

Example 4: Find y_{max}, θ_{max} for the beam shown in Fig. 6.

Solution: With the origin at left end where $y_o = 0$, we determine θ_o by making use of the condition that at $x = L$ the deflection $y_o = 0$:

$$EIy_{x=L} = EI\theta_o L + \frac{P}{2}\frac{(L-0)^3}{6} - \frac{P(L-L/2)^3}{6} = 0$$

and

$$\theta_o = -\frac{PL^2}{16EI} = \theta_A = \theta_{max}$$

$$\theta_B = -\theta_A = \frac{PL^2}{16EI}.$$

The maximum deflection is at $x = L/2$:

$$EIy_{x=L/2} = -\frac{PL^2}{16} \times \frac{L}{2} + \frac{P}{2}\frac{(L/2)^3}{6} = -\frac{PL^3}{48}.$$

It should be noted that the distance from the origin to the reactive force $P/2$ is zero but to the applied load P it is $L/2$.

Since the equation (13) contains y_o and θ_o, it is necessary to choose the location of the origin of coordinates such that either of the two terms (or both) are equal to zero.

Example 5: Find the deflection at point D and the angle of rotation at point A of the beam loaded as shown in Fig. 7.

Solution: We place the origin at the left support where $y_o = 0$. The angle of rotation θ_o then will be determined from the condition that $y = 0$ at $x = 20$ in.

$$EIy_{x=20} = EI\theta_o 20 + \frac{100(20-10)^2}{2} + \frac{20(20-0)^3}{6} - \frac{10(20-0)^4}{24} + \frac{10(20-10)^4}{24} = 0$$

or

$$\theta_o = +\frac{1542}{EI}$$

As was pointed out before, the general equations, developed here, recognize only a distributed load that does not break off before the section where y or θ are determined. Since the uniform load in Fig. 6 breaks off at point B, it is necessary to extend the load to point C. To compensate for this addition, we add a uniform load of opposite direction (upward) in the portion BC of the beam. The last member of above equation reflects this change.

Deflection at point D ($x = 30$ in.):

$$EIy_D = 1542(30) + \frac{100(30-10)^2}{2} + \frac{20(30-0)^3}{6} + \frac{180(30-20)^3}{6} - \frac{10(30-0)^4}{24} + \frac{10(30-10)^4}{24} = -85{,}573.$$

In this case since the uniform load is extended to point D additional upward uniform load is added in the section BD of the beam.

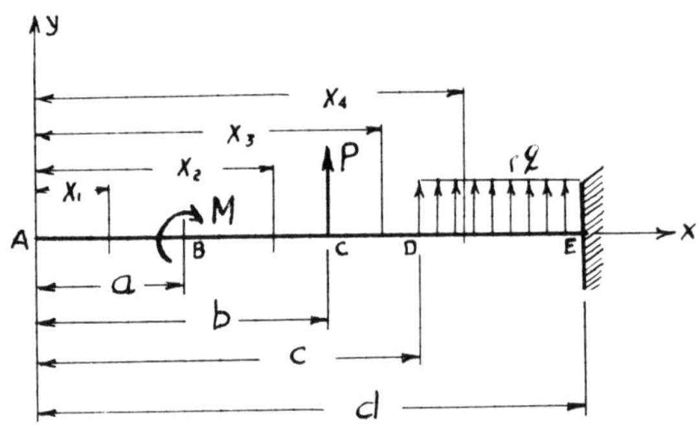

FIG. 2
NOTE: POSITIVE DIRECTION OF M, P, q IS INDICATED

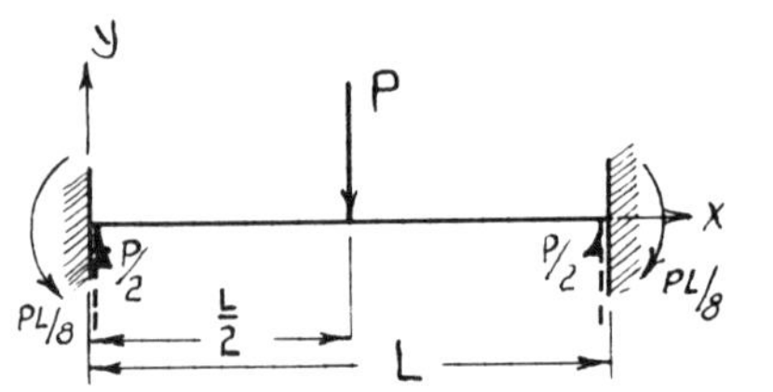

FIG. 3

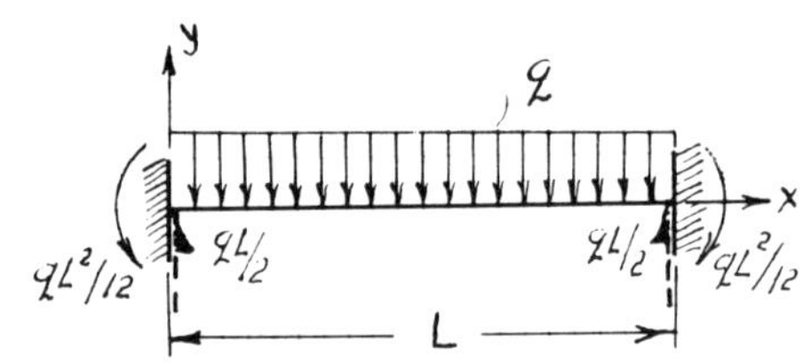

FIG. 4

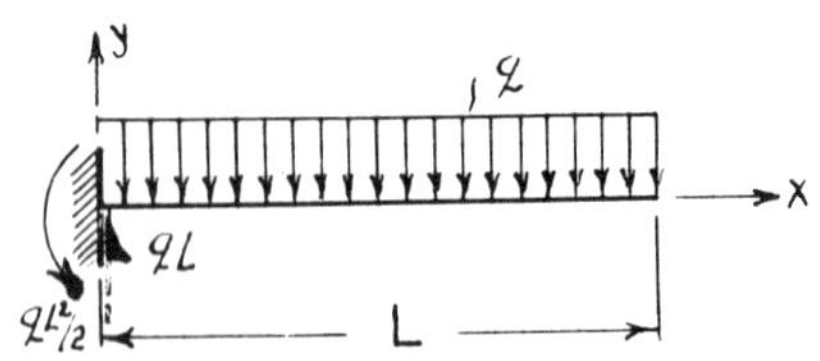

FIG. 5

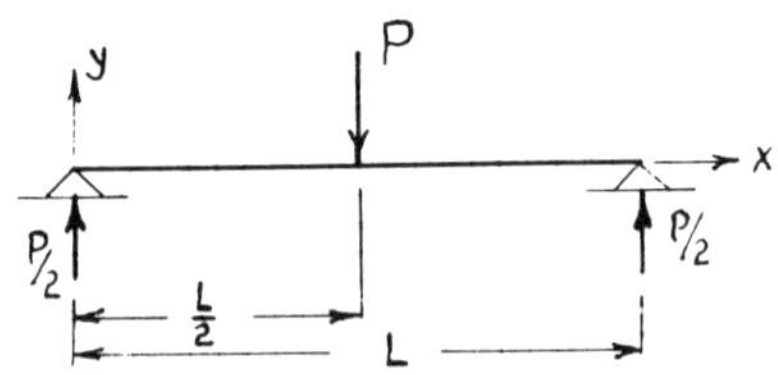

FIG. 6

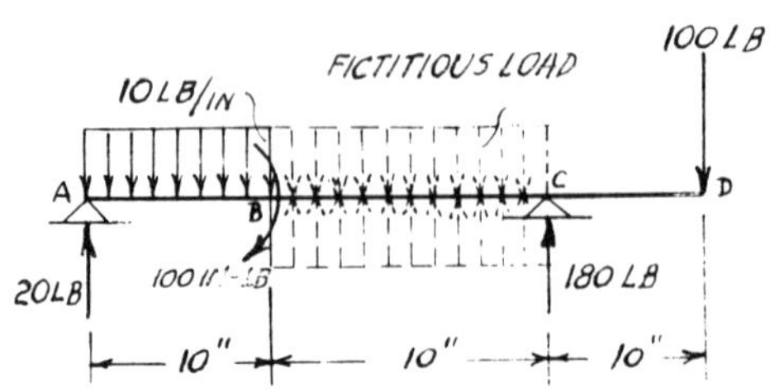

FIG 7

19. TABULATION OF FORMULAS WITH DIAGRAMS

(*Ref. 3*)

REACTION, MOMENT, SHEAR, DEFLECTION, END SLOPE, MAXIMUM RESULTING STRESS IN TERMS OF LOADING, MAXIMUM DEFLECTION IN TERMS OF RESULTING STRESS

NOTATION

W = load (lb.)
w = unit load (lb. per linear in.)
M = bending moment (in.-lb.); positive when clockwise
V = shear (lb.); positive when upward
y = deflection (in.); positive when upward
θ = end slope (radians)
R = reaction (lb.); positive as shown
$2c$ = depth of beam (in.)
S = fiber stress (psi)
I = moment of inertia (in.^{4})
E = modulus of elasticity (psi)

The formulas of this article are based on the following assumptions: (1) The beam is of homogeneous material with same modulus of elasticity in tension and compression. (2) The beam is straight or nearly so; if slightly curved, the curvature is at least 10 times the depth. (3) The cross section is uniform. (4) The beam has at least one longitudinal plane of symmetry. (5) All loads and reactions are perpendicular to the axis of the beam and lie in the same plane, which is a longitudinal plane of symmetry. (6) The beam is long in proportion to its depth. (7) The maximum stress does not exceed the proportional unit.

A *simple beam* rests on supports at its ends which permit rotation. A *cantilever* beam is fixed (no rotation) at one end. When computing reactions and moments, distributed loads may be replaced by their resultant acting at the center of gravity of the distributed-load area.

Reactions R, are the forces and/or couples acting at the supports and holding the beam in place. In general, the weight of the beam should be accounted for.

The bending moment M (in.-lb.) at any section is the algebraic sum of the external forces acting on the beam on one side of the section, $M = Fx$. A bending moment that produces tensile stress on bottom fiber is considered *positive*, while compression on top fiber is *negative*.

The vertical shear V (pounds) effective on a section is the algebraic sum of all the forces acting parallel to and on one side of the section, $V = \sum F$.

Moment and shear diagrams may be constructed by plotting to scale the particular entity as the ordinate for each section of the beam. Such diagrams show in continuous form the variation along the length of the beam.

TABULATED FORMULAS

In Table 2 are given formulas and diagrams for the bending moments, deflections, vertical shear, end reactions, end moments, end slopes and fiber stresses of beams supported and loaded in various ways. The cases listed have been so selected that, by superposition, the formulas may be applied to almost any combination of loading and support likely to occur.

Table 2

1. END SUPPORTS, UNIFORM LOADING

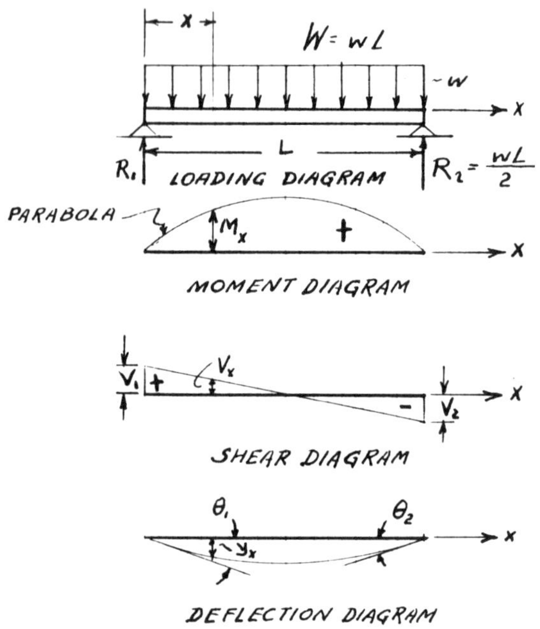

Reactions

$$R_1 = R_2 = \frac{W}{2}$$

Moment

$$M_X = \frac{1}{2}W\left(x - \frac{x^2}{L}\right)$$

$$M_{\text{max}} = \frac{1}{8}WL \text{ at } x = \frac{L}{2}$$

Shear

$$V_X = \frac{1}{2}W\left(1 - 2\frac{x}{L}\right)$$

$$V_1 = \frac{1}{2}W \text{ at } x = 0$$

$$V_2 = -\frac{1}{2}W \text{ at } x = L$$

Deflection

$$y_X = -\frac{1}{24}\frac{Wx}{EIL}(L^3 - 2Lx^2 + x^3)$$

$$y_{\text{max}} = -\frac{5}{384}\frac{WL^3}{EI} \text{ at } x = \frac{L}{2}$$

End Slope

$$\theta_1 = \theta_2 = \frac{1}{24}\frac{WL^2}{EI}$$

Max. Resulting Stress in Terms of Loading

$$S_{\text{max}} = \frac{1}{8}\frac{WLc}{I} \text{ at } x = \frac{L}{2}$$

Max. Deflection in Terms of Resulting Stress

$$y_{\text{max}} = -\frac{5}{48}\frac{SL^2}{Ec} \text{ at } x = \frac{L}{2}$$

2. END SUPPORTS, INTERMEDIATE LOAD

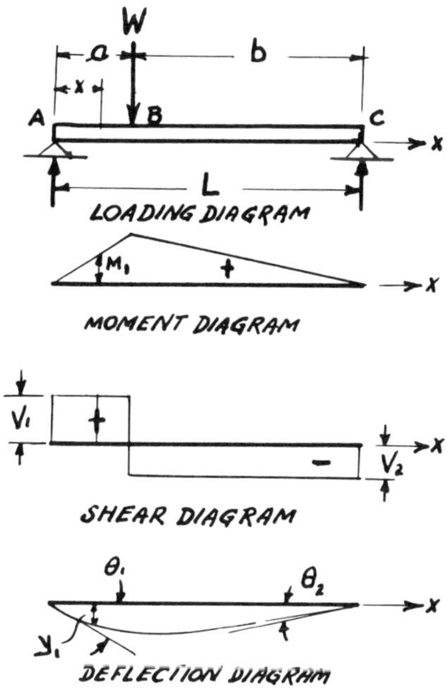

Reactions

$$R_1 = W(L - a)L$$

$$R_2 = Wa/L$$

Moment

$$(A \text{ to } C)\colon M_1 = W\frac{b}{L}x$$

$$(C \text{ to } B)\colon M_2 = W\frac{a}{L}(L - x)$$

$$M_{\max} = W\frac{ab}{L} \text{ at } C$$

Shear

$$(A \text{ to } C)\colon V_1 = W\frac{b}{L}$$

$$(B \text{ to } C)\colon V_2 = -W\frac{a}{L}$$

Deflection

$$(A \text{ to } C)\colon y_1 = -\frac{Wbx}{6EIL}[2L(L - x) - b^2 - (L - x)^2]$$

$$(C \text{ to } B)\colon y_2 = -\frac{Wa(L - x)}{6EIL}[2Lb - b^2 - (L - x)^2]$$

$$y_{\max} = -\frac{Wab}{27EIL}(a + 2b)\sqrt{3a(a + 2b)}$$

$$\text{at } x = \sqrt{\frac{1}{3}a(a + 2b)} \text{ when } a > b$$

End Slope

$$\theta_1 = \frac{1}{6}\frac{W}{EI}\left(bL - \frac{b^3}{L}\right) \text{ at } A$$

$$\theta_2 = \frac{1}{6}\frac{W}{EI}\left(2bL + \frac{b^3}{L} - 3b^2\right) \text{ at } C$$

Max. Resulting Stress in Terms of Loading

$$S_{\max} = \frac{Wabc}{LI} \text{ at } C$$

Max. Deflection in Terms of Resulting Stress

$$y_{\max} = \frac{-SL^2}{27Ec}\left[3\frac{a}{L}\left(2 - \frac{a}{L}\right)^3\right]^{\frac{1}{2}}$$

3. END SUPPORTS, CENTER LOAD

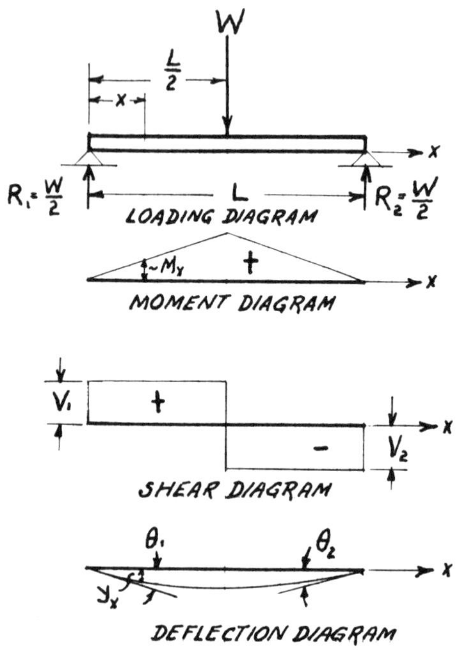

Reactions

$$R_1 = R_2 = \frac{W}{2}$$

Moment

$$(A \text{ to } C)\colon M_X = \frac{1}{2} Wx$$

$$(C \text{ to } B)\colon M_X \doteq \frac{1}{2} W(L - x)$$

$$M_{\max} = \frac{WL}{4} \text{ at } C$$

Shear

$$(A \text{ to } C)\colon V_1 = \frac{W}{2}$$

$$(C \text{ to } B)\colon V_2 = -\frac{W}{2}$$

Deflection

$$(A \text{ to } C)\colon y_X = -\frac{1}{48}\frac{W}{EI}(3L^2x - 4x^3)$$

$$y_{\max} = -\frac{1}{48}\frac{WL^3}{EI} \text{ at } C$$

End Slope

$$\theta_1 = \theta_2 = \frac{WL^2}{16EI}$$

Max. Resulting Stress in Terms of Loading

$$S_{\max} = \frac{WLc}{4I}$$

Max. Deflection in Terms of Resulting Stress

$$y_{\max} = -\frac{SL^2}{12Ec} \text{ at } C$$

4. END SUPPORTS, TRIANGULAR LOAD

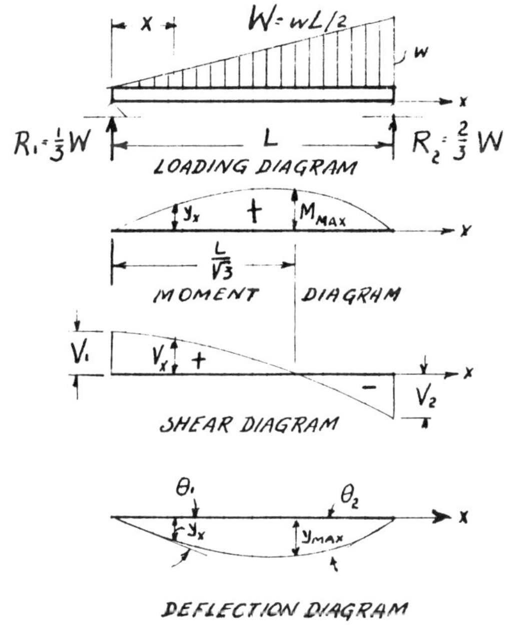

Reactions

$$R_1 = \frac{1}{3}W$$

$$R_2 = \frac{2}{3}W$$

Moment

$$M_X = \frac{1}{3}W\left(x - \frac{x^3}{L^3}\right)$$

$$M_{max} = 0.128WL \text{ at } x = 0.5774L$$

Shear

$$V_X = W\left(\frac{L}{3} - \frac{x^2}{L^2}\right)$$

$$V_1 = \frac{1}{3}W \text{ at } x = 0$$

$$V_2 = \frac{2}{3}W \text{ at } x = L$$

Deflection

$$y_X = \frac{-1}{180}\frac{Wx}{EIL^2}(3x^4 - 10L^2x^2 + 7L^4)$$

$$y_{max} = -0.01304\frac{WL^3}{EI} \text{ at } x = 0.519L$$

End Slope

$$\theta_1 = \frac{7}{180}\frac{WL^2}{EI} \text{ at } x = 0$$

$$\theta_2 = \frac{8}{180}\frac{WL^2}{EI} \text{ at } x = L$$

Max. Resulting Stress in Terms of Loading

$$S_{max} = 0.128\frac{WLc}{I}$$

Max. Deflection in Terms of Resulting Stress

$$y_{max} = \frac{-SL^2}{599Ec}$$

5. END SUPPORTS, TRIANGULAR LOAD

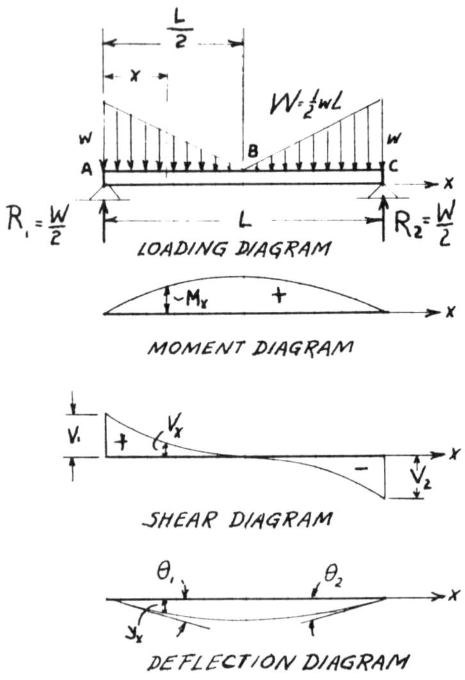

Reactions

$$R_1 = R_2 = \frac{W}{2}$$

Moment

$$(A \text{ to } B)\text{: } M_x = \frac{1}{2}W\left(x - 2\frac{x^2}{L} + \frac{4x^3}{3L^3}\right)$$

$$(B \text{ to } C)\text{: } M_x = \frac{1}{2}W\left[(L-x) - 2\frac{(L-x)^2}{L} + \frac{4}{3}\frac{(L-x)^3}{L^2}\right]$$

$$M_{\max} = \frac{1}{12}WL \text{ at } B$$

Shear

$$(A \text{ to } B)\text{: } V_x = \frac{1}{2}W\left(\frac{L-2x}{L}\right)^2$$

$$(B \text{ to } C)\text{: } V_x = -\frac{1}{2}W\left(\frac{2x-L}{L}\right)^2$$

$$V_1 = \frac{W}{2}; \quad V_2 = -\frac{W}{2}$$

Deflection

$$(A \text{ to } B)\text{: } y_x = \frac{1}{12}\frac{W}{EI}\left(x^3 - \frac{x^4}{L} + \frac{2}{5}\frac{x^5}{L^2} - \frac{3}{8}L^2x\right)$$

$$y_{\max} = -\frac{3}{320}\frac{WL^3}{EI} \text{ at } B$$

End Slope

$$\theta_1 = \theta_2 = \frac{1}{32}\frac{WL^2}{EI}$$

Max. Resulting Stress in Terms of Loading

$$S_{\max} = \frac{1}{12}\frac{WLc}{I}$$

Max. Deflection in Terms of Resulting Stress

$$y_{\max} = \frac{-SL^2}{8.9Ec}$$

6. END SUPPORTS, TRIANGULAR LOAD

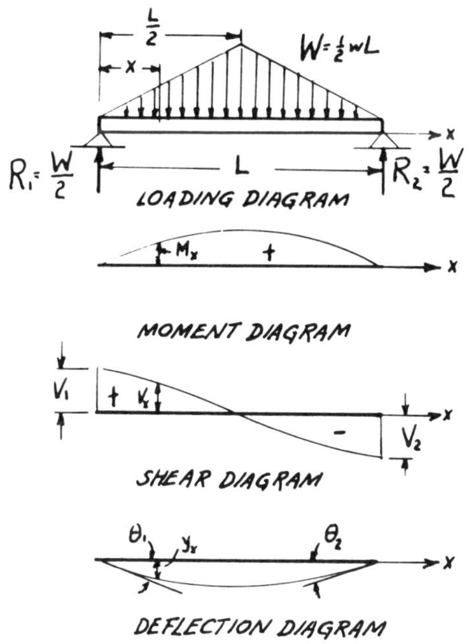

Reactions

$$R_1 = R_2 = \frac{W}{2}$$

Moment

$$(A \text{ to } B)\colon M_x = \frac{1}{6}W\left(3x - 4\frac{x^3}{L^2}\right)$$

$$(B \text{ to } C)\colon M_x = \frac{1}{6}W\left[3(L-x) - 4\frac{(L-x)^3}{L^2}\right]$$

$$M_{\max} = \frac{1}{6}WL \text{ at } B$$

Shear

$$(A \text{ to } B)\colon V_x = \frac{1}{2}W\left(1 - 4\frac{x^2}{L^2}\right)$$

$$(B \text{ to } C)\colon V_x = -\frac{1}{2}W\left[1 - 4\frac{(L-x)^2}{L^2}\right]$$

$$V_1 = \frac{W}{2},\ V_2 = -\frac{W}{2}$$

Deflection

$$(A \text{ to } B)\colon y_x = \frac{1}{6}\frac{W_x}{EIL^2}\left[\frac{1}{2}L^2x^2 - \frac{1}{5}x^4 - \frac{5}{16}L^4\right]$$

$$y_{\max} = -\frac{1}{60}\frac{WL^3}{EI} \text{ at } B$$

End Slope

$$\theta_1 = \theta_2 = \frac{5}{96}\frac{WL^2}{EI}$$

Max. Resulting Stress in Terms of Loading

$$S_{\max} = \frac{1}{6}\frac{WLc}{I}$$

Max. Deflection in Terms of Resulting Stress

$$y_{\max} = \frac{-SL^2}{10Ec}$$

7. END SUPPORTS, END COUPLE

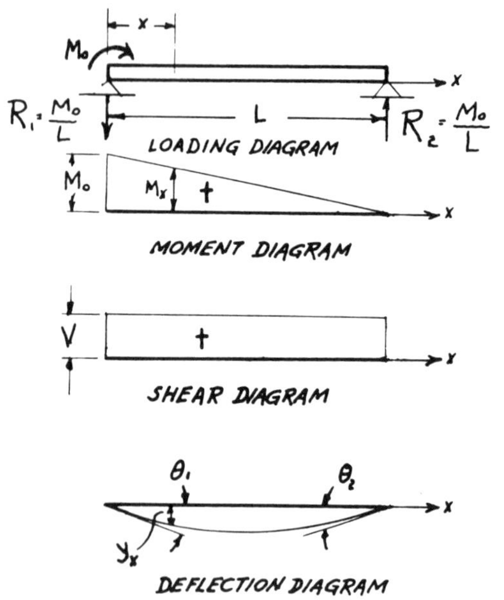

Reactions

$$R_1 = -\frac{M_o}{L}$$

$$R_2 = +\frac{M_o}{L}$$

Moment

$$M_x = M_o\left(1 - \frac{x}{L}\right)$$

$$M_{\max} = M_o$$

Shear

$$V = \frac{M_o}{L}$$

Deflection

$$y_x = \frac{1}{6}\frac{M_o}{EI}\left(3x^2 - \frac{x^3}{L} - 2Lx\right)$$

$$y_{\max} = -0.0642\frac{M_oL^2}{EI} \text{ at } x = 0.422L$$

End Slope

$$\theta_1 = \frac{1}{3}\frac{M_oL}{EI}$$

$$\theta_2 = \frac{1}{6}\frac{M_oL}{EI}$$

Max. Resulting Stress in Terms of Loading

$$S_{\max} = \frac{M_oc}{I}$$

8. END SUPPORTS, INTERMEDIATE COUPLE

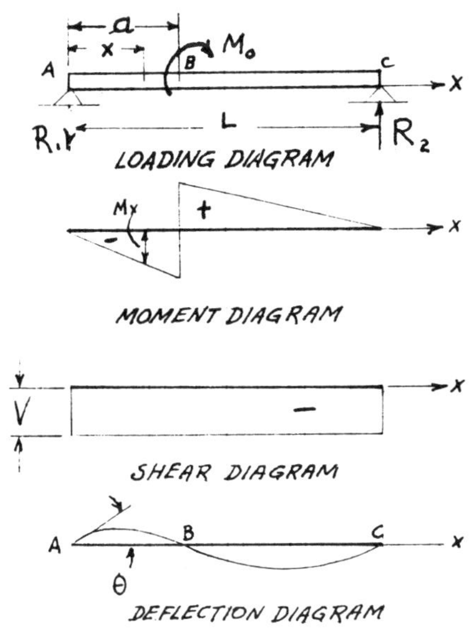

Reactions

$$R_1 = -\frac{M_o}{L},\ R_2 = \frac{M_o}{L}$$

Moment

$$(A \text{ to } B)\colon M_x = R_1 x$$

$$(B \text{ to } C)\colon M_x = R_1 x + M_o$$

$$+M_{\max} = R_1 a + M_o$$

$$-M_{\max} = R_1 a$$

Shear

$$V = R_1 \text{ (constant)}$$

Deflection

$$(A \text{ to } B)\colon y_x = \frac{1}{6}\frac{M_o}{EI}\left[\left(6a - 3\frac{a^2}{L} - 2L\right)x - \frac{x^3}{L}\right]$$

$$(B \text{ to } C)\colon y_x = \frac{1}{6}\frac{M_o}{EI}\left[3a^2 + 3x^2 - \frac{x^3}{L} - \left(2L + 3\frac{a^2}{L}\right)x\right]$$

End Slope

$$\theta = \frac{1}{6}\frac{M_o}{EI}\left(2L - 6a + 3\frac{a^2}{L}\right) \text{ at } A$$

$$\theta = \frac{1}{6}\frac{M_o}{EI}\left(L - 3\frac{a^2}{L}\right) \text{ at } C$$

$$\theta = \frac{M_o}{EI}\left(a - \frac{a^3}{L} - \frac{1}{3}L\right) \text{ at } B$$

Max. Resulting Stress in Terms of Loading

$$S_{\max} = \frac{(R_1 a + M_o)c}{I}$$

9. END SUPPORTS, PARTIAL UNIFORM LOAD

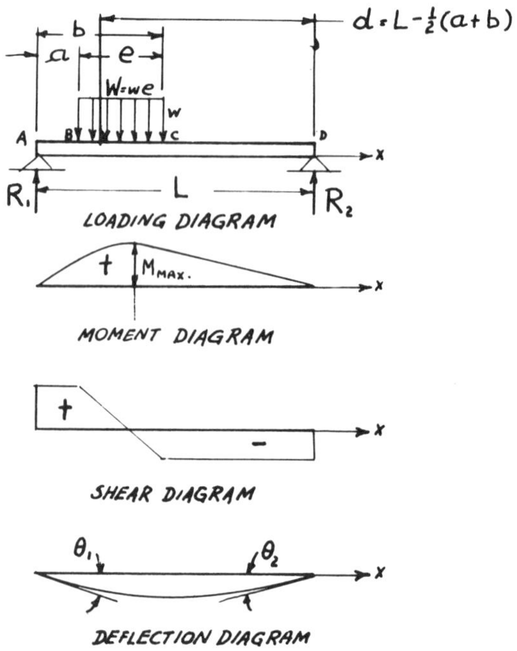

Reactions

$$R_1 = W\frac{d}{L};\quad R_2 = \frac{W}{L}\left(a + \frac{1}{2}e\right)$$

Moment

$$(A \text{ to } B)\text{: } M_x = R_1 x$$

$$(B \text{ to } C)\text{: } M_x = R_1 x - W\frac{(x-a)^2}{2e}$$

$$(C \text{ to } D)\text{: } M_x = R_1 x - W\left(x - \frac{1}{2}a - \frac{1}{2}b\right)$$

$$M_{\max} = W\frac{d}{L}\left(a + \frac{ed}{2L}\right) \text{ at } x = a + \frac{ed}{L}$$

Shear

$$(A \text{ to } B)\text{: } V = R_1$$

$$(B \text{ to } C)\text{: } V = R_1 - W\frac{x-a}{e}$$

$$(C \text{ to } D)\text{: } V = R_1 - W$$

Deflection

$$(A \text{ to } B)\text{: } y_x = \frac{1}{48EI}\left\{8R_1(x^3 - L^2x) + Wx\left[\frac{8d^3}{L} - \frac{2be^2}{L} + \frac{e^3}{L} + 2e^2\right]\right\}$$

$$(B \text{ to } C)\text{: } y_x = \frac{1}{48EI}\left\{8R_1(x^3 - L^2x) + Wx\left[\frac{8d^3}{L} - \frac{2be^2}{L} + \frac{e^3}{L} + 2e^2\right] - 2W\frac{(x-a)^4}{e}\right\}$$

$$(C \text{ to } D)\text{: } y_x = \frac{1}{48EI}\left\{8R_1(x^3 - L^2x) + Wx\left[\frac{8d^3}{L} - \frac{2be^2}{L} + \frac{e^3}{L}\right] - 8W\left(x - \frac{1}{2}a - \frac{1}{2}b\right)^3 + W(2be^2 - e^3)\right\}$$

End Slope

$$\theta_1 = \frac{1}{48EI}\left[-8R_1L^2 + W\left(\frac{8d^3}{L} - \frac{2be^2}{L} + \frac{e^3}{L} + 2e^2\right)\right] \quad \text{at } A$$

$$\theta_2 = \frac{1}{48EI}\left[16R_1L^2 - W\left(24d^2 - \frac{8d^3}{L} + \frac{2be^2}{L} - \frac{e^3}{L}\right)\right] \quad \text{at } B$$

Max. Resulting Stress in Terms of Loading

$$S_{\max} = \frac{Wc}{I}\frac{d}{L}\left(a + \frac{ed}{2L}\right)$$

10. END SUPPORTS, PARTIAL TRIANGULAR LOAD

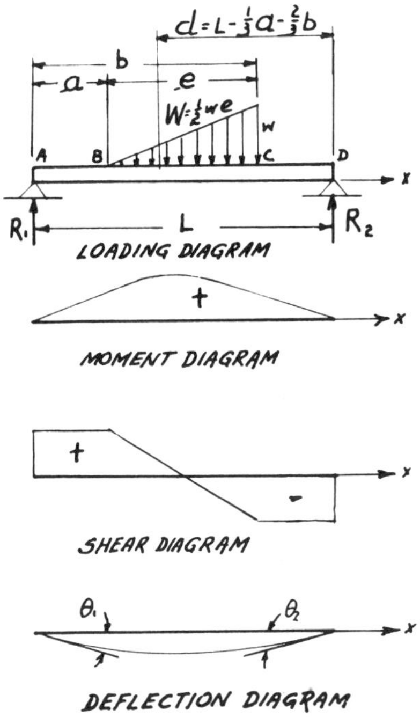

Reactions

$$R_1 = W\frac{d}{L},\quad R_2 = W\frac{L-d}{L}$$

Moment

$$(A \text{ to } B)\colon M_x = R_1 x$$

$$(B \text{ to } C)\colon M_x = R_1 x - W\frac{(x-a)^3}{3e^2}$$

$$(C \text{ to } D)\colon M_x = R_1 x - \frac{1}{3}W(3x - a - 2b)$$

$$M_{\max} = W\frac{d}{L}\left(a + \frac{2}{3}e\sqrt{\frac{d}{L}}\right) \text{ at } x = a + e\sqrt{\frac{d}{L}}$$

Shear

$$(A \text{ to } B)\quad V = R_1$$

$$(B \text{ to } C)\colon V_x = R_1 - \left(\frac{x-a}{e}\right)^2 W$$

$$(C \text{ to } D)\colon V = R_1 - W$$

Deflection

$$(A \text{ to } B)\colon y_x = \frac{1}{6EI}\left\{R_1(x^3 - L^2x) + Wx\left[\frac{d^3}{L} + \frac{1}{6}e^2\left(1 - \frac{b}{L}\right) + \frac{17}{270}\frac{e^3}{L}\right]\right\}$$

$$(B \text{ to } C)\colon y_x = \frac{1}{6EI}\left[R_1(x^3 - L^2x) - \frac{1}{10}W\frac{(x-a)^5}{e^2} + Wx\left(\frac{d^3}{L} + \frac{1}{6}e^2 - \frac{1}{6}e^2\frac{b}{L} + \frac{17}{270}\frac{e^3}{L}\right)\right]$$

$$(C \text{ to } D)\colon y_x = \frac{1}{6EI}\left\{R_1(x^3 - L^2x) - W\left[d^3 - d^3\frac{x}{L} - \frac{1}{6}be^2\left(1 - \frac{x}{L}\right) + \frac{17}{270}e^3\left(1 - \frac{x}{L}\right)\right]\right\}$$

End Slope

$$\theta_1 = \frac{1}{6EI}\left[-R_1L^2 + W\left(\frac{d^3}{L} + \frac{1}{6}e^2 + \frac{17}{270}\frac{e^3}{L} - \frac{1}{6}\frac{e^2b}{L}\right)\right]$$

at A

$$\theta_2 = \frac{1}{6EI}\left[2R_1L^2 + W\left(\frac{d^3}{L} + \frac{17}{270}\frac{e^3}{L} - \frac{1}{6}\frac{e^2b}{L} - 3d^2\right)\right]$$

at D

11. BEAM ON THREE SUPPORTS, UNIFORM LOAD

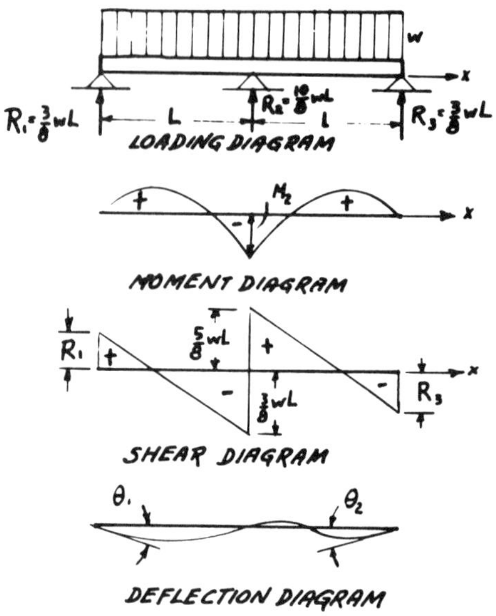

Reactions

$$R_1 = \frac{3}{8}wL$$

$$R_2 = \frac{10}{8}wL$$

$$R_3 = \frac{3}{8}wL$$

Moment

$$M_x = \frac{wLx}{8}\left(3 - 4\frac{x}{L}\right)$$

$$M_1 = \frac{9}{128}wL^2 \text{ at } x = \frac{3}{8}L$$

$$M_2 = \frac{wL^2}{8} \text{ at } x = L$$

Shear

$$V_x = \frac{wL}{8}\left(3 - 8\frac{x}{L}\right)$$

$$V_{\max} = -\frac{5}{8}wL \text{ at } x = L$$

Deflection

$$y_X = -\frac{wL^3x}{48EI}\left(1 - 3\frac{x^2}{L^2} + 2\frac{x^3}{L^3}\right)$$

$$y_{\max} = -\frac{1}{185}\frac{wL^4}{EI} \text{ at } x = 0.4215L$$

End Slope

$$\theta_1 = \theta_2 = \frac{wL^3}{48EI}$$

Max. Resulting Stress in Terms of Loading

$$S_{\max} = \frac{1}{8}\frac{wL^2c}{I}$$

Max. Deflection in Terms of Resulting Stress

$$y_{\max} = -0.0432\frac{SL^2}{Ec}$$

12. SIMPLE SUPPORTS, TWO SYMMETRIC LOADS

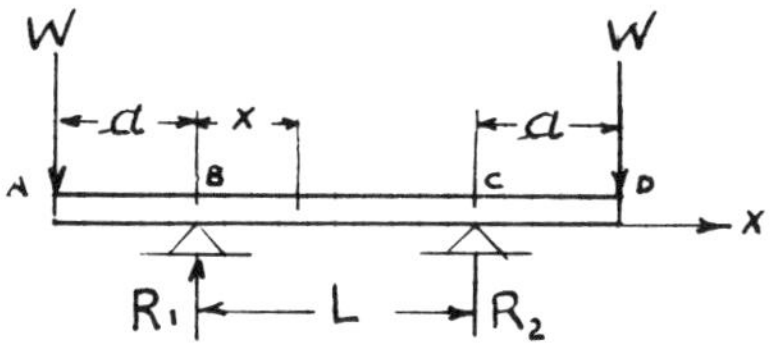

LOADING DIAGRAM

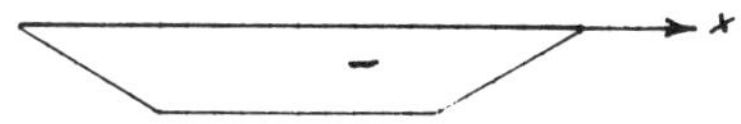

MOMENT DIAGRAM

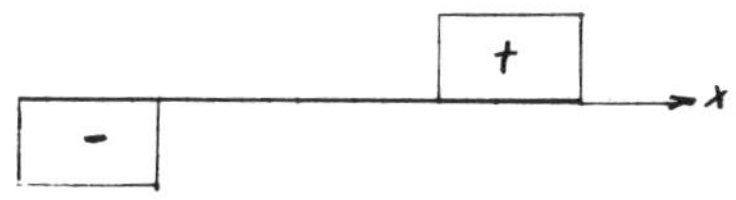

SHEAR DIAGRAM

DEFLECTION DIAGRAM

Reactions

$$R_1 = W$$

$$R_2 = W$$

Moment

$$M_x = -Wa = \text{constant}$$

Shear

$$(A \text{ to } B)\colon V = -W$$

$$(B \text{ to } C)\colon V = 0$$

$$(C \text{ to } D)\colon V = +W$$

Deflection

$$y_1 = \frac{WaL^2}{8EI} (\text{max.})$$

$$y_2 = \frac{W}{EI}\frac{a^2}{3}\left(a + \frac{3L}{2}\right) (\text{max.})$$

Resulting Stress in Terms of Loading

$$S = \frac{Wac}{I}$$

13. SIMPLE SUPPORTS WITH ONE OVERHANG, UNIFORM LOAD

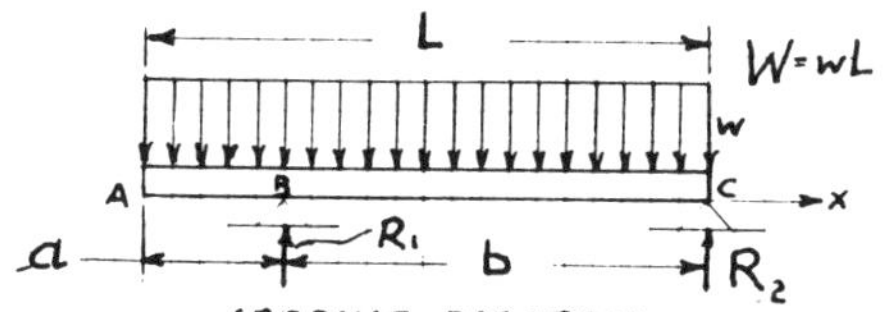

LOADING DIAGRAM

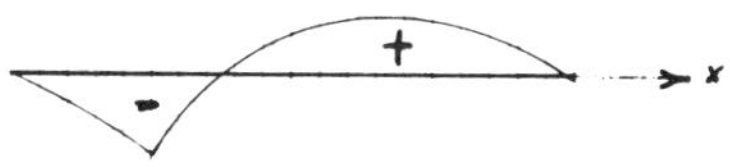

MOMENT DIAGRAM

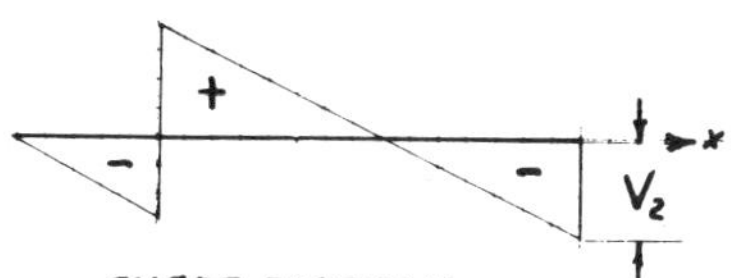

SHEAR DIAGRAM

DEFLECTION DIAGRAM

Moment

$$(A \text{ to } B)\colon M_x = -\frac{W}{L}\frac{x^2}{a}$$

$$(B \text{ to } C)\colon M_x = -\frac{W}{L}\frac{x^2}{2} + R_1(x - a)$$

Shear

$$V_2 = R_2$$

$$(A \text{ to } B)\colon V_x = -\frac{W}{L}x$$

$$(B \text{ to } C)\colon V_x = -\frac{W}{L}x + R_1$$

Deflection

$$(A \text{ to } B)\colon y_x = \frac{1}{24}\frac{W}{EIL}\left[-x^4 + \frac{(L^4 - a^4)x}{b} - 2L^2bx + a^4 - \frac{(L^4 - a^4)a}{b} + 2L^2ab\right]$$

$$(B \text{ to } C)\colon y_x = \frac{1}{24}\frac{W}{EIL}\left[-x^4 + \frac{2L^2(x - a)^3}{b} + \frac{(L^4 - a^4)x}{b} - 2L^2bx + a^4 - \frac{(L^4 - a^4)a}{b} + 2L^2ab\right]$$

Reactions

$$R_2 = \frac{W}{L - a}\left(\frac{L}{2} - a\right)$$

$$R_1 = R_2 + W$$

14. CANTILEVER, UNIFORM LOAD

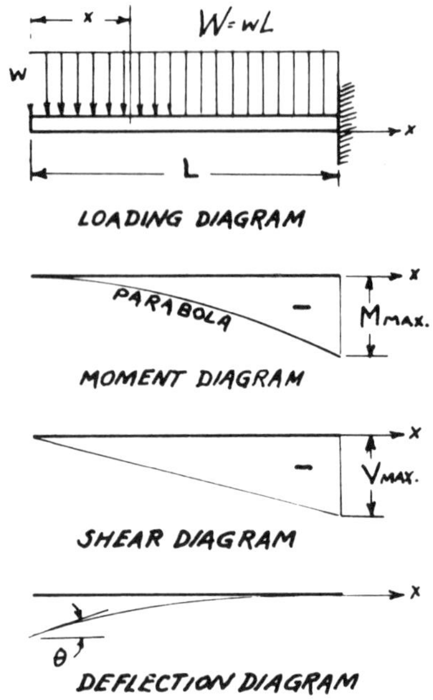

Reaction

$$R_1 = 0$$

$$R_2 = W$$

Moment

$$M_x = -\frac{1}{2}\frac{W}{L}x^2$$

$$M_{\max} = -\frac{1}{2}WL \text{ at } x = L$$

Shear

$$V_x = -\frac{W}{L}x$$

$$V_{\max} = -W \text{ at } x = L$$

Deflection

$$y_x = -\frac{1}{24}\frac{W}{EIL}(x^4 - 4L^3x + L^4)$$

$$y_{\max} = -\frac{1}{8}\frac{WL^3}{EI} \text{ at } x = 0$$

End Slope

$$\theta = \frac{1}{6}\frac{WL^2}{EI} \text{ at } x = 0$$

Max. Resulting Stress in Terms of Loading

$$S_{\max} = \frac{WLc}{2I} \text{ at } x = L$$

Max. Deflection in Terms of Resulting Stress

$$y_{\max} = \frac{-SL^2}{12Ec} \text{ at } x = 0$$

15. CANTILEVER, END LOAD

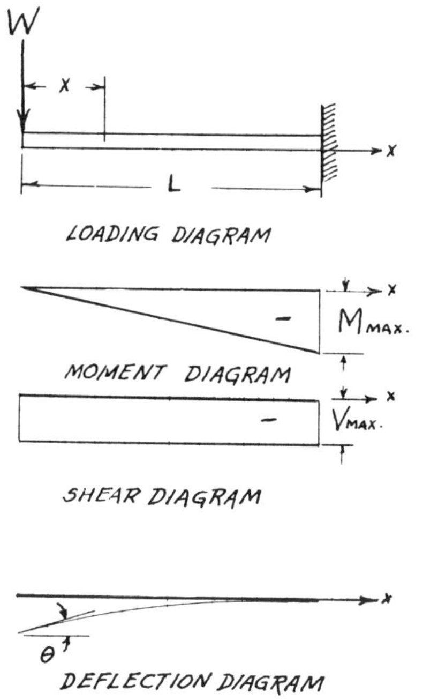

Reactions

$$R_1 = 0$$

$$R_2 = W$$

Moment

$$M_x = -Wx$$

$$M_{\max} = -WL \text{ at } x = L$$

Shear

$$V_{\max} = -W = \text{constant}$$

Deflection

$$y_x = -\frac{1}{6}\frac{W}{EI}(x^3 - 3L^2x + 2L^3)$$

$$y_{\max} = -\frac{1}{3}\frac{WL}{EI} \text{ at } x = 0$$

End Slope

$$\theta = \frac{1}{2}\frac{WL^2}{EI} \text{ at } x = 0$$

Max. Resulting Stress in Terms of Loading

$$S_{\max} = \frac{1}{2}\frac{WLc}{2} \text{ at } x = L$$

Max. Deflection in Terms of Resulting Stress

$$y_{\max} = \frac{1}{12}\frac{SL^2}{Ec} \text{ at } x = 0$$

16. CANTILEVER, INTERMEDIATE LOAD

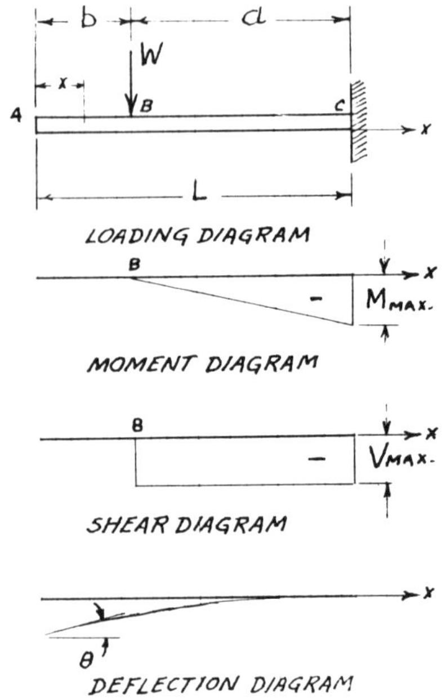

Reactions

$$R_1 = 0$$

$$R_2 = W$$

Moment

$$(A \text{ to } B)\text{: } M_x = 0$$

$$(B \text{ to } C)\text{: } M_x = -W(x - b)$$

$$M_{\max} = -Wa \text{ at } C$$

Shear

$$(A \text{ to } B) \quad V_x = 0$$

$$(B \text{ to } C) \quad V_x = -W = V_{\max}$$

Deflection

$$(A \text{ to } B) \quad y_x = -\frac{1}{6}\frac{W}{EI}(-a^3 + 3a^2L - 3a^2x)$$

$$(B \text{ to } C) \quad y_x = -\frac{1}{6}\frac{W}{EI}[(x - b)^3 - 3a^2(x - b) + 2a^3]$$

$$y_{\max} = -\frac{1}{6}\frac{W}{EI}(3a^2L - a^3) \text{ at } A$$

End Slope

$$(A \text{ to } B) \quad \theta = \frac{1}{2}\frac{Wa^2}{EI}$$

Max. Resulting Stress in Terms of Loading

$$S_{\max} = \frac{Wa\,c}{I} \text{ at } C$$

Max. Deflection in Terms of Resulting Stress

$$y_{\max} = -\frac{1}{6}\frac{SL}{Ec}(2L + b)a$$

17. CANTILEVER, TRIANGULAR LOAD

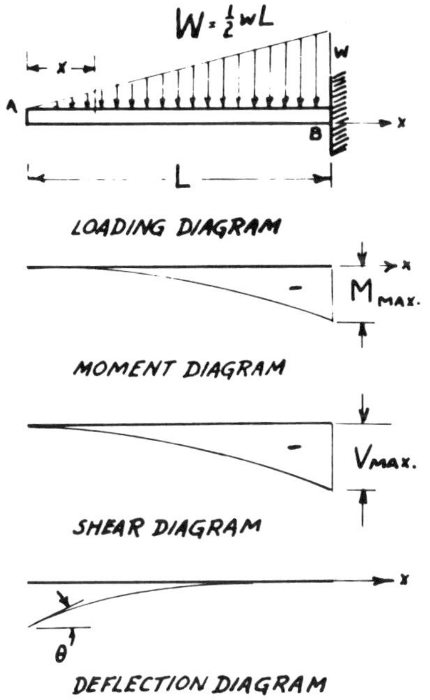

Reactions

$$R_1 = 0$$

$$R_2 = W$$

Moment

$$M_x = -\frac{1}{3}\frac{W}{L^2}x^3$$

$$M_{\max} = -\frac{1}{3}WL \text{ at } B$$

Shear

$$V_x = -\frac{W}{L^2}x^2$$

$$V_{\max} = -W \text{ at } B$$

Deflection

$$y_x = -\frac{1}{60}\frac{W}{EIL^2}(x^5 - 5L^4x + 4L^5)$$

$$y_{\max} = -\frac{1}{15}\frac{WL^3}{EI} \text{ at } A$$

End Slope

$$\theta = \frac{1}{12}\frac{WL^2}{EI} \text{ at } A$$

Max. Resulting Stress in Terms of Loading

$$S_{\max} = \frac{1}{3}\frac{WLc}{I} \text{ at } B$$

Max. Deflection in Terms of Resulting Stress

$$y_{\max} = -\frac{1}{28}\frac{SL^2}{Ec}$$

18. CANTILEVER, TRIANGULAR LOAD

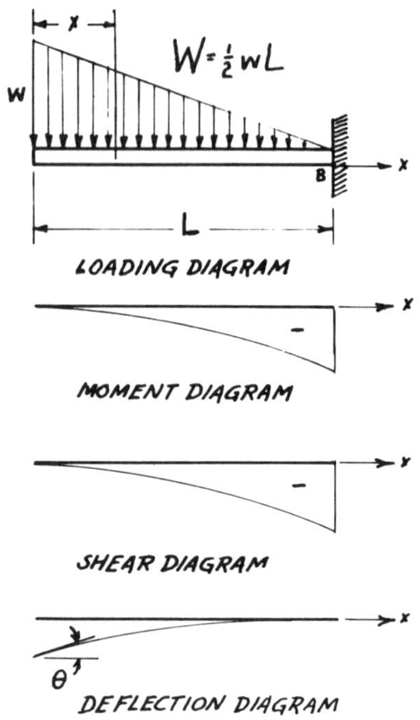

Reactions

$$R_1 = 0$$

$$R_2 = W$$

Moment

$$M_x = -\frac{1}{3}\frac{W}{L^3}(3Lx^2 - x^3)$$

$$M_{\text{max}} = -\frac{2}{3}WL \text{ at } B$$

Shear

$$V_x = -W\left(\frac{2Lx - x^2}{L^2}\right)$$

$$V_{\text{max}} = -W \text{ at } B$$

Deflection

$$y_x = -\frac{1}{60}\frac{W}{EIL^2}(-x^5 - 15L^4x + 5Lx^4 + 11L^5)$$

$$y_{\text{max}} = -\frac{11}{60}\frac{WL^3}{EI} \text{ at } x = 0$$

End Slope

$$\theta = \frac{1}{4}\frac{WL^2}{EI}$$

Max. Resulting Stress in Terms of Loading

$$S_{\text{max}} = \frac{2}{3}\frac{WLc}{I} \text{ at } x = L$$

Max. Deflection in Terms of Resulting Stress

$$y_{\text{max}} = -\frac{1}{3.6}\frac{SL^2}{Ec}$$

19. CANTILEVER, PARTIAL TRIANGULAR LOAD

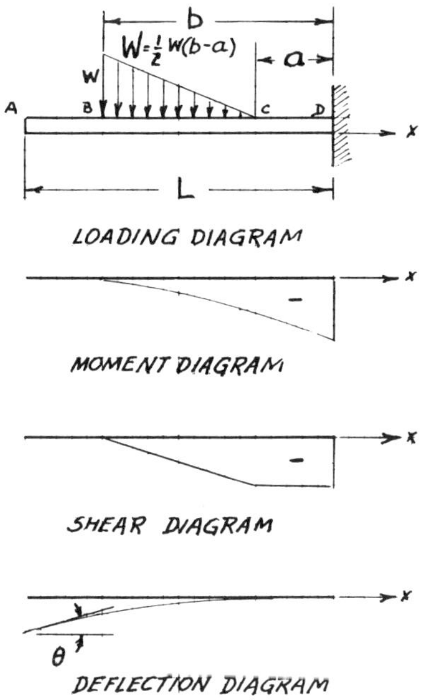

Reactions

$$R_1 = 0$$

$$R_2 = W$$

Moment

(A to B) $M = 0$

$$(B \text{ to } C)\colon M_x = -\frac{1}{3}W\left[\frac{3(x-L+b)^2}{b-a} - \frac{(x-L+b)^3}{(b-a)^2}\right]$$

$$(C \text{ to } D)\colon M_x = -\frac{1}{3}W(-3L+3x+2b+a)$$

$$M_{\max} = -\frac{1}{3}W(2b+a) \text{ at } D$$

Shear

$$(A \text{ to } B)\colon V = 0$$

$$(B \text{ to } C)\colon V_x = -W\left[1 - \frac{(L-a-x)^2}{(b-a)^2}\right]$$

$$V_{\max} = -W$$

Deflection

$$(A \text{ to } B)\colon y_x = -\frac{1}{60}\frac{W}{EI}[(5a^2+10ab+15b^2) \times (L-x) - a^3 - 2a^2b - 3ab^2 - 4b^3]$$

$$(B \text{ to } C)\colon y_x = -\frac{1}{60}\frac{W}{EI}\left[\frac{(L-x-a)^5}{(b-a)^2} - 10(L-x)^3 + (10a+20b)(L-x)^2\right]$$

$$(C \text{ to } D)\colon y_x = -\frac{1}{60}\frac{W}{EI}[(a+2b)(L-x)^2 - (L-x)^3]$$

$$y_{\max} = -\frac{1}{60}\frac{W}{EI}[(5a^2+10ab+15b^2)L - a^3 - 2a^2b - 3ab^2 - 4b^3]$$

End Slope

$$\theta = \frac{1}{12}\frac{W}{EI}(a^2+2ab+3b^2)$$

Max. Resulting Stress in Terms of Loading

$$S_{\max} = \frac{1}{3}\frac{W(2b+a)c}{I} \text{ at } D$$

20. CANTILEVER, END COUPLE

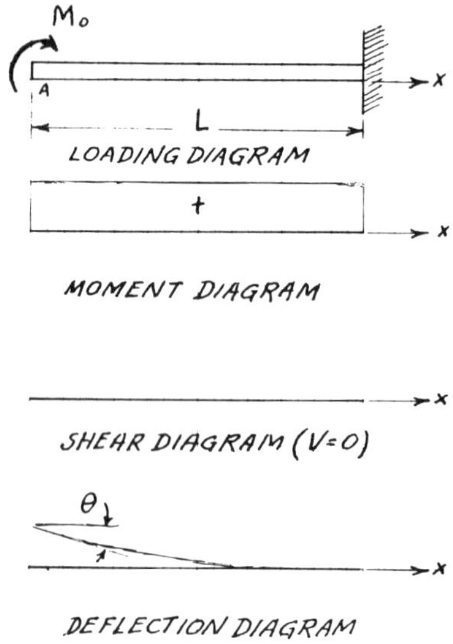

Reactions

$$R_1 = R_2 = 0$$

Moment

$$M_{\max} = M_o \text{ (constant)}$$

Shear

$$V = 0$$

Deflection

$$y_x = \frac{1}{2}\frac{M_o}{EI}(L^2 - 2Lx + x^2)$$

$$y_{\max} = \frac{1}{2}\frac{M_oL^2}{EI} \text{ at } A$$

End Slope

$$\theta = \frac{M_oL}{EI}$$

Max. Resulting Stress in Terms of Loading

$$S_{\max} = \frac{M_oc}{T}$$

21. CANTILEVER, INTERMEDIATE COUPLE

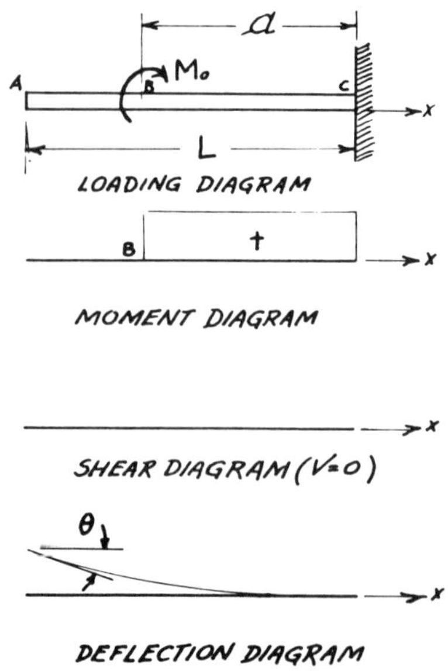

Reactions

$$R_1 = R_2 = 0$$

Moment

$$(A \text{ to } B): M = 0$$

$$(B \text{ to } C): M = M_o$$

$$M_{\max} = M_o$$

Shear

$$V = 0$$

Deflection

$$(A \text{ to } B): y_x = \frac{M_oa}{EI}\left(L - \frac{1}{2}a - x\right)$$

$$(B \text{ to } C): y_x = \frac{1}{2}\frac{M_o}{EI}[(x - L + a)^2 - 2a(x - L + a) + a^2]$$

$$y_{\max} = \frac{M_oa}{EI}\left(L - \frac{1}{2}a\right) \text{ at } A$$

End Slope

$$\theta = \frac{M_oa}{EI}$$

Max. Resulting Stress in Terms of Loading

$$S_{\max} = \frac{M_oc}{I}$$

22. CANTILEVER, PARTIAL TRIANGULAR LOAD

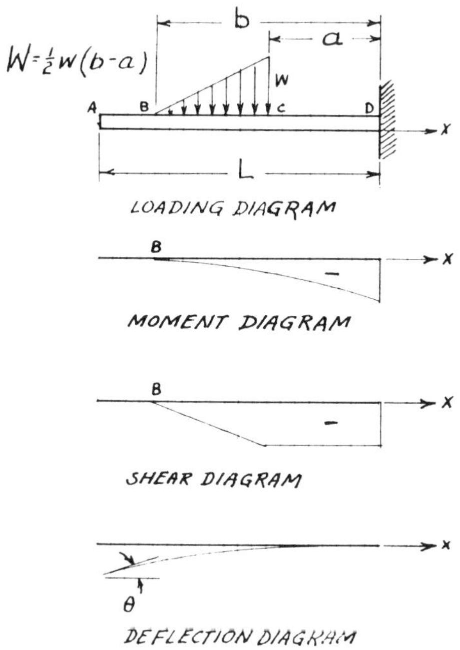

Reactions

$$R_1 = 0$$

$$R_2 = W$$

Moment

$$(A \text{ to } B): M = 0$$

$$(B \text{ to } C): M_x = -\frac{1}{3}\frac{W(x - L + b)^3}{(b - a)^2}$$

$$(C \text{ to } D): M_x = -\frac{1}{3}W(3x - 3L + b + 2a)$$

$$M_{\max} = -\frac{1}{3}W(b + 2a) \text{ at } D$$

Shear

$$(A \text{ to } B): V = 0$$

$$(B \text{ to } C): V_x = -\frac{W(x - L + b)^2}{(b - a)^2}$$

$$V_{\max} = -W$$

Deflection

$$(A \text{ to } B): y_x = -\frac{1}{60}\frac{W}{EI}[(5b^2 + 10ba + 15a^2) \times (L - x) - 4a^3 - 2ab^2 - 3a^2b - b^3]$$

$$(B \text{ to } C): y_x = -\frac{1}{60}\frac{W}{EI}\left[(20a + 10b)(L - x)^2 - 10(L - x)^3 + 5\frac{(L - x - a)^4}{b - a} - \frac{(L - a - x)^5}{(b - a)^2}\right]$$

$$(C \text{ to } D): y_x = -\frac{1}{60}\frac{W}{EI}[(2a + b)(L - x)^2 - (L - x)^3]$$

$$y_{\max} = -\frac{1}{60}\frac{W}{EI}[(5b^2 + 10ba + 15a^2)L - 4a^3 - 2ab^2 - 3a^2b - b^3] \text{ at } A$$

End Slope

$$\theta = \frac{1}{12}\frac{W}{EI}(3a^2 + 2ab + b^2)$$

Max. Resulting Stress in Terms of Loading

$$S_{\max} = \frac{1}{3}\frac{W(b + 2a)c}{I} \text{ at } D$$

23. CANTILEVER, PARTIAL UNIFORM LOAD

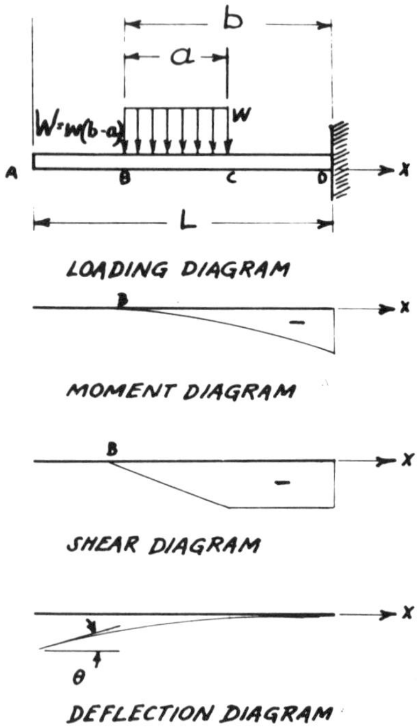

Reactions

$$R_1 = 0$$

$$R_2 = W$$

Moment

$$(A \text{ to } B)\text{: } M = 0$$

$$(B \text{ to } C)\text{: } M_x = -\frac{1}{2}\frac{W}{b-a}(x - L + b)^2$$

$$(C \text{ to } D)\text{: } M_x = -\frac{1}{2}W(2x - 2L + a + b)$$

$$M_{\max} = -\frac{1}{2}W(a + b) \text{ at } D$$

Shear

$$(A \text{ to } B)\text{: } V = 0$$

$$(B \text{ to } C)\text{: } V_x = -\frac{W}{b-a}(x - L + b)$$

$$V_{\max} = -W$$

Deflection

$$(A \text{ to } B)\text{: } y_x = -\frac{1}{24}\frac{W}{EI}[4(a^2 + ab + b^2)(L - x) - a^3 - ab^2 - a^2b - b^3)]$$

$$(B \text{ to } C)\text{: } y_x = -\frac{1}{24}\frac{W}{EI}\left[6(a + b)(L - x)^2 - 4(L - x)^3 + \frac{(L - x - a)^4}{b - a}\right]$$

$$(C \text{ to } D)\text{: } y_x = -\frac{1}{12}\frac{W}{EI}[3(a + b)(L - x)^2 - 2(L - x)^3]$$

$$y_{\max} = -\frac{1}{24}\frac{W}{EI}[4(a^2 + ab + b^2)L - a^3 - ab^2 - a^2b - b^3] \text{ at } A$$

End Slope

$$\theta = \frac{1}{16}\frac{W}{EI}[a^2 + ab + b^2]$$

Max. Resulting Stress in Terms of Loading

$$S_{\max} = \frac{1}{2}\frac{W(a + b)c}{I} \text{ at } D$$

24. ONE END FIXED, ONE END SUPPORTED, UNIFORM LOAD

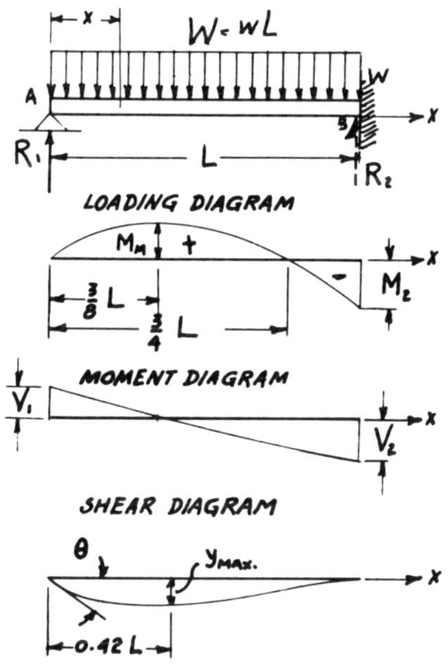

Reactions

$$R_1 = \frac{3}{8}W$$

$$R_2 = \frac{5}{8}W$$

Moment

$$M_x = W\left(\frac{3}{8}x - \frac{1}{2}\frac{x^2}{L}\right)$$

$$M_2 = -\frac{1}{8}WL \text{ at } B$$

$$M_{\max} = \frac{9}{128}WL \text{ at } x = \frac{3}{8}L$$

Shear

$$V_x = W\left(\frac{3}{8} - \frac{x}{L}\right)$$

$$V_1 = \frac{3}{8}W, \; V_2 = -\frac{5}{8}W$$

Deflection

$$y_x = \frac{1}{48}\frac{W}{EIL}(3Lx^3 - 2x^4 - L^3x)$$

$$y_{\max} = -\frac{1}{184}\frac{WL^2}{EI} \text{ at } x = 0.4215L$$

End Slope

$$\theta = \frac{1}{48}\frac{WL^2}{EI}$$

Max. Resulting Stress in Terms of Loading

$$S_{\max} = \frac{1}{8}\frac{WLc}{I}$$

Max. Deflection in Terms of Resulting Stress

$$y_{\max} = -\frac{1}{13.8}\frac{SL^2}{Ec}$$

25. ONE END FIXED, ONE END SUPPORTED, INTERMEDIATE LOAD

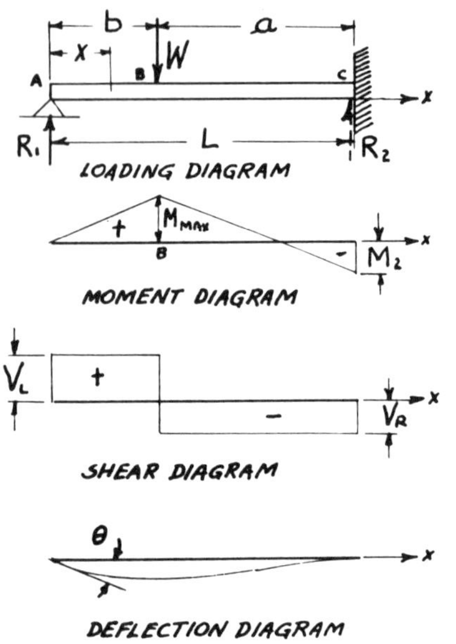

Reactions

$$R_1 = \frac{1}{2}W\left(\frac{3a^2L - a^3}{L^3}\right)$$

$$R_2 = W - R_1$$

Moments

$$(A \text{ to } B)\text{: } M_x \quad R_1 x$$

$$(B \text{ to } C)\text{: } M_x = R_1 x - W(x - L + a)$$

$$+M_{\max} = 0.174WL \text{ at } a = 0.634L$$

$$M_{2(\max)} = -0.1927WL \text{ at } a = 0.422L$$

Shear

$$(A \text{ to } B)\text{: } V_L = R_1$$

$$(B \text{ to } C)\text{: } V_2 = R_1 - W$$

Deflection

$$(A \text{ to } B)\text{: } y_x = \frac{1}{6EI}[R_1(x^3 - 3L^2x) + 3Wa^2x]$$

$$(B \text{ to } C)\text{: } y_x = \frac{1}{6EI}\{R_1(x^3 - 3L^2x) + W[3a^2x - (x - b)^3]\}$$

$$(A \text{ to } B)\text{: } y_{\max}\text{: at 1. } x = L\sqrt{1 - \frac{2L}{3L - a}} \quad \text{for } a < 0.586L$$

$$2.\ x = \frac{L(L^2 + b^2)}{3L^2 - b^2} \text{ for } a > 0.586L$$

$$y_{\max} = -\frac{1}{102}\frac{WL^3}{EI} \text{ at } B, \text{ for } a = 0.586L$$

End Slope

$$\theta = \frac{1}{4}\frac{W}{EI}\left(\frac{a^3}{L} - a^2\right)$$

Max. Resulting Stress in Terms of Loading

$$S_{\max} = \frac{0.1927WLc}{I}$$

Max. Deflection in Terms of Resulting Stress

$$y_{\max} = -\frac{1}{20}\frac{SL^2}{Ec}$$

26. ONE END FIXED, ONE END SUPPORTED, CENTER LOAD

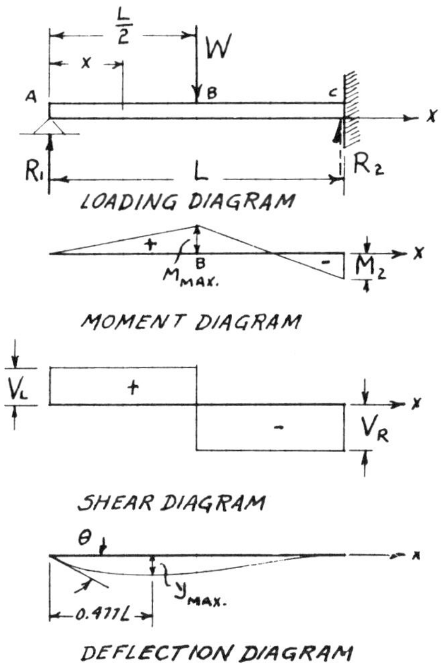

Reactions

$$R_1 = \frac{5}{16}W,\ R_2 = \frac{11}{16}W$$

Moment

$$(A \text{ to } B)\text{: } M_x = \frac{5}{16}Wx$$

$$(B \text{ to } C)\text{: } M_x = W\left(\frac{L}{2} - \frac{11}{16}x\right)$$

$$+M_{\max} = M_c = \frac{5}{32}WL \text{ at } B$$

$$M_2 = -\frac{3}{16}WL \text{ at } C$$

Shear

$$(A \text{ to } B)\text{: } V_L = \frac{5}{16}W$$

$$(B \text{ to } C)\text{: } V_R = -\frac{11}{16}W$$

Deflection

$$(A \text{ to } B)\text{: } y_x = \frac{1}{96}\frac{W}{EI}(5x^3 - 3L^2x)$$

$$(B \text{ to } C)\text{: } y_x = \frac{1}{96}\frac{W}{EI}\left[5x^3 - 16\left(x - \frac{L}{2}\right)^3 - 3L^2x\right]$$

$$y_{\max} = -\frac{1}{107}\frac{WL^3}{EI} \text{ at } x = 0.4472L$$

End Slope

$$\theta = \frac{1}{32}\frac{WL^2}{EI}$$

Resulting Stress in Terms of Loading

$$S_{\max} = \frac{3}{16}\frac{WLc}{I} \text{ at } C$$

Max. Deflection in Terms of Resulting Stress

$$y_{\max} = \frac{1}{12}\frac{SL^2}{Ec}$$

27. ONE END FIXED, ONE END SUPPORTED, TRIANGULAR LOAD

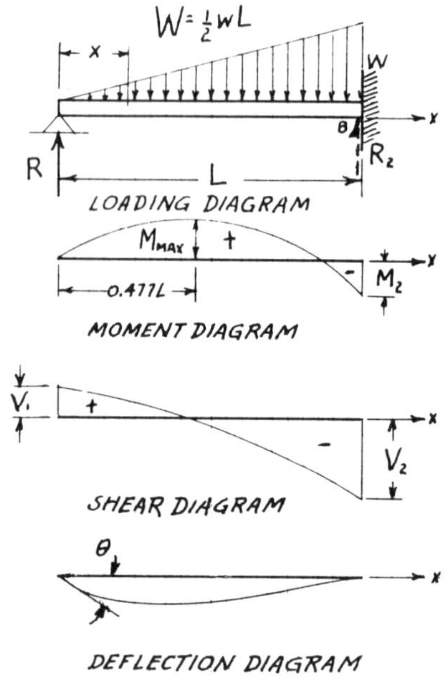

Reaction

$$R_1 = \frac{1}{5}W, \; R_2 = \frac{4}{5}W$$

Moment

$$M_x = W\left(\frac{1}{5}x - \frac{1}{3}\frac{x^3}{L^2}\right)$$

$$+M_{\max} = 0.06WL \text{ at } x = 0.4474L$$

$$-M_{\max} = M_2 = -\frac{2}{15}WL$$

Shear

$$V_x = W\left(\frac{1}{5} - \frac{x^2}{L^2}\right)$$

$$V_1 = \frac{1}{5}W, \; V_2 = \frac{4}{5}W$$

Deflection

$$y_x = \frac{1}{60}\frac{W}{EIL}\left(2Lx^3 - L^3x - \frac{x^5}{L}\right)$$

$$y_{\max} = -\frac{1}{210}\frac{WL^3}{EI} \text{ at } x = L\sqrt{\frac{1}{5}}$$

End Slope

$$\theta = \frac{1}{60}\frac{WL^2}{EI}$$

Max. Resulting Stress in Terms of Loading

$$S_{\max} = \frac{2}{15}\frac{WL}{I} \text{ at } B$$

Max. Deflection in Terms of Resulting Stress

$$y_{\max} = -\frac{1}{28}\frac{SL^2}{Ec}$$

28. ONE END FIXED, ONE END SUPPORTED, TRIANGULAR LOAD

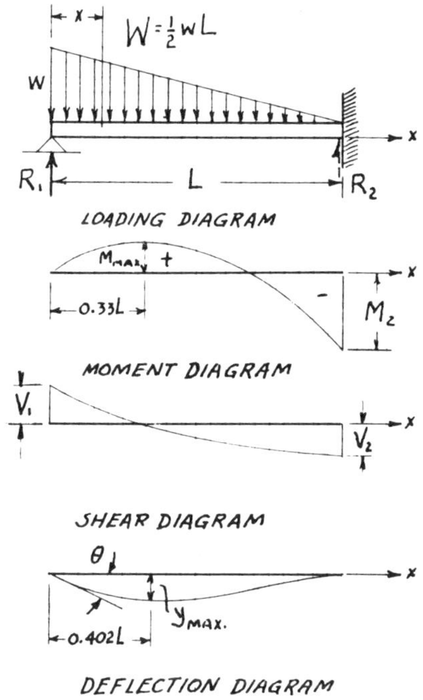

Reaction

$$R_1 = \frac{11}{20}W, \; R_2 = \frac{9}{20}W$$

Moment

$$M_x = W\left(\frac{11}{20}x - \frac{x^2}{L} + \frac{1}{3}\frac{x^3}{L^2}\right)$$

$$+M_{\max} = 0.0846WL \text{ at } x = 0.329L$$

$$-M_{\max} = M_2 = -\frac{7}{60}WL \text{ at } B$$

Shear

$$V_x = W\left(\frac{11}{20} - \frac{2x}{L} + \frac{x^2}{L^2}\right)$$

$$V_1 = \frac{11}{20}W, \; V_2 = -\frac{9}{20}W$$

Deflection

$$y_x = \frac{1}{120}\frac{W}{EIL}\left(11Lx^3 - 3L^3x - 10x^4 + 2\frac{x^5}{L}\right)$$

$$y_{\max} = -\frac{1}{164}\frac{WL^3}{EI} \text{ at } x = 0.402L$$

End Slope

$$\theta = \frac{1}{40}\frac{WL^2}{EI}$$

Max. Resulting Stress in Terms of Loading

$$S_{\max} = \frac{7}{60}\frac{WLc}{I} \text{ at } B$$

Max. Deflection in Terms of Resulting Stress

$$y_{\max} = -\frac{1}{19.1}\frac{SL^2}{Ec}$$

29. ONE END FIXED, ONE END SUPPORTED, END COUPLE

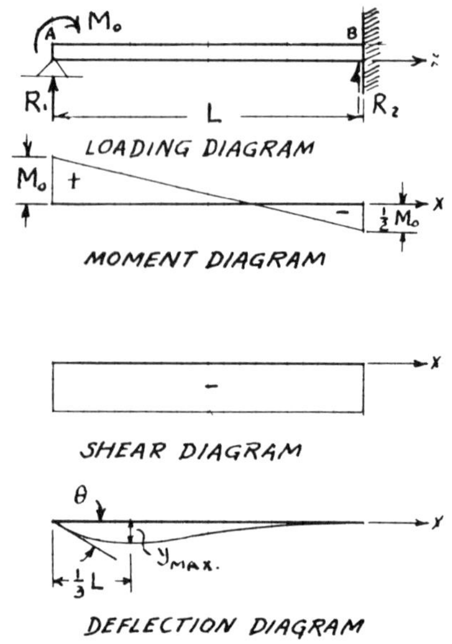

Reactions

$$R_1 = -\frac{3}{2}\frac{M_o}{L}$$

$$R_2 = \frac{3}{2}\frac{M_o}{L}$$

Moment

$$M_x = \frac{1}{2}M_o\left(2 - 3\frac{x}{L}\right)$$

$$+M_{\max} = M_o \text{ at } A$$

$$-M_{\max} = -\frac{1}{2}M_o \text{ at } B$$

Shear

$$V = -\frac{3}{2}\frac{M_o}{L} \text{(constant)}$$

Deflection

$$y_x = \frac{1}{4}\frac{M_o}{EI}\left(2x^2 - \frac{x^3}{L} - Lx\right)$$

$$y_{\max} = -\frac{1}{27}\frac{M_o L^2}{EI} \text{ at } x = \frac{1}{3}L$$

End Slope

$$\theta = \frac{1}{4}\frac{M_o L}{EI} \text{ at } A$$

Max. Resulting Stress in Terms of Loading

$$S_{\max} = \frac{M_o c}{I}$$

30. ONE END FIXED, ONE END SUPPORTED, INTERMEDIATE COUPLE

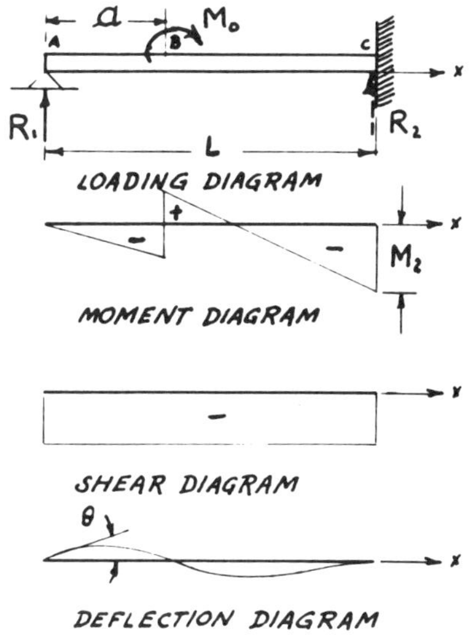

Reactions

$$R_1 = -\frac{3}{2}\frac{M_o}{L}\left(\frac{L^2 - a^2}{L^2}\right)$$

$$R_2 = \frac{3}{2}\frac{M_o}{L}\left(\frac{L^2 - a^2}{L^2}\right)$$

Moment

$$(A \text{ to } B): M_x = R_1 x$$

$$(B \text{ to } C): M_x = R_1 x + M_o$$

$$+M_{\max} = M_o\left[1 - \frac{3a(L^2 - a^2)}{2L^3}\right] \text{ at } B \text{ (to right)}$$

$$M_2 = -\frac{1}{2}M_o\left(1 - 3\frac{a^2}{L^2}\right) \text{ when } a < 0.275L$$

$$-M_{\max} = R_1 a \text{ at } B \text{ (to left) when } a > 0.275L$$

Shear

$$V = R_1 \text{ constant}$$

Deflection

$$(A \text{ to } B): y_x = \frac{M_o}{EI}\left[\frac{L^2 - a^2}{4L^3}(3L^2x - x^3) - (L - a)x\right]$$

$$(B \text{ to } C): y_x = \frac{M_o}{EI}\left[\frac{L^2 - a^2}{4L^3}(3L^2x - x^3) - Lx + \frac{1}{2}(x^2 + a^2)\right]$$

End Slope

$$\theta = \frac{M_o}{EI}\left(a - \frac{1}{4}L - \frac{3}{4}\frac{a^2}{L}\right) \text{ at } A$$

31. ONE END FIXED, ONE END SUPPORTED, PARTIAL UNIFORM LOAD

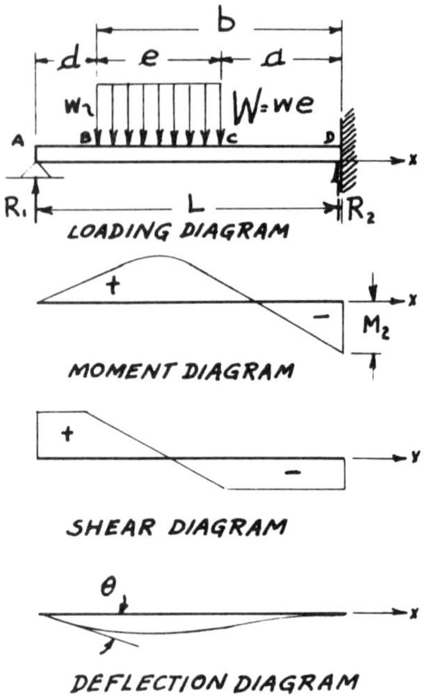

Reactions

$$R_1 = \frac{1}{8}\frac{W}{L^3}[4L(a^2 + ab + b^2) - a^3 - ab^2 - a^2b - b^3]$$

$$R_2 = W - R_1$$

Moment

$$(A \text{ to } B)\text{: } M_x = R_1 x$$

$$(B \text{ to } C)\text{: } M_x = R_1 x - W\frac{(x-d)^2}{2e}$$

$$(C \text{ to } D)\text{: } M_x = R_1 x - W\left(x - d - \frac{1}{2}e\right)$$

$$+M_{\max} = R_1\left(d + \frac{1}{2}\frac{R_1 e}{W}\right) \text{ at } x = d + \frac{R_1}{W}e$$

$$M_2 = -R_1 L + \frac{1}{2}W(a+b)$$

Shear

$$(A \text{ to } B)\text{: } V_x = R_1$$

$$(B \text{ to } C)\text{: } V_x = R_1 - W\left(\frac{x-d}{e}\right)$$

$$(C \text{ to } D)\text{: } V_x = R_1 - W$$

Deflection

$$(A \text{ to } B)\text{: } y_x = \frac{1}{EI}\left[R_1\left(\frac{1}{6}x^3 - \frac{1}{2}L^2x\right) + Wx\left(\frac{1}{2}a^2 + \frac{1}{3}ae + \frac{1}{6}e^2\right)\right]$$

$$(B \text{ to } C)\text{: } = \frac{1}{EI}\left[R_1\left(\frac{1}{6}x^3 - \frac{1}{2}L^2x\right) + Wx\left(\frac{a^2}{2} + \frac{ae}{2} + \frac{e^2}{2}\right) - W\frac{(x-d)^2}{24e}\right]$$

$$(C \text{ to } D)\text{: } y_x = \frac{1}{EI}\left\{R_1\left(\frac{1}{6}x^2 - \frac{1}{2}L^2x + \frac{1}{3}L^3\right) + W\left[\frac{1}{6}\left(a + \frac{1}{2}e\right)^3 - \frac{1}{2}\left(a + \frac{1}{2}e\right)^2 L - \frac{1}{6}\left(x - d - \frac{1}{2}e\right)^3 + \frac{1}{2}\left(a + \frac{1}{2}e\right)^2 x\right]\right\}$$

End Slope

$$\theta = \frac{1}{EI}\left[\frac{1}{2}R_1L^2 - W\left(\frac{1}{2}a^2 + \frac{1}{2}ae + \frac{1}{6}e^2\right)\right] \text{ at } A$$

Max. Resulting Stress in Terms of Loading

$$S_{\max} = \frac{[-R_1L + W\,1/2(a+b)]c}{I}$$

32. ONE END FIXED, ONE END SUPPORTED, PARTIAL TRIANGULAR LOAD

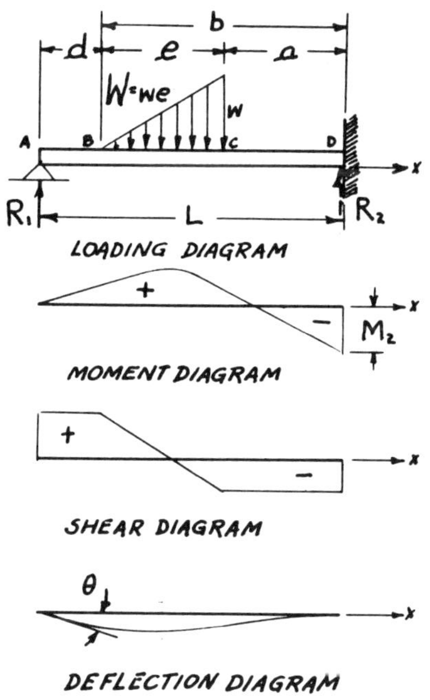

Reactions

$$R_1 = \frac{1}{20}\frac{W}{L^3}\left[(10ab + 15a^2 + 55b^2)L - 4a^3 - 2ab^2 - 3a^2b - b^3\right]$$

$$R_2 = W - R_1$$

Moment

$$(A \text{ to } B)\colon M_x = R_1 x$$

$$(B \text{ to } C)\colon M_x = R_1 x - \frac{1}{3}W\frac{(x-d)^3}{e^2}$$

$$(C \text{ to } D)\colon M_x = R_1 x - W\left(x - d - \frac{2}{3}e\right)$$

$$+M_{\max} = R_1\left(d + \frac{2}{3}e\sqrt{\frac{R_1}{W}}\right) \text{ at } x = d + e\sqrt{\frac{R_1}{W}}$$

$$M_2 = -R_1 L + \frac{1}{3}W(2a + b)$$

Shear

$$(A \text{ to } B)\colon V_x = R_1$$

$$(B \text{ to } C)\colon V_x = R_1 - W\left(\frac{x-d}{e}\right)^2$$

$$(C \text{ to } D)\colon V_x = R_1 - W$$

Deflection

$$(A \text{ to } B)\colon y_x = \frac{1}{EI}\left[R_1\left(\frac{1}{6}x^3 - \frac{1}{2}L^2x\right) + Wx\left(\frac{1}{2}a^2 + \frac{1}{3}ae + \frac{1}{12}e^2\right)\right]$$

$$(B \text{ to } C)\colon y_x = \frac{1}{EI}\left[R_1\left(\frac{x^3}{6} - \frac{L^2x}{2}\right) + Wx\left(\frac{a^2}{2} + \frac{ae}{3} + \frac{e^2}{12}\right) - W\frac{(x-d)^5}{60e^2}\right]$$

$$(C \text{ to } D)\colon y_x = \frac{1}{EI}\left\{R_1\left(\frac{1}{6}x^3 - \frac{1}{2}L^2x + \frac{1}{3}L^3\right) + W\left[\frac{1}{2}\left(a + \frac{1}{3}e\right)^2 x - \frac{1}{6}\left(x - d - \frac{2}{3}e\right)^3 + \frac{1}{6}\left(a + \frac{1}{3}e\right)^3 - \frac{1}{2}\left(a + \frac{1}{3}e\right)^2 L\right]\right\}$$

End Slope

$$\theta = \frac{1}{EI}\left[\frac{1}{2}R_1L^2 - W\left(\frac{1}{12}e^2 + \frac{1}{3}ae + \frac{1}{2}a^2\right)\right] \text{ at } A$$

Max. Resulting Stress in Terms of Loading

$$S_{\max} = \frac{[R_1L + 1/3\,W(2a + b)]c}{I}$$

33. ONE END FIXED, ONE END SUPPORTED, PARTIAL TRIANGULAR LOAD

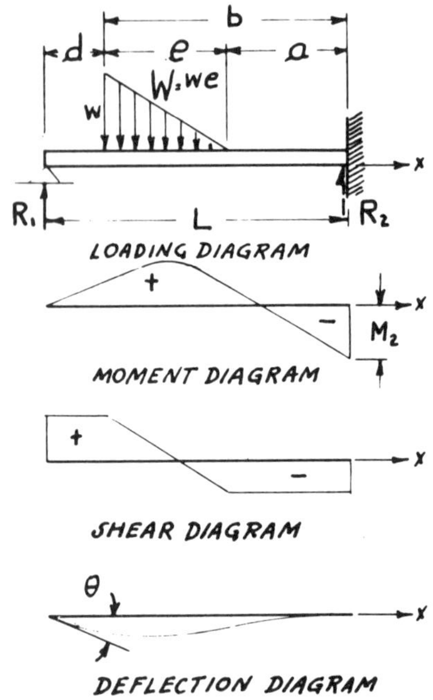

Moment

$(A \text{ to } B)\colon M_x = R_1 x$

$(B \text{ to } C)\colon M_x = R_1 x - \dfrac{(x-d)^2}{e} W + \dfrac{(x-d)^3}{3e^2} W$

$(C \text{ to } D)\colon M_x = R_1 x - W\left(x - d - \dfrac{1}{3}e\right)$

$$+M_{max} = R_1\left(d + e - e\sqrt{1 - \frac{R_1}{W}}\right) - \frac{1}{3}We\left(1 - \sqrt{1 - \frac{R_1}{W}}\right)^2\left(2 + \sqrt{1 - \frac{R_1}{W}}\right)$$

$$\text{at } x = d + e\left(1 - \sqrt{1 - \frac{R_1}{W}}\right)$$

$$M_2 = -R_1 L + \frac{1}{3}W(a + 2b)$$

Deflection

$$(A \text{ to } B)\colon y_x = \frac{1}{EI}\left\{R_1\left(\frac{x^3}{6} - \frac{L^2 x}{2}\right) + Wx\left(\frac{1}{2}a^2 + \frac{2}{3}ae + \frac{1}{4}e^2\right)\right\}$$

$$(B \text{ to } C)\colon y_x = \frac{1}{EI}\left\{R_1\left(\frac{x^3}{6} - \frac{L^2 x}{2}\right) + W\left[\frac{1}{60}\frac{(x-d)^5}{e^2} - \frac{1}{12}\frac{(x-d)^4}{e} + \left(\frac{1}{2}a^2 + \frac{2}{3}ae + \frac{1}{4}e^2\right)x\right]\right\}$$

$$(C \text{ to } D)\colon y_x = \frac{1}{EI}\left\{R_1\left(\frac{x^3}{6} - \frac{L^2 x}{2} + \frac{L^3}{6}\right) - \frac{1}{2}W(L-x)^2\left[\left(a + \frac{2}{3}e\right) - \frac{L-x}{3}\right]\right\}$$

End Slope

$$\theta = \frac{1}{EI}\left[\frac{1}{2}R_1 L^2 - W\left(\frac{1}{2}a^2 + \frac{2}{3}ae + \frac{1}{4}e^2\right)\right] \text{ at } A$$

Max. Resulting Stress in Terms of Loading

$$S_{max} = \frac{[-R_1 L + 1/3\, W(a + b)]c}{I}$$

Reactions

$$R_1 = \frac{1}{20}\frac{W}{L^3}[(10ab + 5a^2 + 15b^2)L - a^3 - 2a^2 b - 3a^2 b - 4b^3]$$

$$R_2 = W - R_1$$

34. BOTH ENDS FIXED, UNIFORM LOAD

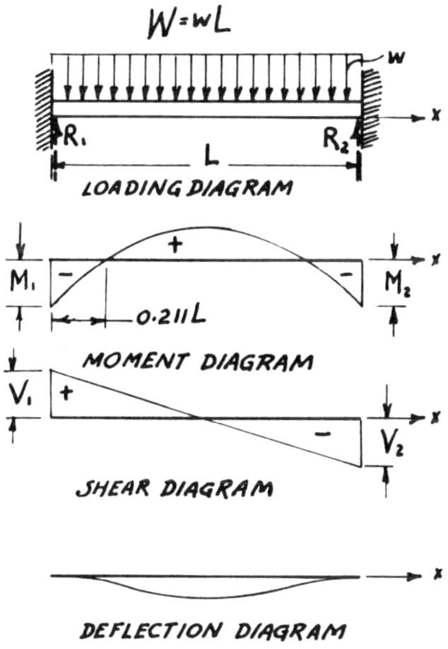

Reactions

$$R_1 = R_2 = \frac{W}{2}$$

Moment

$$M_x = \frac{1}{2}W\left(x - \frac{x^2}{L} - \frac{1}{6}L\right)$$

$$+M_{max} = \frac{1}{24}WL \text{ at } x = \frac{1}{2}L$$

$$-M_{max} = M_2 = -\frac{1}{12}WL = M_1$$

Shear

$$V_x = \frac{1}{2}W\left(1 - \frac{2x}{L}\right)$$

$$V_1 = \frac{1}{2}W, \; V_2 = -\frac{1}{2}W$$

Deflection

$$y_x = \frac{1}{24}\frac{W_x{}^2}{EIL}(2Lx - L^2 - x^2)$$

$$y_{max} = -\frac{1}{384}\frac{WL^3}{EI} \text{ at } x = \frac{L}{2}$$

Max. Resulting Stress in Terms of Loading

$$S_{max} = \frac{1}{12}\frac{WLc}{I} \text{ at } x = \frac{L}{2}$$

Max. Deflection in Terms of Resulting Stress

$$y_{max} = -\frac{SL^2}{32Ec}$$

35. BOTH ENDS FIXED, INTERMEDIATE LOAD

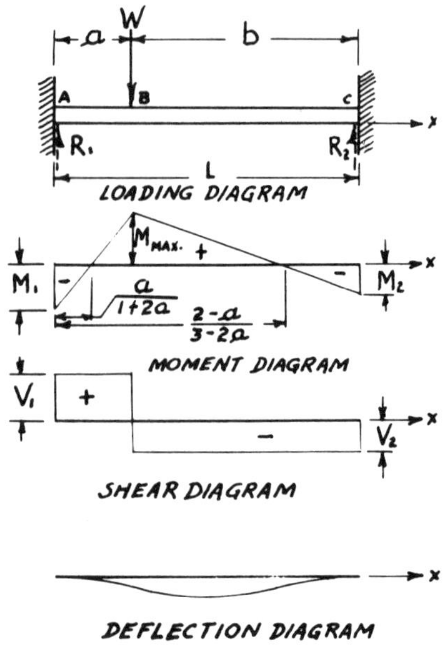

Reactions

$$R_1 = \frac{Wb^2}{L^3}(3a + b); \; R_2 = \frac{Wa^2}{L^3}(3b + a)$$

Moment

$$(A \text{ to } B): M_x = -W\frac{ab^2}{L^2} + R_1x$$

$$(B \text{ to } C): M_x = -W\frac{ab^2}{L^2} + R_1x - W(x - a)$$

$$+M_{\max} = -W\frac{ab^2}{L^2} + R_1a \text{ at } B,$$

$$M_1 = -W\frac{ab^2}{L^2}$$

$$M_2 = -W\frac{a^2b}{L^2}$$

Shear

$$V_1 = R_1, \; V_2 = -R_2$$

Deflection

$$(A \text{ to } B): y_x = \frac{1}{6}\frac{Wb^2x^2}{EIL^3}(3ax + bx - 3aL)$$

$$(B \text{ to } C): y_x = \frac{1}{6}\frac{Wa^2(L - x)^2}{EIL^2}[(3b + a)(L - x) - 3bL]$$

$$y_{\max} = -\frac{2}{3}\frac{W}{EI}\frac{a^3b^2}{(3a + b)^3} \text{ at } x = \frac{3ab}{3a + b} \text{ if } a > b$$

$$y_{\max} = -\frac{2}{3}\frac{W}{EI}\frac{a^2b^3}{(3b + a)^2} \text{ at } x = L - \frac{2bL}{3b + a} \text{ if } a < b$$

36. BOTH ENDS FIXED, CENTER LOAD

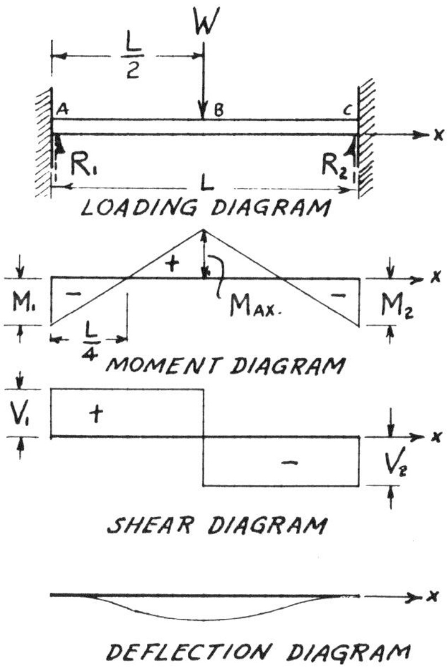

Reactions

$$R_1 = R_2 = \frac{W}{2}$$

Moment

$$(A \text{ to } B)\text{: } M_x = \frac{1}{8} W(4x - L)$$

$$(B \text{ to } C)\text{: } M_x = \frac{1}{8} W(3L - 4x)$$

$$+M_{\max} = \frac{1}{8} WL \text{ at } B$$

$$M_1 = M_2 = -\frac{1}{8} WL$$

Shear

$$(A \text{ to } B)\text{: } V_1 = W$$

$$(B \text{ to } C)\text{: } V_1 = -W$$

Deflection

$$(A \text{ to } B)\text{: } y_x = -\frac{1}{48}\frac{W}{EI}(3Lx^2 - 4x^3)$$

$$y_{\max} = -\frac{1}{192}\frac{WL^3}{EI} \text{ at } B$$

Max. Resulting Stress in Terms of Loading

$$S_{\max} = \frac{1}{8}\frac{WLc}{I}$$

Max. Deflection in Terms of Resulting Stress

$$y_{\max} = -\frac{1}{24}\frac{SL^2}{Ec}$$

37. BOTH ENDS FIXED, TRIANGULAR LOAD

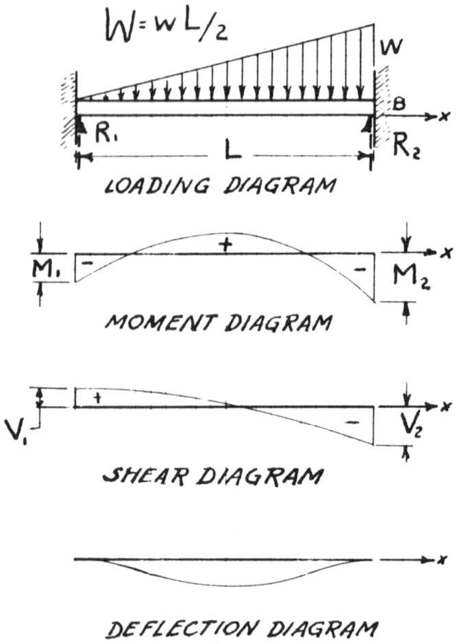

Reactions

$$R_1 = \frac{3}{10} W$$

$$R_2 = \frac{7}{10} W$$

Moment

$$M_x = W\left(\frac{3}{10}x - \frac{1}{15}L - \frac{1}{3}\frac{x^3}{L^2}\right)$$

$$+M_{max} = 0.043\, WL \text{ at } x = 0.548L$$

$$-M_{max} = -\frac{1}{10} WL \text{ at } B$$

$$M_1 = \frac{1}{15} WL$$

$$M_2 = \frac{1}{10} WL$$

Shear

$$V_x = W\left(\frac{3}{10} - \frac{x^2}{L^2}\right)$$

$$V_1 = \frac{3}{10} W$$

$$V_2 = -\frac{7}{10} W$$

Deflection

$$y_x = \frac{1}{60}\frac{W}{EI}\left(3x^3 - 2Lx^2 - \frac{x^5}{2}\right)$$

$$y_{max} = -0.002617 \frac{WL^3}{EI} \text{ at } x = 0.525L$$

Max. Resulting Stress in Terms of Loading

$$S_{max} = \frac{WLc}{10I}$$

Max. Deflection in Terms of Resulting Stress

$$y_{max} = -\frac{SL^2}{38.2\, Ec}$$

38. BOTH ENDS FIXED, PARTIAL UNIFORM LOAD

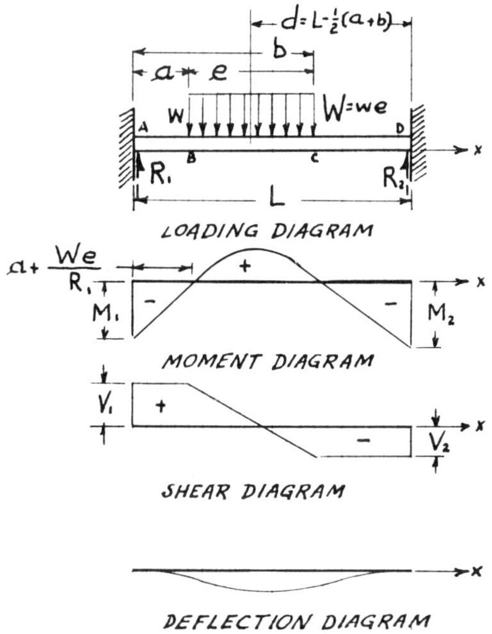

Reactions

$$R_1 = \frac{1}{4}\frac{W}{L^2}\left(12d^2 - 8\frac{d^3}{L} + 2\frac{be^2}{L} - \frac{e^3}{L} - e^2\right)$$

$$R_2 = W - R_1$$

Moment

$$M_1 = -\frac{1}{24}\frac{W}{L}\left(24\frac{d^3}{L} - 6\frac{be^2}{L} + 3\frac{e^3}{L} + 4e^2 - 24d^2\right)$$

$$M_2 = \frac{1}{24}\frac{W}{L}\left(24\frac{d^3}{L} - 6\frac{be^2}{L} + 3\frac{e^3}{L} + 2e^2 - 48d^2 + 24dL\right)$$

$$(A \text{ to } B){:}\ M_x = -M_1 + R_1 x$$

$$(B \text{ to } C){:}\ M_x = -M_1 + R_1 x - \frac{1}{2}W\frac{(x-a)^2}{e}$$

$$(C \text{ to } D){:}\ M_x = -M_1 + R_1 x - W(x - L + d)$$

$$-M_{\max} = -M_1 \text{ when } a < L - b$$

$$-M_{\max} = -M_2 \text{ when } a > L - b$$

Deflection

$$(A \text{ to } B){:}\ y_x = \frac{1}{6EI}(R_1x^3 - 3M_1x^2)$$

$$(B \text{ to } C){:}\ y_x = \frac{1}{6EI}\left(R_1x^3 - 3M_1x^2 - \frac{1}{4}W\frac{(x-a)^4}{e}\right)$$

$$(C \text{ to } D){:}\ y_x = \frac{1}{6EI}[R_1(L-x)^3 - 3M_2(L-x)^2]$$

Shear

$$V_1 = R_1$$

$$V_2 = R_1 - W$$

$$(B \text{ to } C){:}\ V_x = R_1 - W\frac{x-a}{e}$$

39. BOTH ENDS FIXED, PARTIAL TRIANGULAR LOAD

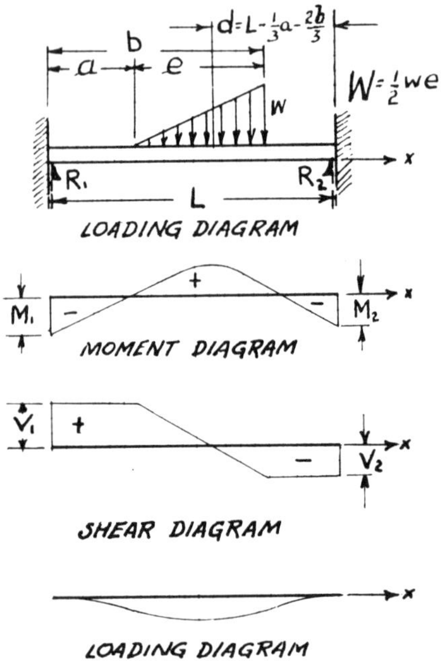

Reactions

$$R_1 = \frac{W}{L^2}\left(3d^2 - \frac{1}{6}e^2 + \frac{1}{3}\frac{be^2}{L} - \frac{17}{135}\frac{e^3}{L} - 2\frac{d^3}{L}\right)$$

$$R_2 = W - R_1$$

Moment

$$M_1 = -\frac{W}{L}\left(\frac{d^3}{L} + \frac{1}{9}e^2 + \frac{51}{810}\frac{e^3}{L} - \frac{1}{6}\frac{e^2 b}{L} - d^2\right)$$

$$M_2 = \frac{W}{L}\left(\frac{d^3}{L} + \frac{1}{18}e^2 + \frac{51}{810}\frac{e^3}{L} - \frac{1}{6}\frac{e^2 b}{L} - 2d^2 + dL\right)$$

$$(A \text{ to } B)\text{: } M_x = -M_1 + R_1 x$$

$$(B \text{ to } C)\text{: } M_x = -M_1 + R_1 x - W\frac{(x-a)^3}{3e^2}$$

$$(C \text{ to } D)\text{: } M_x = -M_1 + R_1 x - W(x - L + d)$$

$$+M_{\max} \text{ at } x = a + c\sqrt{\frac{R_1}{W}}$$

$$-M_{\max} = M_1 \text{ when } d > \frac{L}{2}$$

$$-M_{\max} = -M_2 \text{ when } d < \frac{L}{2}$$

Shear

$$(B \text{ to } C)\text{: } V_x = R_1 - W\frac{(x-a)^2}{e^2}$$

$$V_1 = R_1,\ V_2 = R_1 - W$$

Deflection

$$(A \text{ to } B)\text{: } y_x = \frac{1}{6EI}(R_1 x^3 - 3M_1 x^2)$$

$$(B \text{ to } C)\text{: } y_x = \frac{1}{EI}\left(\frac{1}{6}R_1 x^3 - \frac{1}{2}M_1 x^2 - \frac{1}{60}W\frac{(x-a)^5}{e}\right)$$

$$(C \text{ to } D)\text{: } y_x = \frac{1}{6EI}\{R_1(x^3 - 3L^2 x + 2L^3)$$

$$-3M_1(L-x)^2 + W[3d^2 x + d^3 - 3d^2 L$$

$$-(x - L + d)^3]\}$$

40. BOTH ENDS FIXED, INTERMEDIATE COUPLE

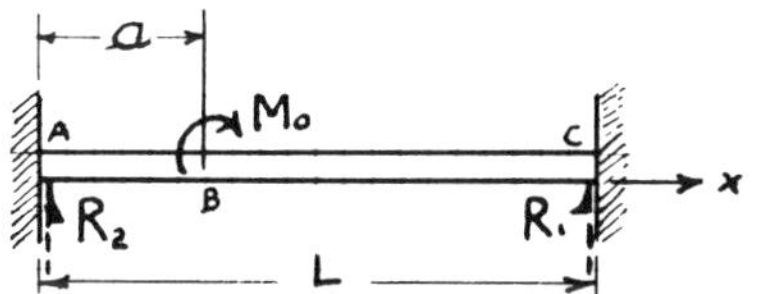

LOADING DIAGRAM

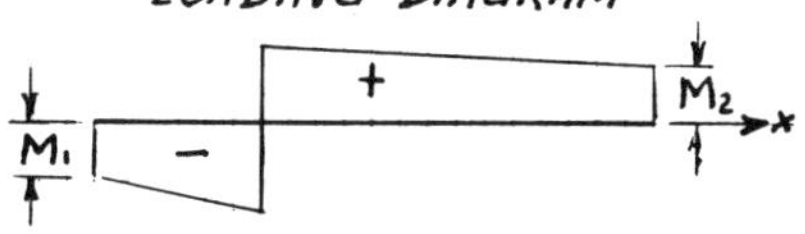

MOMENT DIAGRAM.

SHEAR DIAGRAM

DEFLECTION DIAGRAM

Reactions

$$R_1 = -6\frac{M_o}{L^3}(aL - a^2)$$

$$R_2 = 6\frac{M_o}{L^3}(aL - a^2)$$

Moment

$$M_1 = -\frac{M_o}{L^2}(4La - 3a^2 - L^2)$$

$$M_2 = \frac{M_o}{L^2}(2La - 3a^2)$$

$$(A \text{ to } B)\colon M_x = -M_1 + R_1 x$$

$$(B \text{ to } C)\colon M_x = -M_1 + R_1 x + M_o$$

$$+M_{\max} = M_o\left(4\frac{a}{L} - 9\frac{a^2}{L^2} + 6\frac{a^3}{L^3}\right) \text{ just right of } B$$

$$-M_{\max} = M_o\left(4\frac{a}{L} - 9\frac{a^2}{L^2} + 6\frac{a^3}{L^3} - 1\right) \text{ just left of } B$$

Shear

$$V = R_1$$

Deflection

$$(A \text{ to } B)\colon y_x = -\frac{1}{6EI}(3M_1x^2 - R_1x^3)$$

$$(B \text{ to } C)\colon y_x = \frac{1}{6EI}[(M_o - M_1)(3x^2 - 6Lx + 3L^2) - R_1(3L^2x - x^3 - 2L^3)]$$

$$+y_{\max} \text{ at } x = \frac{2M_1}{R_1} \text{ if } a > \frac{1}{3}L$$

$$-y_{\max} \text{ at } x = L - \frac{2M_2}{R_2} \text{ if } a < \frac{2}{3}L$$

20. MAXIMUM DEFLECTION; GRAPHICAL SOLUTION

(*Ref. 4*)

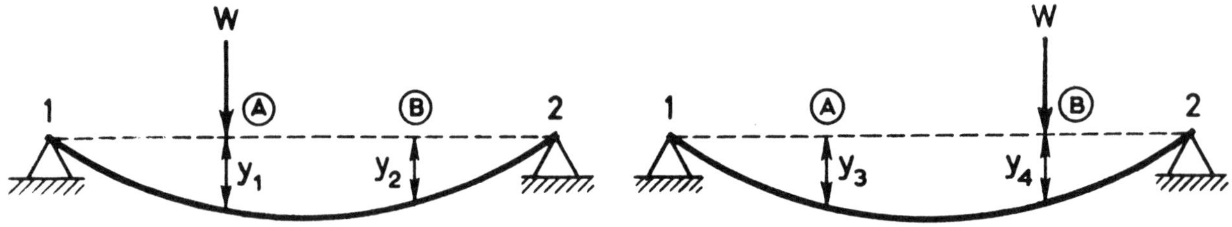

(a) Fig. 8 (b)

Maxwell's theorem of reciprocal displacements establishes a mutual relation between the deflections at any two points in a beam. This theorem states that if the load W at A causes a deflection y_2 at B and the same load W at B causes a deflection y_3 at A, then $y_2 = y_3$ (Fig. 8).

The theorem of reciprocal displacement is very useful while computing the deflection of beams where combinations of concentrated and uniform loading would lead to a long and tedious solution by ordinary methods. The method developed here eliminates all the tedium. For example, assume a simple beam supporting two concentrated loads at A and B. According to the Maxwell's theorem, in order to get the deflection at the center of the beam, O, we place the load W at O and calculate the deflection at A, then we place load W_1 at O and calculate the deflection at B. The sum of these two deflections will give very closely the total deflection produced at O by the two loads. Such calculation, which by ordinary method is lengthy, may readily and rapidly be made by means of graphs and one general equation:

$$y = \frac{WL^3}{6EI}(B), \tag{15}$$

where B is a coefficient which is a function of two constants, K and C. The constant K is a fraction which expresses the distance from one support to the point of deflection in terms of the span length, and C is a corresponding fraction expressing the distance from a support to the load point.

B versus K, for different values of C, were plotted for 16 different most commonly used loading and support conditions.

Example 1: Find the maximum deflection in the beam shown in Fig. 9, ($E = 30 \times 10^6$ psi, $I = 100$ in.^{4}).

Solution: It can be proved that no matter where a concentrated load may be placed on a simple beam, the point of maximum deflection cannot be more remote from the center of the beam than $0.0773L$. It is therefore evident that the mid-span deflection and the maximum deflection for any load, regardless of the position of that on the span, are very nearly equal. Therefore, $K = L/2/L = 0.5$. For the load at A: $C = 4.5/15 = 0.3$. For the load at B: $C = 15\text{–}3/18 = 0.8$.

Thus, from the appropriate curve (Fig. 12)

For $K = 1/2$ and $C = 0.3$, we find: $B = 0.099$.

For $K = 1/2$ and $C = 0.8$, we find: $B = 0.071$.

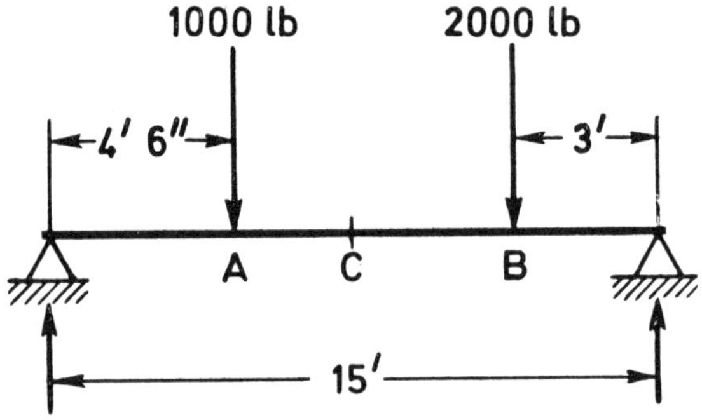

Fig. 9

It should be noted that because of symmetry involved the value of B for $C = 0.8$ is the same as for $C = 0.2$. From equation (15),

$$y = \frac{L^3}{6EI}[1{,}000 \times 0.099 + 2{,}000 \times 0.071]$$

or

$$y = \frac{15^3 \times 1728 \times 241}{6 \times 30 \times 10^6 \times 100} = 0.078 \text{ in.}$$

Example 2: Find the maximum deflection for the beam with a partial uniform loading, shown in Fig. 10 ($E = 30 \times 10^6$ psi, $I = 100$ in.^{4}).

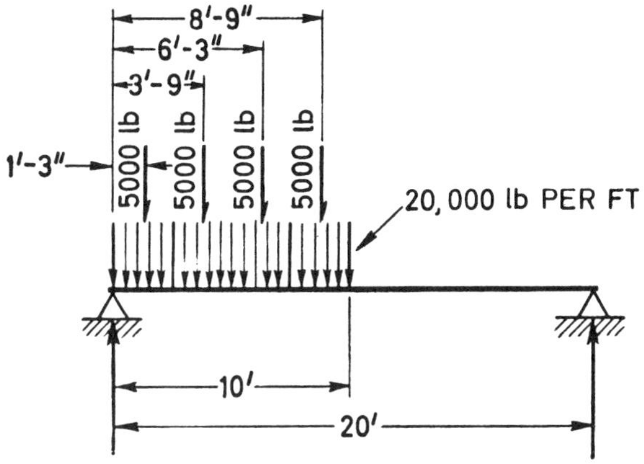

Fig. 10

Solution: When the applied load consists, as in the case of this example, of a uniform load covering a part of the span, the load may be broken up into short sections which in turn may be replaced by the total load in that section as applied at its center of gravity. This was done here by dividing the uniform load into four sections. Since the maximum deflection is at the center of the beam, $K = 0.5$

$$\left.\begin{aligned} C_1 &= \frac{1.25}{20} = 0.0625; \quad B_1 = 0.023 \\ C_2 &= \frac{3.75}{20} = 0.1875; \quad B_2 = 0.068 \\ C_3 &= \frac{6.25}{20} = 0.3125; \quad B_3 = 0.105 \\ C_4 &= \frac{8.75}{20} = 0.4375; \quad B_4 = 0.122 \end{aligned}\right\} \text{From Fig. 12 by interpolation.}$$

Thus

$$y = \frac{L^3}{6EI}[0.023 + 0.068 + 0.105 + 0.122]\ 5{,}000$$

or

$$y = \frac{1590L^3}{6EI} = \frac{1590 \times 20^3 \times 1728}{6 \times 30 \times 10^6 \times 100} = 1.5 \text{ in.}$$

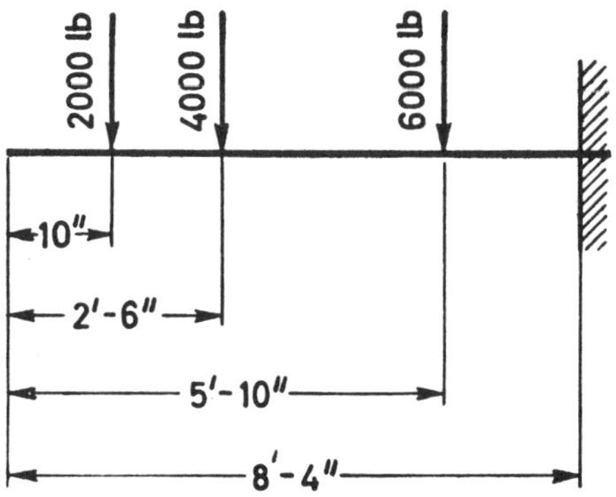

Fig. 11

Example 3: Compute the deflection of the free end of the cantilever beam shown in Fig. 11 ($E = 30 \times 10^6$ psi, $I = 100$ in.^{4}).

Solution: For end deflection $K = 0$

$$\left.\begin{aligned} C_1 &= \frac{10}{100} = 0.1; \quad B_1 = 1.70 \\ C_2 &= \frac{30}{100} = 0.3; \quad B_2 = 1.13 \\ C_3 &= \frac{70}{100} = 0.7; \quad B_3 = 0.24 \end{aligned}\right\} \text{Fig. 17}$$

Thus

$$y = \frac{L^3}{6EI}[2{,}000 \times 1.70 + 4{,}000 \times 1.13 + 6{,}000 \times 0.24]$$

$$y = \frac{9360L^3}{6EI} = \frac{9360 \times 100^3}{6 \times 30 \times 10^6 \times 100} = 0.52 \text{ in.}$$

Fig. 12

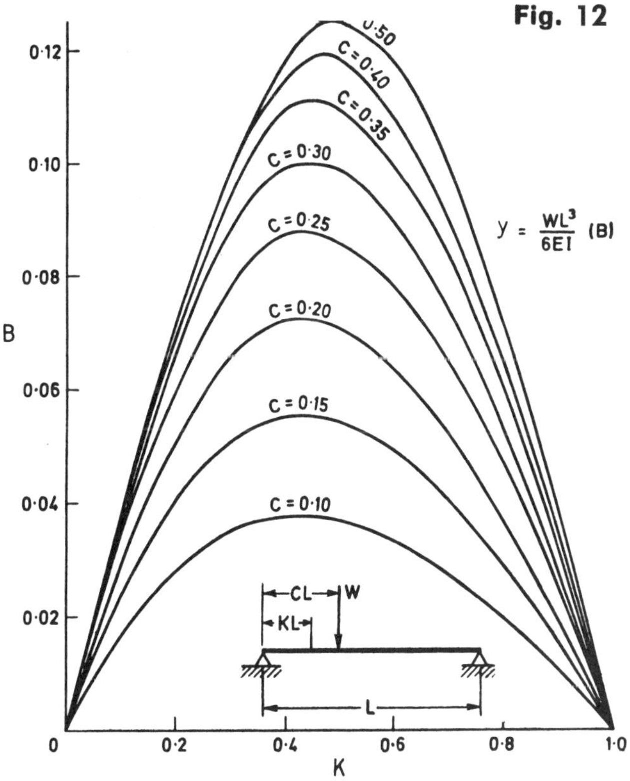

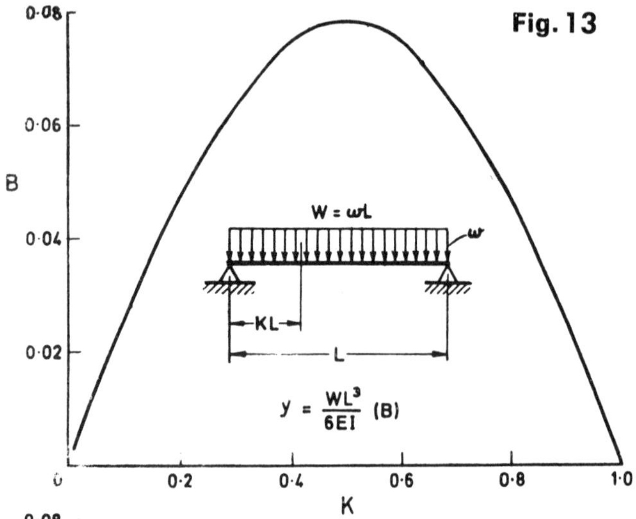

Fig. 13

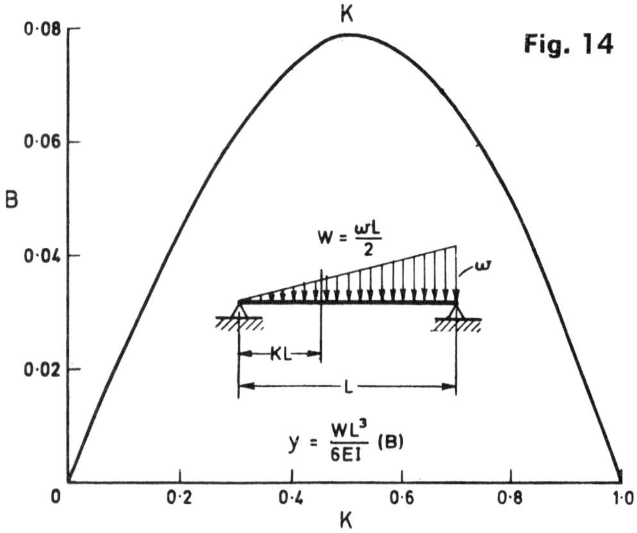

Fig. 14

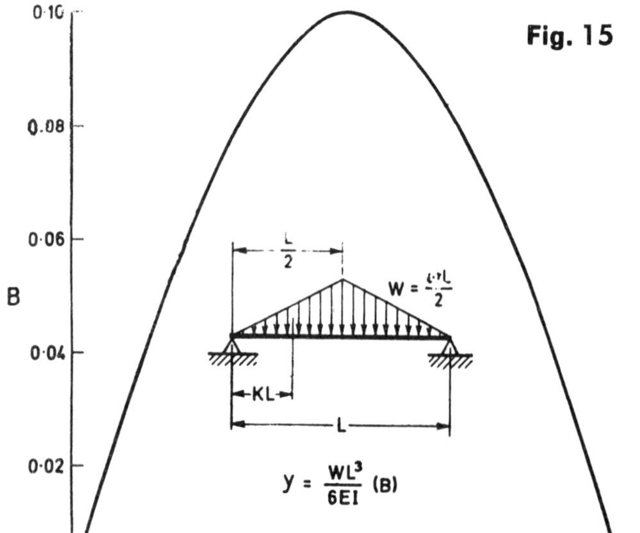

Fig. 15

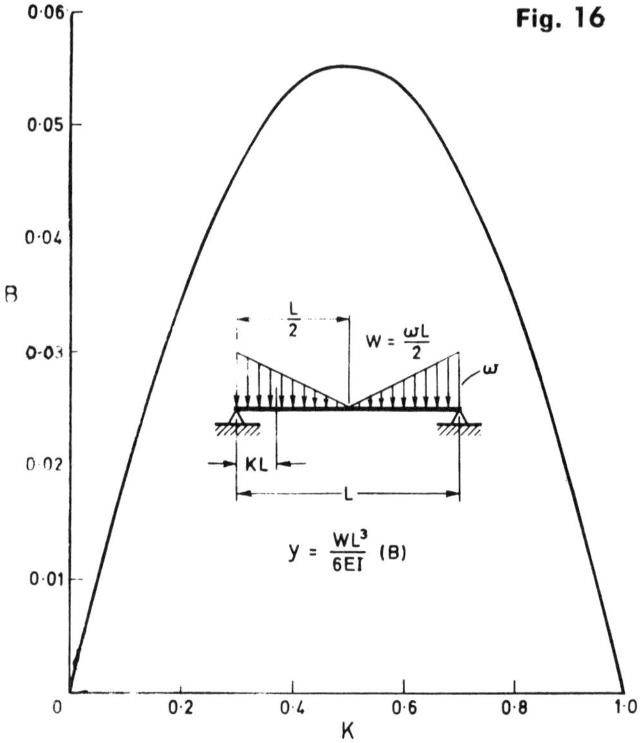

Fig. 16

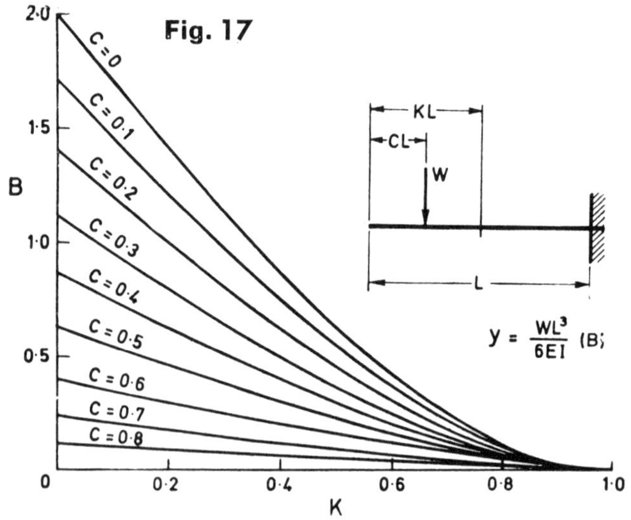

Fig. 17

Fig. 18

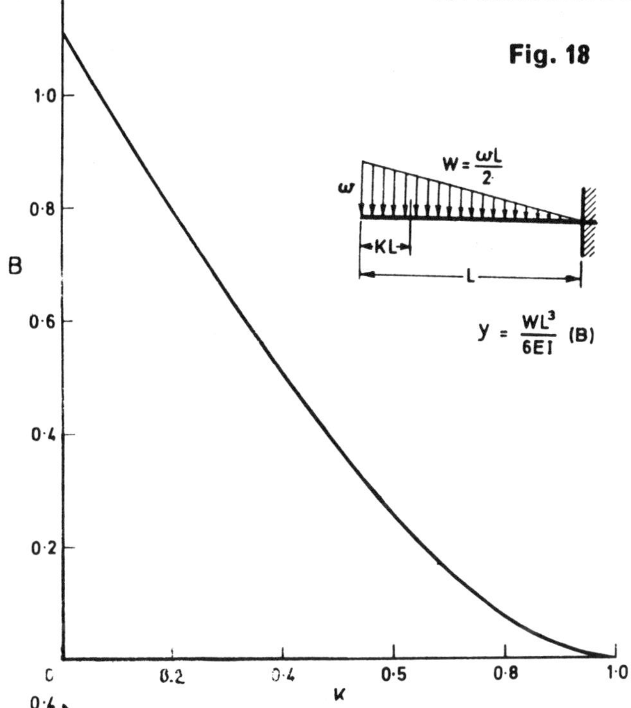

Fig. 21

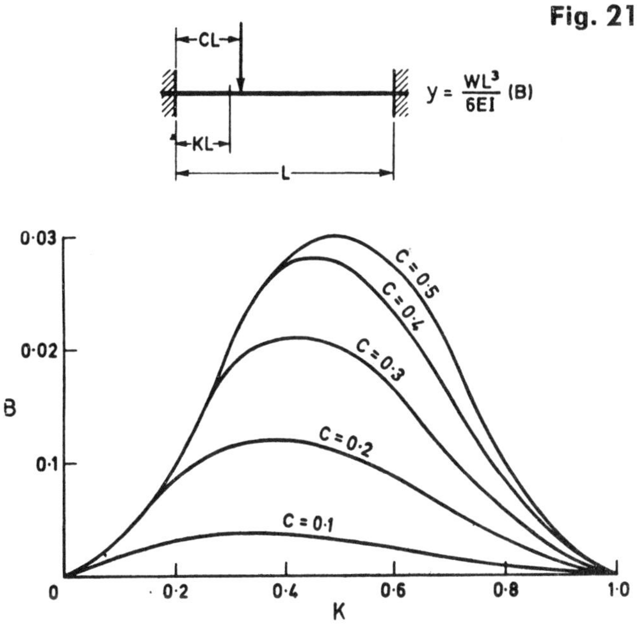

Fig. 19

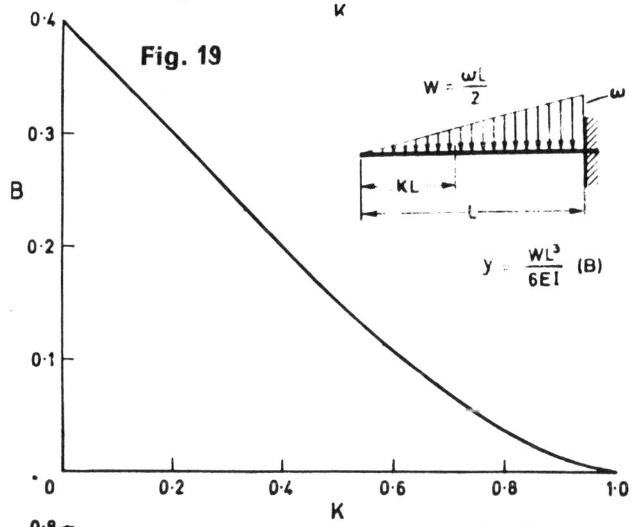

Fig. 22

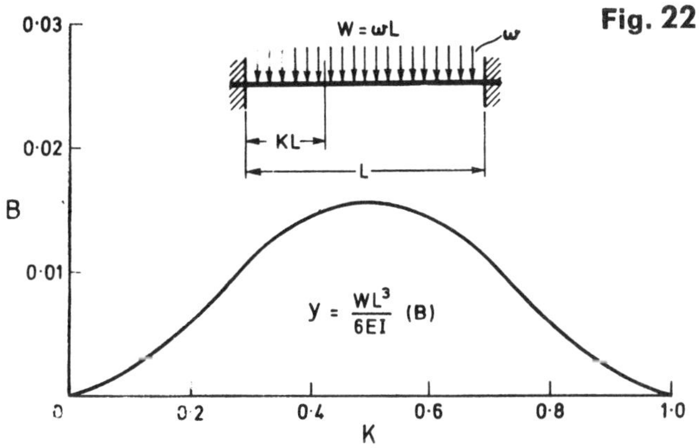

Fig. 20

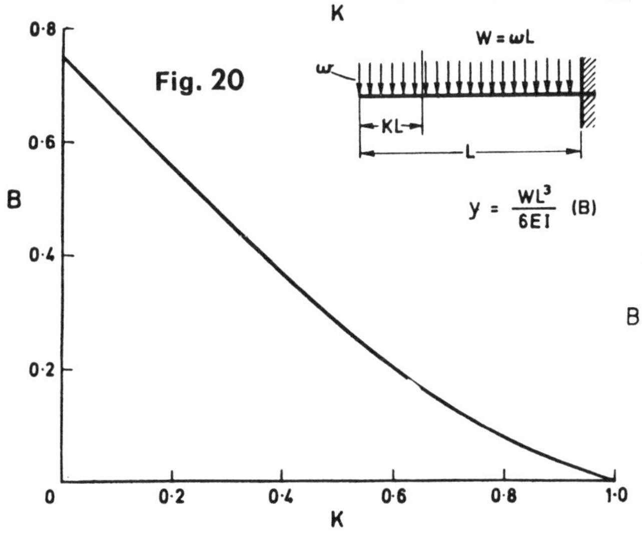

Fig. 23

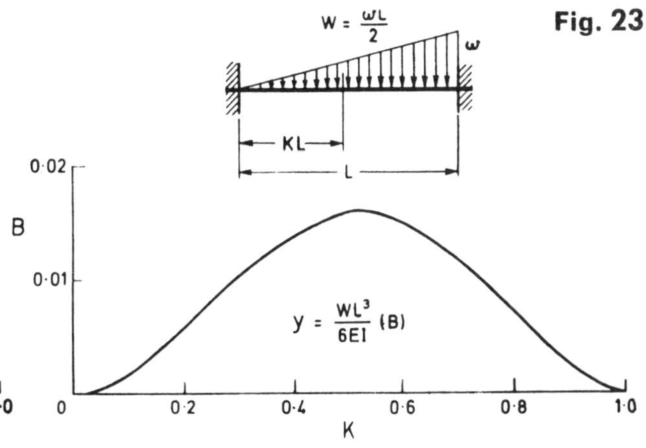

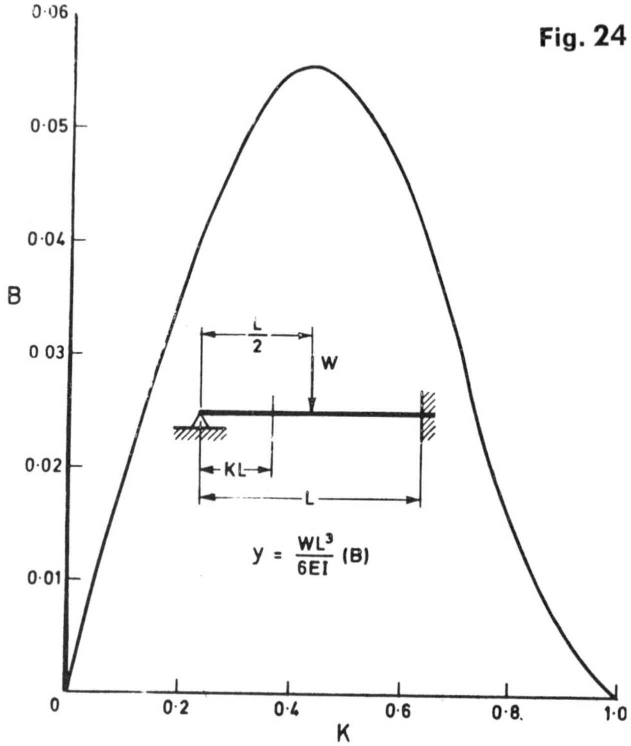

Fig. 24

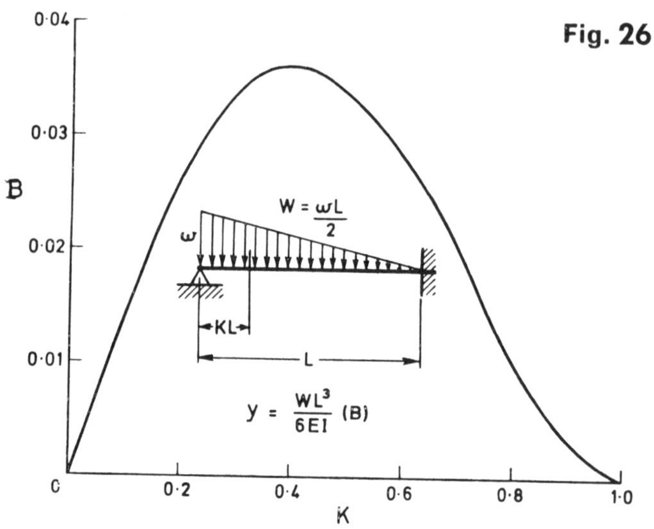

Fig. 26

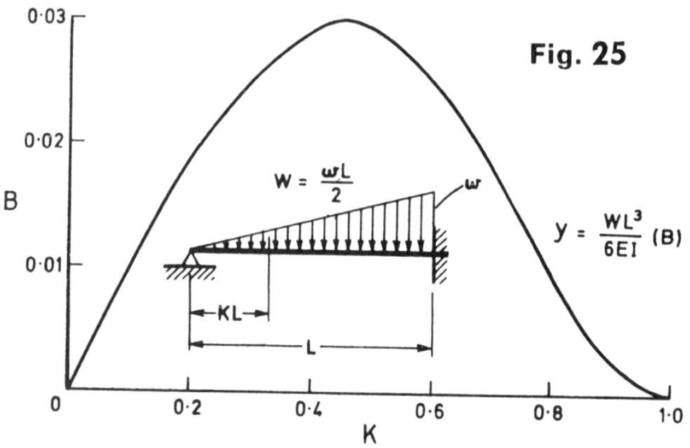

Fig. 25

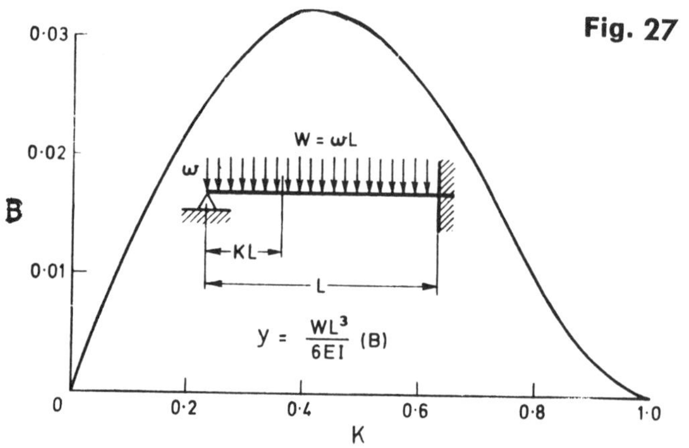

Fig. 27

21. BEAMS OF CONSTANT STRENGTH THROUGHOUT THEIR LENGTH

(*Ref.* 5)

A constant-strength beam has the same maximum bending stress at every section, with enough material at every cross section. Since $S = Mc/I$, this is done by varying the shape so that I/c changes while S is constant, whatever the load on the beam may be.

Suppose a beam fixed at one end and free at the other carries a load P at the free end and no other load, the weight of the beam itself being negligible. We shall show how to make the beam of constant strength.

Measuring x from the load P we find, by taking moments about any section, that

$$M = Px.$$

Since

$$M = IS/c, \qquad I/xc = P/S = \text{Const.}$$

There are many ways of making I proportional to c. Suppose we choose a rectangular section of depth h and width b. Then $c = \frac{1}{2}h$ and $I = bh^3/12$. Therefore, we must make bh^2 proportional to x. We could make h constant and b proportional to x. This would be a wedge-shaped beam with the edge of the wedge along the line of action of P. Or we might make h^2 proportional to x and b constant. This gives a beam whose vertical section, taken along the length of the beam, is a parabola. Or again, we might take b and h, each proportional to $\sqrt[3]{x}$, giving a beam whose cross section has a constant shape but a variable size. In such case if the section of the beam is circular, the radius would have to be proportional to $\sqrt[3]{x}$.

In Table 3 are shown 18 beams all with different loading and support condition; the shape of each beam, required for constant strength, is indicated. It should be noted that the shape given is the theoretical shape required to resist bending only. In order to take care of the vertical shear sufficient cross section of beam must also be added either at the points of support or at the points of application of the load.

The general formula for the carrying capacity of the beam shown in Table 3 is

$$P = K\frac{Sbh^2}{6L},$$

where S is the safe stress per square inch, K is a constant tabulated in Table 3. P is safe load in pounds.

TABLE 3

Type	K		Type	K	
1	1	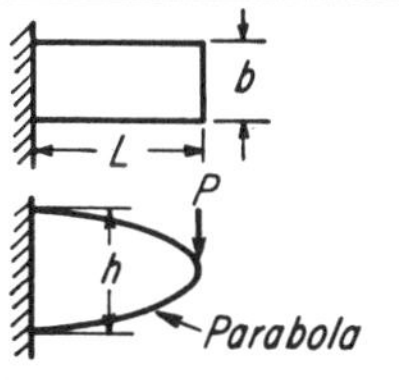	7	3	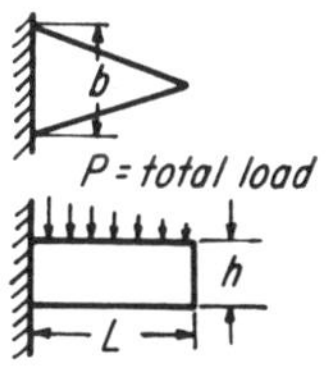
2	1	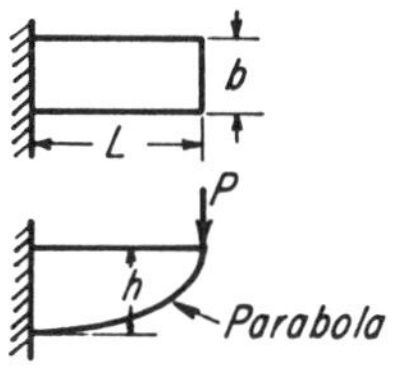	8	4	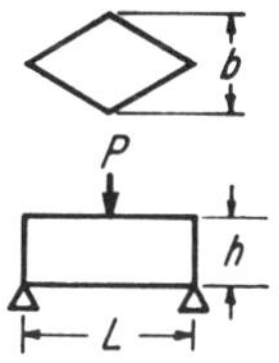
3	1	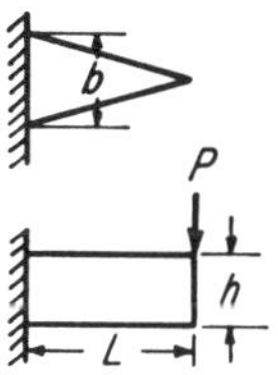	9	4	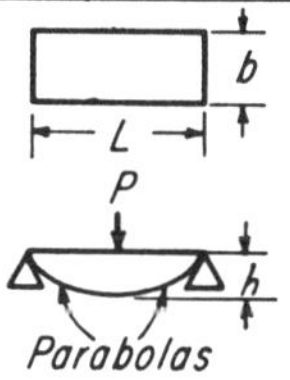
4	1	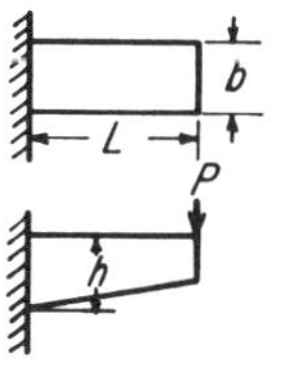	10	5.3	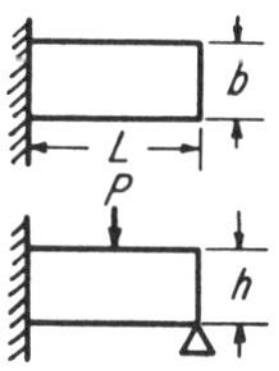
5	2	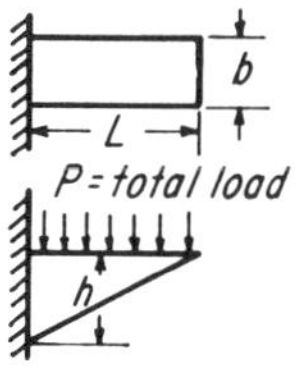	11	6	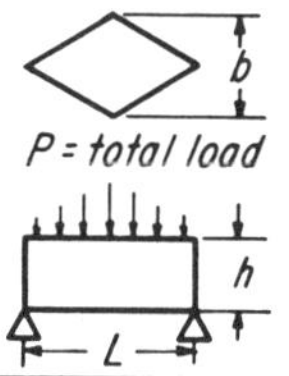
6	2	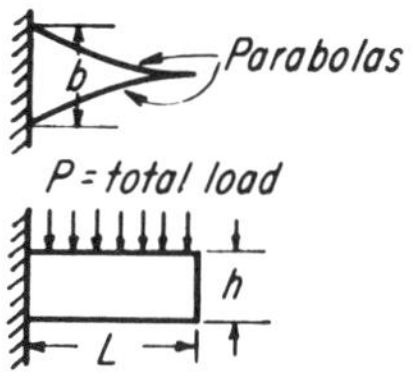	12	7.8	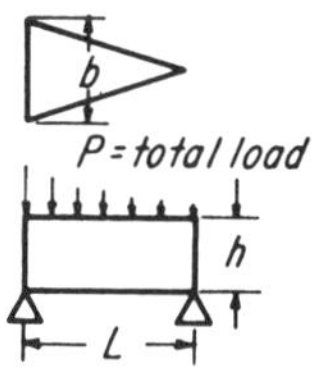

SYMBOLS

b = Maximum width of beam, in.
c = Distance from neutral plane to outer fiber, in.
h = Maximum depth of beam, in.
I = Rectangular moment of inertia about neutral axis, $in.^4$
K = Constant whose value depends on loading and support
L = Length of beam, in.
M = Bending moment, lb-in.
P = Total load on beam, lb
S = Safe stress, psi

Type	K		Type	K	
13	8	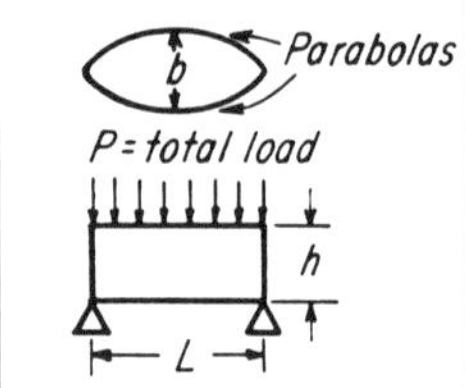	16	8	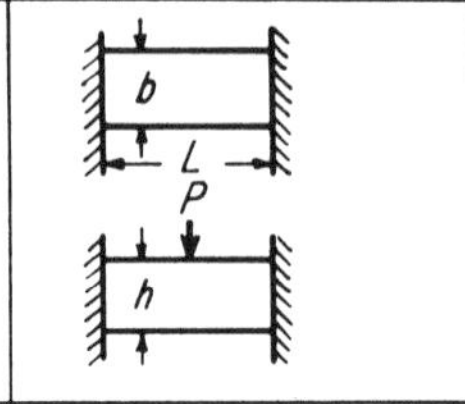
14	8	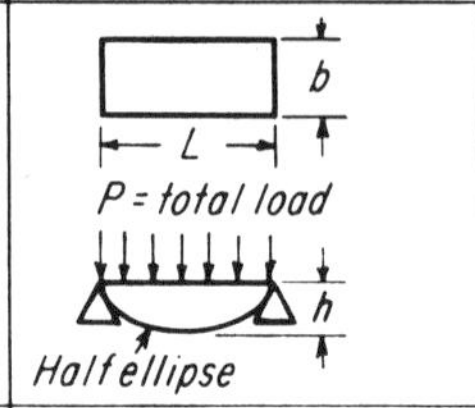	17	12	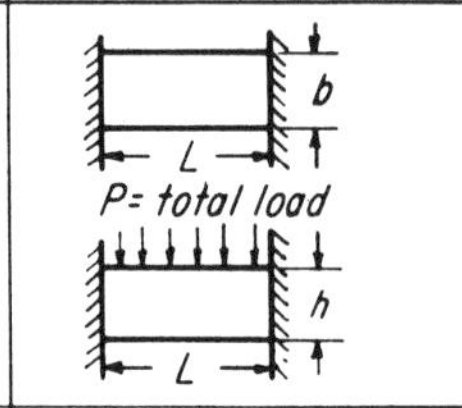
15	8	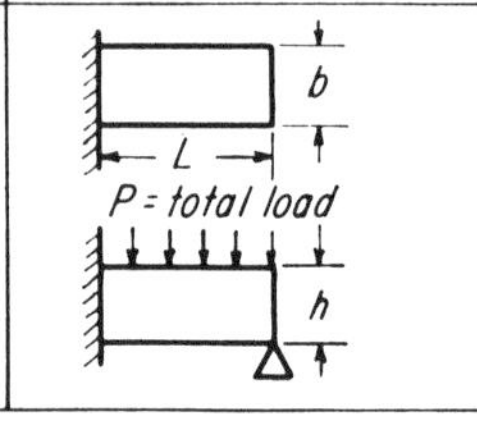	18	12	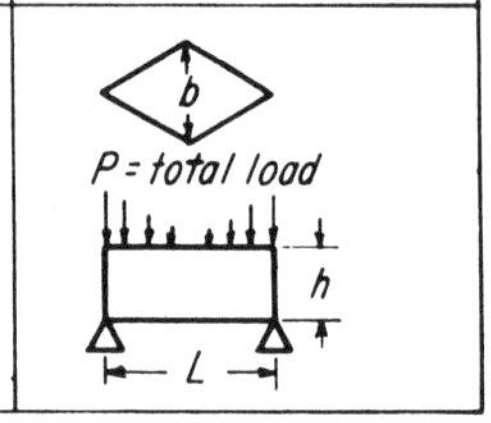

22. NOMOGRAM FOR MAXIMUM DEFLECTION DUE TO ANY NUMBER OF LOADS

(*Ref. 6*)

The basic equations for deflection of the beam illustrated in Fig. 28 are:

$$(A \text{ to } B)\text{: } y = \frac{Wx(L-l)}{6EIL}[2L(L-x) - (L-l)^2 - (L-x)^2] \quad (16)$$

$$(B \text{ to } C)\text{: } y = \frac{Wl(L-x)}{6EIL}[2L^2 - 2Ll - (L-l)^2 - (L-x)^2]. \quad (17)$$

Letting: $l = kL$ and $x = cL$, and substituting into (16, 17):

When $c < k$

$$y = \frac{WL^3}{6EI}\{c[k(2-k) - c^2] \times (1-k)\} \quad (18)$$

$c > k$

$$y = \frac{WL^3}{6EI}\{k[c(2-c) - K^2] \times (1-c)\}. \quad (19)$$

NOTATION

W = total applied load (lb.)
w = unit applied load (lb./in.)
I = moment of inertia (in.^{4})
E = modulus of elasticity (psi)
c = fraction that expresses the distance from one support to the point of deflection ($c < 1$)
k = fraction that expresses the distance from a support to the load point ($k < 1$)
l = the distance of the load from a support (in.)
L = span (in.)
y = deflection (in.)
f = natural frequency (cps)
d = diameter of beam (in.)

NOTE: For convenience, numerical values of f are shown on the scale for deflection.

Similarly, deflection equations for other beam supports have been developed. The nomogram, which solves equations (18) and (19) as well as equations pertaining to other cases, is given in terms of the l/L ratio, which defines the point of application of the load.

The nomogram also allows fast, accurate solution of multiload beams of any material and cross section. For beams of circular cross section, the moment of inertia will be found directly on the nomogram as a function of diameter d.

Example: Find the maximum end deflection of the steel beam shown in Fig. 29. $E = 30 \times 10^6$ psi and $I = 800$ in.^{4}

To find $y_1 \ldots y_4$, first calculate values of the ratio l/L for each load. Then determine each value of y from the nomogram. Determination of the value y_2 is given in detail as follows:

1. Locate the intersection of $l/L = 0.2$ and the curve for Case 1.
2. Project horizontally to the right to Reference line 1. (For end load start at point T.)
3. Align this intersection with $E = 30 \times 10$, intersecting Reference line 2.
4. Align this intersection with $W = 5{,}000$, intersecting Reference line 3 (extended).
5. Align this intersection with $L = 120$, intersecting Reference line 4.
6. Align this intersection with $I = 800$, intersecting $y_2 = 0.083$ in.

Deflections y_1, y_3, y_4 are found in a similar manner. The total deflection is shown below:

Solution of example:

Load (lb.)	Ratio l/L	Deflection (in.)
3000	0	$y_1 = 0.072$
5000	0.2	$y_2 = 0.083$
10000	0.5	$y_3 = 0.065$
6000	0.9	$y_4 = 0.002$
	Total deflection	$y = 0.222$ in.

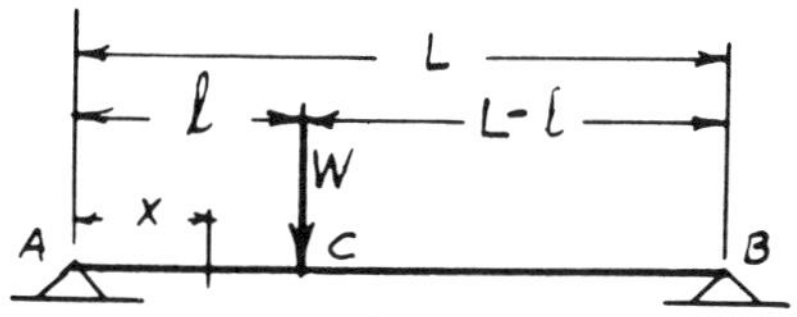

FIG. 28

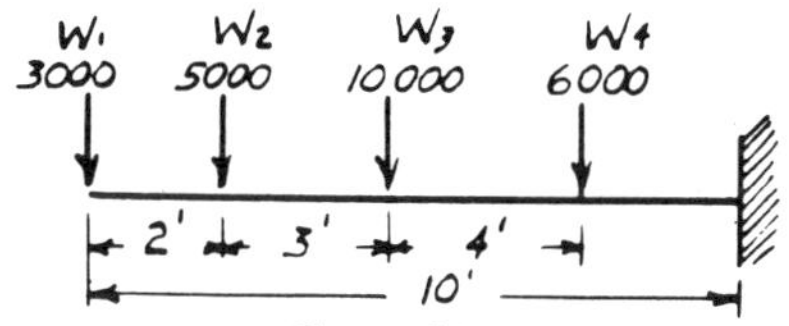

FIG. 29

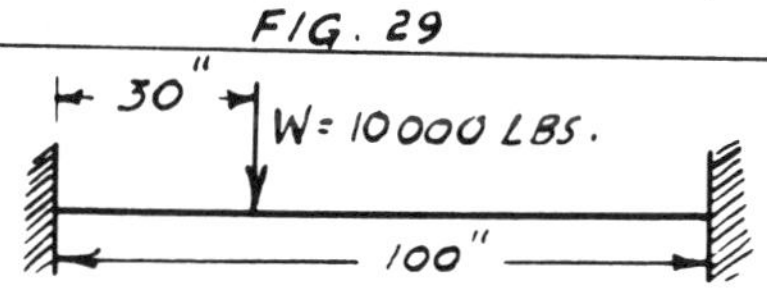

FIG. 30a

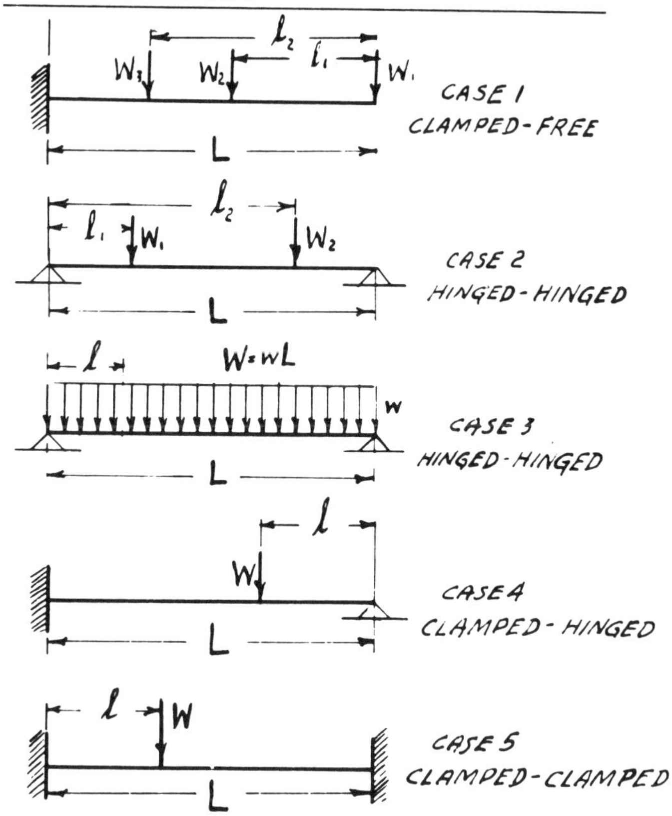

CHART 1

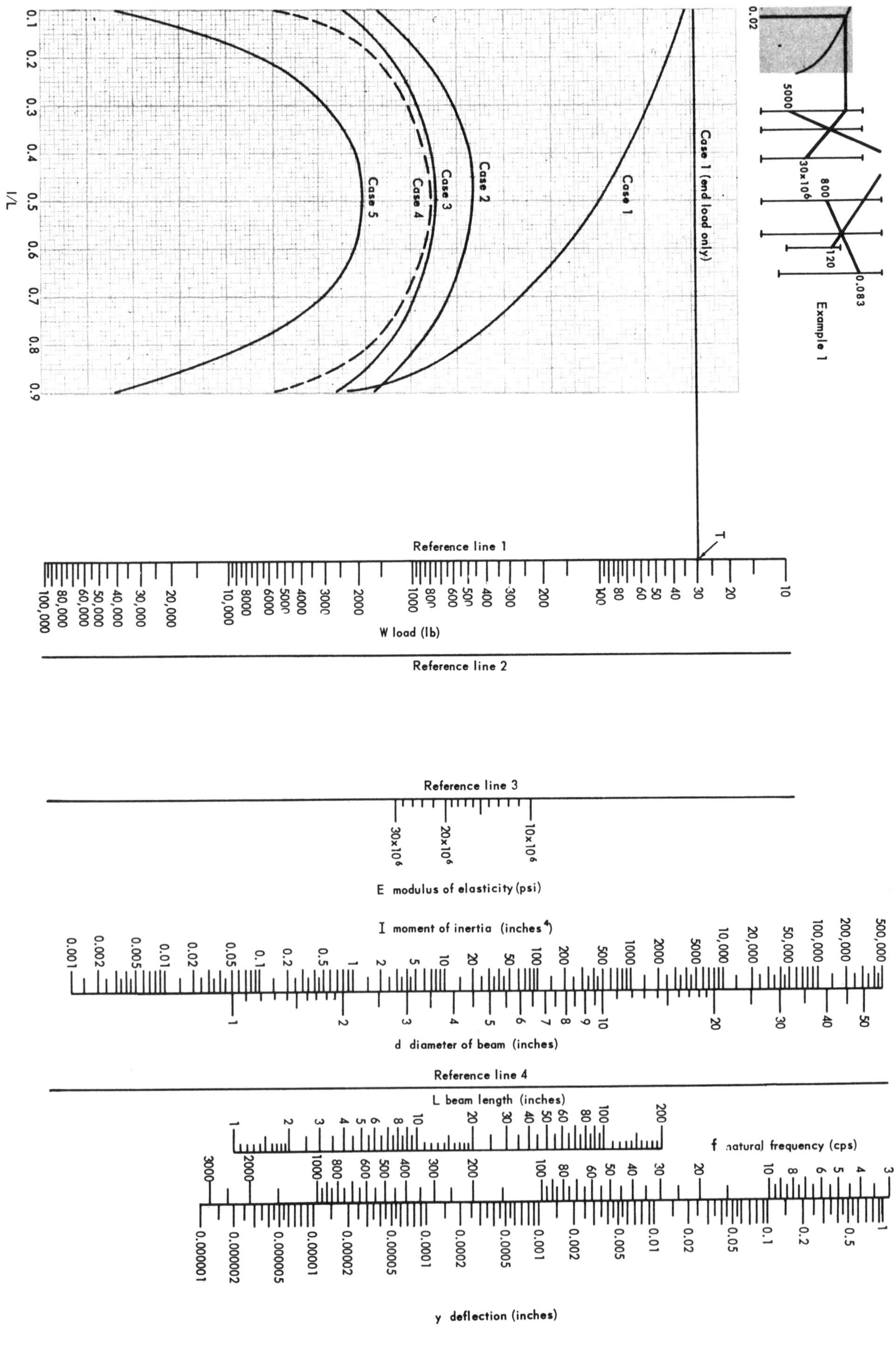
l/L
Case 1 (end load only)
Case 1
Case 2
Case 3
Case 4
Case 5
Example 1
Reference line 1
T
W load (lb)
Reference line 2
Reference line 3
E modulus of elasticity (psi)
I moment of inertia (inches⁴)
d diameter of beam (inches)
Reference line 4
L beam length (inches)
f natural frequency (cps)
y deflection (inches)

23. BEAMS UNDER COMBINED AXIAL AND TRANSVERSE LOADS

(*Ref. 3, 7, 8, 9, 10*)

NOTATION

P = axial load (lb.)
E = modulus of elasticity (psi)
I = moment of inertia (in.^{4})
L = beam length (in.)
S_{max} = maximum stress in beam (psi)
M = maximum bending moment due to combined loading (in.-lb.)
A = cross-sectional area of beam (in.^{2})
Z = section modulus (in.^{3})
M_1 = maximum applied moment (in.-lb.)
M^1 = maximum moment when the axial load is removed (in.-lb.)
y_M = deflection due to applied moment (in.)
y_W = deflection due to applied load (in.)
θ_M = slope of beam due to applied moment (radians)
θ_W = slope of beam due to applied load (radians)
W = transverse load (lb.)
K_y, K_θ, K_M = magnification factors for deflection, slope and moment (nondimensional)
C_y, C_θ, C_M = transverse load support factors for deflection, slope and moment (non-dimensional)

INTRODUCTION

Analysis of the deformation of beams under simultaneous axial and transverse loading can become extremely complex. Axial tension tends to straighten the beam, thus counteracting the bending moments produced by the transverse load. On the other hand, axial compression may greatly increase the bending moment, slope, and deflection of the beam. Any of these effects may also be produced when the transverse load is replaced by an externally applied couple.

A beam under combined loading cannot be analyzed by simply superposing the effects of axial and lateral loads or externally applied moments. The method of solution must take into account the simultaneous effect of these loads.

Two methods of analysis may be used in determining the total fiber stress in such members. One method, which is approximate in nature, assumes that the elastic curve of the deflected member is similar in form to the curve for a similar member under the action of transverse loads. The moment due to the deflection is estimated on this assumption and combined with the moment due to transverse loads. The other method, which is an exact one, makes use of the differential equation of the elastic curve and applies to relatively long and slender members.

The criterion for using either of the two methods is the critical Euler load. The approximate method may be used for cantilever beams if the axial load: $P > 0.125\ EI/L^2$; for beams with end supports if $P > 0.5\ EI/L^2$; for beams with fixed ends if $P > 2\ EI/L^2$. The precise method should be used for cantilevers if $P > 0.8\ EI/L^2$; for beams with end supports if $P > 3\ EI/L^2$; for beams with fixed ends if $P > 4\ EI/L^2$.

THE APPROXIMATE METHOD

As stated above, the approximate method of solution is based on the assumption that the elastic curve for the member with the axial load removed is similar in form to the curve when the axial load is in place. It is also assumed that the deflection and moment under combined loading are proportional to the deflection and moment for a similar member subjected only to transverse loading. All this means that the formula that applies here is appropriate only for beams in which the maximum bending moment and maximum deflection occur at the same section.

For any condition of loading, the maximum stress in the extreme fiber is:

$$S_{max} = \frac{P}{A} \pm \frac{M}{Z}, \tag{20}$$

where P is the axial load, A is the cross-sectional area of the beam, $Z = I/c$ is the section modulus and M is the maximum bending moment due to the combined effect of the axial and transverse loads. The + sign is used for the fibers in which the direct stress P/A and the bending stress M/Z are alike (compression-compression, tension-tension). The minus sign applies to fibers in which direct and bending stresses are unlike.

M in equation (20) is given with sufficient accuracy by the following equation:

$$M = \frac{M^1}{1 \pm (KPL^2)/EI}, \tag{21}$$

where: (1) M^1 is the maximum moment (inch-pounds) in beam when the axial load P is removed; (2) L is the length of the beam (inches); (3) E is the modulus of elasticity (psi); (4) I is the moment of inertia of the section about the central axis normal to the plane of bending (in.[4]); (5) K is a constant depending for its value upon the loading and end conditions for the beam in question. When P is tension, use + sign in the denominator of equation (21) and when P is compression, use a minus sign.

Values of the constant K are as follows:

Cantilever, end load	$K = 1/3$
Cantilever, uniform load	$K = 1/4$
End supports, center load	$K = 1/12$
End supports, uniform load	$K = 5/48$
Equal and opposite end couples	$K = 1/2$
Fixed ends, center load	$K = 1/24$
Fixed ends, uniform load	$K = 1/32$ (for end moments); $K = 1/16$ (for center moments)

THE ACCURATE METHOD (EQUATIONS)

The equations for deflection, slope, and moment, for beams under combined loading, are generally in the form of complex trigonometric and hyperbolic functions.

In what follows these time-consuming equations have been simplified and tabulated; their use is speeded by a table of trigonometric functions for angles in radians and a table of hyperbolic functions.

Here are the important equations involved:

$$y_W = \frac{-WL^3}{EI}K_y \quad (22) \qquad y_M = \frac{-M_1L^2}{EI}K_y, \quad (23)$$

$$\theta_W = \frac{WL^2}{EI}K_\theta \quad (24) \qquad \theta_M = \frac{M_1L}{EI}K_\theta, \quad (25)$$

$$M = WLK_M \quad (26)$$

where y_W, y_M are deflections due to the transverse load W and applied external moment M_1, respectively. θ_W, θ_M, are slopes due to W and M_1. K_y, K_θ, K_M are dimensionless factors the equations of which are tabulated in Table 4. The clue for the solution of these equations is a dimension-quantity:

$$U = L\sqrt{P/EI} \quad (27)$$

Example: A 2 in. × 12 in. steel eye-bar, hinged at the ends and 30 ft. long, is subjected to a compression of 240,000 lb. Find the deflection and fiber stress at the center of the bar due to the combined effect of direct stress and weight of the eye-bar.

Solution: For the conditions stated, $P = 240{,}000$ lb., the cross-sectional are, $A = 24$ in.[2]; $b = 2$ in., $d = 12$ in., $I = bd^3/12 = 288$ in.[4]; $c = d/2 = 6$ in.; $I/c = 288/6 = 48$ in.[3]; $E = 30 \times 10^6$ psi; $L = 360$ in. The weight of a 2 × 12 steel bar is 81.6 lb./ft.

1. Deflection. From equation (27) we find that $U = 1.9$. From Table 4, case 2:

$$Ky = \frac{SEC\,1/2\,U - (1/8)U^2 - 1}{U^4}$$

or by substitution,

$$Ky = \frac{1.719 - 1/8(3.6) - 1}{12.99} = 0.0207.$$

From equation (22):

$$y_W = \frac{-WL^3}{EI}Ky = \frac{-81.6 \times 30 \times 360^3 \times 0.0207}{30 \times 10^6 \times 288}$$
$$= -0.273 \text{ (down) in.}$$

2. Stresses. Again from Table 4, case 2:

$$K_M = \frac{SEC\,1/2\,U - 1}{U} = \frac{1.79 - 1}{3.6} = 0.2.$$

From equation (26): $M = WLK_M = 81.6 \times 30 \times 360 \times 0.2 = 176{,}260$ in.-lb. The maximum compressive stress at the top fiber from equation (20):

$$S_{max} = \frac{P}{A} + \frac{M}{Z} = \frac{240{,}000}{24} + \frac{176{,}260}{48} = 13{,}670 \text{ psi.}$$

THE ACCURATE METHOD (SIMPLIFIED)

To simplify the accurate analysis of beams under combined loading, the complex trigonometric and hyperbolic functions for maximum deflection y, bending moment M, and slope θ may be bypassed. Values of y, M, and θ produced by the transverse load alone (or by the externally applied couple) are multiplied by a *magnification factor K*, representing the effect of the axial force P on the maximum bending moment. Thus:

$$y_W = \frac{-WL^3}{C_yEI}K_y \quad (28) \qquad y_M = \frac{-M_1L^2}{C_yEI}K_y \quad (29)$$

$$\theta_W = \frac{WL^2}{C_\theta EI}K_\theta \quad (30) \qquad \theta_M = \frac{M_1L}{C_\theta EI}K_\theta \quad (31)$$

$$M = \frac{WL}{C_M}K_M. \quad (32)$$

Here again y_W, θ_W are used for a transverse load W, and y_M, θ_M are used for an applied external moment M_1.

The first factor in equation (28), WL^3/C_yEI, represents the deflection produced by the transverse load W

acting alone. The second factor, Ky, indicates in what proportion the deflection produced by W is magnified by the axial compressive force P or decreased by the axial tension.

Transverse load-support factors C_y, C_M, and C_θ are obtained from Table 5 for various loading and support conditions. Also from Table 5, the appropriate K_y, K_M, and K_θ (that is K_1, K_2, $K_3 \ldots$) are found. Numerical values of K are given in Table 6. Magnification factor K depends on the magnitude of nondimensional factor U found from equation (27). The numerical values of K-factors in Table 6 were calculated on the IBM 709 computer.

The following method may be used in analyzing a beam under combined loading:

1. From equation (27), calculate U.
2. Determine, from Table 5, values for C_y, C_M, and C_θ for the applicable case and choose the appropriate $K(K_1, K_2, K_3$, etc.). Where two values of C or K are given, choose the greater value.
3. For the calculated value of U, determine the numerical values of K_y, K_M, K_θ, Table 6.
4. From equations 28, 29, 30, 31, 32 calculate maximum deflection y, moment M and slope θ.
5. If necessary, calculate maximum stress from equation (20).

MULTIPLE TRANSVERSE LOAD

Although the principle of superposition cannot be used to analyze beams under combined loading, a modification of this principle can be used for a beam under a combination of transverse loads, in addition to an axial load. In this case, the deflection, moment, or slope can be found by adding the deflection, moment or slope for each combination of transverse and axial load.

Thus, a beam supported at the ends and subjected to a uniform load, a central load and an axial compressive load would have a maximum y, M and θ equal to the sum of y, M and θ for Cases 1 and 4, the total axial load being used in each Case.

Example: A steel I-beam, 20 ft. long and supported on both ends, is subjected to a transverse load of 50 lb. per ft. (including its own weight) and an axial compressive load of 20,000 lb. Find the maximum deflection, moment, slope and fiber stress.

Solution: $P = 20{,}000$ lb.; $L = 240$ in.; $I = 6$ in.^{4}; $I/C = 3$ in.^{3}; $A = 2.21$ in.^{2}; $W = 50 \times 20 = 1{,}000$ lb.; $E = 30 \times 10^6$ psi.

Step 1: From equation (27), $U = 24(20{,}000/30 \times 10^6)^{1/2} = 2.53$.

Step 2: From Table 5, case 4:

$$C_y = 76.8 \quad C_M = 8 \quad C_\theta = 24 \quad K_y = K_{11} \quad K_M = K_{10} \quad K_\theta = K_{12}$$

Step 3: For $U = 2.53$, from Table 6:

$$K_{11} = 2.8561 \quad K_{10} = 2.9059 \quad K_{12} = 2.8236$$

Step 4: From equation (28)

$$y_W = \frac{-1{,}000 \times 240^3 \times 2.8561}{76.8 \times 30 \times 10^6 \times 6} = -2.85 \text{ in. (down)}$$

Step 5: From equation (32),

$$M = \frac{1{,}000 \times 240 \times 2.9059}{8} = 87{,}200 \text{ in.-lb.}$$

Step 6: Calculate θ from equation (30):

$$\theta = \frac{1{,}000 \times 240^2 \times 2.9059}{24 \times 30 \times 10^6 \times 6} = 0.0388 \text{ radians}$$

Step 7: Finally, from equation (20):

$$S_{max} = \frac{20{,}000}{2.21} + \frac{87{,}200}{3} = 38{,}150 \text{ psi}$$

This is a compressive stress, which occurs in the top fibers at the center of the beam.

ACCURATE METHOD (NOMOGRAMS)

In cases of repetitive solutions of problems involving beams under combined loading it is more expedient to use nomograms; such nomograms are presented here.

$$U = L\sqrt{P/EI} \qquad \text{Nomogram 1}$$

$$\left.\begin{aligned} y_W &= \frac{-WL^3}{C_y EI} K_y \\ y_M &= \frac{-M_1 L^2}{C_y EI} K_y \end{aligned}\right\} \text{Nomogram 2}$$

$$\left.\begin{aligned} \theta_W &= \frac{WL^2}{C_\theta EI} K_\theta \\ \theta_M &= \frac{M_1 L}{C_\theta EI} K_\theta \end{aligned}\right\} \text{Nomogram 3}$$

Example: A 3-in. 5.7-lb. steel I-beam, 10 ft. long, is supported at ends and subjected to a transverse load of 100 lb. per ft. (including its own weight) and an axial compressive load of 30,000 lb. Find the maximum deflection, moment, slope, and fiber stress.

Solution: $P = 30{,}000$ lb., $L = 120$ in., $I = 2.5$ in.^{4}, $E = 30 \times 10^6$ psi, $A = 1.64$ in.^{2}, $W = 100 \times 10 = 1{,}000$ lb.

Step 1: From Nomogram 1, find $U = 2.4$.

Step 2: From Table 5, case 4:

$$C_y = 76.8$$
$$C_M = 8$$
$$C_\theta = 24$$
$$K_y = K_{11}$$
$$K_M = K_{10}$$
$$K_\theta = K_{12}$$

Step 3: For $U = 2.40$, from Table 6:

$$K_{11} = 2.4067 \sim 2.41$$
$$K_{10} = 2.4440 \sim 2.44$$
$$K_{12} = 2.3822 \sim 2.38$$

Step 4: From Nomogram 2, following the key diagram, we obtain $y = 0.72$ in.

Step 5: From Nomogram 3: $\theta_W = 0.02$ radians

Step 6: From equation (32): $M = 36{,}000$ in.-lb.

Step 7:

$$S_{\max} = \frac{P}{A} + \frac{Mc}{I} = \frac{30{,}000}{1.64} + \frac{36{,}000 \times 1.5}{2.5} =$$

$$= 39{,}000 \text{ psi compressive}$$

NOTE: For Cases 5 and 8 (beam loaded by moment M_1) Nomograms 2 and 3 yield y_M/L and θ_M/L respectively.

TABLE 4

MANNER OF LOADING AND SUPPORT	K_y, K_m, K_θ
1. 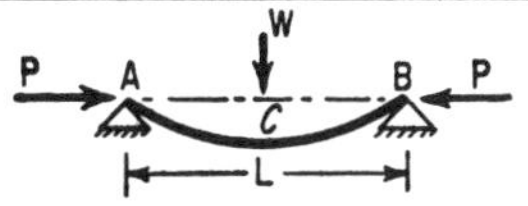	$K_y = \frac{1}{2}\left[\frac{\tan\frac{1}{2}U - \frac{1}{2}U}{U^3}\right]$ at C $K_m = \frac{\tan\frac{1}{2}U}{2U}$ at C $K_\theta = \frac{1-\cos\frac{1}{2}U}{2\cos\frac{1}{2}U}$ at A
2. 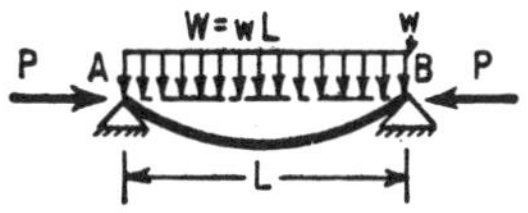	$K_y = \frac{\sec\frac{1}{2}U - \frac{1}{8}U^2 - 1}{U^4}$ at center $K_m = \frac{\sec\frac{1}{2}U - 1}{U^2}$ at center $K_\theta = \frac{1}{U^2}\left[\frac{1-\cos U}{\sin U} - \frac{U}{2}\right]$ at A
3. Beam under axial compression and bent by a couple M_1 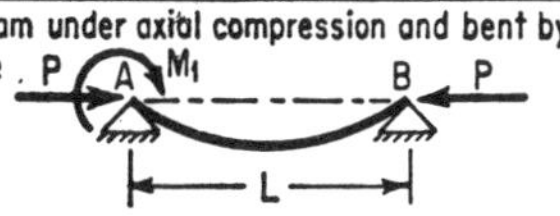	$K_y = \frac{1}{U^2}\left[\frac{\sin 0.422U}{\sin U} - 0.422\right]$ at $0.422L$ $K_{\theta_A} = \frac{1}{U\tan U} - \frac{1}{U^2}$ at A $K_{\theta_B} = \frac{1}{U\sin U} - \frac{1}{U^2}$ at B
4. Beam under axial compression and equal and opposite end couples 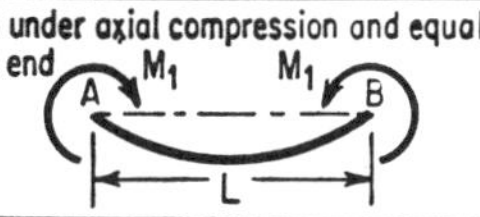	$K_y = \frac{1}{U^2}\left[\frac{1-\cos\frac{1}{2}U}{\cos\frac{1}{2}U}\right]$ at $\frac{L}{2}$ $K_m = \sec\frac{1}{2}U$ at $\frac{L}{2}$ $K_\theta = \frac{1}{U}\tan\frac{1}{2}U$ at A
5. 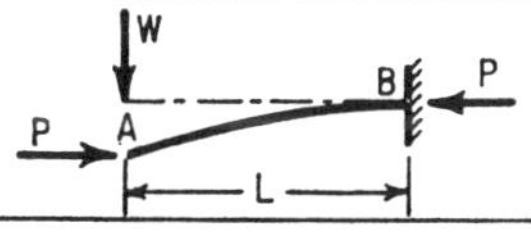	$K_y = \frac{1}{U^3}\left[\tan U - U\right]$ at A $K_m = \frac{1}{U}\tan U$ at B $K_\theta = \frac{1-\cos U}{U^2\cos U}$ at A
6. 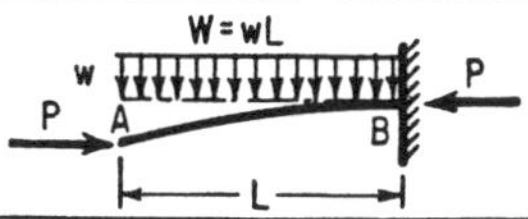	$K_y = \frac{1}{U^4}\left(1+\frac{1}{2}U^2 - \sec U\right) + \frac{1}{U^3}\left(\tan U - U\right)$ $K_m = \frac{1}{U^2}\left(1-\sec U\right) + \frac{1}{U}\left(\tan U\right)$ $K_\theta = \frac{1}{U^2\cos U} - \frac{1}{U^3}\left(\frac{1-\cos 2U}{\sin 2U}\right)$
7. 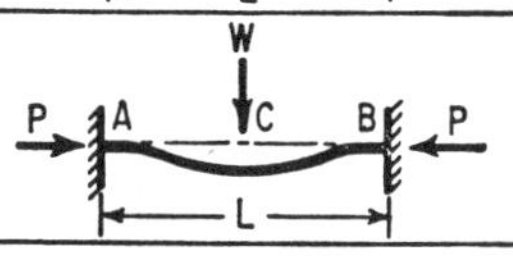	$K_y = \frac{1}{2U^3}\left[\tan\frac{1}{2}U - \frac{(1-\cos\frac{1}{2}U)^2}{\sin\frac{1}{2}U\cos\frac{1}{2}U} - \frac{U}{2}\right]$ at C $K_m = \frac{1}{2U}\left[\tan\frac{1}{2}U - \frac{1-\cos\frac{1}{2}U}{\sin\frac{1}{2}\cos\frac{1}{2}U}\right]$ at C
8. 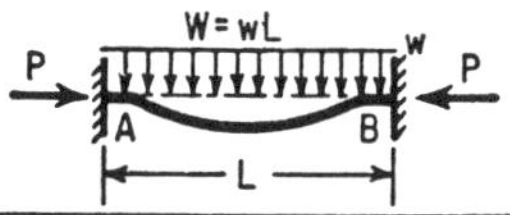	$K_y = \frac{1}{U^4}\left[\sec\frac{1}{2}U - \frac{U^2}{8} - 1 - \left(1 - \frac{\frac{1}{2}U}{\tan\frac{1}{2}U}\right)\left(\frac{1-\cos\frac{1}{2}U}{\cos\frac{1}{2}U}\right)\right]$ at center $K_m = \frac{1}{U^2}\left[\frac{\frac{1}{2}U}{\sin\frac{1}{2}U} - 1\right]$ at center $K_m = \frac{1}{U^2}\left[\frac{\frac{1}{2}U}{\sin\frac{1}{2}U} - 1\right]$ at A and B
9. 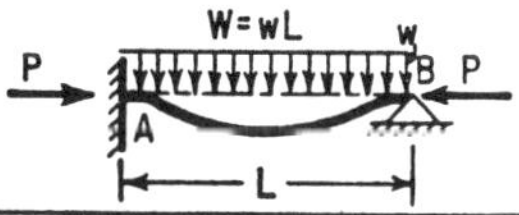	$K_y = \frac{1}{U^3}\left[\frac{\tan U(\tan\frac{1}{2}U - \frac{1}{2}U)}{\tan U - U}\right]\left[\frac{1}{2} + \cot U\sin\frac{1}{2}U - \cos\frac{1}{2}U\right] - \frac{1}{U^4}\left[\cot U\sin\frac{1}{2}U - \frac{\sin\frac{1}{2}U}{\sin U} - \cos\frac{1}{2}U + \frac{U^2}{4} + 1\right]$ at center $K_m = \frac{1}{U}\left[\frac{\tan U(\tan\frac{1}{2}U - \frac{1}{2}U)}{\tan U - U}\right]$
10. 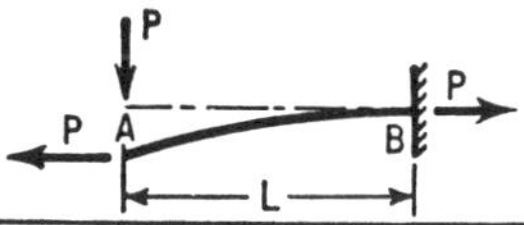	$K_y = \frac{1}{U^2}\left[1 - \frac{1}{U}\tanh U\right]$ at A $K_m = \frac{1}{U}\tanh U$ at B
11. 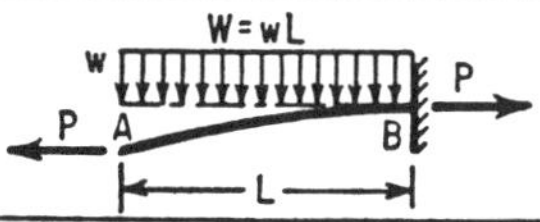	$K_y = \frac{1}{U^3}\left[\frac{1}{U}\left(\frac{3}{2} - \operatorname{sech} U\right) - \tanh U\right]$ at A $K_m = \frac{1}{U}\left[\tanh U - \frac{1}{U}(1 - \operatorname{sech} U)\right]$ at B
12. 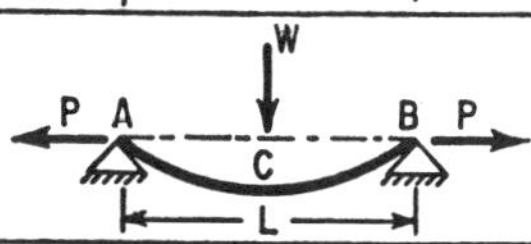	$K_v = \frac{1}{U^2}\left[\frac{1}{4} - \frac{1}{2U}\tanh\frac{1}{2}U\right]$ at C $K_m = \frac{1}{2U}\tanh\frac{1}{2}U$
13. 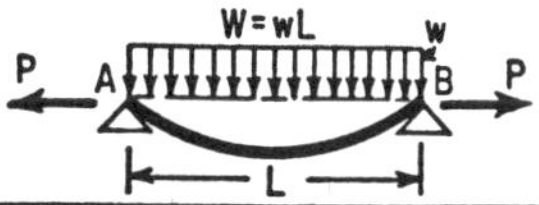	$K_y = \frac{1}{U^2}\left[\frac{1}{8} - \frac{1}{U^2}(1 - \operatorname{sech}\frac{1}{2}U)\right]$ at center $K_m = \frac{1 - \operatorname{sech}\frac{1}{2}U}{U^2}$ at center
14. 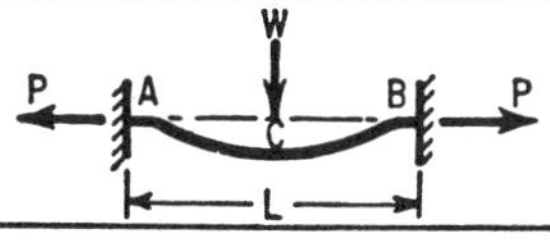	$K_y = \frac{1}{2U^3}\left[\frac{1}{2}U - \tanh U - \frac{(1-\cosh\frac{1}{2}U)^2}{\sinh\frac{1}{2}U\cosh\frac{1}{2}U}\right]$ at C $K_m = \frac{1}{2U}\left[\frac{1-\cosh\frac{1}{2}U}{\sinh\frac{1}{2}U\cosh\frac{1}{2}U} + \tanh\frac{1}{2}U\right]$ at C
15 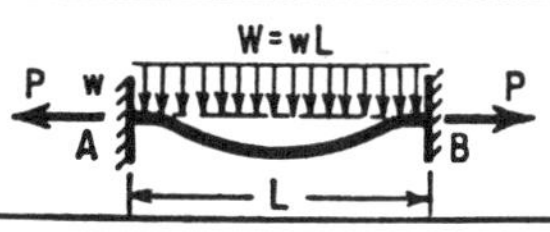	$K_y = \frac{1}{8U^2}\left[\frac{4(1-\cosh\frac{1}{2}U)}{U\sinh\frac{1}{2}U} + 1\right]$ at center $K_m = \frac{1}{U^2}\left[\frac{\frac{1}{2}U - \tanh\frac{1}{2}U}{\tanh\frac{1}{2}U}\right]$ at A and B $K_m = \frac{1}{U^2}\left[1 - \frac{\frac{1}{2}U}{\sinh\frac{1}{2}U}\right]$ at center

TABLE 5. FACTORS FOR VARIOUS LOAD CONDITIONS

Case	C_y	C_M	C_θ	K_y	K_M	K_θ	Max. Values	Case
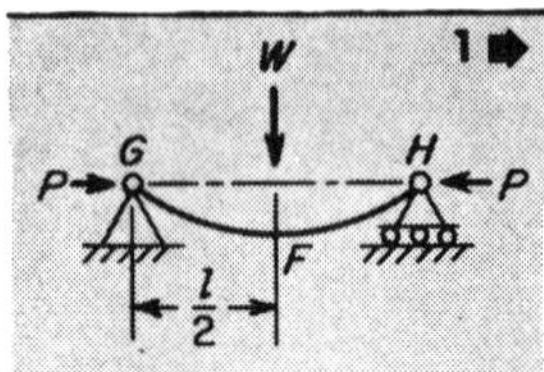 1 ➡	48	4	16	K_1	K_2	K_3	$y_F, M_F, \theta_G (=-\theta_H)$	
	3	1	2	K_5	K_4	K_6	y_G, M_H, θ_G	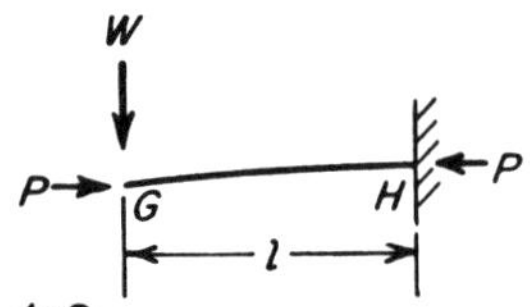 ⬅ 2
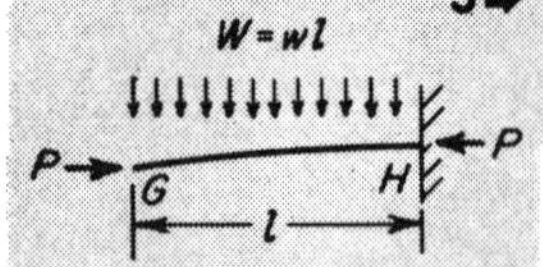 3 ➡	8	2	6	K_8	K_7	K_9	y_G, M_H, θ_G	
	76.8	8	24	K_{11}	K_{10}	K_{12}	$y_F, M_F, \theta_G (=-\theta_H)$	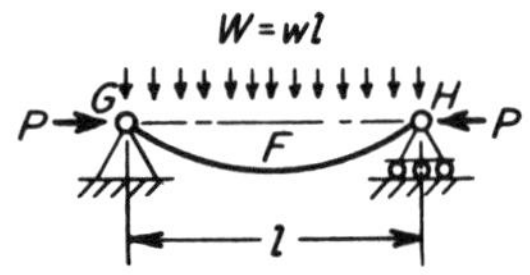 ⬅ 4
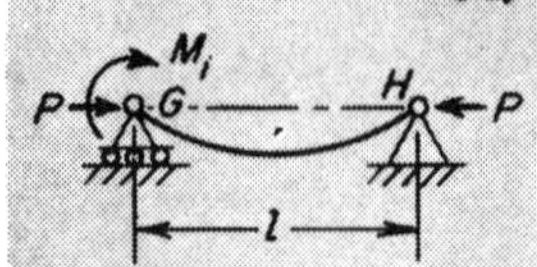 5 ➡	15.6	–	6	K_{15}	–	K_{13}* K_{14}†	$y_{0.422l}$	
	192	8	–	K_{20}	K_{19}	–	y_F, M_F	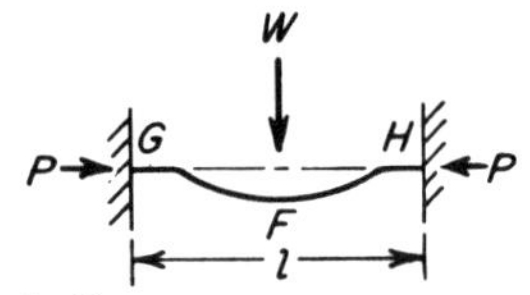 ⬅ 6
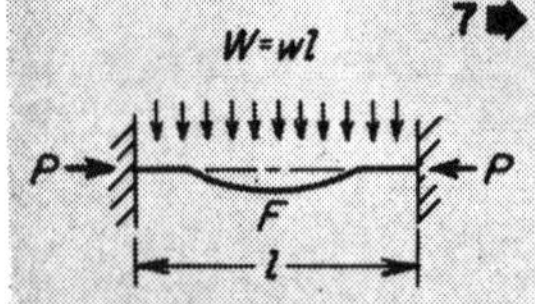 7 ➡	384	24§ 12‡	–	K_{23}	K_{22}§ K_{21}‡	–	y_F	
	8	1	2	K_{16}	K_{18}	K_{17}	$y_F, M_F, \theta_G (=-\theta_H)$	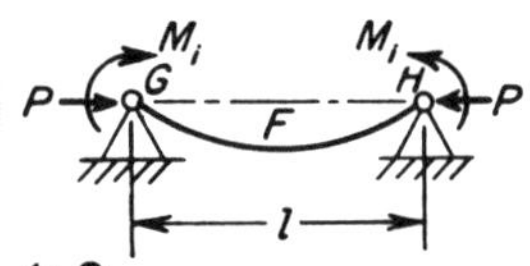 ⬅ 8
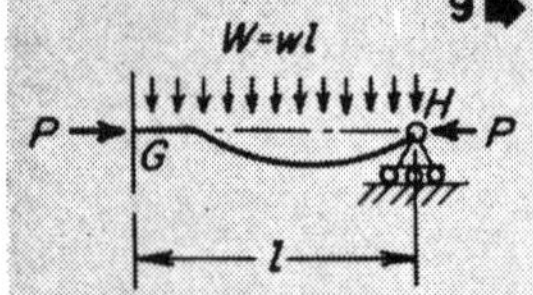 9 ➡	–	8	–	–	K_{24}	–	M_G	
	3	1	–	K_{26}	K_{25}	–	y_G, M_H	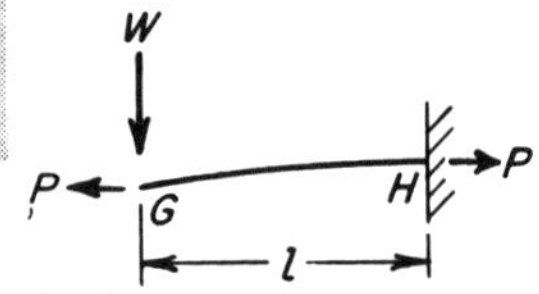 ⬅ 10
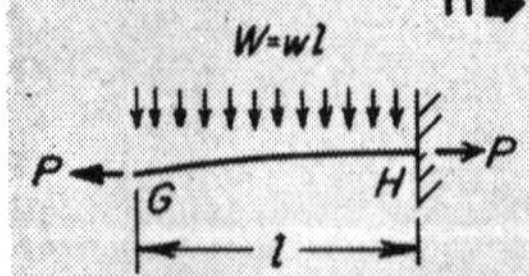 11 ➡	8	2	–	K_{28}	K_{27}	–	y_G, M_H	
	48	4	–	K_{30}	K_{29}	–	y_F, M_F	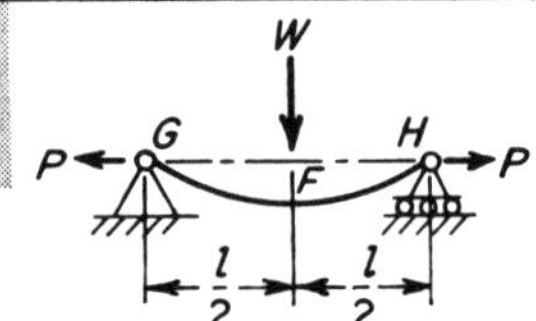 ⬅ 12
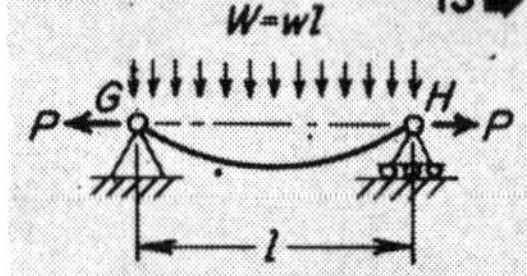 13 ➡	76.8	8	–	K_{32}	K_{31}	–	y_F, M_F	
	192	8	–	K_{34}	K_{33}	–	y_F, M_F	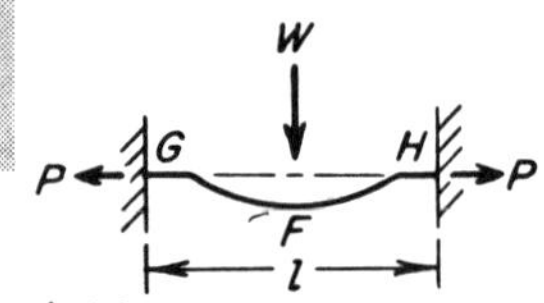 ⬅ 14
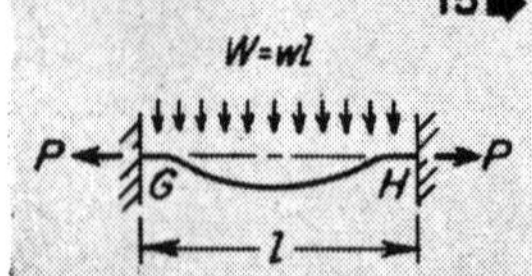 15 ➡	384	24§ 12‡	–	K_{35}	K_{37}§ K_{36}‡	–	y_F	

*For θ_G. †For θ_H. §For M_F. ‡For M_G.

TABLE 6. MAGNIFICATION FACTORS, K

U	K_1	K_2	K_3	K_4	K_5	K_6	K_7	K_8	K_9	K_{10}	K_{11}	K_{12}
0.00	1.0000	1.0000	1.0000	1.0000	1.0000	1.0000	1.0000	1.0000	1.0000	1.0000	1.0000	1.0000
0.05	1.0229	1.0002	1.0003	1.0008	1.0010	1.0010	1.0006	1.0099	1.0000	1.0002	1.0000	1.0229
0.10	1.0017	1.0008	1.0010	1.0033	1.0040	1.0042	1.0025	1.0037	1.0044	1.0010	1.0000	1.0017
0.15	1.0025	1.0019	1.0023	1.0076	1.0091	1.0095	1.0057	1.0086	1.0103	1.0023	1.0022	1.0025
0.20	1.0041	1.0033	1.0042	1.0135	1.0163	1.0169	1.0102	1.0158	1.0183	1.0042	1.0040	1.0041
0.25	1.0063	1.0052	1.0065	1.0214	1.0256	1.0267	1.0160	1.0249	1.0289	1.0066	1.0064	1.0063
0.30	1.0091	1.0076	1.0095	1.0311	1.0374	1.0389	1.0233	1.0363	1.0420	1.0095	1.0092	1.0091
0.35	1.0124	1.0103	1.0129	1.0429	1.0515	1.0537	1.0322	1.0501	1.0580	1.0129	1.0126	1.0124
0.40	1.0163	1.0136	1.0169	1.0570	1.0684	1.0713	1.0427	1.0665	1.0770	1.0169	1.0165	1.0163
0.45	1.0207	1.0172	1.0215	1.0734	1.0882	1.0919	1.0550	1.0857	1.0994	1.0215	1.0210	1.0207
0.50	1.0256	1.0214	1.0267	1.0926	1.1113	1.1159	1.0693	1.1081	1.1253	1.0267	1.0261	1.0256
0.55	1.0312	1.0260	1.0325	1.1147	1.1379	1.1437	1.0358	1.1340	1.1554	1.0325	1.0317	1.0312
0.60	1.0374	1.0311	1.0389	1.1402	1.1686	1.1757	1.1047	1.1638	1.1900	1.0389	1.0380	1.0374
0.65	1.0441	1.0368	1.0460	1.1695	1.2039	1.2125	1.1265	1.1981	1.2299	1.0460	1.0449	1.0441
0.70	1.0515	1.0429	1.0537	1.2033	1.2445	1.2549	1.1516	1.2375	1.2758	1.0537	1.0524	1.0516
0.75	1.0596	1.0497	1.0621	1.2421	1.2913	1.3038	1.1804	1.2831	1.3288	1.0621	1.0606	1.0596
0.80	1.0684	1.0570	1.0713	1.2870	1.3455	1.3604	1.2137	1.3356	1.3901	1.0713	1.0696	1.0684
0.85	1.0779	1.0649	1.0812	1.3392	1.4085	1.4261	1.2523	1.3968	1.4614	1.0312	1.0793	1.0779
0.90	1.0882	1.0734	1.0919	1.4002	1.4821	1.5030	1.2973	1.4682	1.5448	1.0919	1.0897	1.0882
0.95	1.0993	1.0827	1.1035	1.4720	1.5689	1.5937	1.3503	1.5525	1.6432	1.1035	1.1010	1.0993
1.00	1.1113	1.0926	1.1159	1.5574	1.6722	1.7016	1.4132	1.6527	1.7604	1.1159	1.1131	1.1112
1.05	1.1241	1.1033	1.1293	1.6603	1.7967	1.8318	1.4888	1.7735	1.9018	1.1293	1.1262	1.1241
1.10	1.1379	1.1147	1.1437	1.7861	1.9491	1.9911	1.5812	1.9213	2.0750	1.1437	1.1402	1.1379
1.15	1.1527	1.1270	1.1591	1.9430	2.1392	2.1899	1.6962	2.1057	2.2912	1.1592	1.1553	1.1527
1.20	1.1686	1.1402	1.1757	2.1434	2.3822	2.4440	1.8429	2.3413	2.5676	1.1757	1.1714	1.1686
1.25	1.1856	1.1544	1.1935	2.4076	2.7026	2.7793	2.0360	2.6521	2.9326	1.1935	1.1887	1.1856
1.30	1.2039	1.1695	1.2125	2.7708	3.1435	3.2406	2.3010	3.0794	3.4343	1.2125	1.2073	1.2039
1.35	1.2235	1.1858	1.2330	3.3002	3.7863	3.9134	2.6869	3.7025	4.1675	1.2330	1.2273	1.2235
1.40	1.2445	1.2033	1.2549	4.1413	4.8082	4.9831	3.2995	4.6929	5.3331	1.2549	1.2487	1.2445
1.45	1.2671	1.2220	1.2785	5.6814	6.6798	6.9427	4.4201	6.5068	7.4685	1.2785	1.2716	1.2671
1.50	1.2913	1.2421	1.3038	9.4009	11.2012	11.6771	7.1246	10.8883	12.6289	1.3038	1.2963	1.2914
1.55	1.3174	1.2638	1.3311	31.0174	37.4827	39.1986	22.8361	36.3557	42.6304	1.3311	1.3229	1.3174
1.60	1.3455	1.2870	1.3604							1.3604	1.3515	1.3455
1.65	1.3758	1.3121	1.3920							1.3920	1.3823	1.3758
1.70	1.4085	1.3392	1.4261							1.4261	1.4155	1.4085
1.75	1.4438	1.3685	1.4630							1.4630	1.4515	1.4438
1.80	1.4821	1.4002	1.5030							1.5030	1.4904	1.4821
1.85	1.5237	1.4346	1.5464							1.5464	1.5327	1.5237
1.90	1.5689	1.4720	1.5937							1.5937	1.5788	1.5689
1.95	1.6182	1.5128	1.6452							1.6452	1.6290	1.6182
2.00	1.6722	1.5574	1.7016							1.7016	1.6839	1.6722
2.05	1.7315	1.6064	1.7635							1.7635	1.7442	1.7315
2.10	1.7967	1.6603	1.8318							1.8318	1.8106	1.7967
2.15	1.8689	1.7199	1.9072							1.9072	1.8841	1.8689
2.20	1.9491	1.7861	1.9911							1.9911	1.9658	1.9491
2.25	2.0387	1.8601	2.0847							2.0847	2.0569	2.0387
2.30	2.1392	1.9430	2.1899							2.1899	2.1593	2.1392
2.35	2.2529	2.0368	2.3087							2.3087	2.2750	2.2529
2.40	2.3822	2.1434	2.4440							2.4440	2.4067	2.3822
2.45	2.5307	2.2659	2.5993							2.5993	2.5579	2.5307
2.50	2.7027	2.4077	2.7793							2.7793	2.7331	2.7027
2.55	2.9043	2.5737	2.9902							2.9902	2.9383	2.9043
2.60	3.1435	2.7708	3.2406							3.2406	3.1820	3.1435
2.65	3.4319	3.0084	3.5425							3.5425	3.4757	3.4320
2.70	3.7863	3.3002	3.9134							3.9134	3.8366	3.7863
2.75	4.2317	3.6668	4.3796							4.3796	4.2902	4.2317
2.80	4.8082	4.1413	4.9831							4.9831	4.8773	4.8082
2.85	5.5831	4.7791	5.7945							5.7945	5.6666	4.5831
2.90	6.6798	5.6814	6.9427							6.9427	6.7836	6.6798
2.95	8.3499	7.0555	8.6915							8.6915	8.4848	8.3500
3.00	11.2012	9.4009	11.6771							11.6771	11.3889	11.2012
3.05	17.1676	14.3085	17.9249							17.9249	17.4662	17.1676
3.10	37.4835	31.0180	39.1994							39.1994	38.1597	37.4835
3.15	−183.8830	−51.0483	−192.6153							−192.6153	−187.3225	−183.8830
3.20	−26.2447	−21.3956	−27.5371							−27.5371	−26.7535	−26.2448

U	K_{13}	K_{14}	K_{15}	K_{16}	K_{17}	K_{18}	K_{19}	K_{20}	K_{21}	K_{22}	K_{23}	K_{24}
0.00	1.0000	1.0000	1.0000	1.0000	1.0000	1.0000	1.0000	1.0000	1.0000	1.0000	1.0000	1.0000
0.05	1.0001	1.0003	1.0000	1.0003	1.0002	1.0003	1.0000	1.0000	1.0000	1.0001	1.0000	1.0000
0.10	1.0006	1.0012	1.0000	1.0010	1.0008	1.0013	1.0002	1.0002	1.0002	1.0003	1.0000	1.0003
0.15	1.0015	1.0026	1.0000	1.0023	1.0019	1.0028	1.0005	1.0005	1.0004	1.0007	1.0000	1.0007
0.20	1.0026	1.0047	1.0000	1.0042	1.0033	1.0050	1.0008	1.0010	1.0007	1.0012	1.0000	1.0013
0.25	1.0041	1.0073	1.0000	1.0065	1.0052	1.0079	1.0013	1.0016	1.0010	1.0018	1.0013	1.0021
0.30	1.0050	1.0106	1.0000	1.0095	1.0076	1.0114	1.0019	1.0022	1.0015	1.0028	1.0021	1.0030
0.35	1.0032	1.0145	1.0000	1.0129	1.0103	1.0155	1.0026	1.0031	1.0020	1.0036	1.0027	1.0041
0.40	1.0108	1.0190	1.0000	1.0169	1.0135	1.0203	1.0033	1.0040	1.0027	1.0037	1.0037	1.0054
0.45	1.0137	1.0241	1.0000	1.0215	1.0172	1.0259	1.0042	1.0061	1.0034	1.0059	1.0051	1.0068
0.50	1.0170	1.0299	1.0000	1.0267	1.0214	1.0321	1.0052	1.0063	1.0042	1.0073	1.0062	1.0084
0.55	1.0207	1.0365	1.0000	1.0325	1.0260	1.0390	1.0064	1.0076	1.0061	1.0089	1.0076	1.0102
0.60	1.0248	1.0437	1.0000	1.0389	1.0311	1.0468	1.0076	1.0091	1.0061	1.0106	1.0090	1.0122
0.65	1.0293	1.0516	1.0000	1.0460	1.0368	1.0552	1.0089	1.0107	1.0071	1.0125	1.0106	1.0144
0.70	1.0342	1.0603	1.0000	1.0537	1.0429	1.0645	1.0103	1.0124	1.0083	1.0145	1.0124	1.0167
0.75	1.0396	1.0698	1.0000	1.0621	1.0497	1.0747	1.0119	1.0143	1.0095	1.0167	1.0142	1.0192
0.80	1.0454	1.0801	1.0000	1.0713	1.0570	1.0857	1.0136	1.0163	1.0108	1.0190	1.0163	1.0220
0.85	1.0517	1.0912	1.0000	1.0812	1.0649	1.0976	1.0153	1.0184	1.0123	1.0215	1.0184	1.0249
0.90	1.0585	1.1033	1.0000	1.0919	1.0735	1.1106	1.0172	1.0207	1.0138	1.0241	1.0207	1.0281
0.95	1.0658	1.1163	1.0000	1.1035	1.0827	1.1245	1.0192	1.0231	1.0154	1.0270	1.0231	1.0314
1.00	1.0737	1.1304	1.0086	1.1159	1.0926	1.1395	1.0214	1.0256	1.0171	1.0299	1.0256	1.0350
1.05	1.0821	1.1455	1.0211	1.1293	1.1033	1.1556	1.0236	1.0284	1.0189	1.0331	1.0283	1.0387
1.10	1.0912	1.1617	1.0345	1.1437	1.1147	1.1730	1.0260	1.0312	1.0208	1.0365	1.0312	1.0427
1.15	1.1009	1.1792	1.0489	1.1591	1.1270	1.1916	1.0285	1.0342	1.0228	1.0399	1.0342	1.0470
1.20	1.1113	1.1979	1.0644	1.1757	1.1402	1.2116	1.0311	1.0374	1.0249	1.0437	1.0374	1.0515
1.25	1.1225	1.2180	1.0810	1.1935	1.1544	1.2331	1.0339	1.0407	1.0271	1.0475	1.0407	1.0562
1.30	1.1344	1.2396	1.0988	1.2125	1.1695	1.2561	1.0368	1.0441	1.0294	1.0516	1.0441	1.0611
1.35	1.1473	1.2628	1.1179	1.2330	1.1858	1.2809	1.0398	1.0478	1.0318	1.0558	1.0478	1.0664
1.40	1.1610	1.2878	1.1384	1.2549	1.2033	1.3075	1.0429	1.0516	1.0343	1.0603	1.0516	1.0719
1.45	1.1757	1.3146	1.1605	1.2785	1.2220	1.3360	1.0462	1.0555	1.0369	1.0649	1.0555	1.0777
1.50	1.1915	1.3434	1.1842	1.3038	1.2421	1.3667	1.0497	1.0596	1.0396	1.0697	1.0596	1.0838
1.55	1.2084	1.3744	1.2096	1.3311	1.2638	1.3997	1.0533	1.0639	1.0425	1.0748	1.0639	1.0902
1.60	1.2266	1.4079	1.2371	1.3604	1.2870	1.4353	1.0570	1.0684	1.0454	1.0801	1.0684	1.0969
1.65	1.2462	1.4439	1.2667	1.3920	1.3121	1.4737	1.0609	1.0731	1.0485	1.0855	1.0731	1.1040
1.70	1.2673	1.4829	1.2987	1.4261	1.3392	1.5152	1.0649	1.0779	1.0517	1.0912	1.0780	1.1114
1.75	1.2901	1.5252	1.3333	1.4630	1.3685	1.5601	1.0691	1.0830	1.0551	1.0971	1.0830	1.1191
1.80	1.3147	1.5710	1.3707	1.5030	1.4002	1.6087	1.0735	1.0882	1.0585	1.1033	1.0882	1.1273
1.85	1.3414	1.6208	1.4114	1.5464	1.4346	1.6616	1.0780	1.0937	1.0621	1.1097	1.0937	1.1358
1.90	1.3704	1.6750	1.4558	1.5937	1.4720	1.7191	1.0827	1.0993	1.0659	1.1163	1.0993	1.1448
1.95	1.4020	1.7343	1.5041	1.6452	1.5128	1.7820	1.0876	1.1052	1.0697	1.1232	1.1052	1.1542
2.00	1.4364	1.7992	1.5570	1.7016	1.5574	1.8508	1.0926	1.1113	1.0737	1.1304	1.1113	1.1641
2.05	1.4742	1.8706	1.6152	1.7635	1.6064	1.9264	1.0978	1.1176	1.0779	1.1378	1.1176	1.1745
2.10	1.5157	1.9494	1.6793	1.8318	1.6603	2.0098	1.1033	1.1241	1.0822	1.1455	1.1241	1.1854
2.15	1.5616	2.0365	1.7502	1.9072	1.7199	2.1020	1.1089	1.1309	1.0866	1.1535	1.1309	1.1968
2.20	1.6124	2.1336	1.8290	1.9911	1.7861	2.2046	1.1147	1.1379	1.0912	1.1617	1.1379	1.2088
2.25	1.6690	2.2421	1.9170	2.0847	1.8601	2.3192	1.1208	1.1452	1.0960	1.1703	1.1451	1.2215
2.30	1.7325	2.3641	2.0159	2.1899	1.9430	2.4481	1.1270	1.1527	1.1010	1.1792	1.1527	1.2347
2.35	1.8041	2.5021	2.1277	2.3087	2.0368	2.5937	1.1335	1.1605	1.1061	1.1884	1.1605	1.2487
2.40	1.8854	2.6595	2.2551	2.4440	2.1435	2.7597	1.1402	1.1686	1.1114	1.1979	1.1686	1.2635
2.45	1.9786	2.8403	2.4013	2.5993	2.2659	2.9503	1.1472	1.1769	1.1169	1.2078	1.1779	1.2790
2.50	2.0863	3.0502	2.5708	2.7793	2.4077	3.1714	1.1544	1.1856	1.1225	1.2180	1.1856	1.2954
2.55	2.2124	3.2964	2.7695	2.9902	2.5737	3.4305	1.1618	1.1946	1.1284	1.2286	1.1946	1.3127
2.60	2.3617	3.5890	3.0055	3.2406	2.7708	3.7383	1.1695	1.2039	1.1345	1.2396	1.2039	1.3310
2.65	2.5415	3.9422	3.2900	3.5425	3.0084	4.1097	1.1775	1.2135	1.1408	1.2510	1.2135	1.3504
2.70	2.7619	4.3766	3.6397	3.9134	3.3002	4.5661	1.1858	1.2235	1.1473	1.2628	1.2235	1.3709
2.75	3.0386	4.9232	4.0795	4.3796	3.6668	5.1401	1.1944	1.2338	1.1540	1.2751	1.2338	1.3926
2.80	3.3962	5.6315	4.6489	4.9831	4.1413	5.8835	1.2033	1.2445	1.1610	1.2878	1.2445	1.4157
2.85	3.8763	6.5845	5.4146	5.7945	4.7791	6.8832	1.2125	1.2556	1.1682	1.3009	1.2556	1.4403
2.90	4.5550	7.9343	6.4984	6.9427	5.6814	8.2985	1.2220	1.2671	1.1757	1.3146	1.2671	1.4665
2.95	5.5874	9.9915	8.1494	8.6915	7.0555	10.4548	1.2319	1.2790	1.1835	1.3287	1.2790	1.4944
3.00	7.3485	13.5056	10.9684	11.6771	9.4009	14.1368	1.2421	1.2913	1.1915	1.3434	1.2914	1.5243
3.05	11.0313	20.8628	16.8683	17.9249	14.3085	21.8433	1.2527	1.3042	1.1998	1.3586	1.3042	1.5563
3.10	23.5656	45.9228	36.9600	39.1994	31.0180	48.0883	1.2638	1.3174	1.2084	1.3744	1.3174	1.5906
3.15	−112.9817	−227.1810	−181.9753	−192.6153	−151.0482	−237.9031	1.2752	1.3312	1.2174	1.3908	1.3312	1.6275
3.20	−15.7399	−32.7066	−26.0700	−27.5371	−21.3955	−34.2475	1.2870	1.3455	1.2266	1.4078	1.3455	1.6674

TABLE 6. MAGNIFICATION FACTORS, K (cont.)

U	K_{25}	K_{26}	K_{27}	K_{28}	K_{29}	K_{30}	K_{31}	K_{32}	K_{33}	K_{34}	K_{35}	K_{36}	K_{37}
0.00	1.0000	1.0000	1.0000	1.0000	1.0000	1.0000	1.0000	1.0000	1.0000		1.0000	1.0000	1.0000
0.5	0.9242	0.9092	0.9430	0.9117	0.9797	0.9756	0.9746	0.9752	0.9948		0.9939	0.9959	0.9928
1.0	0.7616	0.7152	0.8193	0.7228	0.9242	0.9092	0.9054	0.9077	0.9797		0.9756	0.9837	0.9716
1.5	0.6034	0.5288	0.6958	0.5407	0.8469	0.8167	0.8093	0.8137	0.9556		0.9468	0.9644	0.9381
2.0	0.4820	0.3885	0.5969	0.4031	0.7616	0.7152	0.7039	0.7107	0.9242		0.9092	0.9391	0.8945
2.5	0.3946	0.2906	0.5215	0.3063	0.6786	0.6170	0.6022	0.6110	0.8874		0.8651	0.9092	0.8436
3.0	0.3317	0.2228	0.4632	0.2386	0.6034	0.5288	0.5110	0.5216	0.8469	0.4970	0.8167	0.8762	0.7881
3.5	0.2852	0.1751	0.4170	0.1904	0.5379	0.4526	0.4327	0.4445	0.8045	0.6389	0.7662	0.8415	0.7305
4.0	0.2498	0.1407	0.3792	0.1552	0.4820	0.3885	0.3671	0.3797	0.7616	0.6623	0.7152	0.8060	0.6729
4.5	0.2222	0.1152	0.3478	0.1288	0.4347	0.3350	0.3127	0.3258	0.7194	0.6423	0.6652	0.7707	0.6167
5.0	0.1999	0.0960	0.3210	0.1086	0.3946	0.2906	0.2678	0.2812	0.6786	0.6068	0.6170	0.7363	0.5633
5.5	0.1818	0.0811	0.2980	0.0928	0.3607	0.2536	0.2308	0.2441	0.6399	0.5668	0.5714	0.7032	0.5133
6.0	0.1667	0.0694	0.2780	0.0802	0.3317	0.2228	0.2001	0.2133	0.6034	0.5266	0.5288	0.6716	0.4670
6.5	0.1538	0.0601	0.2605	0.0700	0.3068	0.1969	0.1747	0.1875	0.5694	0.4881	0.4892	0.6418	0.4247
7.0	0.1429	0.0525	0.2450	0.0616	0.2852	0.1750	0.1534	0.1659	0.5379	0.4521	0.4526	0.6138	0.3862
7.5	0.1333	0.0462	0.2312	0.0547	0.2664	0.1565	0.1355	0.1475	0.5088	0.4189	0.4191	0.5876	0.3514
8.0	0.1250	0.0410	0.2188	0.0488	0.2498	0.1407	0.1204	0.1319	0.4820	0.3884	0.3885	0.5630	0.3200
8.5	0.1176	0.0366	0.2076	0.0439	0.2352	0.1270	0.1076	0.1186	0.4574	0.3605	0.3605	0.5401	0.2919
9.0	0.1111	0.0329	0.1975	0.0396	0.2222	0.1152	0.0966	0.1071	0.4347	0.3350	0.3350	0.5187	0.2667
9.5	0.1053	0.0297	0.1884	0.0360	0.2105	0.1050	0.0871	0.0971	0.4139	0.3117	0.3118	0.4987	0.2441
10.0	0.1000	0.0270	0.1800	0.0328	0.1999	0.0960	0.0789	0.0884	0.3946	0.2906	0.2906	0.4801	0.2238
10.5	0.0952	0.0246	0.1723	0.0300	0.1905	0.0881	0.0718	0.0808	0.3770	0.2712	0.2713	0.4626	0.2057
11.0	0.0909	0.0225	0.1653	0.0276	0.1818	0.0811	0.0656	0.0741	0.3607	0.2536	0.2536	0.4463	0.1894
11.5	0.0870	0.0207	0.1588	0.0254	0.1739	0.0750	0.0601	0.0682	0.3456	0.2375	0.2375	0.4310	0.1748
12.0	0.0833	0.0191	0.1528	0.0235	0.1667	0.0694	0.0553	0.0630	0.3317	0.2228	0.2228	0.4167	0.1617
12.5	0.0800	0.0177	0.1472	0.0218	0.1599	0.0645	0.0510	0.0583	0.3188	0.2093	0.2093	0.4032	0.1499
13.0	0.0769	0.0164	0.1420	0.0203	0.1538	0.0601	0.0472	0.0541	0.3068	0.1969	0.1969	0.3905	0.1392
13.5	0.0741	0.0152	0.1372	0.0189	0.1481	0.0561	0.0438	0.0504	0.2956	0.1855	0.1855	0.3786	0.1296
14.0	0.0714	0.0142	0.1327	0.0177	0.1429	0.0525	0.0407	0.0470	0.2852	0.1751	0.1750	0.3673	0.1209
14.5	0.0690	0.0133	0.1284	0.0166	0.1379	0.0492	0.0380	0.0439	0.2755	0.1654	0.1654	0.3567	0.1130
15.0	0.0667	0.0124	0.1244	0.0156	0.1333	0.0462	0.0355	0.0412	0.2664	0.1565	0.1565	0.3467	0.1058
15.5	0.0645	0.0117	0.1207	0.0146	0.1290	0.0435	0.0333	0.0386	0.2578	0.1482	0.1483	0.3371	0.0992
16.0	0.0625	0.0110	0.1172	0.0138	0.1250	0.0410	0.0312	0.0363	0.2498	0.1406	0.1407	0.3281	0.0932
16.5	0.0606	0.0104	0.1139	0.0130	0.1212	0.0387	0.0294	0.0342	0.2423	0.1336	0.1336	0.3196	0.0878
17.0	0.0588	0.0098	0.1107	0.0123	0.1176	0.0366	0.0277	0.0323	0.2352	0.1270	0.1270	0.3114	0.0828
17.5	0.0571	0.0092	0.1078	0.0116	0.1143	0.0347	0.0261	0.0305	0.2285	0.1209	0.1209	0.3037	0.0781
18.0	0.0556	0.0087	0.1049	0.0110	0.1111	0.0329	0.0247	0.0289	0.2222	0.1152	0.1152	0.2963	0.0739
18.5	0.0541	0.0083	0.1023	0.0105	0.1081	0.0313	0.0237	0.0274	0.2162	0.1099	0.1099	0.2893	0.0699
19.0	0.0526	0.0079	0.0997	0.0099	0.1053	0.0297	0.0222	0.0260	0.2105	0.1050	0.1050	0.2825	0.0664
19.5	0.0513	0.0075	0.0973	0.0095	0.1026	0.0283	0.0210	0.0247	0.2051	0.1003	0.1003	0.2761	0.0630
20.0	0.0500	0.0071	0.0950	0.0091	0.1000	0.0270	0.0199	0.0235	0.1999	0.0960	0.0960	0.2700	0.0599
20.5	0.0488	0.0068	0.0928	0.0086	0.0976	0.0258	0.0190	0.0224	0.1951	0.0919	0.0919	0.2641	0.0571
21.0	0.0476	0.0065	0.0907	0.0082	0.0952	0.0246	0.0181	0.0214	0.1905	0.0881	0.0881	0.2585	0.0544
21.5	0.0465	0.0062	0.0887	0.0079	0.0930	0.0235	0.0173	0.0204	0.1860	0.0845	0.0845	0.2531	0.0519
22.0	0.0455	0.0059	0.0868	0.0075	0.0909	0.0225	0.0165	0.0195	0.1818	0.0811	0.0811	0.2479	0.0496
22.5	0.0444	0.0057	0.0849	0.0072	0.0889	0.0216	0.0158	0.0187	0.1778	0.0780	0.0780	0.2430	0.0474
23.0	0.0435	0.0054	0.0832	0.0069	0.0870	0.0207	0.0151	0.0179	0.1739	0.0749	0.0750	0.2382	0.0454
23.5	0.0426	0.0052	0.0815	0.0067	0.0851	0.0199	0.0145	0.0171	0.1702	0.0721	0.0721	0.2336	0.0435
24.0	0.0417	0.0050	0.0799	0.0064	0.0833	0.0191	0.0139	0.0164	0.1667	0.0694	0.0694	0.2292	0.0417
24.5	0.0408	0.0048	0.0783	0.0061	0.0816	0.0183	0.0133	0.0158	0.1633	0.0669	0.0669	0.2249	0.0399
25.0	0.0400	0.0046	0.0768	0.0059	0.0800	0.0177	0.0128	0.0152	0.1599	0.0645	0.0645	0.2208	0.0384
25.5	0.0392	0.0044	0.0754	0.0057	0.0784	0.0170	0.0123	0.0146	0.1569	0.0622	0.0622	0.2168	0.0369
26.0	0.0385	0.0043	0.0740	0.0055	0.0769	0.0164	0.0118	0.0140	0.1538	0.0601	0.0601	0.2130	0.0355
26.5	0.0377	0.0041	0.0726	0.0053	0.0755	0.0158	0.0114	0.0135	0.1509	0.0580	0.0580	0.2093	0.0342
27.0	0.0370	0.0040	0.0713	0.0051	0.0741	0.0152	0.0110	0.0130	0.1481	0.0561	0.0561	0.2058	0.0329
27.5	0.0364	0.0038	0.0701	0.0649	0.0727	0.0147	0.0106	0.0126	0.1455	0.0542	0.0542	0.2023	0.0317
28.0	0.0357	0.0037	0.0689	0.0048	0.0714	0.0142	0.0102	0.0121	0.1429	0.0525	0.0525	0.1990	0.0306
28.5	0.0351	0.0036	0.0677	0.0046	0.0702	0.0137	0.0098	0.0117	0.1404	0.0508	0.0508	0.1958	0.0295
29.0	0.0345	0.0034	0.0666	0.0044	0.0690	0.0133	0.0095	0.0113	0.1379	0.0492	0.0492	0.1926	0.0285
29.5	0.0339	0.0033	0.0655	0.0043	0.0678	0.0129	0.0092	0.0109	0.1356	0.0477	0.0477	0.1896	0.0276
30.0	0.0333	0.0032	0.0644	0.0042	0.0667	0.0124	0.0089	0.0106	0.1333	0.0462	0.0462	0.1867	0.0267
30.5	0.0328	0.0031	0.0634	0.0040	0.0656	0.0121	0.0086	0.0102	0.1311	0.0448	0.0448	0.1838	0.0258
31.0	0.0323	0.0030	0.0624	0.0039	0.0645	0.0117	0.0083	0.0099	0.1290	0.0435	0.0435	0.1811	0.0250
31.5	0.0317	0.0029	0.0615	0.0038	0.0635	0.0113	0.0081	0.0096	0.1270	0.0422	0.0422	0.1784	0.0242
32.0	0.0312	0.0028	0.0605	0.0037	0.0625	0.0109	0.0078	0.0093	0.1250	0.0410	0.0410	0.1758	0.0234
32.5	0.0308	0.0027	0.0596	0.0036	0.0615	0.0107	0.0076	0.0090	0.1231	0.0399	0.0398	0.1733	0.0227
33.0	0.0303	0.0027	0.0588	0.0035	0.0606	0.0104	0.0073	0.0088	0.1212	0.0387	0.0387	0.1708	0.0220
33.5	0.0299	0.0026	0.0579	0.0034	0.0597	0.0101	0.0071	0.0085	0.1194	0.0377	0.0377	0.1684	0.0214
34.0	0.0294	0.0025	0.0571	0.0033	0.0588	0.0097	0.0069	0.0082	0.1176	0.0366	0.0366	0.1661	0.0208
34.5	0.0290	0.0024	0.0563	0.0032	0.0580	0.0095	0.0067	0.0080	0.1159	0.0357	0.0357	0.1638	0.0202
35.0	0.0286	0.0024	0.0555	0.0031	0.0571	0.0092	0.0065	0.0078	0.1143	0.0347	0.0347	0.1616	0.1096
35.5	0.0282	0.0023	0.0547	0.0030	0.0563	0.0090	0.0063	0.0076	0.1127	0.0338	0.0338	0.1595	0.0190
36.0	0.0278	0.0022	0.0540	0.0029	0.0556	0.0087	0.0062	0.0074	0.1111	0.0329	0.0329	0.1574	0.0185
36.5	0.0274	0.0022	0.0533	0.0028	0.0548	0.0085	0.0060	0.0072	0.1096	0.0320	0.0320	0.1554	0.0180
37.0	0.0270	0.0021	0.0526	0.0028	0.0540	0.0083	0.0058	0.0070	0.1081	0.0313	0.0313	0.1534	0.0175
37.5	0.0267	0.0020	0.0519	0.0027	0.0533	0.0081	0.0057	0.0068	0.1067	0.0305	0.0305	0.1515	0.0171
38.0	0.0263	0.0020	0.0512	0.0026	0.0526	0.0079	0.0055	0.0066	0.1053	0.0296	0.0297	0.1496	0.0166
38.5	0.0260	0.0020	0.0506	0.0026	0.0519	0.0077	0.0054	0.0064	0.1039	0.0290	0.0290	0.1477	0.0162
39.0	0.0256	0.0019	0.0499	0.0025	0.0513	0.0075	0.0053	0.0063	0.1026	0.0283	0.0283	0.1460	0.0158
39.5	0.0253	0.0019	0.0494	0.0024	0.0506	0.0073	0.0051	0.0061	0.1013	0.0277	0.0276	0.1442	0.0154
40.0	0.0250	0.0018	0.0488	0.0024	0.0500	0.0071	0.0050	0.0060	0.1000	0.0270	0.0270	0.1425	0.0150
40.5	0.0247	0.0018	0.0482	0.0023	0.0494	0.0069	0.0049	0.0058	0.0988	0.0264	0.0264	0.1408	0.0146
41.0	0.0244	0.0017	0.0476	0.0023	0.0488	0.0068	0.0048	0.0057	0.0976	0.0258	0.0258	0.1392	0.0143
41.5	0.0241	0.0017	0.0470	0.0022	0.0482	0.0066	0.0046	0.0055	0.0964	0.0252	0.0252	0.1376	0.0139
42.0	0.0238	0.0017	0.0465	0.0022	0.0476	0.0065	0.0045	0.0054	0.0952	0.0246	0.0246	0.1361	0.0136
42.5	0.0235	0.0016	0.0460	0.0021	0.0471	0.0063	0.0044	0.0053	0.0941	0.0241	0.0241	0.1345	0.0133
43.0	0.0233	0.0016	0.0454	0.0021	0.0465	0.0062	0.0043	0.0052	0.0930	0.0235	0.0236	0.1330	0.0130
43.5	0.0230	0.0015	0.0449	0.0020	0.0460	0.0061	0.0042	0.0051	0.0919	0.0230	0.0230	0.1316	0.0127
44.0	0.0227	0.0015	0.0444	0.0020	0.0455	0.0059	0.0041	0.0049	0.0909	0.0225	0.0225	0.1302	0.0124
44.5	0.0225	0.0015	0.0439	0.0019	0.0449	0.0058	0.0040	0.0048	0.0899	0.0221	0.0221	0.1288	0.0121
45.0	0.0222	0.0014	0.0435	0.0019	0.0444	0.0057	0.0039	0.0047	0.0889	0.0216	0.0216	0.1274	0.0119
45.5	0.0220	0.0014	0.0430	0.0018	0.0440	0.0055	0.0039	0.0046	0.0879	0.0211	0.0211	0.1261	0.0116
46.0	0.0217	0.0014	0.0425	0.0018	0.0435	0.0054	0.0038	0.0045	0.0870	0.0207	0.0207	0.1248	0.0113
46.5	0.0215	0.0014	0.0421	0.0018	0.0430	0.0053	0.0037	0.0044	0.0860	0.0203	0.0203	0.1235	0.0111
47.0	0.0213	0.0013	0.0416	0.0017	0.0425	0.0052	0.0036	0.0043	0.0851	0.0199	0.0199	0.1222	0.0109
47.5	0.0211	0.0013	0.0412	0.0017	0.0421	0.0051	0.0035	0.0042	0.0842	0.0195	0.0195	0.1210	0.0106
48.0	0.0208	0.0013	0.0408	0.0017	0.0417	0.0050	0.0035	0.0042	0.0833	0.0191	0.0191	0.1198	0.0104
48.5	0.0206	0.0012	0.0404	0.0016	0.0412	0.0049	0.0034	0.0041	0.0825	0.0187	0.0187	0.1186	0.0102
49.0	0.0204	0.0012	0.0399	0.0016	0.0408	0.0048	0.0033	0.0040	0.0816	0.0184	0.0184	0.1175	0.0099
49.5	0.0202	0.0012	0.0396	0.0016	0.0404	0.0047	0.0033	0.0039	0.0808	0.0180	0.0180	0.1163	0.0098
50.0	0.0200	0.0012	0.0392	0.0015	0.0400	0.0046	0.0032	0.0038	0.0800	0.0177	0.0177	0.1152	0.0096

Nomogram 1

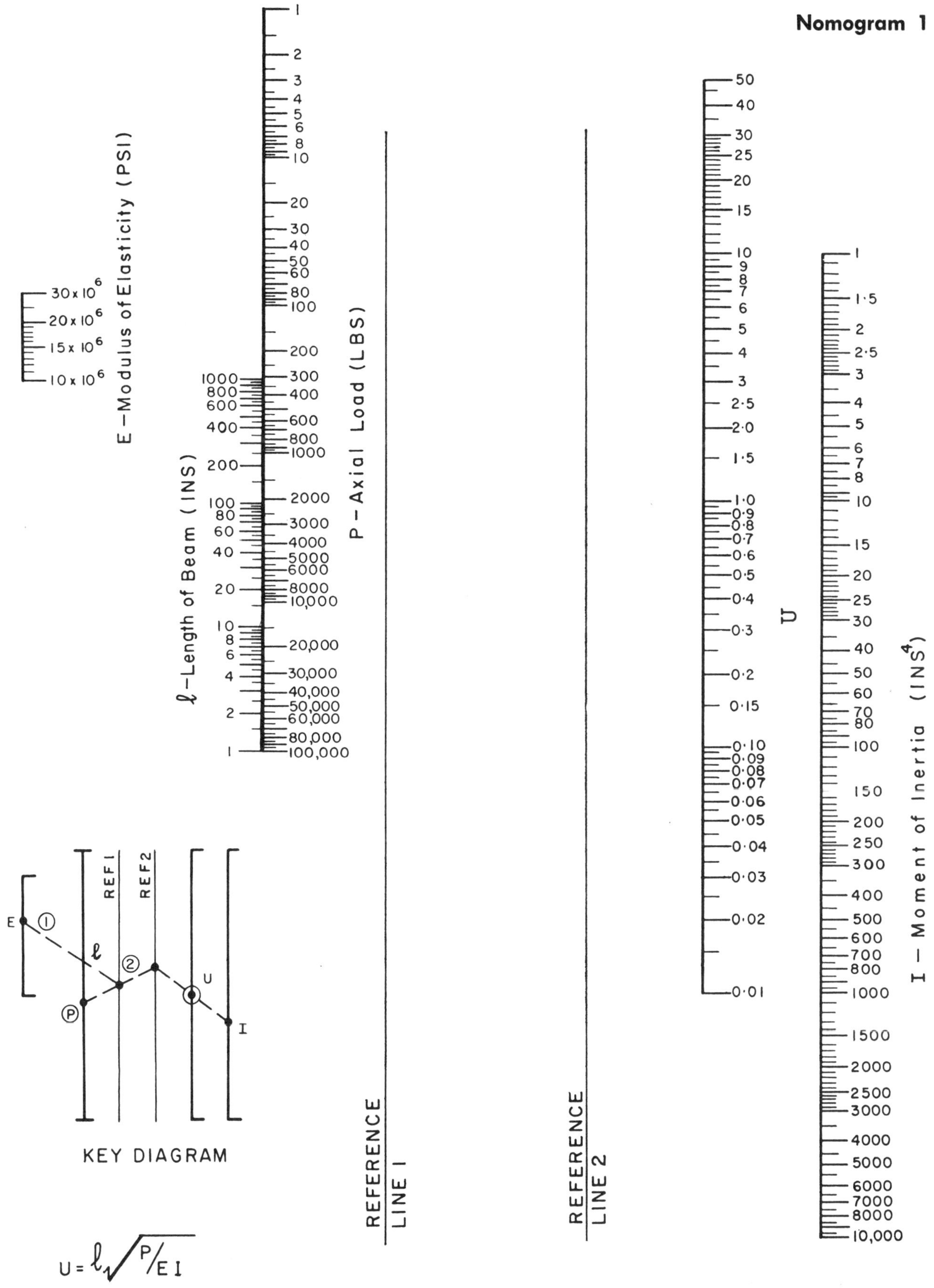
E – Modulus of Elasticity (PSI)
30 x 10^6
20 x 10^6
15 x 10^6
10 x 10^6
ℓ – Length of Beam (INS)
P – Axial Load (LBS)
REFERENCE LINE 1
REFERENCE LINE 2
U
I – Moment of Inertia (INS^4)
KEY DIAGRAM
REF 1
REF 2
E
ℓ
P
U
I
U = ℓ √(P/EI)

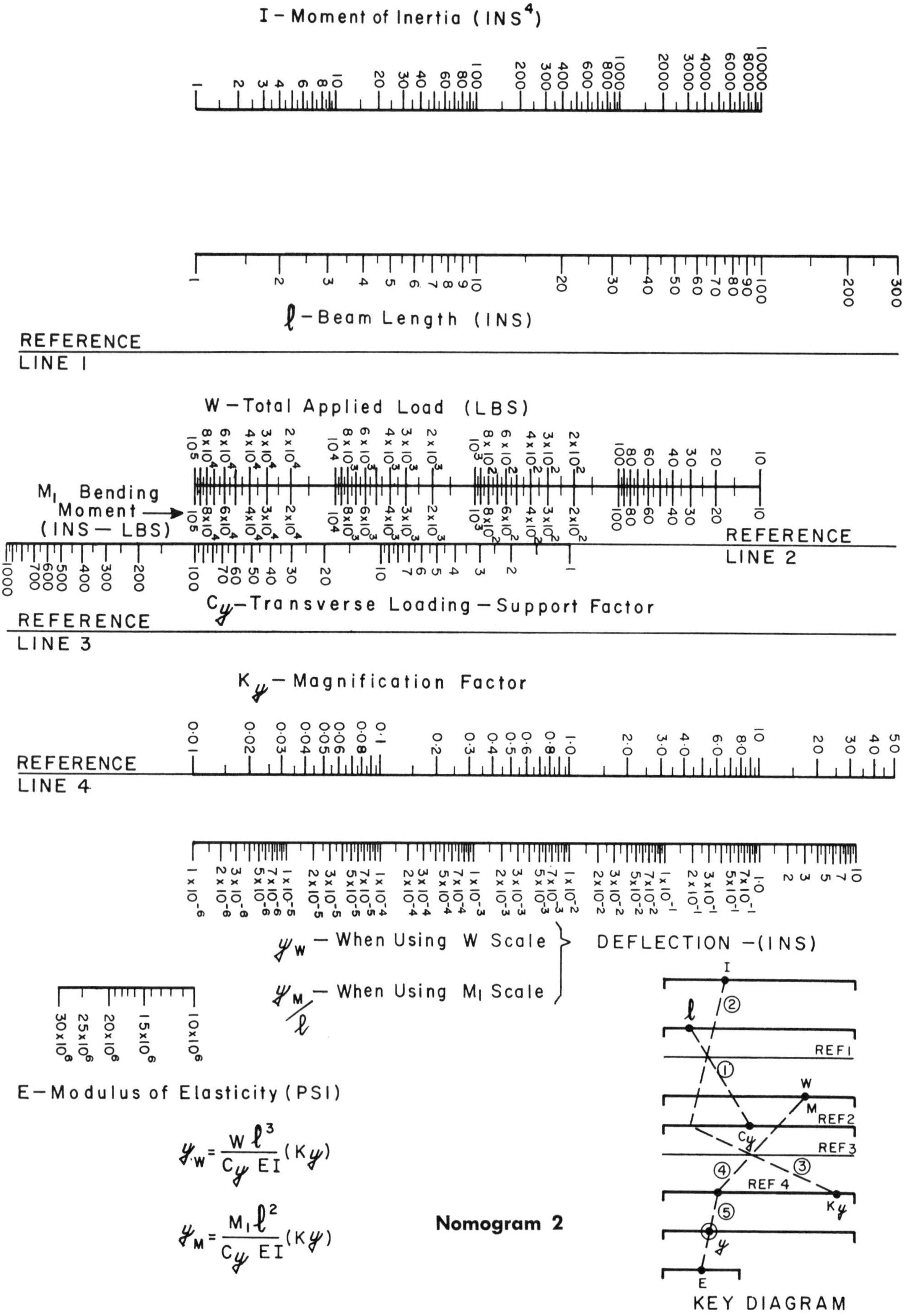

$$y_W = \frac{W \ell^3}{C_y EI}(K_y)$$

$$y_M = \frac{M_1 \ell^2}{C_y EI}(K_y)$$

Nomogram 2

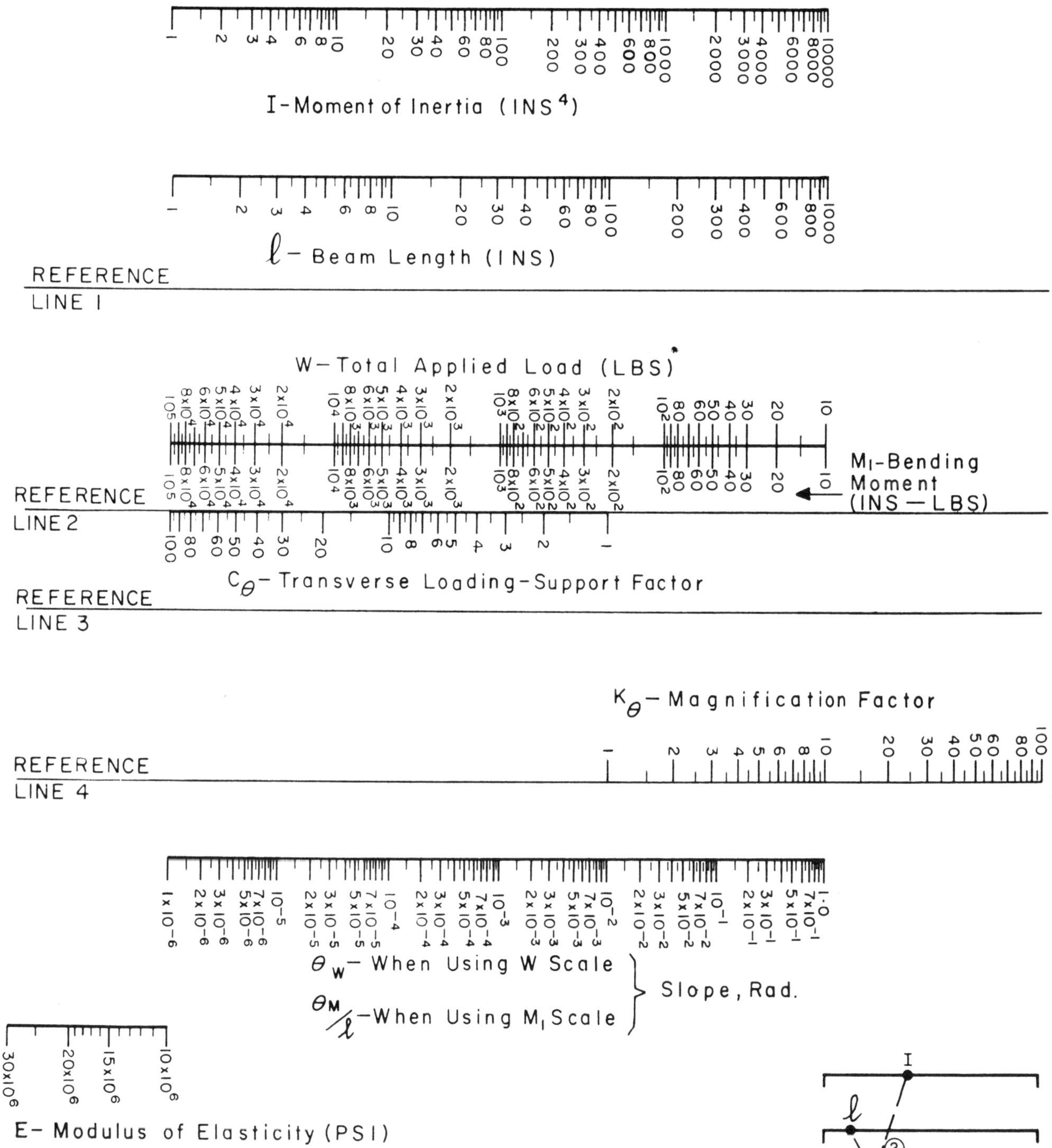

$$\theta_W = \frac{W\ell^2}{C_\theta EI}(K_\theta)$$

$$\theta_M = \frac{M_1\ell}{C_\theta EI}(K_\theta)$$

Nomogram 3

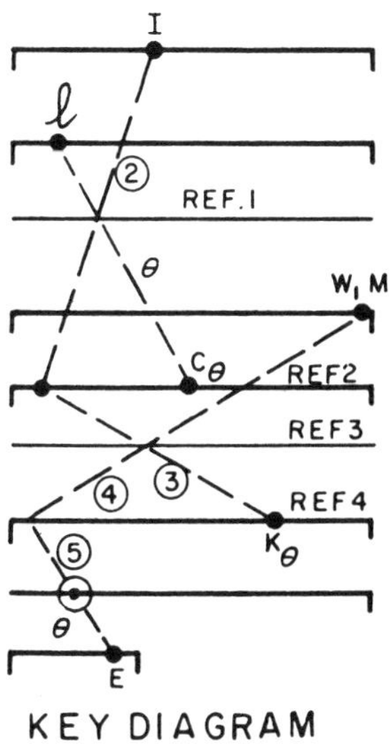

KEY DIAGRAM

24. MAXIMUM SAFE LOAD

(*Ref. 11*)

The calculation of the maximum safe load, in pounds, based on flexural stress $S = Mc/I$ may be tedious when unusual cross sections are involved; the expressions for the moment of inertia usually present a problem.

A simplified formula for the maximum safe load was developed for 15 loading and support conditions. This formula is applicable to the following eight most frequently used beam sections: (1) solid rectangle; (2) solid cylinder; (3) even-legged angle; (4) tee; (5) channel; (6) Z-bar; (7) deck beam; (8) I-beam.

$$W = \frac{1{,}000KAd}{L},$$

where W = maximum safe load (lb.)
A = sectional area of beam (in.^{2})
d = depth of beam (in.)
L = distance between supports (in.)
K = nondimensional factor.

The numerical values for K are tabulated in Table 7. It is important to note that values of K are for steel beams with allowable fiber stress of 16,000 psi. For values of $S > 16{,}000$ psi multiply the safe load, as found from the formula, by $S/16{,}000$.

TABLE 7.

Note:
Values of K are for allowable fiber stress of 16,000 psi. For values of $S > 16,000$, multiply the safe load as found from the formula $S / 16,000$

Loading style and support position		K factor (Section shape) 	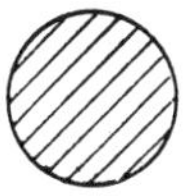	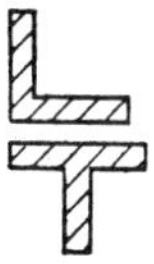	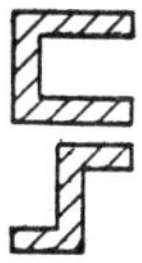		
Cantilever, end load	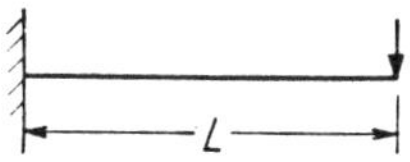	2.67	2.00	2.65	4.57	4.10	5.08
Cantilever, uniform load	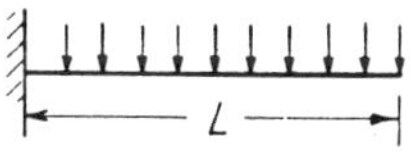	4.34	4.00	5.30	9.14	8.20	10.16
Cantilever, triangular load	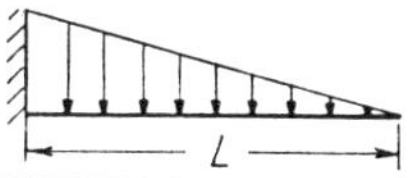	8.01	6.00	7.95	13.71	12.30	15.24
Cantilever, triangular load	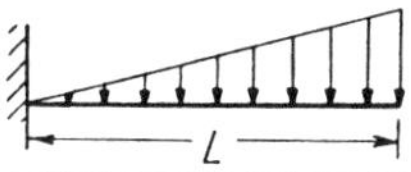	4.00	3.00	3.97	6.85	6.15	7.62
End supports, center load	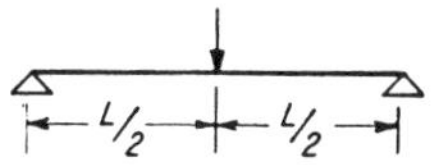	10.68	8.00	10.60	18.28	16.40	20.32
End supports, uniform load	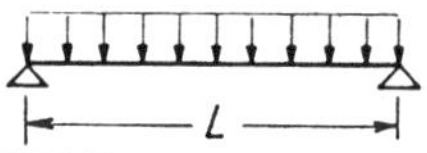	21.36	16.00	21.20	36.56	32.80	40.64
End supports, triangular load	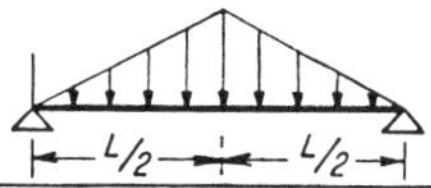	16.02	12.00	15.90	27.42	24.60	30.48
End supports triangular load	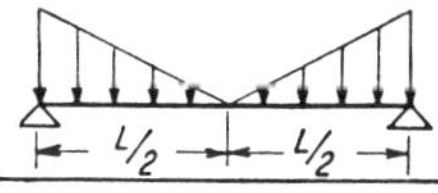	32.04	24.00	31.80	54.84	49.20	60.69
One end fixed, one end supported, center load	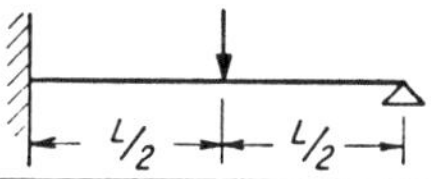	14.24	10.67	14.13	24.37	21.87	27.09
One end fixed, one end supported, uniform load	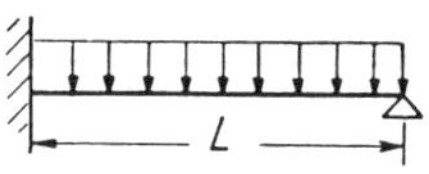	21.36	16.00	21.20	36.56	32.80	40.64
One end fixed, one end supported, triangular load	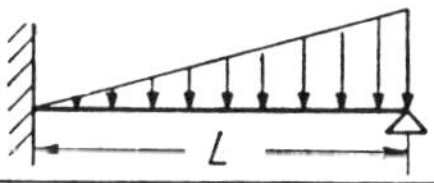	22.2	16.7	22.1	38.1	34.2	40.3
One end fixed, one end supported, triangular load	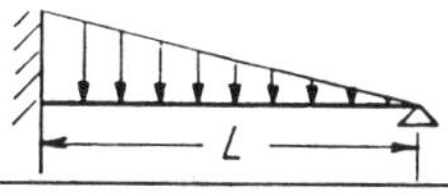	44.50	33.33	44.17	76.17	68.33	84.67
Both ends fixed, center load	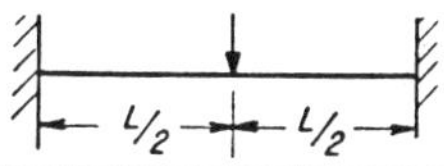	21.36	16.00	21.20	36.56	32.80	40.64
Both ends fixed, uniform load	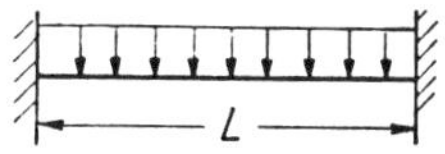	32.04	24.00	31.80	54.84	49.20	60.69
Both ends fixed, triangular load	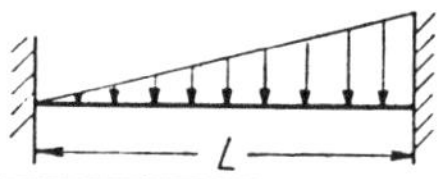	26.70	20.00	26.50	45.70	41.00	50.80

25. UNSYMMETRICAL BENDING

(*Ref. 12*)

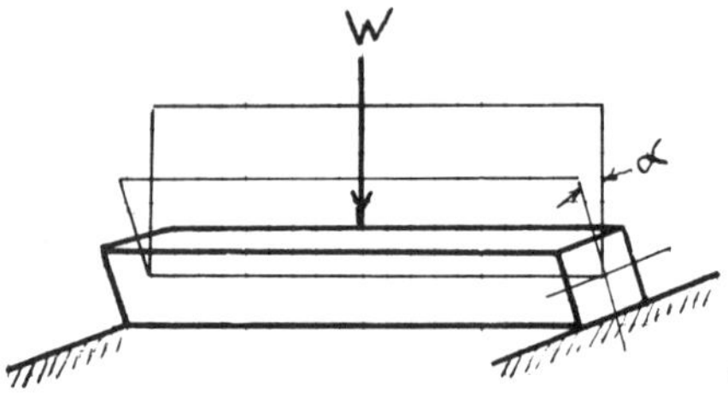

FIG. 30

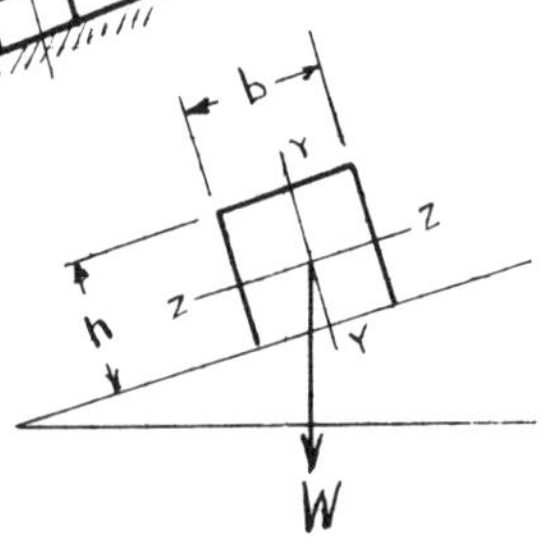

FIG. 31

Sometimes it is necessary to analyze a beam that has its axes of symmetry at an angle to the plane of loading, as shown in Fig. 30. For the condition shown, the principal axes of the section and the plane of loading do not coincide. Bending of the nature shown in Fig. 30 is known as unsymmetrical bending. The brief treatment of the subject given here is confined to cases of pure bending only.

It is assumed that the loads act through and in the direction perpendicular to the bending axis of the beam. Therefore the beam is not subjected to torsion and the bending stresses are not accompanied by direct axial tensile or compressive stresses.

The elastic deflection of a beam subjected to unsymmetrical loading may be found as follows: The bending moment may be resolved into components parallel to principal planes, and the deflection caused by these components of the moment are calculated from the usual equations for deflections of symmetrically loaded beams; the actual deflection of the beam is the vector sum of the deflections found from the component moments. This method gives both the magnitude and the direction of the deflection.

The general equations for the deflection y and stress S are:

$$y = \frac{KWL^3}{E}\left[\frac{\cos^2\alpha}{I_z} + \frac{\sin^2\alpha}{I_y}\right]$$

$$S = M\left[\frac{z\cos\alpha}{I_z} + \frac{y\sin\alpha}{I_y}\right].$$

For a beam of rectangular cross section of depth h and width b the respective moment of inertias are:

$$I_z = bh^3/12;\ I_y = hb^3/12;\ A = bh.$$

Thus

$$y_{\max} = \frac{KWL^3}{AE}\left[\frac{\cos^2\alpha}{h^2} + \frac{\sin^2\alpha}{b^2}\right] \qquad (33)$$

$$S_{\max} = \frac{6M}{A}\left[\frac{\cos\alpha}{h} + \frac{\sin\alpha}{b}\right], \qquad (34)$$

where M = bending moment with respect to the section in question (in.-lb.)

α = angle which the plane of bending moment or the plane of loads makes with Y-axis

K = dimensionless factor for beams of rectangular cross section (Table 8)

$M\cos\alpha$ = the component of bending moment causing bending about the principal axis which has been designated as the Z-axis

$M\sin\alpha$ = the component of bending moment causing bending about the principal axis designated as Y-axis.

The sign of the two terms of unit stress may be determined by inspection by the usual way, and the result will be tension or compression as determined by the algebraic sum of the two terms in equation (34).

In general, it may be stated when the plane of the bending moment coincides with one of the principal axes, then the other axis is the neutral axis. This is the ordinary case, in which the ordinary formula for unit stress may be applied. When the plane of the bending moment does not coincide with one of the principal axes, then the above formulas for oblique loading may be applied.

Example: A steel beam of rectangular cross section carries a uniformly distributed load of intensity w and is supported at the ends in the position shown in Fig. 31. Find the maximum normal stress and the maximum deflection if the length L of the beam is 10 ft., w = 200 lb./ft., h = 8 in., b = 6 in., and $\tan\alpha = 1/3$.

Solution: From Table 8, case 6, $K = 0.156$, $M = WL/8$ = 30,000 in.-lb., $A = b \times h = 48$ in.², $E = 30 \times 10^6$ psi.

By substitution in equations (33) and (34) we obtain:

$$S_{\max} = \frac{6 \times 30{,}000}{48}\left[\frac{0.94869}{8} + \frac{0.31620}{6}\right] = 642 \text{ psi.}$$

$$y_{\max} = \frac{0.156 \times 2{,}000 \times 120^3}{48 \times 30 \times 10^6}\left[\frac{0.94819^2}{64} + \frac{0.31620^2}{36}\right] = 0.006 \text{ in.}$$

TABLE 8. FOR RECTANGULAR CROSS SECTIONS

Deflection and stress factors for unsymmetrical beams				
Beam type		K (for deflection)	M (for stress)	Location of maximum deflection
1	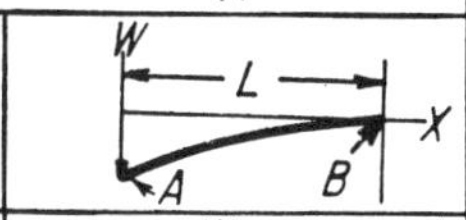	4.00	WL	A
2	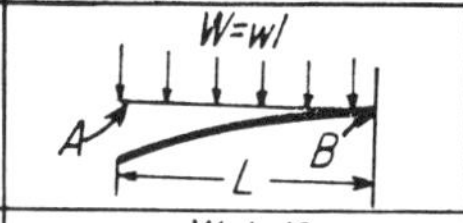	1.50	WL/2	A
3	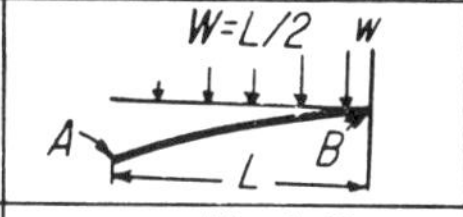	0.80	WL/3	A
4	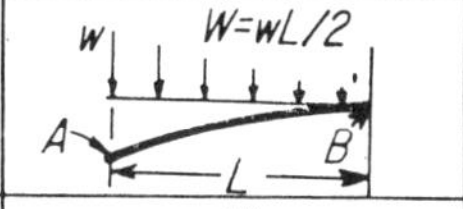	2.20	2WL	A
5	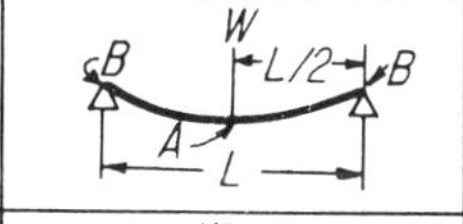	0.25	WL/4	A
6	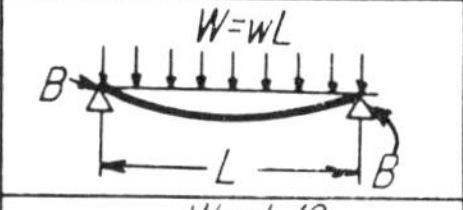	0.156	WL/8	L/2
7	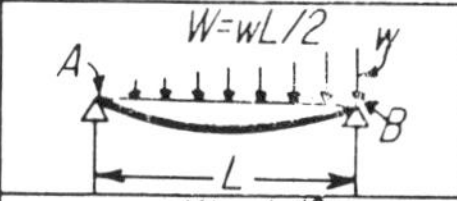	0.156	WL/8	0.519L
8	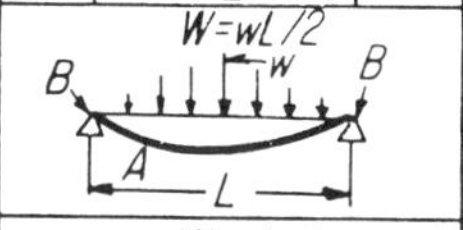	0.20	WL/6	L/2
9	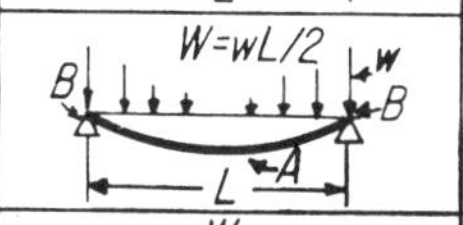	0.112	WL/2	L/2
10	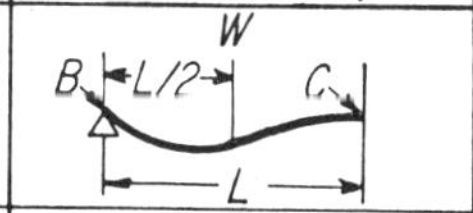	0.112	3WL/16	0.447L
11	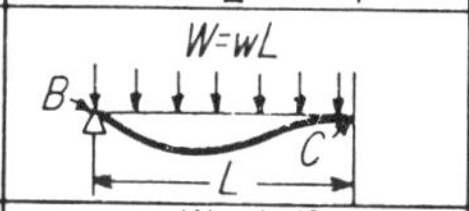	0.065	WL/6	0.421L
12	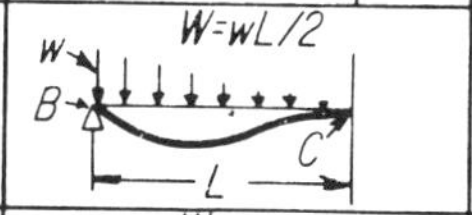	0.073	7WL/60	0.402L
13	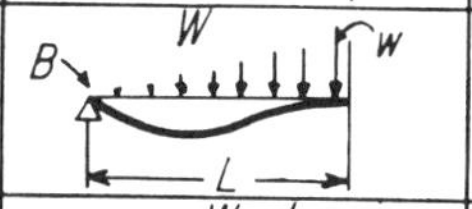	0.057	2WL/15	$L\sqrt{1/5}$
14	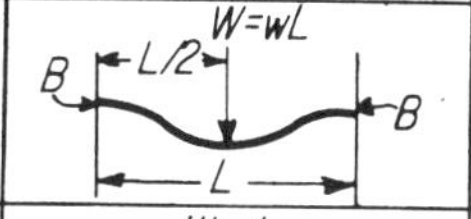	0.062	WL/8	L/2
15	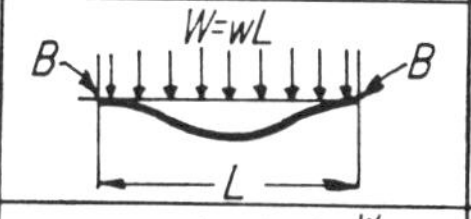	0.031	WL/12	L/2
16	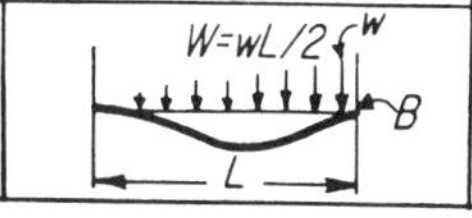	0.031	WL/10	0.525L

26. DEFLECTION CAUSED BY SHEARING STRESS. BEAMS OF RELATIVELY GREAT DEPTH

(*Ref. 13*)

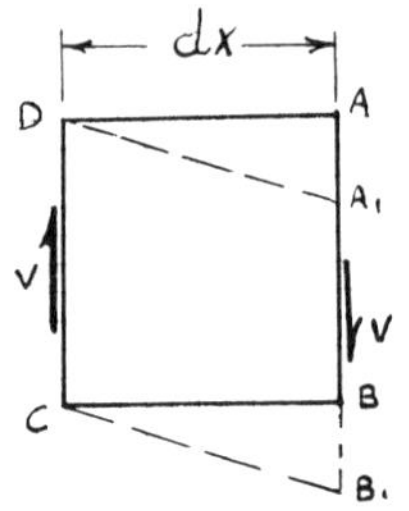

FIG. 32

The deflection of beams due to shearing stress is generally so small, compared to the bending deflection, that it is usually neglected in practice. In short beams, however, with the ratio of section depth to beam length greater than 1:10, the shearing deflection becomes significant. Neglecting it can lead to inaccurate answers.

Figure 32 represents a beam element $ABCD$ acted upon by shearing stresses of intensity v applied to the face AB and CD. Because of the action of these stresses the element is deformed to A_1, B_1, C_1, D_1. The modulus of elasticity in shear, G, is equal to the unit shearing stress divided by the unit shearing strain. In Fig. 32, dy_S represents the shearing strain hence dy_S/dx represents the unit shearing strain.

Thus

$$G = \frac{v}{dy_S/dx} = \frac{\text{unit shearing stress}}{\text{unit shearing strain}}$$

from which $dy_S/dx = v/G$.

If V is shear on section due to external forces and A is the cross-sectional area,

$$v = V/A,$$

and consequently:

$$dy_S/dx = V/AG \text{ and } Y_S = \frac{1}{AG}\int_0^x V\,dx, \qquad (35)$$

where Y_S is the shearing deflection at any point at a distance X from a convenient origin.

In equation (35) $\int_0^x V dx = M$ (the moment), hence:

$$Y_s = \frac{M}{AG} = \frac{WL}{K_S AG} \qquad (36)$$

the maximum deflection due to bending is

$$Y_B = \frac{WL^3}{K_B EI},$$

where K_S and K_B are dimensionless factors.

$$Y_S/Y_B = \frac{K_B}{K_S}\frac{E}{AG}\frac{I}{L^2}.$$

Noting that $I = Ar^2$ where r is the radius of gyration, we obtain

$$Y_S/Y_B = \frac{K_B E}{K_S G}\left(\frac{r}{L}\right)^2. \qquad (37)$$

Tests show that for steel beam $G = 0.4E$: Expressing the radius of gyration, r, as a function of the depths, d, of the beam we obtain from equation (37):

$$Y_S/Y_B = C\left(\frac{d}{L}\right)^2. \qquad (38)$$

C is a factor depending on the form of the cross section.

NOTE: For I-beams and channels: $r = 0.4d$. For rectangular cross sections: $r = 0.289d$. For circular cross section: $r = 0.25d$.

From equations (36) and (38):

$$Y_B = \frac{Y_S}{C}\left(\frac{L}{d}\right)^2 = \frac{WL}{K_S CAG}\left(\frac{L}{d}\right)^2.$$

Finally, the total deflection due to bending and shear:

$$Y_T = Y_B + Y_S = \frac{WL}{K_S AG}\left[1 + \frac{1}{C}\left(\frac{L}{d}\right)^2\right]. \qquad (39)$$

Numerical values for K_S and C, applicable to I-beams, channels, beams with rectangular cross sections, and beams with circular cross section, for 15 most commonly used loading and support conditions are shown in Table 9.

Example: Find the maximum deflection for a simply supported I-beam, 12 ft. long. The beam is 36 in. deep and is subjected to a load of 20,000 lb. per ft. The cross-sectional area of the beam is 31 in². Shear modulus $G = 12.0 \times 10^6$ psi.

Solution: From Table 9, case 6:

$$K_S = 8, C = 3.84.$$

From equation (39), the total deflection:

$$Y_T = \frac{20{,}000 \times 12 \times 144}{8 \times 31 \times 12 \times 10}\left[1 + \frac{1}{3.84}\left(\frac{144}{36}\right)^2\right] = 0.018 \text{ in.}$$

The deflection due to shearing alone would be, again Table 9, case 6,

$$Y_S = \frac{WL}{8AG} = \frac{20{,}000 \times 12 \times 144}{8 \times 31 \times 12 \times 10^6} = 0.012 \text{ in.}$$

This is about 65 per cent of the total deflection and would represent a sizable error if not included in the total deflection of the beam.

TABLE 9

Loading style and support position		Section shape				
		All sections		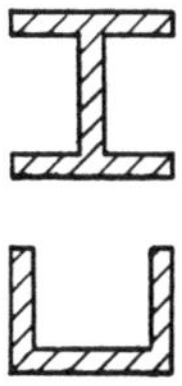	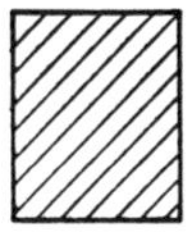	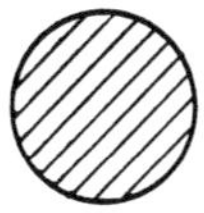
		y_s	K_s	C factor		
1 Cantilever, end load	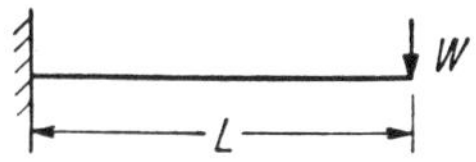	$\frac{WL}{AG}$	1	1.20	0.75	0.46
2 Cantilever, uniform load	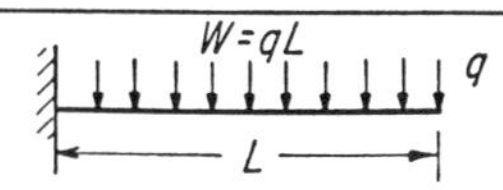	$\frac{WL}{2AG}$	2	1.60	1.00	0.62
3 Cantilever, triangular load	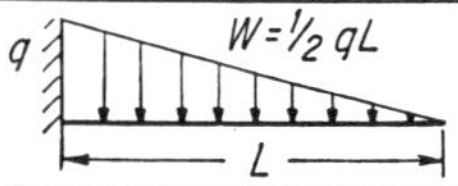	$\frac{WL}{3AG}$	3	2.00	1.25	0.77
4 Cantilever, triangular load	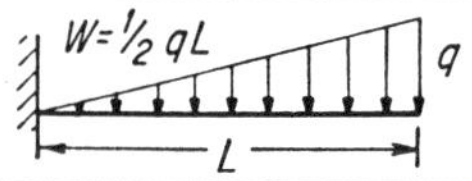	$\frac{WL}{1.5\,AG}$	1.5	1.56	0.91	0.56
5 End supports, center load	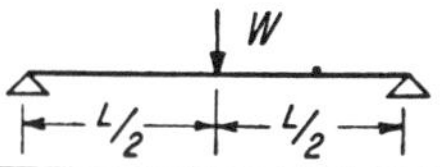	$\frac{WL}{4AG}$	4	4.80	3.00	1.86
6 End supports, uniform load	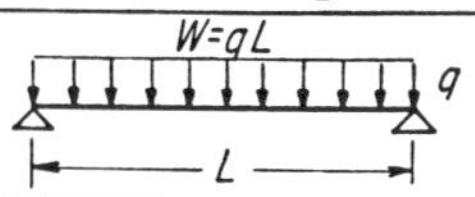	$\frac{WL}{8AG}$	8	3.84	2.40	1.49
7 End supports, triangular load	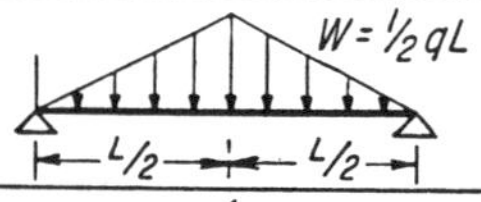	$\frac{WL}{6AG}$	6	4.00	2.50	1.55
8 End supports, triangular load	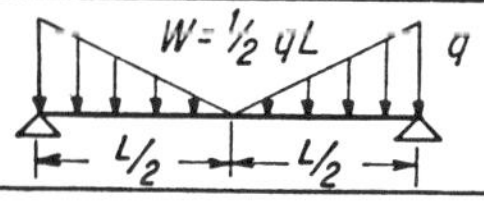	$\frac{WL}{12\,AG}$	12	3.56	2.22	1.38
9 One end fixed, one end supported, center load	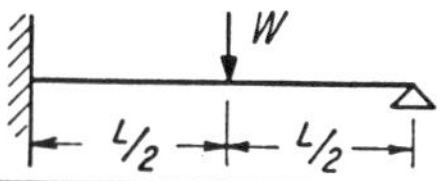	$\frac{WL}{6.4\,AG}$	6.4	6.71	4.19	2.60
10 One end fixed, one end supported, uniform load	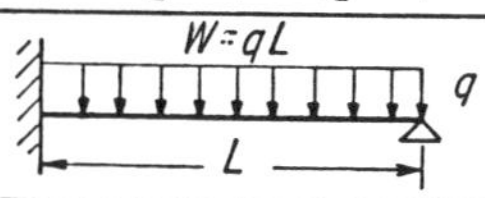	$\frac{WL}{14.2AG}$	14.2	5.21	3.25	2.02
11 One end fixed, one end supported, triangular load	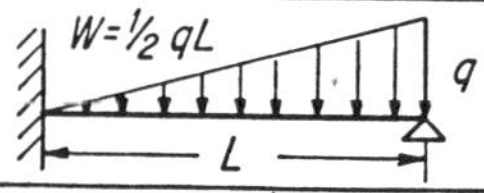	$\frac{WL}{11.8\,AG}$	11.8	5.56	3.47	2.15
12 One end fixed, one end supported, triangular load	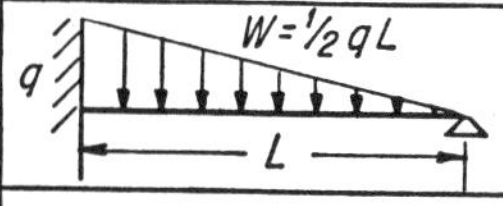	$\frac{WL}{16.7\,AG}$	16.7	5.03	3.14	1.95
13 Both ends fixed, center load	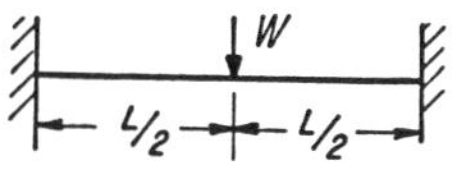	$\frac{WL}{8\,AG}$	8	9.60	6.00	3.72
14 Both ends fixed, uniform load	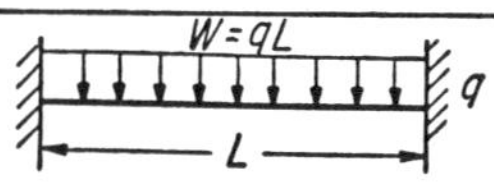	$\frac{WL}{24\,AG}$	24	6.4	4.00	2.48
15 Both ends fixed, triangular load	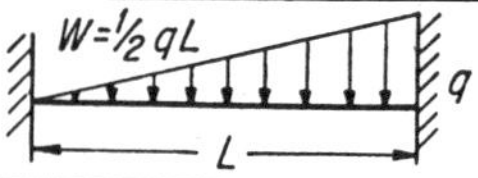	$\frac{WL}{23.2\,AG}$	23.2	6.57	4.11	2.55

27. ENERGY STORAGE CAPACITY IN BENDING AND SPRING RATE OF BEAMS OF RECTANGULAR CROSS SECTIONS

(*Ref. 14*)

The formulas for the resilience U and spring rate p are:

$$U = S^2V/K_RE$$

$$p = K_SEI/L^3,$$

where U = resilience (in.-lb.)
S = maximum stress (psi)
V = volume of beam (cu. in.)
E = modulus of elasticity (psi)
K_R, K_S = loading-support nondimensional factors, (Table 10)
I = moment of inertia (in.[4])
L = beam length (in.)

Numerical values for K_R, K_S, for 16 different loading and support conditions are tabulated in Table 10.

TABLE 10

CASE NO.	LOADING AND SUPPORT	BEAM	K_R	K_S
1	Cantilever, end load	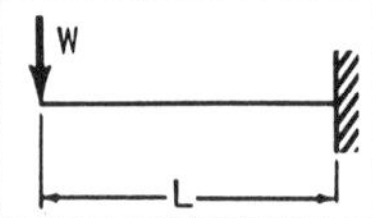	18	3
2	Cantilever, uniform load	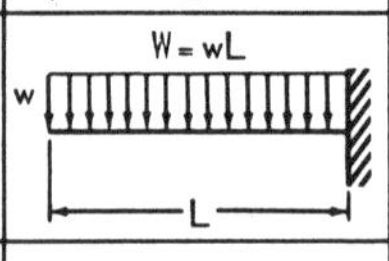	12	8
3	Cantilever, triangular load	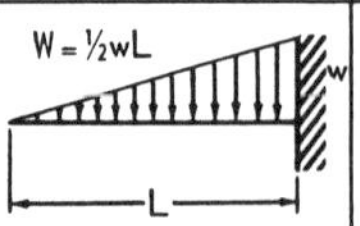	10	15
4	Cantilever, triangular load	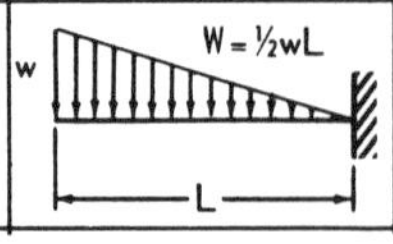	14.5	5.4
5	End supports, center load	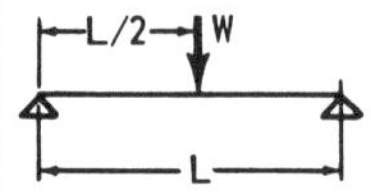	18	48
6	End supports, uniform load	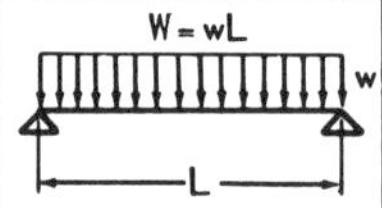	7.2	76.8
7	End supports, triangular load	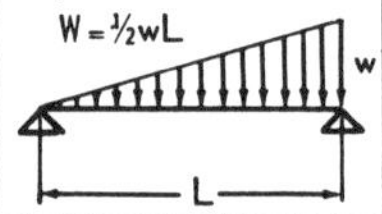	7.5	76.7
8	End supports, triangular load	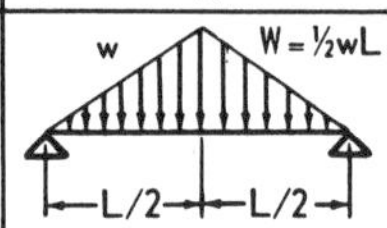	10	60
9	End supports, triangular load	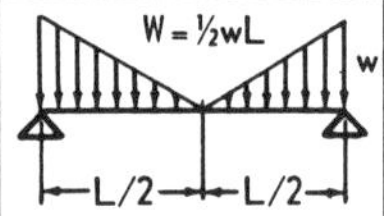	4.4	106.7
10	One end fixed, one end supported, center load	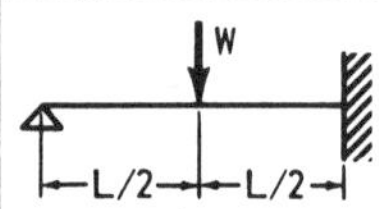	22.7	107.3
11	One end fixed, one end supported, uniform load	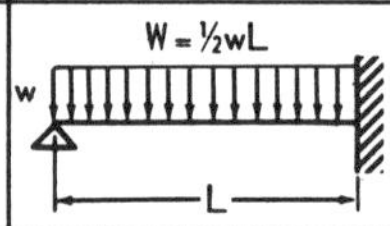	17.3	185
12	One end fixed, one end supported triangular load	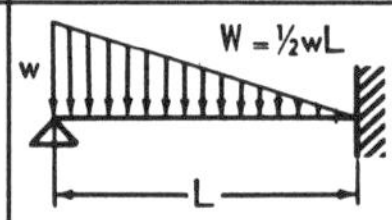	13.5	164
13	One end fixed, one end supported, triangular load	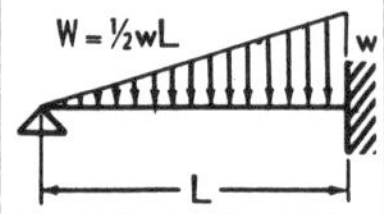	4.5	210
14	Both ends fixed, center load	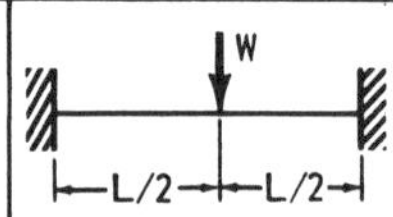	18	192
15	Both ends fixed, uniform load	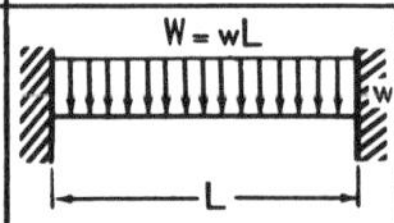	16	384
16	Both ends fixed, triangular load	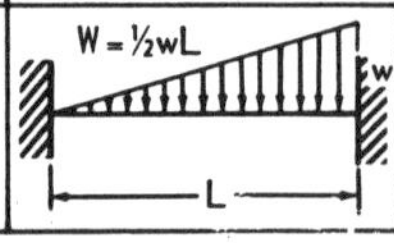	22.9	382

28. CIRCULAR SHAFT UNDER COMBINED BENDING AND TWISTING MOMENT

Simplified Method for Determining Shaft Diameter

(*Ref. 15*)

The expression for diameter of a shaft subjected to combined bending and torsion can be simplified as follows:

$$d = 1.72\sqrt{\frac{M_B + \sqrt{M_B^2 + M^2}}{S}}. \qquad (40)$$

NOTATION

d = diameter of shaft (in.)
S = maximum fiber stress at outer fiber (psi)
M_B = applied bending moment (in.-lb.)
M_T = applied twisting moment (in.-lb.)
$M_E = M_B + \sqrt{M_B{}^2 + M_T{}^2}$, equivalent twisting moment

Equation (40) is presented graphically in Graph I.

It allows: (1) Finding the required diameter of a shaft for a given twisting moment and a given stress; (2) finding the required diameter of a shaft for a given bending moment and a given stress ($M_T = O$ in equation (40), therefore use 2 M_B); (3) finding the equivalent twisting moment for a shaft acted upon by a combined twisting and bending moment when diameter of the shaft and the fiber stress are given; (4) finding the diameter of a shaft required for a given combined twisting and bending moment and for a given fiber stress. As plotted, the chart covers all possible moments and shaft diameters and intensity of stresses up to and including 25,000 psi.

Scale A represents the bending moment in in.-lb. Each numbered division may represent 100, 1,000, or 10,000 in.-lb. as the nature of the problem demands. Scale B represents the intensity of the fiber stress, each numbered division representing 1,000 psi. Scale C represents twisting moment used only to find the equivalent moment. It is important to note that same scale division value for Scales A and C should be used. Three values are given on the diagonal scales. The smallest quantity is used when numbered divisions on the A Scale have been taken as 100 in.-lb. of twisting moment. Intermediate values correspond to 1,000 in.-lb. moment divisions, and largest values correspond to 10,000 in-lb. divisions. If the point of intersection of the perpendiculars does not fall on a diagonal, the values may be obtained by interpolation.

Example: Find the diameter of a shaft if $M_B = 300$ in.-lb., $M_T = 400$ in.-lb., $S = 10{,}000$ psi.

Solution: First find the equivalent twisting moment by laying out the $M_B = 300$ on Scale A, and $M_T = 400$ on Scale C. Set a pair of dividers on the graph, as shown, to the length between these two points. Add this length, to which the dividers have been set, to the length previously plotted on the A Scale. The point thus found will represent the combined twisting and bending moment, 800 in.-lb. From this point on Scale A draw a perpendicular line; from $S = 10{,}000$ on Scale B draw a horizontal line. These two lines intersect at point $d = 0.735$ in. which is the diameter of the shaft.

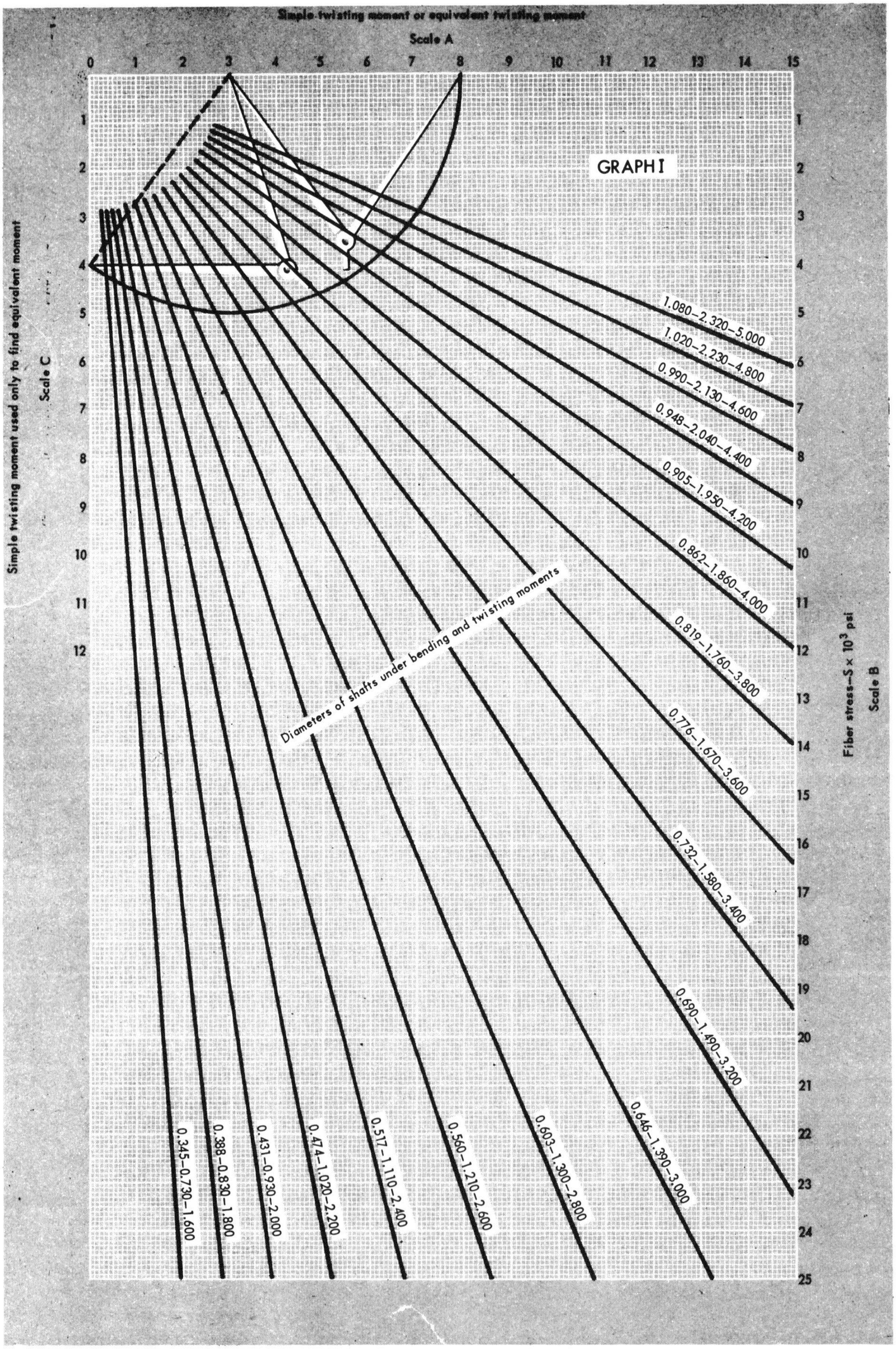
Simple twisting moment or equivalent twisting moment
Scale A
GRAPH I
Simple twisting moment used only to find equivalent moment
Scale C
Fiber stress—S × 10³ psi
Scale B
Diameters of shafts under bending and twisting moments
1.080–2.320–5.000
1.020–2.230–4.800
0.990–2.130–4.600
0.948–2.040–4.400
0.905–1.950–4.200
0.862–1.860–4.000
0.819–1.760–3.800
0.776–1.670–3.600
0.732–1.580–3.400
0.690–1.490–3.200
0.646–1.390–3.000
0.603–1.300–2.800
0.560–1.210–2.600
0.517–1.110–2.400
0.474–1.020–2.200
0.431–0.930–2.000
0.388–0.830–1.800
0.345–0.730–1.600

29. SHEAR CENTER FOR BEAM SECTIONS

WITH ONE AXIS OF SYMMETRY

(*Ref. 16*)

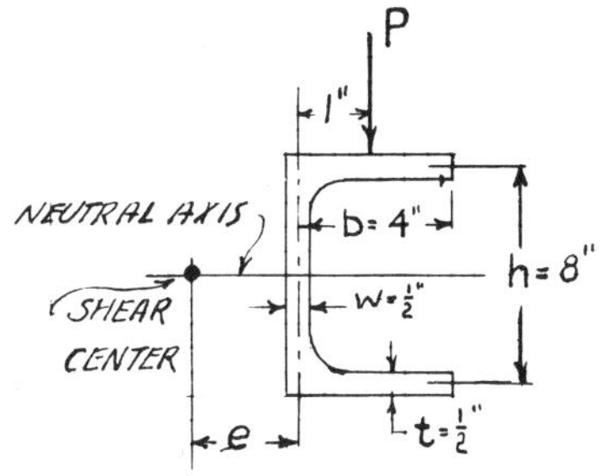

FIG. 33

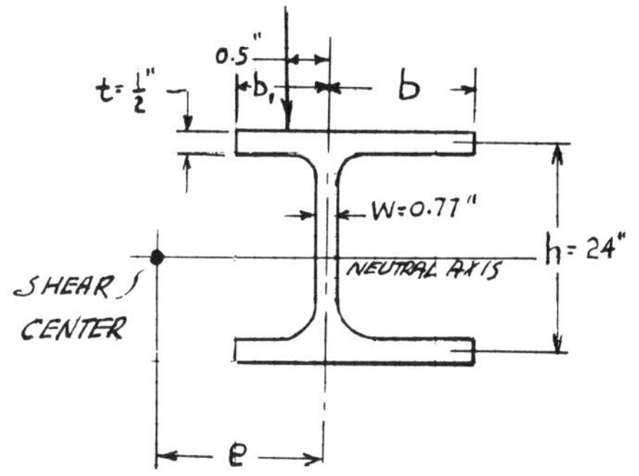

FIG. 34

The beam formula $S = Mc/I$ applies to beams on which the load acts in a plane of symmetry of the beam cross sections. A case of bending sometimes encountered is that in which a beam is subjected to loads acting in a plane parallel to one of the principal planes of bending. The simplest cases are those in which the cross sections of the beam have one axis of symmetry and the loads act in a plane perpendicular to this axis (Figs. 33 and 34).

In order to determine the stresses produced in beams loaded in this manner, it is desirable to locate the position of the load, defined by the point O, which will produce only bending free from rotation or twisting of the cross section. The point of loading for this condition is called the "shear center." Thus, unless the loads pass through the "shear center" the beam will be subjected to torsion as well as bending.

Beams whose cross sections are composed of narrow rectangles such as I and channel sections are designed to offer large resistance to bending in accordance with the flexure formula, but are poorly designed to resist twisting. For such sections, it may be of great importance to prevent twisting by insuring that the transverse bending loads pass through the shear center.

For any section having two or more axes of symmetry and for any section having a point of symmetry (such as equilateral triangle, Z-bar etc. . . .), the shear center O is at the centroid. For any section having but one axis of symmetry, O is on that axis, but in general not at the centroid. For such sections, and for unsymmetrical sections in general, the position of the shear center must be determined by calculations.

Two most commonly used sections with one axis of symmetry are considered here: (1) Channel, with

$$e = \frac{0.5b}{1 + 1/6(wh/bt)},$$

and (2) I-beam with unsymmetrical flanges for which

$$e = \frac{3(b^2 - b_1^2)}{wh/t + 6(b + b_1)}; \quad b_1 < b.$$

The solution of equations for the shear center involve many factors with a rather large spread in the magnitude of the values. To save time and to reduce the likelihood of error it is suggested to use the two nomograms; Nomogram I for channels, Nomogram II for I-beams.

Example 1: How much torsion will force P produce in the channel section shown in Fig. 33; ($b = 4$ in., $t = \frac{1}{2}$ in., $w = \frac{1}{2}$ in., $h = 8$ in.)

Solution: The stresses consist of bending stresses produced by the bending moment $M = PL$, transverse shear stresses produced by the shear force $V = P$ and torsional stresses produced by the torque $T = P(e + 1.0)$.

To find e from the nomogram, align $w/t = 1$ with b/h (Scale 1) $= 0.5$, intersecting the Reference line 1. Align this intersection with $b = 4$, intersecting Reference line 2. Align this intersection with b/h (Scale 2) $= 0.5$, and read $e = 1.5$ in.

$$\therefore T = (1.5 + 1)P = 2.5P.$$

Example 2: How much torsion will force P produce in the section shown in Fig. 34, ($b = 11$ in., $b_1 = 4$ in., $t = \frac{1}{2}$ in., $w = 0.77$ in., $h = 24$ in.)

Solution: Using Nomogram II, align $h = 24$ with $w/t = 0.7$, intersecting Reference line 1. Align this intersection with $b + b_1$ (Scale 1) $= 15$, intersecting Reference line 2. Align this intersection with $b + b_1$ (Scale 2) $= 15$, intersecting Reference line 3. Align this intersection with $b - b_1 = 7$, reading $e = 3$ in.

$$\therefore T = (3 - 0.5)P = 2.5P.$$

NOTATION

b = width of channel flange (in.)
$b + b_1$ = width of I-beam flange (in.)
h = depth (in.)
w = web thickness (in.)
t = flange thickness (in.)
e = location of shear center (in.)

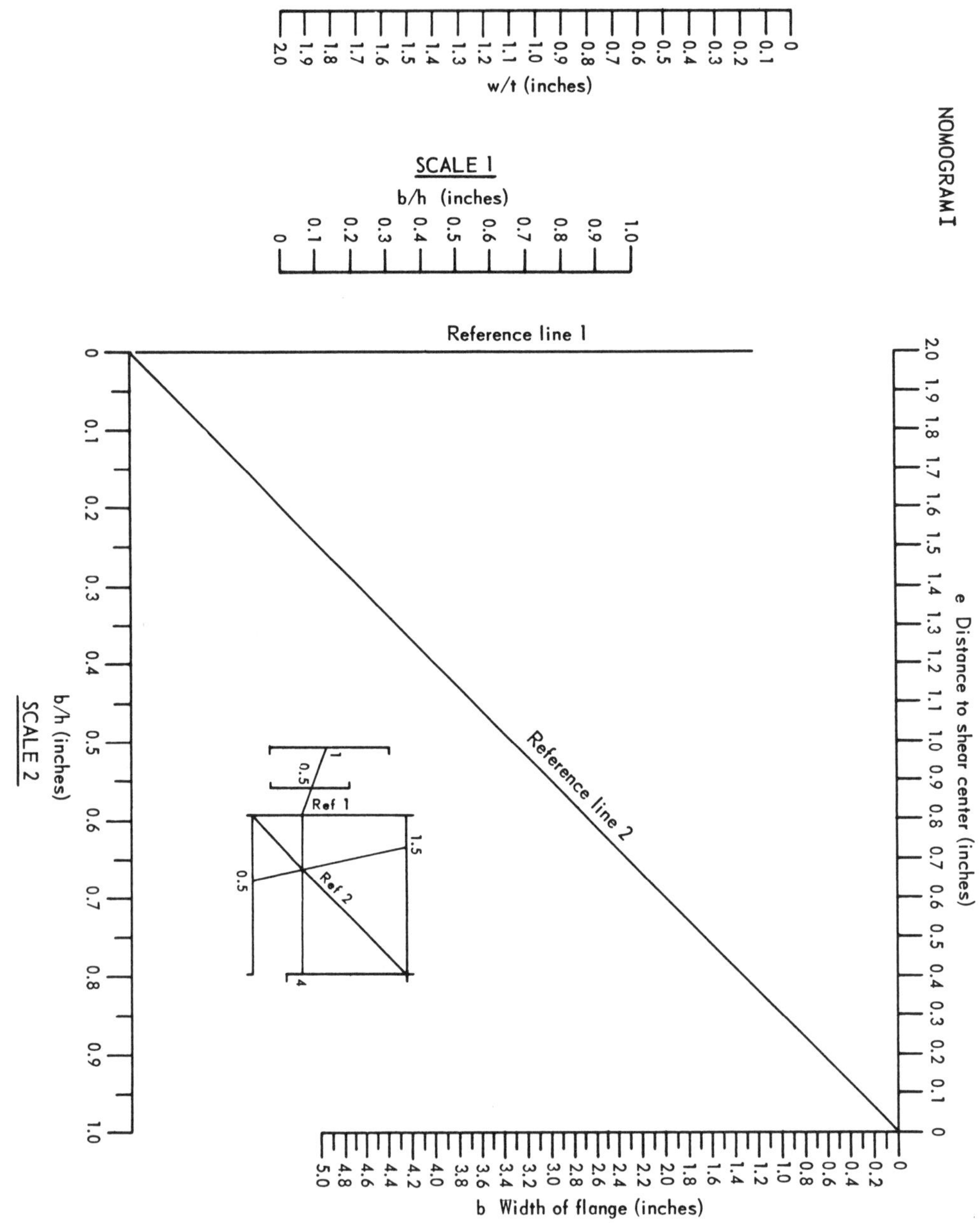
NOMOGRAM I
w/t (inches)
SCALE 1
b/h (inches)
Reference line 1
Reference line 2
b/h (inches)
SCALE 2
e Distance to shear center (inches)
b Width of flange (inches)
Ref 1
Ref 2

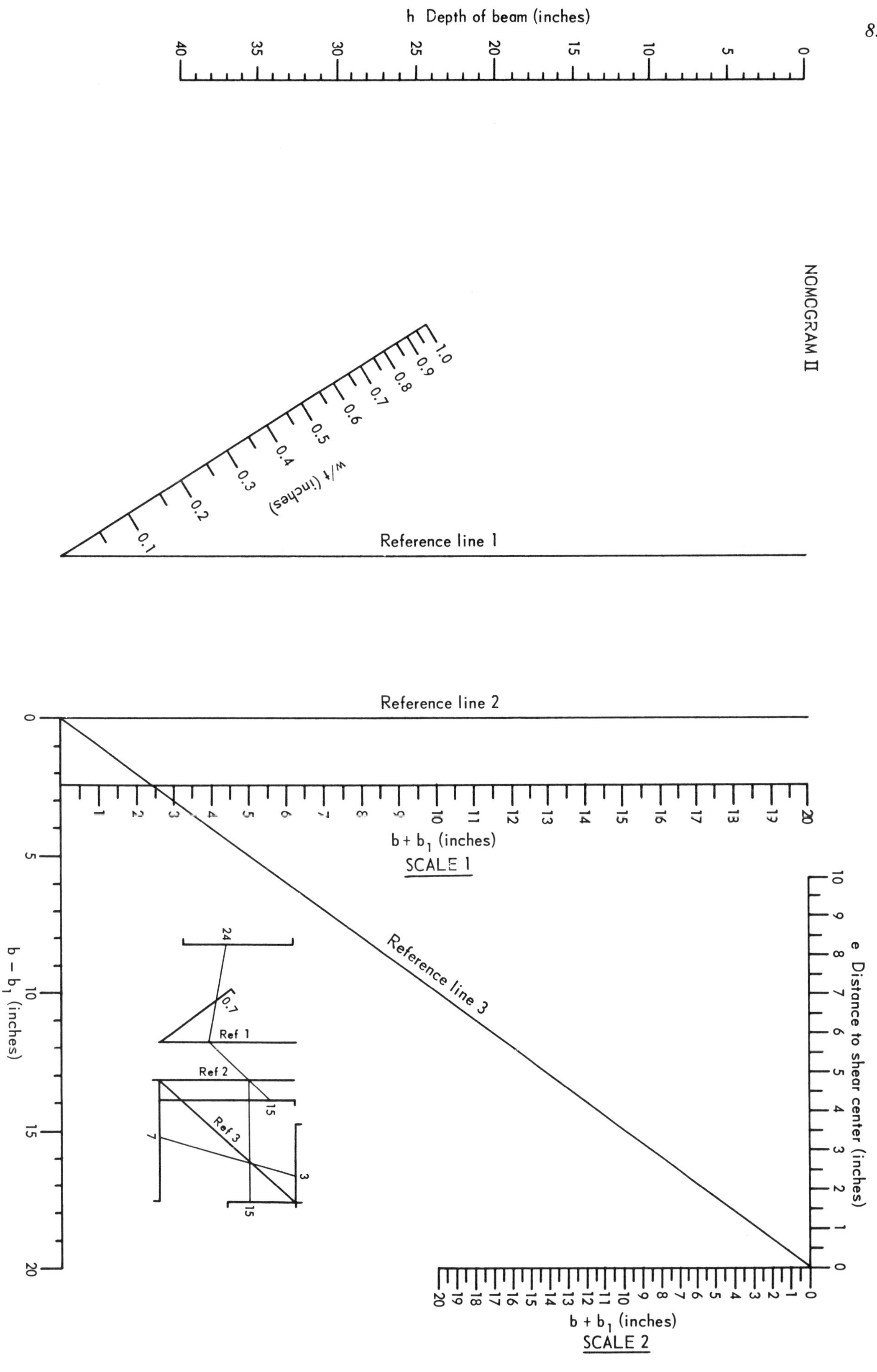
h Depth of beam (inches)
40 35 30 25 20 15 10 5 0
w/t (inches)
0.1 0.2 0.3 0.4 0.5 0.6 0.7 0.8 0.9 1.0
Reference line 1
Reference line 2
b + b1 (inches)
SCALE 1
Reference line 3
b − b1 (inches)
0 5 10 15 20
e Distance to shear center (inches)
0 1 2 3 4 5 6 7 8 9 10
b + b1 (inches)
SCALE 2
24
0.7
Ref 1
Ref 2
Ref 3
15
7
3
15

NOMOGRAM II

30. STEPPED BEAMS

Deflection and Slope

In the previous cases of beams with moment of inertia constant all along the length of the beam, we integrated only M as a function of x. On the other hand, if I also varies with x the differential equation of the deflection curve will be

$$d^2y/dx^2 = \frac{M}{I_o/I}\left(\frac{1}{EI}\right), \tag{41}$$

where I is an arbitrarily chosen basis to which we refer the varying I_o. Integrating $M/(I_o/I)$ function we can obtain the deflection line.

Since I_o/I is dimensionless, $M/(I_o/I)$, having the dimension of bending moment, represents as a function of x, a corrected bending-moment for the beam with uniform moment of inertia I. Hence, it follows that the M moment diagram on the beam of constant I will result in the same deflection curve as the $M/(I_o/I)$ moment diagram on the beam of varying moment of inertia. In other words, a certain $P/(I_o/I)$ loading system which originates the $M/(I_o/I)$ moment diagram, will produce exactly the same deflection curve on the beam of constant cross section as the original P loading would do on the beam of variable moment of inertia. Thus, if a beam or a portion of a beam with a moment of inertia I_o is replaced by a beam with a moment of inertia of $I = kI_o$ and at the same time all the loads and reactions acting upon the beam are changed by a factor of $k = I_o/I$, the elastic curve of the two beams will be the same.

In the case of beam of uniform cross section it was shown that when several moments and forces are applied the general equations for deflection y and slope θ are:

$$EIy = EIy_o + EI\theta_o x + \sum\frac{M(x-a)^2}{2} + \sum\frac{P(x-b)^2}{6} + \sum\frac{q(x-c)^4}{24} \tag{42}$$

$$EI\theta = EI\theta_o + \sum M(x-a) + \sum\frac{P(x-b)^2}{2} + \sum\frac{q(x-c)^3}{6}, \tag{43}$$

where a, b, c, are distances from the origin of coordinates to the points of application of the moment M, the concentrated load P, and the point where the uniform load begins, y_o and θ_o are the deflection and slope at $x = 0$.

In order to apply the general equations (42) and (43) to a beam, the cross section of which varies by steps, it is necessary, by means of simple transformation, to replace the beam of variable rigidity by a beam of constant rigidity. This will be explained on the following numerical example.

Example: Let it be required to determine the end slope θ_o, and the deflection y under the load of a steel shaft shown in Fig. 35a.

Solution: The two-step shaft has a 1-in. diameter with $I_o = 0.049$ in.^{4} and $1\frac{1}{4}$-in. diameter with $I = 0.120$ in.^{4}. It is proposed to replace the stepped shaft by a shaft of constant cross section, $1\frac{1}{4}$-in. in diameter. The resulting transformation of the loads and reactions in each portion will be carried as follows. First, construct a bending moment (Fig. 35b) followed by shear diagram (Fig. 35c). Figure 35d shows the shaft with moments and shears at the discontinuities. In Fig. 35e the stepped shaft is transformed to a constant diameter shaft of $1\frac{1}{4}$ in. with the loads transformed by a factor of $I/I_o = 2.444$. The loads acting upon the middle portion of the shaft remain unchanged, since we do not change the moment of inertia of this portion. Finally, in Fig. 35f we show, schematically, the shaft of uniform rigidity for which the general equations (42) and (43) apply.

Thus, the effect of the change in the cross section being established as identical to that of a certain force system, it is possible to find the deflection in a beam with any number of discontinuities by a simple expedient of using one equation.

There is no approximation involved in the method developed here. It is exact, simple, and devoid of the usual tedium.

Assume the origin of coordinates at the left support. Thus at $x = 26$ in., $y_o = 0$, and equation (42) yields:

$$EIy_{x=26} = EI\theta_o 26 - \frac{2755(26-9)^2}{2} + \frac{3742(26-17)^2}{2} + \frac{517(26-0)^3}{6} - \frac{305(26-9)^3}{6} - \frac{500(26-15)^3}{6} - \frac{415(26-17)^3}{6} = 0$$

thence,

$$\theta_o = -\frac{32{,}948}{EI} = -0.0092 \text{ radians}$$

$$EIy_{x=15} = -32{,}948 \times 15 - \frac{2755(15-9)^2}{2} - \frac{305(15-9)^3}{6} + \frac{517(15-0)^3}{6} = -264{,}518$$

or,

$$y_{x=15} = -0.073 \text{ in.}$$

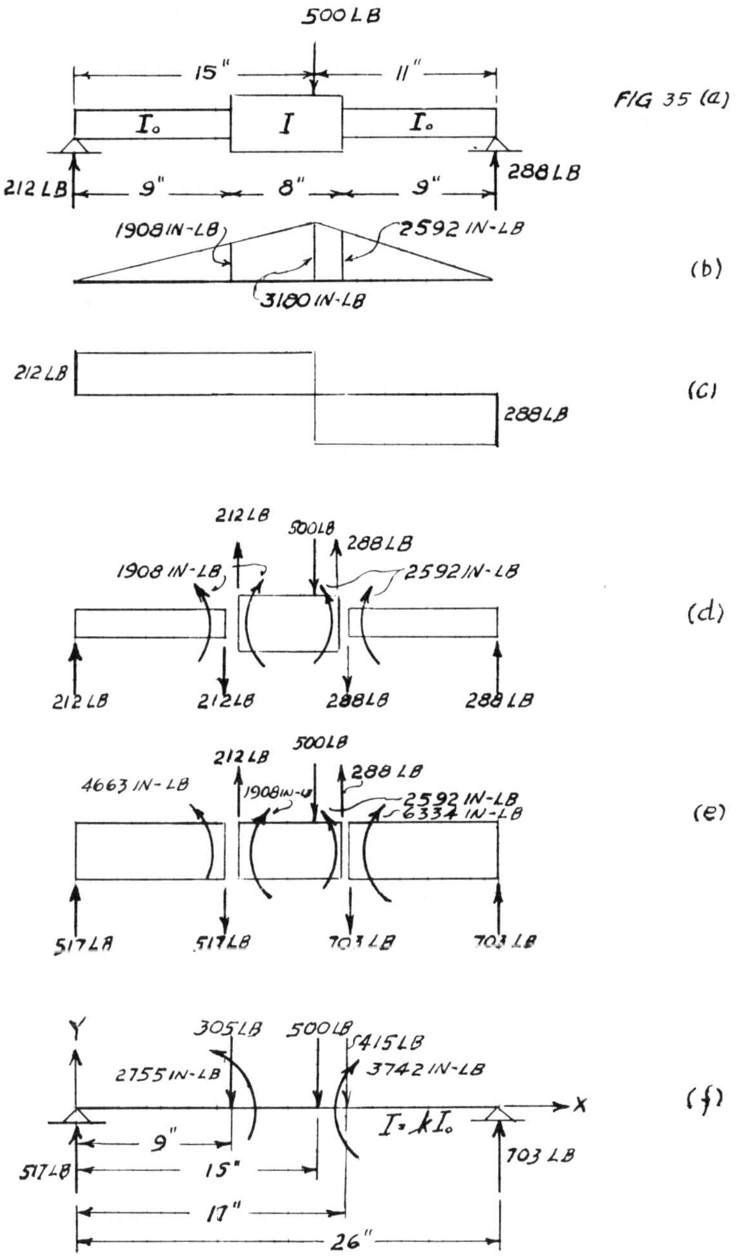

FIG 35 (a)
500 LB
15"
11"
I_0
I
I_0
212 LB
288 LB
9"
8"
9"
(b)
1908 IN-LB
2592 IN-LB
3180 IN-LB
(c)
212 LB
288 LB
(d)
212 LB
500 LB
288 LB
1908 IN-LB
2592 IN-LB
212 LB
212 LB
288 LB
288 LB
(e)
212 LB
500 LB
288 LB
4663 IN-LB
1908 IN-LB
2592 IN-LB
6334 IN-LB
517 LB
517 LB
703 LB
703 LB
(f)
Y
X
305 LB
500 LB
415 LB
2755 IN-LB
3742 IN-LB
$I = kI_0$
517 LB
703 LB
9"
15"
17"
26"

31. BEAMS ON ELASTIC FOUNDATION

A Graphical Presentation of Basic Equations

(*Ref. 10, 17*)

NOTATION

y = deflection of beam (in.)
θ = slope of beam (radian)
M = bending moment of beam (in.-lb.)
Q = shearing force (lb.)
P = applied load (lb.)
q = unit applied load (lb./in.)
E = modulus of elasticity (lb./in.^{2})
I = moment of inertia of beam (in.^{4})
λ = stiffness characteristic (in.$^{-1}$); for beam $\lambda = \sqrt[4]{k/4EI}$
k = modulus of foundation (lb./in.^{2})
K = beam spring constant
K_y = loading-support factor for y
K_θ = loading-support factor for θ
K_M = loading-support factor for M
K_Q = loading-support factor for Q
(K_y to K_Q: Figs. 37 to 50)
L = length of beam (in.)
λL = stiffness parameter (radians)

CLASSIFICATION OF BEAMS ACCORDING TO STIFFNESS

An important factor influencing the shape of the elastic line is the factor $\lambda = \sqrt[4]{(k/4EI)}$, called the characteristic of the system. The dimensional unit of λ is L^{-1}, in which L is the dimension of length. Consequently, λL is dimensionless and is usually thought of as being expressed in *radians*. The λL quantity characterizes the relative stiffness of a beam on an elastic foundation. It determines the magnitude of the curvature of the elastic line and defines the rate at which the effect of a loading force dies out in the form of a damped wave along the length of the beam. According to these λL values we may classify beams into three groups:

I. Short beams: $\lambda L < \pi/4$
II. Beams of medium length: $\pi/4 < \lambda L < \pi$
III. Long beams: $\lambda L > \pi$

For beams belonging to Group I we can neglect, in most practical problems, the bending deformation of the beam; this deformation will be so small as to be negligible compared with the deformation produced in the foundation. Hence, computing beams of $\lambda L < \pi/4$, we can assume them to be absolutely rigid; consequently the computation of the deflections is a statically determinate problem. The deflection then may be calculated by assuming a straight-line distribution of the reaction forces from the elastic supports and applying the equations of equilibrium to the system of forces acting on the beam.

Group II comprises the situations in which accurate computation of the beam is necessary. The characteristic of this group is that a force acting at one end of the beam has a finite, and not negligible, effect at the other end.

Beams belonging to Group III have a λL value such that the countereffect of the end-conditioning forces on each other is a diminishing one. When investigating one end of the beam, we may assume that the other end is infinitely far away. Forces applied at one end will have a negligible effect at the other.

BEAMS OF MEDIUM LENGTH

The basic equations for the deflection, end slope, moment and shear of medium-length beams on elastic foundation sometimes have the form of perplexing combinations of trigonometric and hyperbolic functions for which special tables are required. In order to remove all the tedium that is connected with the use of these equations it was conceived to solve them by a simple expedient of graphs (Figs. 37–50).

Fourteen different loading and support cases, as shown in Table 11, are considered. The tabulated, simplified equations together with the graphs in Figs. 37–50, is all one needs to find y, M, θ, Q at any section of the beam.

The equations in graphs presented here apply to (1) beams on continuous elastic foundation, and (2) beams supported by a number of equidistant elastic supports.

SIGN CONVENTION

Downward acting load (P), downward deflection (y), angular deflection (θ) rotating clockwise shall be regarded as positive. In addition, the bending moment (M) is positive when it causes compression in the top fibers, and the shearing force (Q) is positive when it is an upward force on the left face of an elemental portion of the beam.

The curves in Fig. 37 through 50 were plotted for four different lengths ranging from $L = 2/\lambda$ to $L = 5/\lambda$. If beams on elastic support shorter than $L = 2/\lambda$ are encountered and if approximate values of the bending moment, deflection, slope, and shearing force will suffice, two alternative procedures are possible. First, if the length L is between $1/\lambda$ and $2/\lambda$, the curves for $2/\lambda$ give approximate values for bending moment, etc., which are conservative. Second, if the beam and its elastic support are such that $L = 1/\lambda$ or less, it may be assumed to be rigid, in which case the bending moments are negligible and the computation of the deflection is a statically determined problem.

If the beam length is not equal to one of the given lengths, the K factor at any section may be found for lengths above and below actual length of the beam and the approximate K value found by interpolation.

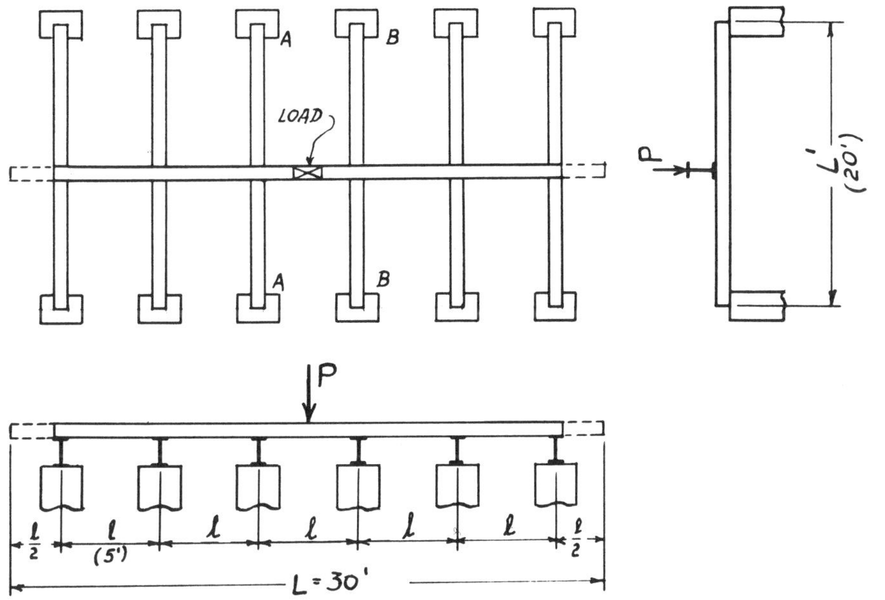

FIG. 36

NOTE: ALL BEAMS ARE 18-IN, 50LB. I-BEAMS

***Example:** A load $P = 50$ tons is supported upon a structure consisting of seven 18 in., 50 lb. steel I beams arranged as shown in Fig. 36. One I beam lies across the other six I beams, which acts as separate elastic supports for the one I beam. Compute the maximum values of the deflection, slope, shearing force, and bending stress in the main I beam. Also, find the maximum deflection and bending stress in the supporting beams AA (or BB) nearest the load P.

Solution: From a steel-maker's handbook the value of the moment of inertia for a 18 in., 50 lb. I beam is 800 in^4. The spring constant K of a separate elastic support such as beam AA is $K = 48EI/L = 48 \times 30 \times 10^6 \times 800/20^3 \times 12^3 = 83{,}360$ lb./in. which comes from the equation $y = (1/48)(P_1 L_1^3/EI)$ for the deflection at the center of a simply supported beam such as AA in Fig. 36 of length L_1 supporting a load P_1 at its center. Remembering Winkler's assumption that the foundation deforms only along the portion directly under loading we find the spring constant k of the *equivalent continuous* elastic foundation:

$$k = \frac{K}{l} = \frac{48EI}{lL_1^3} = \frac{83{,}360}{60} = 1390 \text{ lb./in.}^2$$

Therefore

$$\lambda = \sqrt[4]{\frac{k}{4EI}} = \sqrt[4]{\frac{48EI}{4EIlL_1^3}} = \sqrt[4]{\frac{12}{lL_1^3}} = 0.011 \text{ in.}^{-1}$$

It is important to note that while replacing the separate elastic supports by an equivalent continuous support the span length l should not exceed $l = \pi/4\lambda = 3.14/4 \times 0.011 = 71$ in. Also, note that the value of λ depends only upon the distance l between elastic supports and the span length L of the beam and is independent of the size of the I beams. For the purpose of getting the length L of the one I beam, it is extended beyond the elastic supports at the ends because it is assumed that each separate

*NOTE: This example was taken, by permission, from Ref. 19.

elastic support exerts a uniform pressure on the beam for a distance of $l/2$ on either side of the support.

The load supporting I beam is shown schematically in Table 11, case 14, from which we find

$$y = \frac{P\lambda}{2k}(K_y); \qquad \theta = \frac{P\lambda^2}{k}(K_\theta)$$

$$M = \frac{P}{4\lambda}(K_M); \qquad Q = \frac{P}{2}(K_Q),$$

where $P = 100{,}000$ lb.
$\lambda = 0.011$ in^{-1}., thus $\lambda L = 0.011 \times 30 \times 12 = 4$
$k = 1390$ lb./in.^{2}

From Fig. 50, for $\lambda L = 4$:

$K_y = 1.08$, for $x = 0$ (max. y under load)
$K_\theta = -0.37$, for $x = L/2$ (slope at end points)
$K_M = 1.05$, for $x = 0$ (max. bending moment under the load)
$K_Q = -1.0$, for $x = 0$ (max. shear under the load).

By substitution:

$$y_{max} = \frac{100{,}000 \times 0.011 \times 1.08}{2 \times 1{,}390} = 0.43 \text{ in.}$$

$$M_{max} = \frac{100{,}000 \times 1.05}{4 \times 0.011} = 2{,}390{,}000 \text{ in.-lb.}$$

$$\theta_{max} = \frac{-100{,}000 \times (0.011)^2 \times 0.37}{1{,}390} = -0.0032 \text{ radians}$$

$$Q_{max} = \frac{-100{,}000 \times 1}{2} = -50{,}000 \text{ lb.}$$

The maximum stress in the load supporting I beam is:

$$S_{max} = \frac{M_{max}}{I/c} = \frac{2{,}390{,}000 \times 9}{800} = 27{,}000 \text{ lb./in}^2.$$

MAXIMUM DEFLECTION AND STRESS IN ELASTIC SUPPORT

The beam AA (or BB) in Fig. 36 nearest the load P is subjected to the greatest stresses in the elastic supports. The deflection of beam AA at its midpoint is equal to the deflection of the load carrying main beam at the section where this I beam rests on beam AA. This section is at a distance $x = 2.5/30 = 0.083L$ from the center of the main beam, and the deflection, as before, is $y_{AA} = P\lambda/2k \times (K_y)$; from Fig. 50, for $\lambda L = 4$ and $x = 0.083L$, $K_y = 1.00$ and $y = (100{,}000 \times 0.011 \times 1)/2 \times 1390 = 0.396$ in. This value of the deflection is substituted into the equation

$$y = \frac{P_1 L_1^3}{48EI} = y_{AA} = 0.396 \text{ in.}$$

from which we find that the maximum load at the center of beam AA is:

$$P_1 = \left[\frac{48EI \times 0.396}{L_1^3}\right] \frac{48 \times 30 \times 10^6 \times 800 \times 0.396}{20^3 \times 12^3}$$

$$P_1 = 33{,}000 \text{ lb.}$$

The maximum moment in beam AA is

$$P_1L_1/4 = \frac{33{,}000 \times 20 \times 12}{4} = 1{,}980{,}000 \text{ in.-lb.}$$

The maximum stress in beam AA is

$$S_{max} = \frac{M}{I/c} = \frac{1{,}980{,}000 \times 9}{800} = 22{,}300 \text{ lb./in.}^2$$

LONG BEAMS ($\lambda L > \pi$)

Beams in this group have their length very large compared with the depth and width. In other words, the ends of the beam are removed far enough from the loads to justify the assumption that deflection, slope, shear, and moments at the ends are zero. (Table 12, cases 15 and 16).

In the same category are the so-called semi-infinite beams; the term semi-infinite beam is used for a beam which has unlimited extension in only one direction, having one finite end (Table 12, cases 17, 18, 19, 20).

The formulas for deflection, slope, shear, and moment for beams in this group are shown in Table 12. The formulas are to be used in conjunction with Fig. 51 where values for $A_{\lambda X}, B_{\lambda X}, C_{\lambda X}, D_{\lambda X}$, for specific values of λX, were plotted. Here too, it is important to note the sign convention; the positive sense of x is to the right, positive deflection y is downward, the bending moment M is positive when it causes compression in the top fibers, and the shearing force is positive when it is an upward force on the left face of an elemental portion of the beam.

The formulas in Table 12 will be best explained in the following examples.

Example:

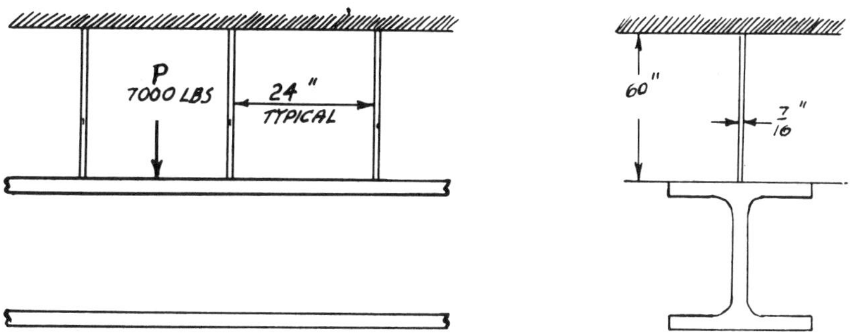

A *long* beam is suspended by a series of steel rods 7/16 in. in diameter and 60 in. long, spaced 24 in. center to center. The beam is a 6-in., 12.5-lb. steel I beam; the moment of inertia of the cross section of the beam is 21.8 in^4., and, hence, its section modulus is 7.2 in^3. If a 7,000-lb. vertical concentrated load is applied to the beam at its mid-point, compute the maximum bending stress in the beam and the maximum tensile stress in the hanger rod. The value of E for steel is 30×10^6 lb./in.^{2}

Solution: The spring constant K of the hanger rod is obtained from the equation for the elongation of a centrally loaded bar under tension. Thus: $y = FL/AE$, where y is the elongation of the rod, F is the axial load on the bar, L is the length of the rod, A is the cross sectional area of the rod, and E is the modulus of elasticity, the spring constant K of the rod is

$$K = \frac{F}{y} = \frac{AE}{L} = \frac{0.15 \times 30 \times 10^6}{60} = 75{,}000 \text{ lb./in.}$$

The value of the constant k of the equivalent continuous elastic support is, as before,

$$k = K/l = 75{,}000/24 = 3{,}120 \text{ lb./in.}^2$$

Hence

$$\lambda = \sqrt[4]{\frac{k}{4EI}} = \sqrt[4]{\frac{3{,}120}{4 \times 30 \times 10^6 \times 21.8}} = 0.033 \text{ in.}^{-1}$$

Also

$$\pi/4\lambda = 3.14/4 \times 0.033 = 24 \text{ in.}$$

and hence the span length l between the suspension rods does not exceed $\pi/4\lambda$. From Table 12, case 1, we find

$$M = \frac{P}{4\lambda}(C_{\lambda x}).$$

From Fig. 50, the maximum value of $C_{\lambda x} = 1$ for $x = 0$. Therefore:

$$M_{\text{max}} = \frac{7{,}000}{4 \times 0.033} = 53{,}000 \text{ in.-lb.}$$

and

$$S_{\text{max}} = \frac{M}{I/c} = \frac{53{,}000}{7.3} = 7{,}260 \text{ lb./in.}^2$$

The stress in the rod is $S_t = F/A = yE/L$. Therefore the maximum stress in the rods will be in the rod nearest the load where the deflection y of the beam (and hence of the rod) is a maximum. The formula for deflection again will be found from Table 12, case 1: $y = P\lambda/2k\ (A_{\lambda x})$, from Fig. 50, the maximum value of $A_{\lambda x} = 1$ for $x = 0$, therefore:

$$y_{\text{max}} = \frac{7{,}000 \times 0.033}{2 \times 3{,}120} = 0.037 \text{ in.}$$

Finally, the maximum stress in the rods is

$$S_{\text{max}} = \frac{y_{\text{max}}E}{L} = \frac{0.037 \times 30 \times 10^6}{60} = 18{,}500 \text{ lb./in.}^2$$

Table 11

EQUATIONS FOR MEDIUM LENGTH BEAMS

LOADING AND CASE No.	DEFLECTION, y	SLOPE, θ	MOMENT, M	SHEAR, Q	FIG. No.
1. HINGED ENDS, CONCENTRATED CENTER LOAD. 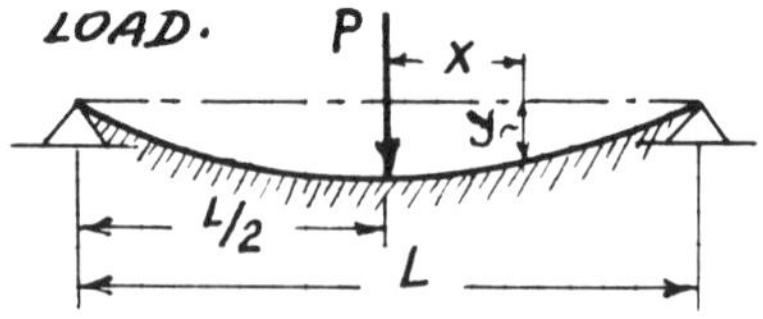	$\frac{P\lambda}{2k}(K_y)$ (a)	$-\frac{P\lambda^2}{k}(K_\theta)$ (b)	$\frac{P}{4\lambda}(K_M)$ (c)	$-\frac{P}{2}(K_Q)$ (d)	37 (a) (b) (c) (d)
2. HINGED ENDS, UNIFORM LOAD. 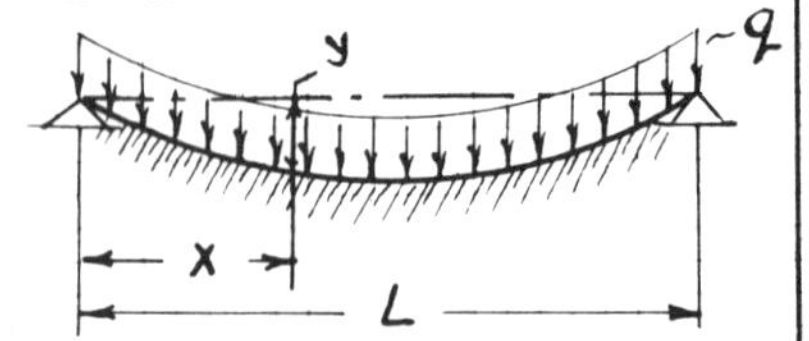	$\frac{q}{k}(K_y)$ (a)	$\frac{q\lambda}{k}(K_\theta)$ (b)	$\frac{q}{2\lambda^2}(K_M)$ (c)	$-\frac{q}{2\lambda}(K_Q)$ (d)	38 (a) (b) (c) (d)
3. HINGED ENDS, EQUAL MOMENTS AT BOTH ENDS. 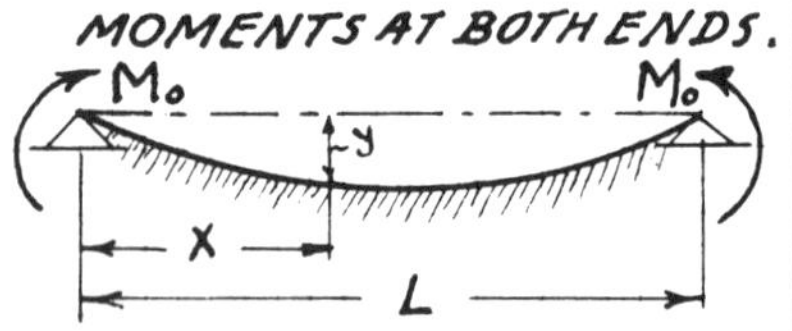	$\frac{2M_0\lambda^2}{k}(K_y)$ (a)	$\frac{2M_0\lambda^3}{k}(K_\theta)$ (b)	$M_0(K_M)$ (c)	$M_0\lambda(K_Q)$ (d)	39 (a) (b) (c) (d)
4. HINGED ENDS, CONCENTRATED MOMENT AT ONE END. 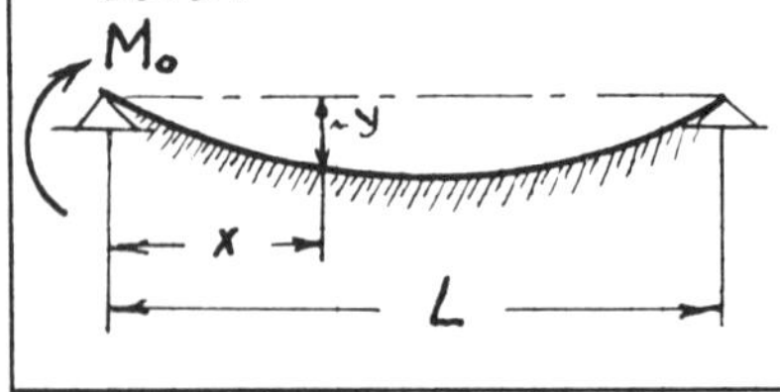	$\frac{2M_0\lambda^2}{k}(K_y)$ (a)	$\frac{2M_0\lambda^3}{k}(K_\theta)$ (b)	$M_0(K_M)$ (c)	$M_0\lambda(K_Q)$ (d)	40 (a) (b) (c) (d)
5. FIXED ENDS, CONCENTRATED CENTER LOAD. 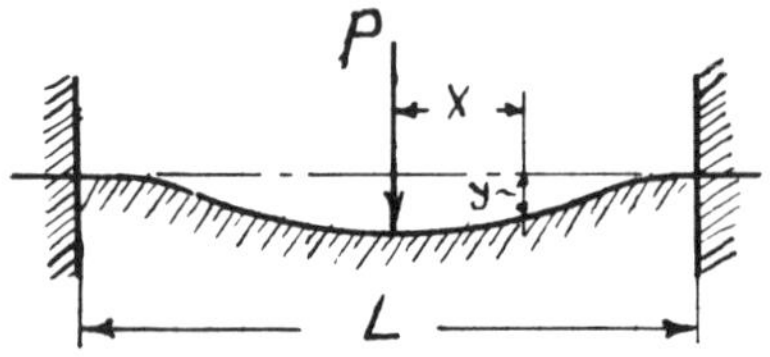	$\frac{P\lambda}{2k}(K_y)$ (a)	$\frac{P\lambda^2}{2k}(K_\theta)$ (b)	$-\frac{P}{4\lambda}(K_M)$ (c)	$-\frac{P}{2}(K_Q)$ (d)	41 (a) (b) (c) (d)

NOTE: FOR TABULATED VALUES OF K_y, K_θ, K_M, K_Q, SEE APPENDIX B

Case					
6. FIXED ENDS, UNIFORM LOAD. 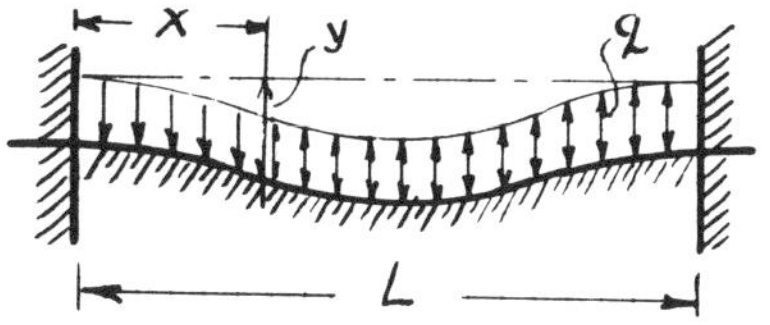	$\frac{q}{k}(K_y)$ (a)	$-\frac{2q\lambda}{k}(K_\theta)$ (b)	$-\frac{q}{2\lambda^2}(K_M)$ (c)	$-\frac{q}{\lambda}(K_\varphi)$ (d)	42 (a) (b) (c) (d)
7. CANTILEVER BEAM, CONCENTRATED END LOAD. 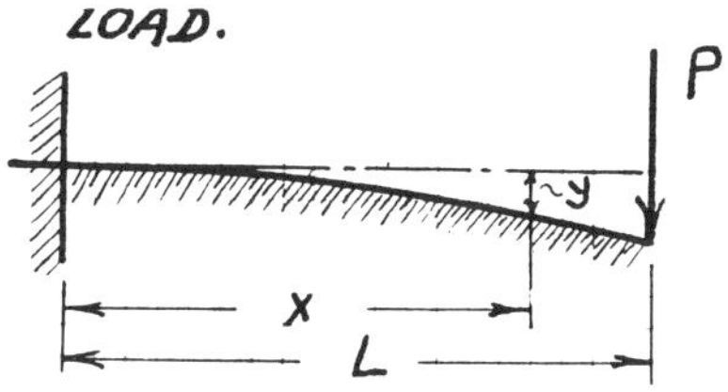	$\frac{2P\lambda}{k}(K_y)$ (a)	$\frac{2P\lambda^2}{k}(K_\theta)$ (b)	$-\frac{P}{\lambda}(K_M)$ (c)	$-P(K_\varphi)$ (d)	43 (a) (b) (c) (d)
8. CANTILEVER BEAM, UNIFORM LOAD. 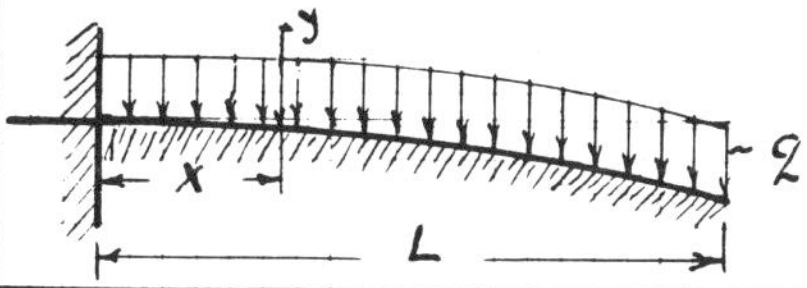	$\frac{q}{k}(K_y)$ (a)	$-\frac{2q\lambda}{k}(K_\theta)$ (b)	$\frac{q}{2\lambda^2}(K_M)$ (c)	$\frac{q}{\lambda}(K_\varphi)$ (d)	44 (a) (b) (c) (d)
9. CANTILEVER BEAM, TRIANGULAR LOAD. 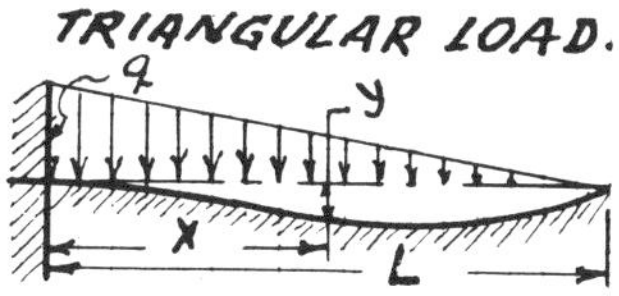	$\frac{q}{k}(1+\frac{x}{L})+$ $+\frac{q}{2\lambda kL}(K_y)$ (a)	$-\frac{q}{kL}+$ $+\frac{q}{2kL}(K_\theta)$ (b)	$-\frac{q}{4\lambda^3 L}(K_M)$ (c)	$-\frac{q}{2\lambda^2 L}(K_\varphi)$ (d)	45 (a) (b) (c) (d)
10. FREE ENDS, EQUAL CONCENTRATED LOADS AT BOTH ENDS. 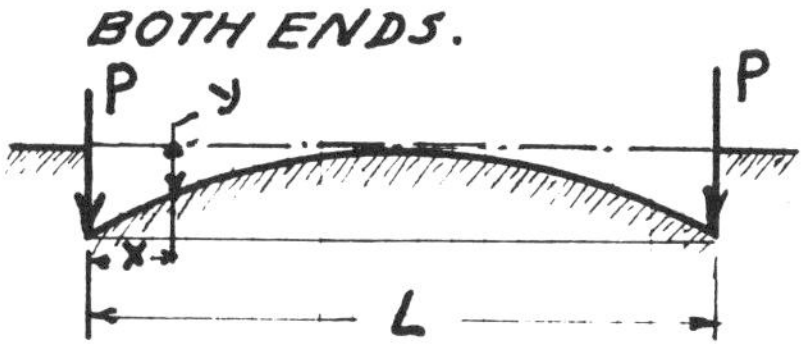	$\frac{2P\lambda}{k}(K_y)$ (a)	$\frac{2P\lambda^2}{k}(K_\theta)$ (b)	$-\frac{P}{\lambda}(K_M)$ (c)	$P(K_\varphi)$ (d)	46 (a) (b) (c) (d)
11. FREE ENDS, EQUAL MOMENTS AT BOTH ENDS. 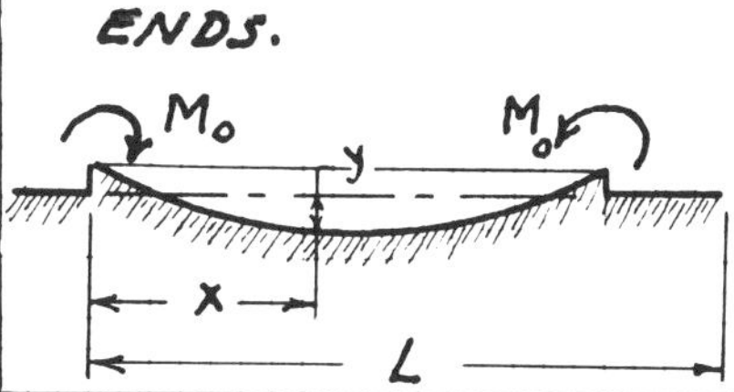	$-\frac{2M_0\lambda^2}{k}(K_y)$ (a)	$-\frac{4M_0\lambda^3}{k}(K_\theta)$ (b)	$M_0(K_M)$ (c)	$2M_0\lambda(K_\varphi)$ (d)	47 (a) (b) (c) (d)

Loading					
12. FREE ENDS, CONCENTRATED LOAD AT ONE END 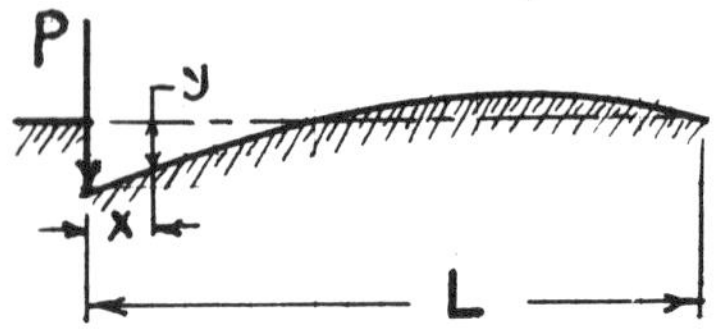	$\frac{2P\lambda}{k}(K_y)$ (a)	$-\frac{2P\lambda^2}{k}(K_\theta)$ (b)	$-\frac{P}{\lambda}(K_M)$ (c)	$-P(K_Q)$ (d)	48 (a) (b) (c) (d)
13. FREE ENDS, CONCENTRATED MOMENT AT ONE END. 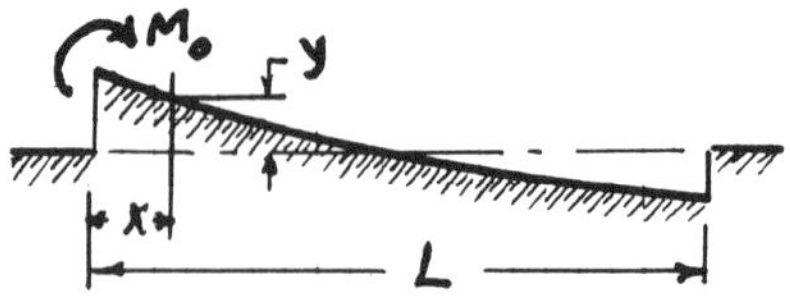	$\frac{2M_0\lambda^2}{k}(K_y)$ (a)	$\frac{4M_0\lambda^3}{k}(K_\theta)$ (b)	$M_0(K_M)$ (c)	$-2M_0\lambda(K_Q)$ (d)	49 (a) (b) (c) (d)
14. FREE ENDS, CENTER LOAD. 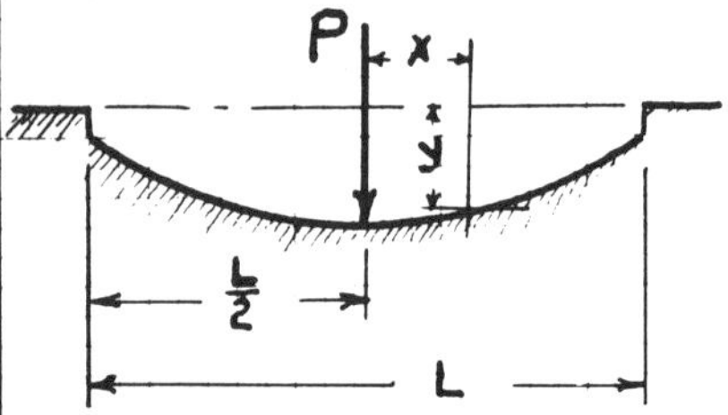	$\frac{P\lambda}{2k}(K_y)$ (a)	$\frac{P\lambda^2}{k}(K_\theta)$ (b)	$\frac{P}{4\lambda}(K_M)$ (c)	$\frac{P}{2}(K_Q)$ (d)	50 (a) (b) (c) (d)

TABLE 12

EQUATIONS FOR UNLIMITED LENGTH BEAMS
(SEE FIG. 51)

LOADING AND CASE NO	DEFLECTION, y	SLOPE, θ	MOMENT, M	SHEAR, Q
INFINITE BEAM, SINGLE CONCENTRATED LOAD. 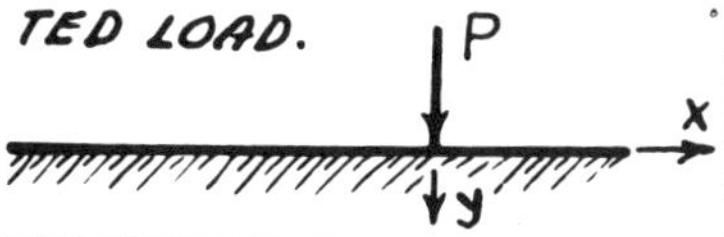	$\frac{P\lambda}{2k}(A_{\lambda x})$	$-\frac{P\lambda^2}{k}(B_{\lambda x})$	$\frac{P}{4\lambda}(C_{\lambda x})$	$-\frac{P}{2}(D_{\lambda x})$
INFINITE BEAM, CONCENTRATED MOMENT. 	$\frac{M_o\lambda^2}{k}(B_{\lambda x})$	$\frac{M_o\lambda^3}{k}(C_{\lambda x})$	$\frac{M_o}{2}(D_{\lambda x})$	$-\frac{M_o\lambda}{2}(A_{\lambda x})$
SEMI-INFINITE BEAM, CONCENTRATED LOAD AT ONE END. 	$\frac{2P\lambda}{k}(D_{\lambda x})$	$-\frac{2P\lambda^2}{k}(A_{\lambda x})$	$-\frac{P}{\lambda}(B_{\lambda x})$	$-P(C_{\lambda x})$
SEMI-INFINITE BEAM, CONCENTRATED MOMENT AT ONE END. 	$-\frac{2M_o\lambda^2}{k}(C_{\lambda x})$	$\frac{4M_o\lambda^3}{k}(D_{\lambda x})$	$M_o(A_{\lambda x})$	$-2M_o\lambda(B_{\lambda x})$
SEMI-INFINITE BEAM, ONE HINGED SUPPORT. END MOMENT. 	$\frac{2M_o\lambda^2}{k}(B_{\lambda x})$	$\frac{2M_o\lambda}{k}(C_{\lambda x})$	$M_o(D_{\lambda x})$	$-M_o\lambda(A_{\lambda x})$
SEMI-INFINITE CANTILEVER BEAM, TRIANGULAR LOAD. 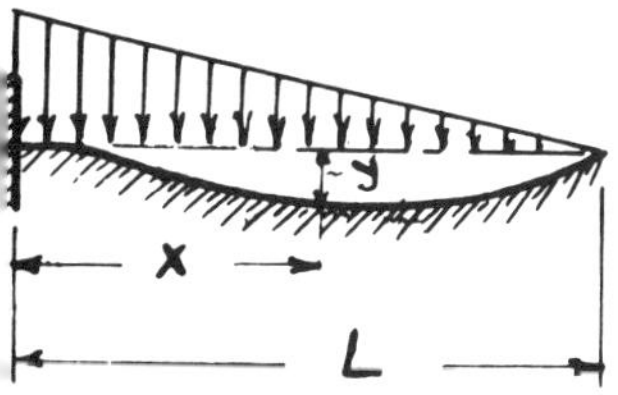	$\frac{q}{\lambda L k}\left[\lambda(L-x) - \lambda L A_{\lambda x} + B_{\lambda x}\right]$	$-\frac{q}{Lk}(1 - 2\lambda L B_{\lambda x} - C_{\lambda x})$	$-\frac{q}{2\lambda^3 L}(\lambda L C_{\lambda x} - D_{\lambda x})$	$-\frac{q}{2\lambda^2 L}(2\lambda L D_{\lambda x} - A_{\lambda x})$

$A_{\lambda x} = e^{-\lambda x}(\cos\lambda x + \sin\lambda x)$; $B_{\lambda x} = e^{-\lambda x}\sin\lambda x$; $C_{\lambda x} = e^{-\lambda x}(\cos\lambda x - \sin\lambda x)$

$D_{\lambda x} = e^{-\lambda x}\cos\lambda x$; SEE FIG. 51

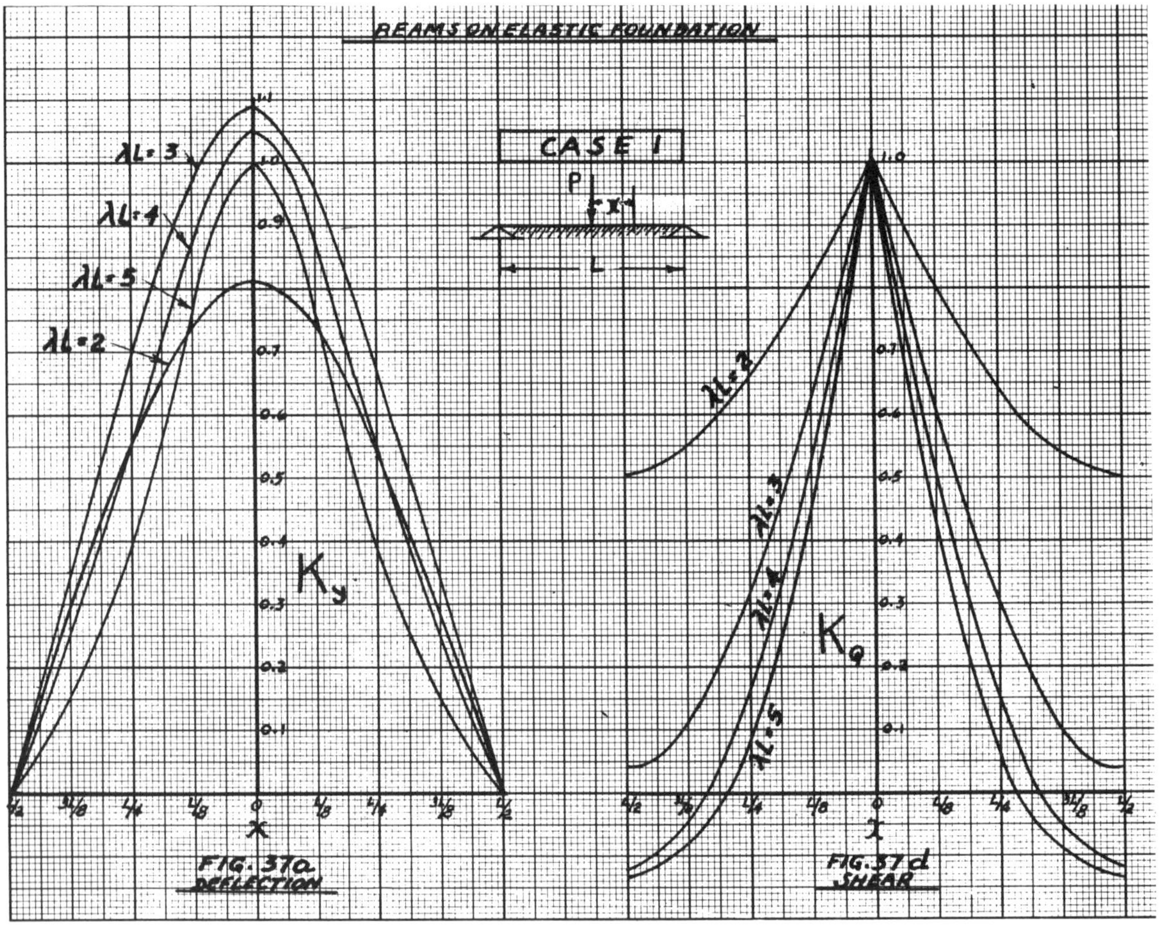
BEAMS ON ELASTIC FOUNDATION
CASE I
P
x
L
λL = 3
λL = 4
λL = 5
λL = 2
K_y
K_Q
FIG. 37a
DEFLECTION
FIG. 37d
SHEAR

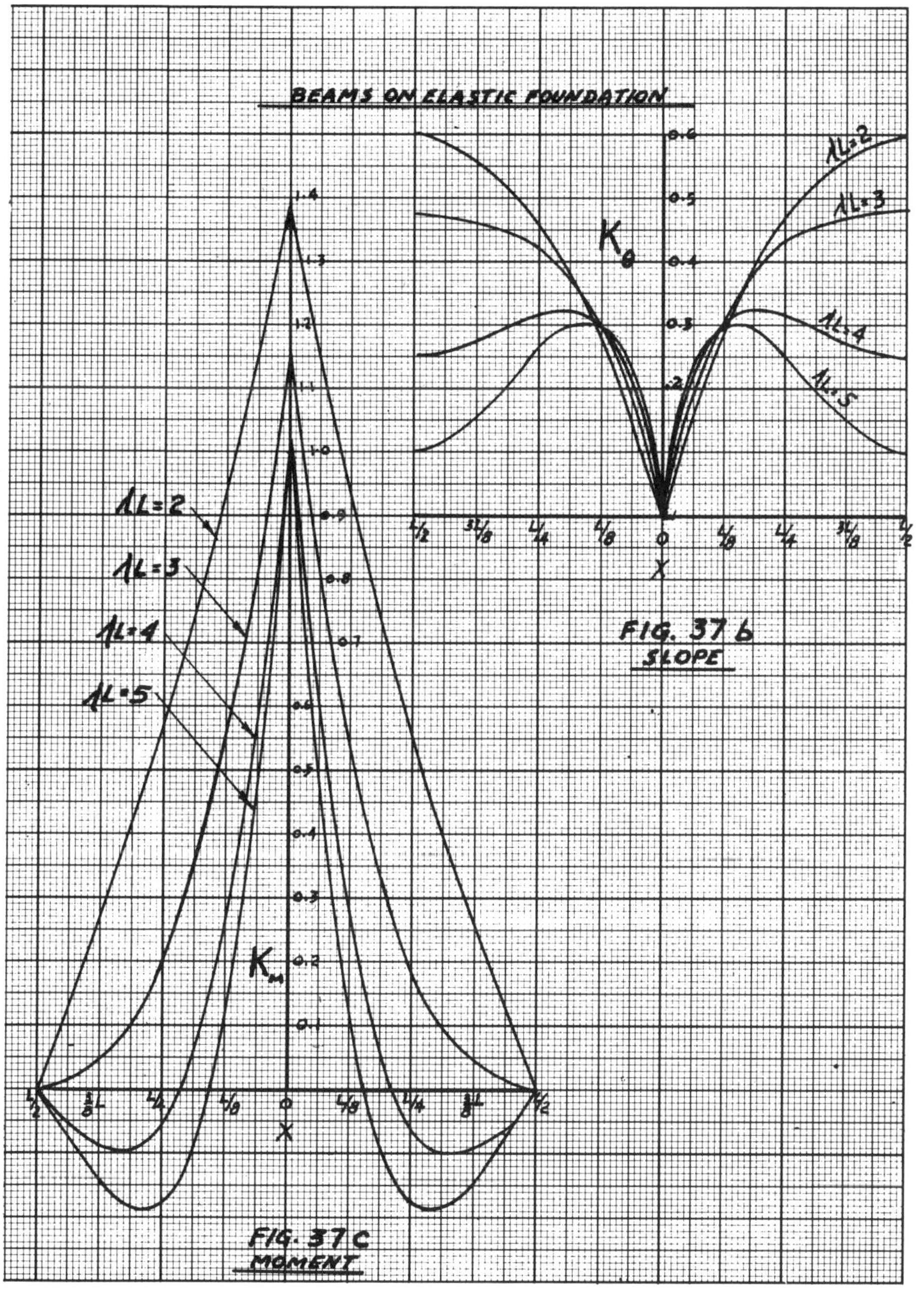
BEAMS ON ELASTIC FOUNDATION
λL=2
λL=3
λL=4
λL=5
K_θ
K_M
X
FIG. 37 b
SLOPE
FIG. 37 C
MOMENT

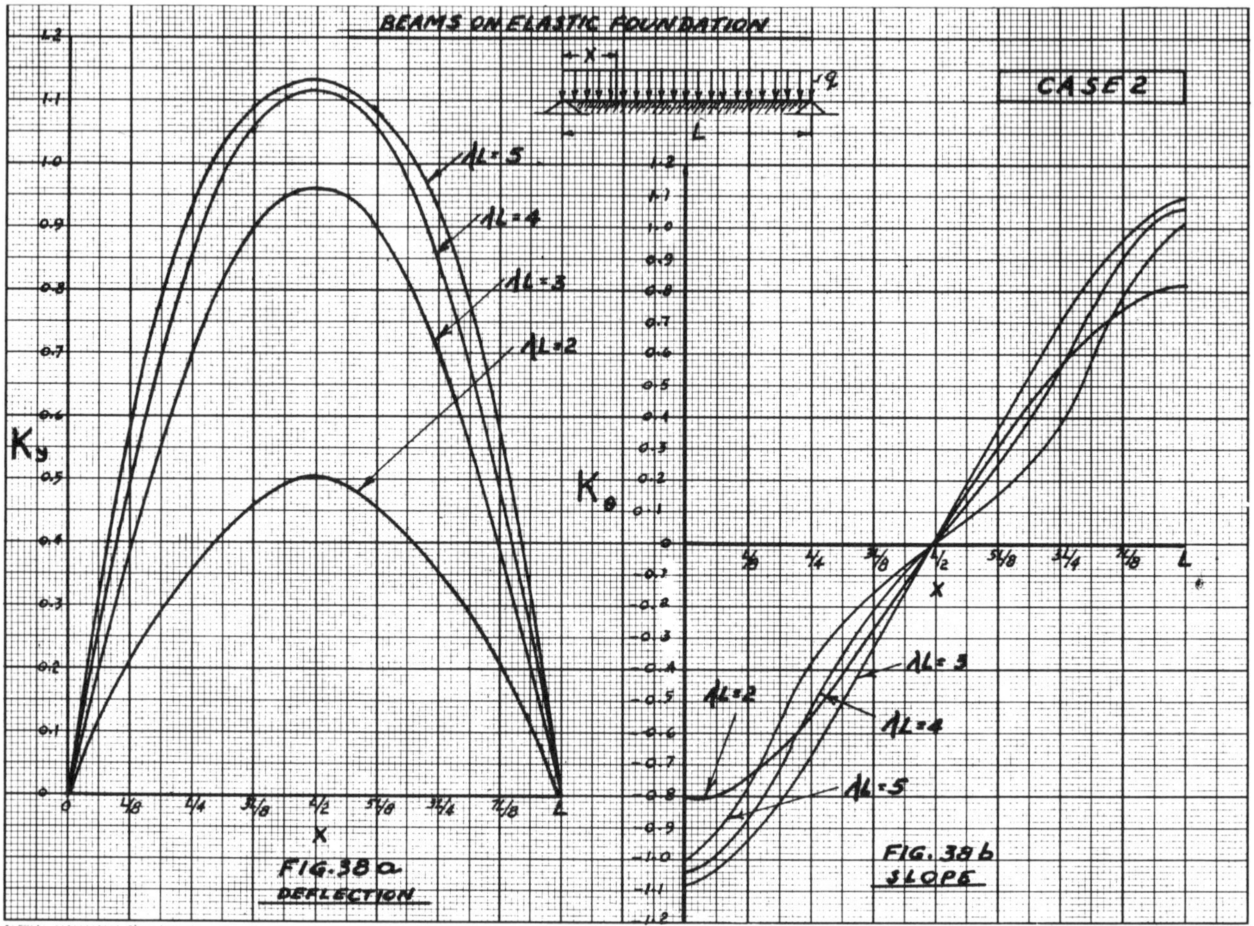
BEAMS ON ELASTIC FOUNDATION
CASE 2
X
q
L
K_y
λL=5
λL=4
λL=3
λL=2
0 L/8 L/4 3L/8 L/2 5L/8 3L/4 7L/8 L
X
FIG. 38 a
DEFLECTION
K_θ
λL=3
λL=2
λL=4
λL=5
FIG. 38 b
SLOPE

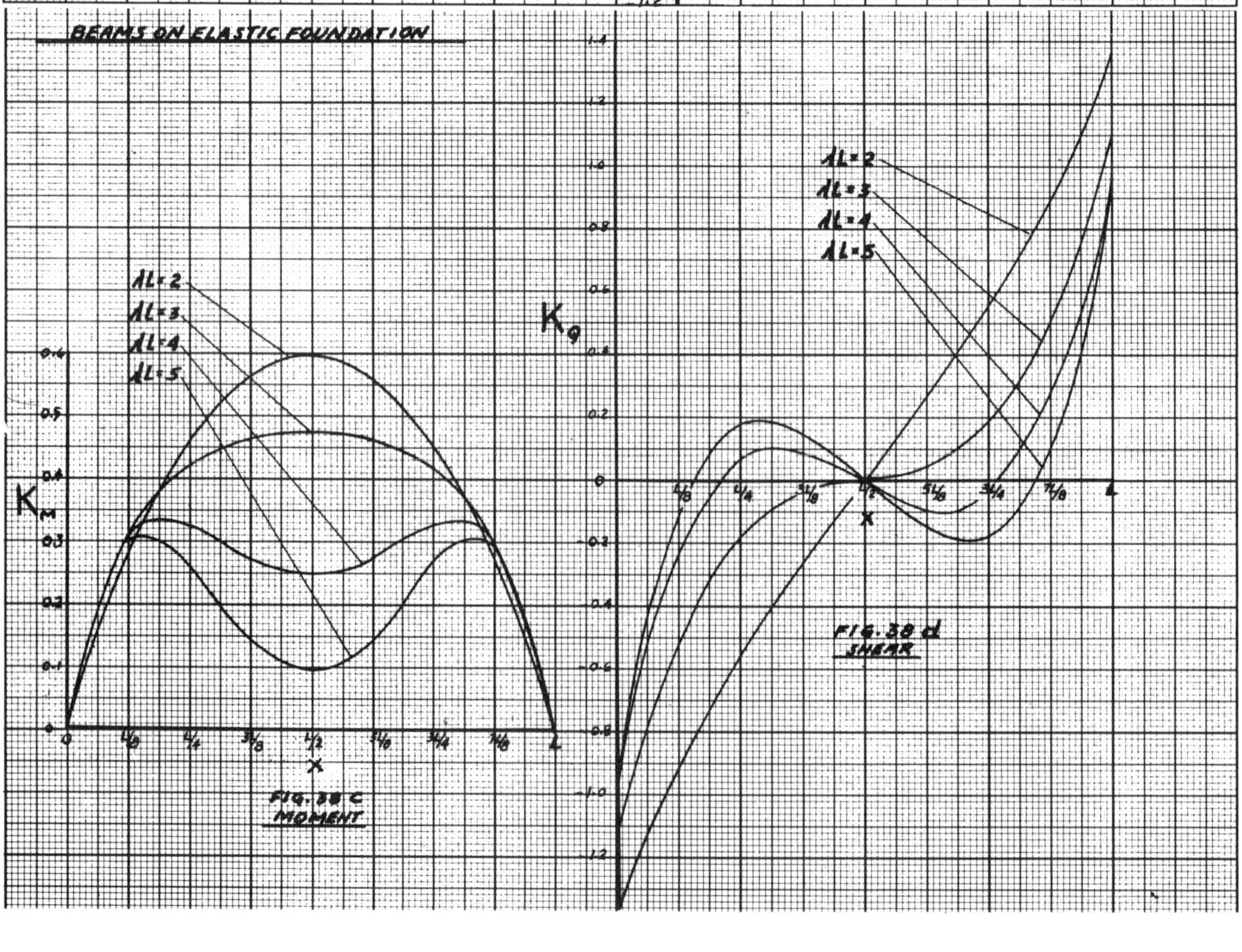
BEAMS ON ELASTIC FOUNDATION
K_M
λL=2
λL=3
λL=4
λL=5
0 L/8 L/4 3L/8 L/2 5L/8 3L/4 7L/8 L
X
FIG. 38 C
MOMENT
K_q
λL=2
λL=3
λL=4
λL=5
FIG. 38 d
SHEAR

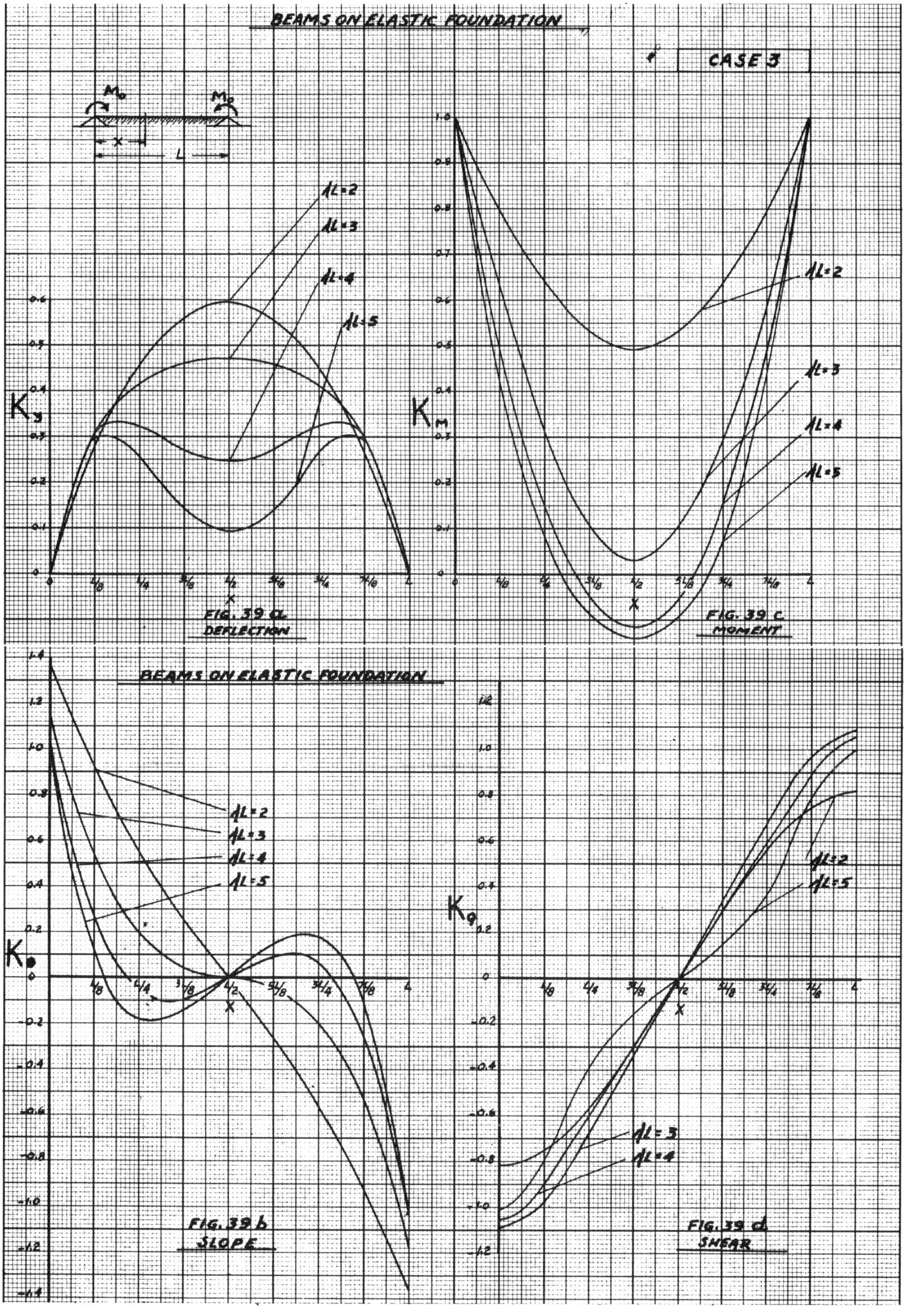
BEAMS ON ELASTIC FOUNDATION
CASE 3
M_o
M_o
x
L
λL=2
λL=3
λL=4
λL=5
K_y
K_M
X
FIG. 39 a
DEFLECTION
FIG. 39 c
MOMENT
BEAMS ON ELASTIC FOUNDATION
λL=2
λL=3
λL=4
λL=5
K_θ
K_Q
λL=2
λL=5
λL=3
λL=4
X
FIG. 39 b
SLOPE
FIG. 39 d
SHEAR

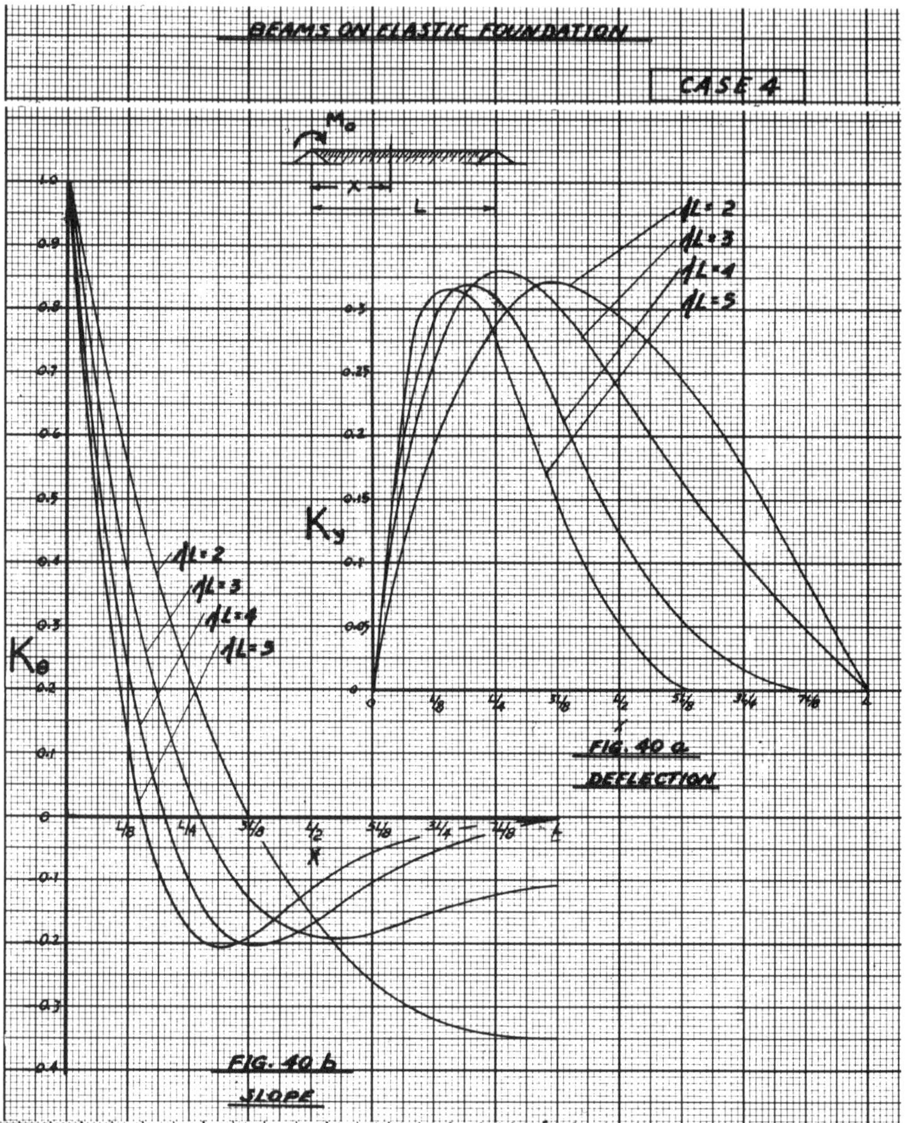
BEAMS ON ELASTIC FOUNDATION
CASE 4
M_o
X
L
$\lambda L = 2$
$\lambda L = 3$
$\lambda L = 4$
$\lambda L = 5$
K_y
K_θ
FIG. 40 a
DEFLECTION
FIG. 40 b
SLOPE

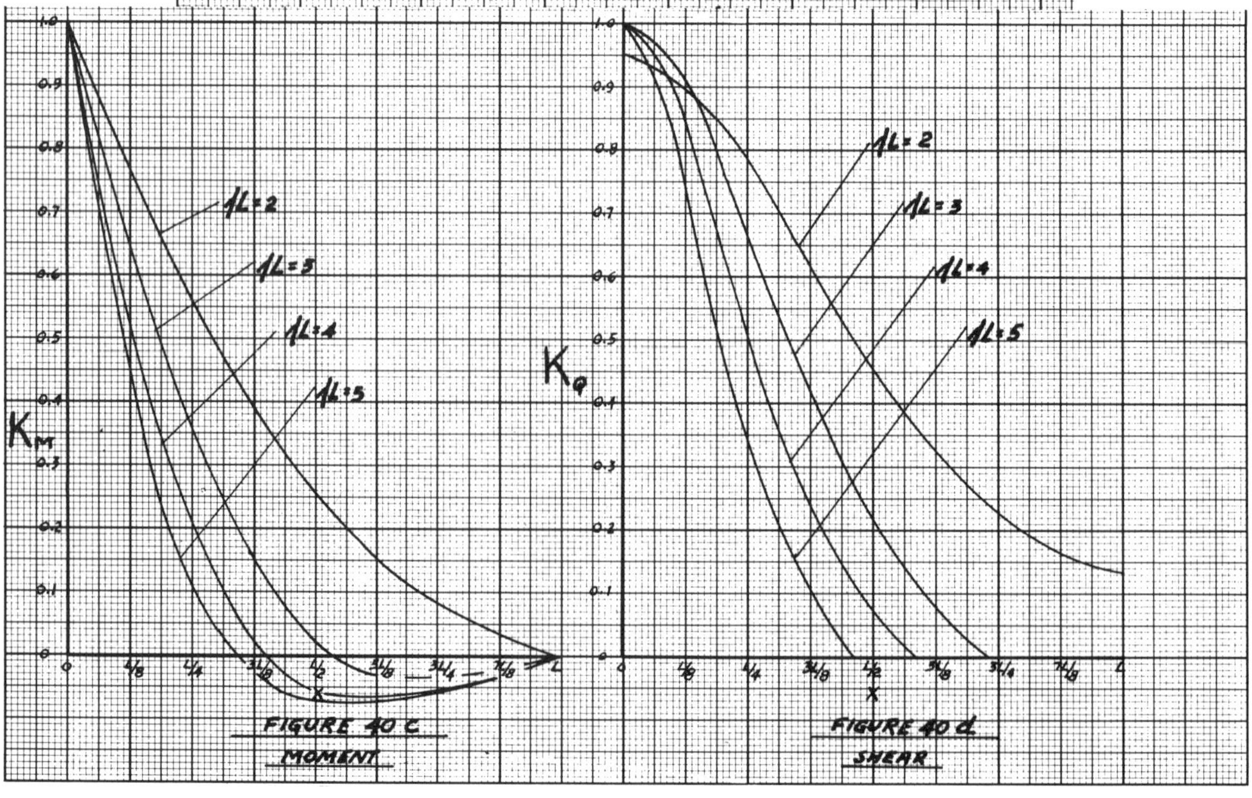
$\lambda L = 2$
$\lambda L = 3$
$\lambda L = 4$
$\lambda L = 5$
K_M
K_Q
FIGURE 40 C
MOMENT
FIGURE 40 d
SHEAR

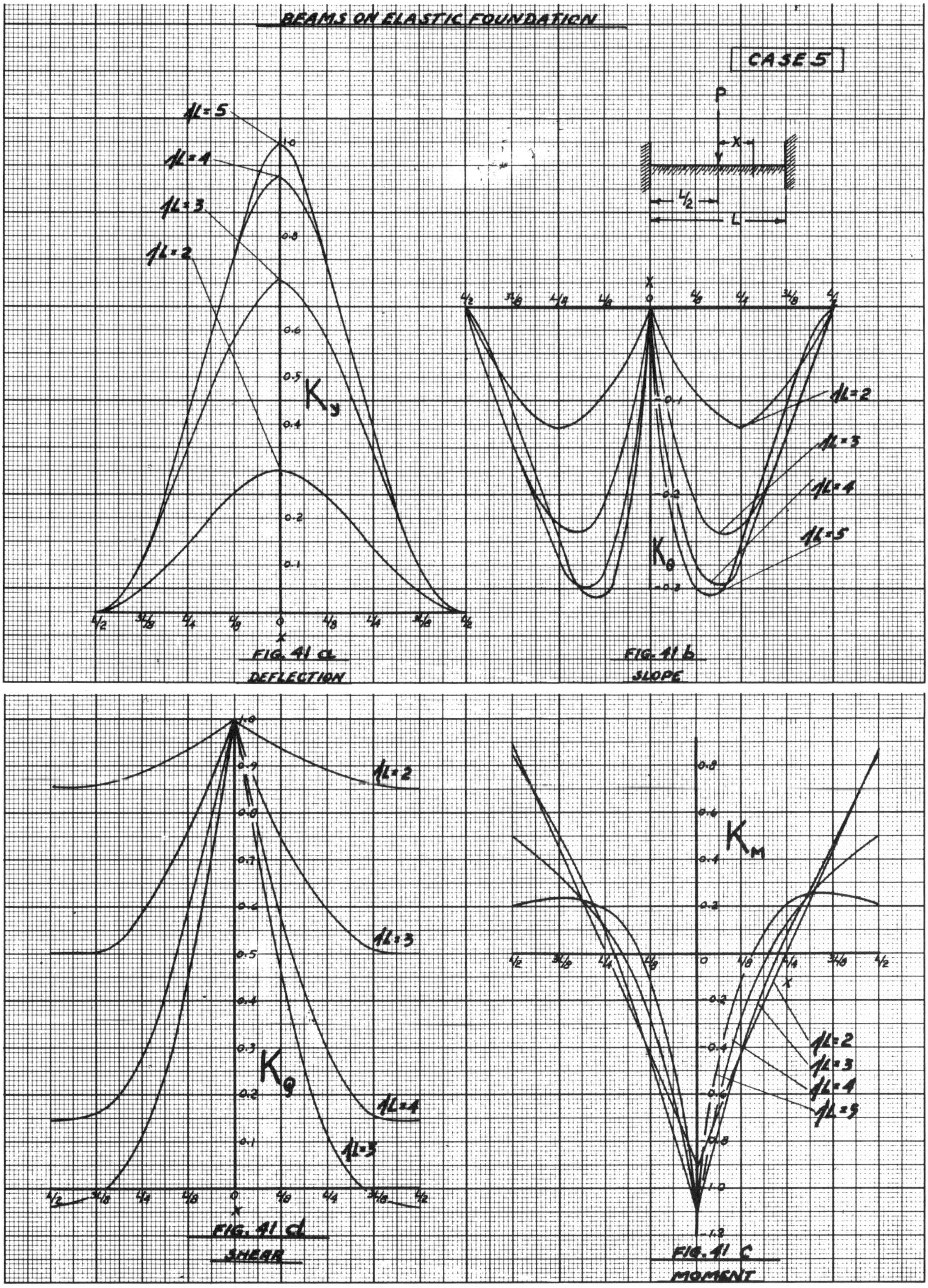
BEAMS ON ELASTIC FOUNDATION
CASE 5
P
X
L/2
L
λL=5
λL=4
λL=3
λL=2
Ky
FIG. 41 a
DEFLECTION
Kθ
FIG. 41 b
SLOPE
KQ
FIG. 41 d
SHEAR
KM
FIG. 41 c
MOMENT

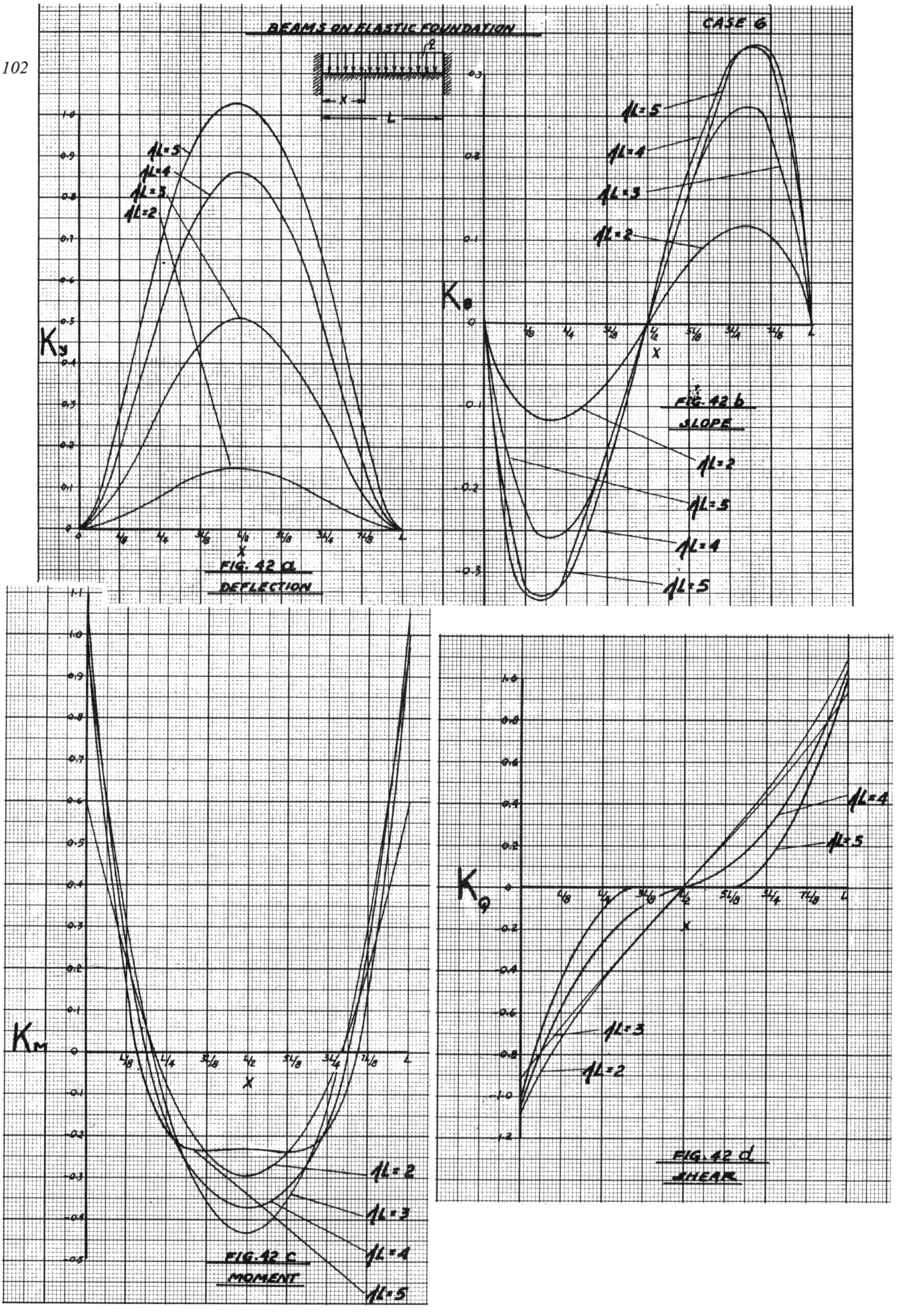

FIG. 42 a
DEFLECTION

FIG. 42 b
SLOPE

FIG. 42 c
MOMENT

FIG. 42 d
SHEAR

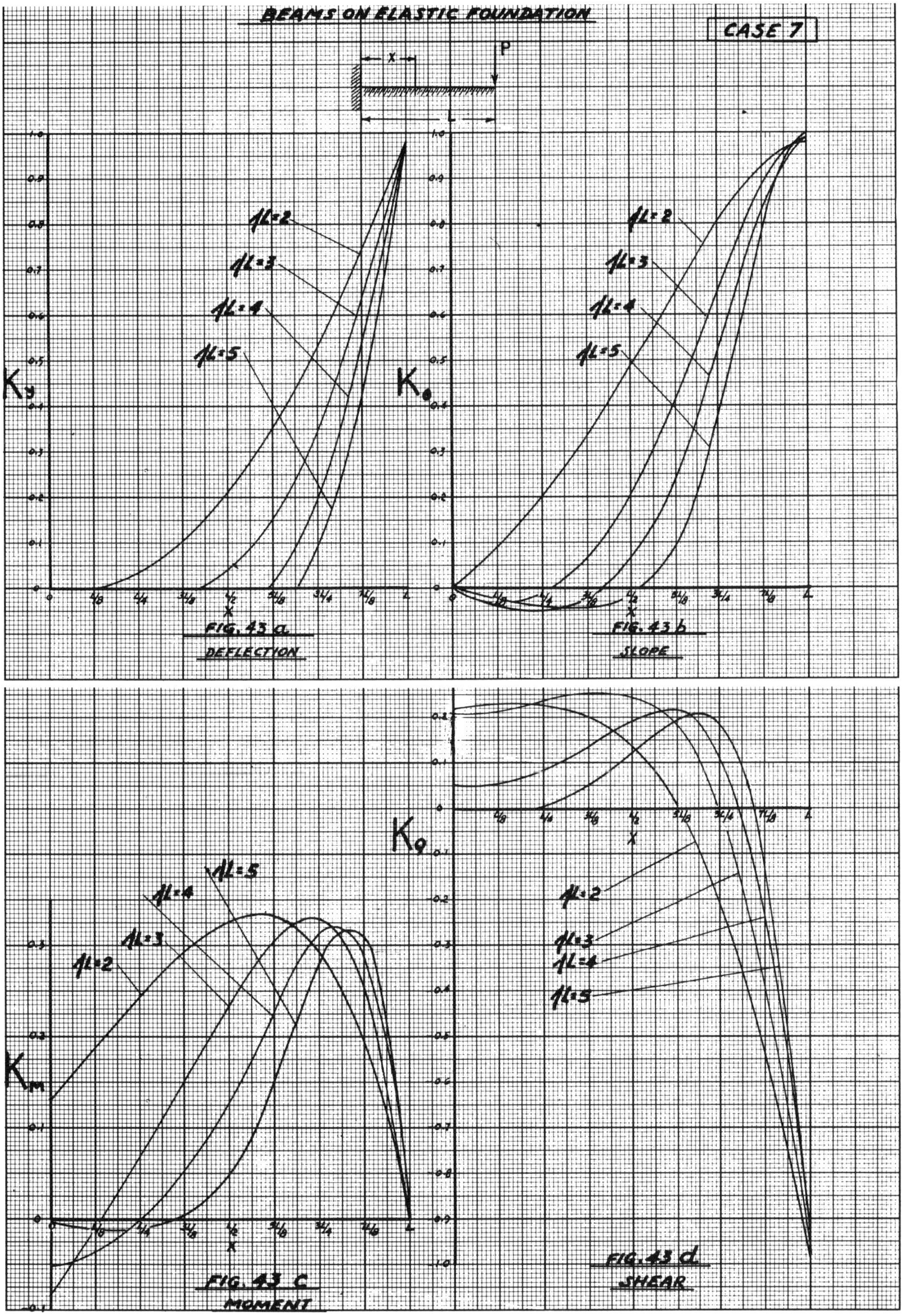
BEAMS ON ELASTIC FOUNDATION
CASE 7
P
X
L
Ky
λL=2
λL=3
λL=4
λL=5
X
FIG. 43 a
DEFLECTION
Kθ
λL=2
λL=3
λL=4
λL=5
X
FIG. 43 b
SLOPE
Km
λL=5
λL=4
λL=3
λL=2
X
FIG. 43 C
MOMENT
Kq
X
λL=2
λL=3
λL=4
λL=5
FIG. 43 d
SHEAR

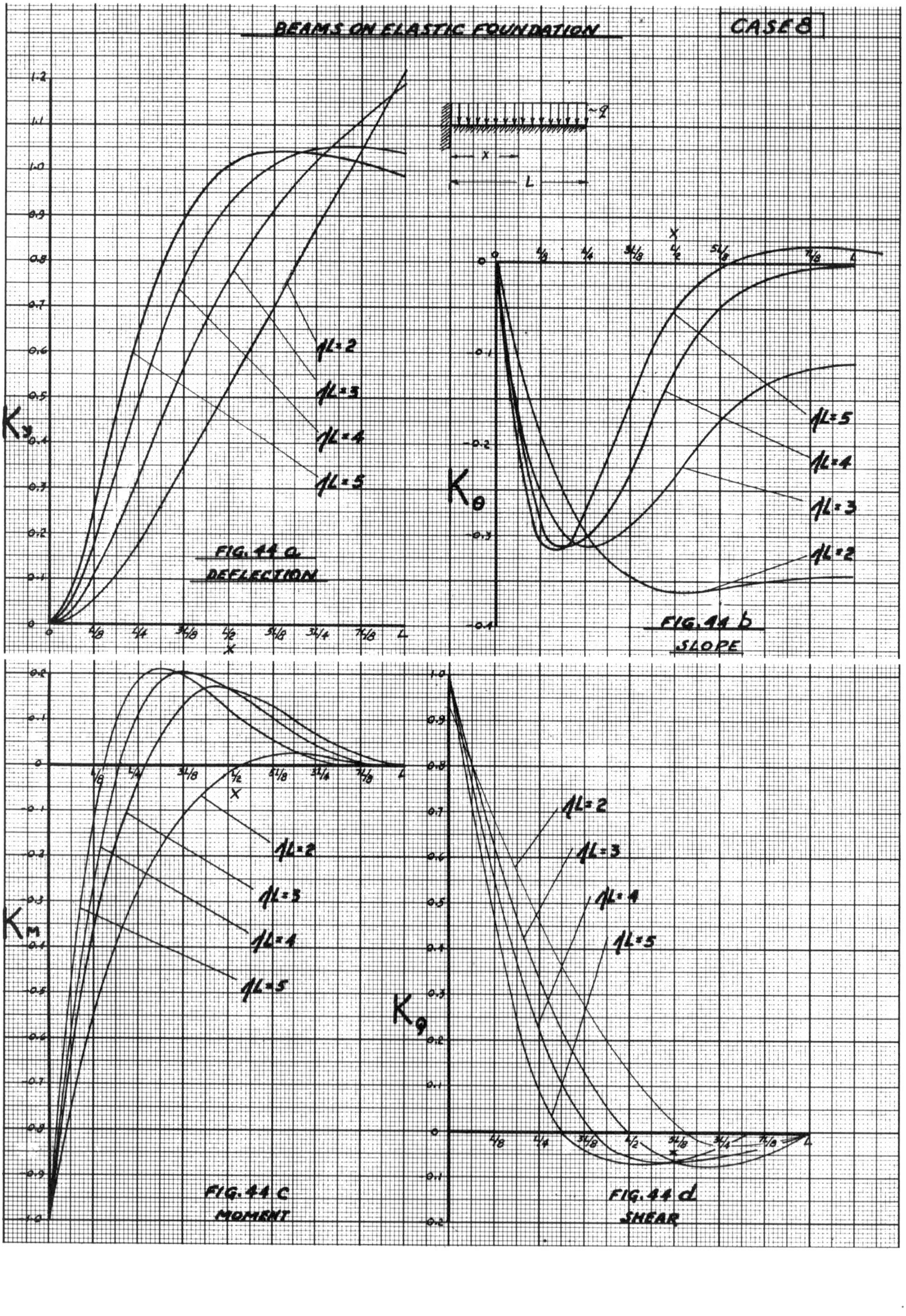
BEAMS ON ELASTIC FOUNDATION
CASE 8
FIG. 44 a
DEFLECTION
FIG. 44 b
SLOPE
FIG. 44 c
MOMENT
FIG. 44 d
SHEAR
λL=2
λL=3
λL=4
λL=5
q
x
L

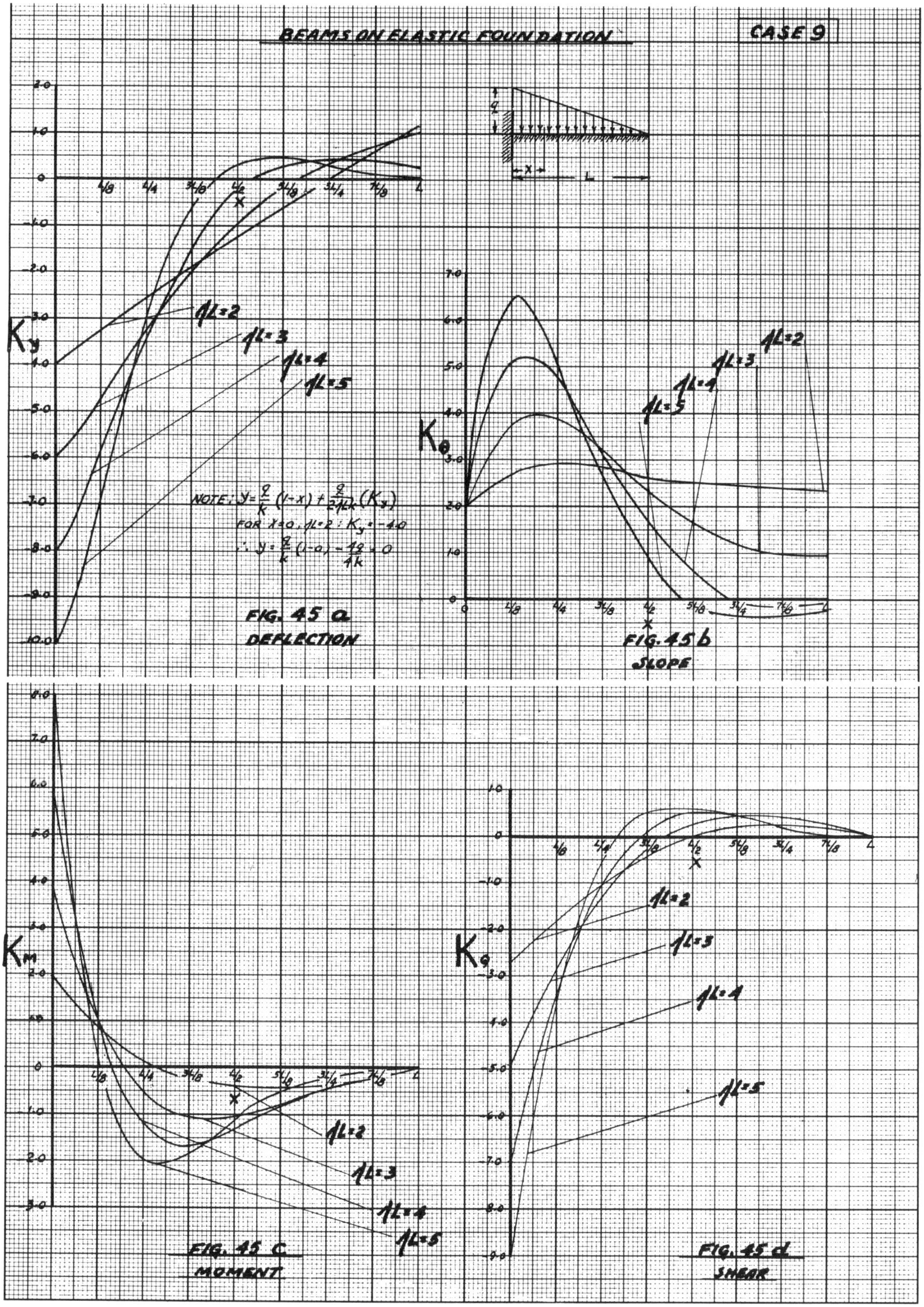
BEAMS ON ELASTIC FOUNDATION
CASE 9
q
x
L
K_y
λL=2
λL=3
λL=4
λL=5
NOTE: y = q/k (l−x) + q/2λ²k (K_y)
FOR X=0, λL=2 : K_y = −4.0
∴ y = q/k (l−0) − 4q/4k = 0
FIG. 45 a
DEFLECTION
K_θ
FIG. 45 b
SLOPE
K_m
FIG. 45 C
MOMENT
K_Q
FIG. 45 d
SHEAR

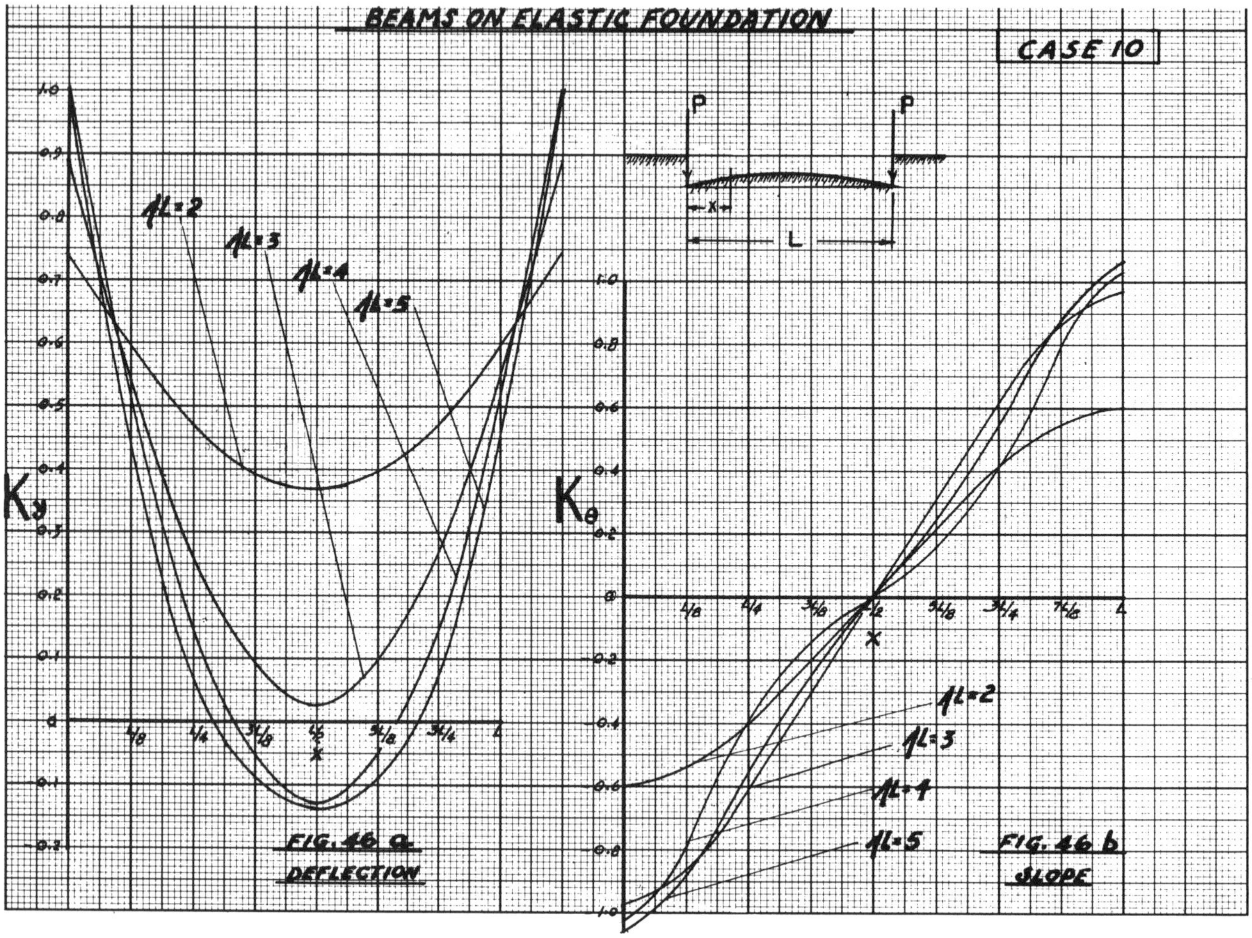

FIG. 46 a
DEFLECTION

FIG. 46 b
SLOPE

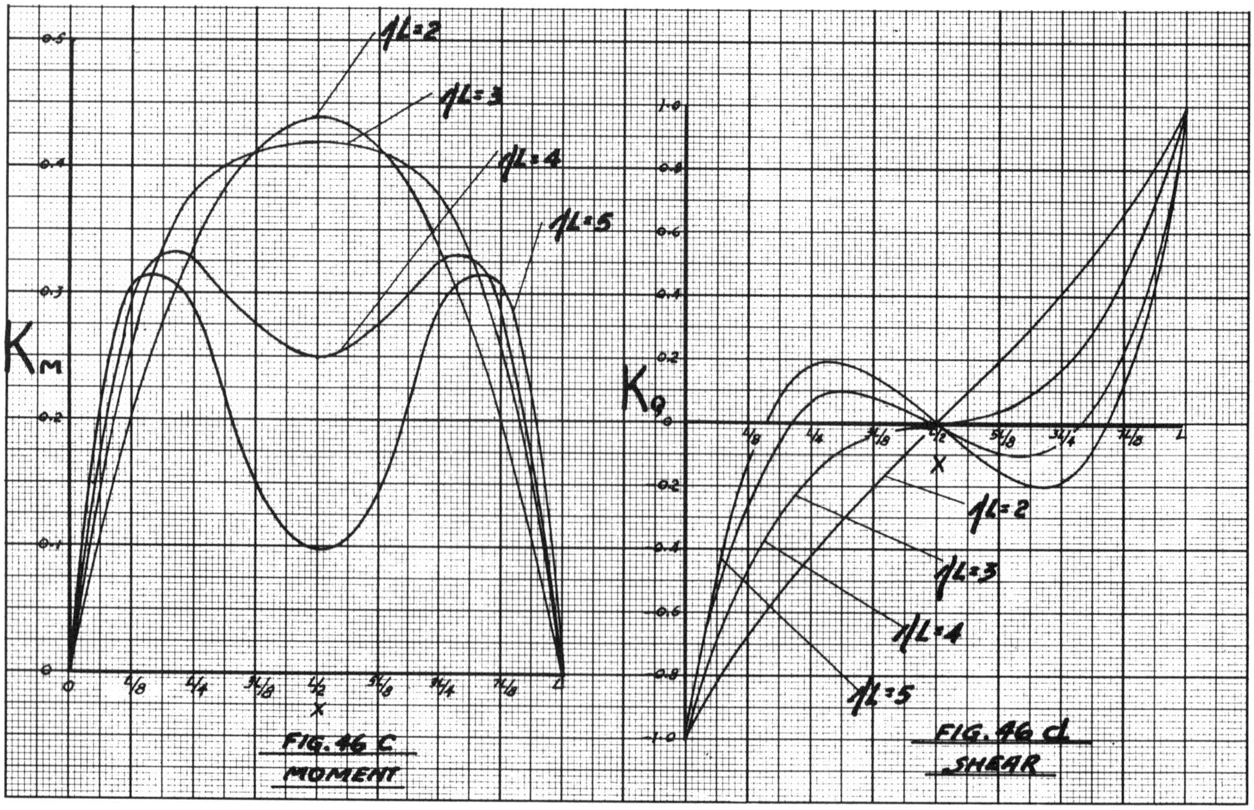

FIG. 46 C
MOMENT

FIG. 46 d
SHEAR

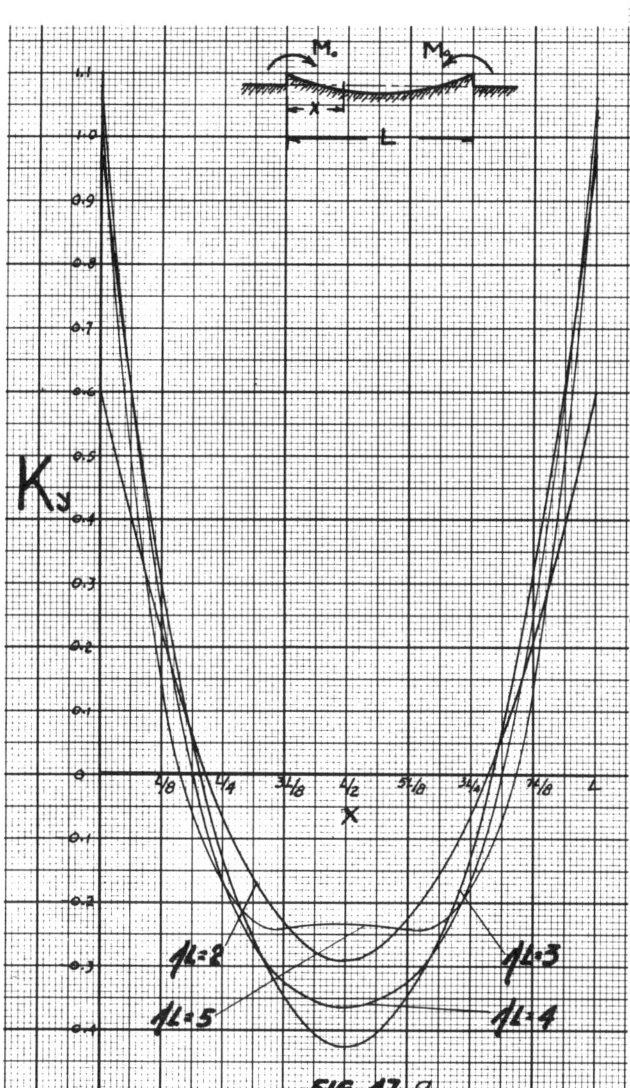

FIG. 47 a
DEFLECTION

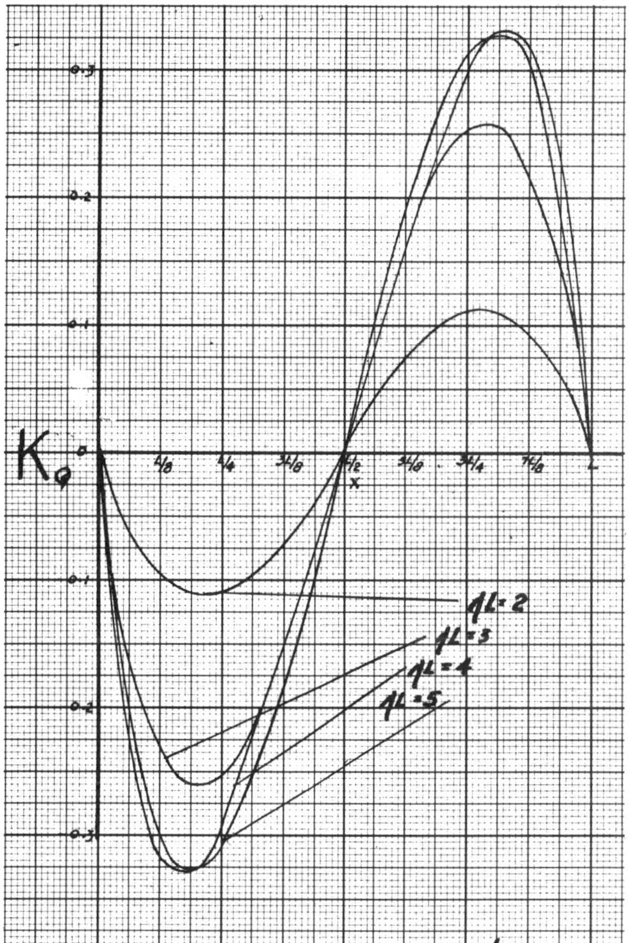

FIG. 47 d
SHEAR

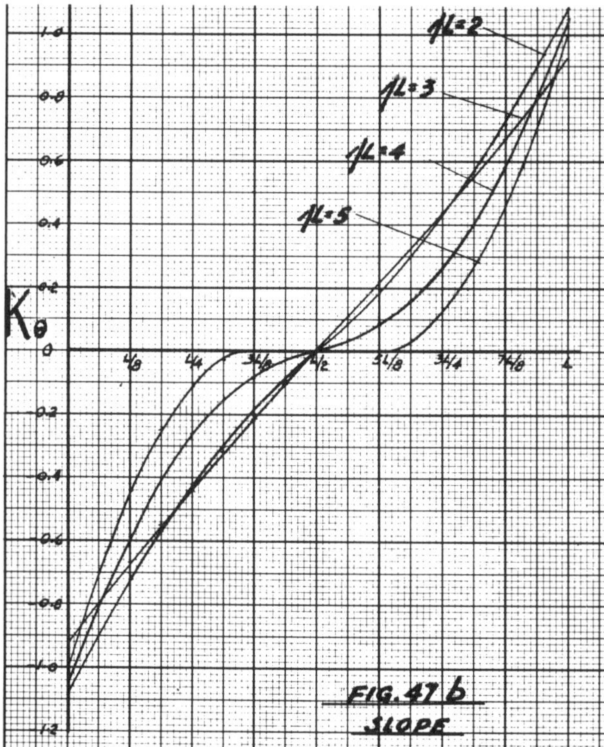

FIG. 47 b
SLOPE

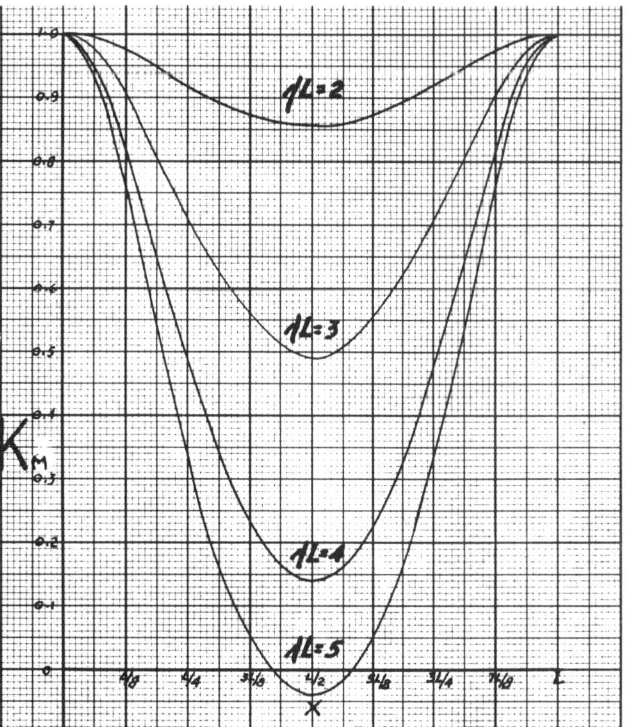

FIG. 47 C
MOMENT

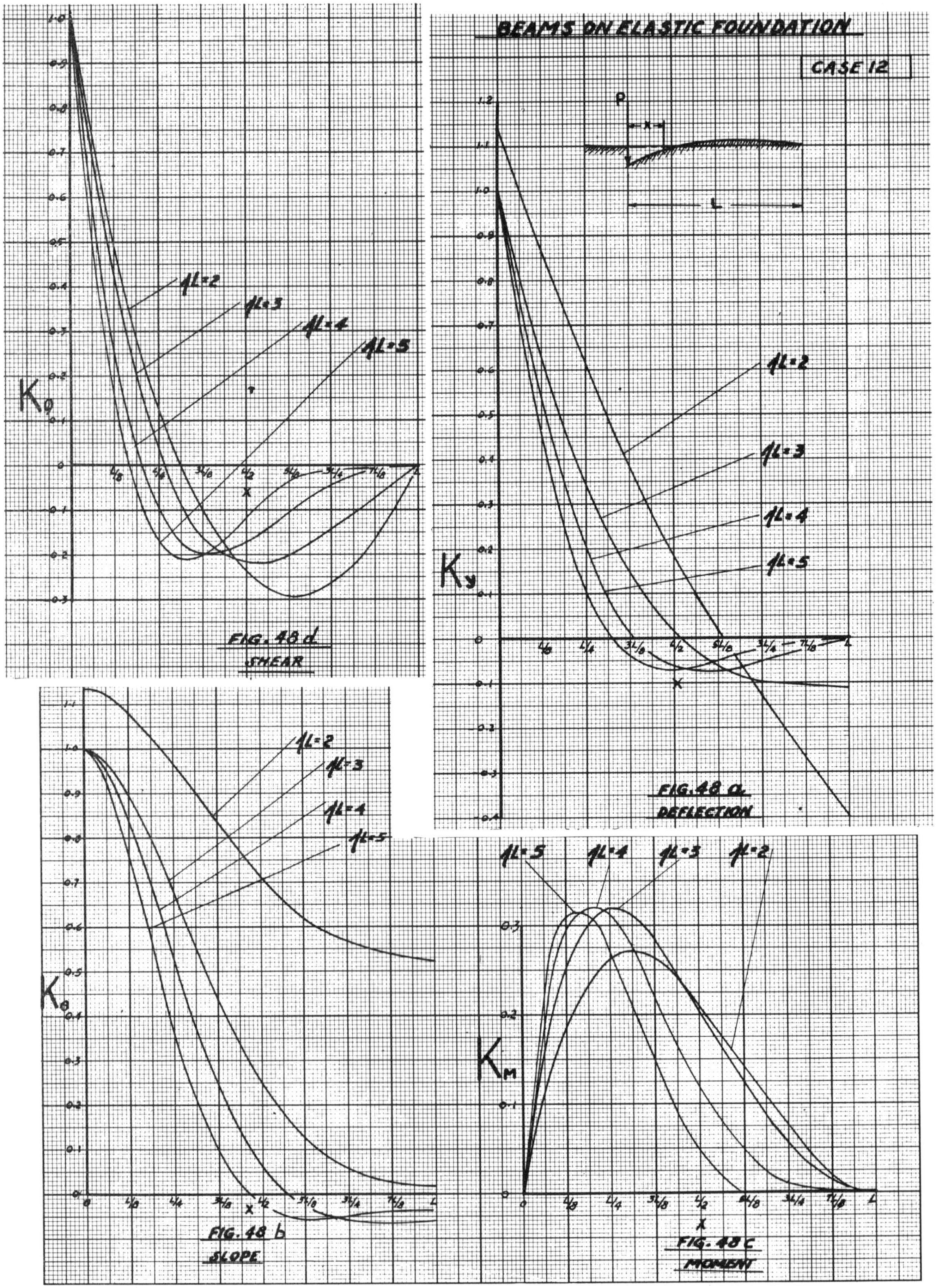
BEAMS ON ELASTIC FOUNDATION
CASE 12
P
x
L
K_Q
K_y
K_θ
K_M
λL=2
λL=3
λL=4
λL=5
FIG. 48 d
SHEAR
FIG. 48 a
DEFLECTION
FIG. 48 b
SLOPE
FIG. 48 c
MOMENT

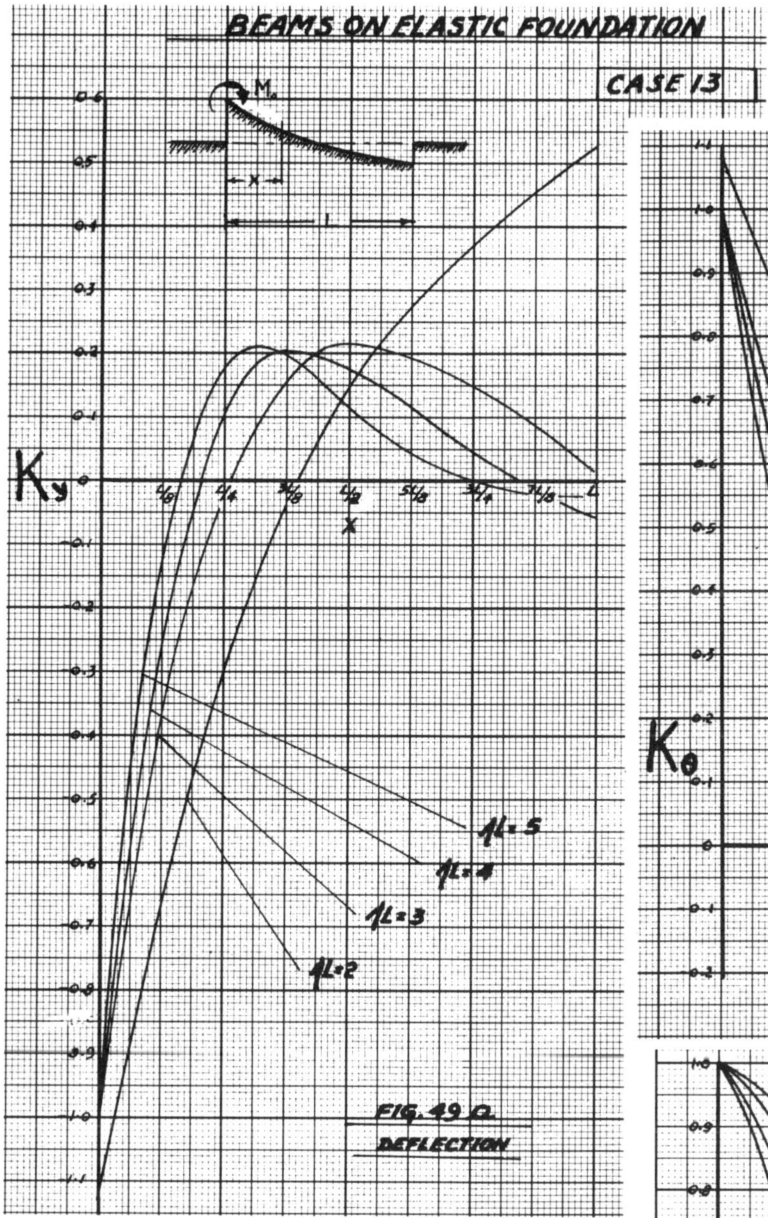

FIG. 49 a
DEFLECTION

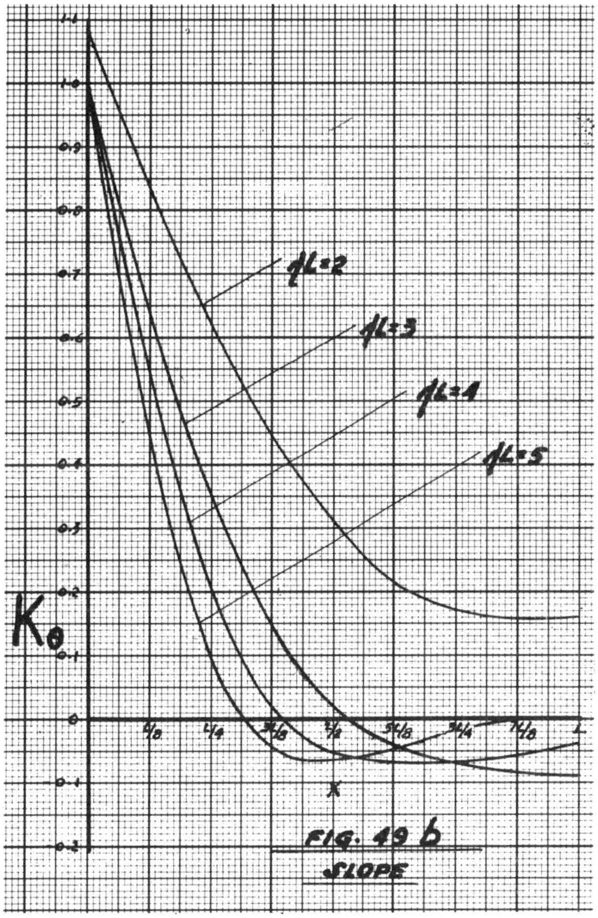

FIG. 49 b
SLOPE

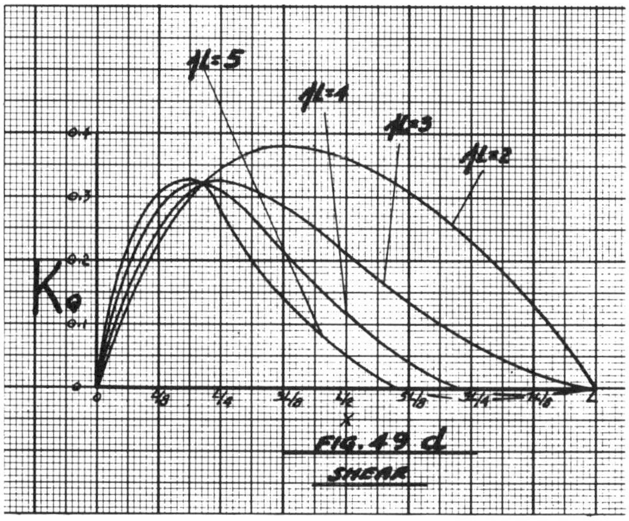

FIG. 49 d
SHEAR

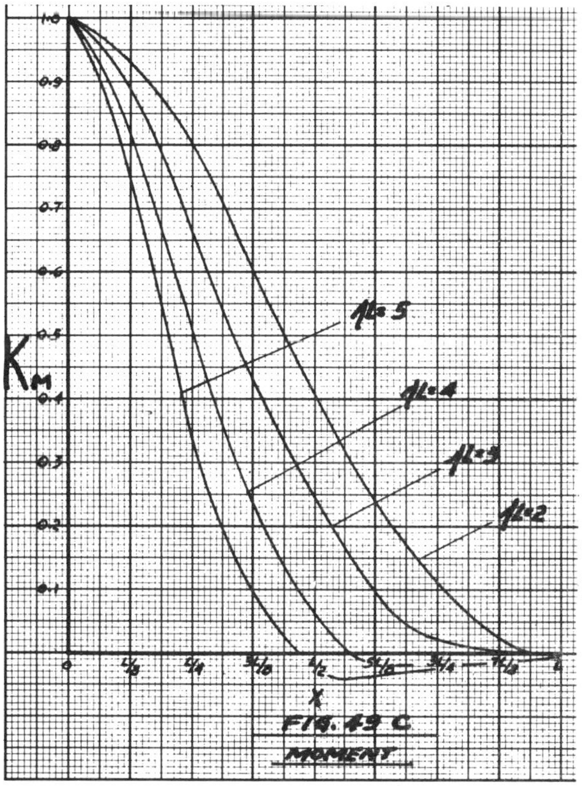

FIG. 49 C
MOMENT

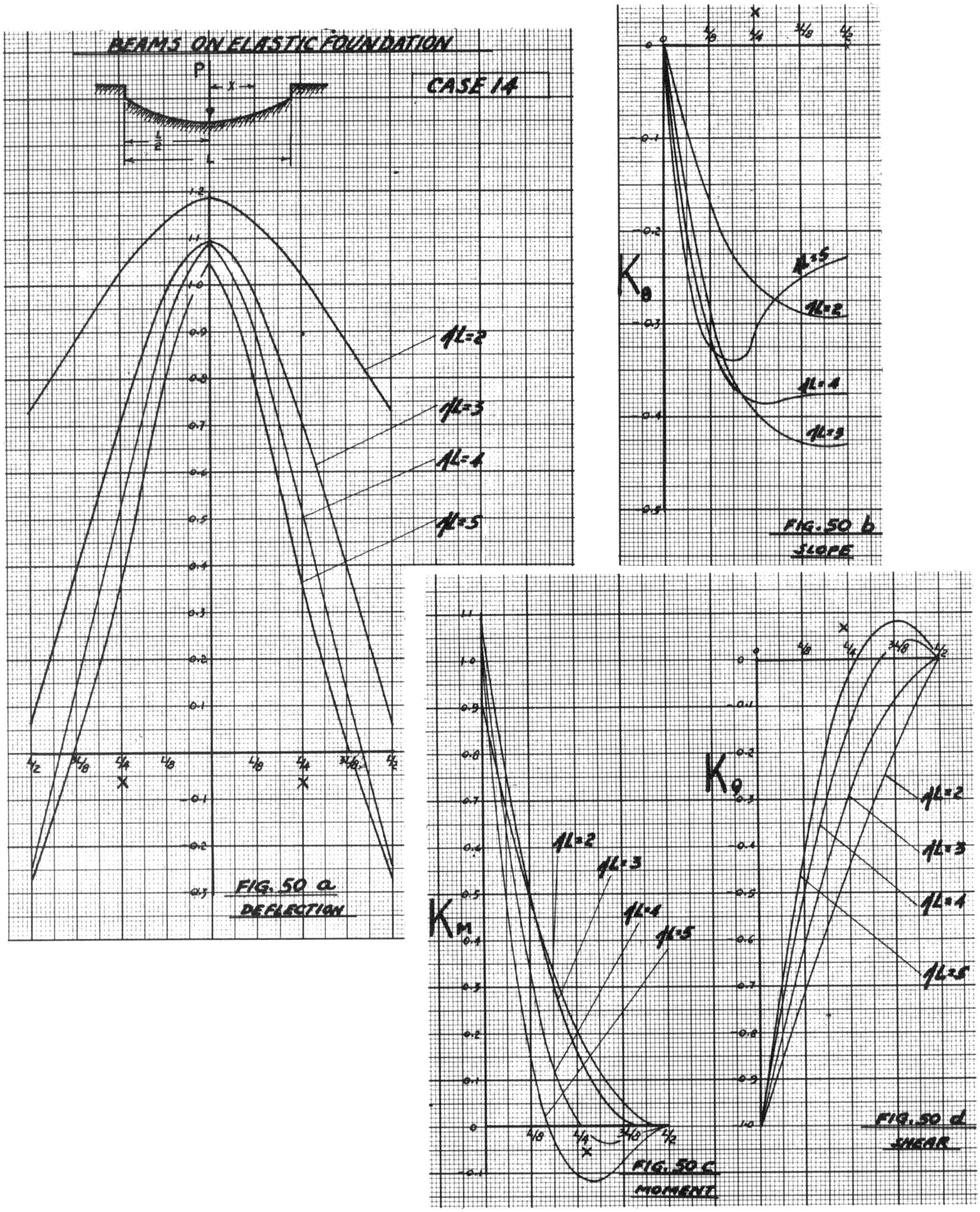
BEAMS ON ELASTIC FOUNDATION
CASE 14
P
λL=2
λL=3
λL=4
λL=5
FIG. 50 a
DEFLECTION
K
θ
FIG. 50 b
SLOPE
K
M
FIG. 50 C
MOMENT
K
Q
FIG. 50 d
SHEAR

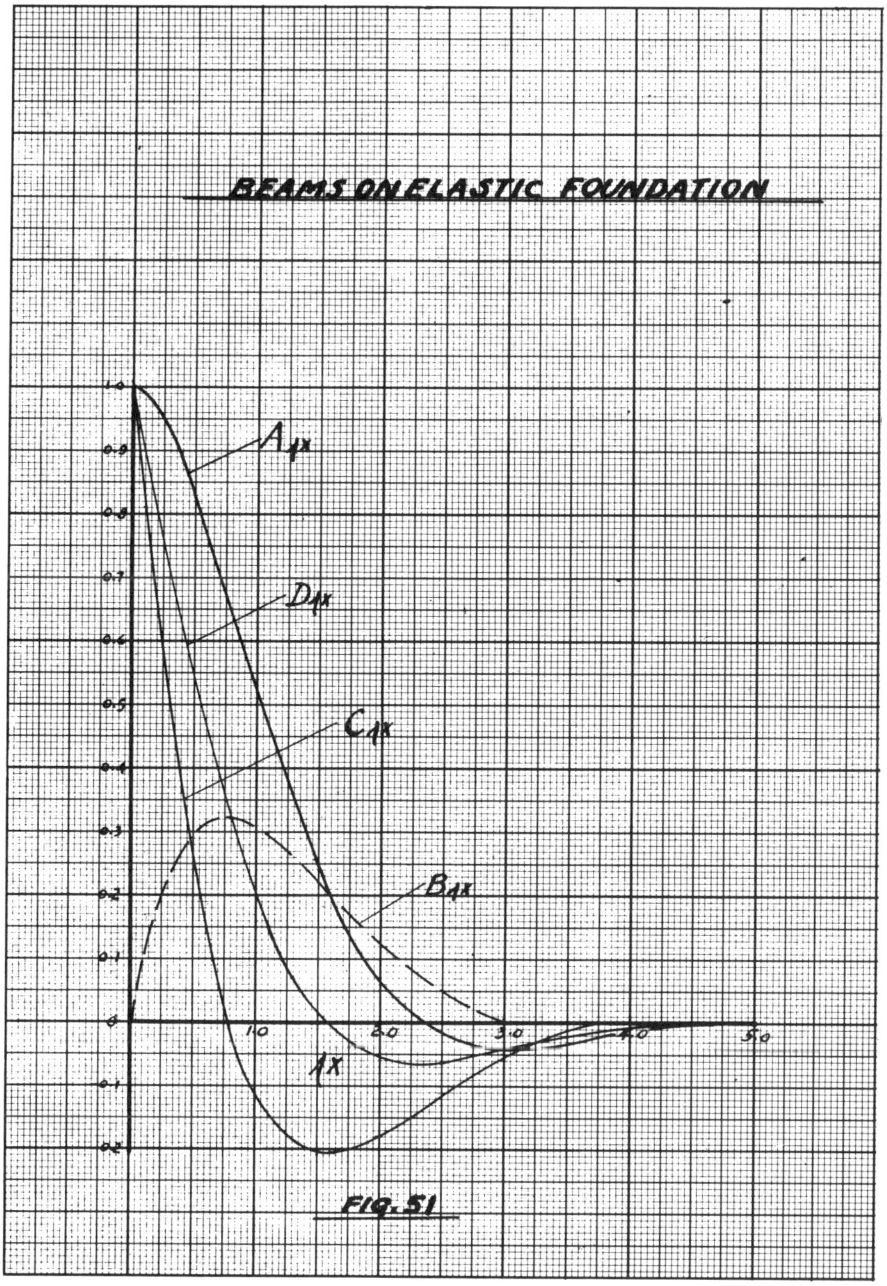
BEAMS ON ELASTIC FOUNDATION
$A_{\lambda x}$
$D_{\lambda x}$
$C_{\lambda x}$
$B_{\lambda x}$
λx
1.0
2.0
3.0
4.0
5.0

FIG. 51

32. BEAMS ON ELASTIC FOUNDATION; A SPECIAL CASE OF A BEAM WITH AN INTERMEDIATE LOAD

MEDIUM-LENGTH BEAM CONSIDERED: $\pi/4 < \lambda L < \pi$

(*Ref. 18, 19*)

In this section we shall discuss medium-length beams subjected to a concentrated load P located at arbitrary distances a and b from the free ends of the beam (Fig. 52). The following equations apply:

$$y = (P\lambda/k)K_y \tag{44}$$

$$M = (P/2\lambda)K_M. \tag{45}$$

The values of K_y and K_M for various load locations are given as ordinates to the curves in Figs. 53–60. These curves have been plotted for four different lengths of beams: $L = 2/\lambda$, $3/\lambda$, $4/\lambda$ and $5/\lambda$.

These curves in conjunction with equations 44 and 45, provide an easy method for finding y and M caused, at any section, by a single concentrated load P located at the left end and at distances $L/6$, $L/3$ and $L/2$ from the left end, as indicated on the curves. If the load is located between the midpoint and the right end of the beam, the curve corresponding to this distance from the left end may be used by exchanging ends of the beam. If the load P does not act exactly at one of the locations for which curves are given, the K factors may be obtained easily by interpolating between the given curves. Also, if the beam length is not equal to one of the given lengths, the K factor at any section may be found for lengths above and below actual length of the beam and the approximate K value found by interpolation.

Fig. 52

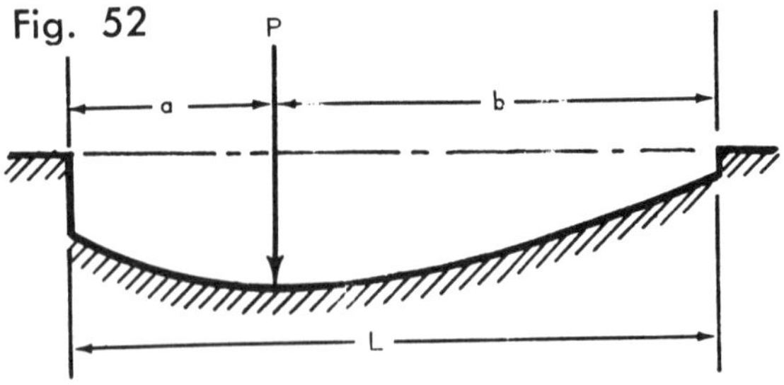

If beams on elastic support shorter than $L = 2/\lambda$ are encountered, values of y and M may be obtained by either of two methods. First, if the length L is between $1/\lambda$ and $2/\lambda$, the curves for $2/\lambda$ give approximate, conservative values for y, and M. Second, if the beam and its elastic supports are such that $L = 1/\lambda$ or less, it may be assumed rigid, in which case the bending moment is negligible and the computation of the deflections is a statically determined problem.

The results of this analysis are applicable to (1) beams on continuous elastic foundation and (2) to beams supported by a number of equidistant elastic supports.

NOTE: The curves for K_y show the actual shape and position the beam will assume as a result of the action of the load P, in comparison to the original position of the beam as shown by the line AB in each diagram.

Example: Assume that the configuration of the beam is such that $L = 2.5/\lambda$ which is between $L = 2/\lambda$ and $L = 3/\lambda$. Also, assume that $a = L/3$.

From Fig. 53, $K_M = 0.375$ and from Fig. 54, $K_M = 0.475$. Therefore, by interpolation, for $L = 2.5/\lambda$, $K_M = 0.42$. Also, from Fig. 57, $K_y = 0.7$ and Fig. 58, $K_y = 0.55$; by interpolation, for $L = 2.5/\lambda$, $K_y = 0.625$.

Thus:

$$M = \left(\frac{P}{2\lambda}\right)0.524$$

$$y = \left(\frac{P\lambda}{k}\right)0.625.$$

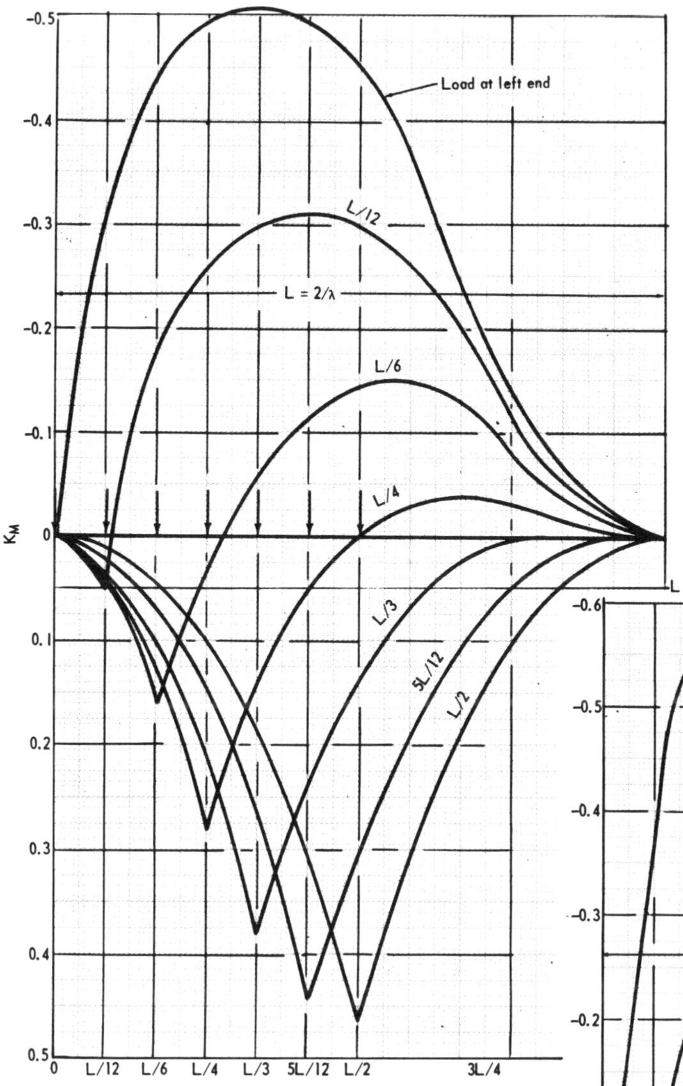
Load at left end
L/12
L = 2/λ
L/6
L/4
L/3
5L/12
L/2
K_M
-0.5
-0.4
-0.3
-0.2
-0.1
0
0.1
0.2
0.3
0.4
0.5
0
L/12
L/6
L/4
L/3
5L/12
L/2
3L/4
L

Fig. 53

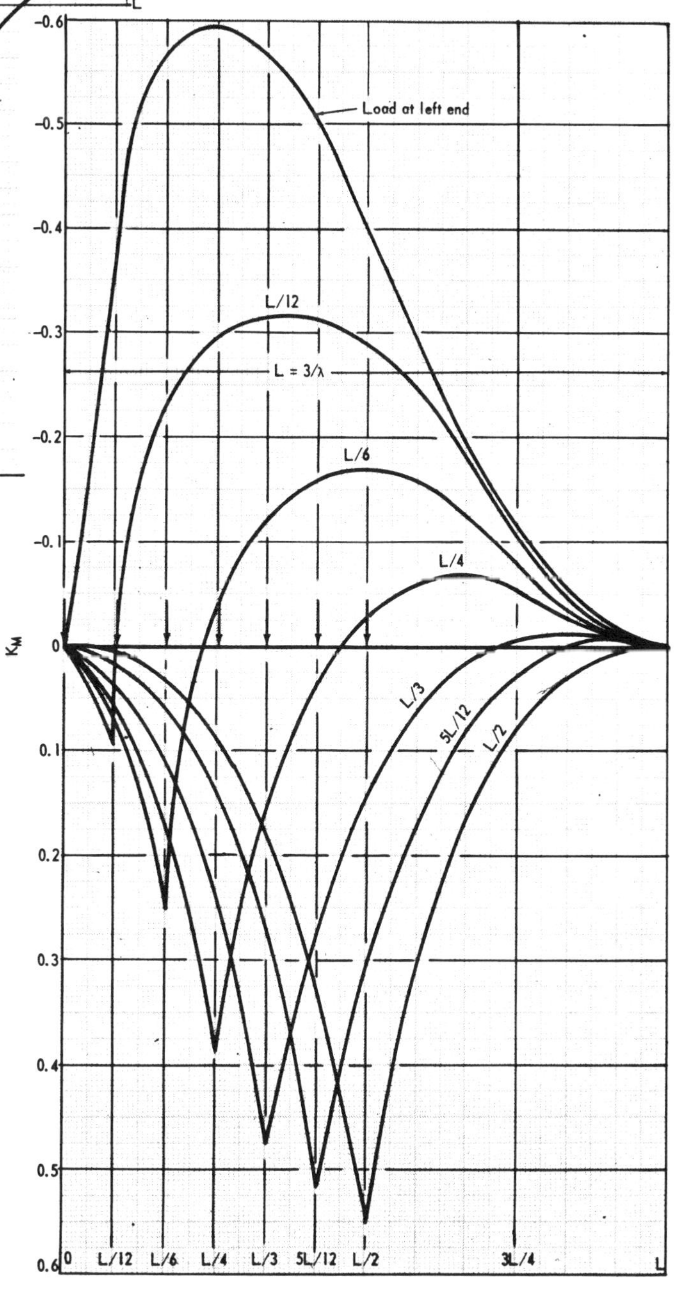
Load at left end
L/12
L = 3/λ
L/6
L/4
L/3
5L/12
L/2
K_M
-0.6
-0.5
-0.4
-0.3
-0.2
-0.1
0
0.1
0.2
0.3
0.4
0.5
0.6
0
L/12
L/6
L/4
L/3
5L/12
L/2
3L/4
L

Fig. 54

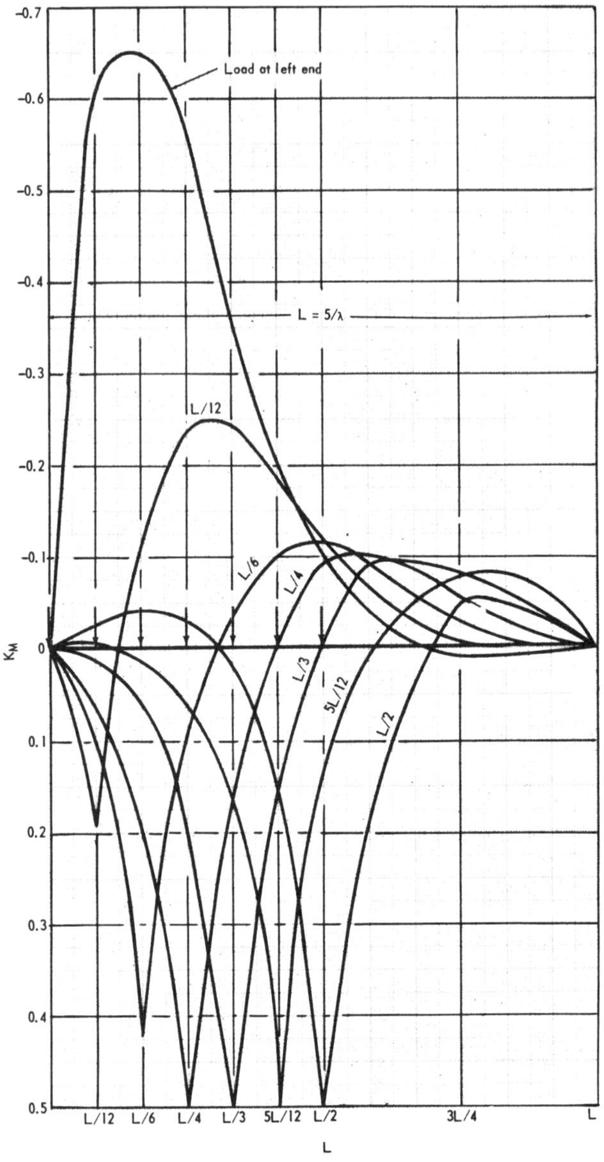
Load at left end
L = 5/λ
L/12
L/6
L/4
L/3
5L/12
L/2
K_M
-0.7
-0.6
-0.5
-0.4
-0.3
-0.2
-0.1
0
0.1
0.2
0.3
0.4
0.5
L/12 L/6 L/4 L/3 5L/12 L/2 3L/4 L
L

Fig. 56

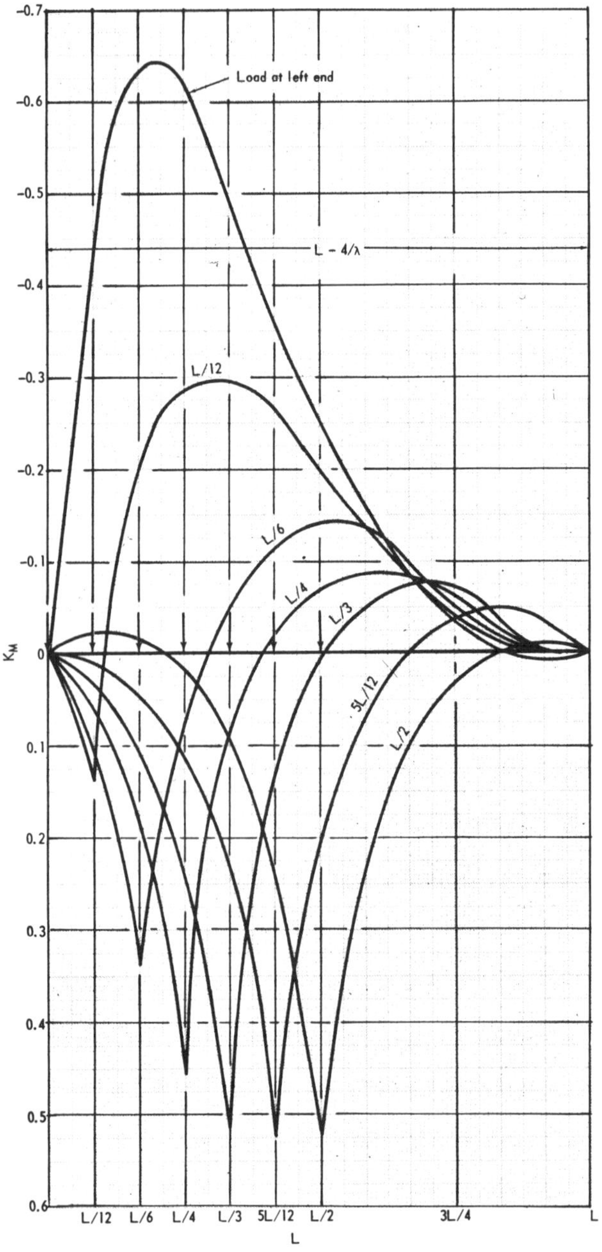
Load at left end
L = 4/λ
L/12
L/6
L/4
L/3
5L/12
L/2
K_M
-0.7
-0.6
-0.5
-0.4
-0.3
-0.2
-0.1
0
0.1
0.2
0.3
0.4
0.5
0.6
L/12 L/6 L/4 L/3 5L/12 L/2 3L/4 L
L

Fig. 55

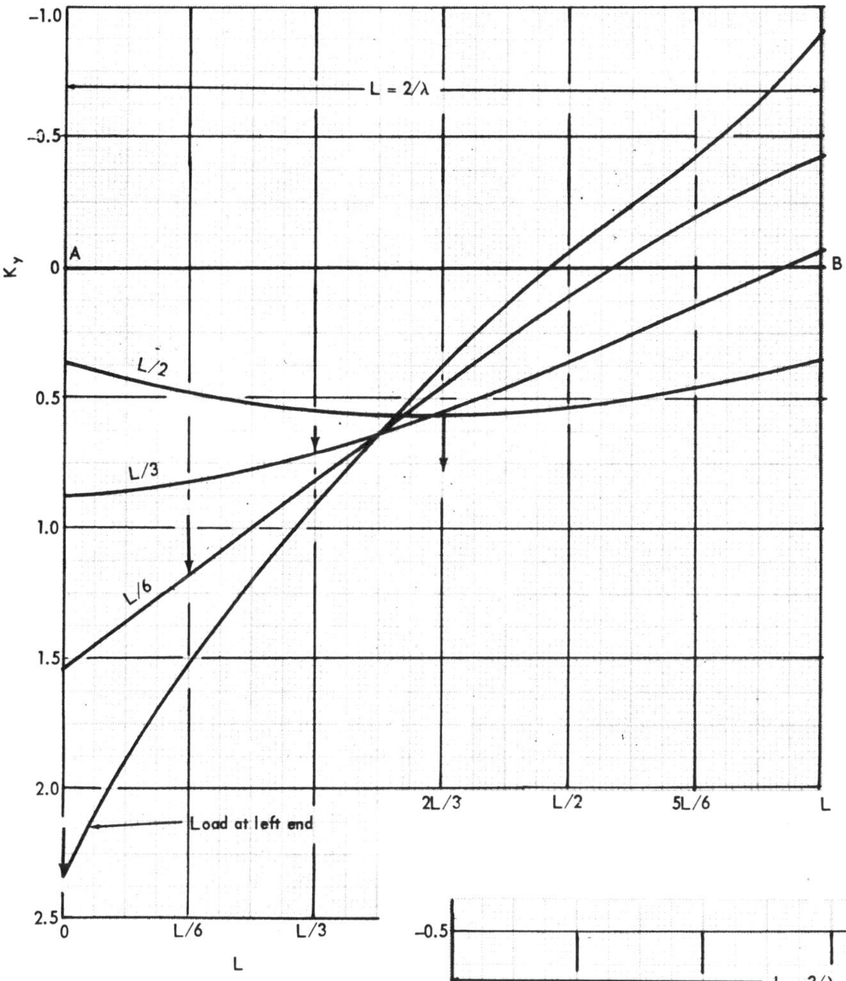

-1.0
-0.5
0
0.5
1.0
1.5
2.0
2.5
K_y
L = 2/λ
A
B
L/2
L/3
L/6
Load at left end
0
L/6
L/3
2L/3
L/2
5L/6
L
L

Fig. 57

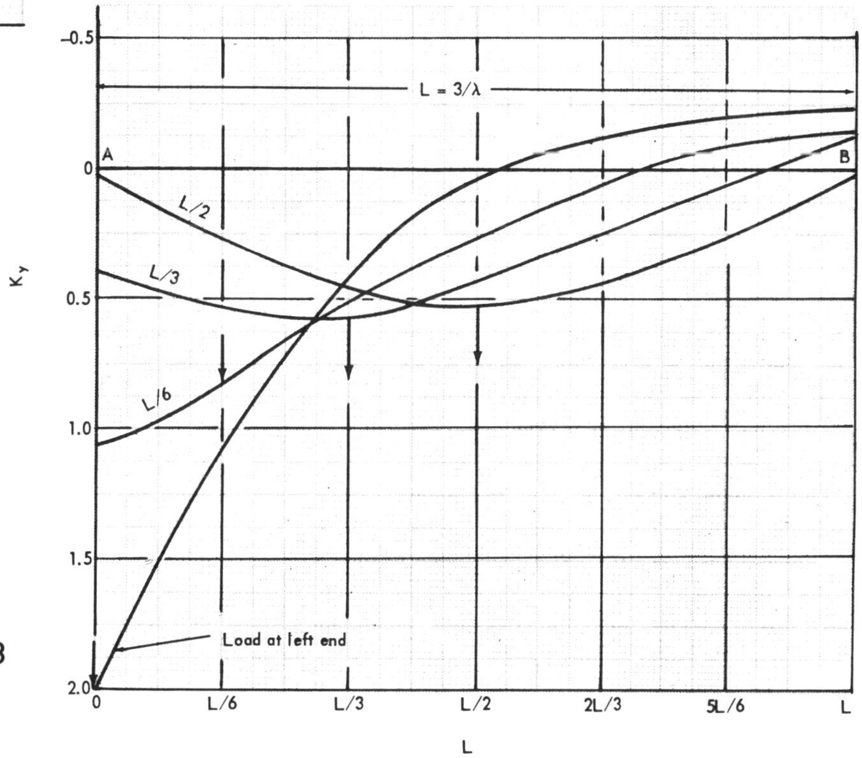

-0.5
0
0.5
1.0
1.5
2.0
K_y
L = 3/λ
A
B
L/2
L/3
L/6
Load at left end
0
L/6
L/3
L/2
2L/3
5L/6
L
L

Fig. 58

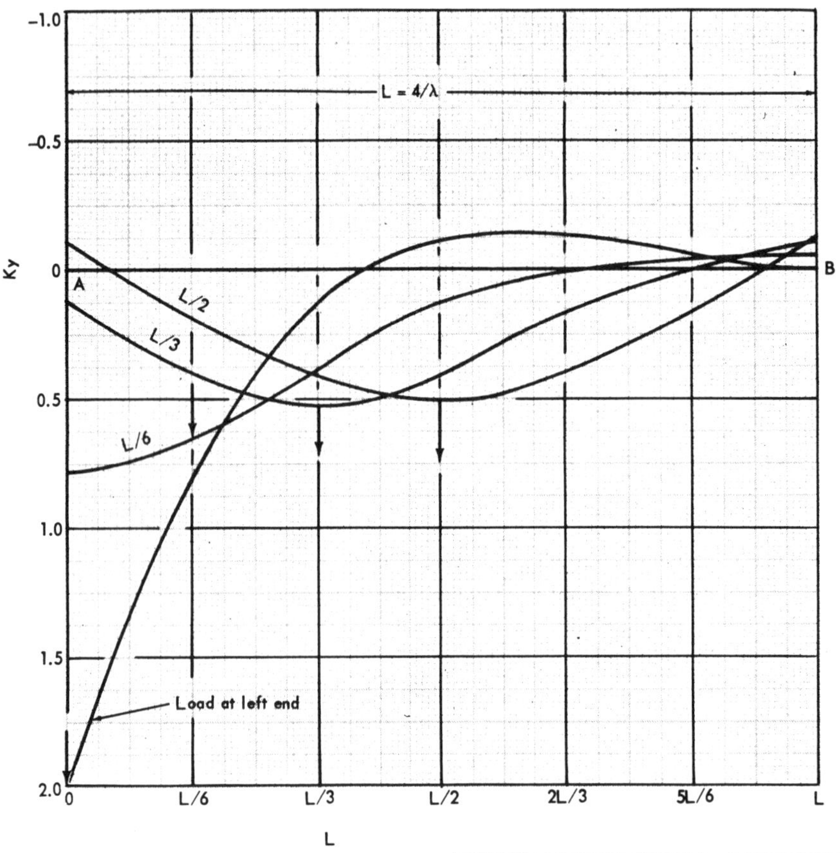
-1.0
-0.5
0
0.5
1.0
1.5
2.0
Ky
L = 4/λ
A
B
L/2
L/3
L/6
Load at left end
0
L/6
L/3
L/2
2L/3
5L/6
L
L

Fig. 59

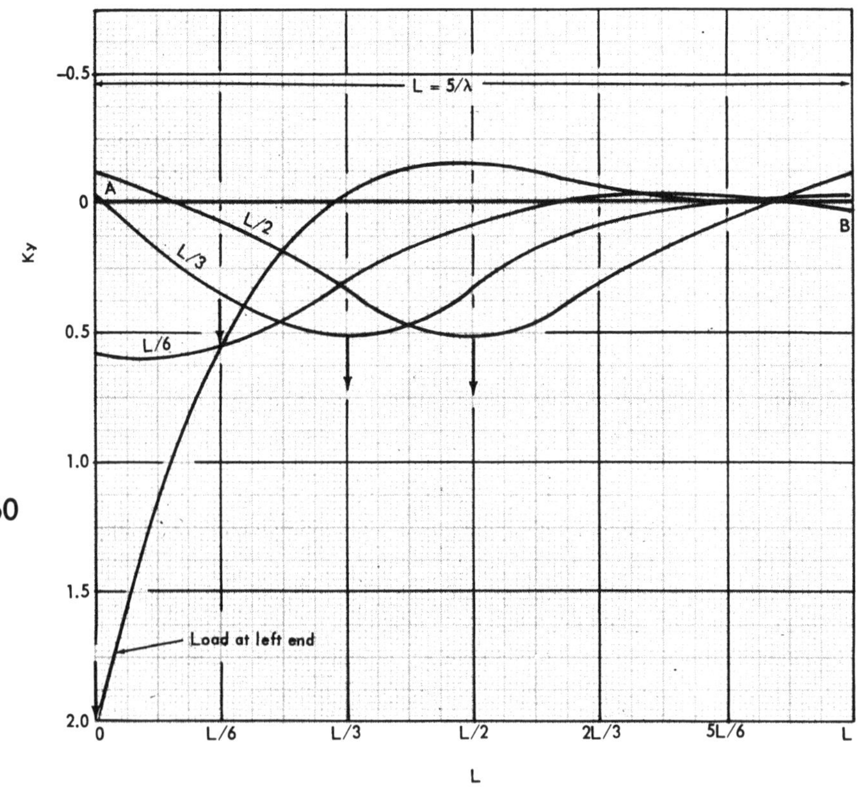
-0.5
0
0.5
1.0
1.5
2.0
Ky
L = 5/λ
A
B
L/2
L/3
L/6
Load at left end
0
L/6
L/3
L/2
2L/3
5L/6
L
L

Fig. 60

33. BEAMS ON ELASTIC FOUNDATION OF UNLIMITED LENGTHS

Beams may be of unlimited lengths in both directions (infinite beams) or of unlimited extension in only one direction, having one finite end (semi-infinite beam). This end may be (1) free, (2) hinged, (3) fixed. Equations for y, θ, M, Q for infinite beams are tabulated in Table 12, cases 15, 16, and for semi-infinite beams, Cases 17, 18, 19, 20. The parameters $A_{\lambda x}$, $B_{\lambda x}$, $C_{\lambda x}$, $D_{\lambda x}$ are plotted in Fig. 51.

The use of equations in Table 13 and graphs, Fig. 51 will be demonstrated in the following example.

Example: A circular tube of unlimited length has a reinforcing ring at *one* place and the tube is subjected to an internal pressure p, find the increase of the bending stress which occurs under the reinforcing ring in the longitudinal fibers of the tube over the circumferential stress existing before the application of the ring.

Solution: From consideration of statics we find that, in the wall of the tube, there will be a uniform circumferential stress $S_c = pR/t$, where R denotes the mean radius and t the wall thickness of the tube. Because of this stress the tube will expand radially by the amount

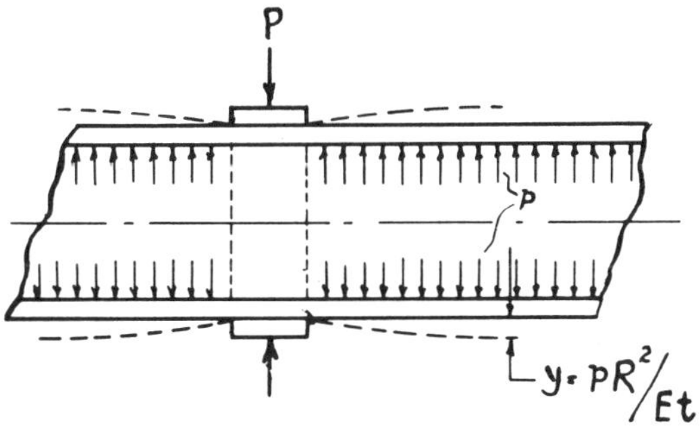

$y = pR^2/Et$ on an infinite beam supported on elastic foundation having the values of k and λ defined as before:

$$k = \frac{Et}{R^2};\ \lambda = \sqrt[4]{3(1-\nu^2)}\,\frac{1}{\sqrt{Rt}}.$$

For steel

$$\nu = 0.3 \text{ and } \lambda = \frac{1.285}{Rt}.$$

The maximum bending moment and maximum deflection due to this action can be determined from Table 2–13 case 15:

$$M_{\max} = \frac{P}{4\lambda} \text{ for } C_{\lambda x} = 1(x = 0)$$

$$Y_{\max} = \frac{P\lambda}{2k} \text{ for } A_{\lambda x} = 1(x = 0).$$

The problem before us is to find the unknown concentrated load P. As was stated before the deflection due to P is equal to the radial expansion of the tube due to the pressure p:

$$Y = \frac{PR^2}{Et} = \frac{P\lambda}{2k} \text{ from which } P = \frac{2p\sqrt{Rt}}{1.285};$$

by substitution:

$$M_{\max} = \frac{P}{4\lambda} = \frac{pRt}{3.3022}.$$

Finally the maximum bending stress:

$$S_{\max} = \frac{6M_{\max}}{t^2} = 1.82\,\frac{pR}{t}.$$

Conclusion: The bending stress caused by the reinforcing ring is 82 per cent larger than the circumferential stress existing before the application of the ring.

34. BEAMS ON ELASTIC FOUNDATIONS

SIMPLIFIED METHOD FOR CALCULATION OF DEFLECTION, END SLOPE AND MOMENT

(Ref. 20, 17)

As indicated earlier, beams supported continuously along their length by an elastic surface can be classified in three categories, defined by the stiffness parameter λl, where $\lambda = (k/4\ EI)^{1/4}$. Modulus of foundation, k, is defined by the equation $p = ky$, where p is the pressure (lb./in.) required to produce vertical deflection, y, of the foundation. The three categories are:

1. Short beams: $\lambda l < \pi/4$
2. Beams of medium length: $\pi/4 < \lambda l < \pi$
3. Long beams: $\lambda l > \pi$

NOTATION

a, b = beam dimensions (in.)
E = modulus of elasticity (psi)
I = moment of inertia of beam (in.^{4})
k = modulus of foundation (psi)
l = length of beam (in.)
M = maximum bending moment of beam (lb.-in.*)
M_o = applied moment of beam (lb.-in.)
P = concentrated load (lb.)
w = uniformly distributed load (lb./in.)
y = maximum deflection of beam (in.*)
θ = end slope of beam (radians*)
λ = stiffness characteristics (in.$^{-1}$)
$= (k/4\ EI)^{1/4}$
λl = stiffness parameter.

*M', y', and θ' represent values of m, y, and θ computed by standard beam formulas.

Beams in Group 1 are considered completely rigid. Therefore, bending deformation may be neglected, since the deflection caused by bending is usually negligibly small in comparison with the deflection of the elastic foundation.

The beams in Group 2 must be treated as beams of finite length. Beam deflections are significant, and should be calculated. This is the case discussed here.

Beams in Group 3 are considered infinitely long, so that a force applied at one end has no effect at the other end.

The following method provides a quick means of calculating deflection, end slope, and bending moment of elastically supported beams which fall into Group 2. Seven different load conditions are covered. The general method is:

1. Calculate moment of inertia, I, of the beam.
2. Determine $\lambda = (k/4\ EI)^{1/4}$
3. Find λl.
4. Determine which of the seven loading conditions applies to the problem. From the equations in Table 13, calculate y', θ', and M'.
5. Choose the proper graph for the load condition. Find y/y', θ/θ', and M/M'.
6. Calculate $y = y'(y/y')$; $\theta = \theta'(\theta/\theta')$; $M = M'(M/M')$.

Table 13 –Standard Equations Deflections, Slope, and Moment

Case	Deflection, y'	Slope, θ'	Moment, M'	Maximum Values
1. Hinged ends, center load	$\frac{Pl^3}{48\ EI}$	$\frac{Pl^2}{16\ EI}$	$\frac{Pl}{4}$	y_C, $\theta_A(=-\theta_B)$, M_C
2. Hinged ends, uniform load	$\frac{5\ wl^4}{384\ EI}$	$\frac{wl^3}{24\ EI}$	$\frac{wl^2}{8}$	y_C, $\theta_A(=-\theta_B)$, M_C
3. Hinged ends, equal moments at both ends	$\frac{M_o l^2}{8\ EI}$	$\frac{M_o l}{2\ EI}$	M_o	y_C, $\theta_A(=-\theta_B)$, M_C
4. Fixed ends, center load	$\frac{Pl^3}{192\ EI}$		$\frac{Pl}{8}$	y_C, M_C
5. Fixed ends, uniform load	$\frac{wl^4}{384\ EI}$		$-\frac{wl^2}{12}$	y_C, M_C
6. Cantilever beam, end load	$\frac{Pl^3}{3\ EI}$	$\frac{Pl^2}{2\ EI}$	$-Pl$	y_B, θ_B, M_A
7. Cantilever beam. uniform load	$\frac{wl^4}{8\ EI}$	$\frac{wl^3}{6\ EI}$	$-\frac{wl^2}{2}$	y_B, θ_B, M_A

Example: An aluminum-alloy beam of rectangular cross section rests with its larger face (side a) on an elastic foundation having a modulus of foundation, k = 100 psi. Length of beam, l = 15 in.; a = 1 in.; b = 0.25 in. Find the maximum deflection, maximum bending moment, and end slope caused by a center load, P = 100 lb. Modulus of elasticity of the beam, $E = 10 \times 10^6$ psi and the ends of the beam are hinged (case 1).

1. Calculate $I = [1(0.25)^3]/12 = 0.0013$ in.^{4}
2. Find $\lambda = (k/4\ EI)^{1/4} = [100/4(10 \times 10^6)(0.0013)]^{1/4} = 0.209$ in.$^{-1}$
3. Then, $\lambda l = 0.209(15) = 3.13$.
4. From Table 13, case 1, $y' = Pl^3/48\ EI = [100(15)^3]/48\ (10 \times 10^6)\ (0.0013) = 0.544$ in. Also, $\theta' = 11.25$ rad.; $M' = 375$ lb.-in.
5. From the chart for case 1, $y/y' = 0.22$; $\theta/\theta' = 0.17$; $M/M' = 0.34$.
6. Finally, $y = 0.544(0.22) = 0.12$ in.; $\theta = 11.25(0.17) = 1.91$ rad.; $M = 375(0.34) = 127.5$ lb.-in.

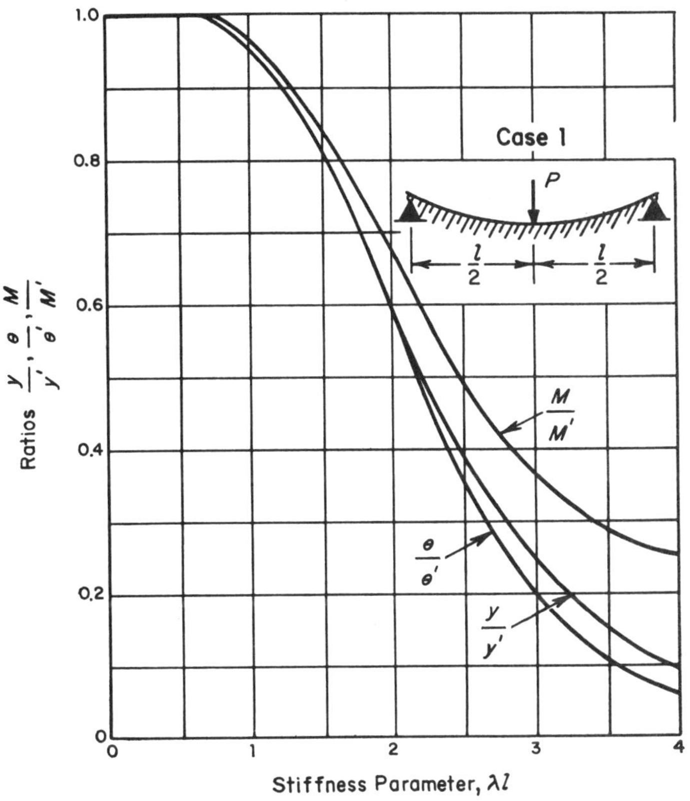

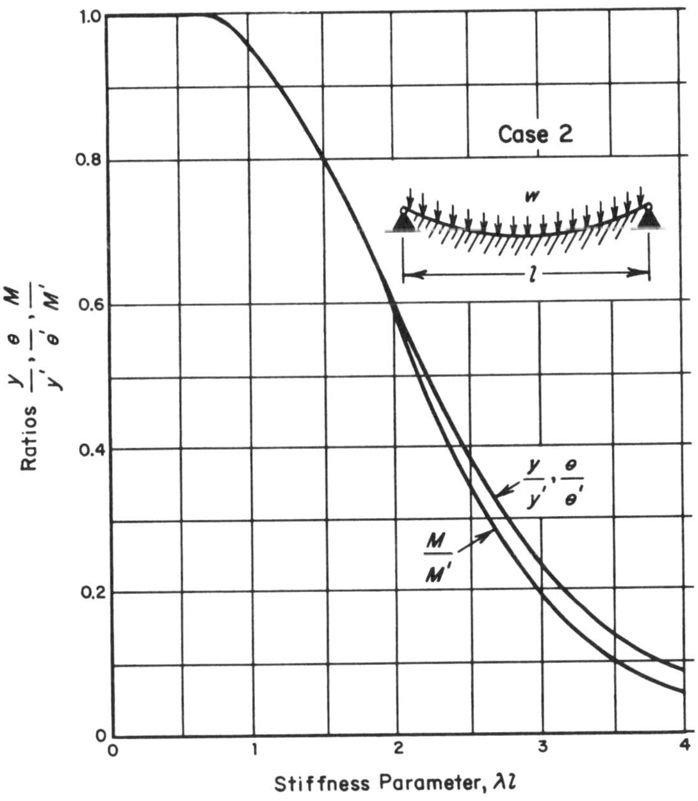

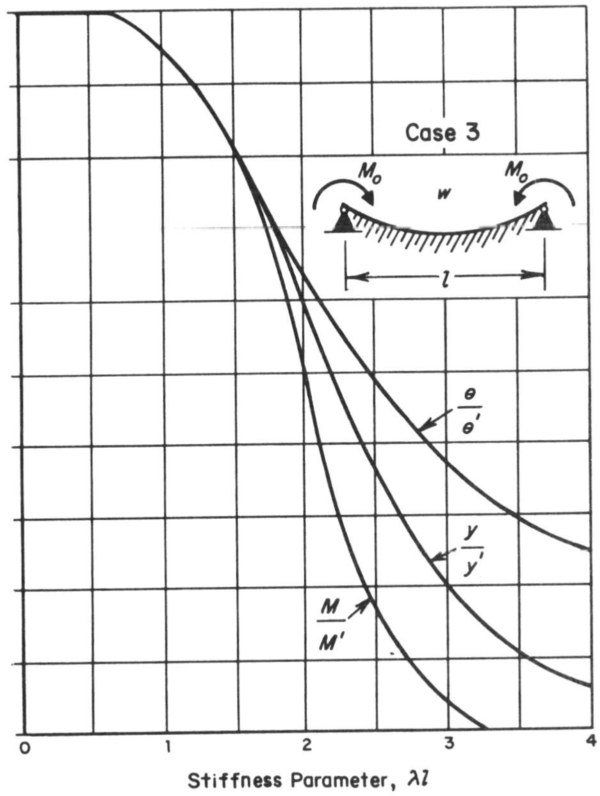

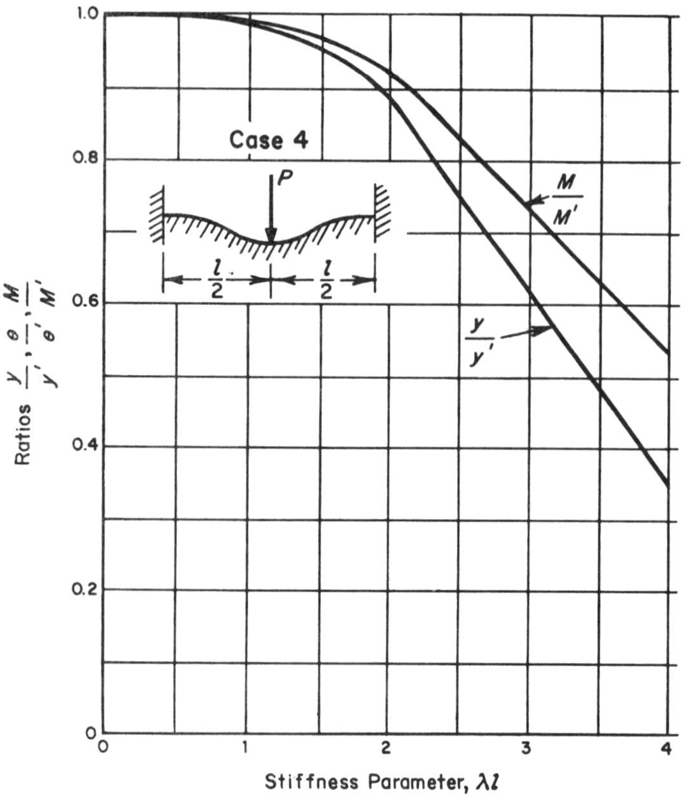
Case 4
P
l/2
l/2
M/M'
y/y'
Ratios y/y', θ/θ', M/M'
1.0
0.8
0.6
0.4
0.2
0
0
1
2
3
4
Stiffness Parameter, λl

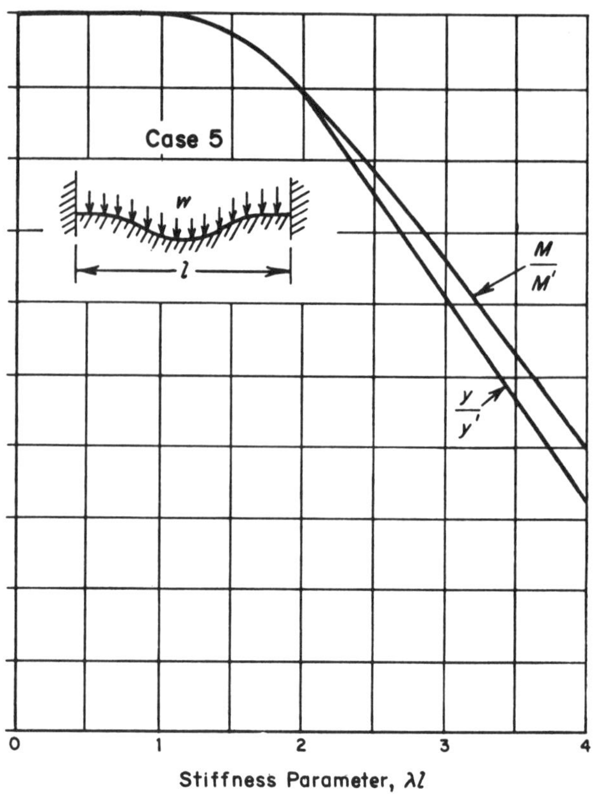
Case 5
w
l
M/M'
y/y'
0
1
2
3
4
Stiffness Parameter, λl

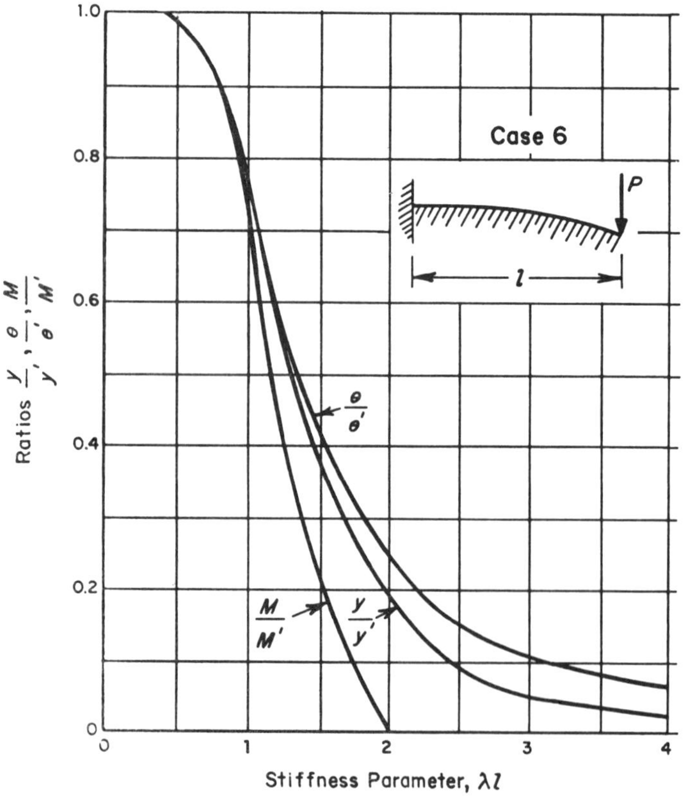
Case 6
P
l
θ/θ'
M/M'
y/y'
Ratios y/y', θ/θ', M/M'
1.0
0.8
0.6
0.4
0.2
0
0
1
2
3
4
Stiffness Parameter, λl

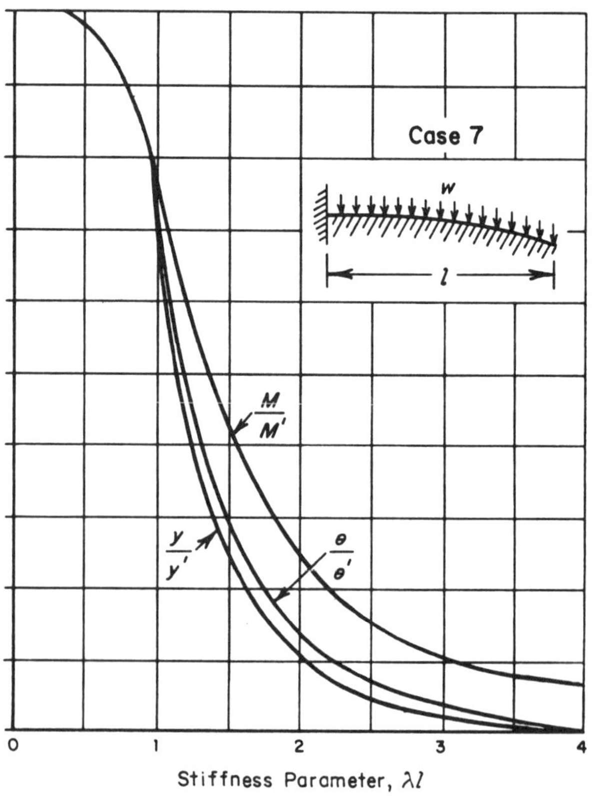
Case 7
w
l
M/M'
y/y'
θ/θ'
0
1
2
3
4
Stiffness Parameter, λl

35. APPLICATION OF THE THEORY OF BEAMS ON ELASTIC FOUNDATION TO PRESSURE VESSELS

(*Ref. 17*)

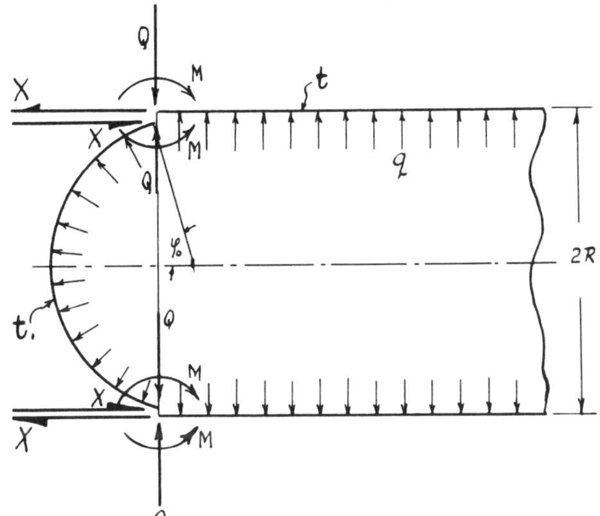

FIG. 61

A longitudinal element of a cylindrical shell under axially symmetrical loading can be regarded as a beam on elastic foundation, in which event the modulus of the foundation k and the characteristic λ are

$$k = \frac{Et}{R^2}, \lambda = \sqrt[4]{3(1-\nu^2)} \times \frac{1}{\sqrt{Rt}},$$

where R is the mean radius, t the thickness of the cylinder, E the modulus of elasticity, and ν Poisson's Ratio for the material. If steel ($\nu = 0.3$) is the material of the vessel we have

$$\lambda = 1.285 \frac{1}{\sqrt{Rt}}.$$

When computing stresses in a pressure vessel the first step is to separate the head and the cylindrical part and consider each part a membrane which expands because of the internal pressure q, but does not have resistance against bending deformation. The membrane expansion will have a different value for each of these parts, and consequently a discontinuity in displacements will result.

In order to reassemble the head and the cylinder we must apply at the adjoining edges shearing forces Q and bending moments M, uniformly distributed along the circle of juncture, which will cancel the difference in displacements and slopes. At the same time the Q and M will produce, by bending, the so-called *discontinuity stresses* which have primary importance in the design and strength of containers and pressure vessels. Since a meridional strip of the cylinder as well as of the head can be regarded as a beam on elastic foundation, the problem can be analyzed on this basis.

At this time it should be mentioned that in addition to the loading Q and M, there will also be a distributed axial force X (*see* Fig. 61) around the juncture circle due to the reaction of the head, but the effect of this axial force on the displacement due to Q and M in the cylinder is quite negligible in most engineering structures.

COMPUTATION OF Q and M

Let Y_c, Y_h, θ_c, θ_h, be the deflection and slope for the cylinder and head along the juncture circle. Then we shall have discontinuities in displacement $Y_c - Y_h$ and in slope $\theta_c - \theta_h$.

Displacements and rotations of the end of the cylinder due to unit values of Q and M may be obtained from the applicable equations of Table 11 and will be denoted by Y_c^Q, θ_c^Q, Y_c^M, θ_c^M respectively. Corresponding values for the spherical head Y_h^Q, θ_h^Q, Y_h^M, θ_h^M will be computed from the following equations:

$$Y_h^Q = \frac{2\lambda_1 r \sin^2 \varphi_o}{Et_1} \tag{46}$$

$$\theta_h^Q = Y_h^M = \frac{2\lambda_1^2 \sin \varphi_o}{Et_1} \tag{47}$$

$$\theta_h^M = \frac{4\lambda_1^3}{Ert_1}, \tag{48}$$

where $\lambda_1^4 = 3(1 - r^2)\, r^2/t_1^2$.

Thus, the equations for continuity in deflection and slope along the juncture are:

$$(Y_c^Q + Y_h^Q)\, Q + (Y_c^M + Y_h^M)\, M = Y_c - Y_h \tag{49}$$

$$(\theta_c^Q + \theta_h^Q)\, Q + (\theta_c^M + \theta_h^M)\, M = \theta_c - \theta_h, \tag{50}$$

from which the unknown Q and M quantities can be determined.

Let us solve equations (49) and (50) for the particular case of a pressure vessel as shown in Fig. 61 assuming a hemispherical head: $\varphi_o = 90$, $r = R$.

MEMBRANE ANALYSIS

The slope of the cylinder and spherical head, caused by the internal pressure q, is zero; therefore $\theta_c - \theta_h = 0$.

In Chapter 7 we find the radial displacement of a cylinder $Y_c = (q\ R^2/Et)\ (1 - \nu/2)$ and sphere $Y_h = (q\ R^2/2Et_1)\ (1 - \nu)$.

BENDING ANALYSIS

A meridional strip of the cylinder shown in Fig. 61 may be regarded as a beam on elastic foundation having free ends and loaded by an end force Q and end moment M. This corresponds to Cases 12 and 13 in Table 11 from which we find the deflection (radial displacement) and the slope due to unit load and unit moment ($P = 1, M = 1$):

$$\left.\begin{aligned} Y_c^Q &= \frac{2\lambda}{k}(K_y) \\ \theta_c^Q &= -\frac{2\lambda^2}{k}(K_\theta) \end{aligned}\right\}\ \text{Case 12,}$$

$$\left.\begin{aligned} Y_c^M &= \frac{2\lambda^2}{k}(K_y) \\ \theta_c^M &= \frac{4\lambda^3}{k}(K_\theta) \end{aligned}\right\}\ \text{Case 13.}$$

From equations 46, 47, 48 we find the expressions for deflection and slope for the hemispherical head ($\varphi_0 = 90$ deg., $r = R$, $t_1 = t$).

$$Y_h^Q = \frac{2\lambda_1 R}{Et_1}$$

$$\theta_h^Q = \frac{2\lambda_1^2}{Et_1}$$

$$Y_h^M = \frac{2\lambda_1^2}{Et_1}$$

$$\theta_h^M = \frac{4\lambda_1^3}{E\ Rt_1}$$

Note that:

$$k = \frac{Et}{R^2}$$

$$\lambda = \sqrt[4]{3\,(1-\nu^2)} \times \frac{1}{\sqrt{Rt}}$$

$$\lambda_1 = \sqrt[4]{3(1-\nu^2)}\quad \sqrt{\frac{R}{t_1}}.$$

By substitution, equations 49 and 50 yield:

$$\left[\frac{2\lambda}{k}(K_y) + \frac{2\lambda_1 R}{Et}\right] Q + \left[\frac{2\lambda^2}{k}(K_y) + \frac{2\lambda_1^2}{Et}\right] M = y_c - y_h$$

$$\left[-\frac{2\lambda^2}{k}(K_\theta) + \frac{2\lambda_1^2}{Et}\right] Q + \left[\frac{4\lambda^3}{k}(K_\theta) + \frac{4\lambda_1^3}{EtR}\right] M = 0$$

K_y, K_θ will be obtained from appropriate curves of Figs. 48 and 49 thus affording the solution of the above equations.

Note: If $t = t_1$ and $\varphi_o = 90$ deg., $M = 0$; $Q = q/8\lambda_1$

36. STRESSES IN CURVED BARS*

Concentration Factors and True Bending Stresses for a Wide Variety of Cross Sections

(*Ref. 21*)

Bars with curvatures ranging from large-radius arcs to sharp-angle bends are common components in machines and products. Yet many designers are hazy about how to evaluate the true stresses when such bars are subjected to bending loads.

An analytical method for determining such stresses has been developed by S. Timoshenko in his book *Strength of Materials—Part I* (Princeton, N.J.: Van Nostrand, 1956). We have verified the accuracy of Timoshenko's method by subjecting various curved bars to bending loads and measuring the strains in the fibers with electric strain gages. The measurements indicate close agreement with theory—even when the cross sections are complex and the bend radius is much smaller than the height of the section being tested.

Timoshenko's method, however, requires evaluation of an integral relating to the cross section of the bar. Integration can be cumbersome. Hence, in addition to presenting his method as a series of step-by-step equations, we have tabulated the solution of the integral for 12 common geometric shapes. Three problems are included to illustrate the use of the equations.

CURVED-BAR THEORY

In a straight bar, during bending, the fibers are elongated on the convex side and shortened on the concave side. Those on the neutral axis remain at their original length. The strains of compression and tension fibers equidistant from the gravity axis are equal in magnitude because the elemental lengths between two plane

*Reprinted, by permission, from *Product Engineering*, Dec. 24, 1962. Copyright by McGraw-Hill.

1 . . STRAIGHT-BAR STRESS DISTRIBUTION. Fibers equidistant from neutral axis are equally stressed.

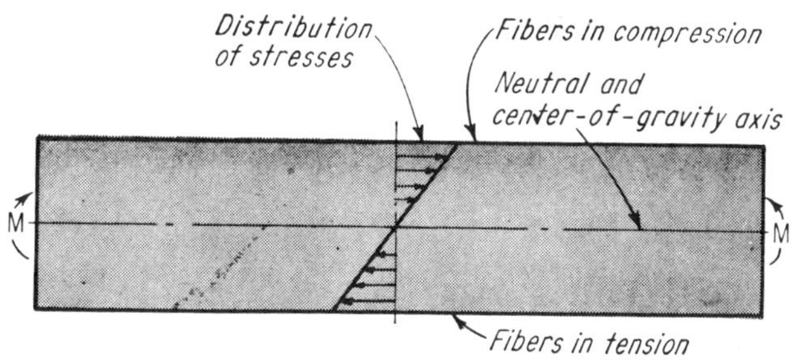

sections before bending are equal. Strain is determined as the quotient of the length of deformation divided by the original length.

In a curved bar, the deformations at equidistant points above and below the neutral axis are equal, but the elemental lengths between two radial sections vary. Thus the strains due to bending for compression and tension fibers equidistant from the neutral axis are not equal. Specifically, the elemental arc at the concave surface is minimum and the deformation is maximum. Therefore, the strain on this surface is maximum.

To meet the condition of equilibrium on a cross section, the neutral axis (the axis where there is no stress) cannot coincide with the gravity axis. The neutral axis shifts toward the concave surface, so that the summation of the distributed normal forces on the cross section is equal to zero.

The key equation in curved bar analysis is given below. It gives the location of the neutral surface with respect to the center of curvature for any cross section (refer to the list of symbols):

$$r = \frac{A}{\int_A dA/v}, \tag{51}$$

where

$$dA = L\,dv.$$

For a rectangle, dimension r becomes (See Table 14)

$$r = \frac{A}{b \ln(c/a)}. \tag{52}$$

Starting with equation (51), the procedure and equations for determining the bending stress with the stress concentration due to the curvature are as follows:

Distance e between center-of-gravity and neutral axes is obtained from the equation

$$e = R - r, \tag{53}$$

where the value for R is known or calculated from the relationship $R = a + d$, and the value for r is obtained from equation (51) for the particular cross section.

Factor d is the distance from the base of the section to its gravity axis. (It is calculated as shown in Problem II). Hence

$$h_B = d - e \tag{54}$$

and

$$h_A = h - d + e. \tag{55}$$

The *true* bending stress—the normal bending stress with the additional stress concentration—on the concave surface is

$$\sigma_B = \frac{Mh_B}{Aea} \tag{56}$$

and on the convex surface is

$$\sigma_A = \frac{Mh_A}{Aec}. \tag{57}$$

The *nominal* bending stress—*without* the stress concentration—can be obtained from the familiar expression

$$S_B = \frac{M}{Z_B} \qquad S_A = \frac{M}{Z_A}, \tag{58}$$

where Z is the section modulus equal to

$$Z_B = I_{CG}/d \qquad Z_A = I_{CG}/(h - d) \tag{59}$$

and where

$$I_{CG} = I_{xx} - Ad^2.$$

The stress concentration factor is the ratio of σ_B/S_B or σ_A/S_A. Usually the first of these factors will be greater than unity, and the other less than unity. The factor that is greater than unity is usually of greatest interest.

To simplify the analysis of curved bars, twelve of the most common shapes are illustrated in Table 14 with equations for the integral which appears in the denominator of equation (51).

NOTATION

a = inner radius of curvature (in.)
A = area of section (in.^{2})
b = width of section (in.)
c = outer radius of curvature (in.)
d = distance from base of section to gravity axis (in.)
e = distance of the neutral axis from gravity axis (in.)

h = height of section (in.) ($h = h_B + h_A$)
h_B = distance from inner radius of curvature to neutral axis (in.)
h_A = distance from neutral axis to outer radius of curvature (in.)
I_{xx} = moment of inertia about the base of section (in.^{4})
I_{CG} = moment of inertia about the center of gravity of section (in.^{4})
L = length of differential area (in.)
ln = natural logarithm (to base 2.718)
M = bending moment with respect to gravity axis (in.-lb.)
r = distance from center of curvature to neutral axis (in.)
R = distance from center of curvature to gravity axis (in.) ($R = a + d$)
S_B = nominal bending stress, lower fiber (psi)
S_A = nominal bending stress, upper fiber (psi)
v = distance from center of curvature to a differential area (dA) of the section (in.)
Z_B = section modulus, lower fiber (in.^{3})
Z_A = section modulus, upper fiber (in.^{3})
σ_B = bending stress with stress concentration at the concave surface (psi)
σ_A = bending stress with stress concentration at the convex surface (psi)

PROBLEM I—RECTANGULAR CURVED BAR

Given: Rectangular-section curved bar with these dimensions in inches: $h = 2\frac{1}{2}, b = 1\frac{1}{4}, R = 1\frac{3}{4}, a = \frac{1}{2}$.

Solution: From the geometry of the figure

$$A = 3.125 \text{ in.}^2;\ d = h/2 = 1.250 \text{ in.}$$
$$h - d = 1.250 \text{ in.};\ I_{CG} = 1.628 \text{ in.}^4;\ Z_B = 1.302 \text{ in.}^3$$

From equation (51)

$$r = \frac{3.125}{1.25 \ln (3/0.5)} = 1.395 \text{ in.}$$

From equations (53, 54, 55)

$$e = 1.750 - 1.395 = 0.355 \text{ in.}$$
$$h_B = 1.250 - 0.355 = 0.895 \text{ in.}$$
$$h_A = 1.250 + 0.355 = 1.605 \text{ in.}$$

From equations (56, 57)

$$\sigma_B = \frac{0.895M}{3.125 \times 0.355 \times 0.500} = 1.614M.$$

$$\sigma_A = \frac{1.605M}{3.125 \times 0.355 \times 3.00} = 0.482M.$$

From equation (58)

$$S_B = S_A = \frac{M}{1.302} = 0.768M.$$

Therefore the magnitude of the stress concentration factor is $\sigma_B/S_B = 1.614/0.768 = 2.10$.

PROBLEM II—C-CLAMP

The C-clamp illustrated here, has a constant cross section around the $\frac{3}{8}$-in. radius curvature. The stress concentration factor due to the curvature will remain constant, but the nominal stress will vary directly as the bending moment. Therefore, the maximum bending stress including stress concentration will occur at the inner fiber of a section (section A-B) taken vertically through the center of the radius.

A systematic approach to determine the area, neutral axis, and moment of inertia of section A-B is tabulated with the illustration. These properties are ob-

2 . . CURVED-BAR GEOMETRY. Neutral axis does not coincide with center-of-gravity axis.

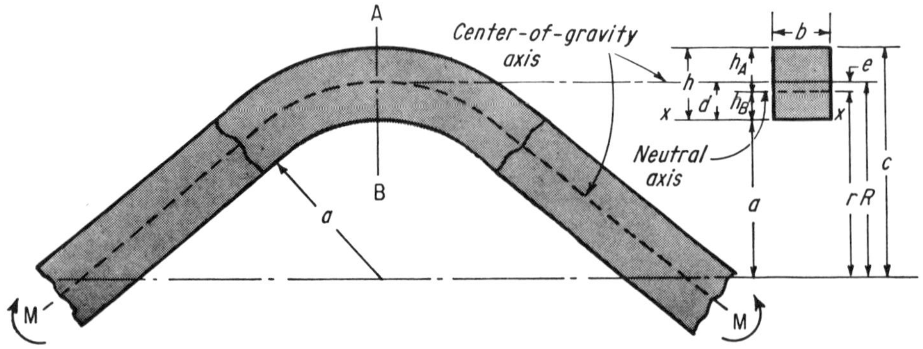

3 . . ELEMENTAL SECTION of curved bar with varying cross section.

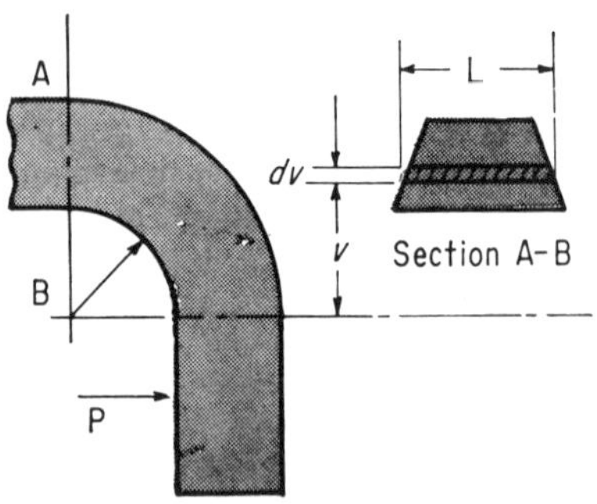

tained by adding and subtracting rectangles. For computing the moment of inertia, each rectangle is taken from the base line x—x. This method is especially convenient when calculating machines are used in the numerical work. Thus from the table: $\sum A = 0.2265$; $\sum I_{xx} = 0.0574$.

The nominal vertical bending stress in tension and compression is now calculated. Subscripts A and B refer to the outer and inner fibers, respectively.

$$d = 0.0911/0.2265 = 0.402 \text{ in.}$$
$$h - d = 0.875 - 0.402 = 0.473 \text{ in.}$$
$$R = 0.375 + 0.402 = 0.777 \text{ in.}$$
$$I_{CG} = 0.0574 - (0.0911)(0.402)^2 = 0.0208 \text{ in.}^4$$
$$Z_B = 0.0208/0.402 = 0.052 \text{ in.}^3$$
$$Z_A = 0.0208/0.473 = 0.044 \text{ in.}^3$$
$$M = 300 \times 2.527 = 760 \text{ lb.-in.}$$

TABLE 14. CURVED-BAR EQUATIONS FOR 12 SHAPES

RECTANGLE

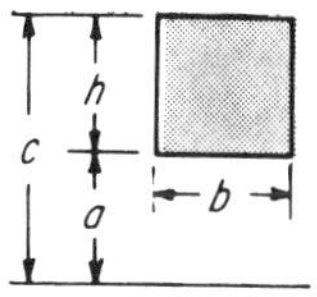

$$\int \frac{dA}{v} = b \ln \frac{c}{a}$$

TRAPEZOID

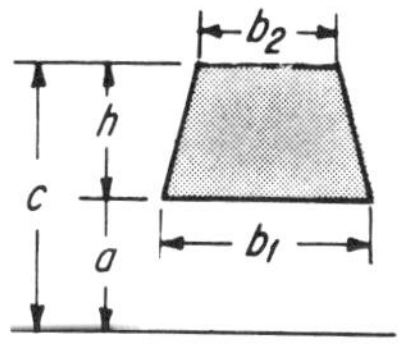

$$\int \frac{dA}{v} = \left[\frac{b_1c - b_2a}{h} \ln \frac{c}{a}\right] - (b_1 - b_2)$$

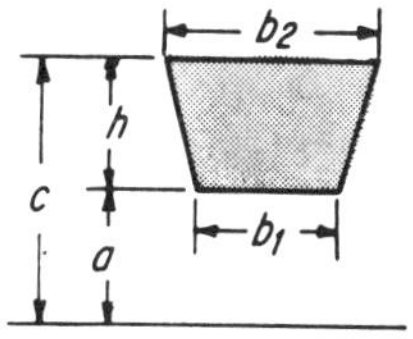

$$\int \frac{dA}{v} = \left[\frac{b_1c - b_2a}{h} \ln \frac{c}{a}\right] + (b_2 - b_1)$$

TRIANGLE

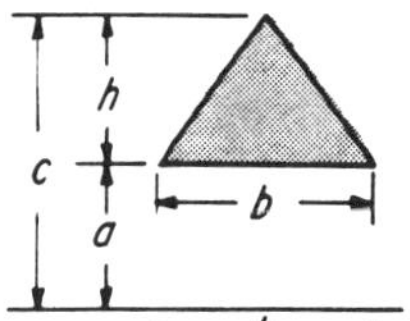

$$\int \frac{dA}{v} = \left[\frac{bc}{h} \ln \frac{c}{a}\right] - b$$

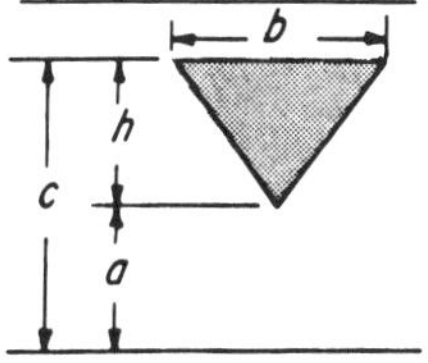

$$\int \frac{dA}{v} = b - \left[\frac{ba}{h} \ln \frac{c}{a}\right]$$

CIRCLE

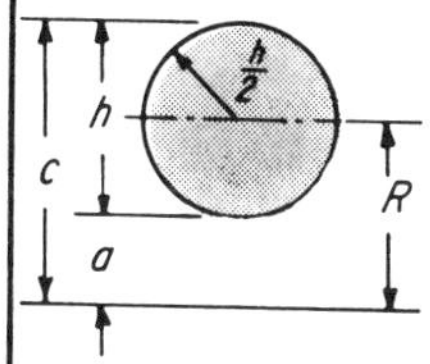

$$\int \frac{dA}{v} = 2\pi\left[R - \left(R^2 - \frac{h^2}{4}\right)^{1/2}\right]$$

SEMI-CIRCLE

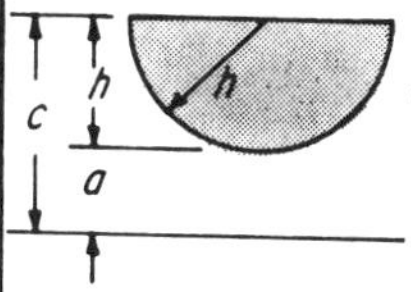

$$\int \frac{dA}{v} = 2h + \pi c - \frac{\pi}{45}(\theta + \beta)(c^2 - h^2)^{1/2}$$

where:

$$\theta = \text{arc tan} \frac{c - h}{(c^2 - h^2)^{1/2}}, \text{ deg}$$

$$\beta = \text{arc tan} \frac{h}{(c^2 - h^2)^{1/2}}, \text{ deg}$$

SEMI-CIRCLE

When $h < a$

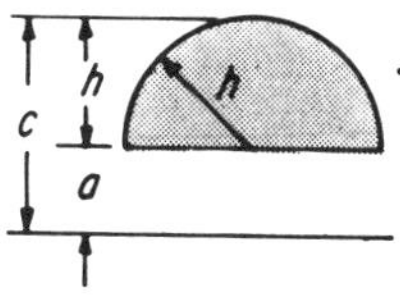

$$\int \frac{dA}{v} = a\pi - 2h - \frac{\pi}{45}(\theta - \beta)(a^2 - h^2)^{1/2}$$

$$\theta = \text{arc tan} \frac{a + h}{(a^2 - h^2)^{1/2}}, \text{ deg}$$

$$\beta = \text{arc tan} \frac{h}{(a^2 - h^2)^{1/2}}, \text{ deg}$$

When $a < h$

$$\int \frac{dA}{v} = a\pi - 2h + 2(h^2 - a^2)^{1/2} \ln\left[\frac{a + h + (h^2 - a^2)^{1/2}}{a + h - (h^2 - a^2)^{1/2}}\right]$$

When $a = h$ $\quad \int \frac{dA}{v} = 1.1416h$

QUARTER-CIRCLE

$$\int \frac{dA}{v} = h + \frac{\pi c}{2} - \frac{\pi}{90}(\theta + \beta)(c^2 - h^2)^{1/2}$$

where:

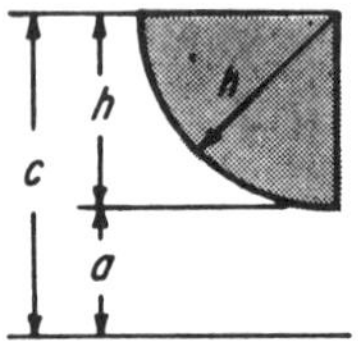

$$\theta = \text{arc tan}\ \frac{c - h}{(c^2 - h^2)^{1/2}}, \text{ deg}$$

$$\beta = \text{arc tan}\ \frac{h}{(c^2 - h^2)^{1/2}}, \text{ deg}$$

QUARTER-CIRCLE

When $h < a$

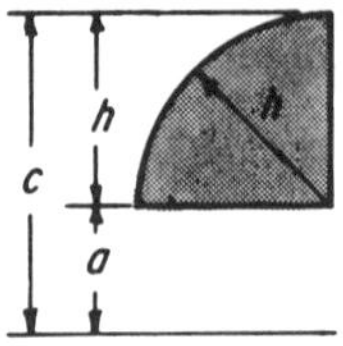

$$\int \frac{dA}{v} = \frac{a\pi}{2} - h - \frac{\pi}{90}(\theta - \beta)(a^2 - h^2)^{1/2}$$

When $a < h$

$$\int \frac{dA}{v} = \frac{a\pi}{2} - h + (h^2 - a^2)^{1/2} \ln\left[\frac{a + h + (h^2 - a^2)^{1/2}}{a + h - (h^2 - a^2)^{1/2}}\right]$$

When $a = h$ where:

$$\int \frac{dA}{v} = 0.5708h \qquad \theta = \text{arc tan}\ \frac{a + h}{(a^2 - h^2)^{1/2}}, \text{ deg}$$

$$\beta = \text{arc tan}\ \frac{h}{(a^2 - h^2)^{1/2}}, \text{ deg}$$

REVERSE-FILLET

$$\int \frac{dA}{v} = \left[h \ln \frac{c}{a}\right] - h - \frac{\pi}{2}c + \frac{\pi}{90}(\theta + \beta)(c^2 - h^2)^{1/2}$$

Where $\theta = \text{arc tan}\ \dfrac{c - h}{(c^2 - h^2)^{1/2}}$, deg

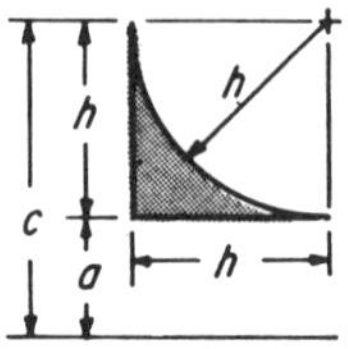

$$\beta = \text{arc tan}\ \frac{h}{(c^2 - h^2)^{1/2}}, \text{ deg}$$

REVERSE-FILLET

When $h < a$

$$\int \frac{dA}{v} = \left[h \ln \frac{c}{a}\right] - \frac{a}{2}\pi + h + \frac{\pi}{90}(\theta - \beta)(a^2 - h^2)^{1/2}$$

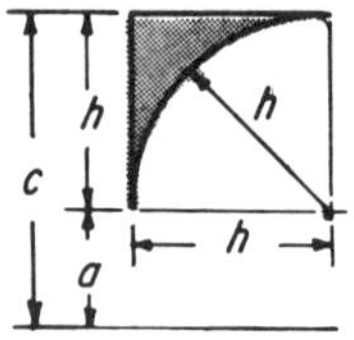

$$\theta = \text{arc tan}\ \frac{a + h}{(a^2 - h^2)^{1/2}}, \text{ deg}$$

$$\beta = \text{arc tan}\ \frac{h}{(a^2 - h^2)^{1/2}}, \text{ deg}$$

When $a < h$

$$\int \frac{dA}{v} = \left[h \ln \frac{c}{a}\right] + h - \frac{\pi}{2}a - (h^2 - a^2)^{1/2} \ln\left[\frac{a + h + (h^2 - a^2)^{1/2}}{a + h - (h^2 - a^2)^{1/2}}\right]$$

When $a = h$

$$\int \frac{dA}{v} = 0.1223h$$

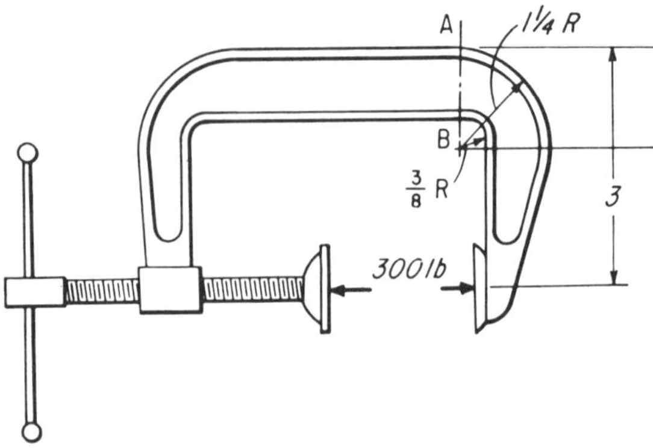

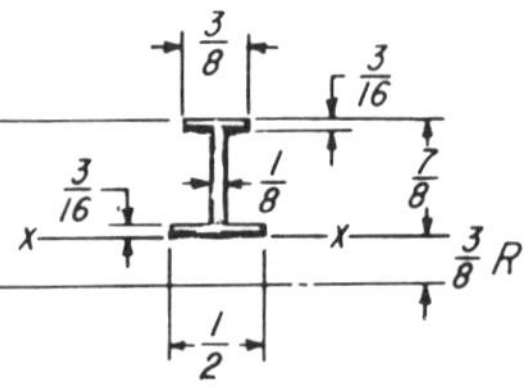

Section A – B

CROSS-SECTION PROPERTIES

Rectangle	Area, A	d	$A \times d$	I_{xx}
$3/8 \times 7/8$	+0.3281	0.4375	+0.1436	+0.0837
$3/8 \times 3/16$	+0.0703	0.0938	+0.0066	+0.0008
$1/4 \times 11/16$	−0.1719	0.3438	−0.0591	−0.0271
Summation, $\sum$	=0.2265		=0.0911	=0.0574

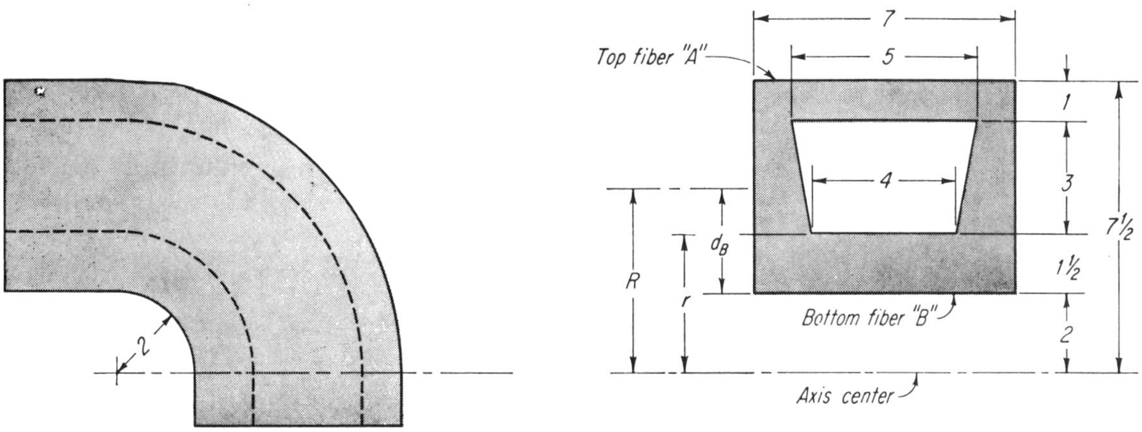

HOLLOW CURVED BEAM (for Problem III)

Thus the nominal stresses are

S_A = 760/0.044 = 17,300 psi (compression)
S_B = 760/0.052 = 14,600 psi (tension)

To compute the effect of the bend, section A-B is divided into three rectangles—lower, center, and top. The bend integral for a rectangle is calculated for each:

Lower rectangle

$$\int \frac{dA}{v} = \frac{1}{2}\left(\ln \frac{9/16}{3/8}\right) = 0.2027 \text{ in.}$$

Inner rectangle

$$\int \frac{dA}{v} = \frac{1}{8} \ln\left(\frac{1\ 1/16}{9/16}\right) = 0.0795 \text{ in.}$$

Upper rectangle

$$\int \frac{dA}{v} = \frac{3}{8} \ln\left(\frac{1\ 1/4}{1\ 1/16}\right) = 0.0611 \text{ in.}$$

Sum of the integrals = 0.3433 in. Hence

$$r = \frac{A}{\int \frac{d_A}{v}} = \frac{0.2265}{0.3433} = 0.660 \text{ in.}$$

$e = 0.777 - 0.660 = 0.117$ in.
$h_B = 0.402 - 0.117 = 0.285$ in.
$h_A = 0.473 + 0.117 = 0.590$ in.

The corrected (true) bending stresses are then

$$\sigma_B = \frac{(760)\ (0.285)}{(0.227)\ (0.117)\ (0.375)} = 21{,}700 \text{ psi (tension)}$$

$$\sigma_A = \frac{(760)\ (0.590)}{(0.227)\ (0.117)\ (1.250)} = 13{,}500 \text{ psi (compression)}$$

Thus the nominal bending stress in tension should be multiplied by a factor of 21,700/14,600 = 1.49 to obtain the true bending stress in tension. Also, the true compression stress in bending is 13,500/17,300 = 0.78 of the nominal value. Hence, it is reduced by 22%. These figures indicate that stress concentration must be considered to avoid large errors in evaluating curved-bar bending stress, especially when hardened material or fatigue loading is involved.

Problem III—Hollow sections

Frequently hollow irregular sections, above, are employed as machine members. From the geometry of the section (calculated as described in Problem II): A = 25.00 in.^{2}, d = 2.585 in., R = 4.585 in., I_{CG} = 85.03 in.^{4}, z_B = 32.89 in.^{3}

Outer rectangle

$$\int \frac{d_A}{v} = 7 \ln \frac{7.5}{2.0} = 9.252 \text{ in.}$$

Inner trapezoid

$$\int \frac{d_A}{v} = \left(\frac{4 \times 6\frac{1}{2} - 5 \times 3\frac{1}{2}}{3}\right)\left(\ln \frac{6\frac{1}{2}}{3\frac{1}{2}}\right) + (5 - 4)$$

$$= 2.754 \text{ in.}$$

$$r = \frac{25.00}{9.252 - 2.754} = 3.847 \text{ in.}$$

$e = 4.585 - 3.847 = 0.738$ in.
$h_B = 2.585 - 0.738 = 1.847$ in.

$$\sigma_B = \frac{1.847\ M}{(25.00\ (0.738)\ (2.000)} = 0.0501\ M$$

$$s_B = M/32.89 = 0.0304\ M \text{ (tension)}$$

Stress concentration factor is 0.0501/ 0.0304 = 1.65.

37. CORRECTION FACTORS, K, FOR USE IN THE STRAIGHT BEAM FORMULA

(*Ref. 22, 19*)

NOTATION

R = radius of curvature measured to centroid of section (in.)
c = distance from centroidal axis to extreme fiber on concave side of beam (in.)
A = area of section (in.^{2})
P = axial load (lb.)
S_A = circumferential stress in extreme fiber on concave side (psi)
S_B = circumferential stress in extreme fiber on convex side (psi)
M = bending moment (in.-lb.)
I = moment of inertia (in.^{4})
K_A = correction fiber for inside fiber (nondimensional)
K_B = correction fiber for outside fiber (nondimensional)
Z = section property (nondimensional)
h = distance from centroidal axis to neutral axis, measured toward center of curvature (in.)

If a curved beam is subjected to an axial load P passing through the centroid of an area, the stress at the outer fiber of the section of the curved beam would be given by the formula

$$S = K\left(\frac{P}{A} + \frac{Mc}{I}\right), \qquad (60)$$

where:

$$K = \frac{\frac{M}{AR}\left(1 + \frac{1}{2}\frac{c}{R + c}\right)}{Mc/I}$$

Here Z is an abstract number defined by the equation:

$$Z = -\frac{1}{A}\int \frac{y}{R + y}\, dA$$

(y is the distance from the centroidal axis to a given section).

Values of correction factors K_A, K_B for the inside and outside fiber respectively, for 14 common sections, each having 10 different degrees of curvature, as found by Wilson and Quereau, are given in Table 15.

Example: Find the circumferential stress S_A, inside the fiber of a 10-ton crane hook.

Solution: Assume the centroid G at $c = 1.85$ in. therefore, $R = 2.5 + 1.85 = 4.35$ in. In the case of a traperoidal cross section, as shown, $I = 15$ in.^{4}; $A = 9.84$ in.^{2}

From Table 15, case 14, for $R/c = 2.35$ we obtain $K_A = 1.52$. The bending moment $M = 20{,}000 \times 4.35 = 87{,}000$ in.-lb. Thus, from equation (60) by substitution:

$$S_A = 19{,}400 \text{ psi.}$$

Note: In Table 15, R = radius of curvature measured to centroid of section (in.); h = distance from centroidal axis to neutral axis when beam is subjected to pure bending.

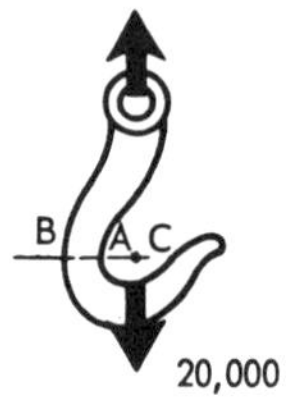

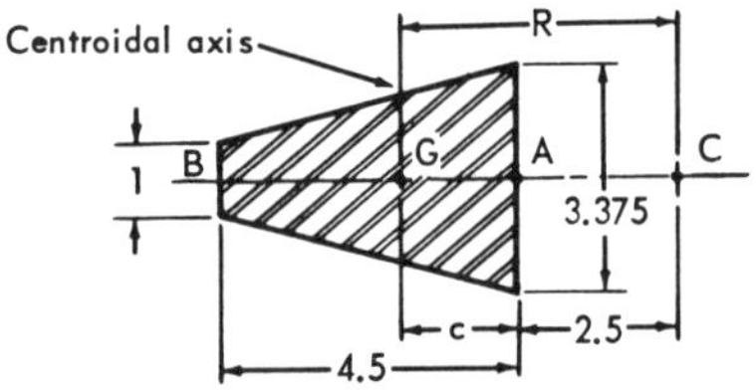

TABLE 15

1. Solid Rectangular Section

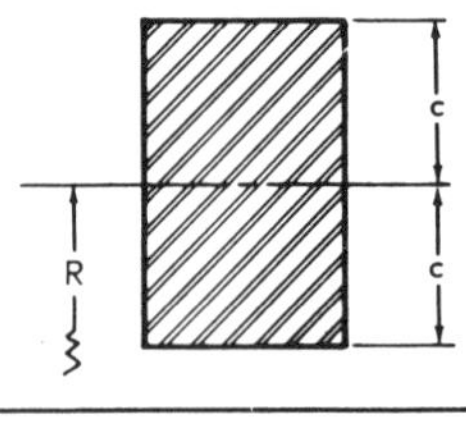

R/C	K_A	K_B	h/r
1.2	2.89	0.57	0.305
1.4	2.13	0.63	0.204
1.6	1.79	0.67	0.149
1.8	1.63	0.70	0.112
2.0	1.52	0.73	0.90
3.0	1.30	0.81	0.041
4.0	1.20	0.85	0.021
6.0	1.12	0.90	0.0093
8.0	1.09	0.92	0.0052
10.0	1.07	0.94	0.0033

2. Solid Circular Section

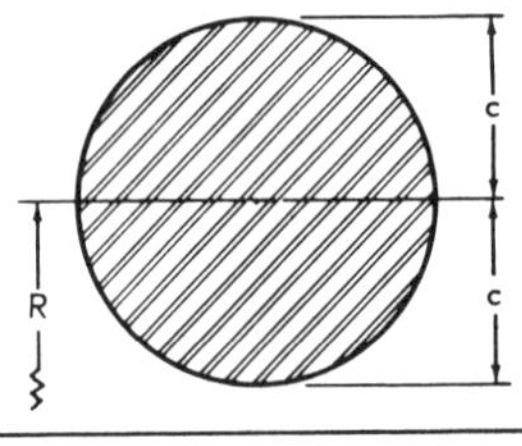

R/C	K_A	K_B	h/r
1.2	3.41	0.54	0.224
1.4	2.40	0.60	0.151
1.6	1.96	0.65	0.108
1.8	1.75	0.68	0.084
2.0	1.62	0.71	0.069
3.0	1.33	0.79	0.03
4.0	1.23	0.84	0.016
6.0	1.14	0.89	0.007
8.0	1.10	0.91	0.0039
10.0	1.08	0.93	0.0025

TABLE 15 (Continued)

3. Hollow Circular Section

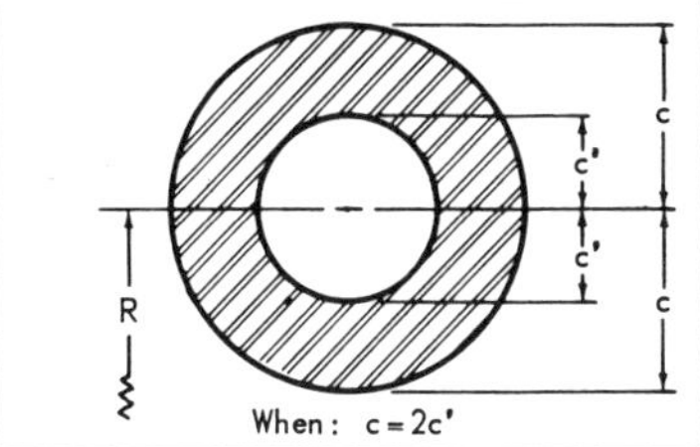

R/C	K_A	K_B	h/r
1.2	3.28	0.58	0.269
1.4	2.31	0.64	0.182
1.6	1.89	0.68	0.134
1.8	1.70	0.71	0.104
2.0	1.57	0.73	0.083
3.0	1.31	0.81	0.038
4.0	1.21	0.85	0.20
6.0	1.13	0.90	0.0087
8.0	1.10	0.92	0.0049
10.0	1.07	0.93	0.0031

4. Solid Elliptical Section

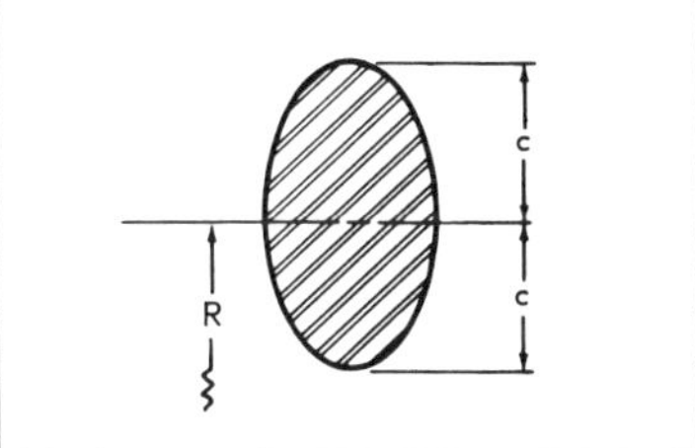

R/C	K_A	K_B	h/r
1.2	3.41	0.54	0.224
1.4	2.40	0.60	0.151
1.6	1.96	0.65	0.108
1.8	1.75	0.68	0.084
2.0	1.62	0.71	0.069
3.0	1.33	0.79	0.03
4.0	1.23	0.84	0.016
6.0	1.14	0.89	0.007
8.0	1.10	0.91	0.0039
10.0	1.08	0.93	0.0025

5. Hollow Elliptical Section

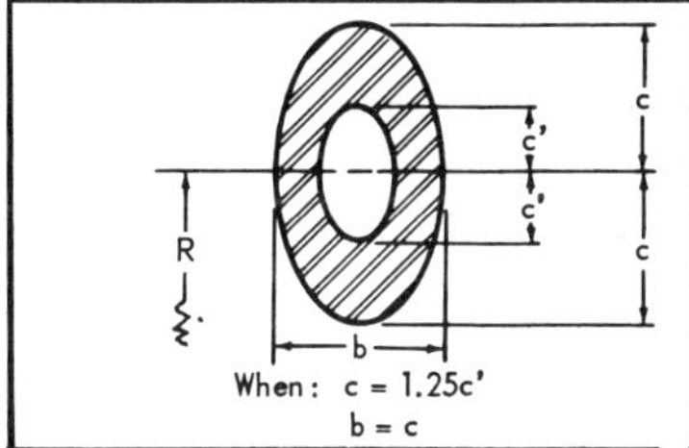

R/C	K_A	K_B	h/r
1.2	3.03	0.58	0.295
1.4	2.15	0.64	0.195
1.6	1.82	0.68	0.146
1.8	1.65	0.71	0.114
2.0	1.53	0.73	0.085
3.0	1.29	0.81	0.041
4.0	1.20	0.85	0.021
6.0	1.12	0.90	0.0088
8.0	1.09	0.92	0.005
10.0	1.07	0.93	0.0032

6. Triangular Section

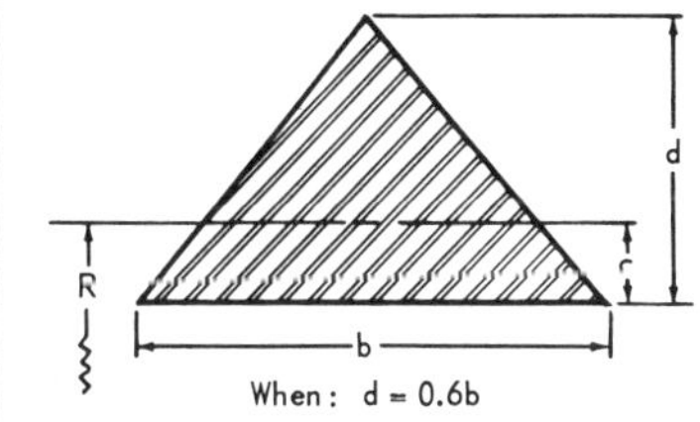

R/C	K_A	K_B	h/r
1.2	3.26	0.44	0.361
1.4	2.39	0.50	0.251
1.6	1.99	0.54	0.186
1.8	1.78	0.57	0.144
2.0	1.66	0.60	0.116
3.0	1.37	0.70	0.052
4.0	1.27	0.75	0.029
6.0	1.16	0.82	0.013
8.0	1.12	0.86	0.006
10.0	1.09	0.88	0.0039

7. Trapezoidal Section

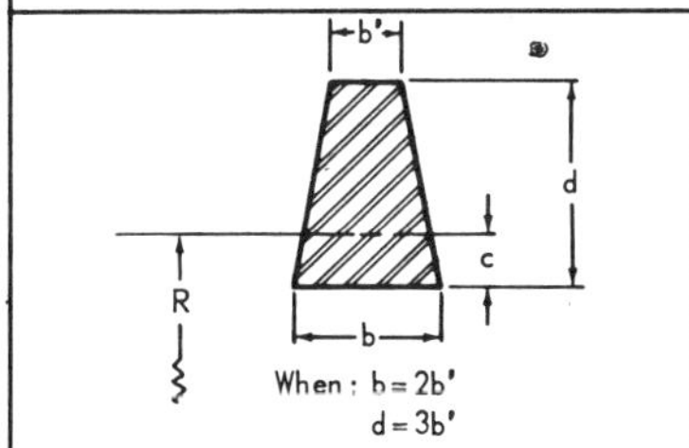

R/C	K_A	K_B	h/r
1.2	3.09	0.56	0.336
1.4	2.25	0.62	0.229
1.6	1.91	0.66	0.168
1.8	1.73	0.70	0.128
2.0	1.61	0.73	0.102
3.0	1.37	0.81	0.046
4.0	1.26	0.86	0.24
6.0	1.17	0.91	0.011
8.0	1.13	0.94	0.006
10.0	1.11	0.95	0.0039

8. T-Beam

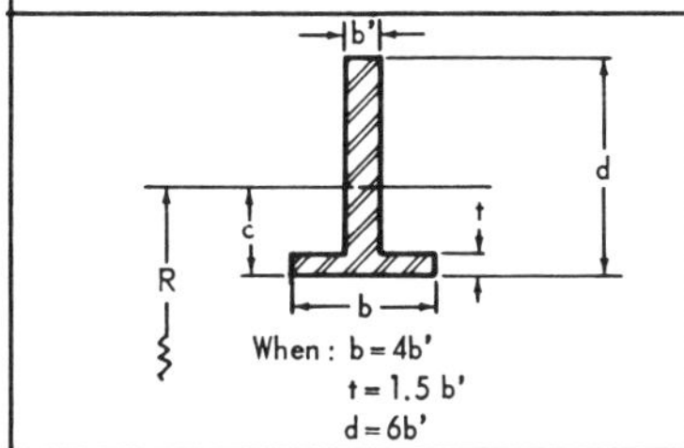

R/C	K_A	K_B	h/r
1.2	3.63	0.58	0.418
1.4	2.54	0.63	0.299
1.6	2.14	0.67	0.229
1.8	1.89	0.70	0.183
2.0	1.73	0.72	0.149
3.0	1.41	0.79	0.069
4.0	1.29	0.83	0.04
6.0	1.18	0.88	0.018
8.0	1.13	0.91	0.01
10.0	1.10	0.92	0.0065

9. Channel Section

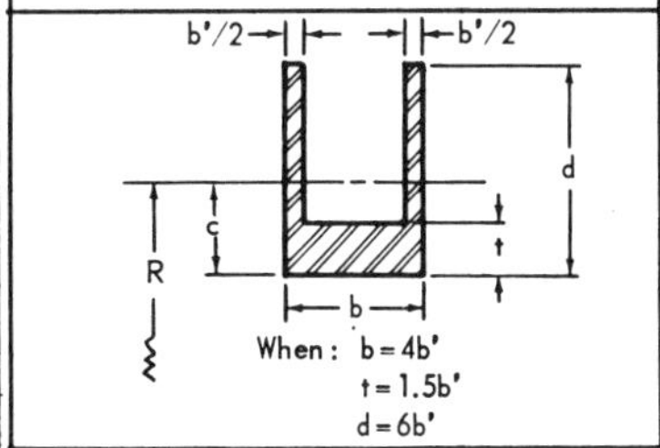

R/C	K_A	K_B	h/r
1.2	3.63	0.58	0.418
1.4	2.54	0.63	0.299
1.6	2.14	0.67	0.229
1.8	1.89	0.70	0.183
2.0	1.73	0.72	0.149
3.0	1.41	0.79	0.069
4.0	1.29	0.83	0.04
6.0	1.18	0.88	0.018
8.0	1.13	0.91	0.01
10.0	1.10	0.92	0.0065

10. Symmetrical I-Beam

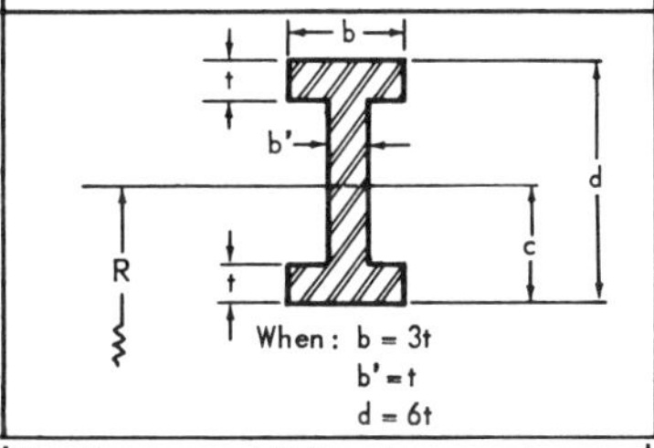

R/C	K_A	K_B	h/r
1.2	2.52	0.67	0.408
1.4	1.90	0.71	0.285
1.6	1.63	0.75	0.208
1.8	1.50	0.77	0.160
2.0	1.41	0.79	0.127
3.0	1.23	0.86	0.058
4.0	1.16	0.89	0.03
6.0	1.10	0.92	0.013
8.0	1.07	0.94	0.0076
10.0	1.05	0.95	0.0048

11. Symmetrical Hollow Rectangular Section

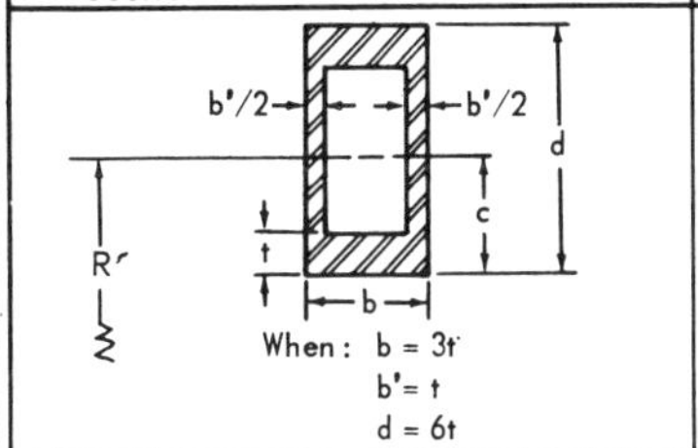

R/C	K_A	K_B	h/r
1.2	2.52	0.67	0.408
1.4	1.90	0.71	0.285
1.6	1.63	0.75	0.208
1.8	1.50	0.77	0.160
2.0	1.41	0.79	0.127
3.0	1.23	0.86	0.058
4.0	1.16	0.89	0.03
6.0	1.10	0.92	0.013
8.0	1.07	0.94	0.0076
10.0	1.05	0.95	0.0048

12. Unsymmetrical I-Beam

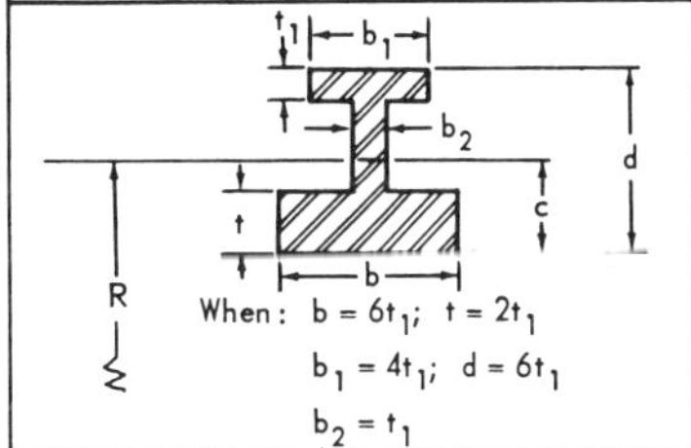

R/C	K_A	K_B	h/r
1.2	3.55	0.67	0.409
1.4	2.48	0.72	0.292
1.6	2.07	0.76	0.224
1.8	1.83	0.78	0.178
2.0	1.69	0.80	0.144
3.0	1.30	0.86	0.067
4.0	1.26	0.89	0.038
6.0	1.15	0.92	0.018
8.0	1.10	0.94	0.010
10.0	1.08	0.95	0.0065

13. Trapezoidal Section

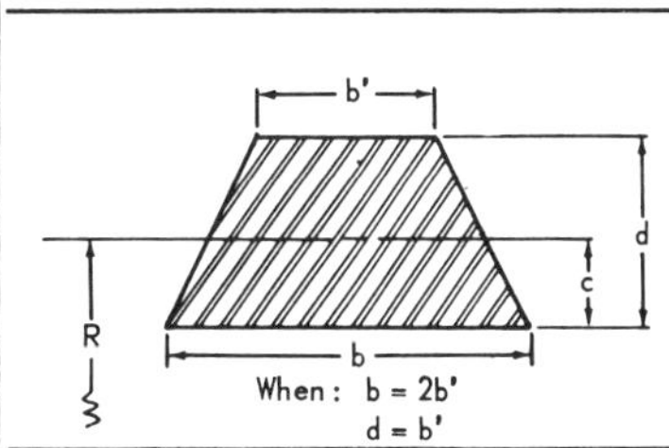

R/C	K_A	K_B	h/r
1.2	3.01	0.54	0.336
1.4	2.18	0.60	0.229
1.6	1.87	0.65	0.168
1.8	1.69	0.68	0.128
2.0	1.58	0.71	0.102
3.0	1.33	0.80	0.046
4.0	1.23	0.84	0.024
6.0	1.13	0.88	0.011
8.0	1.10	0.91	0.006
10.0	1.08	0.93	0.0039

14. Trapezoidal Section

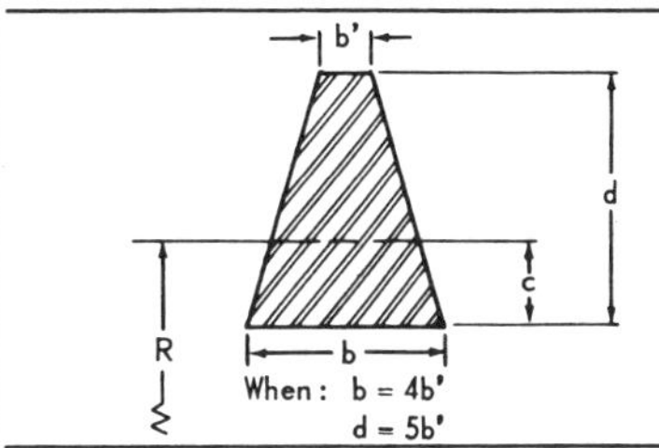

R/C	K_A	K_B	h/r
1.2	3.14	0.52	0.352
1.4	2.29	0.54	0.243
1.6	1.93	0.62	0.179
1.8	1.74	0.65	0.138
2.0	1.61	0.68	0.11
3.0	1.34	0.76	0.05
4.0	1.24	0.82	0.028
6.0	1.15	0.87	0.012
8.0	1.12	0.91	0.006
10.0	1.10	0.93	0.0039

38. CIRCULAR CANTILEVER BEAMS

OUT OF PLANE LOADING (ARBITRARY CONCENTRATED LOAD, TRANSVERSE UNIFORM LOAD, END MOMENT) (*Ref. 23, 24*)

Frequently a member is subjected to loads that produce at any cross section a bending moment and a twisting moment. A typical example is a circular cantilever beam with a concentrated load acting at any point along the center line of the beam and perpendicular to its plane of curvature (Fig. 62).

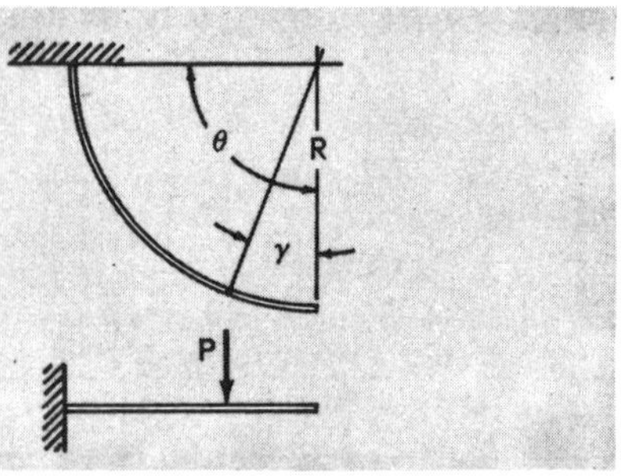

Fig. 62

Despite its importance, the subject of circular beams with this type loading has a meager treatment in engineering literature. Because of the complexity of expressions involved, engineers have been content largely to employ rough approximation in the solution of these problems.

The circular cantilever beam with an out-of-plane load has a wide application in a number of common design elements in structural engineering and electrical machinery, such as: springs, retaining brackets, chains, and steam piping subject to temperature changes.

We present here equations for the deflection y, angle of slope S, angle of twist ϕ of the free end of a circular cantilever beam for the following three loading conditions: (1) concentrated load acting at any point along the center line of the beam and perpendicular to its plane of curvature, (2) end moment, and (3) transverse uniform load.

For concentrated load (Table 16)

$$y = \frac{PR^3}{EI} K_1$$

$$S = \frac{PR^2}{EI} K_2$$

$$\phi = \frac{PR^2}{EI} K_3.$$

For transverse uniform load (Table 17)

$$y = \frac{WR^3}{EI} K_4$$

$$S = \frac{WR^2}{EI} K_5$$

$$\phi = \frac{WR^2}{EI} K_6$$

For end bending moment (Table 18)

$$y = \frac{MR^2}{EI} K_7$$

$$S = \frac{MR}{EI} K_8$$

$$\phi = \frac{MR}{EI} K_9,$$

where

$$K_1 = \frac{1-C}{2}[\cos\gamma(\theta-\gamma-\sin\theta\cos\theta)+\sin\gamma\cos^2\theta] + C[(\theta-\gamma)(1+\cos\gamma)-\sin(\theta-\gamma) - \sin\theta+\sin\gamma]$$

$$K_2 = \frac{1-C}{2}[\cos\gamma\sin^2\theta-\sin\gamma(\theta-\gamma+\sin\theta\cos\theta)] + C[\cos\gamma-\cos\theta-(\theta-\gamma)\sin\gamma]$$

$$K_3 = \frac{1-C}{2}[\cos\gamma(\theta-\gamma-\sin\theta\cos\theta)+\sin\gamma\cos^2\theta] + C[(\theta-\gamma)\cos\gamma-\sin\theta+\sin\gamma]$$

$$K_4 = \frac{1}{2\theta}[(1-\cos\theta)^2+C(\theta-\sin\theta)^2]$$

$$K_5 = \frac{1}{2\theta}\{(1+C)[\sin\theta(2-\cos\theta)-\theta] - - 2C[\cos\theta(\theta-\sin\theta)]\}$$

$$K_6 = \frac{1}{2\theta}[(1+C)(1-\cos\theta)^2-2C\sin\theta(\theta-\sin\theta)]$$

$$K_7 = \frac{1-C}{2}(1+\cos\theta)+C(1-\cos\theta)$$

$$K_8 = \frac{1+C}{2}(\theta+\sin\theta\cos\theta)-C\sin\theta\cos\theta$$

$$K_9 = \frac{1-C}{2}\sin^2\theta,$$

and where $C = EI/GJ$.

NOTATION

y = vertical end deflection (in.)
S = angle of slope (radians)
ϕ = angle of twist (radians)
P = concentrated load (lb.)
w = intensity of loading of a distributed load (lb./in.)
W = total transverse load (lb.)
R = radius of center line of curved beam (in.)
E = modulus of elasticity in direct stress (psi)
G = modulus of elasticity in shear stress (psi)
I = principal moment of inertia of the cross-sectional area with respect to the axis of bending (in.^{4})
J = torsional effect of the cross-sectional area (in.^{4})
EI = flexural rigidity
GJ = torsional rigidity
C = EI/GJ
M = bending moment (lb.-in.)
T = twisting moment (lb.-in.)
θ = angle subtended by a circular arc (radians)
γ = variable angle of arc denoting position of the applied load P (radians)
γ_o = running coordinate
$K_1, K_2 \ldots K_9$ = dimensionless factors for deflection, angle of slope and angle of twist tabulated in Tables 16, 17, 18.

For steel, (Poisson's Ratio = 0.3), the ratio of modulus of elasticity E to the modulus of rigidity G is assumed to be 2.6· Therefore, $C = 2.6(I/J)$. The ratio I/J depends upon the cross-sectional form of the beam. In the case of the circular cross section, J is twice I with respect to the diameter. Hence, the ratio of the flexural rigidity EI to torsional rigidity GJ is 1.30. Similarly: for rectangular cross section with a width three halves the height, $C = 1.109$; for a rectangular cross section with a height three halves the width, $C = 2.493$; for elliptical cross sections with a semimajor axes ratio $a/b = 2$, $C = 3.25$.

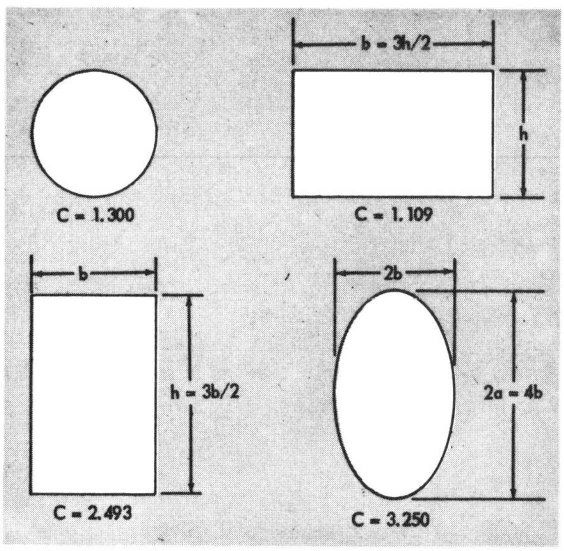

Values of K_1 through K_9 for different values of angle subtended by the circular arc for the typical cross sections specified above were calculated by computer and selected results tabulated in Tables 16, 17, 18. Also, Table 16 shows values for K_1, K_2, K_3 for various ratios of γ/θ, which determines the location of the concentrated load. Values of K_n (n = 1 to 9) for intermediate values of θ and γ/θ may be obtained by interpolation.

Example: Find the deflection, angle of slope, and angle of twist at the free end of the cantilever shown in Fig. 62.

Solution: θ = 60 deg., γ = 60 deg., $\gamma/\theta = 0.1$, P = 100 lb. and R = 10 in. The beam has a circular cross section with r = 0.25 in., and $E = 30 \times 10^6$ psi (steel). Thus, $I = \pi r^4/4 = 0.00306$ in.^{4} and $C = 1.300$. From Table 16, for $C = 1.300$, θ = 60 deg. and interpolating between $\gamma/\theta = 0$ and $\gamma/\theta = 0.2$: $K_1 = 0.3222$, $K_2 = 0.4311$, $K_3 = 0.1430$.
Thus:

$$y = \frac{PR^3}{EI} K_1 = 0.351 \text{ in.}$$

$$S = \frac{PR^2}{EI} K_2 = 0.047 \text{ radians}$$

$$\phi = \frac{PR^2}{EI} K_3 = 0.015 \text{ radians.}$$

DERIVATIONS OF EQUATIONS

The derivations were based on the concept of virtual work in which the total strain energy was partially differentiated with respect to a given fictitious load applied at a point and in the direction of the required deflection.

The method of finding the deflection and angle of twist or change of slope is the same, regardless of what combination of loads is applied to the member. Hence, only the deflection is discussed here.

For purposes of illustrating the method, let it be required to find the deflection of the free end of a cylindrical rod bent into a circular arc of radius R as shown in Fig. 63. The rod is fixed at one end, B, and loaded at point C by a concentrated force P that is perpendicular to the plane AOB of the circular arc.

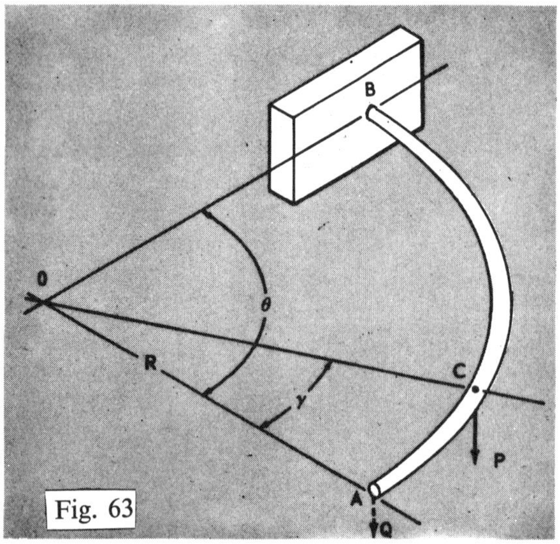

Fig. 63

Because it is required to find the vertical elastic deflection of the point A by means of Castigliano's theorem, it is assumed that a fictitious load Q is applied in the direction of the desired deflection y of point A. From Castigliano's theorem: $y = \partial U/\partial Q$, where U is equal to the total elastic strain energy in the beam due to the work done by loads Q and P.
Or:

$$y = \int \frac{M\partial M}{EI\partial Q}\,ds + \int \frac{T\partial T}{GJ\partial Q}\,ds \tag{61}$$

The bending moment M and twisting moment T will be determined by considering two arcs of the beam AC and CB. Angle γ is subtended by AC and angle $\theta - \gamma$ by CB. Assume a running coordinate γ_o, which varies in the sector OAC from 0 to γ and in the sector OCB from γ to θ (Fig. 64). Hence at any section:

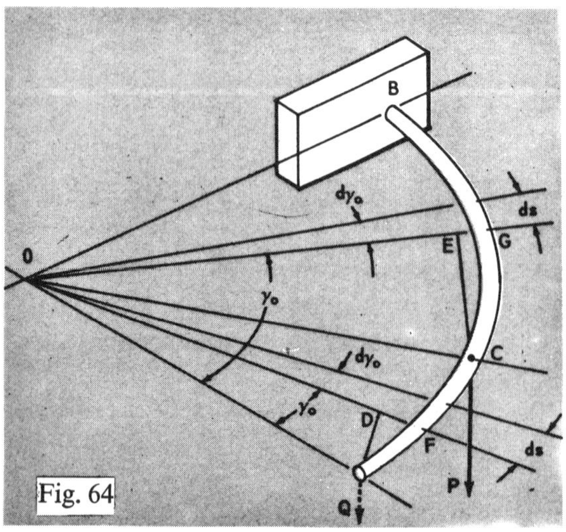

Fig. 64

Due to load Q:

$$M_1 = Q(DA) = QR\sin\gamma_o$$

$$T_1 = Q(DF) = QR(1 - \cos\gamma_o)$$

$$\frac{\partial M_1}{\partial Q} = R\sin\gamma_o$$

$$\frac{\partial T_1}{\partial Q} = R(1 - \cos\gamma_o).$$

Due to loads Q and P:

$$M_2 = M_1 + P(CE) = QR\sin\gamma_o + PR\sin(\gamma_o - \gamma)$$

$$T_2 = T_1 + P(EG) = QR(1 - \cos\gamma_o) + PR[1 - \cos(\gamma_o - \gamma)]$$

$$\frac{\partial M_2}{\partial Q} = R\sin\gamma_o$$

$$\frac{\partial T_2}{\partial Q} = R(1 - \cos\gamma_o).$$

For either section: $ds = Rd\gamma_o$

By substitution in equation (61):

$$y = \frac{QR^3}{EI}\int_0^\gamma \sin^2\gamma_o d\gamma_o + \frac{1}{EI}\int_\gamma^\theta [QR^3\sin^2\gamma_o + PR^3\sin\gamma_o\sin(\gamma_o - \gamma)]d\gamma_o + \frac{QR^3}{GJ}\int_0^\gamma (1 - \cos\gamma_o)^2 d\gamma_o + \frac{1}{GJ}\int_\gamma^\theta \{QR^3(1 - \cos\gamma_o) + PR^3(1 - \cos\gamma_o)[1 - \cos(\gamma_o - \gamma)]\}d\gamma_o.$$

Letting $Q = 0$ and $EI/GJ = C$, then

$$y = \frac{PR^3}{EI}\int_\gamma^\theta \sin\gamma_o\sin(\gamma_o - \gamma)d\gamma_o + C\int_\gamma^\theta (1 - \cos\gamma_o)[1 - \cos(\gamma_o - \gamma)]\,d\gamma_o.$$

In the derivation of equations presented here it was assumed that the effect of direct shear was negligible compared to the effects of bending and torsion. Hence, strain energy expressions contain only the terms due to flexure and torsion. It was further assumed that the members are of relatively small thickness and of constant cross section within the elastic limit of the material.

TABLE 16

K₁ K₂ K₃ FOR CONCENTRATED LOAD

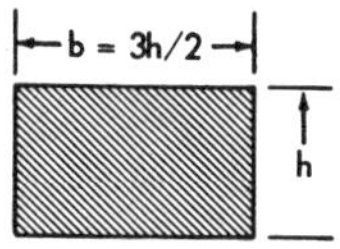

C = 1.109 $\gamma/\theta = 0$

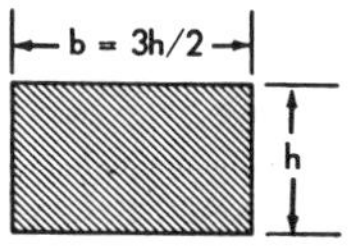

C = 1.109 $\gamma/\theta = 0.2$

θ deg	K_1	K_2	K_3	K_1	K_2	K_3
0	0.0000	0.0000	0.0000	0.0000	0.0000	0.0000
30	0.0474	0.1349	0.0212	0.0334	0.0852	0.0199
60	0.3683	0.5136	0.1674	0.2594	0.3115	0.1545
90	1.1804	1.0545	0.5474	0.8328	0.5934	0.4939
120	2.5867	1.6226	1.2245	1.8328	0.8020	1.0776
150	4.5313	2.0558	2.1825	3.2339	0.7930	1.8717
180	6.7966	2.2180	3.3127	4.8991	0.4485	2.7638
210	9.0620	2.0558	4.4429	6.6055	−0.2786	3.5844
240	11.0067	1.6227	5.4010	8.1014	−1.3391	4.1547
270	12.4131	1.0546	6.0781	9.1754	−2.5961	4.3429
300	13.2252	0.5137	6.4581	9.7115	−3.8660	4.1059
330	13.5462	0.1350	6.6044	9.7098	−4.9750	3.4971
360	13.5937	0.0000	6.6256	9.2701	−5.8072	2.6410

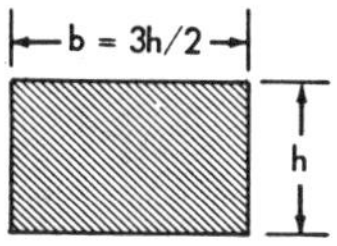

C = 1.109 $\gamma/\theta = 0.4$

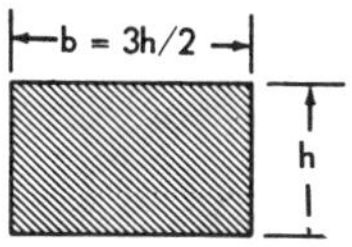

C = 1.109 $\gamma/\theta = 0.6$

θ deg	K_1	K_2	K_3	K_1	K_2	K_3
0	0.0000	0.0000	0.0000	0.0000	0.0000	0.0000
30	0.0204	0.0470	0.0147	0.0098	0.0204	0.0081
60	0.1569	0.1614	0.1119	0.0733	0.0639	0.0599
90	0.4949	0.2689	0.3469	0.2226	0.0838	0.1776
120	1.0633	0.2669	0.7245	0.4519	0.0220	0.3470
150	1.8199	0.0532	1.1869	0.7145	−0.1674	0.5136
180	2.6527	−0.4386	1.6171	0.9323	−0.4939	0.5934
210	3.4032	−1.2078	1.8616	1.0190	−0.9150	0.4961
240	3.9056	−2.1693	1.7703	0.9110	−1.3342	0.1558
270	4.0353	−3.1613	1.2425	0.5971	−1.6171	−0.4386
300	3.7485	−3.9779	0.2646	0.1378	−1.6227	−1.2244
330	3.1010	−4.4190	−1.0740	−0.3327	−1.2443	−2.0635
360	2.2365	−4.3429	−2.5960	−0.6285	−0.4486	−2.7638

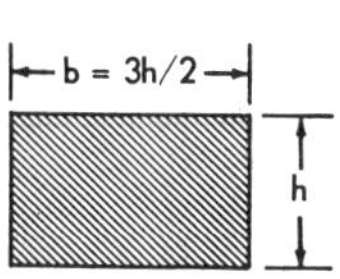

C = 1.109 $\gamma/\theta = 0.8$

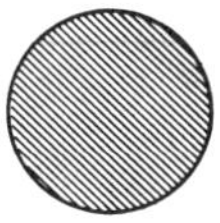

C = 1.300 $\gamma/\theta = 0$

θ deg	K_1	K_2	K_3	K_1	K_2	K_3
0	0.0000	0.0000	0.0000	0.0000	0.0000	0.0000
30	0.0026	0.0049	0.0024	0.0478	0.1367	0.0171
60	0.0189	0.0136	0.0172	0.3789	0.5375	0.1434
90	0.0538	0.0108	0.0481	1.2484	1.1500	0.5064
120	0.0987	−0.0199	0.0852	2.8145	1.8375	1.2177
150	0.1324	−0.0859	0.1063	5.0489	2.3883	2.2956
180	0.1288	−0.1776	0.0838	7.6966	2.6000	3.6127
210	0.0683	−0.2668	−0.0026	10.3444	2.3884	4.9298
240	−0.0496	−0.3115	−0.1545	12.5789	1.8376	6.0078
270	−0.1989	−0.2690	−0.3469	14.1452	1.1501	6.7192
300	−0.3276	−0.1118	−0.5285	15.0148	0.5376	7.0822
330	−0.3679	0.1565	−0.6323	15.3460	0.1367	7.2086
360	−0.2546	0.4939	−0.5934	15.3938	0.0000	7.2257

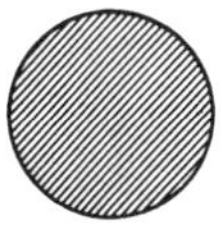

C = 1.300 $\gamma/\theta = 0.2$

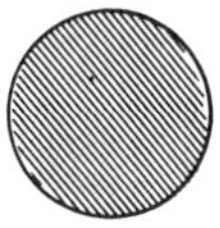

C = 1.300 $\gamma/\theta = 0.4$

θ deg	K_1	K_2	K_3	K_1	K_2	K_3
0	0.0000	0.0000	0.0000	0.0000	0.0000	0.0000
30	0.0336	0.0862	0.0178	0.0205	0.0475	0.0138
60	0.2655	0.3247	0.1425	0.1598	0.1674	0.1071
90	0.8733	0.6471	0.4761	0.5144	0.2933	0.3410
120	1.9738	0.9247	1.0885	1.1331	0.3224	0.7359
150	3.5701	0.9824	1.9733	1.9911	0.1365	1.2491
180	5.5172	0.6529	3.0141	2.9775	−0.3598	1.7635
210	7.5400	−0.1487	3.9986	3.9092	−1.1926	2.1021
240	9.3126	−1.3708	4.6861	4.5756	−2.2838	2.0725
270	10.5649	−2.8312	4.9000	4.8055	−3.4476	1.5317
300	11.1665	−4.2840	4.5954	4.5294	−4.4287	0.4455
330	11.1465	−5.5122	3.8638	3.8111	−4.9722	−1.0829
360	10.6510	−6.3957	2.8802	2.8338	−4.9001	−2.8311

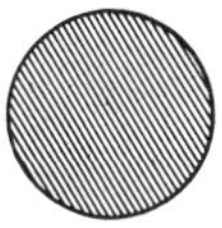

C = 1.300 $\gamma/\theta = 0.6$

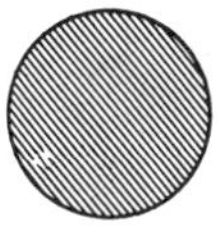

C = 1.300 $\gamma/\theta = 0.8$

θ deg	K_1	K_2	K_3	K_1	K_2	K_3
0	0.0000	0.0000	0.0000	0.0000	0.0000	0.0000
30	0.0098	0.0205	0.0078	0.0026	0.0050	0.0024
60	0.0743	0.0657	0.0585	0.0190	0.0139	0.0170
90	0.2291	0.0914	0.1764	0.0547	0.0118	0.0480
120	0.4755	0.0389	0.3525	0.1020	−0.0178	0.0862
150	0.7730	−0.1434	0.5375	0.1405	−0.0831	0.1098
180	1.0444	−0.4761	0.6471	0.1441	−0.1764	0.0914
210	1.1945	−0.9276	0.5816	0.0919	−0.2708	0.0088
240	1.1418	−1.4050	0.2566	−0.0195	−0.3247	−0.1425
270	0.8542	−1.7635	−0.3598	−0.1675	−0.2933	−0.3409
300	0.3792	−1.8375	−1.2176	−0.3016	−0.1446	−0.5372
330	−0.1459	−1.4879	−2.1747	−0.3528	0.2353	−0.6763
360	−0.5110	−0.6531	−3.0141	−0.2499	0.4761	−0.6472

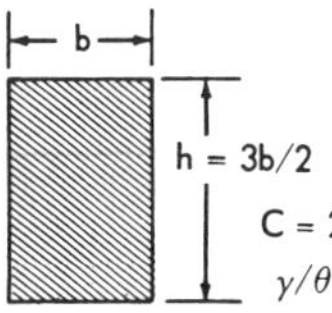

C = 2.493 $\gamma/\theta = 0$

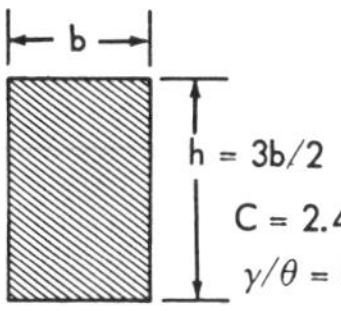

C = 2.493 $\gamma/\theta = 0.2$

θ deg	K_1	K_2	K_3	K_1	K_2	K_3
0	0.0000	0.0000	0.0000	0.0000	0.0000	0.0000
30	0.0500	0.1474	−0.0088	0.0349	0.0921	0.0046
60	0.4448	0.6866	−0.0068	0.3037	0.4074	0.0679
90	1.6733	1.7464	0.2504	1.1264	0.9828	0.3647
120	4.2377	3.1795	1.1755	2.8547	1.6907	1.1570
150	8.2823	4.4653	3.0024	5.6700	2.1658	2.6078
180	13.3183	4.9860	5.4865	9.3775	1.9299	4.5775
210	18.3543	4.4655	7.9707	13.3770	0.6629	6.5857
240	22.3992	3.1798	9.7978	16.8774	1.5687	8.0053

θ deg	K_1	K_2	K_3	K_1	K_2	K_3
270	24.9639	1.7467	10.7231	19.2434	-4.2997	8.3800
300	26.1926	0.6868	10.9804	20.2542	-6.8951	7.6528
330	26.5875	0.1474	10.9824	20.1205	-8.8675	6.1545
360	26.6376	0.0000	10.9736	19.2762	-10.0716	4.3742

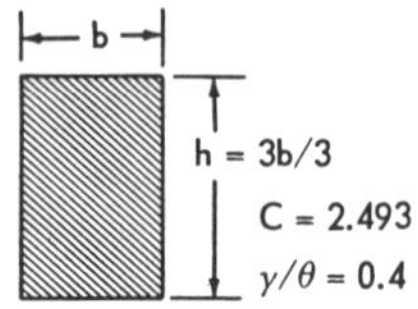

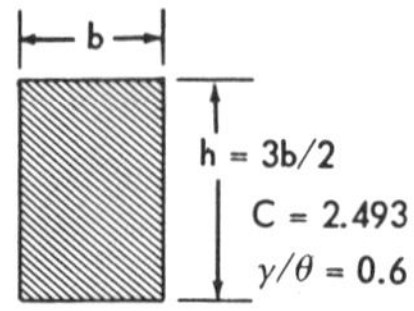

θ deg	K_1	K_2	K_3	K_1	K_2	K_3
0	0.0000	0.0000	0.0000	0.0000	0.0000	0.0000
30	0.0211	0.0501	0.0083	0.0100	0.0213	0.0062
60	0.1779	0.2046	0.0769	0.0803	0.0774	0.0500
90	0.6367	0.4454	0.3040	0.2699	0.1388	0.1689
120	1.5690	0.6688	0.8072	0.6226	0.1449	0.3867
150	3.0606	0.6564	1.6376	1.1382	0.0068	0.6866
180	5.0063	0.1325	2.6782	1.7446	-0.3646	0.9828
210	7.0696	-1.0980	3.6042	2.2909	-1.0065	1.1154
240	8.7602	-2.9992	3.9601	2.5834	-1.8473	0.8858
270	9.6144	-5.2358	3.3382	2.4607	-2.6782	0.1326
300	9.4067	-7.2444	1.5750	1.8868	-3.1796	-1.1754
330	8.2467	-8.4276	-1.1386	1.0208	-3.0093	-2.8698
360	6.5640	-8.3801	-4.2995	0.2227	-1.9302	-4.5774

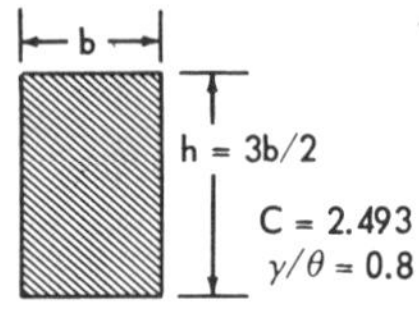

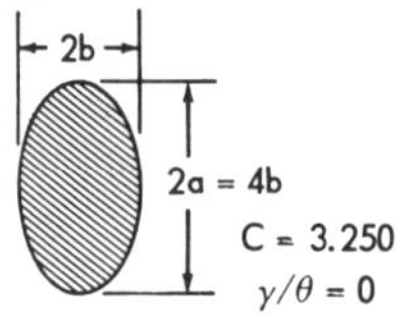

θ deg	K_1	K_2	K_3	K_1	K_2	K_3
0	0.0000	0.0000	0.0000	0.0000	0.0000	0.0000
30	0.0026	0.0051	0.0022	0.0515	0.1542	-0.0252
60	0.0198	0.0154	0.0160	0.4866	0.7812	-0.1021
90	0.0604	0.0179	0.0475	1.9429	2.1249	0.0879
120	0.1223	-0.0046	0.0921	5.1408	4.0311	1.1488
150	0.1909	-0.0661	0.1320	10.3341	5.7832	3.4509
180	0.2398	-0.1689	0.1388	16.8854	6.5000	6.6755
210	0.2393	-0.2959	0.0800	23.4368	5.7835	9.9003
240	0.1680	-0.4074	-0.0678	28.6305	4.0315	12.2027
270	0,0287	-0.4454	-0.3040	31.8288	2.1253	13.2637
300	-0.1395	-0.3493	-0.5912	33.2853	0.7814	13.4539
330	-0.2583	-0.0779	-0.8525	33.7206	0.1542	13.3770
360	-0.2210	0.3646	-0.9828	33.7721	-0.0000	13.3518

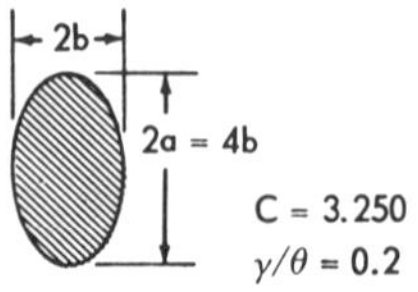

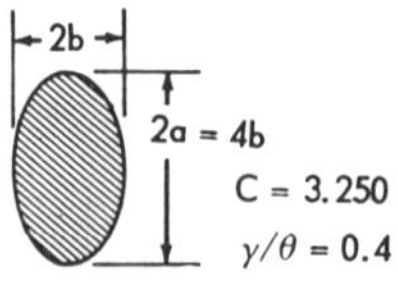

θ deg	K_1	K_2	K_3	K_1	K_2	K_3
0	0.0000	0.0000	0.0000	0.0000	0.0000	0.0000
30	0.0357	0.0958	-0.0038	0.0215	0.0518	0.0048
60	0.3280	0.4598	0.0205	0.1894	0.2283	0.0577
90	1.2870	1.1958	0.2939	0.7143	0.5419	0.2806
120	3.1436	2.1768	1.2005	1.8455	0.8886	0.8524

θ deg	K_1	K_2	K_3	K_1	K_2	K_3
150	7.0024	2.9167	3.0104	3.7392	0.9864	1.8842
180	11.8271	2.7402	5.5695	6.2937	0.4449	3.2586
210	17.0807	1.1778	8.2272	9.0750	-1.0379	4.5573
240	21.6776	-1.6943	10.1114	11.4155	-3.4531	5.1579
270	24.7503	-5.2315	10.5881	12.6690	-6.3704	4.4845
300	26.0207	-8.5519	9.5929	12.5015	-9.0311	2.2917
330	25.8148	-10.9965	7.6080	11.0613	-10.6202	-1.1739
360	24.7492	-12.4041	5.3222	8.9309	-10.5883	-5.2312

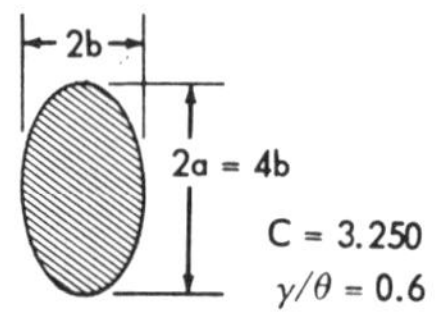

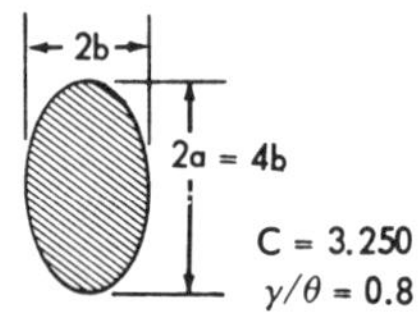

θ deg	K_1	K_2	K_3	K_1	K_2	K_3
0	0.0000	0.0000	0.0000	0.0000	0.0000	0.0000
30	0.0101	0.0219	0.0051	0.0026	0.0051	0.0020
60	0.0841	0.0848	0.0446	0.0204	0.0164	0.0154
90	0.2958	0.1689	0.1641	0.0639	0.0217	0.0472
120	0.7159	0.2122	0.4084	0.1352	0.0038	0.0958
150	1.3700	0.1022	0.7812	0.2228	-0.0552	0.1461
180	2.1889	-0.2939	1.1958	0.3006	-0.1641	0.1689
210	2.9866	-1.0566	1.4542	0.3328	-0.3119	0.1251
240	3.4982	-2.1280	1.2850	0.2870	-0.4598	-0.0205
270	3.4800	-3.2586	0.4450	0.1532	-0.5420	-0.2805
300	2.8435	-4.0312	-1.1485	-0.0367	-0.4791	-0.6254
330	1.7611	-3.9748	-3.3109	-0.1983	-0.2061	-0.9730
360	0.6882	-2.7405	-5.5693	-0.2027	0.2938	-1.1958

TABLE 17 Factors K_4, K_5 and K_6 for uniformly distributed transverse load

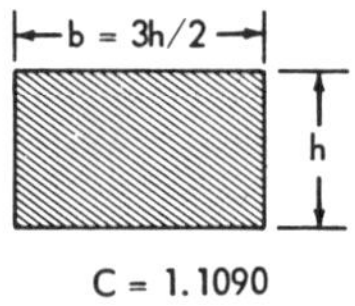

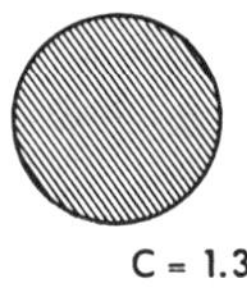

θ deg	K_4	K_5	K_6	K_4	K_5	K_6
0	0.0000	0.0000	0.0000	0.0000	0.0000	0.0000
30	0.0177	0.0441	0.0112	0.0178	0.0446	0.0101
60	0.1367	0.1577	0.0856	0.1397	0.1641	0.0798
90	0.4333	0.2881	0.2683	0.4531	0.3142	0.2597
120	0.9366	0.3608	0.5695	1.0054	0.4200	0.5751
150	1.6151	0.2997	0.9539	1.7787	0.3903	1.0037
180	2.3786	0.0545	1.3426	2.6786	0.1500	1.4642
210	3.0996	-0.3753	1.6319	3.5516	-0.3201	1.8312
240	3.6509	-0.903	1.7254	4.2334	-0.9599	1.9763
270	3.9458	-1.5020	1.5682	4.6071	-1.6380	1.8200
300	3.9671	-1.9623	1.1697	4.6462	-2.1927	1.3670
330	3.7739	-2.2021	0.6060	4.4236	-2.4868	0.7101
360	3.4841	-2.1635	0.0001	4.0841	-2.4500	0.0001

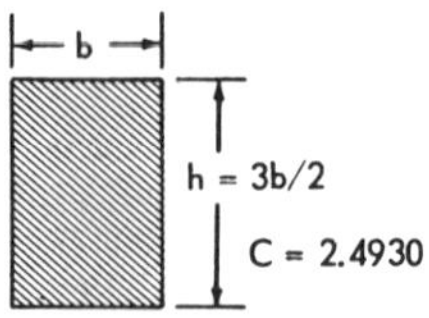

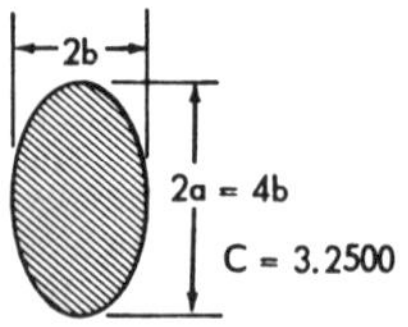

θ deg	K_4	K_5	K_6	K_4	K_5	K_6
0	0.0000	0.0000	0.0000	0.0000	0.0000	0.0000
30	0.0185	0.0474	0.0037	0.0189	0.0492	−0.0004
60	0.1584	0.2044	0.0434	0.1703	0.2299	0.0204
90	0.5768	0.4772	0.2059	0.6553	0.5806	0.1718
120	1.4351	0.7900	0.6100	1.7078	1.0247	0.6321
150	2.8008	0.9561	1.3144	3.4493	1.3152	1.5116
180	4.5524	0.7465	2.2236	5.7415	1.1251	2.7055
210	6.3750	0.0243	3.0757	8.1665	0.2429	3.8654
240	7.8719	−1.1448	3.5435	10.1807	−1.2622	4.5379
270	8.7375	−2.4875	3.3927	11.3585	−3.0266	4.3907
300	8.8881	−3.6323	2.5997	11.5797	−4.5458	3.3818
330	8.4816	−4.2648	1.3604	11.0566	−5.3930	1.7730
360	7.8321	−4.2396	0.0003	10.2103	−5.3751	0.0003

TABLE 18 Factors K_7, K_8 and K_9 for end bending moment

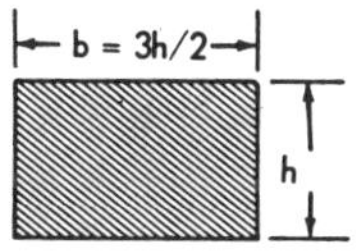

C = 1.1090

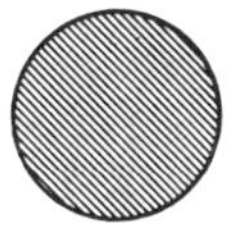

C = 1.3000

θ deg	K_7	K_8	K_9	K_7	K_8	K_9
0	0.0000	0.0000	0.0000	0.0000	0.0000	0.0000
30	0.1349	0.5285	−0.0136	0.1367	0.5372	−0.0375
60	0.5136	1.0807	−0.0409	0.5375	1.1393	−0.1125
90	1.0545	1.6564	−0.0545	1.1500	1.8064	−0.1500
120	1.6226	2.2321	−0.0409	1.8375	2.4735	−0.1125
150	2.0558	2.7842	−0.0136	2.3883	3.0756	−0.0375
180	2.2180	3.3128	−0.0000	2.6000	3.6128	−0.0000
210	2.0558	3.8413	−0.0136	2.3884	4.1500	−0.0375
240	1.6227	4.3934	−0.0409	1.8376	4.7521	−0.1125
270	1.0546	4.9691	−0.0545	1.1501	5.4191	−0.1500
300	0.5137	5.5449	−0.0409	0.5376	6.0862	−0.1125
330	0.1350	6.0970	−0.0136	0.1367	6.6884	−0.0375
360	0.0000	6.6255	−0.0000	0.0000	7.2256	−0.0000

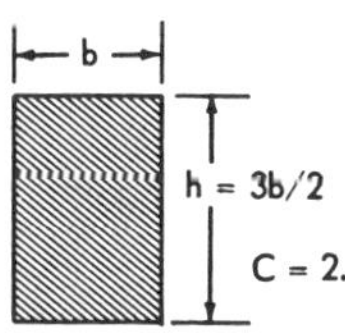

C = 2.4930

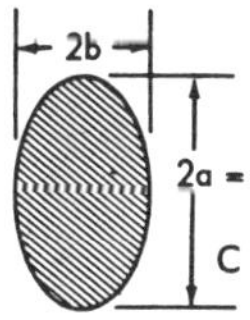

C = 3.2500

θ deg	K_7	K_8	K_9	K_7	K_8	K_9
0	0.0000	0.0000	0.0000	0.0000	0.0000	0.0000
30	0.1474	0.5912	−0.1866	0.1542	0.6255	−0.2812
60	0.6866	1.5056	−0.5599	0.7812	1.7381	−0.8437
90	1.7464	2.7433	−0.7465	2.1249	3.3379	−1.1250
120	3.1795	3.9810	−0.5599	4.0311	4.9376	−0.8438
150	4.4653	4.8955	−0.1867	5.7832	6.0503	−0.2813
180	4.9860	5.4867	−0.0000	6.5000	6.6758	−0.0000
210	4.4655	6.0779	−0.1866	5.7835	7.3013	−0.2812
240	3.1798	6.9923	−0.5598	4.0315	8.4139	−0.8437
270	1.7467	8.2300	−0.7465	2.1253	10.0136	−1.1250
300	0.6868	9.4677	−0.5599	0.7814	11.6134	−0.8438
330	0.1474	10.3822	−0.1867	0.1542	12.7261	−0.2813
360	0.0000	10.9735	−0.0000	0.0000	13.3517	−0.0000

39. CIRCULAR BEAMS

One End Supported, One End Free

Out-of-plane Loading

(Arbitrary Concentrated Load, Transverse Uniform Load, End Moment)

CIRCULAR CANTILEVER BEAMS

Out-of-plane Loading

(End Twisting Moment)

(Ref. 25, 26)

This section presents a simplified method of calculating deflection, angle of slope and angle of twist of the free end of a circular cantilever beam with a twisting moment acting at the free end. Also, a method is given for calculating the values of the reaction at the simple support for a circular beam rigidly supported at one end, simply supported at the other, and loaded (1) with a bending moment applied at the simply supported end or (2) twisting moment applied at the simply supported end.

Despite its importance, the subject of circular beams loaded normal to the plane of curvature has a very meager treatment in the engineering literature. The problem, being a combination of flexure and torsion, is complicated by expressions containing a large number of terms involving transcendental functions. Further complication of the problem results from the fact that in order to obtain sufficient accuracy it is necessary to carry the decimal point to about seven places.

The results of this analysis are summarized in Tables 19, 20, and 21. Because of the complexity of the subject, however, the equations for deflection, angle of slope, angle of twist, and reaction force are developed below to provide a method for solution of similar problems but with different types of out-of-plane loads.

ASSUMPTIONS

(1) The beam is of uniform cross section. (2) It is of such large radius compared to its radial thickness that the deflection theory of straight beams does not apply. (3) Its deflection is due solely to bending, the effect of shear being negligible. (4) At no point is it stressed beyond the elastic limit. (5) It is not so severely deformed that it loses its essential circular shape.

METHOD OF UNIT LOADS

During the static loading of an elastic system, the external work done by a constant force acting thereon is equal to the internal work done by the stresses due to that constant force. This relationship is the basis of the following method for finding the deflection of any given point of an elastic system: A unit force is imagined to act at the point in question and in the direction of the deflection that is to be found. The stresses produced by such a unit force would do a certain amount of internal work during the application of the actual loads. This work, which can be readily found, is equal to the work done by the unit force. This work, since the unit force is constant, is equal to the deflection sought.

NOTATION

y = vertical deflection of the free end of circular beam (in.)
α = angle of slope of the free end (radians)
ϕ = angle of twist of the free end (radians)
θ = angle subtended by a circular arc (radians)
γ = running coordinate
F = reaction of supports (lb.)
M_o = applied bending moment perpendicular to the plane of beam (lb.-in.)
M = bending moment at section located by angle γ (lb.-in.)
T = twisting moment at section located by angle γ (lb.-in.)
T_o = applied twisting moment, perpendicular to the plane of beam (lb.-in.)
m = bending moment at any section of the beam caused by a dummy load of unity
t = twisting moment at any section of the beam caused by a dummy load of unity
m' = bending moment at any section of the beam caused by a dummy couple of unity
t' = twisting moment at any section of the beam caused by a dummy couple of unity
E = modulus of elasticity in direct stress (psi)
G = modulus of elasticity in shear stress (psi)
I = principal moment of inertia of the cross sectional area with respect to the axis of bending (in.^{4})
J = torsional effect of the cross sectional area (in.^{4})
EI = flexural rigidity
GJ = torsional rigidity
$C = EI/GJ$ = rigidity ratio
R = radius of center line of curved beam (in.)
K_1, K_2, K_3, K_4, dimensionless factors, Tables 19 and 20.
Z = section modulus (in.^{3})
S_B = bending stress (psi)
S_S = shear stress (psi)
$d = 2r$, radial thickness of circular cross section (in.)

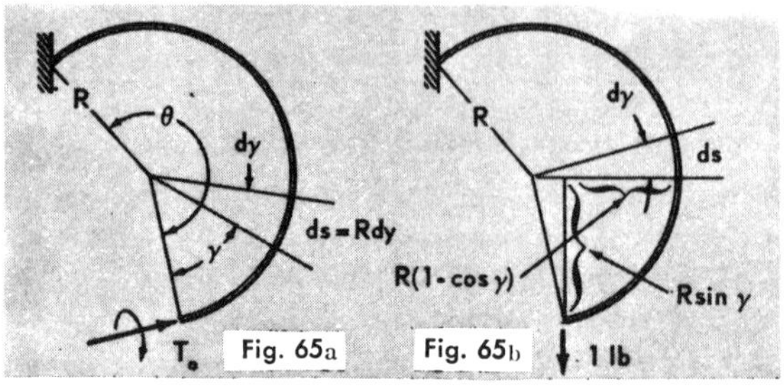

Fig. 65a Fig. 65b

VERTICAL DEFLECTION OF THE FREE END

Fig. 65a shows a circular cantilever beam with a twisting moment T_o applied at the free end. A part of the vertical deflection of the free end is due to bending moment and a part is due to a torsional moment. Therefore:

$$y = \frac{Mm}{EI}\,ds + \frac{Tt}{GJ}\,ds. \tag{62}$$

Resolving the torsional vector at the free end into its torsion and bending moment components, we obtain at any section located by angle γ:

$$\left.\begin{aligned} \text{Bending moment:}\quad & M = T_o \sin\gamma \\ \text{Twisting moment:}\quad & T = T_o \cos\gamma \end{aligned}\right\} \tag{63}$$

By applying a unit load at the free end and normal to the curvature of the beam, we find from Fig. 65b:

$$\left.\begin{aligned} m &= R\sin\gamma \\ t &= R(1 - \cos\gamma) \end{aligned}\right\} \tag{64}$$

By substitution in equation (62):

$$y = \frac{T_oR^2}{EI}\int_0^\theta \sin^2\gamma d\gamma + \frac{T_oR^2}{GJ}\int_0^\theta \cos^2\gamma(1\text{-}\cos\gamma)d\gamma. \tag{65}$$

By integration, and assuming $C = EI/GJ$, we obtain:

$$y = \frac{T_oR^2}{EI}\left[\frac{\theta - \sin\theta\cos\theta}{2} + C\left(\frac{\theta + \sin\theta\cos\theta}{2} - \sin\theta\right)\right].$$

ANGLE OF SLOPE

Instead of unit load, a dummy couple of unity (1 lb.-in.) will be assumed acting at the free end. Again, resolving the unit moment vector at the free end into its torsion and bending moment components, we obtain at any section located by angle γ:

$$m' = \cos\gamma$$

$$t' = -\sin\gamma.$$

The angle of slope of the free end is:

$$\alpha = \int_0^\theta \frac{Mm'}{EI}\,ds + \int_0^\theta \frac{Tt'}{GJ}\,ds. \tag{66}$$

Note: M and T are given by equation (63).

By substitution and integration we obtain:

$$\alpha = \frac{T_oR}{EI}\left[\frac{1}{2}\sin^2\theta - C\frac{1}{2}\sin^2\theta\right].$$

ANGLE OF TWIST

$$\phi = \int_0^\theta \frac{M}{EI}\frac{\partial M}{\partial T_o}\,ds + \int_0^\theta \frac{T}{GJ}\frac{\partial T}{\partial T_o}\,ds \tag{67}$$

From equation (63):

$$\frac{\partial M}{\partial T_o} = \sin\gamma$$

$$\frac{\partial T}{\partial T_o} = \cos\gamma$$

or by substitution in equation (67):

$$\phi = \frac{RT_o}{EI}\left[\int_0^\theta \sin^2\gamma\,d\gamma + C\int_0^\theta \cos^2\gamma\,d\gamma\right]. \tag{68}$$

Finally, by integration:

$$\phi = \frac{T_oR}{EI}\left[\frac{(1+C)}{2}(\theta - \sin\theta\cos\theta) + C\sin\theta\cos\theta\right].$$

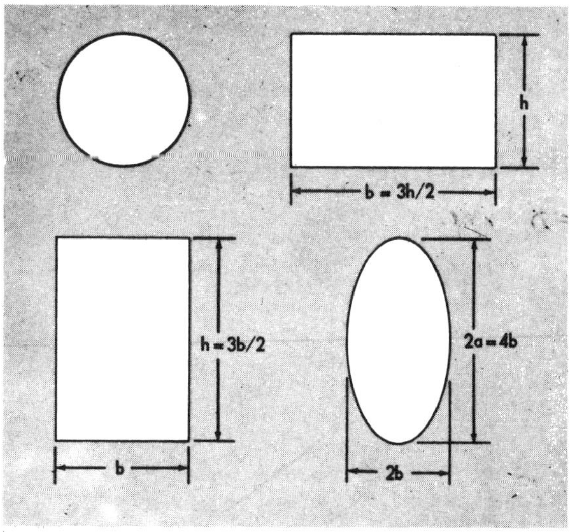

The equations for y, α and ϕ contain a constant C that is the ratio of the flexural rigidity EI to torsional rigidity GJ. For steel, (Poisson's Ratio = 0.30), the ratio $E/G = 2.6$. Therefore, $C = 2.6\,(I/J)$. The ratio I/J depends upon the cross sectional form of the beam. In the case of the circular cross section, J is twice I with respect to the diameter. Hence, $EI/GJ = 1.30$. Similarly: for rectangular cross section with a width three halves the height, $C = 1.109$; for a rectangular cross section with a height three halves the width, $C = 2.493$: for elliptical cross sections with semimajor axes ratio $a/b = 2$, $C = 3.25$.

Values of EIy/T_oR^2, $EI\alpha/T_oR^2$, $EI\phi/T_oR^2$ for values of θ from 0 to 360 deg. for the typical cross sections specified above were calculated by computer and selected results tabulated in Table 19.

The bending and shear stresses at any section of the beam

$$S_B = \frac{M}{Z}$$

$$S_S = \frac{2T}{\pi r^3}$$

for circular cross section and where the bending moment M and twisting moment T are determined from equation (63).

REACTION FORCE

Fig. 66 shows a circular beam rigidly supported at one end, simply supported at the other and loaded with a bending moment applied at the simply supported end.

The reaction force F at the support will be obtained by assuming that the deflection y_M of the free end, due to moment M_o alone (with supports removed), is equal to the deflection y_F of the free end due to the force F alone (with supports removed).

Fig. 66

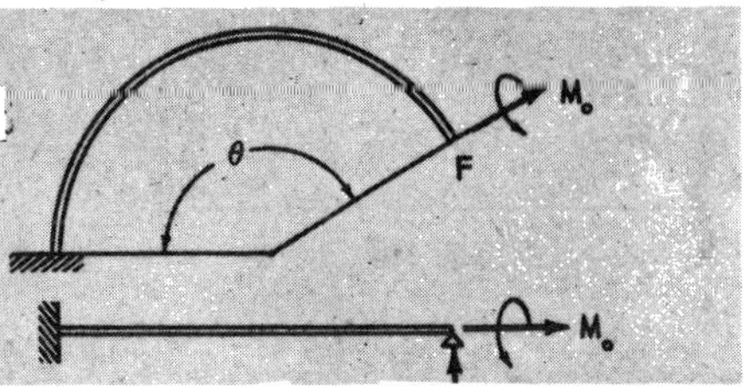

Thus,

$$y_M = y_F. \tag{69}$$

Following the method of unit load we obtain:

$$y_M = \frac{M_oR^2}{EI}\left[\frac{1-\cos^2\theta}{2} + C\left(\frac{1+\cos^2\theta}{2} - \cos\theta\right)\right], \tag{70}$$

where, as before, $C = EI/GJ$.

For the computation of the deflection y_F of the free end due to the reaction force F alone (with supports removed), it is convenient to use Castigliano's theorem, which states that when the total strain energy U expressed in terms of external loads is partially differentiated with respect to a given external load, the result gives the displacement of that load in its own line of action.

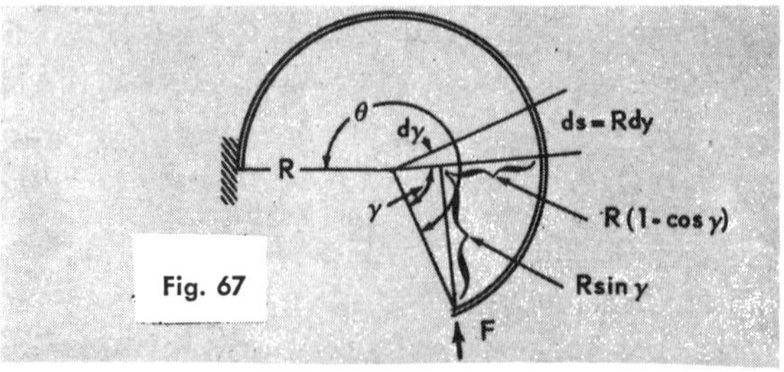

Fig. 67

Thus:

$$y_F = \frac{\partial U}{\partial F}$$

or

$$y_F = \int \frac{M}{EI}\frac{\partial M}{\partial F}\,ds + \int \frac{T}{GJ}\frac{\partial T}{\partial F}\,ds. \quad (71)$$

From Fig. 67 we obtain:

$$\text{Bending moment } M = FR\sin\gamma$$

$$\text{Twisting moment } T = FR(1-\cos\gamma)$$

$$\frac{\partial M}{\partial F} = R\sin\gamma$$

$$\frac{\partial T}{\partial F} = R(1-\cos\gamma)$$

Substituting in equation 71 and integrating:

$$y_F = \frac{FR^3}{EI}\left[\frac{1+3C}{2}(\theta - \sin\theta\cos\theta) - 2C\sin\theta(1-\cos\theta)\right]. \quad (72)$$

Let the bracket expressions of equation (70) be K_1 and of equation (72) be K_2:

$$y_M = \frac{M_oR^2}{EI}K_1 \text{ and } y_F = \frac{FR^3}{EI}K_2. \quad (73)$$

Because $y_M = y_F$, we obtain from equation (73):

$$F = \frac{M_o}{R}\frac{K_1}{K_2}$$

or

$$F = \frac{M_o}{L}\frac{\theta K_1}{K_2},$$

where $L = R\theta$ is the length of the curved beam.

In the case of a circular beam rigidly supported at one end, simply supported at the other and loaded with a twisting moment T_o, applied at the supported end,

$$F = \frac{T_o}{L}\frac{\theta K_3}{K_4},$$

where

$$K_3 = \frac{\theta - \sin\theta\cos\theta}{2} + C\left(\frac{\theta + \sin\theta\cos\theta}{2} - \sin\theta\right)$$

and

$$K_4 = \frac{1+3C}{2}(\theta - \sin\theta\cos\theta) - 2C\sin\theta(1-\cos\theta).$$

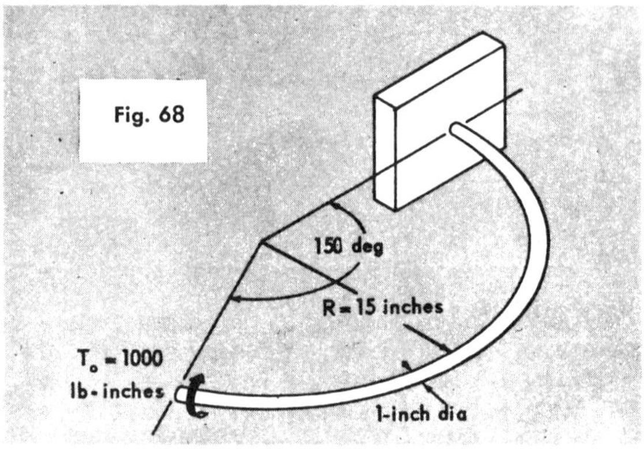

Fig. 68

The ratios $\theta K_1/K_2$ and $\theta K_3/K_4$ for values of θ from 0 to 360 deg. for typical cross sections specified above were calculated by computer and selected results tabulated in Tables 20 and 21.

Example 1: Calculate the deflection, angle of twist, and angle of slope at the free end of the steel rings shown in Fig. 68. Also, find the magnitude of the stresses at the fixed end.

Solution: The rigidity ratio, is

$C = 1.30. \quad I = (0.785)(0.5^4) = 0.049 \text{ in.}^4.$

$$Z = \frac{0.049}{0.5} = 0.098 \text{ in.}^3.\ E = 30\times10^6 \text{ psi} \quad \theta = 150\,\text{deg.}$$

$$\frac{EI}{T_oR^2} = \frac{30\times10^6(0.049)}{(1000)(15^2)} = 6.5333.$$

From Table 19 for $C = 1.30$ and $\theta = 150$ deg.:

$$\frac{EI}{T_oR^2}y = 2.2956;\ \frac{EI}{T_oR^2}\alpha = -0.0375;\ \frac{EI}{T_oR^2}\theta = 2.9457.$$

Hence,

$y = 0.351$ in.; $\alpha = 0.006$ radian; $\phi = 0.451$ radian.

STRESSES AT FIXED END

From equation (63) for $\gamma = 150$ deg.: $M = 500$ lb.-in. and $T = -866$ lb.-in.

$$S_B = \frac{M}{Z} = \frac{500}{0.098} = 5{,}100 \text{ psi},$$

top surface in compression

$$S_S = \frac{2T}{\pi r^3} = -\frac{(2)(866)}{(3.14)(0.5^3)} = -4{,}400 \text{ psi},$$

shear in clockwise direction.

Example 2: Assume the beam in Fig. 68 has the free end simply supported. Find the reaction of the supports.

Solution: From Table 21: $\theta K_3/K_4 = 1.1903$

$$\therefore F = \frac{T_o}{L}(1.1903)$$

$$L = R\theta = (15)(2.618) = 39.27 \text{ in.}$$

$$F = \frac{1{,}000}{39.27}(1.1903) = 30.3 \text{ lb.}$$

TABLE 19

Circular cantilever beam with one free end.

Twisting moment applied at free end.

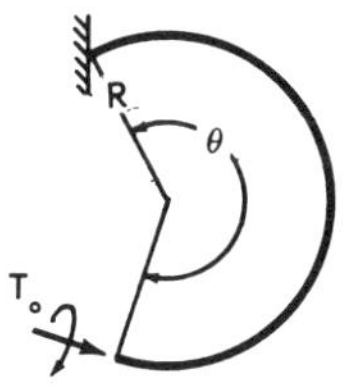

θ deg	b = 3h/2, h, C = 1.1090: EIy/T_oR^2	$EI\gamma/T_oR^2$	$EI\phi/T_oR^2$	C = 1.300: EIy/T_oR^2	$EI\gamma/T_oR^2$	$EI\phi/T_oR^2$
0	0.0000	0.0000	0.0000	0.0000	0.0000	0.0000
30	0.0212	-0.0136	0.5757	0.0171	-0.0375	0.6671
60	0.1674	-0.0409	1.1279	0.1434	-0.1125	1.2692
90	0.5474	-0.0545	1.6564	0.5064	-0.1500	1.8064
120	1.2245	-0.0409	2.1849	1.2177	-0.1125	2.3436
150	2.1825	-0.0136	2.7370	2.2956	-0.0375	2.9457
180	3.3127	-0.0000	3.3128	3.6127	-0.0000	3.6128
210	4.4429	-0.0136	3.8885	4.9298	-0.0375	4.2798
240	5.4010	-0.0409	4.4406	6.0078	-0.1125	4.8820
270	6.0781	-0.0545	4.9691	6.7192	-0.1500	5.4192
300	6.4581	-0.0409	5.4977	7.0822	-0.1125	5.9563
330	6.6044	-0.0136	6.0498	7.2086	-0.0375	6.5585
360	6.6256	-0.0000	6.6255	7.2257	-0.0000	7.2255

θ deg	b, h = 3b/2, C = 2.4930: EIy/T_oR^2	$EI\gamma/T_oR^2$	$EI\phi/T_oR^2$	2b, 2a = 4b, C = 3.2500: EIy/T_oR^2	$EI\gamma/T_oR^2$	$EI\phi/T_oR^2$
0	0.0000	0.0000	0.0000	0.0000	0.0000	0.0000
30	-0.0088	-0.1866	1.2377	-0.0252	-0.2812	1.5998
60	-0.0068	-0.5599	2.1522	-0.1021	-0.8437	2.7124
90	0.2504	-0.7465	2.7434	0.0879	-1.1250	3.3379
120	1.1755	-0.5599	3.3346	1.1488	-0.8438	3.9634
150	3.0024	-0.1867	4.2490	3.4509	-0.2813	5.0760
180	5.4865	-0.0000	5.4867	6.6755	-0.0000	6.6757
210	7.9707	-0.1866	6.7244	9.9003	-0.2812	8.2755
240	9.7978	-0.5598	7.6389	12.2027	-0.8437	9.3882
270	10.7231	-0.7465	8.2301	13.2637	-1.1250	10.0137
300	10.9804	-0.5599	8.8213	13.4539	-0.8438	10.6392
330	10.9824	-0.1867	9.7357	13.3770	-0.2813	11.7517
360	10.9736	-0.0000	10.9733	13.3518	-0.0000	13.3514

TABLE 20

Circular cantilever beam with one end supported.

Bending moment, M_o, applied at simply supported end.

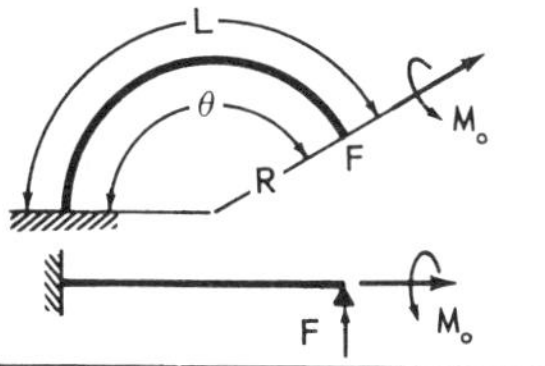

θ deg	b = 3h/2, h, C = 1.1090: $\theta K_1/K_2$	C = 1.3000: $\theta K_1/K_2$	b, h = 3b/2, C = 2.4930: $\theta K_1/K_2$	2b, 2a = 4b, C = 3.2500: $\theta K_1/K_2$
0	0.0000	0.0000	0.0000	0.0000
30	1.4906	1.4980	1.5420	1.5679
60	1.4601	1.4855	1.6164	1.6811
90	1.4032	1.4469	1.6394	1.7179
120	1.3138	1.3673	1.5714	1.6423
150	1.1877	1.2384	1.4114	1.4651
180	1.0252	1.0612	1.1761	1.2093
210	0.8315	0.8462	0.8917	0.9044
240	0.6175	0.6119	0.5946	0.5898
270	0.4003	0.3831	0.3297	0.3146
300	0.2034	0.1875	0.1373	0.1229
330	0.0574	0.0513	0.0319	0.0263
360	0.0000	0.0000	0.0000	0.0000

TABLE 21

Circular cantilever beam with one end supported.

Twisting moment, T_o, applied at simply supported end.

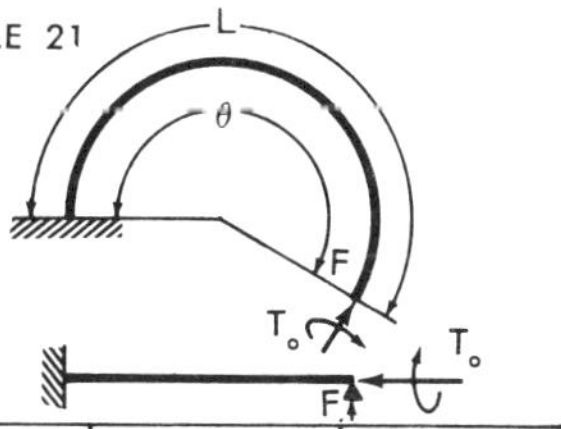

θ deg	b = 3h/2, h, C = 1.1090: $\theta K_3/K_4$	C = 1.3000: $\theta K_3/K_4$	b, h = 3b/2, C = 2.4930: $\theta K_3/K_4$	2b, 2a = 4b, C = 3.2500: $\theta K_3/K_4$
0	0.0000	0.0000	0.0000	0.0000
30	0.2345	0.1873	-0.0920	-0.2564
60	0.4760	0.3963	-0.0161	-0.2198
90	0.7284	0.6372	0.2350	0.0711
120	0.9914	0.9061	0.5810	0.4680
150	1.2609	1.1903	0.9490	0.8742
180	1.5312	1.4746	1.2942	1.2420
210	1.7969	1.7467	1.5917	1.5482
240	2.0554	2.0006	1.8322	1.7853
270	2.3074	2.2384	2.0241	1.9637
300	2.5568	2.4697	2.1950	2.1163
330	2.8080	2.7054	2.3790	2.2848
360	3.0624	2.9492	2.5884	2.4840

40. INTERNAL REACTIONS FOR OUT-OF-PLANE LOADING OF CIRCULAR BEAMS

Computer Analysis Permits Rapid Determination of Internal Reactions, *Mo*, *To*, *Vo* for Bending and Shear Stresses on Circular Beams

(*Ref.* 27)

NOTATION

P = concentrated load (lb.)
R = radius of center line of curved beam (in.)
E = modulus of elasticity in direct stress (psi)
G = modulus of elasticity in shear stress (psi)
I = principal moment of inertia of the cross-sectional area with respect to the axis of bending (in.^{4})
J = torsional effect of the cross-sectional area (in.^{4})
C = (EI/GJ), rigidity constant
θ = angle subtended by a circular arc
γ = variable angle of arc denoting position
M_o = bending moment at cross section of symmetry (lb.-in.)
T_o = twisting moment at cross section of symmetry (lb.-in.)
V_o = shear force at cross section of symmetry (lb.)
K_{Mo}, K_{To}, K_{Vo} = factors for M_o, T_o, V_o (*see* Table 22)
S_B = bending stress (psi)
S_S = shearing stress (psi)
Z = section modulus (in.^{3})
r = radius of circular cross section (in.)

This presentation yields a simplified method for finding three internal reactions: bending moment, twisting moment, and shear existing at the plane of symmetry of a circular beam built in at both ends with an arbitrarily placed, concentrated load acting perpendicular to the plane of the arc and its center line.

Fig. 69 shows a circular beam of arc 2θ, built in at both ends and loaded by the force P at the angle γ from the axis of symmetry of the arc. The positive direction of V_o, M_o, T_o is indicated in the free body diagram. The shear force V_o is acting in the same direction as P on the right-hand side and the opposite direction on the left-hand side.

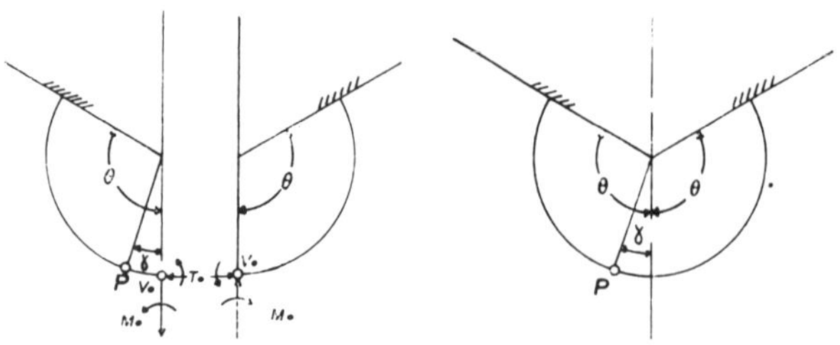

Fig. 69

The derivation of the results represented here was accomplished by use of Castigliano's theorem and expressions for the strain energy of the members. It was assumed that the beams were sufficiently flexible and that the effect of direct shear was negligible compared with the effects of bending and torsion.

$$M_o = \frac{PR}{2D_2}[(1+c)\{\sin\theta\sin(\theta-\gamma) - (\theta-\gamma)\sin\gamma\} - 2c\{\sin\theta\sin(\theta-\gamma) + \cos\theta - \cos\gamma\}] \quad \text{(a)}$$

$$T_o = \frac{PR}{2D_3}\left[\frac{(1+c)}{2}\{\theta(1-\cos\theta)\sin(\theta-\gamma) - (\theta-\gamma)[\theta(1-\cos\gamma) + \sin\theta(\cos\gamma-\cos\theta]\} + c\sin\theta\{\sin\theta - \sin\gamma - \sin(\theta-\gamma) + \theta[\cos(\theta-\gamma) - \cos\theta] - \gamma(1-\cos\theta)\}\right] \quad \text{(b)}$$

$$V_o = \frac{P}{2D_3}\left[\frac{(1+c)}{2}\{(\theta-\gamma) \times [\theta + \sin\theta(\cos\gamma - \cos\theta)] - \theta\sin(\theta-\gamma)\} - c\sin\theta\{\sin\theta - \sin\gamma - (\theta-\gamma)\cos\theta\}\right] \quad \text{(c)}$$

where

$$c = \frac{EI}{GJ}$$

$$D_2 = (1+c)(\theta + \sin\theta\cos\theta) - 2c\sin\theta\cos\theta$$

$$D_3 = \frac{1+c}{2}\theta(\theta - \sin\theta\cos\theta) - c\sin\theta(\sin\theta - \theta\cos\theta).$$

Substituting for the transcendental functions in above equations factors K_{Mo}, K_{To}, K_{Vo}, we obtain:

$$M_o = PR(K_{Mo}) \quad (74)$$

$$T_o = PR(K_{To}) \quad (75)$$

$$V_o = P(K_{Vo}) \quad (76)$$

The factors K_{Mo}, K_{To}, K_{Vo}, for various values of θ and γ/θ were calculated on a digital computer and the results are tabulated in Table 22.

Equations (a), (b), (c) contain a constant C which is the ratio of the flexural rigidity EI to the torsional

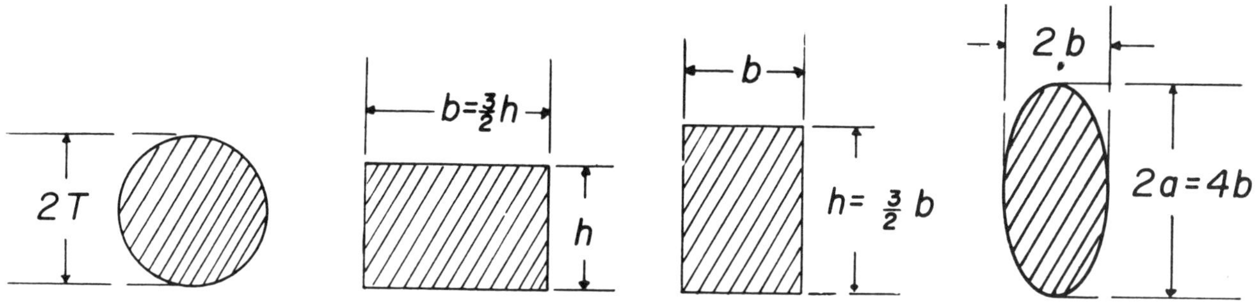

$C = 1.30$ $S_S = 2T_0/\pi T^3$ | $C = 1.109$ $S_S = T_0/0.231bh^2$ | $C = 2.493$ | $C = 3.25$ $S_S = 2T_0/\pi ab^2$

Fig. 70

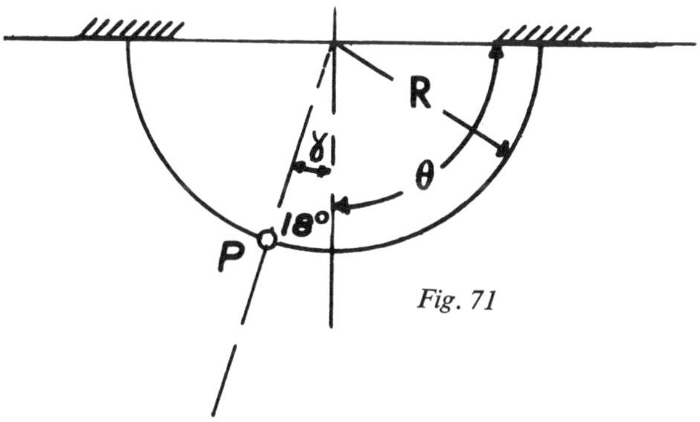

Fig. 71

rigidity GJ. Assuming material such as steel, for which the Poisson's Ratio is 0.30, the ratio $E/G = 2.6$, therefore $C = 2.6I/J$. The ratio I/J depends upon the cross-sectional form of the beam. In the case of the circular cross section, for example, the area has a polar moment of inertia J equal to twice the moment of inertia I with respect to the diameter. Hence $EI/GJ = 1.30$.

Similarly:

1. For a rectangular cross section having a width of three halves the height, $C = 1.109$.
2. For a rectangular cross section having a height three halves the width, $C = 2.493$.
3. For an elliptical cross section with semimajor axes ratio $a/b = 2$, $C = 3.25$.

The values of K_{Mo}, K_{To}, K_{Vo}, shown in Table 22 are for the four typical cross sections shown in Fig. 70.

The use of Table 22 in conjunction with equations (74), (75) and (76) should simplify the engineer's task by affording, at the same time, an exact solution of bending and torsion problems.

Knowing the bending moment M_o, the bending stresses will be obtained from: $S_B = M_o/Z$.

For a given twisting moment T_o, the equations for shearing stress are as specified in Fig. 70.

Example: Assume a steel cylindrical bar 1.0-in. diameter, in the form of a semicircular bar fixed at both ends and subjected to a concentrated load P, perpendicular to the plane of curvature and located as shown in Fig. 71, find: M_o, T_o, V_o, S_B, S_S at the cross section of symmetry of the beam.

$r = 0.5$ in., $R = 20$ in., $\theta = 90$ deg., $\gamma = 18$ deg.
$\therefore \gamma/\theta = 0.2$, $c = 1.30$ (*see* Fig. 70), $P = 1{,}000$ lb.

From Table 22:

$$K_{Vo} = 0.3343$$
$$K_{To} = 0.0381$$
$$K_{Mo} = 0.1791.$$

Hence:

$$V_o = PK_{Vo} = 334 \text{ lb.}$$
$$T_o = PR(K_{To}) = 762 \text{ lb.-in.}$$
$$M_o = PR(K_{Mo}) = 3582 \text{ lb.-in.}$$

STRESSES

$$I = 0.785 \times 0.5^4 = 0.049 \text{ in.}^4$$

$$Z = \frac{0.049}{0.5} = 0.098 \text{ in.}^3$$

Bending stress

$$S_B = \frac{M_o}{Z} = 36{,}550 \text{ psi}$$

Shearing stress, from Fig.70 :

$$S_S = \frac{2T_o}{\pi T^3} = 3{,}900 \text{ psi}$$

Table 22—Factors K_{Vo}, K_{To}, K_{Mo} for four typical cross sections as defined in Fig. 70.

Notes:

1. Data is applicable to any material with Poisson's Ratio of 0.30.
2. Interpolate for intermediate values of θ and γ/θ.

C	θ°	γ/θ	K_{V0}	K_{T0}	K_{M0}	C	θ°	γ/θ	K_{V0}	K_{T0}	K_{M0}
1.109	0.	0.	0.	—0.	—0.	1.300	0.	0.	0.	—0.	—0.
1.109	10.	0.	0.5000	0.0000	0.0435	1.300	10.	0.	0.5000	0.0000	0.0435
1.109	20.	0.	0.5000	—0.	0.0863	1.300	20.	0.	0.5000	—0.	0.0861
1.109	30.	0.	0.5000	—0.	0.1277	1.300	30.	0.	0.5000	—0.	0.1272
1.109	40.	0.	0.5000	0.0000	0.1670	1.300	40.	0.	0.5000	0.0000	0.1661
1.109	50.	0.	0.5000	0.0000	0.2038	1.300	50.	0.	0.5000	0.0000	0.2024
1.109	60.	0.	0.5000	—0.	0.2376	1.300	60.	0.	0.5000	—0.	0.2359
1.109	70.	0.	0.5000	—0.	0.2682	1.300	70.	0.	0.5000	—0.	0.2664
1.109	80.	0.	0.5000	—0.	0.2951	1.300	80.	0.	0.5000	—0.	0.2939
1.109	90.	0.	0.5000	—0.0000	0.3183	1.300	90.	0.	0.5000	—0.0000	0.3183
1.109	100.	0.	0.5000	0.0000	0.3375	1.300	100.	0.	0.5000	0.0000	0.3395
1.109	110.	0.	0.5000	—0.	0.3526	1.300	110.	0.	0.5000	—0.	0.3573
1.109	120.	0.	0.5000	—0.0000	0.3635	1.300	120.	0.	0.5000	—0.0000	0.3714
1.109	130.	0.	0.5000	—0.0000	0.3699	1.300	130.	0.	0.5000	—0.0000	0.3816
1.109	140.	0.	0.5000	—0.0000	0.3718	1.300	140.	0.	0.5000	—0.0000	0.3873
1.109	150.	0.	0.5000	—0.0000	0.3692	1.300	150.	0.	0.5000	—0.0000	0.3883
1.109	160.	θ.	0.5000	—0.0000	0.3620	1.300	160.	0.	0.5000	—0.0000	0.3841
1.109	170.	0.	0.5000	—0.0000	0.3505	1.300	170.	0.	0.5000	—0.0000	0.3746
1.109	180.	0.	0.5000	—0.0000	0.3348	1.300	180.	0.	0.5000	—0.0000	0.3598
1.109	0.	0.2	0.	—0.	—0.	1.300	0.	0.2	0.	—0.	—0.
1.109	10.	0.2	0.3518	0.0005	0.0278	1.300	10.	0.2	0.3518	0.0005	0.0278
1.109	20.	0.2	0.3511	0.0019	0.0549	1.300	20.	0.2	0.3511	0.0019	0.0548
1.109	30.	0.2	0.3500	0.0044	0.0806	1.300	30.	0.2	0.3499	0.0044	0.0802
1.109	40.	0.2	0.3485	0.0078	0.1044	1·300	40.	0.2	0.3483	0.0078	0.1036
1.109	50.	0.2	0.3466	0.0121	0.1257	1.300	50.	0.2	0.3463	0.0121	0.1244
1.109	60.	0.2	0.3442	0.0174	0.1441	1.300	60.	0.2	0.3439	0.0173	0.1425
1.109	70.	0.2	0.3415	0.0236	0.1593	1.300	70.	0.2	0.3410	0.0234	0.1577
1.109	80.	0.2	0.3384	0.0306	0.1711	1.300	80.	0.2	0.3378	0.0304	0.1699
1.109	90.	0.2	0.3350	0.0384	0.1793	1.300	90.	0.2	0.3343	0.0381	0.1791
1.109	100.	0.2	0.3313	0.0469	0.1831	1.300	100.	0.2	0.3306	0.0464	0.1851
1.109	110.	0.2	0.3273	0.0560	0.1836	1.300	110.	0.2	0.3267	0.0553	0.1878
1.109	120.	0.2	0.3233	0.0654	0.1797	1.300	120.	0.2	0.3227	0.0646	0.1869
1.109	130.	0.2	0.3193	0.0750	0.1715	1.300	130.	0.2	0.3188	0.0740	0.1822
1.109	140.	0.2	0.3154	0.0844	0.1591	1.300	140.	0.2	0.3150	0.0832	0.1732
1.109	150.	0.2	0.3120	0.0932	0.1424	1.300	150.	0.2	0.3117	0.0920	0.1597
1.109	160.	0.2	0.3091	0.1009	0.1215	1.300	160.	0.2	0.3090	0.0999	0.1415
1.109	170.	0.2	0.3072	0.1069	0.0965	1.300	170.	0.2	0.3071	0.1063	0.1183
1.109	180.	0.2	0.3065	0.1107	0.0677	1.300	180.	0.2	0.3065	0.1107	0.0904
1.109	0.	0.4	0.	—0.	—0.	1.300	0.	0.4	0.	—0.	—0.
1.109	10.	0.4	0.2157	0.0005	0.0156	1.300	10.	0.4	0.2156	0.0005	0.0156
1.109	20.	0.4	0.2147	0.0022	0.0306	1.300	20.	0.4	0.2146	0.0022	0.0305
1.109	30.	0.4	0.2130	0.0049	0.0445	1.300	30.	0.4	0.2128	0.0049	0.0442
1.109	40.	0.4	0.2107	0.0087	0.0567	1.300	40.	0.4	0.2104	0.0087	0.0561
1.109	50.	0.4	0.2077	0.0135	0.0669	1.300	50.	0.4	0.2073	0.0135	0.0659
1.109	60.	0.4	0.2042	0.0193	0.0747	1.300	60.	0.4	0.2036	0.0192	0.0734
1.109	70.	0.4	0.2000	0.0260	0.0798	1.300	70.	0.4	0.1993	0.0258	0.0785
1.109	80.	0.4	0.1954	0.0336	0.0820	1.300	80.	0.4	0.1945	0.0332	0.0811
1.109	90.	0.4	0.1902	0.0418	0.0812	1.300	90.	0.4	0.1893	0.0413	0.0812
1.109	100.	0.4	0.1847	0.0506	0.0773	1.300	100.	0.4	0.1837	0.0499	0.0786
1.109	110.	0.4	0.1789	0.0597	0.0701	1.300	110.	0.4	0.1779	0.0587	0.0733
1.109	120.	0.4	0.1729	0.0689	0.0598	1.300	120.	0.4	0.1720	0.0676	0.0652
1.109	130.	0.4	0.1671	0.0776	0.0462	1.300	130.	0.4	0.1663	0.0761	0.0540
1.109	140.	0.4	0.1615	0.0856	0.0294	1.300	140.	0.4	0.1609	0.0839	0.0397
1.109	150.	0.4	0.1565	0.0920	0.0096	1.300	150.	0.4	0.1561	0.0904	0.0222
1.109	160.	0.4	0.1524	0.0964	—0.0132	1.300	160.	0.4	0.1522	0.0950	0.0013
1.109	170.	0.4	0.1497	0.0978	—0.0386	1.300	170.	0.4	0.1496	0.0969	—0.0227
1.109	180.	0.4	0.1486	0.0954	—0.0662	1.300	180.	0.4	0.1486	0.0954	—0.0498
1.109	0.	0.6	0.	—0.	—0.	1.300	0.	0.6	0.	—0.	—0.
1.109	10.	0.6	0.1037	0.0004	0.0069	1.300	10.	0.6	0.1037	0.0004	0.0069
1.109	20.	0.6	0.1028	0.0015	0.0135	1.300	20.	0.6	0.1028	0.0015	0.0134
1.109	30.	0.6	0.1014	0.0033	0.0193	1.300	30.	0.6	0.1012	0.0033	0.0191
1.109	40.	0.6	0.0994	0.0057	0.0241	1.300	40.	0.6	0.0991	0.0057	0.0237
1.109	50.	0.6	0.0969	0.0089	0.0276	1.300	50.	0.6	0.0965	0.0088	0.0270
1.109	60.	0.6	0.0938	0.0126	0.0296	1.300	60.	0.6	0.0933	0.0125	0.0289
1.109	70.	0.6	0.0903	0.0169	0.0299	1.300	70.	0.6	0.0896	0.0167	0.0292
1.109	80.	0.6	0.0863	0.0215	0.0285	1.300	80.	0.6	0.0856	0.0212	0.0280
1.109	90.	0.6	0.0820	0.0265	0.0253	1.300	90.	0.6	0.0812	0.0261	0.0253
1.109	100.	0.6	0.0774	0.0317	0.0203	1.300	100.	0.6	0.0765	0.0310	0.0210
1.109	110.	0.6	0.0726	0.0367	0.0135	1.300	110.	0.6	0.0718	0.0359	0.0152
1.109	120.	0.6	0.0677	0.0415	0.0049	1.300	120.	0.6	0.0670	0.0404	0.0079
1.109	130.	0.6	0.0630	0.0456	—0.0053	1.300	130.	0.6	0.0624	0.0443	—0.0010
1.109	140.	0.6	0.0586	0.0486	—0.0170	1.300	140.	0.6	0.0581	0.0473	—0.0114
1.109	150.	0.6	0.0546	0.0503	—0.0301	1.300	150.	0.6	0.0543	0.0489	—0.0233
1·109	160.	0.6	0.0515	0.0499	—0.0442	1.300	160.	0.6	0.0513	0.0487	—0.0365
1.109	170.	0.6	0.0494	0.0470	—0.0592	1.300	170.	0.6	0.0493	0.0462	—0.0508
1.109	180.	0.6	0.0486	0.0409	—0.0745	1.300	180.	0.6	0.0486	0.0409	—0.0659
1.109	0.	0.8	0.	—0.	—0.	1.300	0.	0.8	0.	—0.	—0.
1.109	10.	0.8	0.0279	0.0001	0.0017	1.300	10.	0.8	0.0279	0.0001	0.0017
1.109	20.	0.8	0.0275	0.0005	0.0033	1.300	20.	0.8	0.0275	0.0005	0.0033
1,109	30.	0.8	0.0269	0.0011	0.0047	1.300	30.	0.8	0.0268	0.0011	0.0046
1.109	40.	0.8	0.0261	0.0019	0.0057	1.300	40.	0.8	0.0260	0.0019	0.0056
1.109	50.	0.8	0.0250	0.0029	0.0062	1.300	50.	0.8	0.0249	0.0029	0.0061
1.109	60.	0.8	0.0237	0.0041	0.0063	1.300	60.	0.8	0.0235	0.0040	0.0061
1.109	70.	0.8	0.0223	0.0054	0.0058	1.300	70.	0.8	0.0220	0.0053	0.0056
1.109	80.	0.8	0.0207	0.0068	0.0048	1.300	80.	0.8	0.0204	0.0067	0.0047
1.109	90.	0.8	0.0190	0.0083	0.0033	1.300	90.	0.8	0.0186	0.0081	0.0033
1.109	100.	0.8	0.0171	0.0097	0.0012	1.300	100.	0.8	0.0168	0.0094	0.0014
1.109	110.	0.8	0.0153	0.0109	—0.0014	1.300	110.	0.8	0.0149	0.0106	—0.0009
1.109	120.	0.8	0.0134	0.0120	—0.0045	1.300	120.	0.8	0.0131	0.0116	—0.0036
1.109	130.	0.8	0.0116	0.0127	—0.0079	1.300	130.	0.8	0.0114	0.0122	—0.0066
1.109	140.	0.8	0.0100	0.0130	—0.0115	1.300	140.	0.8	0.0098	0.0124	—0.0100
1.109	150.	0.8	0.0085	0.0126	—0.0154	1.300	150.	0.8	0.0084	0.0121	—0.0135
1.109	160.	0.8	0.0074	0.0114	—0.0193	1.300	160.	0.8	0.0074	0.0110	—0.0172
1.109	170.	0.8	0.0067	0.0094	—0.0232	1.300	170.	0.8	0.0067	0.0091	—0.0209
1.109	180.	0.8	0.0065	0.0062	—0.0268	1.300	180.	0.8	0.0065	0.0062	—0.0244
1.109	0.	1.0	0.	—0.	—0.	1.300	0.	1.0	0.	—0.	—0.
1.109	10.	1.0	—0.0000	—0.0000	0.0000	1.300	10.	1.0	—0.0000	—0.0000	0.0000
1.109	20.	1.0	—0.0000	0.0000	0.0000	1.300	20.	1.0	—0.0000	0.0000	0.0000
1.109	30.	1.0	—0.0000	—0.0000	0.0000	1.300	30.	1.0	—0.0000	—0.0000	0.0000
1.109	40.	1.0	—0.0000	—0.0000	—0.0000	1.300	40.	1.0	—0.0000	—0.0000	—0.0000
1.109	50.	1.0	0.0000	—0.0000	—0.0000	1.300	50.	1.0	0.0000	—0.0000	—0.0000
1.109	60.	1.0	0.0000	—0.0000	0.0000	1.300	60.	1.0	0.0000	—0.0000	0.0000
1.109	70.	1.0	0.0000	—0.0000	0.0000	1.300	70.	1.0	0.0000	—0.0000	0.0000
1.109	80.	1.0	—0.0000	—0.0000	0.0000	1.300	80.	1.0	—0.0000	—0.0000	0.0000
1.109	90.	1.0	0.0000	0.0000	0.0000	1.300	90.	1.0	0.0000	0.0000	0.0000
1.109	100.	1.0	0.0000	—0.0000	—0.0000	1.300	100.	1.0	0.0000	—0.0000	—0.0000
1.109	110.	1.0	0.0000	0.0000	—0.0000	1.300	110.	1.0	0.0000	0.0000	—0.0000
1.109	120.	1.0	0.0000	0.0000	0.0000	1.300	120.	1.0	0.0000	0.0000	0.0000
1.109	130.	1.0	0.0000	—0.0000	—0.0000	1.300	130.	1.0	0.0000	—0.0000	—0.0000
1.109	140.	1.0	—0.0000	—0.0000	—0.0000	1.300	140.	1.0	—0.0000	—0.0000	—0.0000
1.109	150.	1.0	0.0000	0.0000	—0.0000	1.300	150.	1.0	0.0000	0.0000	—0.0000
1·109	160.	1.0	—0.0000	—0.0000	—0.0000	1.300	160.	1.0	—0.0000	—0.0000	—0.0000
1.109	170.	1.0	0.0000	—0.0000	—0.0000	1.300	170.	1.0	0.0000	—0.0000	—0.0000
1.109	180.	1.0	0.0000	—0.	—0.0000	1.300	180.	1.0	0.0000	—0.	—0.0000

C	$\theta°$	γ/θ	K_{V0}	K_{T0}	K_{M0}	C	$\theta°$	γ/θ	K_{V0}	K_{T0}	K_{M0}
2.493	0.	0.	0.	—0.	—0.	3.250	0.	0.	0.	—0.	—0.
2.493	10.	0.	0.5000	0.0000	0.0434	3.250	10.	0.	0.5000	0.0000	0.0433
2.493	20.	0.	0.5000	—0.	0.0852	3.250	20.	0.	0.5000	—0.	0.0847
2.493	30.	0.	0.5000	—0.	0.1246	3,250	30.	0.	0.5000	—0.	0.1232
2.493	40.	0.	0.5000	0.0000	0.1613	3.250	40.	0.	0.5000	0.0000	0.1590
2.493	50.	0.	0.5000	0.0000	0.1956	3.250	50.	0.	0.5000	0.0000	0.1925
2.493	60.	0.	0.5000	—0.	0.2280	3.250	60.	0.	0.5000	—0.	0.2247
2.493	70.	0.	0.5000	—0.	0.2590	3.250	70.	0.	0.5000	—0.	0.2562
2.493	80.	0.	0.5000	—0.	0.2891	3.250	80.	0.	0.5000	—0.	0.2873
2.493	90.	0.	0.5000	—0.0000	0.3183	3.250	90.	0.	0.5000	—0.0000	0.3183
2.493	100.	0.	0.5000	0.0000	0.3467	3.250	100.	0.	0.5000	0.0000	0.3490
2.493	110.	0.	0.5000	—0.	0.3738	3.250	110.	0.	0.5000	—0.	0.3792
2.493	120.	0.	0.5000	—0.0000	0.3993	3.250	120.	0.	0.5000	—0.0000	0.4082
2.493	130.	0.	0.5000	—0.0000	0.4223	3.250	130.	0.	0.5000	—0.0000	0.4352
2.493	140.	0.	0.5000	—0.0000	0.4417	3.250	140.	0.	0.5000	—0.0000	0.4590
2.493	150.	0.	0.5000	—0.0000	0.4561	3.250	150.	0.	0.5000	—0.0000	0.4779
2.493	160.	0.	0.5000	—0.0000	0.4640	3.250	160.	0.	0.5000	—0.0000	0.4902
2.493	170.	0.	0.5000	—0.0000	0.4638	3.250	170.	0.	0.5000	—0.0000	0.4938
2.493	180.	0.	0.5000	—0.0000	0.4544	3.250	180.	0.	0.5000	—0.0000	0.4868
2.493	0.	0.2	0.	—0.	—0.	3.250	0.	0.2	0.	—0.	—0.
2.493	10.	0.2	0.3517	0.0005	0.0277	3.250	10.	0.2	0.3516	0.0005	0.0276
2.493	20.	0.2	0.3508	0.0019	0.0539	3.250	20.	0.2	0.3505	0.0019	0.0534
2.493	30.	0.2	0.3492	0.0043	0.0779	3.250	30.	0.2	0.3488	0.0043	0.0766
2.493	40.	0.2	0.3472	0.0077	0.0992	3.250	40.	0.2	0.3465	0.0076	0.0970
2.493	50.	0.2	0.3446	0.0118	0.1182	3.250	50.	0.2	0.3437	0.0117	0.1153
2.493	60.	0.2	0.3417	0.0169	0.1353	3.250	60.	0.2	0.3404	0.0166	0.1323
2.493	70.	0.2	0.3384	0.0226	0.1510	3.250	70.	0.2	0.3370	0.0222	0.1484
2.493	80.	0.2	0.3349	0.0291	0.1655	3.250	80.	0.2	0.3333	0.0285	0.1639
2.493	90.	0.2	0.3312	0.0362	0.1791	3.250	90.	0.2	0.3296	0.0353	0.1791
2.493	100.	0.2	0.3274	0.0439	0.1917	3.250	100.	0.2	0.3259	0.0428	0.1938
2.493	110.	0.2	0.3236	0.0521	0.2029	3.250	110.	0.2	0.3223	0.0507	0.2078
2.493	120.	0.2	0.3200	0.0607	0.2124	3.250	120.	0.2	0.3188	0.0590	0.2204
2.493	130.	0.2	0.3165	0.0695	0.2192	3.250	130.	0.2	0.3156	0.0677	0.2310
2.493	140.	0.2	0.3134	0.0785	0.2226	3.250	140.	0.2	0.3127	0.0766	0.2383
2.493	150.	0.2	0.3106	0.0874	0.2212	3.250	150.	0.2	0.3102	0.0856	0.2410
2.493	160.	0.2	0.3084	0.0960	0.2138	3.250	160.	0.2	0.3082	0.0945	0.2376
2.493	170.	0.2	0.3070	0.1039	0.1990	3.250	170.	0.2	0.3069	0.1030	0.2262
2.493	180.	0.2	0.3065	0.1107	0.1759	3.250	180.	0.2	0.3065	0.1107	0.2052
2.493	0.	0.4	0.	—0.	—0.	3.250	0.	0.4	0.	—0.	—0.
2.493	10.	0.4	0.2155	0.0005	0.0155	3.250	10.	0.4	0.2154	0.0005	0.0154
2.493	20.	0.4	0.2141	0.0022	0.0299	3.250	20.	0.4	0.2138	0.0022	0.0295
2.493	30.	0.4	0.2118	0.0049	0.0424	3.250	30.	0.4	0.2111	0.0048	0.0414
2.493	40.	0.4	0.2086	0.0085	0.0528	3.250	40.	0.4	0.2076	0.0084	0.0511
2.493	50.	0.5	0.2047	0.0131	0.0612	3.250	50.	0.4	0.2033	0.0129	0.0590
2.493	60.	0.4	0.2003	0.0185	0.0680	3.250	60.	0.4	0.1984	0.0181	0.0657
2.493	70.	0.4	0.1953	0.0246	0.0734	3.250	70.	0.4	0.1932	0.0240	0.0715
2.493	80.	0.4	0.1900	0.0313	0.0778	3.250	80.	0.4	0.1877	0.0304	0.0766
2.493	90.	0.4	0.1845	0.0386	0.0812	3.250	90.	0.4	0.1821	0.0372	0.0812
2.493	100.	0.4	0.1789	0.0462	0.0835	3.250	100.	0.4	0.1766	0.0444	0.0851
2.493	110.	0.4	0.1733	0.0540	0.0846	3.250	110.	0.4	0.1713	0.0518	0.0882
2.493	120.	0.4	0.1680	0.0618	0.0840	3.250	120.	0.4	0.1662	0.0594	0.0900
2.493	130.	0.4	0.1629	0.0695	0.0813	3.250	130.	0.4	0.1616	0.0669	0.0900
2.493	140.	0.4	0.1584	0.0769	0.0759	3.250	140.	0.4	0.1574	0.0741	0.0875
2.493	150.	0.4	0.1545	0.0836	0.0670	3.250	150.	0.4	0.1539	0.0810	0.0815
2.493	160.	0.4	0.1514	0.0892	0.0539	3.250	160.	0.4	0.1511	0.0871	0.0712
2.493	170.	0.4	0.1494	0.0933	0.0357	3.250	170.	0.4	0.1493	0.0921	0.0554
2.493	180.	0.4	0.1486	0.0954	0.0121	3.250	180.	0.4	0.1486	0.0954	0.0333
2.493	0.	0.6	0.	—0.	—0.	3.250	0.	0.6	0.	—0.	—0.
2.493	10.	0.6	0.1036	0.0004	0.0069	3.250	10.	0.6	0.1035	0.0004	0.0068
2.493	20.	0.6	0.1023	0.0014	0.0130	3.250	20.	0.6	0.1021	0.0014	0.0128
2.493	30.	0.6	0.1003	0.0032	0.0181	3.250	30.	0.6	0.0998	0.0032	0.0175
2.493	40.	0.6	0.0976	0.0056	0.0218	3.250	40.	0.6	0.0967	0.0055	0.0208
2.493	50.	0.6	0.0943	0.0085	0.0243	3.250	50.	0.6	0.0930	0.0083	0.0230
2.493	60.	0.6	0.0904	0.0119	0.0257	3.250	60.	0.6	0.0888	0.0116	0.0244
2.493	70.	0.6	0.0862	0.0156	0.0263	3.250	70.	0.6	0.0844	0.0151	0.0252
2.493	80.	0.6	0.0817	0.0196	0.0261	3.250	80.	0.6	0.0797	0.0188	0.0255
2.493	90.	0.6	0.0771	0.0237	0.0253	3.250	90.	0.6	0.0751	0.0226	0.0253
2.493	100.	0.6	0.0725	0.0279	0.0238	3.250	100.	0.6	0.0706	0.0264	0.0247
2.493	110.	0.6	0.0679	0.0319	0.0214	3.250	110.	0.6	0.0662	0.0301	0.0235
2.493	120.	0.6	0.0636	0.0356	0.0182	3.250	120.	0.6	0.0621	0.0335	0.0215
2.493	130.	0.6	0.0596	0.0389	0.0138	3.250	130.	0.6	0.0584	0.0366	0.0185
2.493	140.	0.6	0.0560	0.0415	0.0081	3.250	140.	0.6	0.0552	0.0393	0.0143
2.493	150.	0.6	0.0530	0.0433	0.0007	3.250	150.	0.6	0.0525	0.0413	0.0084
2.493	160.	0.6	0.0507	0.0441	—0.0086	3.250	160.	0.6	0.0504	0.0424	0.0006
2.493	170.	0.6	0.0492	0.0434	—0.0199	3.250	170.	0.6	0.0491	0.0424	0.0095
2.493	180.	0.6	0.0486	0.0409	—0.0332	3.250	180.	0.6	0.0486	0.0409	—0.0220
2.493	0.	0.8	0.	—0.	—0.	3.250	0.	0.8	0.	—0.	—0.
2.493	10.	0.8	0.0278	0.0001	0.0017	3.250	10.	0.8	0.0278	0.0001	0.0017
2.493	20.	0.8	0.0273	0.0005	0.0032	3.250	20.	0.8	0.0272	0.0005	0.0031
2.493	30.	0.8	0.0265	0.0011	0.0043	3.250	30.	0.8	0.0262	0.0010	0.0041
2.493	40.	0.8	0.0253	0.0018	0.0049	3.250	40.	0.8	0.0249	0.0018	0.0046
2.493	50.	0.8	0.0239	0.0028	0.0052	3.250	50.	0.8	0.0234	0.0027	0.0048
2.493	60.	0.8	0.0223	0.0038	0.0051	3.250	60.	0.8	0.0217	0.0037	0.0047
2.493	70.	0.8	0.0206	0.0049	0.0047	3.250	70.	0.8	0.0199	0.0047	0.0044
2.493	80.	0.8	0.0188	0.0060	0.0041	3.250	80.	0.8	0.0180	0.0057	0.0039
2.493	90.	0.8	0.0170	0.0071	0.0033	3.250	90.	0.8	0.0162	0.0066	0.0033
2.493	100.	0.8	0.0151	0.0081	0.0022	3.250	100.	0.8	0.0144	0.0075	0.0025
2.493	110.	0.8	0.0134	0.0090	0.0009	3.250	110.	0.8	0.0127	0.0083	0.0015
2.493	120.	0.8	0.0118	0.0097	—0.0006	3.250	120.	0.8	0.0112	0.0088	0.0004
2.493	130.	0.8	0.0103	0.0101	—0.0023	3.250	130.	0.8	0.0098	0.0092	—0.0010
2.493	140.	0.8	0.0090	0.0102	—0.0044	3.250	140.	0.8	0.0087	0.0093	—0.0026
2.493	150.	0.8	0.0079	0.0099	—0.0067	3.250	150.	0.8	0.0077	0.0091	—0.0046
2.493	160.	0.8	0.0071	0.0092	—0.0094	3.250	160.	0.8	0.0070	0.0086	—0.0068
2.493	170.	0.8	0.0066	0.0080	—0.0123	3.250	170.	0.8	0.0066	0.0076	—0.0094
2.493	180.	0.8	0.0065	0.0062	—0.0154	3.250	180.	0.8	0.0065	0.0062	—0.0123
2.493	0.	1.0	0.	—0.	—0.	3.250	0.	1.0	0.	—0.	—0.
2.493	10.	1.0	—0.0000	—0.0000	0.0000	3.250	10.	1.0	—0.0000	—0.0000	0.0000
2.493	20.	1.0	—0.0000	0.0000	0.0000	3.250	20.	1.0	—0.0000	0.0000	0.0000
2.493	30.	1.0	—0.0000	—0.0000	0.0000	3.250	30.	1.0	—0.0000	—0.0000	0.0000
2.493	40.	1.0	—0.0000	—0.0000	—0.0000	3.250	40.	1.0	—0.0000	—0.0000	—0.0000
2.493	50.	1.0	0.0000	—0.0000	—0.0000	3.250	50.	1.0	0.0000	—0.0000	—0.0000
2.493	60.	1.0	0.0000	—0.0000	0.0000	3.250	60.	1.0	0.0000	—0.0000	0.0000
2.493	70.	1.0	0.0000	—0.0000	0.0000	3.250	70.	1.0	0.0000	—0.0000	0.0000
2.493	80.	1.0	—0.0000	—0.0000	0.0000	3.250	80.	1.0	—0.0000	—0.0000	0.0000
2.493	90.	1.0	0.0000	0.0000	0.0000	3.250	90.	1.0	0.0000	0.0000	0.0000
2.493	100.	1.0	0.0000	—0.0000	—0.0000	3,250	100.	1.0	0.0000	—0.0000	—0.0000
2.493	110.	1.0	0.0000	0.0000	—0.0000	3.250	110.	1.0	0.0000	0.0000	—0.0000
2.493	120.	1.0	0.0000	0.0000	0.0000	3.250	120.	1.0	0.0000	0.0000	0.0000
2.493	130.	1.0	0.0000	—0.0000	—0.0000	3.250	130.	1.0	0.0000	—0.0000	—0.0000
2.493	140.	1.0	—0.0000	—0.0000	—0.0000	3.250	140.	1.0	—0.0000	—0.0000	—0.0000
2.493	150.	1.0	0.0000	0.0000	—0.0000	3.250	150.	1.0	0.0000	0.0000	—0.0000
2.493	160.	1.0	—0.0000	—0.0000	—0.0000	3.250	160.	1.0	—0.0000	—0.0000	—0.0000
2.493	170.	1.0	0.0000	—0.0000	—0.0000	3.250	170.	1.0	0.0000	—0.0000	—0.0000
2.493	180.	1.0	0.0000	—0.	—0.0000	3.250	180.	1.0	0.0000	—0.	—0.0000

41. CIRCULAR RINGS

BENDING MOMENT, TENSION, SHEAR, HORIZONTAL AND VERTICAL DEFLECTIONS OF CIRCULAR RINGS LOADED AND SUPPORTED IN EIGHTEEN DIFFERENT WAYS

(*Ref. 28*)

NOTATION

M = bending moment (in.-lb.), T = circumferential tension (lb.), V = radial shear (lb.) — at angular distance X from bottom of ring

M_1 = bending moment (in.-lb.), T_1 = circumferential tension (lb.) — at bottom of ring ($X = 0$)

D_x = change in horizontal diameter (in.) (+ means increase) (— means decrease)

D_y = change in vertical diameter (in.) (+ means increase) (— means decrease)

ΔR = change in upper half of vertical diameter (in.) (+ means increase) (— means decrease)

E = modulus of elasticity (psi)

I = moment of inertia of ring cross section (in.^{4})

W = applied total load (lb.)

F = reaction (lb.)

w = applied unit load (lb. per linear inch)

k = weight of contained liquid (lb./in.^{3})

M_o = applied moment (in.-lb.)

R = outside radius of ring (in.)

X = angular distance from bottom of ring

θ = angular location of applied load

K_x = dimensionless load support factor for D_x

K_y = dimensionless load support factor for D_y

K_R = dimensionless load support factor for ΔR

K_T = dimensionless load support factor for T

K_V = dimensionless load support factor for V

K_M = dimensionless load support factor for M

K_S = dimentionless load support factor for radial displacement of load point

K_S^1 = dimentionless load support factor for radial displacement at $x = 0, 2\theta, 4\theta$, etc.

A circular ring is an important structural element used in tanks, aircraft, and large pipelines. The formulas for stresses and deflection produced in such rings are in the form of complicated and involved trigonometric functions. (For Case 16, for example,)

$$\Delta R = \frac{kR^5}{2EI}\left[1 + \cos\theta + \sin\theta\left(\theta + \frac{\sin\theta}{2}\right) - \frac{\pi}{2}\left(\frac{\pi}{8} + \theta\sin\theta + \cos\theta\right)\right].$$

The simplification of these equations resulted in calculation of dimensionless load-support factors. The results are tabulated in Table 23. In the table are given formulas for the bending moments, tension, shears, and horizontal and vertical deflections of rings loaded and supported in 18 different ways. By superposition, these formulas can be combined so as to cover almost any condition of loading and support likely to occur.

The ring formulas are based on the following assumptions: (1) The ring is of uniform cross section; (2) it is of such large radius in comparison with its radial thickness that the deflection theory of straight beam is applicable; (3) its deflection is due solely to bending, the effect of direct axial tension or compression, the effect of shear being negligible; (4) it is nowhere stressed beyond the elastic limit; (5) it is not so severely deformed as to lose its essentially circular shape.

The use of the formulas and the way in which they may be combined by superposition are illustrated in the following example.

Example: Assume a pipe supported at sides, at intervals of 44 ft., by rings. The diameter of the pipe is 10 ft. and the thickness is $\frac{1}{4}$ in. Find the bending moments in the ring. Also, find the bending moment when the pipe is filled with water.

Solution: Based on the density of water 62.4 lb. per cubic foot and the weight of the pipe as 10.2 lb. per square foot of surface area, we obtain the total weight W of 44 ft. of pipe carried by one ring as follows:

Weight of pipe $10.2 \times \pi D \times 44 = 14{,}092$ lb.

Weight of water $\frac{\pi D^2}{4} \times 44 \times 62.4 = 215{,}530$ lb.

Total weight $W = 229{,}622$ lb.

From Table 23, case 17:

$$M = K_M W R$$

From Table 24: For,

$$X = 0, \quad K_M = -0.0113$$

$$X = \frac{\pi}{2}, \quad K_M = 0$$

$$X = \pi, \quad K_M = 0.0113$$

Maximum positive value of $K_M = 0.0145$ at $X = 70$ deg.

Maximum minimum value of $K_M =$
-0.0145 at $X = 115$ deg.

Thus:

(1) At the bottom of the ring: $M = -0.0113 \times 229{,}622 \times 5 \times 12 = -155{,}680$ in.-lb.

(2) At the top of the ring: $M = 155{,}680$ in.-lb.

(3) At $X = 70$ deg.: $+M_{max} = 0.0145 \times 229{,}622 \times 5 \times 12 = 199{,}740$ in.-lb.

(4) At $X = 115$ deg.: $-M_{max} = -199{,}740$ in.-lb.

TABLE 23
CIRCULAR RINGS

CASE NUMBER	FORMULAS FOR: D_x, D_y, ΔR, M, T, V
1. 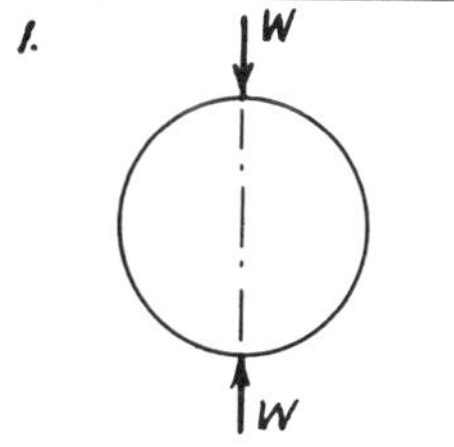	$D_x = 0.137\, WR^3/EI$ $D_y = -0.149\, WR^3/EI$ MAX. $+M = 0.3183\, WR$, AT BOTTOM ($X = 0°$) MAX. $-M = -0.1817\, WR$, AT SIDE ($X = \frac{\pi}{2}$) $T_1 = 0$ $V_1 = -\frac{1}{2} W$
2. 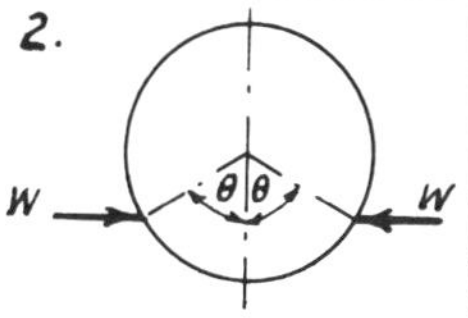	$D_x = K_x WR^3/EI$ $D_y = K_y WR^3/EI$ $\Delta R = K_R WR^3/EI$ $M = K_M WR$ $T = K_T W$ $V = K_V W$
3. 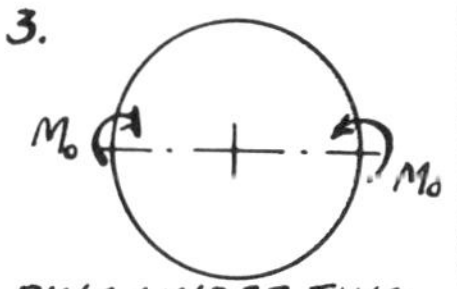 RING UNDER TWO EQUAL OPPOSITE COUPLES.	$D_x = 0$ $D_y = 0$ $\Delta R = 0.0329\, M_0 R^2/EI$ MAX. $+M - \frac{1}{2} M_0 R^2/EI$ MAX. $-M = -\frac{1}{2} M_0$ $T = K_T M_0/R$ $V = -K_V M_0/R$
4. 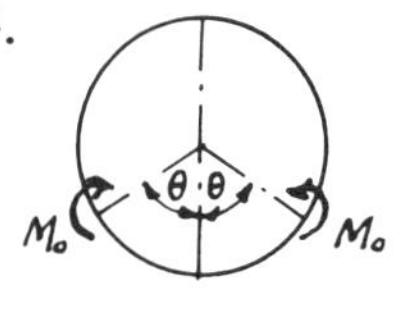	$D_x = K_x M_0 R^2/EI$ $D_y = K_y M_0 R^2/EI$ $\Delta R = K_R M_0 R^2/EI$ $M = K_M M_0$ $T = K_T M_0/R$ $V = K_V M_0/R$
5. 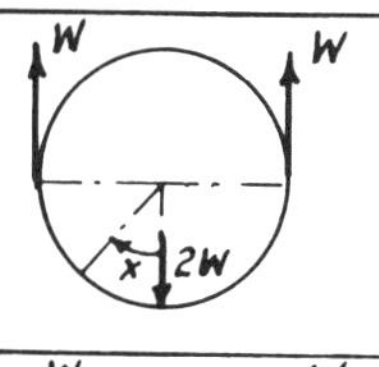	$D_x = -0.1366\, WR^3/EI$ $D_y = 0.1488\, WR^3/EI$ $\Delta R = 0.0554\, WR^3/EI$ $X = 0$ TO $X = \pi/2$: $M = K_M WR$ $T = K_T W$ $V = K_V W$
6. 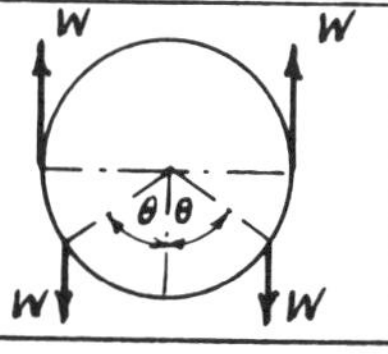	$D_x = K_x WR^3/EI$ $D_y = K_y WR^3/EI$ $\Delta R = K_R WR^3/EI$ $M = K_M WR$ $T = K_T W$ $V = -K_V W$ (AT θ)
7. 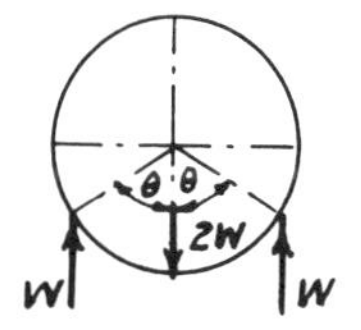	$D_x = K_x WR^3/EI$ $D_y = K_y WR^3/EI$ $\Delta R = K_R WR^3/EI$ AT $\theta°$: $M = K_M WR$ $T = K_T W$ $V = K_V W$
8. 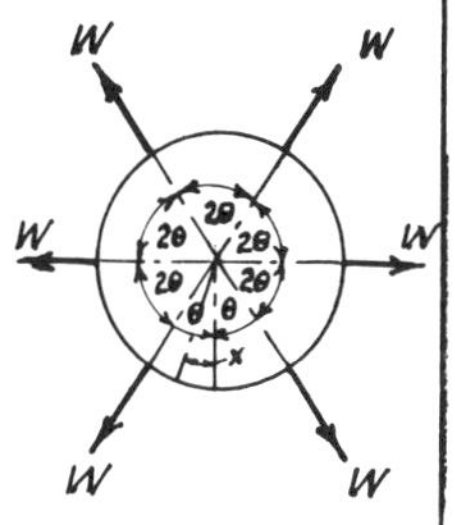	RADIAL DISPLACEMENT OF EACH LOAD POINT: $K_S WR^3/2EI$ (OUTWARD) RADIAL DISPLACEMENT AT $X = 0, 2\theta, 4\theta$ etc.,: $K'_S WR^3/4EI$ (INWARD) AT $X = 0, 2\theta, 4\theta$ etc.,: MAX. $+M = K_M WR/2$ AT EACH LOAD: MAX. $-M = -K'_M WR/2$ AT $X = 0, 2\theta, 4\theta$ etc.,: $T = K_T W/2$ AT EACH LOAD: $T = K'_T W/2$
9. 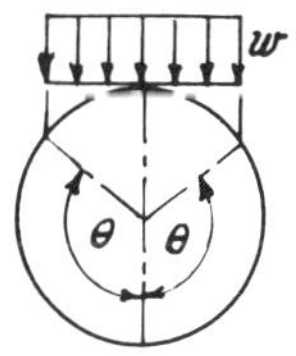	$D_x = K_x 2wR^4/EI$ $D_y = K_y 2wR^4/EI$ $\Delta R = K_R wR^4/EI$ $M_1 = K_M wR^2$, AT BOTTOM $T = -K_T wR$ $V = -K_V wR$ (AT $\theta°$)
10. 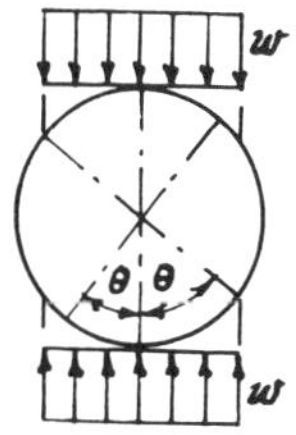	$D_x = -K_x wR^4/EI$ $D_y = -K_y wR^4/EI$ $M_1 = K_M wR^2$, AT BOTTOM AT $\theta°$: $T = -K_T wR$ $V = -K_V wR$
11. 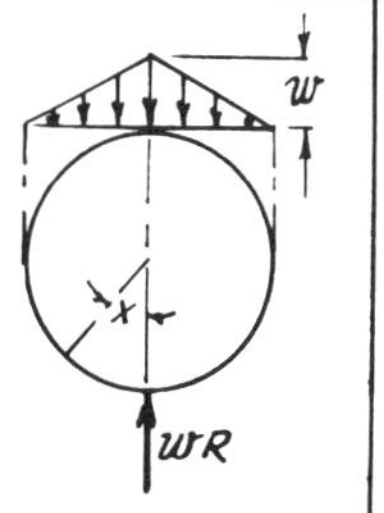	$D_x = 0.1228\, wR^4/EI$ $D_y = -0.1220\, wR^4/EI$ AT BOTTOM ($X = 0$): $M_1 = 0.305\, wR^2$ $T_1 = -0.0265\, wR$ $V_1 = -0.5000\, wR$ AT $X = \frac{\pi}{2}$: $M = -0.165\, wR^2$, $T = -0.500\, wR$ $V = 0.0265\, wR$ AT $X = \pi$: $M = 0.1914\, wR^2$, $T = 0.0265\, wR$, $V = 0$

Case	Loading	Formulas
12.	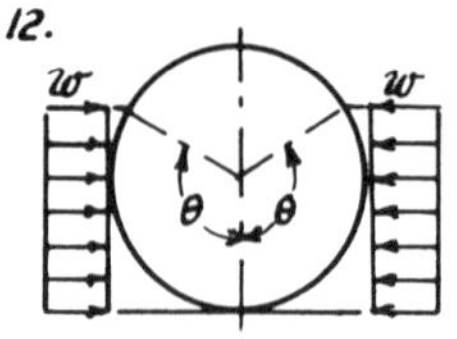	$M_1 = K_M wR^2$, AT BOTTOM, $x=0$ $T = K_T wR$, AT $x=\theta$ $V = K_V wR$, AT $x=\theta$
13.	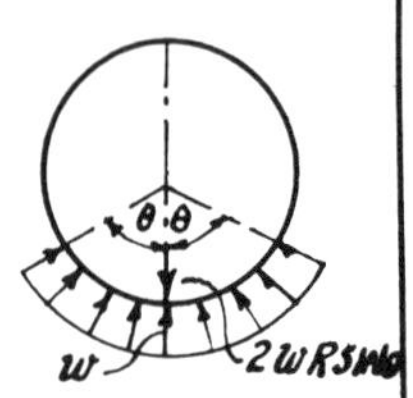	$D_x = K_x 2wR^4/EI$ $\quad D_y = K_y 2wR^4/EI$ $M_1 = K_M wR^2$, AT BOTTOM, $x=0$ $T = K_T wR$, AT $x=\theta$ $V = K_V wR$, AT $x=\theta$
14.	 RING SYMMETRICALLY SUPPORTED AND LOADED BY ITS OWN WEIGHT w (LB. PER LINEAR INCH.	$D_x = K_x 2wR^4/EI$ $\quad D_y = K_y wR^4/EI$ $M_1 = K_M wR^2$, AT BOTTOM, $x=0$ $T = K_T wR$, AT $x=\theta$ $V = K_V wR$, AT $x=\theta$
15.	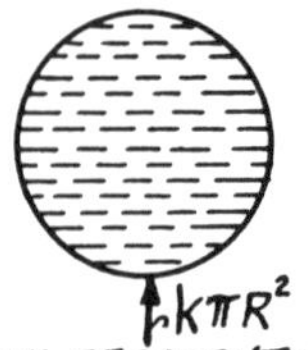 1-INCH SEGMENT OF PIPE FILLED WITH LIQUID OF SPECIFIC WEIGHT k LB. PER CUBIC INCH AND SUPPORTED AT BASE.	$D_x = 0.2146\, kR^5/EI$ $\quad D_y = -0.2337\, kR^5/EI$ $\Delta R = -0.0938\, kR^5/EI$ MAX. $+M = M_1 = 0.750\, kR^3$, AT BOTTOM, $x=0$ MAX. $-M = 0.321\, kR^3$, AT $x=75°$ $M = K_M kR^3$ $T = K_T kR^2$ $V = K_V kR^2$
16.	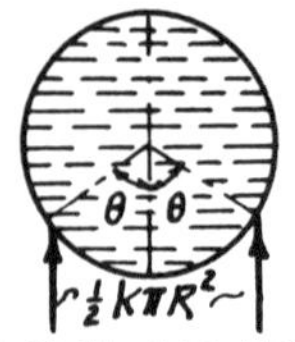 1-INCH SEGMENT OF PIPE FILLED WITH LIQUID OF SPECIFIC WEIGHT k LB. PER CUBIC INCH AND SUPPORTED SYMMETRICALLY ALONG TWO ELEMENTS	$D_x = K_x kR^5/EI$ $\quad D_y = -K_y kR^5/EI$ $\Delta R = K_R kR^5/2EI$ $M_1 = K_M kR^3$, AT BOTTOM, $x=0$ $T = K_T kR^2$, AT $x=\theta$ $V = K_V kR^2$, AT $x=\theta$
17.	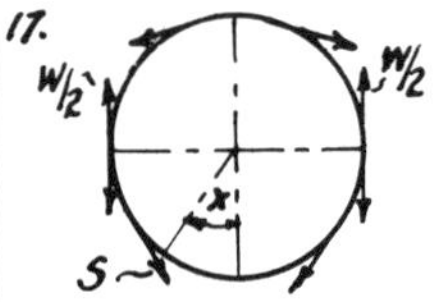 BULKHEAD OR SUPPORTING RING IN PIPE, SUPPORTED AT SIDES AND CARRYING A LOAD W TRANSFERRED BY TANGENTIAL SHEAR OF S LB. PER LINEAR INCH DISTRIBUTED AS SHOWN	$D_x = 0$ $\quad D_y = 0$ $M = K_M WR$ $T = K_T W$ $V = K_V W$ $M_1 = -0.0113\, WR$, AT BOTTOM, $x=0$ MAX. $+M = 0.0146\, WR$, AT $x=66.8°$ MAX. $-M = -0.0146\, WR$, AT $113.2°$ $T_1 = -0.0796\, W$, AT $x=0$ $S = W \sin x/\pi R$
18.	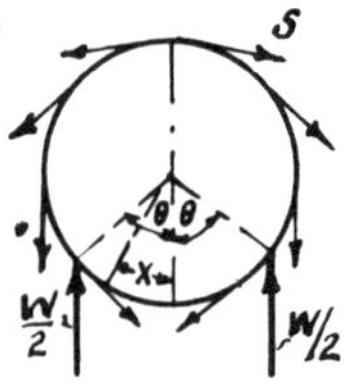 SAME AS CASE 17 EXCEPT SUPPORTED AS SHOWN.	$D_x = K_x WR^3/EI$ $\quad D_y = K_y WR^3/EI$ $\Delta R = K_R WR^3/EI$ AT $x=\theta$: $M = K_M WR$ $T = K_T W$ $V = K_V W$ $S = W \sin x/\pi R$

TABLES 24

CASE 2

θ Degrees	θ Radians	K_x	K_y	K_M	K_T	K_V	K_R
0	0.	-0.	-0.0000	0.	-1.0000	0.	-0.
5	0.0873	-0.0001	0.0001	0.0002	-0.9961	-0.0000	0.0000
10	0.1746	-0.0006	0.0010	0.0017	-0.9837	-0.0002	0.0002
15	0.2619	-0.0021	0.0032	0.0055	-0.9623	-0.0010	0.0008
20	0.3492	-0.0049	0.0071	0.0127	-0.9314	-0.0030	0.0019
25	0.4365	-0.0094	0.0129	0.0241	-0.8908	-0.0072	0.0036
30	0.5238	-0.0157	0.0207	0.0398	-0.8409	-0.0144	0.0060
35	0.6111	-0.0239	0.0303	0.0601	-0.7822	-0.0258	0.0092
40	0.6984	-0.0342	0.0414	0.0846	-0.7157	-0.0422	0.0131
45	0.7857	-0.0461	0.0538	0.1126	-0.6426	-0.0643	0.0179
50	0.8730	-0.0596	0.0668	0.1432	-0.5647	-0.0928	0.0233
55	0.9603	-0.0741	0.0801	0.1751	-0.4838	-0.1279	0.0292
60	1.0476	-0.0892	0.0930	0.2069	-0.4019	-0.1695	0.0356
65	1.1349	-0.1041	0.1051	0.2371	-0.3211	-0.2170	0.0421
70	1.2222	-0.1180	0.1158	0.2643	-0.2436	-0.2696	0.0485
75	1.3095	-0.1303	0.1246	0.2870	-0.1712	-0.3259	0.0546
80	1.3968	-0.1401	0.1312	0.3041	-0.1055	-0.3844	0.0601
85	1.4841	-0.1465	0.1352	0.3147	-0.0481	-0.4433	0.0648

CASE 4

θ Degrees	θ Radians	K_x	K_y	K_R	K_M	K_T	K_V
0	0.	0.	0.0000	0.	-1.0000	0.	0.
5	0.0873	-0.0316	0.0517	0.0118	-0.9169	0.0552	0.0048
10	0.1746	-0.0625	0.0959	0.0235	-0.8355	0.1089	0.0192
15	0.2619	-0.0921	0.1326	0.0348	-0.7574	0.1592	0.0426
20	0.3492	-0.1198	0.1619	0.0454	-0.6841	0.2046	0.0745
25	0.4365	-0.1448	0.1841	0.0552	-0.6171	0.2439	0.1137
30	0.5238	-0.1667	0.1993	0.0640	-0.5575	0.2757	0.1592
35	0.6111	-0.1847	0.2080	0.0715	-0.5063	0.2991	0.2095
40	0.6984	-0.1983	0.2104	0.0777	-0.4642	0.3134	0.2631
45	0.7857	-0.2071	0.2070	0.0823	-0.4316	0.3183	0.3184
50	0.8730	-0.2105	0.1982	0.0852	-0.4086	0.3134	0.3737
55	0.9603	-0.2080	0.1846	0.0863	-0.3953	0.2990	0.4273
60	1.0476	-0.1993	0.1665	0.0853	-0.3910	0.2755	0.4776
65	1.1349	-0.1840	0.1447	0.0823	-0.3951	0.2436	0.5231
70	1.2222	-0.1618	0.1196	0.0770	-0.4066	0.2043	0.5623
75	1.3095	-0.1324	0.0919	0.0695	-0.4243	0.1588	0.5941
80	1.3968	-0.0956	0.0623	0.0596	-0.4468	0.1085	0.6175
85	1.4841	-0.-514	0.0313	0.0474	-0.4726	0.0549	0.6318

CASE 5

x Degrees	x Radians	K_M	K_T	K_V
0	0.	-0.5000	0.3183	1.0000
5	0.0873	-0.4140	0.4042	0.9684
10	0.1746	-0.3311	0.4871	0.9295
15	0.2619	-0.2519	0.5663	0.8834
20	0.3492	-0.1770	0.6412	0.8307
25	0.4365	-0.1070	0.7112	0.7716
30	0.5238	-0.0425	0.7757	0.7067
35	0.6111	0.0161	0.8344	0.6363
40	0.6984	0.0684	0.8867	0.5612
45	0.7857	0.1140	0.9323	0.4817
50	0.8730	0.1524	0.9707	0.3986
55	0.9603	0.1835	1.0018	0.3124
60	1.0476	0.2069	1.0252	0.2239
65	1.1349	0.2225	1.0408	0.1336
70	1.2222	0.2302	1.0485	0.0424
75	1.3095	0.2299	1.0482	-0.0491
80	1.3968	0.2217	1.0400	-0.1403
85	1.4841	0.2055	1.0238	-0.2305
90	1.5708	0.1816	-0.0000	0.3183
95	1.6581	0.1539	-0.0277	0.3170
100	1.7454	0.1264	-0.0552	0.3134
105	1.8327	0.0992	-0.0824	0.3074
110	1.9200	0.0727	-0.1089	0.2990
115	2.0073	0.0471	-0.1345	0.2884
120	2.0946	0.0224	0.1592	0.2756
125	2.1819	-0.0009	-0.1826	0.2606
130	2.2692	-0.0229	-0.2046	0.2437
135	2.3565	-0.0434	-0.2251	0.2250
140	2.4438	-0.0622	-0.2439	0.2045
145	2.5311	-0.0791	-0.2608	0.1824
150	2.6184	-0.0940	-0.2757	0.1590
155	2.7057	-0.1068	-0.2885	0.1343
160	2.7930	-0.1174	-0.2991	0.1087
165	2.8803	-0.1257	-0.3074	0.0822
170	2.9676	-0.1317	-0.3134	0.0551
175	3.0549	-0.1354	-0.3171	0.0275

CASE 6

θ Degrees	θ Radians	K_x	K_y	K_R	K_T	K_V	K_M
0	0.	-0.1366	0.1488	0.0555	0.3183	0.	-0.5000
5	0.0873	-0.1352	0.1465	0.0549	0.3147	0.0275	-0.4176
10	0.1746	-0.1311	0.1400	0.0534	0.3040	0.0536	-0.3454
15	0.2619	-0.1245	0.1303	0.0509	0.2868	0.0769	-0.2833
20	0.3492	-0.1157	0.1180	0.0476	0.2641	0.0962	-0.2309
25	0.4365	-0.1050	0.1040	0.0435	0.2369	0.1105	-0.1875
30	0.5238	-0.0929	0.0891	0.0389	0.2067	0.1194	-0.1522
35	0.6111	-0.0800	0.0741	0.0339	0.1749	0.1225	-0.1237
40	0.6984	-0.0667	0.0595	0.0287	0.1430	0.1200	-0.1007
45	0.7857	-0.0536	0.0461	0.0235	0.1124	0.1125	-0.0821
50	0.8730	-0.0413	0.0341	0.0186	0.0844	0.1007	-0.0667
55	0.9603	-0.0302	0.0239	0.0140	0.0600	0.0857	-0.0536
60	1.0476	-0.0206	0.0157	0.0100	0.0397	0.0688	-0.0420
65	1.1349	-0.0128	0.0094	0.0066	0.0240	0.0514	-0.0314
70	1.2222	-0.0070	0.0049	0.0040	0.0127	0.0349	-0.0218
75	1.3095	-0.0032	0.0021	0.0020	0.0055	0.0205	-0.0133
80	1.3968	-0.0010	0.0007	0.0008	0.0017	0.0094	-0.0064
85	1.4841	-0.0001	0.0001	0.0002	0.0002	0.0024	-0.0017

CASE 7

Θ Degrees	Θ Radians	K_x	K_y	K_R	K_M	K_T	K_V
0	0.	-0.0000	0.0000	0.4292	-0.0000	0.	1.0000
5	0.0873	-0.0014	0.0023	0.3381	0.0036	0.0896	0.9960
10	0.1746	-0.0055	0.0088	0.2635	0.0143	0.1832	0.9831
15	0.2619	-0.0121	0.0185	0.2043	0.0313	0.2795	0.9604
20	0.3492	-0.0209	0.0308	0.1590	0.0538	0.3772	0.9269
25	0.4365	-0.0316	0.0448	0.1261	0.0805	0.4743	0.8822
30	0.5238	-0.0437	0.0597	0.1037	0.1097	0.5691	0.8261
35	0.6111	-0.0566	0.0747	0.0901	0.1398	0.6596	0.7589
40	0.6984	-0.0699	0.0893	0.0834	0.1692	0.7438	0.6813
45	0.7857	-0.0830	0.1027	0.0819	0.1962	0.8199	0.5943
50	0.8730	-0.0953	0.1147	0.0839	0.2192	0.8863	0.4993
55	0.9603	-0.1064	0.1249	0.0882	0.2371	0.9419	0.3982
60	1.0476	-0.1160	0.1331	0.0934	0.2489	0.9856	0.2928
65	1.1349	-0.1238	0.1394	0.0987	0.2540	1.0169	0.1851
70	1.2222	-0.1296	0.1439	0.1033	0.2521	1.0359	0.0773
75	1.3095	-0.1334	0.1467	0.1069	0.2433	1.0428	-0.0286
80	1.3968	-0.1356	0.1481	0.1093	0.2281	1.0384	-0.1310
85	1.4841	-0.1365	0.1487	0.1105	0.2072	1.0236	-0.2281

CASE 8

Θ Degrees	D Radians	K_S	K'_S	K_M	K'_M	K_T	K'_T
0	0.	0.	0.	0.	0.	0.	0.
5	0.0873	-11.3677	11.3532	0.0146	0.0291	11.4693	11.4256
10	0.1746	-5.5545	5.5262	0.0292	0.0583	5.7566	5.6691
15	0.2619	-3.5623	3.5213	0.0440	0.0877	3.8623	3.7305
20	0.3492	-2.5283	2.4765	0.0590	0.1174	2.9227	2.7463
25	0.4365	-1.8811	1.8210	0.0744	0.1474	2.3654	2.1436
30	0.5238	-1.4307	1.3654	0.0902	0.1779	1.9993	1.7312
35	0.6111	-1.0959	1.0294	0.1065	0.2090	1.7429	1.4274
40	0.6984	-0.8364	0.7736	0.1234	0.2407	1.5552	1.1911
45	0.7857	-0.6299	0.5763	0.1410	0.2734	1.4138	0.9994
50	0.8730	-0.4628	0.4250	0.1596	0.3069	1.3050	0.8385
55	0.9603	-0.3263	0.3117	0.1791	0.3417	1.2205	0.6997
60	1.0476	-0.2144	0.2313	0.1999	0.3777	1.1544	0.5768
65	1.1349	-0.1223	0.1799	0.2220	0.4154	1.1032	0.4658
70	1.2222	-0.0466	0.1549	0.2458	0.4548	1.0640	0.3634
75	1.3095	0.0159	0.1539	0.2715	0.4962	1.0351	0.2674
80	1.3968	0.0677	0.1747	0.2994	0.5401	1.0153	0.1758
85	1.4841	0.1114	0.2153	0.3300	0.5869	1.0038	0.0869
90	1.5714	0.1490	0.2737	0.3636	0.6370	1.0000	-0.0006
95	1.6587	0.1827	0.3475	0.4010	0.6910	1.0039	-0.0881
100	1.7460	0.2144	0.4343	0.4428	0.7498	1.0155	-0.1770
105	1.8333	0.2459	0.5312	0.4900	0.8142	1.0355	-0.2687
110	1.9206	0.2786	0.6351	0.5438	0.8855	1.0645	-0.3648
115	2.0079	0.3142	0.7423	0.6057	0.9653	1.1038	-0.4672
120	2.0952	0.3536	0.8484	0.6780	1.0557	1.1552	-0.5784
125	2.1825	0.3980	0.9482	0.7633	1.1596	1.2215	-0.7015
130	2.2698	0.4481	1.0353	0.8658	1.2812	1.3064	-0.8406
135	2.3571	0.5043	1.1012	0.9912	1.4261	1.4155	-1.0018
140	2.4444	0.5670	1.1347	1.1484	1.6031	1.5575	-1.1940
145	2.5317	0.6361	1.1194	1.3509	1.8261	1.7459	-1.4311
150	2.6190	0.7114	1.0296	1.6217	2.1179	2.0035	-1.7361
155	2.7063	0.7925	0.8215	2.0020	2.5198	2.3715	-2.1503
160	2.7936	0.8786	0.4096	2.5745	3.1146	2.9324	-2.7567
165	2.8809	0.9688	-0.4019	3.5326	4.0957	3.8797	-3.7486
170	2.9682	1.0622	-2.1987	5.4593	6.0462	5.7962	-5.7093
175	3.0555	1.1576	-7.9309	11.3024	11.9139	1.6297	11.5866

CASE 9

Θ Degrees	Θ Radians	K_x	K_y	K_R	K_M	K_T	K_V
90	1.5708	0.1100	-0.1161	-0.1041	0.5872	1.0000	0.1061
95	1.6581	0.1097	-0.1158	-0.1039	0.5853	0.9833	0.1914
100	1.7454	0.1089	-0.1149	-0.1034	0.5796	0.9522	0.2709
105	1.8327	0.1076	-0.1135	-0.1024	0.5701	0.9082	0.3424
110	1.9200	0.1058	-0.1115	-0.1009	0.5567	0.8528	0.4042
115	2.0073	0.1033	-0.1089	-0.0991	0.5395	0.7879	0.4547
120	2.0946	0.1002	-0.1057	-0.0967	0.5184	0.7154	0.4928
125	2.1819	0.0964	-0.1017	-0.0937	0.4934	0.6373	0.5177
130	2.2692	0.0918	-0.0970	-0.0901	0.4645	0.5559	0.5290
135	2.3565	0.0864	-0.0914	-0.0857	0.4318	0.4732	0.5265
140	2.4438	0.0801	-0.0849	-0.0804	0.3953	0.3913	0.5104
145	2.5311	0.0728	-0.0774	-0.0741	0.3551	0.3123	0.4812
150	2.6184	0.0646	-0.0690	-0.0667	0.3116	0.2382	0.4394
155	2.7057	0.0554	-0.0595	-0.0581	0.2648	0.1710	0.3861
160	2.7930	0.0455	-0.0490	-0.0483	0.2154	0.1127	0.3225
165	2.8803	0.0348	-0.0376	-0.0374	0.1636	0.0650	0.2500
170	2.9676	0.0235	-0.0255	-0.0255	0.1099	0.0294	0.1706
175	3.0549	0.0118	-0.0128	-0.0129	0.0551	0.0074	0.0863

CASE 10

Θ Degrees	Θ Radians	K_x	K_y	K_M	K_T	K_V
0	0.	0.	-0.0000	0.	0.	0.
5	0.0873	-0.0237	0.0258	0.0518	0.0076	0.0869
10	0.1746	-0.0468	0.0507	0.0961	0.0302	0.1711
15	0.2619	-0.0686	0.0737	0.1332	0.0670	0.2501
20	0.3492	-0.0887	0.0944	0.1636	0.1171	0.3215
25	0.4365	-0.1065	0.1123	0.1879	0.1787	0.3831
30	0.5238	-0.1218	0.1273	0.2068	0.2502	0.4331
35	0.6111	-0.1346	0.1393	0.2211	0.3292	0.4699
40	0.6984	-0.1448	0.1486	0.2315	0.4134	0.4925
45	0.7857	-0.1525	0.1554	0.2387	0.5003	0.5000
50	0.8730	-0.1581	0.1601	0.2436	0.5872	0.4923
55	0.9603	-0.1619	0.1632	0.2466	0.6714	0.4697
60	1.0476	-0.1643	0.1650	0.2484	0.7503	0.4328
65	1.1349	-0.1656	0.1660	0.2493	0.8217	0.3827
70	1.2222	-0.1663	0.1665	0.2498	0.8833	0.3210
75	1.3095	-0.1665	0.1667	0.2499	0.9333	0.2496
80	1.3968	-0.1666	0.1667	0.2500	0.9700	0.1705
85	1.4841	-0.1666	0.1667	0.2500	0.9925	0.0863

CASE 12

Θ Degrees	Θ Radians	K_M	K_T	K_V
0	0.	0.	0.	0.
5	0.0873	-0.0000	0.0000	0.0000
10	0.1746	-0.0001	0.0000	0.0000
15	0.2619	-0.0005	0.0000	0.0000
20	0.3492	-0.0015	0.0002	0.0001
25	0.4365	-0.0034	0.0006	0.0003
30	0.5238	-0.0066	0.0013	0.0008
35	0.6111	-0.0114	0.0027	0.0019
40	0.6984	-0.0180	0.0048	0.0040
45	0.7857	-0.0265	0.0076	0.0076
50	0.8730	-0.0370	0.0113	0.0135
55	0.9603	-0.0493	0.0156	0.0223
60	1.0476	-0.0634	0.0201	0.0348
65	1.1349	-0.0788	0.0241	0.0517
70	1.2222	-0.0953	0.0267	0.0735
75	1.3095	-0.1123	0.0269	0.1006
80	1.3968	-0.1296	0.0234	0.1331
85	1.4841	-0.1466	0.0148	0.1706
90	1.5714	-0.1629	-0.0001	0.2125
95	1.6587	-0.1782	-0.0227	0.2575
100	1.7460	-0.1921	-0.0539	0.3043
105	1.8333	-0.2046	-0.0943	0.3508
110	1.9206	-0.2153	-0.1441	0.3951
115	2.0079	-0.2243	-0.2032	0.4349
120	2.0952	-0.2316	-0.2707	0.4680
125	2.1825	-0.2374	-0.3453	0.4923
130	2.2698	-0.2418	-0.4253	0.5059
135	2.3571	-0.2449	-0.5085	0.5076
140	2.4444	-0.2471	-0.5924	0.4962
145	2.5317	-0.2485	-0.6746	0.4714
150	2.6190	-0.2493	-0.7522	0.4333
155	2.7063	-0.2497	-0.8227	0.3826
160	2.7936	-0.2499	-0.8839	0.3206
165	2.8809	-0.2500	-0.9336	0.2490
170	2.9682	-0.2501	-0.9702	0.1699
175	3.0555	-0.2501	-0.9926	0.0857

CASE 13

Θ Degrees	Θ Radians	K_x	K_y	K_M	K_T	K_V
0	0.	0.	0.0000	-0.0000	-0.0000	0.
5	0.0873	-0.0000	0.0000	-0.0037	-0.0037	-0.0003
10	0.1746	-0.0002	0.0003	-0.0144	-0.0146	-0.0026
15	0.2619	-0.0005	0.0009	-0.0313	-0.0322	-0.0088
20	0.3492	-0.0013	0.0019	-0.0536	-0.0559	-0.0206
25	0.4365	-0.0024	0.0036	-0.0807	-0.0851	-0.0396
30	0.5238	-0.0041	0.0060	-0.1117	-0.1192	-0.0671
35	0.6111	-0.0064	0.0090	-0.1458	-0.1577	-0.1038
40	0.6984	-0.0094	0.0128	-0.1821	-0.1997	-0.1505
45	0.7857	-0.0130	0.0173	-0.2198	-0.2448	-0.2073
50	0.8730	-0.0174	0.0225	-0.2581	-0.2921	-0.2739
55	0.9603	-0.0224	0.0282	-0.2963	-0.3412	-0.3496
60	1.0476	-0.0280	0.0344	-0.3335	-0.3912	-0.4334
65	1.1349	-0.0341	0.0410	-0.3691	-0.4418	-0.5238
70	1.2222	-0.0407	0.0478	-0.4023	-0.4921	-0.6188
75	1.3095	-0.0476	0.0547	-0.4325	-0.5418	-0.7165
80	1.3968	-0.0547	0.0616	-0.4592	-0.5904	-0.8144
85	1.4841	-0.0617	0.0682	-0.4819	-0.6372	-0.9100

CASE 14

θ Degrees	θ Radians	K_x	K_y	K_R	K_M	K_T	K_V
0	0.	0.2146	-0.4674	-0.1876	1.5000	-0.5000	0.
5	0.0873	0.2124	-0.4601	-0.1860	1.2375	-0.4829	0.1299
10	0.1746	0.2060	-0.4399	-0.1812	0.9996	-0.4323	0.2536
15	0.2619	0.1957	-0.4092	-0.1734	0.7873	-0.3504	0.3651
20	0.3492	0.1818	-0.3705	-0.1629	0.6013	-0.2403	0.4591
25	0.4365	0.1650	-0.3265	-0.1501	0.4413	-0.1066	0.5314
30	0.5238	0.1460	-0.2798	-0.1356	0.3067	0.0457	0.5785
35	0.6111	0.1257	-0.2326	-0.1199	0.1963	0.2108	0.5985
40	0.6984	0.1048	-0.1870	-0.1036	0.1084	0.3828	0.5905
45	0.7857	0.0843	-0.1447	-0.0874	0.0408	0.5560	0.5552
50	0.8730	0.0649	-0.1070	-0.0719	-0.0087	0.7249	0.4941
55	0.9603	0.0474	-0.0750	-0.0575	-0.0426	0.8851	0.4101
60	1.0476	0.0324	-0.0491	-0.0449	-0.0639	1.0325	0.3066
65	1.1349	0.0202	-0.0293	-0.0342	-0.0751	1.1646	0.1875
70	1.2222	0.0111	-0.0154	-0.0258	-0.0791	1.2796	0.0572
75	1.3095	0.0050	-0.0066	-0.0198	-0.0783	1.3770	-0.0803
80	1.3968	0.0016	-0.0020	-0.0159	-0.0753	1.4571	-0.2211
85	1.4841	0.0002	-0.0002	-0.0140	-0.0722	1.5212	-0.3621
90	1.5714	-0.0000	-0.0000	-0.0135	-0.0708	1.5711	-0.5009
95	1.6587	-0.0002	0.0003	-0.0137	-0.0727	1.6091	-0.6360
100	1.7460	-0.0020	0.0020	-0.0139	-0.0789	1.6374	-0.7668
105	1.8333	-0.0069	0.0067	-0.0130	-0.0903	1.6582	-0.8936
110	1.9206	-0.0170	0.0156	-0.0098	-0.1072	1.6732	-1.0176
115	2.0079	-0.0343	0.0296	-0.0032	-0.1296	1.6833	-1.1406
120	2.0952	-0.0609	0.0494	0.0084	-0.1572	1.6888	-1.2649
125	2.1825	-0.0993	0.0754	0.0262	-0.1892	1.6890	-1.3927
130	2.2698	-0.1516	0.1075	0.0517	-0.2248	1.6822	-1.5263
135	2.3571	-0.2199	0.1452	0.0862	-0.2629	1.6659	-1.6676
140	2.4444	-0.3064	0.1876	0.1310	-0.3020	1.6367	-1.8177
145	2.5317	-0.4127	0.2332	0.1874	-0.3410	1.5910	-1.9768
150	2.6190	-0.5404	0.2805	0.2562	-0.3782	1.5246	-2.1442
155	2.7063	-0.6906	0.3272	0.3382	-0.4125	1.4333	-2.3181
160	2.7936	-0.8641	0.3711	0.4341	-0.4424	1.3134	-2.4953
165	2.8809	-1.0612	0.4097	0.5440	-0.4670	1.1615	-2.6718
170	2.9682	-1.2817	0.4403	0.6680	-0.4852	0.9753	-2.8426
175	3.0555	-1.5248	0.4603	0.8056	-0.4963	0.7535	-3.0018

CASE 15

x Degrees	x Radians	K_M	K_T	K_V
0	0.	0.	1.2500	-1.5708
5	0.0873	0.6159	1.1159	-1.4995
10	0.1746	0.4885	0.9885	-1.4175
15	0.2619	0.3687	0.8687	-1.3260
20	0.3492	0.2572	0.7572	-1.2264
25	0.4365	0.1547	0.6547	-1.1200
30	0.5238	0.0618	0.5618	-1.0084
35	0.6111	-0.0212	0.4788	-0.8928
40	0.6984	-0.0940	0.4060	-0.7748
45	0.7857	-0.1565	0.3435	-0.6559
50	0.8730	-0.2085	0.2915	-0.5373
55	0.9603	-0.2503	0.2497	-0.4204
60	1.0476	-0.2820	0.2180	-0.3066
65	1.1349	-0.3040	0.1960	-0.1970
70	1.2222	-0.3166	0.1834	-0.0929
75	1.3095	-0.3204	0.1796	0.0049
80	1.3968	-0.3159	0.1841	0.0952
85	1.4841	-0.3040	0.1960	0.1773
90	1.5714	-0.2853	0.2147	0.2505
95	1.6587	-0.2605	0.2395	0.3141
100	1.7460	-0.2307	0.2693	0.3678
105	1.8333	-0.1966	0.3034	0.4112
110	1.9206	-0.1592	0.3408	0.4441
115	2.0079	-0.1194	0.3806	0.4665
120	2.0952	-0.0781	0.4219	0.4784
125	2.1825	-0.0362	0.4638	0.4801
130	2.2698	0.0055	0.5055	0.4719
135	2.3571	0.0460	0.5460	0.4542
140	2.4444	0.0845	0.5845	0.4278
145	2.5317	0.1204	0.6204	0.3932
150	2.6190	0.1529	0.6529	0.3512
155	2.7063	0.1815	0.6815	0.3028
160	2.7936	0.2056	0.7056	0.2488
165	2.8809	0.2248	0.7248	0.1904
170	2.9682	0.2388	0.7388	0.1285
175	3.0555	0.2472	0.7472	0.0644

CASE 16

Θ Degrees	Θ Radians	K_x	K_y	K_R	K_M	K_T	K_V
0	0.	0.2146	0.2337	-0.1877	0.7500	1.2500	0.
5	0.0873	0.2124	0.2301	-0.1860	0.6187	1.2566	0.0649
10	0.1746	0.2060	0.2200	-0.1812	0.4998	1.2762	0.1268
15	0.2619	0.1957	0.2046	-0.1734	0.3937	1.3078	0.1825
20	0.3492	0.1818	0.1853	-0.1629	0.3007	1.3496	0.2296
25	0.4365	0.1650	0.1633	-0.1501	0.2207	1.3998	0.2657
30	0.5238	0.1460	0.1399	-0.1356	0.1534	1.4558	0.2893
35	0.6111	0.1257	0.1163	-0.1199	0.0982	1.5149	0.2992
40	0.6984	0.1048	0.0935	-0.1036	0.0542	1.5743	0.2953
45	0.7857	0.0843	0.0723	-0.0874	0.0204	1.6314	0.2776
50	0.8730	0.0649	0.0535	-0.0719	-0.0043	1.6837	0.2471
55	0.9603	0.0474	0.0375	-0.0575	-0.0213	1.7292	0.2051
60	1.0476	0.0324	0.0245	-0.0449	-0.0319	1.7661	0.1533
65	1.1349	0.0202	0.0147	-0.0342	-0.0376	1.7934	0.0938
70	1.2222	0.0111	0.0077	-0.0258	-0.0395	1.8106	0.0286
75	1.3095	0.0050	0.0033	-0.0198	-0.0392	1.8177	-0.0401
80	1.3968	0.0016	0.0010	-0.0159	-0.0377	1.8151	-0.1106
85	1.4841	0.0002	0.0001	-0.0140	-0.0361	1.8039	-0.1811
90	1.5714	-0.0000	0.0000	-0.0135	-0.0354	1.7852	-0.2505
95	1.6587	-0.0002	-0.0001	-0.0137	-0.0363	1.7606	-0.3180
100	1.7460	-0.0020	-0.0010	-0.0139	-0.0395	1.7316	-0.3834
105	1.8333	-0.0069	-0.0034	-0.0130	-0.0452	1.6994	-0.4468
110	1.9206	-0.0170	-0.0078	-0.0098	-0.0536	1.6652	-0.5088
115	2.0079	-0.0343	-0.0148	-0.0032	-0.0648	1.6300	-0.5703
120	2.0952	-0.0609	-0.0247	0.0084	-0.0786	1.5941	-0.6324
125	2.1825	-0.0993	-0.0377	0.0262	-0.0946	1.5574	-0.6963
130	2.2698	-0.1516	-0.0538	0.0517	-0.1124	1.5194	-0.7632
135	2.3571	-0.2199	-0.0726	0.0862	-0.1314	1.4791	-0.8338
140	2.4444	-0.3064	-0.0938	0.1310	-0.1510	1.4350	-0.9088
145	2.5317	-0.4127	-0.1166	0.1874	-0.1705	1.3857	-0.9884
150	2.6190	-0.5404	-0.1402	0.2562	-0.1891	1.3290	-1.0721
155	2.7063	-0.6906	-0.1636	0.3382	-0.2062	1.2633	-1.1590
160	2.7936	-0.8641	-0.1855	0.4341	-0.2212	1.1867	-1.2476
165	2.8809	-1.0612	-0.2048	0.5440	-0.2335	1.0976	-1.3359
170	2.9682	-1.2817	-0.2201	0.6680	-0.2426	0.9951	-1.4213
175	3.0555	-1.5248	-0.2302	0.8056	-0.2482	0.8786	-1.5009
180	3.1428	-1.7892	-0.2337	0.9561	-0.2500	0.7481	-1.5717

CASE 17

x Degrees	x Radians	K_T	K_M	K_V
0	0.	-0.0113	-0.0796	0.
5	0.0873	-0.0110	-0.0781	0.0069
10	0.1746	-0.0101	-0.0735	0.0135
15	0.2619	-0.0087	-0.0661	0.0197
20	0.3492	-0.0067	-0.0558	0.0250
25	0.4365	-0.0043	-0.0427	0.0293
30	0.5238	-0.0016	-0.0272	0.0324
35	0.6111	0.0013	-0.0094	0.0340
40	0.6984	0.0043	0.0105	0.0340
45	0.7857	0.0072	0.0322	0.0321
50	0.8730	0.0098	0.0553	0.0283
55	0.9603	0.0121	0.0796	0.0224
60	1.0476	0.0137	0.1047	0.0144
65	1.1349	0.0145	0.1301	0.0041
70	1.2222	0.0143	0.1556	-0.0084
75	1.3095	0.0130	0.1808	-0.0230
80	1.3968	0.0103	0.2052	-0.0399
85	1.4841	0.0060	0.2284	-0.0588
90	1.5708	-0.0000	-0.2500	-0.0796
95	1.6581	-0.0060	-0.2283	-0.0587
100	1.7454	-0.0103	-0.2050	-0.0398
105	1.8327	-0.0130	-0.1806	-0.0229
110	1.9200	-0.0144	-0.1555	-0.0083
115	2.0073	-0.0145	-0.1300	0.0042
120	2.0946	-0.0137	-0.1045	0.0144
125	2.1819	-0.0121	-0.0794	0.0225
130	2.2692	-0.0098	-0.0552	0.0283
135	2.3565	-0.0072	-0.0320	0.0321
140	2.4438	-0.0043	-0.0104	0.0340
145	2.5311	-0.0013	0.0095	0.0340
150	2.6184	0.0016	0.0273	0.0324
155	2.7057	0.0043	0.0428	0.0293
160	2.7930	0.0067	0.0558	0.0250
165	2.8803	0.0087	0.0661	0.0196
170	2.9676	0.0101	0.0736	0.0135
175	3.0549	0.0110	0.0781	0.0069
180	3.1422	0.0113	0.0796	-0.0000

CASE 18

Θ Degrees	Θ Radians	K_x	K_y	K_R	K_M	K_T	K_V
0	0.	0.0683	-0.0744	-0.0299	0.2387	-0.2387	0.
5	0.0873	0.0676	-0.0732	-0.0296	0.1978	-0.2354	0.0207
10	0.1746	0.0656	-0.0700	-0.0288	0.1626	-0.2255	0.0404
15	0.2619	0.0623	-0.0651	-0.0276	0.1330	-0.2095	0.0581
20	0.3492	0.0579	-0.0590	-0.0259	0.1087	-0.1878	0.0731
25	0.4365	0.0525	-0.0520	-0.0239	0.0894	-0.1612	0.0846
30	0.5238	0.0465	-0.0445	-0.0216	0.0745	-0.1305	0.0921
35	0.6111	0.0400	-0.0370	-0.0191	0.0631	-0.0968	0.0953
40	0.6984	0.0334	-0.0298	-0.0165	0.0546	-0.0610	0.0940
45	0.7857	0.0268	-0.0230	-0.0139	0.0482	-0.0240	0.0884
50	0.8730	0.0207	-0.0171	-0.0114	0.0432	0.0131	0.0786
55	0.9603	0.0151	-0.0120	-0.0092	0.0388	0.0496	0.0653
60	1.0476	0.0103	-0.0078	-0.0071	0.0347	0.0848	0.0488
65	1.1349	0.0064	-0.0047	-0.0055	0.0302	0.1182	0.0298
70	1.2222	0.0035	-0.0025	-0.0041	0.0252	0.1493	0.0091
75	1.3095	0.0016	-0.0011	-0.0032	0.0197	0.1780	-0.0128
80	1.3968	0.0005	-0.0003	-0.0025	0.0135	0.2043	-0.0352
85	1.4841	0.0001	-0.0001	-0.0022	0.0068	0.2283	-0.0576

Chap. II

REFERENCES

Section	*Ref. No.*	
16	1	GRIFFEL, W.: "Maximum Deflection and Slope," *Product Engineering*, Dec. 25, 1961
17	2	GRIFFEL, W.: "Beam Deflection," *Design News*, Nov. 10, 1961
19	3	ROARK, R. J.: *Formulas for Stress and Strain*, McGraw-Hill, N.Y., 1954
20	4	HOOL, A. G. and KINNE, W. S.: *Structural Members and Connections*, McGraw-Hill, N.Y., 1923
21	5	GRIFFEL, W.: "Single Nomogram for 18 Constant-strength Beams," *Product Engineering*, May 8, 1961

22 6 GRIFFEL, W.: "Maximum Deflection and Natural Frequency of Beams," *Design News*, Feb. 6, 1963

23 7 GRIFFEL, W.: "Beam Columns," *Product Engineering*, Sept. 14, 1964

23 8 GRIFFEL, W.: "Beams under Combined Axial and Transverse Loads," *Machine Design*, July 19, 1962

23 9 GRIFFEL, W.: "Beam Columns, Deflection and Slope," *Product Design and Value Engineering*, Canada, Nov.–Dec., 1963

23 10 TIMOSHENKO, S.: *Strength of Materials*, Part II, 3rd ed., Van Nostrand, Princeton, N.J., 1956

24 11 GRIFFEL, W.: "Maximum Safe Load for a Steel Beam," *Product Engineering*, Oct. 1, 1962

25 12 GRIFFEL, W.: "Unsymmetrical Bending of Beams," *Product Engineering*, July 22, 1963

26 13 GRIFFEL, W.: "Beam Deflection Caused by Shearing Stress," *Product Engineering*, Jan. 7, 1963

27 14 GRIFFEL, W.: "Beams, Energy Storage Capacity in Bending and Spring Rate," *Design News*, Jan. 24, 1962

28 15 GRIFFEL, W.: "Circular Shaft Under Combined and Twisting Moment; Simplified Method for Determining Shaft Diameter and Angle of Twist," *Design News*, Nov. 14, 1962

29 16 GRIFFEL, W.: "Shear Center for Beam Sections," *Design News*, June 13, 1962

31 17 HETÉNYI, M.: *Beams on Elastic Foundation*, University of Michigan Press, Ann Arbor, Mich., 1958

32 18 GRIFFEL, W.: "Beams on Elastic Foundation," *Design News*, Jan. 23, 1963

32 19 SEELY, F. B. and SMITH, S. O.: *Advanced Mechanics of Materials*, John Wiley, N.Y., 1962.

34 20 GRIFFEL, W.: "Beams on Elastic Foundation, Charts," *Machine Design*, March 1, 1962

36 21 SPARING, W. H., and RADWILL, R. P.: "Stresses in Curved Bars," *Product Engineering*, Dec. 24, 1962

37 22 GRIFFEL, W.: "Bending of Curved Beams," *Design News*, April 4, 1962

38 23 GRIFFEL, W.: "Circular Cantilever Beams; Load Normal to the Plane of Curvature," *Design News*, Sept. 11, 1964

38 24 HOGAN, MERVIN B.: "Circular Beams Loaded Normal to the Plane of Curvature," *A.S.M.E. Transaction*, 1938, Vol. 60, p. A-81

39 25 GRIFFEL, W.: "Circular Cantilever Beams, With Out-of-Plane Loading," *Design News*, Aug. 5, 1964

39 26 HOGAN, M. B.: "Circular Beams Loaded Normal to the Plane of Curvature—2," *Design Data and Methods*; *Applied Mechanics A.S.M.E.*, 1953

40 27 GRIFFEL, W.: "How to Predict Internal Reactions for Out-of-Plane Loading of Circular Beams," *Product Design Engineering*, Canada, Sept.-Oct., 1963

41 28 ROARK, R. J.: *Formulas For Stress and Strain*, McGraw-Hill, N.Y., 3rd. Ed., 1954

Chapter III

TORSION

42. STRAIGHT BAR OF UNIFORM CIRCULAR SECTION UNDER PURE TORSION

(*Ref. 1*)

Let: T = twisting moment (in.-lb.)
L = length of the member (in.)
d = diameter of the section (in.)
J = polar moment of inertia of the section (in.^{4})
c = distance from the center of the section to any point (in.)
S = shear stress (psi)
θ = angle of twist (radians)
G = modulus of rigidity of the material (psi)

If a bar having a constant circular cross section of diameter d is subjected to twisting moments T at its ends, the shearing stress at any point on a transverse cross section may be expressed by the equation:

$$S = Tc/J \qquad (1)$$

in which $J = \pi\, d^4/32$.
Also,

$$\theta = TL/JG = \frac{32TL}{\pi d^4 G} = \frac{10.19TL}{d^4 G} \qquad (2)$$

The above equations apply only to bars or shafts having constant cross sections, for which plane cross sections remain plane as the bar is twisted, and are valid only when the shearing stress in the shaft does not exceed the shearing elastic strength of the material.

Expressing θ in *degrees* and assuming $G = 11 \times 10^6$ psi, T = 1,000 in.-lb., and L = 1 in., equation (2) becomes:

$$\theta = \frac{0.05315}{d^4} \qquad (3)$$

The solution of equation (3) may be obtained graphically from Fig. 1, in which the quantity $0.05315/d^4$ has been plotted against the diameter d. The use of this

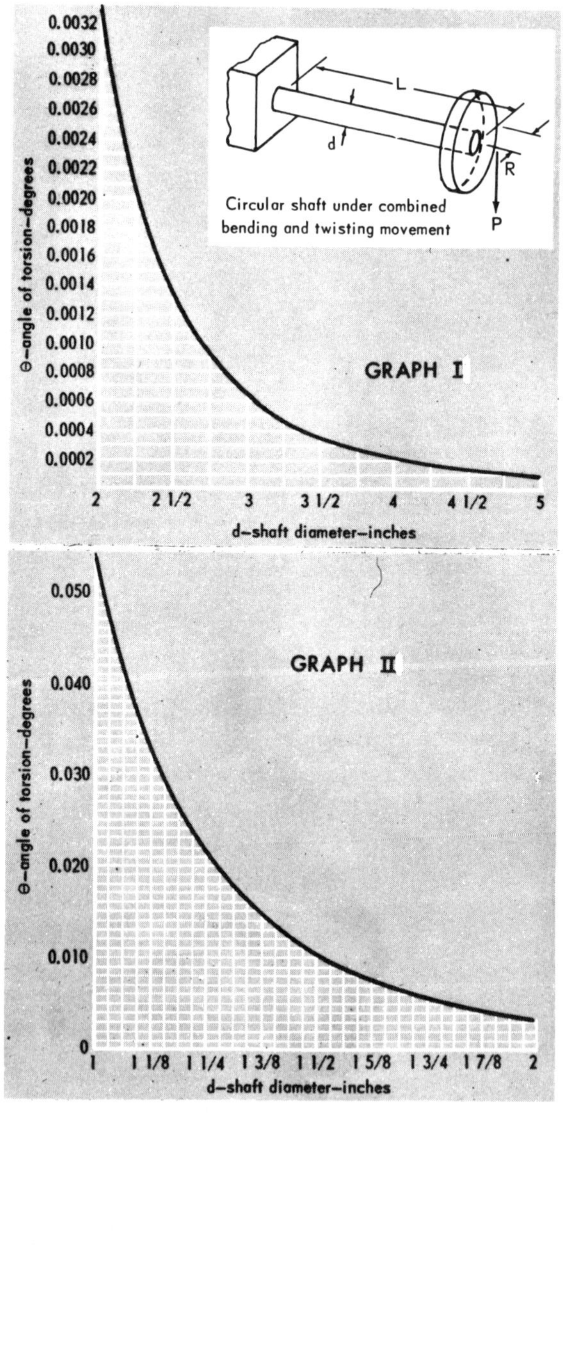

chart will be best demonstrated on the following example.

NOTE: For values of G greater than 11×10^6 psi multiply the results obtained from Fig. 1 by the ratio $11 \times 10^6/G$.

Example: Determine the angle of twist, in degrees, of a shaft, if $d = 2.25$ in., $R = 12$ in., $P = 1{,}000$ lb. and $L = 60$ in.

Solution: $T = PR = 12{,}000$ in.-lb. The ordinate to the curve at $d = 2.25$ in. is 0.00207 which, multiplied by 12, the twisting moment in thousands of in.-lb., and by 60, the length of the shaft in inches, gives 1.49 deg., the angle through which the shaft will twist.

43. TORSION IN SHAFTS OF NONUNIFORM SECTION*

(*Ref. 2*)

When a uniform-section shaft is twisted, its angular deflection is given by the familiar formula

$$\theta = \frac{TL}{JG}$$

where θ = the angular deflection, radians; T = torque, in.-lb.; L = length of the shaft, in.; G = shear modulus; J = polar moment of inertia, in.[4] ($= \pi/32\ D^4$ for a solid shaft of diameter D).

If the shaft is not uniform, however, the solution is not readily available; we must use the equation

$$\theta = \frac{T}{G}\int_0^L \frac{dx}{J}$$

In the case of a shaft with a constant taper (Fig. 1) the diameter D at any point is given by

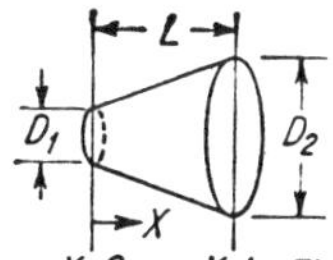

Fig. 1

$$D = \left[\frac{D_2 - D_1}{L}\right] x + D_1$$

then

$$\theta = \frac{T}{G}\int_0^L \frac{dx}{\frac{\pi D^4}{32}} = \frac{32T}{\pi G}\int_0^L \frac{dx}{\left[\left(\frac{D_2 - D_1}{L}\right) x + D_1\right]^4}$$

$$= \frac{32TL}{3\pi G D_1 D_2}\left[\frac{1}{D_1^2} + \frac{1}{D_1 D_2} + \frac{1}{D_2^2}\right]$$

Using the same procedure, a table was prepared for several sections giving the equation for angular deflection.

Table 1

ANGULAR DEFLECTION OF SHAFTS

Section	Angular deflection
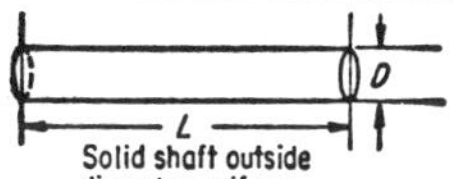 Solid shaft outside diameter uniform	$\theta = \frac{32TL}{\pi G D^4}$
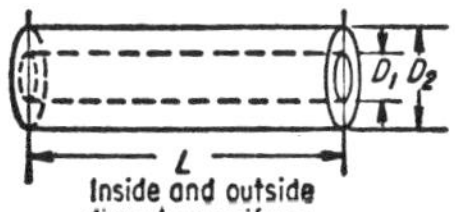 Inside and outside diameters uniform	$\theta = \frac{32TL}{\pi G(D_2^4 - D_1^4)}$

*Reprinted, by permission, from *Product Engineering*, Aug. 17, 1964. Copyright 1964 by McGraw-Hill.

ANGULAR DEFLECTION OF SHAFTS

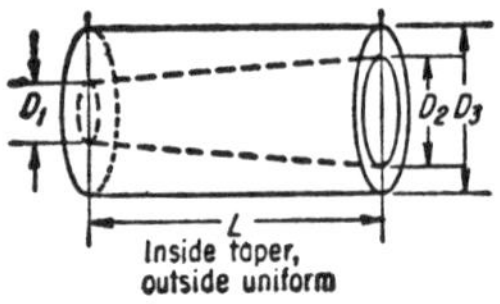

$$\theta = \frac{8TL}{\pi G(D_2 - D_1)}\,\frac{1}{D_3^3} \times \left\{2\arctan\left(\frac{D_2}{D_3}\right) - 2\arctan\left(\frac{D_1}{D_3}\right) + \mathrm{Log}_e\left[\left(\frac{D_3 + D_2}{D_3 - D_2}\right)\left(\frac{D_3 - D_1}{D_3 + D_1}\right)\right]\right\}$$

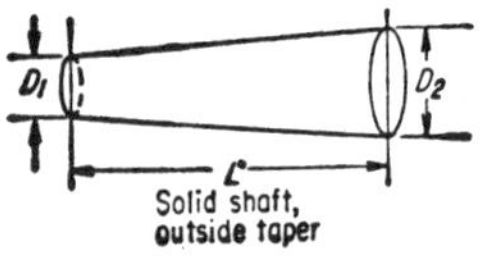

$$\theta = \frac{32TL}{3\pi G D_1 D_2}\left[\frac{1}{D_1^2} + \frac{1}{D_1 D_2} + \frac{1}{D_2^2}\right]$$

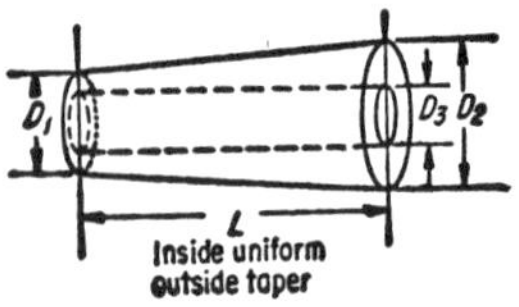

$$\theta = \frac{8TL}{\pi G(D_2 - D_1)}\,\frac{1}{D_3^3} \times \left\{\mathrm{Log}_e\left[\left(\frac{D_2 - D_3}{D_1 - D_3}\right)\left(\frac{D_1 + D_3}{D_2 + D_3}\right)\right] - 2\left[\arctan\left(\frac{D_2}{D_3}\right) - \arctan\left(\frac{D_1}{D_3}\right)\right]\right\}$$

FOR THIN SECTIONS

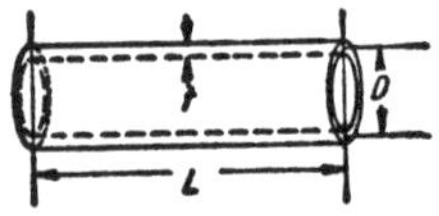

$$\theta = \frac{4TL}{\pi G D^3 t}$$

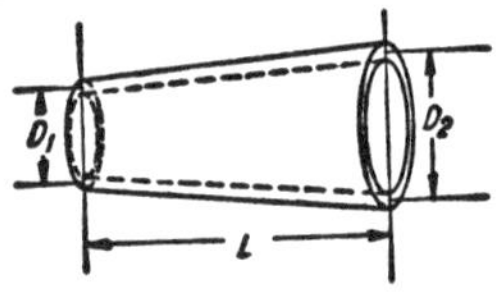

$$\theta = \frac{2TL(D_1 + D_2)}{\pi G t D_1^2 D_2^2}$$

44. NON-CIRCULAR, UNIFORM SECTION

(*Ref. 3*)

QUICK WAY TO ADJUST MOMENT OF INERTIA AND TORSIONAL EFFECT K

Very often, when designing a member for maximum strength and minimum weight, it is necessary to adjust the moment of inertia. In the case of a hollow rectangular cross section this can result in quite involved and time-consuming operations.

The purpose of the following is to show how to determine I_x, I_y and the torsional effect factor K without any effort and while maintaining the desired accuracy.

The moments of inertia of the hollow rectangular cross section shown in Fig. 2 are:

$$I_x = \frac{bh^3}{12} - \frac{(b-2t)(h-2t)^3}{12} \quad (4)$$

$$I_y = \frac{hb^3}{12} - \frac{(h-2t)(b-2t)^3}{12} \quad (5)$$

The torsional effect K used in equation $\theta = T/KG$ is

$$K = \frac{2t^2(h-t)^2(b-t)^2}{t(h+b)-2t^2} \quad (6)$$

Note: θ = Angle of twist per unit length (radians).
T = Twisting moment (lb.-in.)
G = Modulus of rigidity of the material (psi)

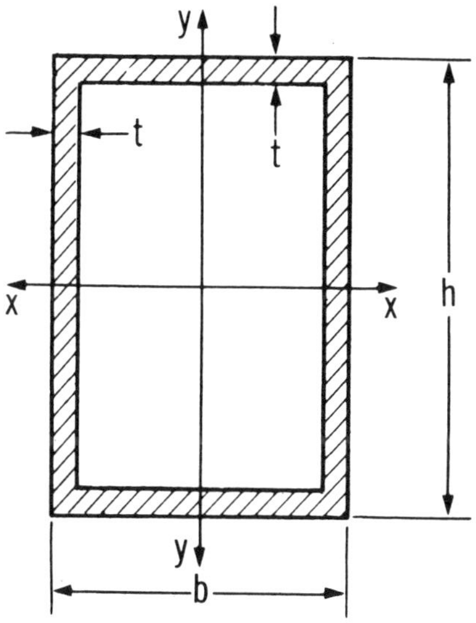

Fig. 2 Cross-section under consideration.

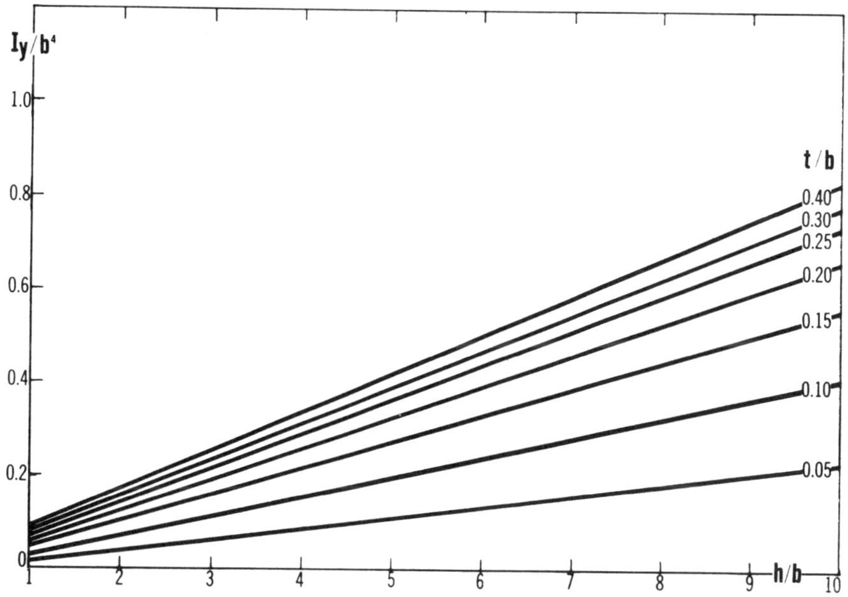

Fig. 4

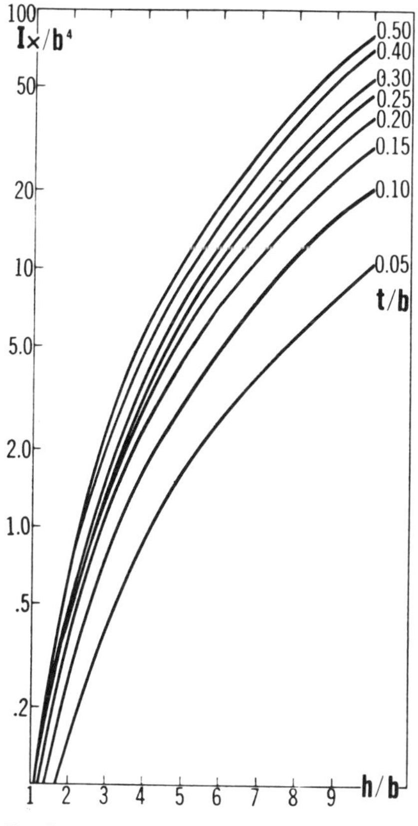

Fig. 3

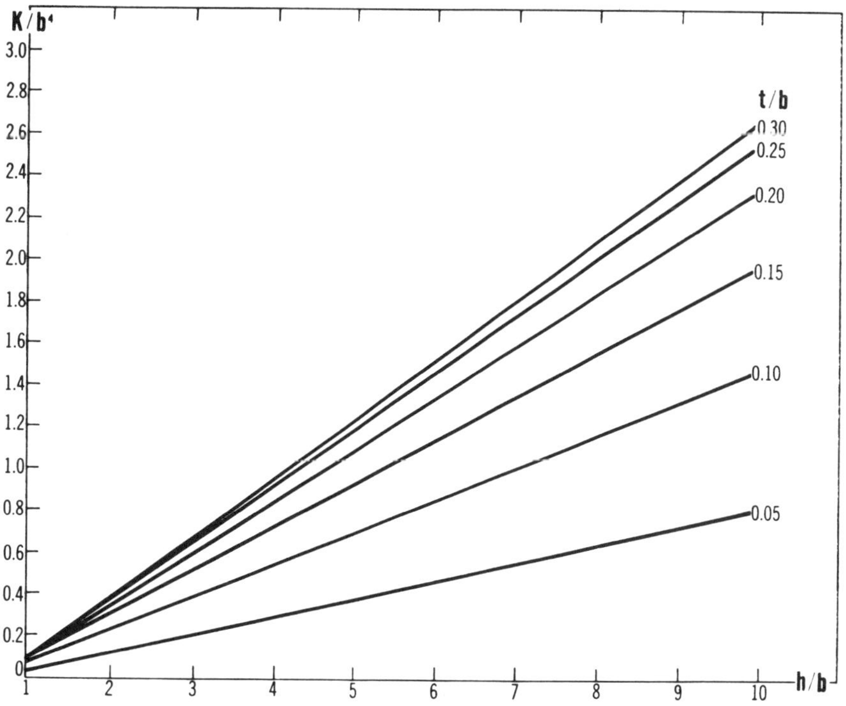

Fig. 5

Equations (1), (2), (3) may be simplified as follows:

$$I_x/b^4 = \frac{(h/b)^3}{12} - \frac{[1 - 2(t/b)]\,[h/b - 2(t/b]^3}{12} \qquad (7)$$

$$I_y/b^4 = \frac{h/b}{12} - \frac{[h/b - 2(t/b)]\,[1 - 2(t/b)]^3}{12} \qquad (8)$$

$$K/b^4 = \frac{2(t/b)^2\,(h/b - t/b)^2\,(1 - t/b)^2}{t/b(h/b + 1) - 2(t/b)^2} \qquad (9)$$

Equations (7), (8), (9), which actually represent I_x, I_y, K for $b = 1$, are plotted in Figs. 2, 3, 4, where the values of these parameters, for given values of h/b and t/b, were calculated on a digital computer.

The use of Figs. 2, 3, 4 will be best illustrated by the following example:

Let $b = 3$ in., $h = 15$ in., $t = 0.3$ in., $h/b = 5$, $t/b = 0.1$.

From Fig. 2: $I_x/b^4 = 3.05 \therefore I_x = 3.05 \times 3^4 = 247$ in.^{4}

From Fig. 3: $I_y/b^4 = 0.21 \therefore I_y = 0.21 \times 3^4 = 17.01$ in.^{4}

From Fig. 4: $K/b^4 = 0.67 \therefore K = 0.67 \times 3^4 = 54.27$ in.^{4}

45. MINIMUM WEIGHT SECTIONS IN TORSION

Of two cross sections having the same section modulus Z and the same allowable twisting moment, the section with the smaller cross-sectional area, A, will have the least of material. In order to compare various shaped cross sections it is convenient to employ a dimensionless quantity: $Z/\sqrt{A^3}$.

In the following table are given values of $Z/\sqrt{A^3}$ for some typical cross sections.

Type of Section	$Z/\sqrt{A^3}$
Channel	0.04 — 0.05
I-beam	0.05 — 0.07
Rectangular section with $a/b = 10$	0.1
Rectangular section with $a/b = 2$	0.18
Square	0.21
Solid circular section	0.28
Circular ring with $c = d/D = 0.5$	0.37
Circular ring with $c = 0.9$	1.16

As can be seen, channels, I-beams, and narrow rectangular sections are least economical in torsion, and circular ring sections are the most economical especially when the wall thickness is small, i.e., the values of $c = d/D$ is close to unity.

Also of interest to a designer is a weight comparison of bars of tubular section, W_T, with the weight of bars of solid section, W_S, for various values of $c = d/D$, assuming an equal strength.

By equating the polar section moduli of solid and ring sections we obtain:

$$\frac{\pi d_1^3}{16} = \frac{\pi D^3}{16}(1 - c^3).$$

from which, assuming an equal strength,

$$D = \frac{d_1}{\sqrt[3]{1 - c^3}}$$

The ratio of weight:

$$\frac{W_T}{W_S} = \frac{\pi}{4}(D^2 - d^2)\frac{4}{\pi d_1^2} = \frac{D_2^2 - d^2}{d_1^2} = \frac{D_2(1 - c^2)}{d_1^2}$$

Substituting here the value of D obtained from the condition of equal strength, we obtain:

$$\frac{W_T}{W_S} = \frac{1 - c^2}{\sqrt[3]{(1 - c^3)^2}}$$

In the following table are given values of W_T/W_S for various values of $c = d/D$

C	0	0.1	0.2	0.3	0.4	0.5	0.6	0.7	0.8	0.9
W_T/W_S	1	0.99	0.96	0.92	0.85	0.79	0.70	0.61	0.51	0.39

It is seen from this tabulation that greatest saving of metal will result from the use of tubular thin-walled bars.

Chap. III

REFERENCES

Section	Ref. No.	
42	1	HALSEY, A. F.: *Handbook for Machine Designers*, McGraw-Hill, N.Y., 1916
43	2	KRAVITZ, S.: "Your Turn," *Product Engineering*, Aug. 17, 1964
44	3	GRIFFEL, W.: "Quick Way to Adjust Moment of Inertia," *Design Engineering*, Canada, Jan., 1964

Chapter IV

THE THEORY OF ELASTIC ENERGY

46. BASIC CONSIDERATIONS

(*Ref. 1, 5*)

In the hands of a capable engineer, the *theory of elastic energy* is a powerful tool. Easy to grasp, the theory is basically simple, and yet methods based on it will accomplish all that any other theory or method for calculating stresses in structures will do. It is equally applicable to straight beams, curved beams, and trusses. The only assumptions are: The structure to be analyzed behaves as a homogeneous and elastic body; deformations of the structure are small enough to maintain the geometric relation of structure parts one to another; and the law of superposition holds.

In what follows, the basic theory will be developed and its application to some of the problems of out-of-plane loading of curved cantilever beams will be shown in detail.

Work is defined as the product of force times distance. For a constant force P acting over a finite distance δ, the work done is given by the expression $W = P\delta$. On the other hand if the force varies over the distance, work done by this varying force P in deforming a body an amount of δ is: $W = \int_0^\delta P d\delta$ and is represented by the area under the load deformation curve as shown in Fig. 1.

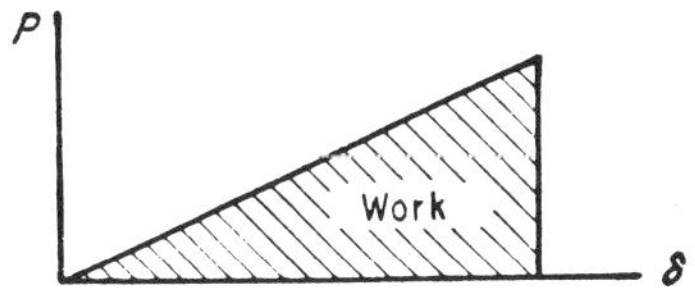

Fig. 1

If the deformed body is perfectly elastic, the energy stored in the body may be completely recovered during a gradual unloading. This energy is called the elastic strain energy and is denoted by U. Thus, $U = \int_0^\delta P d\delta$. The straight line in Fig. 1 assumes the body to be linearly elastic; its equation is $P = k\delta$. Therefore

$$U = \frac{k\delta^2}{2} = \frac{P^2}{2k} = \frac{P\delta}{2}$$

STRAIN ENERGY EXPRESSIONS FOR VARIOUS LOADINGS

(1) *Tension:* In an axially loaded slender bar of a constant cross-sectional area A, length L and modulus of elasticity E, the deformation $\delta = PL/AE$, and

$$U = \int^\delta P d\delta = \frac{P^2 L}{2AE} \tag{1}$$

For a tensile bar with varying AE and/or P

$$U = \frac{1}{2}\int \frac{P^2\,dx}{AE} \tag{2}$$

(2) *Flexure:* Assume a uniform beam of length L loaded as shown in Fig. 2

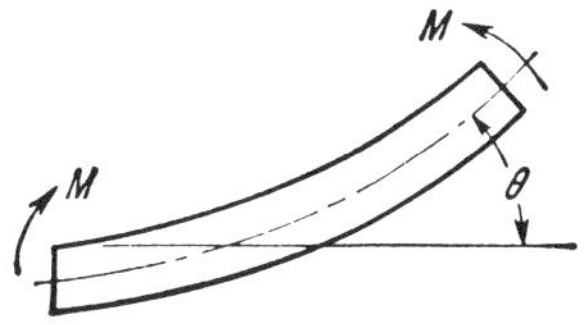

Fig. 2

This beam, under the action of a pure couple, undergoes an angular rotation θ proportional to the couple. Thus, from elementary strength of material we obtain: $\theta = ML/EI$, where EI, the product of modulus of elasticity in tension, and section moment of inertia, is called the bending stiffness. Since the rotation θ builds up linearly with M the strain energy stored is

$$U = \frac{M\theta}{2} \text{ or } U = \frac{1}{2}\left(\frac{M^2 L}{EI}\right) \tag{3}$$

For a beam with varying EI and/or varying M:

$$U = \frac{1}{2}\int \frac{M^2\,dx}{EI} \tag{4}$$

(3) *Torsion:* A uniform circular shaft carrying a torque T undergoes a total twist ϕ in length L proportional to the torque. From the elementary strength of materials

$$\phi = \frac{L}{GJ}T \tag{5}$$

where GJ, the product of the modulus of elasticity in shear and the cross section polar moment of inertia, is called the torsional stiffness. Since the twist ϕ builds up linearly with T, the strain energy stored is

$$U = \frac{1}{2}T\phi \text{ or } U = \frac{1}{2}\frac{T^2 L}{GJ} \tag{6}$$

For a shaft with nonuniform properties and varying loading:

$$U = \frac{1}{2}\int \frac{T^2\,dx}{GJ} \tag{7}$$

(4) *Shear:* Through a similar reasoning as before the reader will find that the strain energy in shear is

$$U = \frac{1}{2}\int \frac{V^2\,dx}{AG} \tag{8}$$

where V is the transverse shear loading.

TOTAL ELASTIC STRAIN ENERGY OF A STRUCTURE

$$U_{\text{total}} = \int \frac{P^2\,dx}{2A\,E} + \int \frac{M^2\,dx}{2EI} + \int \frac{T^2\,dx}{2JG} + \int \frac{V^2\,dx}{2AG} \tag{9}$$

It is of interest to note that it is seldom that all the terms of equation (9) are used in a calculation. Many of the loadings, if actually present, may be of a localized or secondary nature and their energy contribution may be neglected. A few examples illustrating the above outlined methods for calculating the strain energy follow:

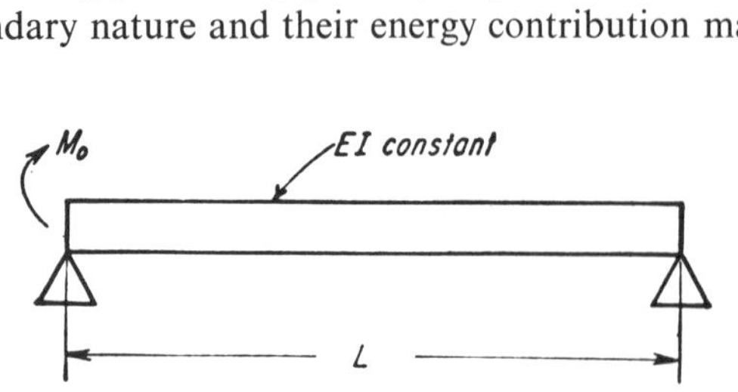

Fig. 3

(1) Find the expression of the strain energy as a function of applied moment M_o (Fig. 3).

Solution: The bending moment at any section of the beam is: $M = M_o[1 - (x/L)]$; then,

$$U = \int_0^L \frac{M^2\,dx}{2EI} = \frac{1}{2EI}\int_0^L M_0^2(1 - x/L)^2\,dx = \frac{M_0^2 L}{6EI}$$

(2) A linearly tapered aluminum bar is under an axial load of 1,500 lb. as shown in Fig. 4. Find U:

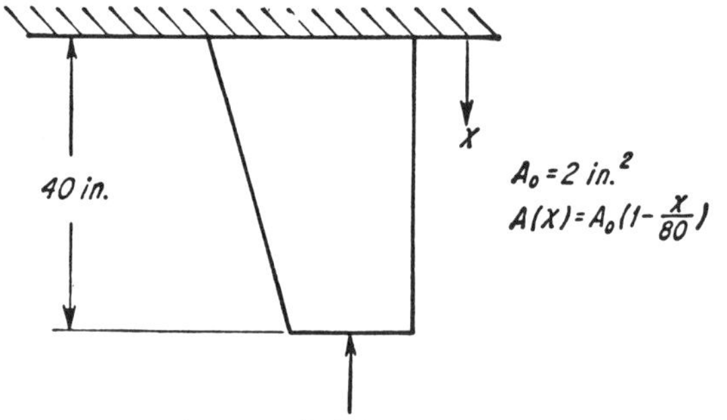

Fig. 4*

$$U = \int_0^L \frac{P^2\,dx}{2EA} = \int_0^{40} \frac{1{,}500^2\,dx}{4(1 - x/80)\,10 \times 10^6} = 4.5 \ln 2 \text{ in.-lb.}$$

$= 1.35$ in.-lb.

NOTATION

A = area (in.^{2})
E = elastic modulus (psi)
G = shear modulus (psi)
I = moment of inertia (in.^{4})
J = polar moment of inertia (in.^{4})
k = constant
L = length (in.)
M = bending moment (in.-lb.)
M_o = applied moment (in.-lb.)
P = force (lb.)
T = torque (in.-lb.)
U = elastic strain energy (in.-lb.)
V = shear force (lb.)
W = work (lb.-in.)
x = coordinate
δ = deflection (in.)
θ = slope (radians)

---

*Example from Ref. 1.

47. CASTIGLIANO'S THEOREM

(*Ref.* 2)

It was previously established that the elastic energy is $U = PL^2/2AE$, from which can be obtained

$$\frac{dU}{dP} = \frac{PL}{AE} = \delta$$

where δ = the displacement of the particles of a given structural member under the influence of the load P., In other words, this equation indicates that the rate of change of strain energy with respect to the load is equal to associated deflection; this is Castigliano's theorem. For a body under the simultaneous action of several loads, the theorem is written so as to apply individually to each load and its associated deflection, thus

$$\delta = \frac{\partial U}{\partial P} \tag{10}$$

This relation states in mathematical terms that if the partial derivative (∂) of U with respect to a given force is taken, the value of this derivative is equal to the deflection in the direction of the force and at the point of application of the force. Similarly, if the expression for U incorporates a couple, M_1, the rotation or slope in the direction of couple is given by $\theta = \partial U/\partial M_1$

Note that any generalizations of the meanings of "force and deflection" are possible only so long as the units are such that their product yields the units of strain energy (in.-lb.). For example: (1) force (lb.) and translation (in.); (2) moment (in.-lb.) and rotation (radians); (3) torque (in.-lb.) and rotation (radians); (4) pressure (psi) and volume (in.[3]). Once again, for emphasis, it is repeated that Castigliano's theorem is restricted to linearly elastic structures.

USE OF FICTITIOUS LOADS, Q

Castigliano's theorem allows us to calculate the deflection under the load. This results from the definition of the theorem that states that the deflection is equal to the rate of change of strain energy with respect of load, or $\delta = \partial U/\partial P$. As such rate exists even though the load itself is zero, it is possible to calculate a deflection at a point where no load is applied. In such cases a fictitious load will be added acting at the point and in the direction of desired deflection. The rate of change of strain energy with respect to this fictitious load will be found, after which the load will be set to zero. Let us demonstrate this technique on the following example.

Example: Fig. 5 shows a cantilever round rod formed in a quarter circle of radius R and acted upon by a torque T_o Find the vertical movement of the free end.

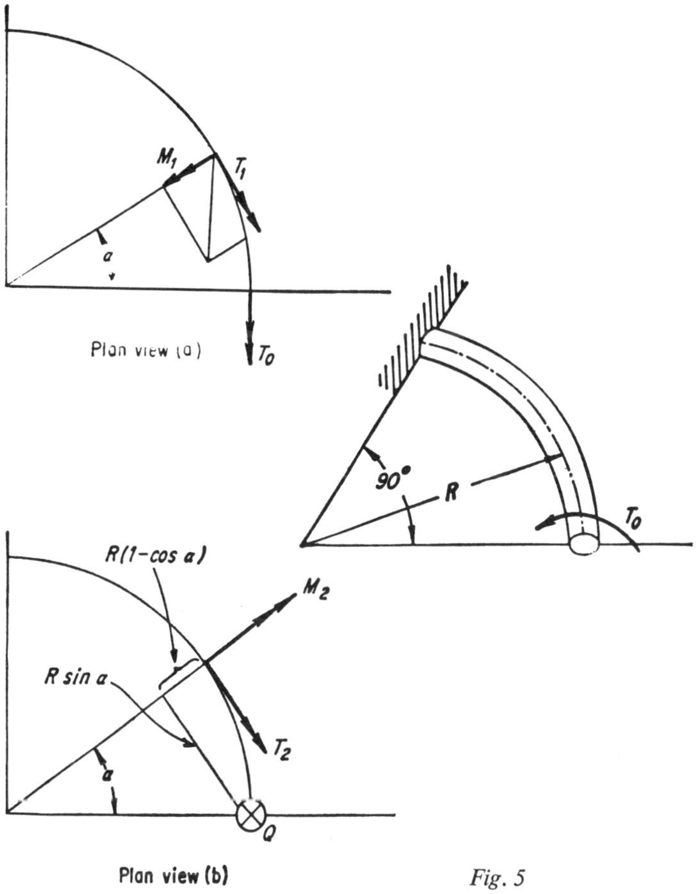

Fig. 5

Solution: The vector resolutions of the applied torque, shown in Fig. 5a, yield

$$T_1 = T_o \cos\alpha;\ M_1 = T_o \sin\alpha$$

Now let us apply, at the point of desired deflection, a fictitious load Q (down). The resulting loadings, shown in Fig. 5b, yield $M_2 = QR\sin\alpha$; $T_2 = QR(1 - \cos\alpha)$; thus the total loadings are: $M = M_1 - M_2 = (T_o - QR) \times \sin\alpha$; $T = T_1 + T_2 = T_o\cos\alpha + QR(1 - \cos\alpha)$. The strain energy expressions that apply in the case of this example are the ones for *flexure* and *torsion*, therefore

$$U = \int \frac{M^2\,dx}{2EI} + \int \frac{T^2\,dx}{2GJ}$$

by substitution we obtain

$$U = \frac{1}{2EI}\int_0^{\pi/2}(T_o - QR)^2 \sin^2\alpha \, Rd\alpha + \frac{1}{2GJ}\int_0^{\pi/2}[T_o \cos\alpha + QR\,(1 - \cos\alpha)]^2 \, Rd\alpha$$

where $Rd\alpha = dx$.

Differentiating under the integral sign:

$$\frac{\partial U}{\partial Q} = -\frac{(T_o - QR)R^2}{EI}\int_0^{\pi/2}\sin^2\alpha \, d\alpha + \frac{R^2}{GJ}\int_0^{\pi/2}[T_o\cos\alpha + QR\,(1-\cos\alpha)]\,(1-\cos\alpha)\,d\alpha$$

Putting Q, the fictitious load, equal to zero and integrating:

$$\delta = \frac{\partial U}{\partial Q} = \frac{T_o R^2}{4}\left(\frac{4-\pi}{GJ} - \frac{\pi}{EI}\right)$$

Since $J = 2I$ and $G = E/2.6$, $\delta = -2.71/EI$. The minus sign indicates that the deflection in this case is *upwards*.

APPLICATION OF CASTIGLIANO'S THEOREM TO MORE ADVANCED PROBLEMS

Beginning with the general expression for strain energy

$$U = \frac{1}{2}\int \frac{P^2}{AE}dx + \frac{1}{2}\int\frac{M^2}{EI}dx + \frac{1}{2}\int\frac{T^2}{GJ}dx + \frac{1}{2}\int\frac{V^2}{AG}dx \quad (11)$$

Differentiating under the integral sign with respect to a fictitious load Q, we obtain:

$$\frac{\partial U}{\partial Q} = \delta = \frac{1}{AE}\int P\frac{\partial P}{\partial Q}dx + \frac{1}{EI}\int M\frac{\partial M}{\partial Q}dx + \frac{1}{GJ}\int T\frac{\partial T}{\partial Q}dx + \frac{1}{AG}\int V\frac{\partial V}{\partial Q}dx \quad (12)$$

It should be noted that in equation (12) the integral represents the sum of the partial derivative of the energy in each element, dx, of the structural member and that, since differentiation is with respect to Q instead of x, the differentiation of equation (12) does not remove the integral sign. The partial derivative in equation (12) is the rate at which the load, bending moment, torque, and shear at the section change with changes in Q. Here too, it should be emphasized that it is seldom that all of the terms in equation (12) are used.

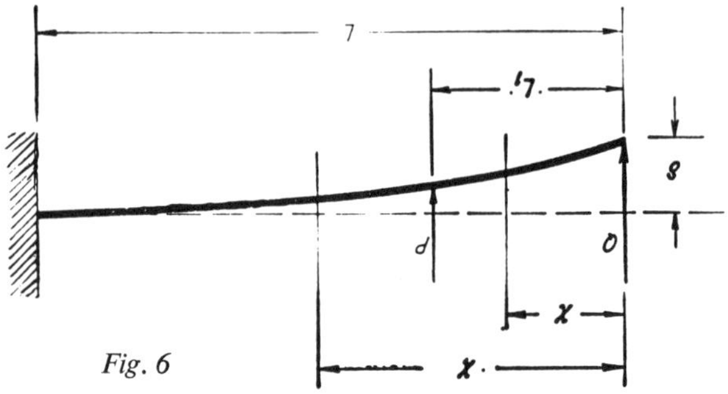

Fig. 6

Example: Find the vertical deflection of the free end of a cantilever beam of length L subjected to a single concentrated load P at a distance L_1 from the free end, as shown in Fig. 6.

Solution: No load is acting at the free end; hence a fictitious vertical load Q is assumed to act. Let δ be the deflection of the free end. Then we have by equation (12):

$$\delta = \int \frac{M}{EI} \times \frac{\partial M}{\partial Q}dx$$

Note that in the case of this example, bending is the primary cause of the deflection, as it is assumed that the length of the beam is large relative to the depth. If the beam is relatively short and deep then the shear would be another contributing factor in the calculation of the deflection.

EXPRESSIONS FOR M

$$0 \leqslant x \leqslant L_1; \quad M = QX; \quad \partial M/\partial Q = x.$$

$$L_1 \leqslant x \leqslant L; \quad M = Qx + P(x - L_1); \quad \partial M/\partial Q = x$$

Thus

$$\delta = \int_0^{L_1}\frac{QX^2}{EI}dx + \int_{L_1}^{L}\frac{1}{EI}[Qx^2 + P\,(x - L_1)\,x]\,dx$$

The fictitious load Q has served its purpose; therefore we will make it equal to zero; then all terms containing Q will vanish, and hence

$$\delta = \frac{1}{EI}\int_{L_1}^{L}P\,(x - L_1)\,x\,dx = \frac{1}{EI}\left[\frac{1}{3}P\,(L^3 - L_1^3) - \frac{PL_1}{2}(L^2 - L_1^2)\right]$$

NOTATION

Q = Fictitious load
P = force (lb.)
R = radius of ring (in.)
T_o = applied torque (in.-lb.)
W = work (lb.-in.)
∂ = partial differential
α = subtended angle (deg.)

48. UNIT LOAD METHOD

(*Ref. 3*)

A strict application of the calculus of Castigliano's theorem leads to a number of cumbersome techniques ill-suited to the solution of complex structures. A more flexible approach is the method of unit loads. While Castigliano's theorem applies to structures in which the loads and deflections are linearly related, the unit load method is not restricted to the linear relationship between load and deflection. However, the deflection must be relatively small.

As was shown in the previous section the equation for the deflection under a load, P, is

$$\delta = \frac{1}{EA}\int P\frac{\partial P}{\partial P_1}dx + \frac{1}{EI}\int M\frac{\partial M}{\partial P_1}dx + \frac{1}{GJ}\int T\frac{\partial T}{\partial P_1}dx + \frac{1}{AG}\int V\frac{\partial V}{\partial P_1}dx \qquad (13)$$

Let

$$\mu = \frac{\partial P}{\partial P_1}; \quad m = \frac{\partial M}{\partial P_1}dx; \quad t = \frac{\partial T}{\partial P_1}; \quad v = \frac{\partial V}{\partial P_1}$$

Hence

$$\delta = \int\frac{P\mu}{AE}dx + \int\frac{Mm}{EI}dx + \int\frac{Tt}{GJ}dx + \int\frac{Vv}{AG}dx \qquad (14)$$

In applying equation (14), the labor of a deflection calculation divides conveniently into several steps: (1) calculation of the real (actual) load distribution (P, M, T, V); (2) calculation of the *unit* (virtual) load distribution (u, m, t, v) due to a unit load applied at the point of desired deflection and reacted at the reference or points; (3) calculation of flexibilities $1/AE$, $1/EI$, $1/GJ$, $1/AG$; (4) summation and/or integration.

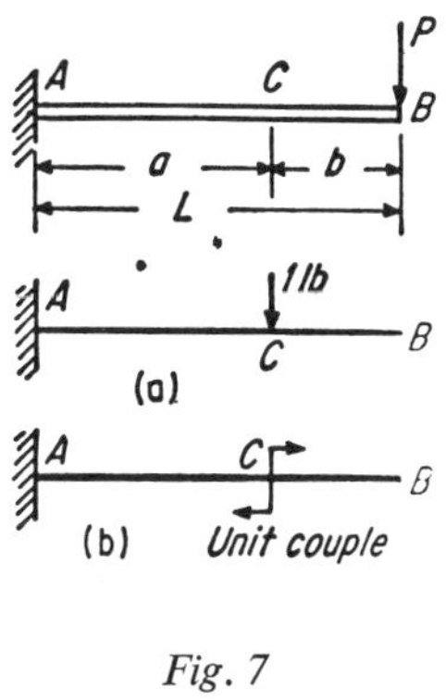

Fig. 7

Despite the preference of the unit load method over Castigliano's theorem, which may require tedious partial differentiation, the two methods are frequently used to establish independently the basic integral as a check.

The following examples apply the method of unit load:

Example 1: Find the vertical deflection at point C for the cantilever beam of Fig. 7 carrying a concentrated load P at its end. Also find the slope of the elastic curve at C.

Solution: (a) The deflection, assuming the shear negligible, is from equation (14):

$$\delta = \int (Mm/EI)\,(dx)$$

Assume origin at B: $M = Px$. For virtual loading of 1 lb. (Fig. 7a), $m = 0$, for $x < b$, $m = 1(x - b)$, for $x > b$; hence $Mm = Px(x - b) = P(x^2 - bx)$.

Therefore

$$\delta = \frac{P}{EI}\int_b^L (x^2 - bx)\,dx = \frac{P}{EI}\left[\frac{x^3}{3} - \frac{bx^2}{2}\right]_b^L = \frac{P}{3EI}\left[L^3 - \frac{3L^2 b}{2} + \frac{b^3}{2}\right]$$

If $b = 0$ we obtain the well known formula $\delta_B = PL^3/3EI$.

(b) The slope $\theta_c = \int Mm/EI\ dx$; for virtual loading see Fig. 7b: $m = 0$, $x < b$; $m = 1$, $x > b$. Hence $Mm = Px(1) = Px$;

therefore $\theta_c = \int Mm/EI\ dx = P/EI\int_b^L xdx$
$= P/2EI\,(L^2 - b^2)$; if $b = 0$, $\theta_B = PL^2/2EI$.

Example 2: A closely coiled helical spring is concentrically loaded with a force P. The dimensions of the spring are shown in Fig. 8. Find the elongation of the spring as a function of P.

Solution: Fig. 8b represents a free-body sketch of a portion of the spring under the action of load P. At the point where the spring is cut the reaction consists of the force P and couple PR. In order to obtain the deflection we conceive an auxiliary load of 1 lb., being applied along the vertical axis prior to application of P load. The 1-lb. loading and the P loading will be qualitatively identical. Fig. 8b would show a free-body diagram of part of the spring subject to the auxiliary loading merely by replacing P by 1 lb.

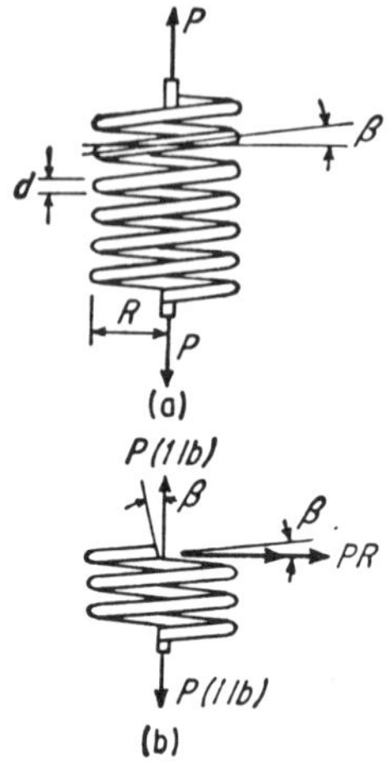

Fig. 8

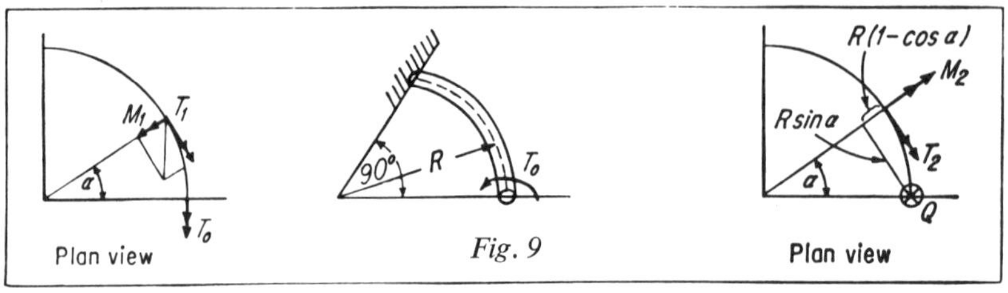

Fig. 9

Referring to the vector resolution, Fig. 9, we obtain: $m = 1 \times R \sin \beta$; $t = 1 \times R \cos \beta$; $M = PR \sin \beta$; $T = PR \cos \beta$. From equation (14),

$$\delta = \int_o^L \frac{Mm}{EI} dx + \int_o^L \frac{Tt}{GJ} dx$$

by substitution we get:

$\delta = PR^2L\,[(\sin^2 \beta / EI) + (\cos^2 \beta / GJ)]$, where L is the length of the wire. $L = 2\pi RN/\cos \beta$ (N representing the number of coils), therefore:

$$\delta = \frac{8PD^3\,N}{d^4}\left(\frac{2 \sin^2 \beta}{E \cos \beta} + \frac{\cos \beta}{G}\right)$$

For a closely coiled spring $\beta = 0$ and $\delta = 8PD^3N/d^4G$, where $D = 2R$ is the mean diameter of the coil.

In the analysis of stresses in a large variety of problems, we have introduced an auxiliary force assumed acting while the actual loading is being applied. Thus, it was possible to formulate the expression for the elastic energy stored in the structure. With reference to the analysis of redundant structures, a large number of theories appear to be available. Just to mention a few: theory of the elastic curve, theory of elastic energy, theory of least work, elastic weights, area moment, conjugate beams, slope deflection theory, kinetic theory of structures.

Within the assumption of elastic behavior of materials all the theories mentioned above may be derived from the theory of elastic energy. In the ensuing sections we shall analyze, in detail, circular cantilever beams with out-of-plane loading, using the material covered.

NOTATION

μ = load
m = moment
t = torque
v = shear

} all caused by 1-lb. force

49. OUT-OF-PLANE LOADING OF CIRCULAR BEAMS

(*Ref. 4*)

Continuing our discussion on Castigliano's theorem, let us solve a rather involved case of out-of-plane loading of circular cantilever beams of various uniform cross sections. The load in this case is perpendicular to the plane of the circular arc, therefore it produces, at any cross section of the member, a combination of an axial force, a bending moment, and a twisting moment. The deflection of the free end δ_A, the angle of slope θ_A, and angle of twist ϕ_A are given by

$$\int \frac{M}{EI}\frac{\partial M}{\partial Q} ds + \int \frac{T}{JG}\frac{\partial T}{\partial Q} ds$$

The shear is assumed to be negligible.

The method of finding the deflection, angle of slope, and angle of twist will be the same, regardless of what combinations of loads are applied to the member, and hence only deflection is discussed here.

For the purpose of illustrating the method, suppose we want to find the deflection of the free end of a cylindrical rod bent into a circular arc of radius R. The rod is fixed at one end, B (Fig. 10), and loaded at point C by a concentrated load P that is perpendicular to the plane AOB of the circular arc.

In order to find the deflection of the free end, A (Fig. 10a), we assume that a fictitious load Q is acting at point A and perpendicular to the plane of curvature

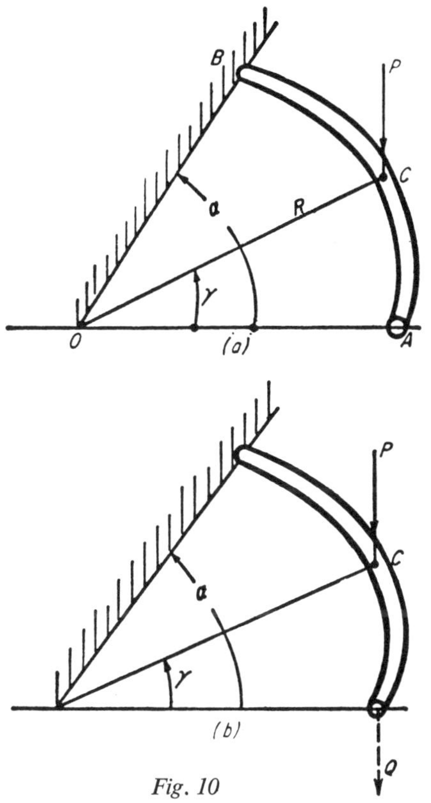

Fig. 10

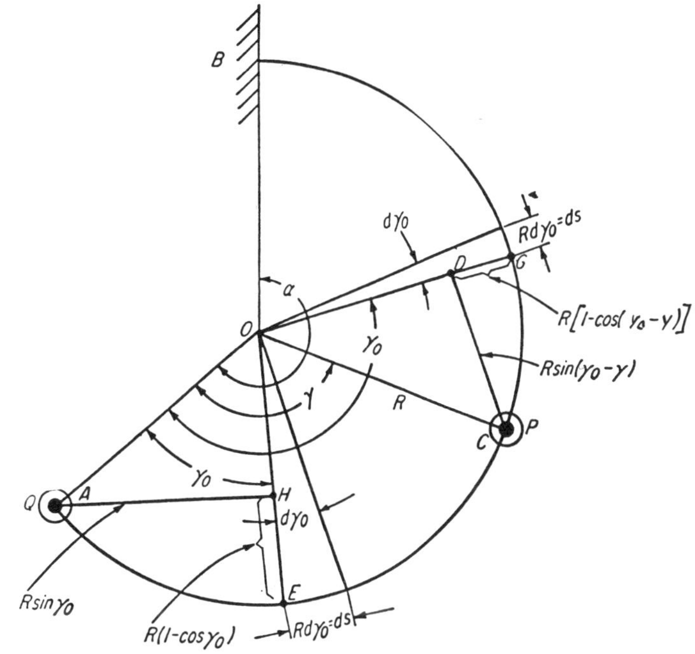

Fig. 11

(Fig. 10b). Referring to the plane view of the ring, shown in Fig. 11, let E be any cross section of the rod, let γ_o be the angle subtended by the arc AE, and let a differential length ds of the rod at the section E subtend the angle $d\gamma_o$. The cross section E was chosen on the arc AC which is the influence area of the fictitious load Q. Similarly, another cross section G was chosen on the arc CB, the influence area of $P + Q$. Since the strain energy is absorbed by the element of length ds (which is here considered as a straight cylindrical element because the radius r of its cross section is small in comparison with R), the deflection of the free end is

$$\delta_A = \int_0^\alpha \frac{M}{EI}\frac{\partial M}{\partial Q}\, ds + \int_0^\alpha \frac{T}{GJ}\frac{\partial T}{\partial Q}\, ds \qquad (15)$$

where α is the angle subtended by the arc AB.

Thus the problem at hand is to determine the expressions for the bending moment M and twisting moment, T, differentiate with respect to the auxiliary load Q, and by substitution in equation (15) find δ_A

Before determining M and T the first question that comes to mind is: What are the moment arms? From Fig. 11 we see that the moment arms at point E, the location of which varies from zero to γ, are

$$AH = R\sin\gamma_o \qquad \rightarrow M_1$$

$$HE = R(1 - \cos\gamma_o) \qquad \rightarrow T_1$$

Similarly at point G, the location of which varies from γ to α

$$DC = R\sin(\gamma_o - \gamma) \qquad \rightarrow M_2$$

$$DG = R[1 - \cos(\gamma - \gamma_o)] \rightarrow T_2$$

Hence at any section between zero and γ

$$M_1 = QR\sin\gamma_o \qquad (16)$$

$$T_1 = QR(1 - \cos\gamma_o). \qquad (17)$$

At any section between γ and α

$$M_2 = QR\sin\gamma_o + PR\sin(\gamma_o - \gamma) \qquad (18)$$

$$T_2 = QR(1 - \cos\gamma_o) + PR[1 - \cos(\gamma_o - \gamma)] \qquad (19)$$

$$\frac{\partial M_1}{\partial Q} = \frac{\partial M_2}{\partial Q} = R\sin\gamma_0; \; \frac{\partial T_1}{\partial Q} = \frac{\partial T_2}{\partial Q} = R(1 - \cos\gamma_0)$$

Now we can rewrite equation (15) as follows

$$\delta_A = \int_0^\gamma \frac{M_1}{EI}\frac{\partial M_1}{\partial Q}\, ds + \int_\gamma^\alpha \frac{M_2}{EI}\frac{\partial M_2}{\partial Q}\, ds + \int_0^\gamma \frac{T_1}{GJ}\frac{\partial T_1}{\partial Q}\, ds + \int_\gamma^\alpha \frac{T_2}{GJ}\frac{\partial T_2}{\partial Q}\, ds$$

By substitution, and remembering that $ds = Rd\gamma_o$, we obtain

$$\delta_A = \frac{QR^3}{EI}\int_0^{\gamma}\sin^2\gamma_0\,d\gamma_0 + \frac{1}{EI}\int_{\gamma}^{\alpha}[QR^3\sin^2\gamma_0$$

$$+ PR^3\sin\gamma_0\sin(\gamma_0-\gamma)]\,d\gamma_0 + \frac{QR^3}{GJ}\int_0^{\gamma}(1-\cos\gamma_0)^2 d\gamma_0$$

$$+ \frac{1}{GJ}\int_{\gamma}^{\alpha}\{QR^3(1-\cos\gamma_0) + PR^3(1-\cos\gamma_0)[1 - \cos(\gamma_0-\gamma)]\}\,d\gamma_0$$

Now that the load Q has served its purpose we shall make it equal to zero; also let $EI/GJ = C$.

$$\delta_A = \frac{PR^3}{EI}\left\{\int_{\gamma}^{\alpha}\sin\gamma_0\sin(\gamma_0-\gamma)\,d\gamma_0 + C\int_{\gamma}^{\alpha}(1-\cos\gamma_0)[1-\cos(\gamma_0-\gamma)]\,d\gamma_0\right\}$$

The solution of this integral yields

$$\delta_A = \frac{PR^3}{EI}K_\delta$$

where $K_\delta = \frac{1-c}{2}[\cos\gamma(\alpha-\gamma-\sin\alpha\cos\alpha) + \sin\gamma \times \cos^2\alpha] + C[(\alpha-\gamma)(1+\cos\gamma) - \sin(\alpha-\gamma) - \sin\alpha + \sin\gamma]$

NOTATION

δ_A = deflection of free end (in.)
θ_A = angle of slope (radians)
ϕ_A = angle of twist (radians)
α = subtended angle (deg.)
γ = angle locating actual load P (deg.)
γ_o = angle locating a given cross section of the beam (deg.)
$ds = Rd\gamma_o$ = differential length
r = radius of cross section (in.)
$C = EI/GJ$

The method of finding the angle of twist or change of slope is the same as finding the deflection, with one exception. Instead of a fictitious load, a fictitious twisting couple is applied at the free end and in the plane of cross section. This was done with the following result: The angle of slope, $\theta_A = (PR^2/EI)K_\theta$, the angle of twist $\phi_A = (PR^2/EI)K_\phi$. Using the same nomenclature and beam configurations as in Fig. 12 we obtain

$$K_\theta = [(1-C)/2)][\cos\gamma\sin^2\alpha - \sin\gamma(\alpha-\gamma + \sin\alpha\cos\alpha)] + C[\cos\gamma - \cos\alpha - (\alpha-\gamma)\sin\gamma]$$

$$K_\phi = [(1-C)/2][\cos\gamma(\alpha-\gamma-\sin\alpha\cos\alpha) + \sin\gamma\cos^2\alpha] + C[(\alpha-\gamma)\cos\gamma - \sin\alpha + \sin\gamma]$$

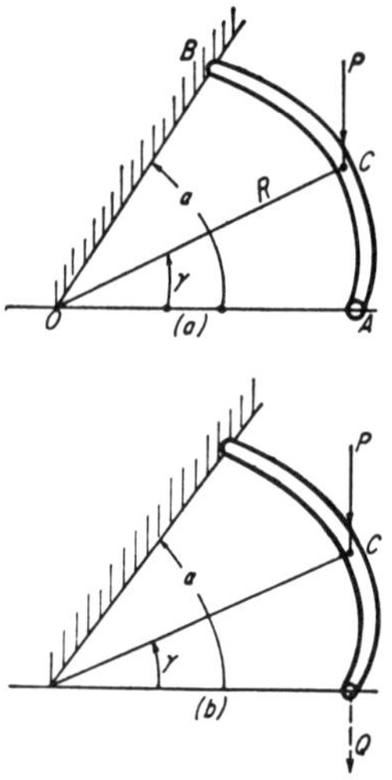

Fig. 12

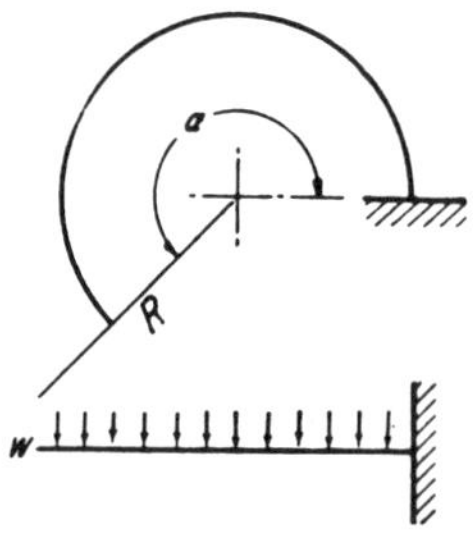

Fig. 13

TRANSVERSE UNIFORM LOAD

The results for the above case of general loading are useful to obtain similar values for any type of loading. It is only necessary to substitute a differential load for the concentrated load, P, in the expressions for δ_A, ϕ_A, and θ_A and integrate. Such integrations have been made for the uniform load (Fig. 13) with the following result: $\delta_A = (WR^3/EI)K_\delta, \theta_A = (WR^2/EI)K_\theta, \phi_A = (WR^2/EI)K_\theta$, where

$$K_\delta = (1/2\alpha)[(1-\cos\alpha)^2 + C(\alpha-\sin\alpha)^2]$$

$$K_\theta = (1/2\alpha)\{(1+C)[\sin\alpha(2-\cos\alpha)-\alpha] - 2C[\cos\alpha(\alpha-\sin\alpha)]\}$$

$$K_\phi = (1/2\alpha)[(1+C)(1-\cos\alpha)^2 - 2C\sin\alpha(\alpha-\sin\alpha)]$$

where, as before, $C = EI/GJ$; since $E/G = 2.6$ (for Poisson's Ratio of 0.3) therefore $C = 2.6\ I/J$, for a circular cross section $I/J = 0.5$, and $C = 1.3$; similarly for rectangular cross section with (1) width equal 1.5 times the height: $C = 1.109$ (2) height equal 1.5 times the width: $C = 2.493$

For useful reference, the results of this study are summarized in Tables 1 and 2, the use of which will be shown in the following example:

Example: Find the deflection δ, angle of slope θ, angle of twist ϕ of the free end of the steel cantilever circular beam, loaded as shown in Fig. 13. The beam is of circular cross section ($r = 0.25$ in.); $R = 10$ in.; $w =$ 1 lb./in. $I = \pi r^4/64 = 0.00306$ in.^{4}; assume $\alpha = 180$ deg. $= 3.14$ radians; $E = 30 \times 10^6$ psi.

Table 1. Circular beams with concentrated and uniform loads acting normal to the plane of curvature. (Cases 1 to 36 are for concentrated loads, Cases 37 to 44 are for uniform loads.)

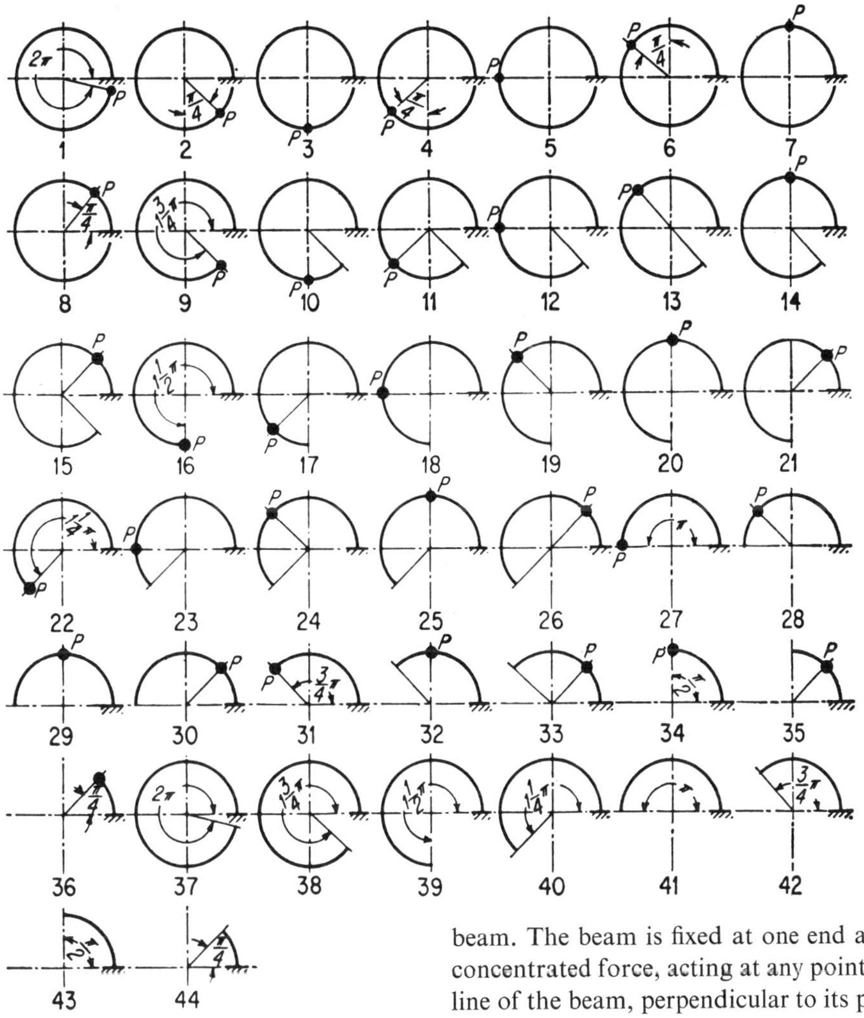

Solution: $W = wR\alpha = 1 \times 10 \times 3.14 = 31.4$ lb. The configuration of our beam corresponds to case 41 of Table 1. From Table 2, our case 41, circular cross section, obtain $K_\delta = 2.679$; $K_\theta = 0.15$; $K_\phi = 1.464$ Thus $\delta = (WR^3/EI)K_\delta = 0.915$ in.; $\theta = (WR^2/EI)K_\theta = 0.005$ radians; and $\phi = (WR^2/EI)K_\phi = 0.047$ radians.

NOTATION

w = unit applied load (lb./in.)
$W = wR\alpha$ = total applied load (lb.)
r = radius of cross section (in.)

The Castigliano theorem can simplify calculations for the *reaction force* at the simple support of a circular beam. The beam is fixed at one end and loaded with a concentrated force, acting at any point along the centerline of the beam, perpendicular to its plane of curvature as in the diagram (*a*). Similar results are also given for three different cases of distributed loading, shown in diagrams *b*, *c*, and *d*.

Following the method outlined in the previous sections, the equations for reaction F were solved as follows. For the case illustrated (*a*), for example, we know that the equation for the end deflection (assuming only the load P is acting, and the free support is removed) is $\delta_P = PR^3K_\delta/EI$. Similarly, when only the force F is acting, with the free support removed, we obtain $\delta_F = FR^3K_F/EI$. Since $\delta_P = \delta_F$, we get $F = PK_\delta/K_F = PK_1$. Values for K_1, K_2, K_3, K_4 for four typical cross sections and various beam configurations (Table 3) are shown in Table 4. These data apply to materials with the Poisson's Ratio of 0.3. The use of Tables 3 and 4 is shown in the following example.

Table 2. *K* factors for circular beams

Case no.	C=1.109 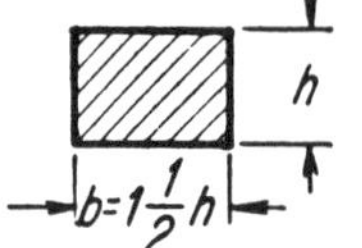			C=1.3 			C=2.493 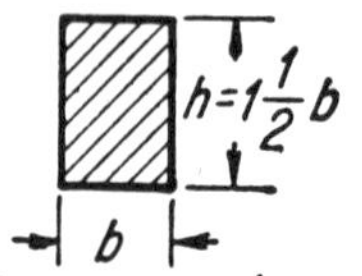		
	K_δ	K_θ	K_ϕ	K_δ	K_θ	K_ϕ	K_δ	K_θ	K_ϕ
1	13.594	0.000	6.625	15.394	0.000	7.226	26.638	0.000	10.973
2	11.276	−4.424	8.372	13.350	−4.851	9.418	23.493	−7.520	15.953
3	7.389	−6.078	5.171	8.576	−6.719	5.976	15.987	−10.723	11.001
4	2.957	−4.821	0.113	3.644	−5.413	0.310	7.938	−9.105	1.545
5	0.1712	−2.218	−3.313	0.471	−2.600	−3.613	2.345	−4.986	−5.487
6	−0.674	−0.136	−3.566	−0.585	−0.303	−3.976	−0.034	−1.346	−6.536
7	−0.421	0.547	−1.687	−0.408	0.506	−1.892	−0.323	0.250	−3.169
8	−0.073	0.261	0.008	−0.073	0.258	0.023	−0.070	0.240	0.113
9	13.436	0.298	10.081	15.233	0.306	11.301	26.460	0.357	18.920
10	11.196	−3.689	10.777	12.790	−4.089	12.299	22.747	−6.591	21.805
11	8.183	−4.864	8.535	9.472	−5.405	9.884	17.526	−8.778	18.317
12	2.194	−3.625	1.948	2.784	−4.080	2.495	6.466	−6.928	5.912
13	0.405	−2.168	−0.304	0.724	−2.479	−0.108	2.716	−4.420	1.122
14	−0.497	−0.460	−1.179	−0.413	−0.570	−1.213	0.106	−1.261	−1.427
15	−0.230	0.070	−0.273	−0.222	0.055	−0.273	−0.172	−0.034	−0.270
16	12.413	1.054	10.195	14.145	1.150	11.545	24.964	1.750	19.978
17	12.001	−0.077	10.714	13.693	−0.084	12.184	24.262	−0.127	21.369
18	6.580	−3.271	7.816	7.669	−3.567	9.117	14.47	−5.417	17.247
19	1.965	−2.502	2.704	2.447	−2.729	3.313	5.457	−4.145	7.118
20	0.186	−1.188	0.239	0.359	−1.295	0.421	1.443	−1.967	1.563
21	−0.199	−0.269	−0.373	−0.167	−0.293	−0.372	0.029	−0.445	−0.363
22	10.092	1.866	7.245	11.535	2.144	8.200	20.547	3.882	14.154
23	7.395	−0.774	7.270	8.477	−0.716	8.330	15.237	−0.354	14.956
24	3.695	−1.673	4.433	4.288	−1.715	5.154	7.994	−1.979	9.654
25	0.992	−1.133	1.254	1.197	−1.171	1.504	2.481	−1.412	3.070
26	0.016	−0.298	−0.240	0.043	−0.306	−0.256	0.213	−0.357	−0.360
27	6.797	2.218	3.313	7.697	2.600	3.613	13.319	4.986	5.487
28	4.331	0.136	3.566	4.873	0.303	3.976	9.378	1.93	6.795
29	1.688	−0.547	1.688	1.892	−0.506	1.892	3.170	−0.250	3.170
30	0.247	−0.261	−0.008	0.276	−0.258	−0.023	0.460	−0.240	−0.113
31	3.502	1.866	0.610	3.859	2.144	0.469	6.090	3.883	−0.412
32	1.944	0.460	1.179	2.110	0.570	1.213	3.147	1.262	1.427
33	0.430	−0.070	0.273	0.457	−0.055	0.273	0.622	0.034	0.270
34	1.180	1.055	−0.088	1.248	1.150	−0.236	1.673	1.747	−1.173
35	0.348	0.160	0.331	0.360	0.175	0.340	0.435	0.265	0.396
36	0.158	0.298	0.240	0.161	0.306	0.256	0.177	0.357	0.360
37	3.484	−2.163	0.000	4.084	−2.450	0.000	7.832	−4.240	0.00
38	3.890	−2.110	0.900	4.557	−2.374	1.053	8.731	−4.017	2.012
39	3.946	−1.502	1.568	4.607	−1.638	1.820	8.737	−2.487	3.393
40	3.400	−0.644	1.705	3.992	−0.627	1.935	7.180	−0.526	3.368
41	2.379	0.054	1.343	2.679	0.150	1.464	4.552	0.746	2.224
42	1.260	0.349	0.756	1.371	0.425	0.780	2.064	0.902	0.930
43	0.433	0.288	0.268	0.453	0.314	0.260	0.577	0.477	0.206
44	0.061	0.095	0.038	0.062	0.098	0.035	0.067	0.112	0.016

Example: A steel circular beam, loaded as shown in the diagram (*b*), has $R = 10$ in., $\alpha = 180$ deg. $= 3.141$ radians, $w = 50$ lb./in. The beam has a circular cross section. Find the reaction force *F*.

Solution: $F = (wr\alpha)K_2$. The configuration of the beam corresponds to case 21 in Table 3. From Table 4 for circular cross section case 21, we obtain $K_2 = 0.348$, thus $F = 50 \times 10 \times 3.141 \times 0.348 = 547$ lb.

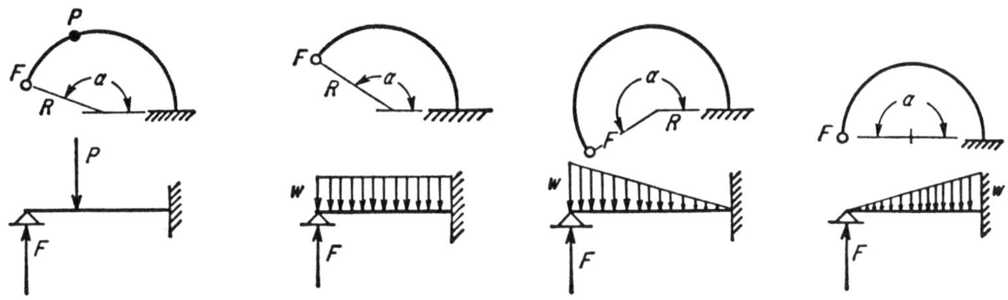

Concentrated load $F = PK_1$ (a)–cases 1 to 16

Uniform load $F = (wR\alpha)K_2$ (b)–cases 17 to 24

Uniformly decreasing load $F = (\frac{1}{2} wR\alpha)K_3$ (c)–cases 25 to 32

Uniformly increasing load $F = (\frac{1}{2} wR\alpha)K_4$ (d)–cases 33 to 40

Fig. 14

Table 3. Circular-beam loading cases (Cases 1 to 16 are for concentrated loads, Cases 17 to 40 are for uniform loads)

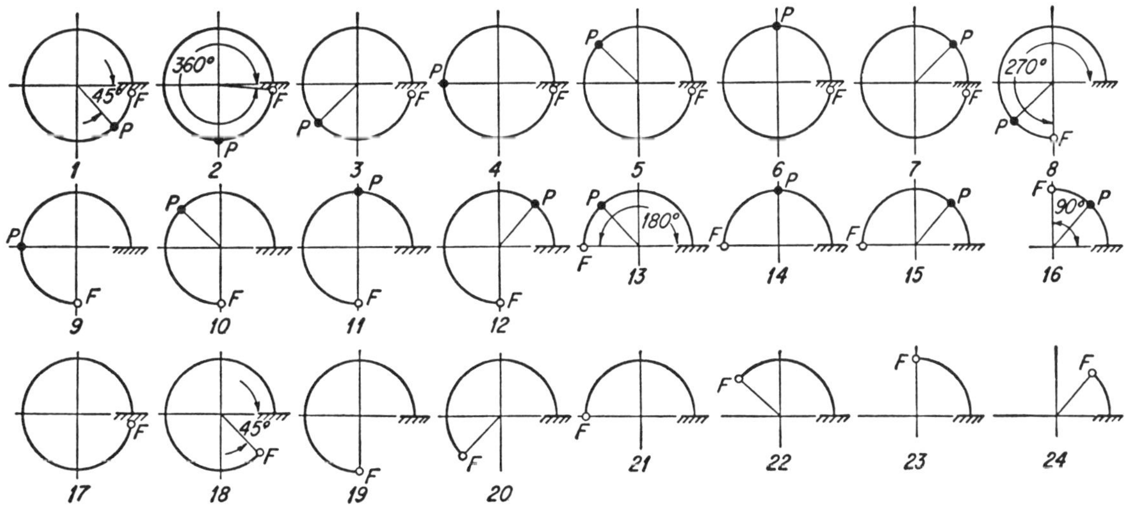

Uniform load

25 26 27 28 29 30 31 32

Uniform decreasing load

33 34 35 36 37 38 39 40

Uniform increasing load

Table 4. *K* factors for circular-beam sections and loadings

Case numbers for:	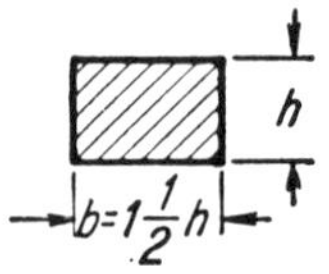							
K_1, K_2, K_3, K_4	K_1	K_2	K_3	K_4	K_1	K_2	K_3	K_4
1, 17, 25, 33	0.859	0.256	0.442	0.071	0.865	0.265	0.452	0.079
2, 18, 26, 34	0.545	0.290	0.480	0.096	0.558	0.300	0.494	0.104
3, 19, 27, 35	0.221	0.318	0.511	0.124	0.240	0.326	0.520	0.132
4, 20, 28, 36	0.014	0.337	0.529	0.148	0.032	0.342	0.530	0.150
5, 21, 29, 37	−0.048	0.350	0.534	0.166	−0.035	0.348	0.532	0.164
6, 22, 30, 38	−0.031	0.360	0.539	0.180	−0.027	0.354	0.533	0.175
7, 23, 31, 39	−0.006	0.369	0.544	0.190	−0.005	0.363	0.539	0.187
8, 24, 32, 40	0.785	0.373	0.548	0.197	0.792	0.371	0.546	0.196
9	0.477				0.490			
10	0.159				0.174			
11	0.015				0.026			
12	−0.014				−0.015			
13	0.637				0.635			
14	0.248				0.246			
15	0.036				0.036			
16	0.295				0.288			

Case numbers for:	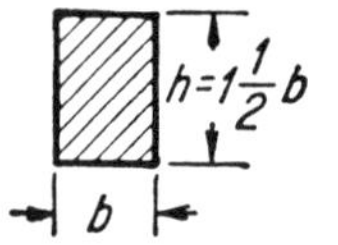				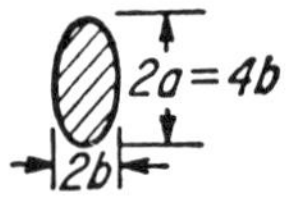			
K_1, K_2, K_3, K_4	K_1	K_2	K_3	K_4	K_1	K_2	K_3	K_4
1, 17, 25, 33	0.879	0.294	0.484	0.104	0.883	0.302	0.494	0.111
2, 18, 26, 34	0.601	0.330	0.527	0.130	0.614	0.340	0.536	0.140
3, 19, 27, 35	0.301	0.350	0.546	0.154	0.319	0.351	0.553	0.160
4, 20, 28, 36	0.089	0.348	0.538	0.160	0.106	0.352	0.540	0.164
5, 21, 29, 37	−0.008	0.342	0.523	0.160	0.020	0.340	0.521	0.159
6, 22, 30, 38	−0.012	0.338	0.513	0.162	−0.008	0.332	0.507	0.157
7, 23, 31, 39	−0.003	0.345	0.518	0.172	−0.005	0.337	0.509	0.166
8, 24, 32, 40	0.812	0.363	0.536	0.190	0.817	0.360	0.531	0.186
9	0.530				0.054			
10	0.219				0.232			
11	0.044				0.052			
12	0.000				0.000			
13	0.620				0.616			
14	0.238				0.236			
15	0.034				0.034			
16	0.260				0.249			

Chap. IV

REFERENCES

Section	*Ref. No.*	
46	1	BRUHN, E. F. and SCHMITT, A. F.: *Analysis and Design of Aircraft Structures*, Vol. I., Tri-State Offset Co., 1958
47	2	SEELY, F. B., and SMITH, J. O.: *Advanced Mechanics of Materials*, John Wiley, N.Y., 1962
48	3	VAN DEN BROOK, S. A.: *Elastic Energy Theory*, John Wiley, N.Y., 1950
49	4	HOGAN, B. M.: "Circular Beams Loaded Normal to the Plane of Curvature," *A.S.M.E. Transaction*, Vol. 60, 1938
49	5	GRIFFEL, W.: "The Theory of Elastic Energy," *Product Engineering*, March 2, 1964, March 16, 1964, March 30, 1964, April 13, 1964

Chapter V

FLAT PLATES

50. FLAT-PLATE DESIGN

(*Ref. 1, 2, 3*)

NOTATION

W = total applied load (lb.)
ω = unit applied load (psi)
w = unit applied load (lb./in.)
E = modulus of elasticity (psi)
a, b = plate dimensions (in.)
K = loading-support factor for deflection, also edge force factor (dimensionless)
K_1 = loading-support factor for stress, also corner force factor (dimensionless)
K_2 = loading-support factor for slope (dimensionless)
R = corner force (lb.)
V_x = edge force distributed along X axis (lb./in.)
V_y = edge force distributed along Y axis (lb./in.)
y = vertical deflection (in.)
θ = slope of plate measured from horizontal (radians)
L = length (in.)
M = end moment (in.-lb./in.)
r_o = radius of a circular area on which load is acting (in.)
S = unit stress at surface of plate (psi)
S_a = stress in the direction of dimension a (psi)
S_b = stress in the direction of dimension b (psi)
S_r = unit stress at surface of plate in radial direction (psi)
S_t = unit stress at surface of plate in tangential direction (psi)
t = plate thickness (in.)
ν = Poisson's Ratio
$m = 1/\nu$

ASSUMPTIONS

The equations of this section apply only if: (1) the plate is flat, of uniform thickness, and of homogenous isotropic material; (2) the thickness is not more than about one-quarter of the least transverse dimension, and the maximum deflection is not more than about one-half the thickness; (3) all forces—loads and reactions—are normal to the plane of the plate; (4) the plate is nowhere stressed beyond the elastic limit.

EQUATIONS

The presentation of equations for deflection, stress, and edge slope for plates of various forms, loaded and supported in various ways, follows the established pattern used throughout this book, namely, an introduction, into the basic equation, of a dimensionless load-support factor K. The text book equations for the deformation of thin plates contain the Poisson's Ratio and are rather long, and tedious. They are based on certain assumptions as to properties of materials, regularity of form, and boundary conditions that are only approximately true. Also, they are derived by mathematical procedures that often involve further approximation. In general, therefore, great precision in numerical work is not justified as the result it yields does not correspond to a real condition. Therefore, it is suggested that the equations for one particular value of Poisson's Ratio, $\nu = 0.3$, a value used for steel, aluminum, magnesium be presented. A considerable change of the Poisson's Ratio will only slightly change the stress and deflection of the plate. (*See* Appendix) Thus, the tabulated values of K could be used for any structural material of plate. All that is needed is to select from the tables the loading condition of the plate and to find the numerical values of K, K_1, K_2, which will be used with any of the following equations:

Table 1—*Circular Plates*

$$y = \frac{KWa^2}{Et^3} \qquad S = \frac{K_1 W}{t^2} \qquad \theta = \frac{K_2 Wa}{Et^3}$$

Table 2—*Rectangular, Square, Triangular and Elliptical Plates*

$$y = \frac{K\omega L^4}{Et^3} \qquad S = \frac{K_1 \omega L^2}{t^2}$$

Table 3—*Circular Plates with End Moments*

$$y = \frac{KMa^2}{Et^3} \qquad S = \frac{K_1 M}{t^2} \qquad \theta = \frac{K_2 Ma}{Et^3}$$

$$\left.\begin{aligned} S &= \frac{K_1 M}{at^2} \\ \theta &= \frac{M}{K_2 Et^3} \end{aligned}\right\} \text{for Cases 6 and 7 only}$$

Table 4—*Partially Loaded Rectangular Plates with all Edges Supported*

$$y = \frac{KwL^3}{Et^3} \qquad S = \frac{K_1 W}{t^2}$$

Table 5—*Corner and Edge Forces for Simply Supported Rectangular Plates*

$$R = K_1\, \omega\, ab \qquad V = K\, \omega\, L$$

Note: A rectangular plate supported in some way along the edges and loaded laterally will usually induce not only reactions, distributed along the boundary of the plate, (edge forces) but also concentrated reactions at the corners (corner forces). Since the corners of the plate have a tendency to rise up under the action of the applied load the concentrated force R must be applied to prevent it.

EDGE CONDITIONS

Quite often it is difficult to decide whether a plate should be calculated as freely supported or fixed, whether a load should be assumed uniformly or otherwise distributed. In any such case it is good practice to calculate the desired quantity on the basis of each of two assumptions representing limits between which the actual conditions must lie.

Example 1: A circular steel plate, 0.2 in. thick and 20 in. in diameter is supported along the edge and loaded with a uniformly distributed load of 3 psi. It is required to determine the deflection and the maximum stress.

Solution: This comes under Table 1, case 1, from where we obtain for $a/r_o = 1$

$$K = 0.212 \text{ and } K_1 = 0.398$$

The total applied load $W = 3 \times 3.14 \times 100 = 942$ lb.

Taking $E = 30 \times 10^6$ psi we obtain:

$$y_{\max} = K\frac{Wa^2}{Et^3} = 0.212\frac{942 \times 100}{30 \times 10^6 \times 0.008} = 0.083 \text{ in.}$$

$$S_{\max} = K_1\frac{W}{t^2} = 0.398\left(\frac{942}{0.04}\right) = 9{,}370 \text{ psi}$$

Example 2: The plate of Example 1 is *partially* fixed at the edges, so that when the uniform load of 3 psi is applied, the plate, instead of remaining horizontal at the edges, assumes a slope there of 0.25 deg. It is required to determine the stress at the center and at the edge under these conditions.

Solution: The principle of superposition is used. The stresses at the edge and at the center are first found on the assumption of true fixity. Then the uniform edge moment (Table 3, case 1) necessary to cause an edge slope of 0.25 deg. is found. Finally the stresses produced by such an edge moment are superposed on the stresses formerly found; the results represent the true stresses. For fixed edges (Table 1, case 2)

At edge: For $a/r_o = 1$, $K_1 = 0.239$ and $S = 0.239 \times (942/.04) = 5{,}630$ psi.

At center: For $a/r_o = 1$, $K_1 = 0.159$ and $S = 0.159 \times (942/.04) = 3{,}740$ psi.

From Table 3, case 1: $K_2 = 8$, $K_1 = 6$.

For $\theta = 0.25$ deg. $= 0.00436$ radian:

$$0.00436 = \frac{8 \times M \times 10}{30 \times 10^6 \times 0.008} = 333 \times 10^{-6} M$$

and

$$M = 13.08 \text{ in.-lb., per in.}$$

therefore

$$S = \frac{6 \times 13.08}{0.04} = 1{,}960 \text{ psi}$$

The resultant stress at the edge:

$$S = 5{,}630 - 1{,}960 = 3{,}670 \text{ psi}$$

The resultant stress at the center:

$$S = 3{,}740 + 1{,}960 = 5{,}700 \text{ psi}$$

TABLE I . . Circular Plates

	LOADING	 CIRCULAR SOLID PLATE  CIRCULAR PLATE WITH CONCENTRIC HOLE OR CIRCULAR FLANGE.	a/b or a/r_0: 1	1.25	1.5	2	3	4	5	Factors: K for deflection K_1 for stress
1	Edges supported. Uniform load over concentric circular area of radius r_0	$W=\omega\pi r_0^2$ 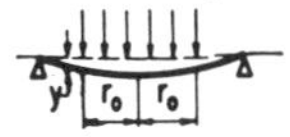	0.212	0.295	0.350	0.413	0.469	0.492	0.503	K
			0.398	0.568	0.700	0.900	1.167	1.353	1.500	K_1 (S_r at center)
2	Edges fixed. Uniform load over concentric circular area of radius r_0	$W=\omega\pi r_0^2$ 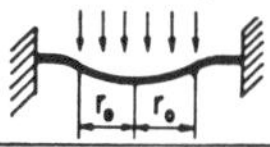	0.053	0.080	0.103	0.135	0.168	0.183	0.192	K
			0.159	0.244	0.328	0.481	0.717	0.892	1.031	K_1 (S_r, S_t, at center)
3	Outer edge supported. Uniform load over entire actual surface	$W=\omega\pi(a^2-b^2)$ 	—	0.163	0.237	0.290	0.295	0.297	0.269	K
			—	0.524	0.559	0.612	0.673	0.744	0.725	K_1 (S_t at inner edge)
4	Outer edge supported. Uniform load along inner edge	W 	—	0.341	0.519	0.672	0.734	0.724	0.704	K
			—	1.10	1.26	1.48	1.88	2.17	2.34	K_1 (S_t at inner edge)
5	Inner edge supported. Uniform load over entire actual surface	$W=\omega\pi(a^2-b^2)$ 	—	0.179	0.281	0.383	0.437	0.444	0.434	K (at outer edge)
			—	0.584	0.682	0.867	1.196	1.458	1.688	K_1 (S_t at inner edge)
6	Outer edge fixed and supported. Uniform load over entire actual surface	$W=\omega\pi(a^2-b^2)$ 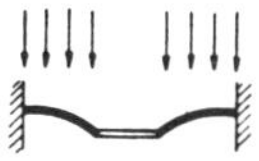	—	0.0018	0.0035	0.024	0.046	0.055	0.0576	K
			—	0.0072	0.025	0.069	0.145	0.192	0.222	K_1 (S_t at inner edge)
			—	0.093	0.148	0.204	0.235	0.240	0.242	K_1 (S_r at outer edge)
7	Outer edge fixed and supported. Uniform load along inner edge	W 	—	0.005	0.024	0.081	0.171	0.215	0.237	K
			—	0.025	0.087	0.269	0.670	1.018	1.300	K_1 (S_t at inner edge)
			—	0.195	0.320	0.455	0.539	0.538	0.532	K_1 (S_r at outer edge)
8	Outer edge fixed and supported, inner edge fixed. Uniform load over entire actual surface	$W=\omega\pi(a^2-b^2)$ 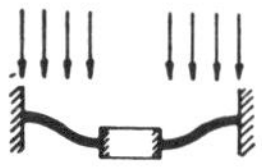	—	0.0004	0.003	0.010	0.023	0.031	0.037	K
			—	0.036	0.065	0.104	0.151	0.176	0.192	K_1 (S at inner edge)
			—	0.062	0.105	0.153	0.195	0.212	0.221	K_1 (S at outer edge)
9	Outer edge fixed and supported, inner edge fixed. Uniform load along inner edge	W 	—	0.0013	0.0064	0.024	0.062	0.092	0.114	K
			—	0.115	0.220	0.405	0.703	0.933	1.130	K_1 (S_r at inner edge)
			—	0.098	0.168	0.257	0.347	0.390	0.415	K_1 (S_r at outer edge)
10	Inner edge fixed and supported. Uniform Load over entire actual surface	$W=\omega\pi(a^2-b^2)$ 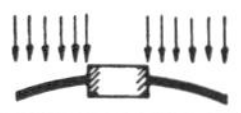	—	0.002	0.010	0.040	0.105	0.152	0.187	K (at outer edge)
			—	0.119	0.235	0.442	0.770	1.014	1.22	K_1 (S_r at inner edge)
11	Inner edge fixed and supported. Uniform load along outer edge	W 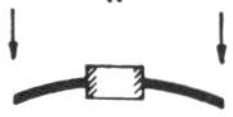	—	0.0051	0.025	0.088	0.209	0.293	0.350	K (at outer edge)
			—	0.227	0.428	0.753	1.205	1.514	1.745	K_1 (S_r at inner edge)
12	Outer edge supported, inner edge fixed. Uniform load over entire actual surface	$W=\omega\pi(a^2-b^2)$ 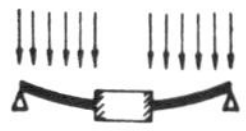	—	0.003	0.018	0.053	0.104	0.141	0.163	K
			—	0.108	0.192	0.314	0.433	0.491	0.526	K_1 (S_r at inner edge)
13	Both edges fixed. Balanced loading (piston)	$W=\omega\pi(a^2-b^2)$ 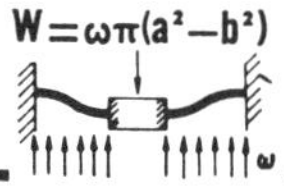	—	0.0007	0.003	0.014	0.039	0.061	0.077	K
			—	0.080	0.156	0.302	0.551	0.756	0.927	K_1 (S_r at inner edge)

TABLE II . . Rectangular, square, triangular and

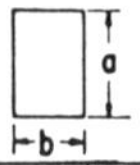

1 . . RECTANGULAR, SOLID

	a/b	1.0		1.1		1.2		1.3		1.4		1.5	
	MANNER OF LOADING AND CASE NUMBER	K	K_1	K	K_1	K	K_1	K	K_1	K	K_1	K	K_1
1	All edges supported Uniform load over entire surface	0.044	0.29	0.053	0.33	0.062	0.38	0.070	0.42	0.077	0.45	0.084	0.49
2	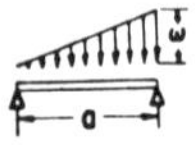 All edges supported Distributed load varying linearly along length	0.022	0.16	0.028	0.18	0.030	0.21	0.035	0.23	0.040	0.25	0.043	0.26
3	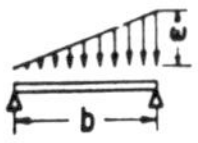 All edges supported Distributed load varying linearly along breadth	0.022	0.16	0.026	0.18	0.031	0.21	0.035	0.23	0.038	0.25	0.042	0.26
4	All edges fixed Uniform load over entire surface	0.014	0.31	0.016	0.35	0.019	0.38	0.021	0.41	0.023	0.44	0.024	0.45
5	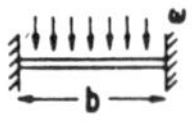 Long edges fixed, short edges supported Uniform load over entire surface	0.021	0.420	0.023	0.44	0.24	0.46	0.025	0.47	0.026	0.49	0.027	0.50
6	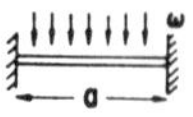 Short edges fixed, long edges supported Uniform load over entire surface	0.021	0.42	0.027	0.47	0.034	0.52	0.042	0.56	0.050	0.60	0.058	0.63
7	One long edge clamped, other three edges supported Uniform load over entire surface	0.030	0.50	0.033	0.54	0.038	0.58	0.040	0.61	0.042	0.63	0.046	0.66
8	One short edge clamped, other three edges supported Uniform load over entire surface	0.030	0.500	0.040	0.54	0.050	0.58	0.059	0.61	0.068	0.63	0.071	0.67
9	One short edge free, other three edges supported Uniform load over entire surface	0.140	0.67	0.146	0.72	0.15	0.74	0.15	0.75	0.16	0.76	0.16	0.77
10	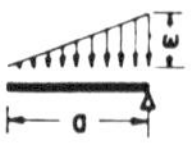 One short edge free, other three edges supported Distributed load varying linearly along length	0.040	0.20	0.041	0.22	0.045	0.24	0.047	0.26	0.048	0.27	0.050	0.28
11	 One long edge free, other three edges supported Uniform load over entire surface	0.14	0.67	0.13	0.62	0.12	0.57	0.12	0.52	0.11	0.48	0.11	0.45
12	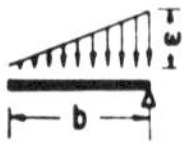 One long edge free, other three edges supported Distributed load varying linearly along length	0.040	0.20	0.039	0.19	0.036	0.18	0.035	0.18	0.033	0.17	0.033	0.16
13	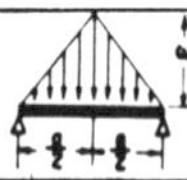 All edges supported Distributed load in form of a triangular prism	0.029	0.204	0.034	0.234	0.040	0.262	0.045	0.287	0.050	0.311	0.054	0.332

2 . . RECTANGULAR SOLID

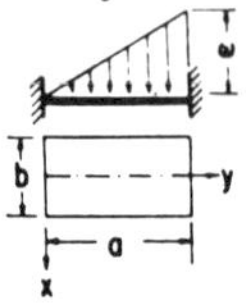

b/a		0.6		0.8		1		1.2		1.4		1.6		1.8		2.0
Manner of loading and case number		K	K_1	K	K_1	K	K_1	K	K_1	K	K_1	K	K_1	K	K_1	K
All edges fixed, distributed load varying linearly along length	14	0.0016	0.13	0.0047	0.14	0.0074	0.17	0.0097	0.18	0.011	0.18	0.013	0.19	0.013	0.19	0.014
	15	0.0016	0.064	0.0047	0.069	0.0074	0.076	0.0097	0.071	0.011	0.061	0.013	0.051	0.013	0.041	0.014
	16	0.0016	0.083	0.0047	0.18	0.0074	0.24	0.0097	0.26	0.011	0.30	0.013	0.31	0.013	0.31	0.014
	17	0.0016	0.021	0.0047	0.050	0.0074	0.090	0.0097	0.12	0.011	0.15	0.013	0.16	0.013	0.17	0.014
	18	0.0016	0.041	0.0047	0.063	0.0074	0.090	0.0097	0.10	0.011	0.11	0.013	0.12	0.013	0.11	0.014

4 . . ELLIPTICAL, SOLID

	a/b	1.0		1.1		1.2		1.3		1.4		1.5		1.6	
	Manner of loading and case number	K	K_1	K	K_1	K	K_1	K	K_1	K	K_1	K	K_1	K	K_1
22	Edge supported Uniform load over entire surface	0.70	1.24	0.84	1.42	0.95	1.57	1.06	1.69	1.17	1.82	1.26	1.92	1.34	2.04
23	Edge fixed Uniform load over entire surface	0.171	0.75	0.20	0.90	0.25	1.04	0.28	1.14	0.30	1.25	0.30	1.34	0.33	1.41

1.6		1.7		1.8		1.9		2.0		2.5		3.0		3.5		4.0			
K	K_1	K	K_1	K	K_1	K	K_1	K	K_1	K	K_1	K	K_1	K	K_1	K	K_1	L	Locations of stress and deflection
0.091	0.52	0.096	0.54	0.10	0.57	0.11	0.59	0.11	0.61	0.12	0.61	0.13	0.61	0.14	0.61	0.14	0.61	b	S max. at center Y max. at center
0.048	0.28	0.050	0.30	0.053	0.31	0.058	0.32	0.060	0.34	0.070	0.38	0.078	0.43	0.086	0.470	0.091	0.49	b	
0.045	0.28	0.048	0.30	0.051	0.31	0.053	0.32	0.056	0.32	0.063	0.35	0.067	0.37	0.069	0.38	0.070	0.38	b	
0.025	0.47	0.026	0.48	0.027	0.49	0.027	0.49	0.028	0.50	0.028	0.50	0.028	0.50	0.028	0.50	0.028	0.50	b	S max. at center of long edges Y max. at center
0.027	0.50	0.028	0.50	0.028	0.50	0.028	0.50	0.028	0.50	0.028	0.50	0.028	0.50	0.028	0.50	0.028	0.50	b	S max. at center of fixed edges Y max. at center
0.066	0.65	0.073	0.67	0.080	0.69	0.086	0.70	0.099	0.71	0.13	0.71	0.13	0.71	0.13	0.71	0.13	0.71	b	S max. at center of fixed edges Y max. at center
0.047	0.68	0.048	0.70	0.050	0.71	0.052	0.72	0.054	0.740	0.056	0.74	0.057	0.74	0.058	0.75	0.058	0.75	b	S max. at center of fixed edge Y max. at center
0.080	0.68	0.083	0.70	0.090	0.71	0.095	0.72	0.10	0.74	0.122	0.74	0.132	0.75	0.137	0.75	0.139	0.75	b	S max. at center of fixed edge Y max. at center
0.16	0.77	0.16	0.78	0.16	0.78	0.16	0.79	0.16	0.79	0.17	0.80	0.17	0.80	0.17	0.80	0.17	0.80	b	S max. at center of free edge Y max. at center
0.051	0.29	0.052	0.30	0.053	0.31	0.056	0.31	0.058	0.32	0.054	0.35	0.067	0.36	0.069	0.37	0.070	0.37	b	S max. at center of free edge Y max. at center
0.10	0.42	0.094	0.40	0.089	0.38	0.84	0.37	0.080	0.036	0.80	0.36	0.080	0.36	0.080	0.36	0.080	0.36	a	S max. at center of free edge Y max. at center
0.030	0.15	0.030	0.14	0.028	0.13	0.025	0.12	0.025	0.11	—	—	—	—	—	—	—	—	a	S max. at center of free edge Y max. at center
0.058	0.352	0.062	0.369	0.065	0.385	0.068	0.398	0.071	0.411	0.082	0.450	0.085	0.476	0.085	0.476	$a/b = \infty$ 0.091	0.500	b	S max. at center Y max. at center

K_1	Locations of stress and deflection	L
0.19	Max. S_b is at $x = \pm \frac{1}{2}b$; $y = 0.55a$	a
0.036	For S_b at $x = 0$; $y = 0.6a$	a
0.30	Max. S_a is at $x = 0$; $y = a$	a
0.17	For S_a at $x = 0$; $y = 0$	a
0.11	For S_x at $x = 0$; $y = 0.6a$	a

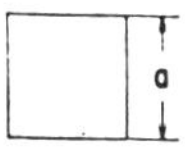

	3 . . SQUARE, SOLID (L = a)	K	K_1	
19	Edges supported above and below (corners held down) Uniform load over entire surface	0.043	0.29	S maximum at center Y maximum at center
20	Edges supported below only (corners free to rise) Uniform load over entire surface	0.044	0.28	S max. at corners on diagonal section Y maximum at center
21	All edges fixed Uniform load over entire surface	0.014	0.31	S maximum at center of each edge Y maximum at center

1.7		1.8		1.9		2.0		2.5		3.0		3.5		4.0			
K	K_1	K	K_1	K	K_1	K	K_1	K	K_1	K	K_1	K	K_1	K	K_1	L	Locations of stress and deflection
1.41	2.09	1.47	2.16	1.53	2.22	1.58	2.26	1.75	2.45	1.88	2.60	1.96	2.70	2.02	2.78	b	S maximum at center Y maximum at center
0.35	1.49	0.36	1.54	0.370	1.59	0.379	1.63	0.40	1.75	0.42	1.84	0.43	1.89	0.43	1.9	b	S maximum at end of shorter principal axis Y maximum at center

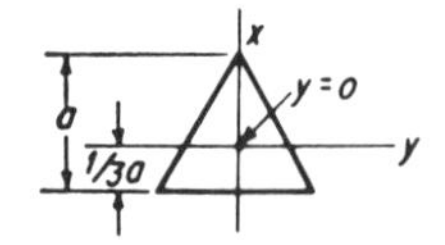

5 . . EQUILATERAL TRIANGLE, SOLID

Manner of loading and case number		K	K_1	Location of stress and deflection	L
Edges supported Distributed load over entire surface	24	0.01	0.149	S_x maximum at y = 0, x = −0.062a Y maximum at center	a
	25	0.01	0.155	S_y maximum at y = 0, x = 0.129a Y maximum at center	a

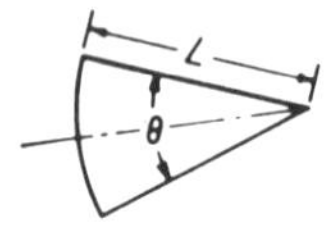

7 . . CIRCULAR SECTOR, SOLID

θ		45		60°		90°		180		
Manner of loading and case number		K	K_1	K	K_1	K	K_1	K	K_1	L
Edges supported Distributed load over entire surface	27	0.0054	0.11	0.01	0.15	0.025	0.22	0.087	0.31	a
	28	0.0054	0.10	0.01	0.15	0.025	0.24	0.087	0.52	a

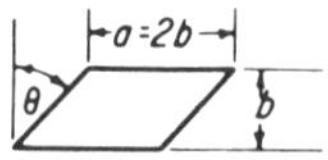

8 . . PARALLELOPIPED (SKEW SLAB)

θ		0°		30		45°		60°		75°	
Manner of loading and case number		K	K_1	K	K_1	K	K_1	K	K_1	K	K_1
All edges supported Distributed load over entire surface	29	—	0.50	—	0.50	—	0.45	—	0.40	—	0.16
Edges b supported Edges a free Distributed load over entire surface	30	—	0.76	—	0.61	—	0.44	—	0.25	—	—

TABLE III . . Circular plates with end moments

Manner of loading and case number	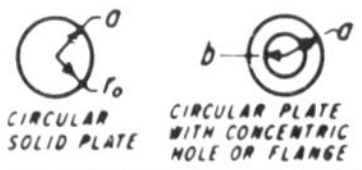	a/b or a/r: 1	1.25	1.50	2	3	4	5	K (deflection) K_1 (stress)
1 No support Uniform edge moment		4	—	—	—	—	—	—	K max. at center
		1	—	—	—	—	—	—	K_1 (S_r = S_t at any point)
2 Outer edge fixed Uniform moment along inner edge		—	0.20	0.48	0.85	0.94	0.80	0.66	K max. at inner edge
		—	0.37	0.86	2.44	4.10	4.84	5.11	K_1 max. at inner edge
3 Inner edge fixed Uniform moment along outer edge	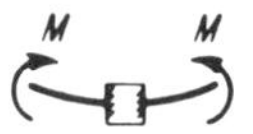	—	0.23	0.66	1.49	2.55	3.10	3.41	K max. at outer edge
		—	6.87	7.50	8.14	8.71	8.94	9.04	K_1 max. at inner edge
4 Outer edge supported Uniform moment along inner edge		—	8.87	6.92	4.65	2.58	1.69	1.21	K max. at inner
		—	27.36	15.60	10.0	7.50	6.80	6.50	K_1 max. S_t at inner
5 Inner edge supported Uniform moment along outer edge		—	10.37	9.23	7.80	6.31	5.62	5.23	K max. at outer
		—	33.3	21.6	16.0	13.5	12.8	12.5	K_1 max. S_t at inner

Manner of loading and case number		r_0/a: 0.1	0.15	0.20	0.25	0.30	0.35	0.40	0.45	0.50	0.55	0.60	0.65	0.70	0.75	0.80	K_2 K_1 stress
6 Edges supported Central couple, trunion loading		0.71	0.97	1.22	1.60	2.0	2.50	3.53	5.60	8.54	12.00	16.30	24.10	41.4	82.0	162.0	K_2 max. at ce
		4.36	3.80	3.27	2.80	2.37	2.10	1.84	1.58	1.41	1.16	1.07	0.90	0.78	0.70	0.57	K_1 max. S at c
7 Edges fixed Central couple, trunion loading		0.87	1.23	1.68	2.31	3.10	4.00	5.45	8.20	12.40	18.0	28.5	44.0	77.9	156.0	314.0	K_2 Max. at ce
		6.06	5.40	4.52	3.40	2.65	2.15	1.75	1.45	1.14	0.89	0.68	0.57	0.47	0.35	0.26	K_1 max. S at c

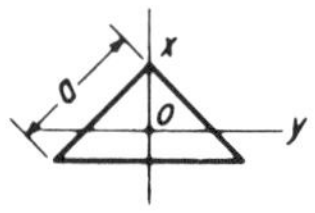

. . RIGHT ANGLE ISOSCELES TRIANGLE, SOLID	Manner of loading and case number		K	K₁	Location of stress and deflection	L
	Edges supported Distributed load over entire surface	26	0.0095	0.13	S_x maximum at center Y maximum at center	a

ocations of stress and deflection
$_t$ maximum (tangential) at midpoint of circular edge maximum at midpoint of circular edge
maximum (radial) on line to midpoint of circular edge maximum at midpoint of circular edge

. maximum at center o data available for deflection
maximum at center o data availble for deflection

TABLE IV 1 . . All edges supported, partially loaded rectangular plates

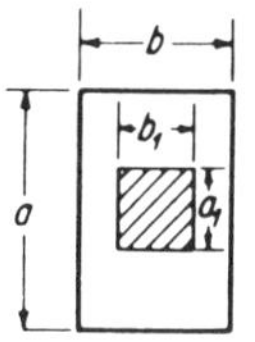

edges supported.
iform load over
ntral rectangular
ea shown shaded.

K₁ factor for maximum stress at center ($S = S_b$)

a_1/b	a/b = 1						a/b = 1.4						a/b = 2					
b_1/b	0	0.2	0.4	0.6	0.8	1.0	0	0.2	0.4	0.8	1.2	1.4	0	0.4	0.8	1.2	1.6	2.0
0	—	1.82	1.38	1.12	0.93	0.76	—	2.0	1.55	1.12	0.84	0.75	—	1.64	1.20	0.97	0.78	0.64
0.2	1.82	1.28	1.08	0.90	0.76	0.63	1.78	1.43	1.23	0.95	0.74	0.64	1.73	1.31	1.03	0.84	0.68	0.57
0.4	1.39	1.07	0.84	0.72	0.62	0.52	1.39	1.13	1.00	0.80	0.62	0.55	1.32	1.08	0.88	0.74	0.60	0.50
0.6	1.12	0.90	0.72	0.60	0.52	0.43	1.10	0.91	0.82	0.68	0.53	0.47	1.04	0.90	0.76	0.64	0.54	0.44
0.8	0.92	0.76	0.62	0.51	0.42	0.36	0.90	0.76	0.68	0.57	0.45	0.40	0.87	0.76	0.63	0.54	0.44	0.38
1.0	0.76	0.63	0.52	0.42	0.35	0.30	0.75	0.62	0.57	0.47	0.38	0.33	0.71	0.61	0.53	0.45	0.38	0.30

Note: Total load $W = \omega a_1 b_1$

2 . . ALL EDGES SUPPORTED, PARTIALLY LOADED RECTANGULAR PLATES

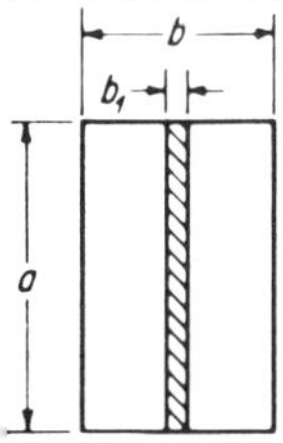

edges supported.
iform load along
e axis of symmetry
rallel to the dimension a
(b_1 very small)

K factor for maximum deflection

b > a	b/a	2	1.5	1.4	1.3	1.2	1.1	1.0
L = a	K	0.108	0.099	0.096	0.092	0.087	0.081	0.074
b < a	a > b	1.1	1.2	1.3	1.4	1.5	2.0	∞
L = b	K	0.088	0.101	0.114	0.126	0.137	0.178	0.227

Note: Use unit applied load w in this case (LB/IN)

TABLE V.. Uniformly loaded and simply supported rectangular plate.

1.. Corner and edge forces for simply supported rectangular plates

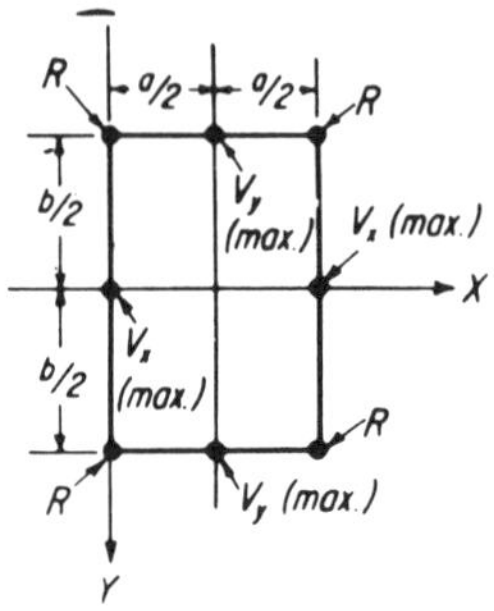

	b/a															
	1	1.1	1.2	1.3	1.4	1.5	1.6	1.7	1.8	1.9	2.0	3.0	4.0	5.0	∞	L
K for V_x (max.)	0.420	0.440	0.455	0.468	0.478	0.486	0.491	0.496	0.499	0.502	0.503	0.505	0.502	0.501	0.500	a
K for V_y (max.)	0.420	0.440	0.453	0.464	0.471	0.480	0.485	0.488	0.491	0.494	0.496	0.498	0.500	0.500	0.500	a
K_1 for R	0.065	0.070	0.074	0.079	0.083	0.085	0.086	0.088	0.090	0.091	0.092	0.093	0.094	0.095	0.095	

Simply supported rectangular plate under hydrostatic pressure.

2.. Corner and edge forces for simply supported rectangular plates

a < b

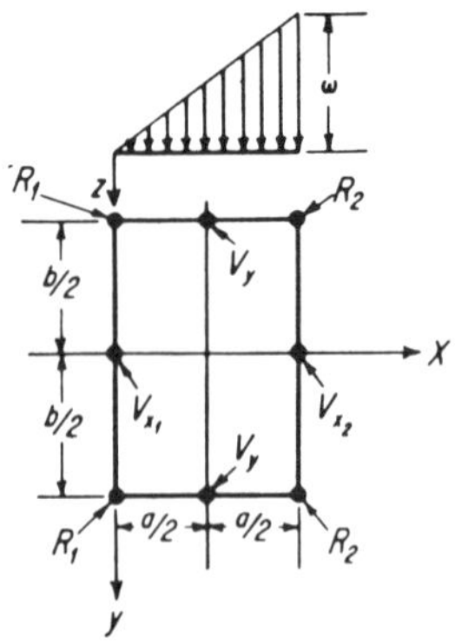

At any point along a dimension $w' = w\frac{x}{a}$

	K			K_1		
b/a	V_{x1}	V_{x2}	V_y	R_1	R_2	Remarks
1.0	0.126	0.294	0.210	0.026	0.039	Use L = a for V_{x1}, V_{x2} Use L = b for V_y Because the load is not symmetrical, the reactions R_1 are different from the reactions R_2, also V_{x1} is different than V_{x2}. The same applies to case V-3.
1.1	0.136	0.304	0.199	0.026	0.038	
1.2	0.144	0.312	0.189	0.026	0.037	
1.3	0.150	0.318	0.178	0.026	0.036	
1.4	0.155	0.323	0.169	0.025	0.035	
1.5	0.159	0.327	0.160	0.024	0.033	
1.6	0.162	0.330	0.151	0.023	0.032	
1.7	0.164	0.332	0.144	0.022	0.030	
1.8	0.166	0.333	0.136	0.021	0.029	
1.9	0.167	0.334	0.130	0.021	0.028	
2.0	0.168	0.335	0.124	0.020	0.026	
3.0	0.169	0.336	0.083	0.014	0.018	
4.0	0.168	0.334	0.063	0.010	0.014	
5.0	0.167	0.334	0.050	0.008	0.011	
∞	0.167	0.333	—	—	—	

NOTES:
1. In this case only, the formula (V) for the corner force R can be used when substituting a for b
2. V_x (max.) and V_y (max.) are at the middle of sides b and a respectively as shown in the figure for this table.

Simply supported rectangular plate under hydrostatic pressure

3 . . Corner and edge forces for simply supported rectangular plates

a > b

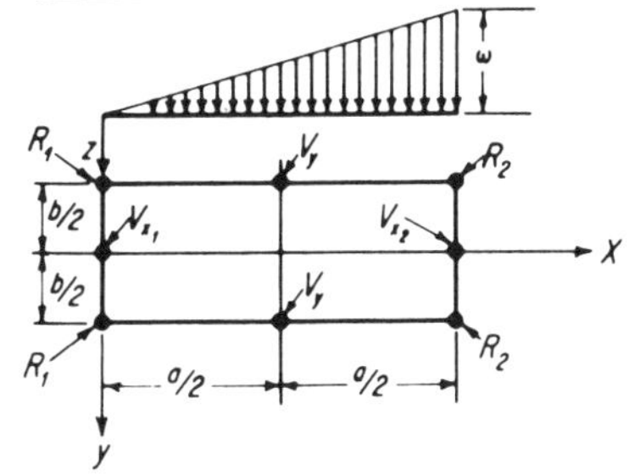

	K			K_1		
a/b	V_{x1}	V_{x2}	V_y	R_1	R_2	Remarks
∞	—	—	0.250	—	—	Use L = a for V_{x1}, V_{x2}
5.0	0.008	0.092	0.250	0.002	0.017	
4.0	0.013	0.112	0.251	0.004	0.020	Use L = b for V_y
3.0	0.023	0.143	0.252	0.006	0.025	
2.0	0.050	0.197	0.251	0.013	0.033	
1.9	0.055	0.205	0.251	0.014	0.034	
1.8	0.060	0.213	0.249	0.016	0.035	
1.7	0.066	0.221	0.248	0.017	0.036	
1.6	0.073	0.230	0.245	0.018	0.037	
1.5	0.080	0.240	0.243	0.020	0.037	
1.4	0.088	0.250	0.239	0.021	0.038	
1.3	0.097	0.260	0.234	0.023	0.039	
1.2	0.106	0.271	0.227	0.024	0.039	
1.1	0.116	0.282	0.220	0.025	0.039	
1.0	0.126	0.294	0.210	0.026	0.039	

Simply supported rectangular plate under a load in the form of a triangular prism

4 . . Corner and edge forces for simply supported rectangular plates

a < b

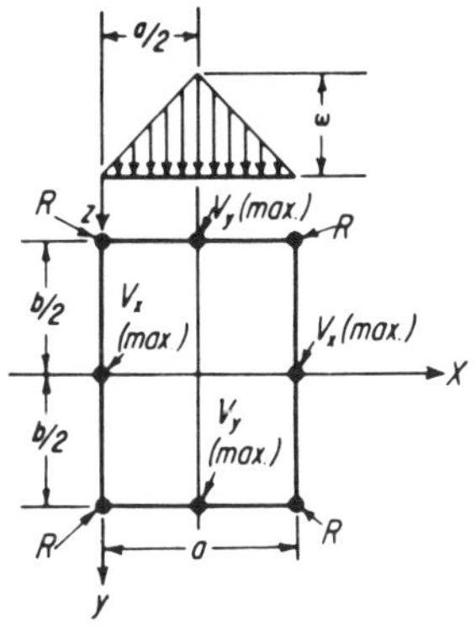

	K		K_1	
b/a	V_x (max.)	V_y (max.)	R	Remarks
1.0	0.147	0.250	0.038	Use L = a for V_x
1.1	0.161	0.232	0.038	Use L = b for V_y
1.2	0.173	0.216	0.037	
1.3	0.184	0.202	0.036	
1.4	0.193	0.189	0.035	
1.5	0.202	0.178	0.034	
1.6	0.208	0.108	0.033	
1.7	0.214	0.158	0.031	
1.8	0.220	0.150	0.030	
1.9	0.224	0.142	0.029	
2.0	0.228	0.135	0.028	
3.0	0.245	0.090	0.019	
∞	0.250	—	—	

Simply supported rectangular plate under a load in the form of a triangular prism

5 . . Corner and edge forces for simply supported rectangular plates

a > b

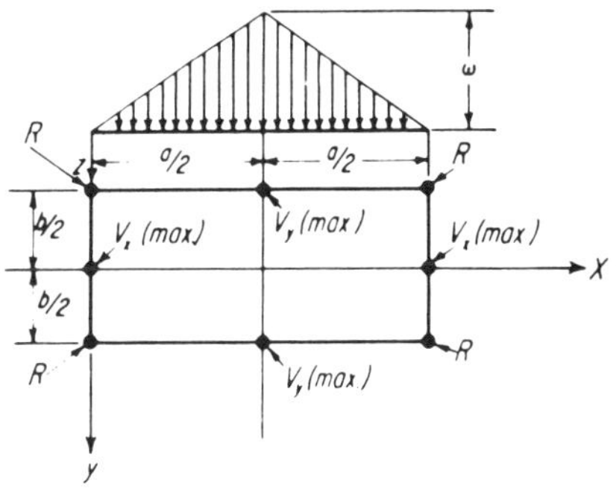

	K		K_1	
a/b	V_x (max.)	V_y (max.)	R	Remarks
∞	—	0.50	—	Use L = a for V_x
3.0	0.027	0.410	0.010	Use L = b for V_y
2.0	0.057	0.365	0.023	
1.9	0.062	0.358	0.024	
1.8	0.098	0.350	0.026	
1.7	0.074	0.342	0.028	
1.6	0.081	0.332	0.029	
1.5	0.090	0.322	0.031	
1.4	0.099	0.311	0.033	
1.3	0.109	0.298	0.035	
1.2	0.120	0.284	0.036	
1.1	0.133	0.268	0.037	
1.0	0.147	0.250	0.038	

51. GRAPHICAL PRESENTATION OF STRESSES AND DEFLECTIONS IN PLATES

NOTATION

W = total applied load (lb.)
ω = unit applied load (psi)
S = maximum unit stress at surface of plate (psi)
y = vertical maximum deflection of plate from original position (in.)
a, b = plate dimensions (in.)
R = outside radius of plate (in.)
r = radius of concentric hole or flange in plate (in.) (For Table 6, cases 1 and 2, radius of a circular area on which load is acting)
t = plate thickness (in.)
E = modulus of elasticity (psi)
L = length (in.). (*See* Table 6)
C = stress loading-support factor for rectangular plates (nondimensional)
K = deflection loading-support factor for rectangular plates (nondimensional)
λ = stress loading-support factor for circular plates
β = deflection loading-support factor for circular plates

The tabulation of constants in Tables 1–5 afford the calculation of stress and deflection of plates of *given* dimensions. This section presents, graphically, data on stress and deflection of rectangular and circular plates for 27 types of loading and edge conditions, for a wide range of plate dimensions. Here too, in all cases considered, it was assumed that the plate is flat, of uniform thickness, not more than about one-quarter of the least transverse dimension and the maximum deflection is not more than about one-half the thickness. All forces-loads are normal to the plane of the plate; the plate is nowhere stressed beyond the elastic limit. In all cases, Poisson's Ratio was taken as 0.3, a value used for steel. A considerable change of this value will only slightly change the stress and deflection. As the thickness of plate is small the additional deflection due to shear is negligible.

The cases of loading of rectangular plates which are considered here are shown in Table 6 and of circular plates in Table 7.

STRESS AND DEFLECTION FORMULAS

1. Rectangular plates (Table 6):

$$S = \frac{C\omega L^2}{t^2} \qquad \text{Cases 1–13}$$

$$S = \frac{C_{14}\omega a_1 b_1}{t^2} \qquad \text{Case 14}$$

$$y = \frac{K\omega L^4}{Et^3} \qquad \text{Case 1–13}$$

2. Circular plates (Table 7):

$$S = \frac{\lambda W}{t^2}$$

$$Y = \frac{\beta W R^2}{Et^3}$$

L may be obtained from Table 6; C and K are numerical factors depending on the magnitude of a/b and may be obtained from the curves of Figs. 1 and 2. a_1 and b_1 are the dimensions of the uniformly loaded portion of the plate. (*See* Table 6, case 14). Values of C_{14} for $a = b$, $a = 1.4b$, $a = 2b$ may be obtained from Figs. 3, 4, and 5, respectively; λ, β are numerical factors depending on the magnitude of R/r and may be obtained from the curves of Figs. 6, 7, 8 and 9.

NUMERICAL EXAMPLES

1. It is required to find the maximum deflection and maximum stress in a rectangular plate simply supported along its four edges and with a uniformly distributed load $\omega = 6$ psi over the entire surface of the plate. The dimensions of the plate and Young's Modulus of Elasticity are: $a = 50$ in., $b = 36$ in., $t = \frac{1}{2}$ in., $E = 30 \times 10^6$ psi.

The loading and edge conditions correspond to those of Table 6, case 7. Therefore for $a/b = 1.39$ we find from curves of Figs. 1 and 2:

$$C_7 = 0.46, K_7 = 0.079$$

From Table 6, case 7: $L = b$

From the data given:

$$S = \frac{0.46 \times 6 \times 36^2}{(1/2)^2} = 14{,}980 \text{ psi}$$

$$Y = \frac{0.079 \times 6 \times 36^4}{30 \times 10^6 \times (1/2)^3} = 0.200 \text{ in.}$$

2. Find the maximum deflection and maximum stress in a circular plate simply supported along the edge and with a uniformly distributed load $\omega = 3$ psi over the entire surface of the plate. The dimensions of the

LOADING AND EDGE CONDITIONS —CASE NUMBER—			STRESS		DEFLECTION		L
			C	FIG.	K	FIG.	
1	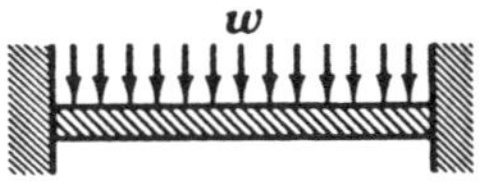	ALL EDGES FIXED. UNIFORM LOAD OVER ENTIRE SURFACE.	C_1	1	K_1	2	b
2	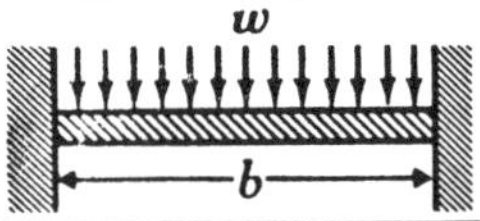	LONG EDGES FIXED, SHORT EDGES SUPPORTED. UNIFORM LOAD OVER ENTIRE SURFACE.	C_2	1	K_2	2	b
3	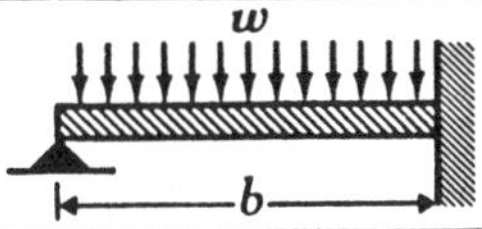	ONE LONG EDGE CLAMPED, OTHER THREE EDGES SUPPORTED. UNIFORM LOAD OVER ENTIRE SURFACE.	C_3	1	K_3	2	b
4	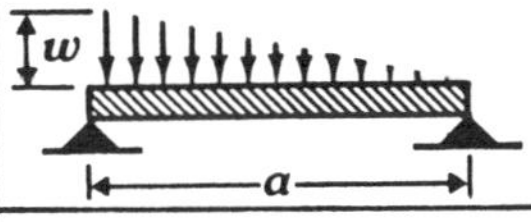	ALL EDGES SUPPORTED. HYDROSTATIC PRESSURE VARYING ALONG LENGTH.	C_4	1	K_4	2	b
5	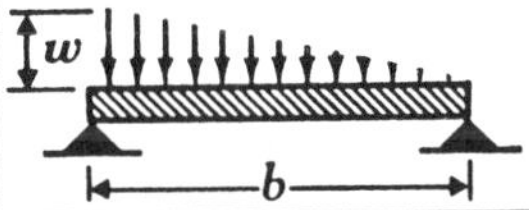	ALL EDGES SUPPORTED. HYDROSTATIC PRESSURE VARYING ALONG BREADTH.	C_5	1	K_5	2	b
6	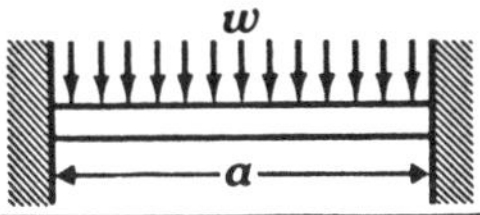	SHORT EDGES FIXED, LONG EDGES SUPPORTED UNIFORM LOAD OVER ENTIRE SURFACE.	C_6	1	K_6	2	b
7	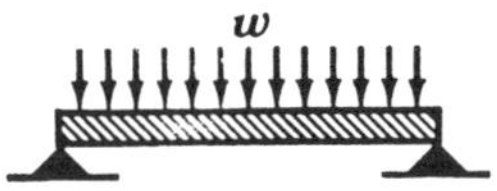	ALL EDGES SUPPORTED. UNIFORM LOAD OVER ENTIRE SURFACE.	C_7	1	K_7	2	b
8	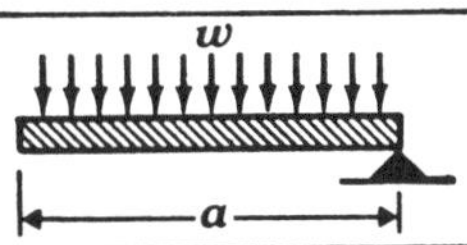	ONE SHORT EDGE FREE, OTHER THREE EDGES SUPPORTED. UNIFORM LOAD OVER ENTIRE SURFACE.	C_8	1	K_8	2	b
9	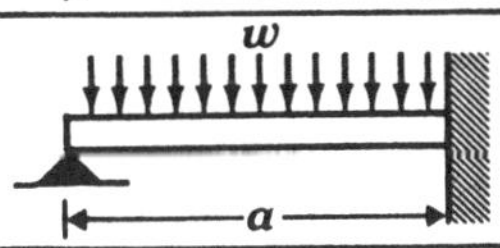	ONE SHORT EDGE CLAMPED, OTHER THREE EDGES SUPPORTED. UNIFORM LOAD OVER ENTIRE SURFACE.	C_9	1	K_9	2	b
10	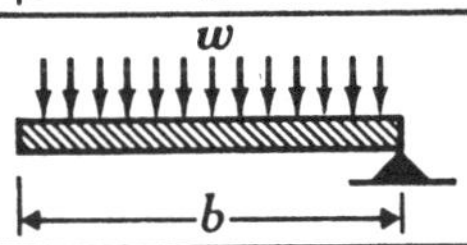	ONE LONG EDGE FREE, OTHER THREE EDGES SUPPORTED. UNIFORM LOAD OVER ENTIRE SURFACE.	10	1	10	2	a
11	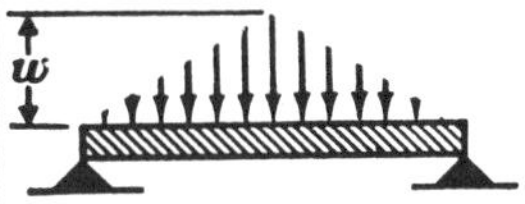	ALL EDGES SUPPORTED. DISTRIBUTED LOAD IN FORM OF A TRIANGULAR PRISM.	11	1	11	2	b
12	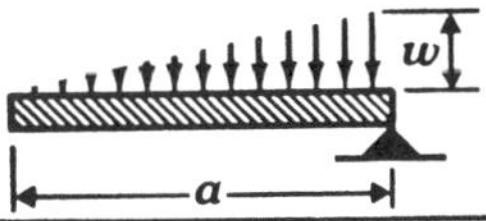	ONE SHORT EDGE FREE, OTHER THREE EDGES SUPPORTED. DISTRIBUTED LOAD VARYING LINEARLY ALONG LENGTH.	12	1	12	2	b
13	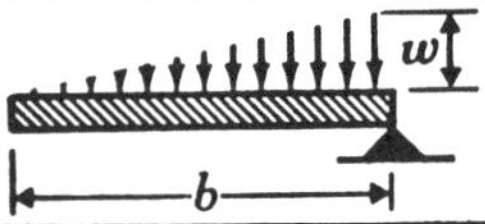	ONE LONG EDGE FREE, OTHER THREE EDGES SUPPORTED. DISTRIBUTED LOAD VARYING LINEARLY ALONG BREADTH.	13	1	13	2	a
14	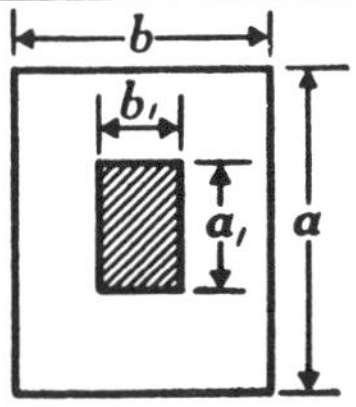	ALL EDGES SUPPORTED, UNIFORMLY DISTRIBUTED LOAD OVER SHADED PORTION.	$S = \frac{C_{14} \times w a_1 b_1}{t^2}$ FIGURES 3, 4, 5				

LOADING AND EDGE CONDITIONS —CASE NUMBER—			TOTAL APPLIED LOAD W (LB)	DEFLECTION		STRESS	
				β	FIG.	λ	FIG.
1	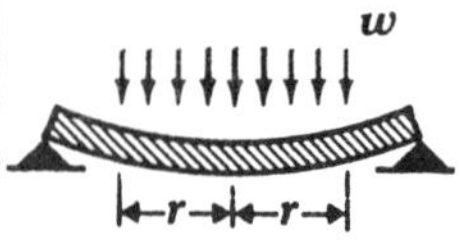	EDGES SUPPORTED. UNIFORM LOAD OVER CONCENTRIC CIRCULAR AREA OF RADIUS r.	$w\pi r^2$	β_6	6	λ_6	8
2	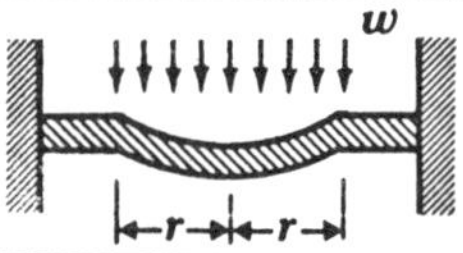	EDGES FIXED. UNIFORM LOAD OVER CONCENTRIC CIRCULAR AREA OF RADIUS r.	$w\pi r^2$	β_2	6	λ_2	8
3	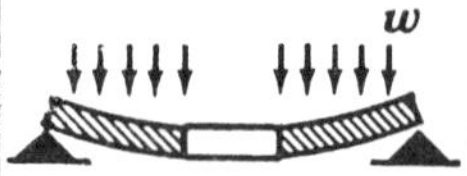	OUTER EDGE SUPPORTED. UNIFORM LOAD OVER ENTIRE ACTUAL SURFACE.	$w\pi(R^2-r^2)$	β_4	6	λ_4	8
4	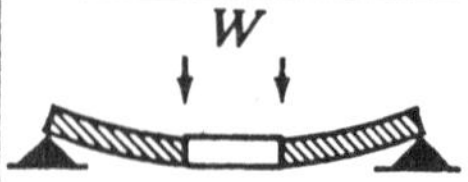	OUTER EDGE SUPPORTED. LOADED ALONG INNER EDGE.	W	β_7	6	λ_7	9
5		INNER EDGE SUPPORTED. UNIFORM LOAD OVER ENTIRE ACTUAL SURFACE.	$w\pi(R^2-r^2)$	β_1	6	λ_1	8
6	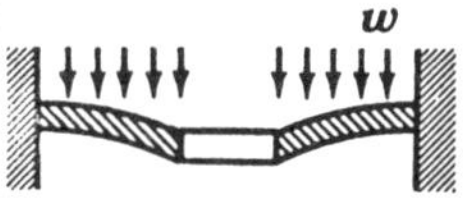	OUTER EDGE FIXED AND SUPPORTED, UNIFORM LOAD OVER ENTIRE ACTUAL SURFACE.	$w\pi(R^2-r^2)$	β_3	6	λ_3	8
7	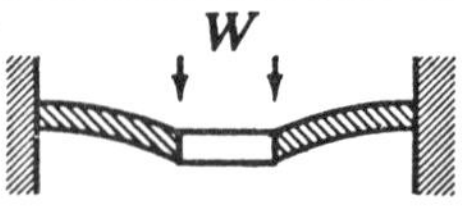	OUTER EDGE FIXED AND SUPPORTED, LOADED ALONG INNER EDGE.	W	β_5	6	λ_5	8
8	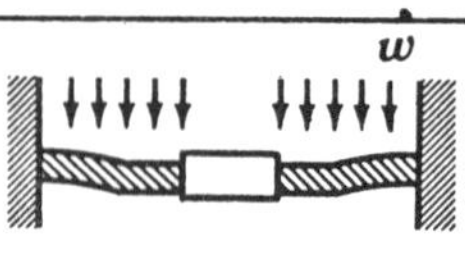	OUTER EDGE FIXED AND SUPPORTED. INNER EDGE FIXED. UNIFORM LOAD OVER ENTIRE ACTUAL SURFACE.	$w\pi(R^2-r^2)$	β_8	7	λ_8	9
9	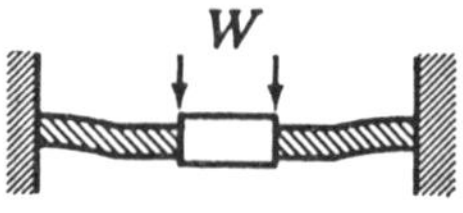	OUTER EDGE FIXED AND SUPPORTED, INNER EDGE FIXED. LOADED ALONG INNER EDGE.	W	β_{10}	7	λ_{10}	9
10	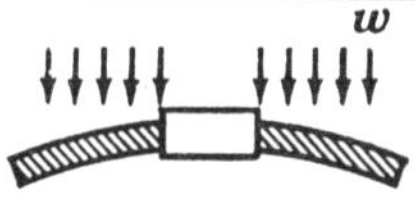	INNER EDGE FIXED AND SUPPORTED. UNIFORM LOAD OVER ENTIRE ACTUAL SURFACE.	$w\pi(R^2-r^2)$	β_{12}	7	λ_{12}	9
11		INNER EDGE FIXED AND SUPPORTED LOADED ALONG OUTER EDGE.	W	β_{13}	7	λ_{13}	9
12	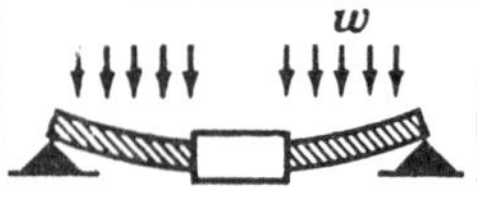	OUTER EDGE SUPPORTED, INNER EDGE FIXED. UNIFORM LOAD OVER ENTIRE ACTUAL SURFACE.	$w\pi(R^2-r^2)$	β_{11}	7	λ_{11}	9
13	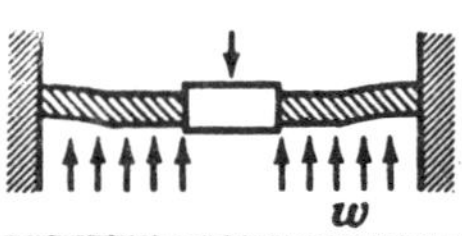	BOTH EDGES FIXED. BALANCED LOADING (PISTON).	$w\pi(R^2-r^2)$	β_9	7	λ_9	9

plate and Young's Modulus are: $R = 10$ in., $t = 0.2$ in., $E = 30 \times 10^6$ psi.

The loading and edge conditions correspond to those of Table 7, case 1, where $\beta = \beta_6$, $\lambda = \lambda_6$, $W = \omega\pi R^2 = 942$ lb.

Since $R/r = 1$ (load uniformly distributed over entire surface of the plate) we find from curves of Figs. 6 and 8:

$$\beta_6 = 0.210, \lambda_6 = 0.40.$$

Finally:

$$S = \frac{0.40 \times 942}{(0.2)^2} = 9{,}400 \text{ psi}$$

$$Y = \frac{0.21 \times 942 \times 10^2}{30 \times 10^6 \times (0.2)^3} = 0.082 \text{ in.}$$

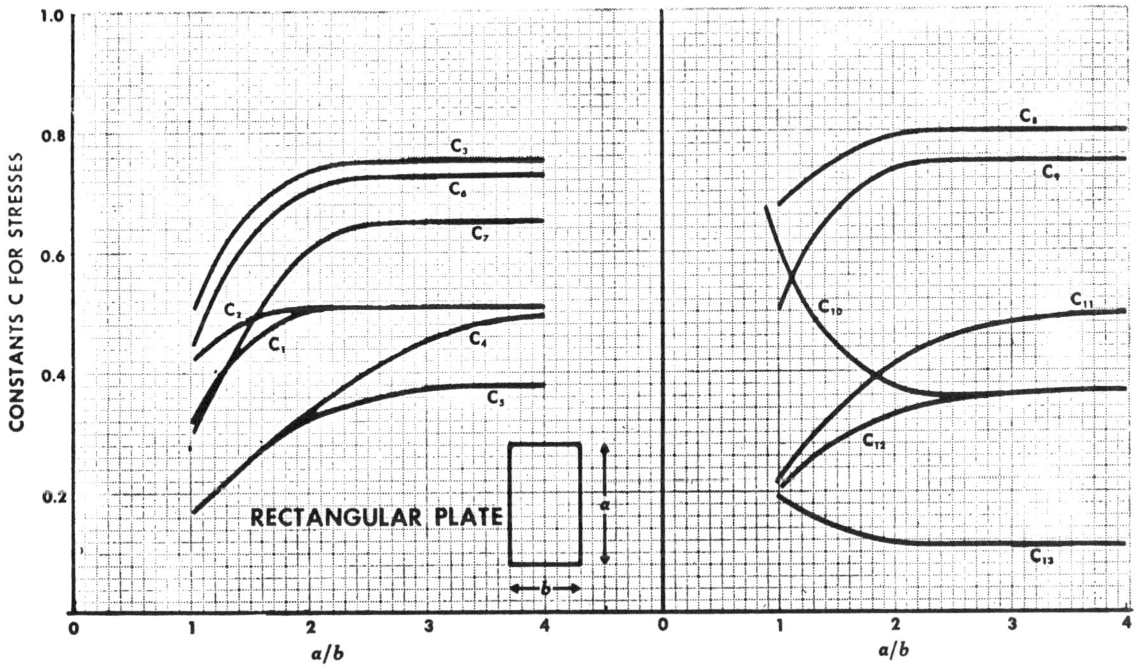

Figure 1

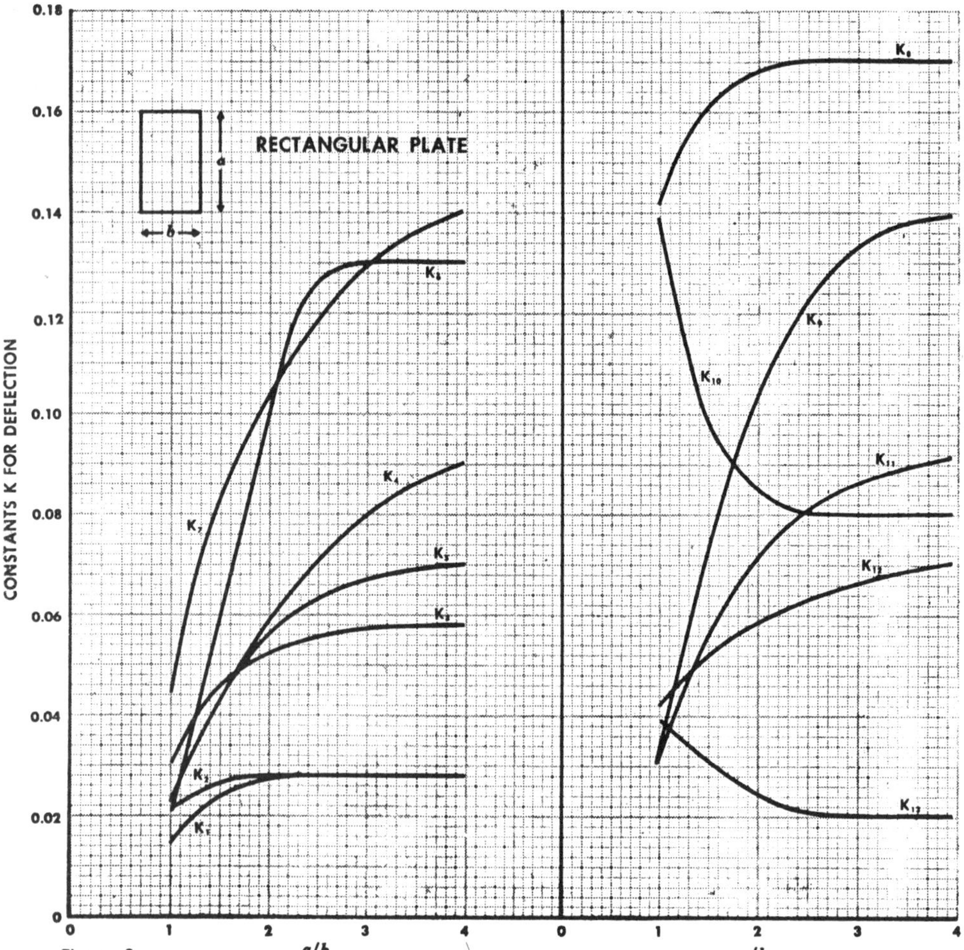

Figure 2

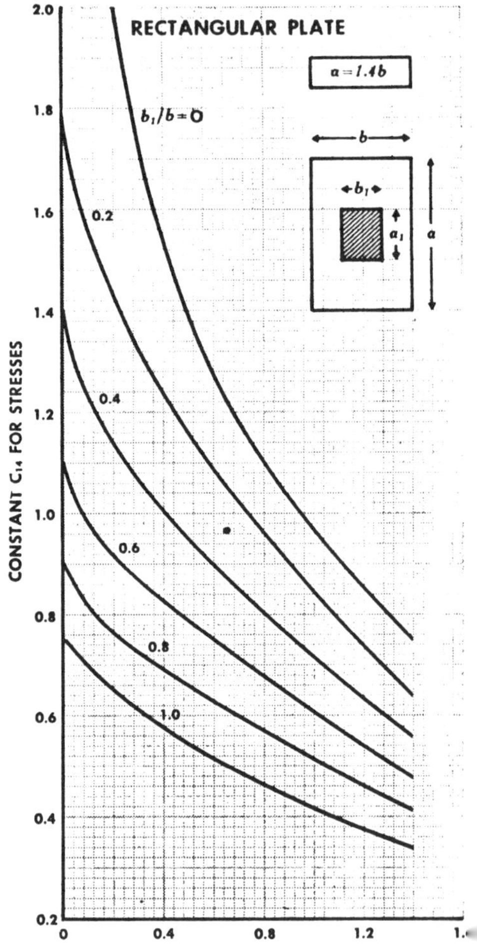

Figure 4

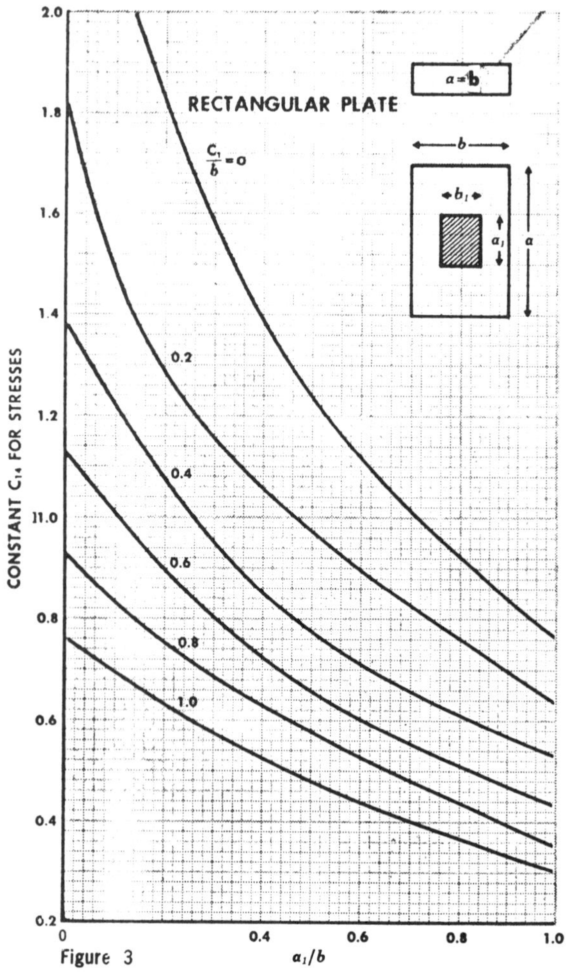

Figure 3

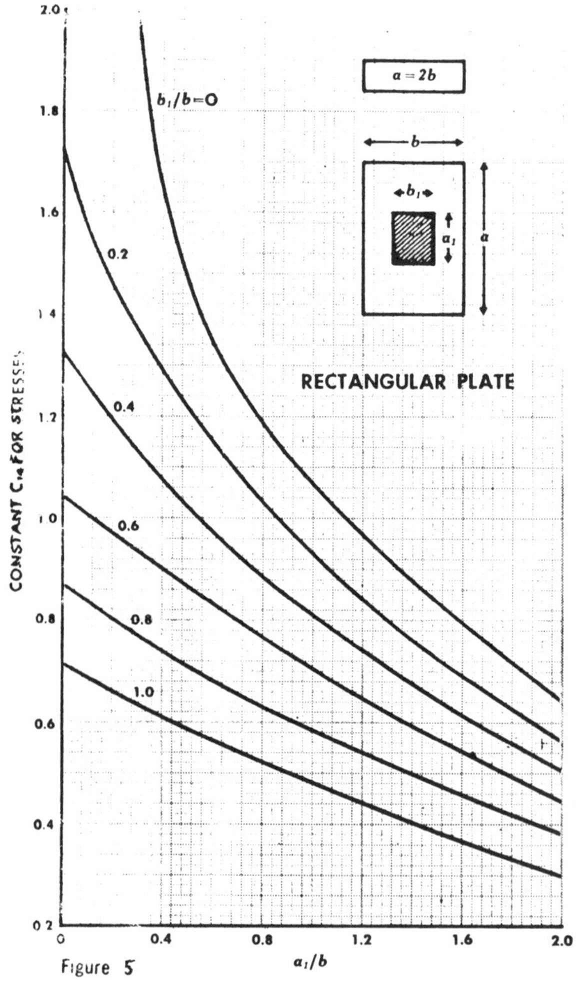

Figure 5

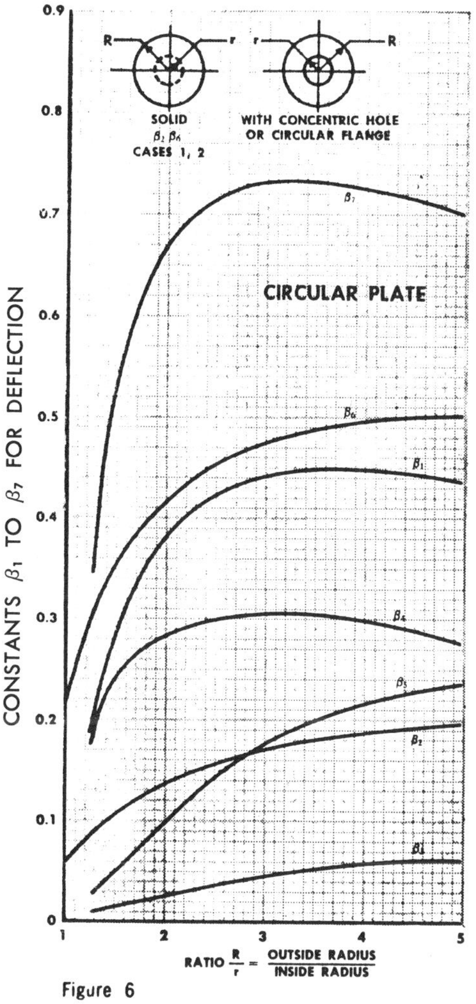

Figure 6

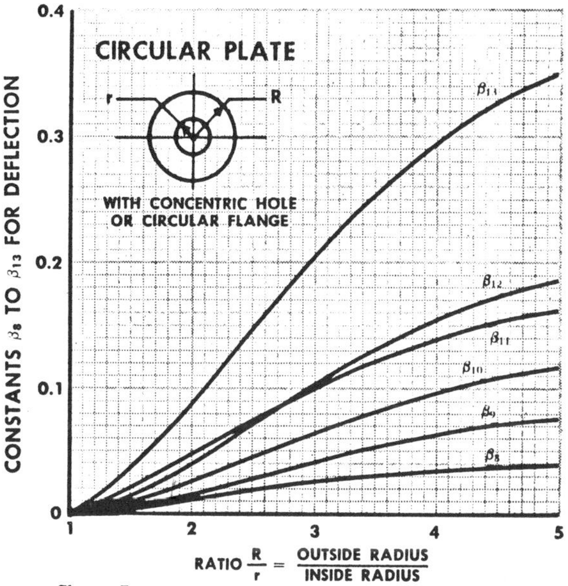

Figure 7

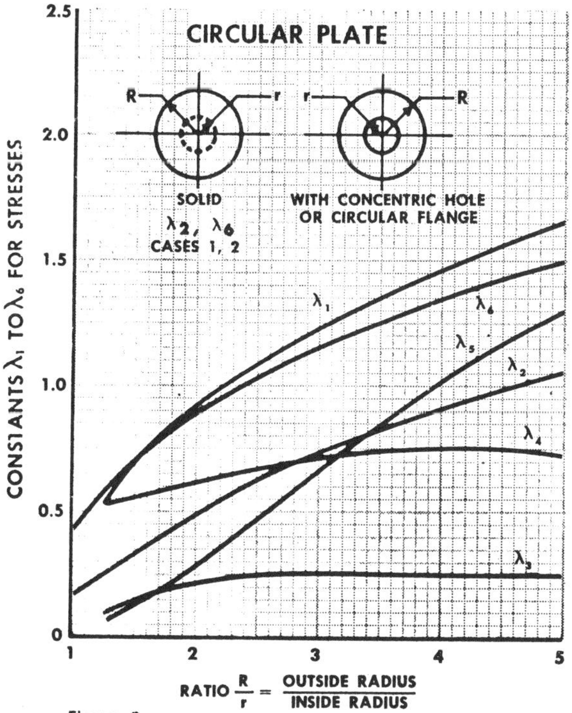

Figure 8

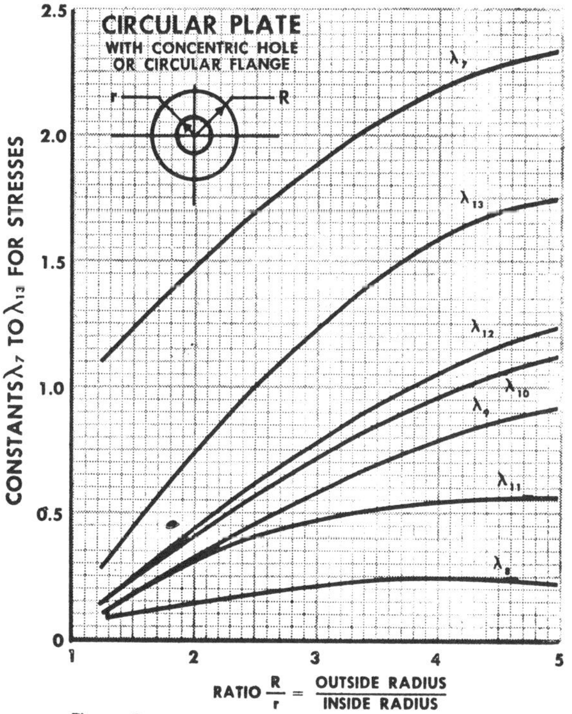

Figure 9

52. SLOPE FOR CIRCULAR PLATES

MEASURED FROM THE HORIZONTAL

(*Ref. 4*)

$$\theta = CWa/Et^3 \tag{1}$$

$$\theta = C_1Ma/Et^3 \tag{2}$$

Tables 8, 9, 10 and 11 give C and C_1 factors for slopes for circular plates loaded and supported in 17 different manners. By superposition these factors may be made to cover a wide variety of loadings not specifically considered. In order to find the slope for a given circular plate, the C and C_1 factors are used with equations (1), (2). Equation (1) solves for slopes of plates loaded with a total load W (lb.). In cases where a plate is loaded with a uniformly distributed load w (psi), compute as follows:

$$W = w\pi a^2 \text{ or}$$

$$W = w\pi(a^2 - b^2)$$

Equation (2) solves for slopes of plates with end moments.

NOTATION

- M = end moment (in.-lb. per in.)
- θ = slope measured from horizontal (radians)
- C = load-support factor (Tables 8, 9)
- C_1 = load-support factor (Tables 10, 11)
- w = unit applied load (psi)
- W = total applied load (lb.)
- E = modulus of elasticity (psi)
- t = plate thickness (in.)
- a = outside radius of plate (in.)
- b = radius of concentric hole or flange in plate (in.)
- r = radius of concentric trunnion (in.)

The C and C_1 factors in the tables are based on a Poisson's Ratio of 0.3, a value generally used for steel and aluminum. For other values, however, there will be little change in slope.

The same assumptions, as specified before, apply here: (1) The plate is flat, of uniform thickness, and of homogeneous isotropic material; (2) the thickness is not more than about one-quarter of the least transverse dimension; (3) all forces (loads and reactions) are normal to the plane of the plate; and (4) the plate is not stressed at any point beyond the elastic limit.

Example 1: A circular, aluminum plate 20 in. in diameter with a concentric hole 4 in. in diameter and a thickness of 0.12 in is fixed and supported along the outer edge and loaded with a uniformly distributed load of 3 psi over the entire surface. Determine the slope.

Solution: a = 10 in., b = 2 in., t = 0.12 in., E = 10×10^6 psi, w = 3 psi, a/b = 10/2 = 5. From Table 8, case 4, C = 0.089, $W = w\pi(a^2 - b^2)$ = 904 lb. From equation (1), θ = 0.05 radians.

Example 2: A circular, solid steel, plate 20 in. in diameter and 0.20 in. thick has no supports. A uniform edge moment of 12.45 in.-lb. per in. is applied. Find the slope of the plate.

Solution: a = 10 in., t = 0.20 in., E = 30×10^6 psi, M = 12.45 in.-lb. per in. From Table 11, case 15, C = 8. From equation (2), θ = 0.004 radians.

TABLE 8 "C" factors (use with Nomogram I)

Manner of loading and case number	b, a	1.25	1.5
1. Outer edge supported. Uniform load over entire actual surface.	$W = w\pi(a^2-b^2)$	0.786 0.919	0.679 0.741
2. Outer edge supported. Uniform load along inner edge.	W	1.646 1.758	1.470 1.650
3. Inner edge supported. Uniform load over entire actual surface.	$W = w\pi(a^2-b^2)$	0.884 1.012	0.794 0.910
4. Outer edge fixed and supported. Uniform load over entire actual surface.	$W = w\pi(a^2-b^2)$	0.013	0.035
5. Outer edge fixed and supported. Uniform load along inner edge.	W	0.045	0.115
6. Inner edge fixed and supported. Uniform load over entire actual surface.	$W = w\pi(a^2-b^2)$	0.012	0.040
7. Inner edge fixed and supported. Uniform load along outer edge.	W	0.046	0.105
8. Outer edge supported, inner edge fixed. Uniform load over entire actual surface.	$W = w\pi(a^2-b^2)$	0.091	0.123

a/b				
2	3	4	5	Location of slope
0.640	0.441	0.401	0.379	At outer edge
0.612	0.447	0.355	0.290	At inner edge
1.237	1.006	0.895	0.832	At outer edge
1.475	1.238	1.082	0.932	At inner edge
0.606	0.565	0.492	0.448	At outer edge
0.862	0.791	0.726	0.642	At inner edge
0.069	0.096	0.096	0.089	At inner edge
0.269	0.448	0.510	0.522	At inner edge
0.097	0.177	0.223	0.252	At outer edge
0.239	0.403	0.499	0.544	At outer edge
0.197	0.260	0.297	0.306	At outer edge

TABLE 9 "C" factors (use with Nomogram I)

Manner of loading and case number	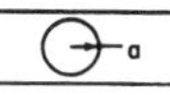	C	Note
9. Edges supported. Uniform load over entire surface.	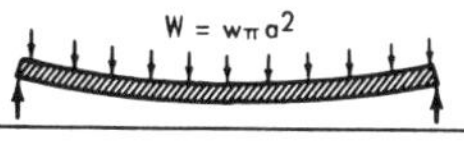	0.318	For cases 1–10: only the total load "W" should be used with Nomogram I. (See example I)
10. Edges supported. Uniform load over concentric circular area of radius r_o.	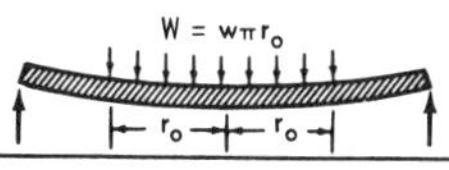	0.636	

TABLE 10 "C_1" factors (use with Nomogram II)

Manner of loading and case number				a/b				
		1.25	1.50	2	3	4	5	Location of slope
11. Outer edge fixed. Uniform moment along inner edge.	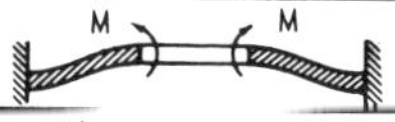	2.20	3.16	3.88	3.31	3.12	2.72	At inner edge
12. Inner edge fixed. Uniform moment along outer edge.		2.30	3.84	5.67	6.94	7.82	8.17	At outer edge
13. Inner edge supported. Uniform moment along outer edge.		51.00	28.00	16.40	11.60	10.23	9.61	At outer edge
		53.30	27.80	15.60	8.78	6.24	4.86	At inner edge
14. Outer edge supported. Uniform moment along inner edge.	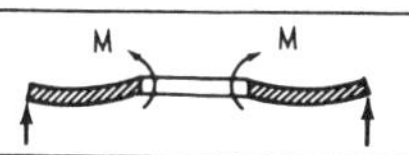	42.70	18.75	7.81	2.93	1.56	0.94	At outer edge
		44.90	22.35	11.27	5.52	4.08	3.17	At inner edge

TABLE 11 "C_1" factors (use with Nomogram II)

Manner of loading and case number	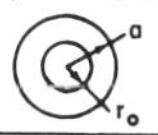	0	0.10	0.15	0.20	0.25	0.30	0.35	0.40	0.45	0.50	0.55	0.60	0.65	0.70	0.75	0.80	Location of slope
									r_o/a									
15. No supports. Uniform edge moment.		8.0	—	—	—	—	—	—	—	—	—	—	—	—	—	—	—	At edge
16. Edges supported. Central couple (trunnion loading)	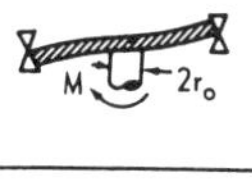	—	0.71	0.92	1.22	1.60	2.00	2.50	3.53	5.60	8.54	12.00	16.30	24.00	41.40	82.00	162.00	At center
17. Edges fixed. Central couple (trunnion loading)	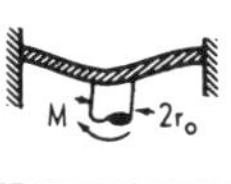	—	0.87	1.23	1.68	2.31	3.10	4.00	5.45	8.20	12.40	18.00	28.50	44.00	77.90	156.00	314.00	At center

Note for cases 16 and 17: multiply Θ, as obtained from Nomogram II by $C_1{}^2 a$

53. BEAM-SUPPORTED PLATES

(*Ref. 5, 6*)

NOTATION

a, b = plate dimensions (in.)
c = half depth of beam (in.)
E = modulus of elasticity (psi)
H = rigidity ratio of beam to plate
I = moment of inertia (in.^{4})
$K_1, K_2 \ldots K_8$ = factors for use in nomograms
P = total concentrated load (lb.)
S = bending stress (psi)
t = plate thickness (in.)
w = uniformly distributed load (psi)
y = deflection (in.)
ν = Poisson's Ratio

Subscripts:

b = beam
p = plate

Most design applications involving load-carrying plates require determining of stress and deflection in the plate and in its supporting members. Such calculations are based, of course, on the rigidity of both the plate and the beams that support it.

This section presents tables that incorporate these rigidity factors, and nomograms that provide a direct method for determining maximum stress and deflection in both plate and beams. Examples are given for the common arrangement shown in Fig. 10.

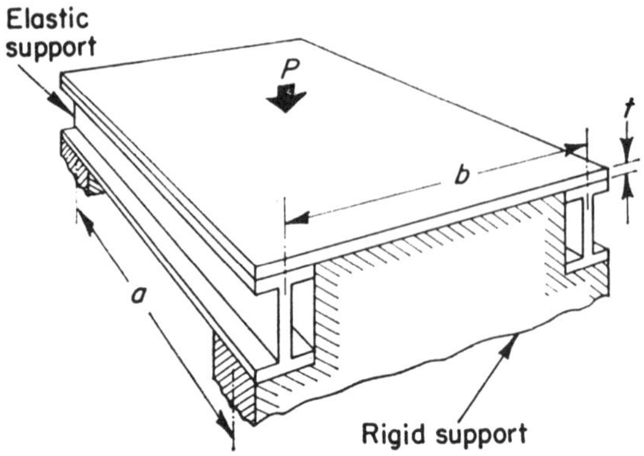

Fig. 10. Plate supported on two rigid and two elastic members.

The ratio of beam rigidity to plate rigidity, H, is often represented simply by E_bI_b/E_pI_p, where $I_p = at^3/12$. However, bending of a loaded plate is not confined to one direction. Therefore, for greater accuracy, Poisson's Ratio is introduced, and the rigidity ratio becomes:

$$H = \frac{(1 - \nu^2)E_bI_b}{E_pI_p} \tag{3}$$

The value of H can vary from zero (support beams not present) to infinity (rigid supports on all four plate edges).

Values of dimensionless factors K_1, K_2, K_3, and K_4 are determined from ratio H and from physical-size ratio b/a, Table 12. These factors, along with the plate and beam dimensions, are used in the nomogram shown in Fig. 11 to find deflections y_p and y_b. For maximum plate stress S_p, factors K_5 and K_7 from Table 13 are used with

Table 12—Factors for Determining

	$H = 0$				$H = 0.05$				$H = 0.1$				$H = 0.2$			
b/a	K_1	K_2	K_3	K_4	K_1	K_2	K_3	K_4	K_1	K_2	K_3	K_4	K_1	K_2	K_3	K_4
0.33	0.1591	0.146	0.7593	0.7641	0.1214	0.1265	0.5810	0.5805	0.0983	0.1013	0.4718	0.4681	0.0714	0.0724	0.3449	0.3375
0.5	0.1579	0.170	0.5076	0.5008	0.1310	0.1381	0.4250	0.4104	0.1122	0.1162	0.3677	0.3477	0.0879	0.0882	0.2932	0.2664
0.6	0.1573	0.172	0.4264	0.4110	0.1345	0.1425	0.3693	0.3454	0.1180	0.1220	0.3280	0.2979	0.0956	0.0947	0.2720	0.2336
0.8	0.1563	0.174	0.3305	0.2937	0.1394	0.1482	0.3004	0.2541	0.1264	0.1296	0.2774	0.2240	0.1080	0.1037	0.2446	0.1810
1.0	0.1557	0.175	0.2789	0.2189	0.1428	0.1515	0.2619	0.1923	0.1327	0.1341	0.2485	0.1715	0.1179	0.1090	0.2290	0.1409
1.2	0.1554	0.175	0.2491	0.1665	0.1455	0.1534	0.2391	0.1475	0.1317	0.1366	0.2312	0.1324	0.1258	0.1121	0.2195	0.1098
1.4	0.1553	0.176	0.2312	0.1278	0.1476	0.1546	0.2253	0.1137	0.1415	0.1382	0.2206	0.1024	0.1322	0.1140	0.2135	0.0855
1.6	0.1554	0.176	0.2202	0.0985	0.1495	0.1553	0.2167	0.0879	0.1446	0.1391	0.2139	0.0793	0.1374	0.1150	0.2095	0.0664
1.8	0.15[illegible]5	0.176	0.2135	0.0760	0.1509	0.1557	0.2114	0.0679	0.1473	0.1396	0.2097	0.0614	0.1416	0.1156	0.2071	0.0515
2.0	0.1557	0.176	0.2093	0.0596	0.1521	0.1559	0.2081	0.0534	0.1493	0.1399	0.2071	0.0484	0.1449	0.1160	0.2055	0.0408
3.0	0.1565	0.189	0.2031	0.0156	0.1557	0.1562	0.2030	0.0140	0.1551	0.1403	0.2030	0.0126	0.1543	0.1165	0.2029	0.0106

INSTRUCTIONS—TABLE 12

1. For a uniformly distributed load, w, use K_1 for plate and K_2 for beam calculations.
2. For concentrated load, P, at center of plate, use K_3 for plate and K_4 for beam calculations.
3. Nomogram of Fig. 11 is to be used with these K values to determine maximum deflection (at center) of plate and beam.

Fig. 12 and for maximum beam stress S_b, K_6 and K_8, also from Table 13, are used with Fig. 13.

The dimensionless factors in both tables are computed for $\nu = 0.3$, a value generally used for steel. Variations of ν have little effect on stress and deflection values however, and these tables can be used for a wide range of ν.

For H values between those shown in the tables, linear interpolation may be used to determine values of the K factors.

Although load P is generally considered to be concentrated at the center of the plate (or in a circular area of diameter $a/4$), these nomograms may also be used to determine stresses and deflections produced by a uniformly distributed load. For a uniform load, the value of wa^2 is used in place of P. Instructions on the tables indicate which K's are to be used for each condition.

Summary—Example 1

To Find:	Use Table:	Constant, K	Fig. No. (nomogram)	Solution
S_p	13	$K_7 = 0.917$	12	8,200 psi
y_p	12	$K_3 = 0.0847$	11	0.020 in.
S_b	13	$K_8 = 0.090$	13	5,800 psi
y_b	12	$K_4 = 0.0390$	11	0.0095 in.

Example 1: A steel plate is supported as shown in Fig. 10. Beams are standard 3-in., 7.5 lb. per ft. I-beams ($I_b = 2.9$ in.^{4}). Plate dimensions are $a = 25$ in., $b = 12.5$ in., $t = 0.75$ in. The plate supports a concentrated load, $P = 5{,}000$ lb., at its center. $E_p = E_b = 30 \times 10^6$ psi, and $\nu = 0.3$. Find S_p, S_b, y_p, and y_b

1. Calculate $I_p = at^3/12 = 0.88$ in.^{4}
2. From equation (1), $H = 3.0$.
3. Using plate ratio $b/a = 0.5$ and $H = 3$, determine values of K_3 and K_4 from Table 12.
4. Similarly, determine constants K_7 and K_8 from Table 13
5. Using the nomograms, Figs. 11, 12 and 13, determine stresses and deflections in the plate and beam.

Example 2: A 0.75-in. thick, 72-in. square plate ($a = b = 72$) is supported as shown in Fig. 10. Support members at two edges are 3-in. I-beams ($I_b = 2.9$ in.^{4}). The plate supports a uniformly distributed load, $w = 4$ psi; $E_p = E_b = 30 \times 10^6$; $\nu = 0.25$. Find S_p, S_b, y_p, and y_b

1. Calculate $I_p = ar^3/12 = 2.52$ in.^{4}
2. From equation (3), $H = 1.07$.
3. Using plate ratio $b/a = 1$ and $H = 1.07$, determine values of K_1 and K_2 from Table 12 and K_5 and K_6 from Table 13.
4. Calculate $P = wa^2 = 20{,}700$ lb.
5. From the nomograms, Fig. 11, 12, and 13, determine stresses and deflection in the plate and beam.

Maximum Plate and Beam Deflection

$H = 0.5$				$H = 1.0$				$H = 2.0$				$H = 4.0$				$H = \infty$			
K_1	K_2	K_3	K_4	K_1	K_2	K_3	K_4	K_1	K_2	K_3	K_4	K_1	K_2	K_3	K_4	K_1	K_2	K_3	K_4
0.0398	0.0390	0.1957	0.1837	0.0234	0.0221	0.1187	0.1044	0.0135	0.0118	0.0718	0.0588	0.0079	0.0061	0.0456	0.0291	0.0019	0.0049	0.0174	0
0.0549	0.0512	0.1925	0.1565	0.0358	0.0301	0.1340	0.0928	0.0233	0.0165	0.0958	0.0511	0.0160	0.0087	0.0736	0.0269	0.0079	0.0070	0.0489	0
0.0636	0.0567	0.1910	0.1412	0.0439	0.0340	0.1429	0.0857	0.0307	0.0189	0.1098	0.0479	0.0228	0.0100	0.0902	0.0254	0.0140	0.0081	0.0689	0
0.0796	0.0648	0.1942	0.1150	0.0609	0.0398	0.1610	0.0715	0.0476	0.0225	0.1374	0.0407	0.0395	0.0120	0.1230	0.0219	0.0301	0.0098	0.1063	0
0.0940	0.0698	0.1975	0.0919	0.0776	0.0436	0.1758	0.0582	0.0656	0.0249	0.1600	0.0335	0.0581	0.0134	0.1501	0.0182	0.0492	0.0109	0.1385	0
0.1064	0.0728	0.2000	0.0727	0.0927	0.0460	0.1863	0.0465	0.0825	0.0265	0.1761	0.0271	0.0761	0.0143	0.1696	0.0150	0.0684	0.0117	0.1619	0
0.1167	0.0747	0.2016	0.0571	0.1057	0.0474	0.1930	0.0368	0.0974	0.0274	0.1866	0.0215	0.0921	0.0149	0.1826	0.0117	0.0856	0.0121	0.1776	0
0.1253	0.0757	0.2024	0.0446	0.1165	0.0482	0.1972	0.0288	0.1098	0.0280	0.1932	0.0169	0.1056	0.0152	0.1907	0.0092	0.1004	0.0124	0.1877	0
0.1321	0.0763	0.2028	0.0347	0.1252	0.0487	0.1996	0.0235	0.1200	0.0283	0.1972	0.0142	0.1165	0.0154	0.1957	0.0082	0.1125	0.0125	0.1938	0
0.1375	0.0767	0.2029	0.0279	0.1321	0.0490	0.2010	0.0194	0.1280	0.0285	0.1996	0.0112	0.1253	0.0155	0.1887	0.0066	0.1222	0.0126	0.1975	0
0.1528	0.0772	0.2028	0.0072	0.1517	0.0494	0.2027	0.0047	0.1509	0.0287	0.2027	0.0027	0.1503	0.0156	0.2026	0.0015	0.1496	0.0127	0.2026	0

Summary—Example 2

To Find:	Use Table:	Constant, K	Fig. No. (nomogram)	Solution
S_p	13	$K_5 = 0.4030$	12	15,000 psi
y_p	12	$K_1 = 0.0768$	11	0.70 in.
S_b	13	$K_6 = 0.0361$	13	29,000 psi
y_b	12	$K_2 = 0.0423$	11	0.38 in.

Table 13—Factors for Determining

	$H = 0$				$H = 0.05$				$H = 0.1$				$H = 0.2$			
b/a	K_5	K_6	K_7	K_8	K_5	K_6	K_7	K_8	K_5	K_6	K_7	K_8	K_5	K_6	K_7	K_8
0.33	0.732	0	4.09	0	0.551	0.0033	3.17	0.026	0.468	0.0052	2.62	0.042	0.339	0.0076	1.99	0.060
0.5	0.733	0	2.86	0	0.615	0.0040	2.47	0.018	0.532	0.0067	2.21	0.030	0.424	0.0103	1.86	0.046
0.6	0.732	0	2.48	0	0.634	0.0055	2.23	0.015	0.562	0.0095	2.03	0.026	0.464	0.0150	1.79	0.040
0.8	0.735	0	2.06	0	0.662	0.0057	1.93	0.011	0.607	0.0101	1.83	0.019	0.528	0.0164	1.69	0.030
1.0	0.735	0	1.83	0	0.667	0.0058	1.76	0.008	0.640	0.0104	1.70	0.014	0.577	0.0172	1.61	0.023
1.2	0.738	0	1.69	0	0.697	0.0059	1.66	0.006	0.664	0.0106	1.63	0.011	0.615	0.0177	1.58	0.018
1.4	0.739	0	1.63	0	0.708	0.0059	1.60	0.005	0.682	0.0107	1.57	0.008	0.645	0.0179	1.55	0.014
1.6	0.742	0	1.57	0	0.717	0.0060	1.57	0.004	0.697	0.0108	1.55	0.006	0.624	0.0181	1.54	0.011
1.8	0.748	0	1.55	0	0.724	0.0060	1.54	0.003	0.709	0.0108	1.54	0.005	0.687	0.0182	1.52	0.008
2.0	....	0	1.54	..	....	0.0060	1.53	....	....	0.0108	1.52	....	....	0.0182	1.52	....
3.0	....	0	1.51	..	....	0.0060	1.51	....	....	0.0108	1.51	....	....	0.0182	1.51	....
∞	....	0	...	..	....	0.0060	...	....	....	0.0109	...	....	....	0.0183	...	....

INSTRUCTIONS—TABLE 13

1. For a uniformly distributed load, w, use K_5 for plate and K_6 for beam calculations.
2. For a concentrated load, P, at center of plate, use K_7 for plate and K_8 for beam calculations.
3. Use K_5 and K_7 with nomogram of Fig. 12 to determine maximum stress in plates.
4. Use K_6 and K_8 with nomogram of Fig. 13 to determine maximum stress in beams.

Maximum Plate and Beam Bending Stress

$H = 0.5$				$H = 1.0$				$H = 2.0$				$H = 4.0$				$H = \infty$			
K_5	K_6	K_7	K_8	K_5	K_6	K_7	K_8	K_5	K_6	K_7	K_8	K_5	K_6	K_7	K_8	K_5	K_6	K_7	K_8
0.196	0.0103	1.25	0.082	0.131	0.0177	0.864	0.093	0.079	0.0126	0.636	0.100	0.054	0.0131	0.510	0.103	0.027	0.0134	0.360	0.106
0.279	0.0152	1.40	0.068	0.167	0.0180	1.120	0.080	0.139	0.0199	0.966	0.088	0.106	0.0210	0.858	0.093	0.070	0.0217	0.750	0.096
0.324	0.0228	1.43	0.060	0.238	0.0276	1.210	0.073	0.179	0.0308	1.060	0.081	0.145	0.0328	0.996	0.086	0.106	0.0349	0.876	0.092
0.406	0.0261	1.46	0.048	0.325	0.0325	1.310	0.060	0.268	0.0370	1.220	0.068	0.233	0.0398	1.150	0.073	0.193	0.0430	1.090	0.079
0.475	0.0282	1.48	0.038	0.407	0.0357	1.390	0.048	0.356	0.0412	1.330	0.056	0.325	0.0446	1.280	0.060	0.287	0.0487	1.240	0.066
0.534	0.0294	1.49	0.030	0.478	0.0376	1.440	0.038	0.435	0.0438	1.400	0.045	0.408	0.0477	1.370	0.049	0.375	0.0524	1.330	0.053
0.583	0.0301	1.50	0.023	0.536	0.0388	1.470	0.030	0.502	0.0454	1.440	0.035	0.480	0.0496	1.430	0.039	0.454	0.0546	1.450	0.042
0.619	0.0305	1.51	0.018	0.583	0.0395	1.480	0.024	0.556	0.0463	1.470	0.028	0.538	0.0507	1.460	0.030	0.517	0.0560	1.450	0.033
0.648	0.0308	1.51	0.014	0.620	0.0399	1.490	0.018	0.599	0.0469	1.480	0.022	0.586	0.0514	1.470	0.024	0.568	0.0568	1.470	0.026
....	0.0309	1.51	....	....	0.0402	1.500	....	....	0.0472	1.490	....	....	0.0518	1.490	....	....	0.0573	1.480	....
....	0.0310	1.51	....	....	0.0404	1.510	....	....	0.0475	1.510	....	....	0.0520	1.510	....	....	0.0575	1.510	....
....	0.0311	...	....	....	0.0405	....	....	....	0.0477	....	....	....	0.0523	....	....	....	0.0579	....	....

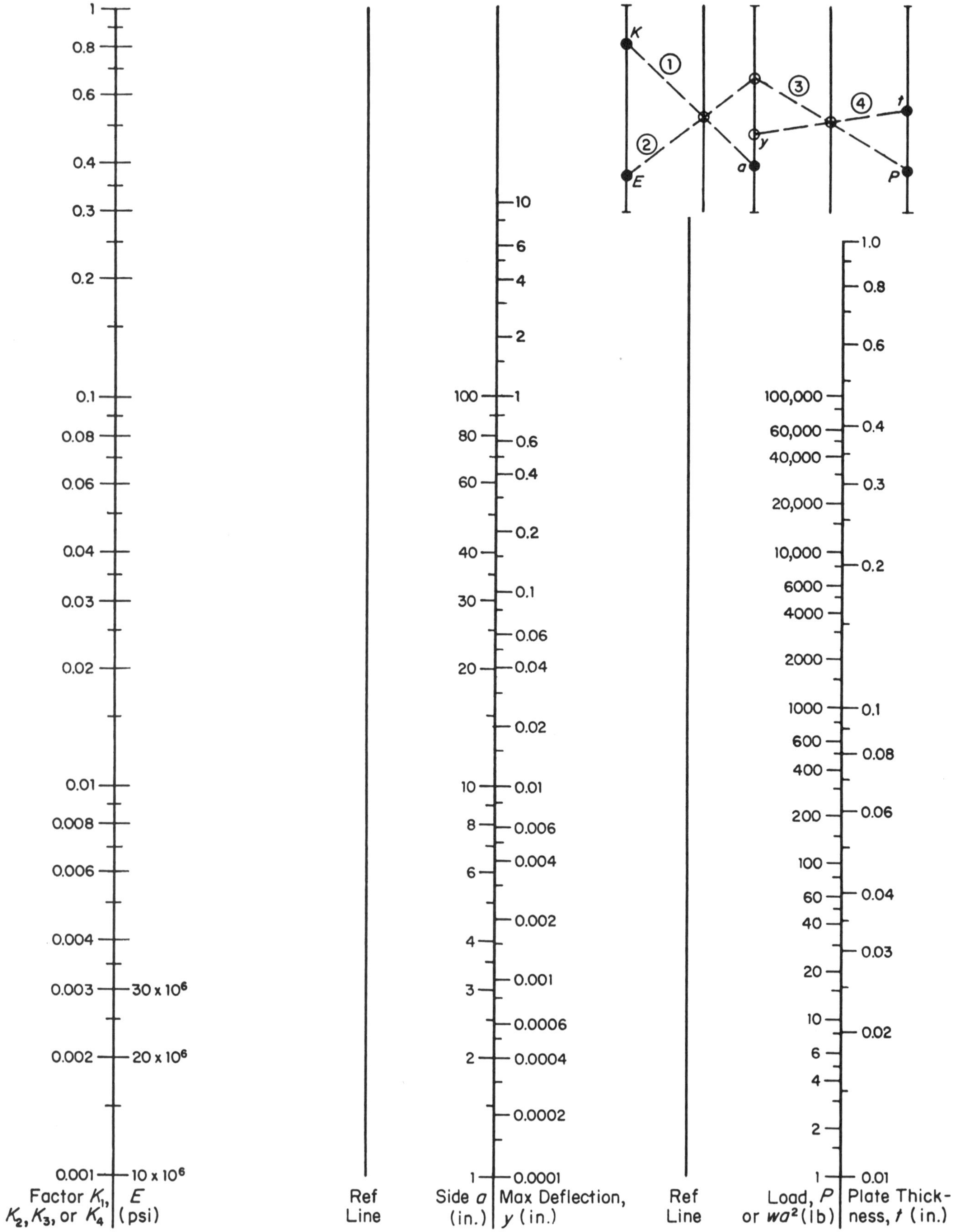

Fig. 11 — Nomogram for determining plate deflections y_a and y_b
Actual deflections are about 91 per cent of those indicated by nomogram.

Fig. 12 — Nomogram for determining plate stress S_a

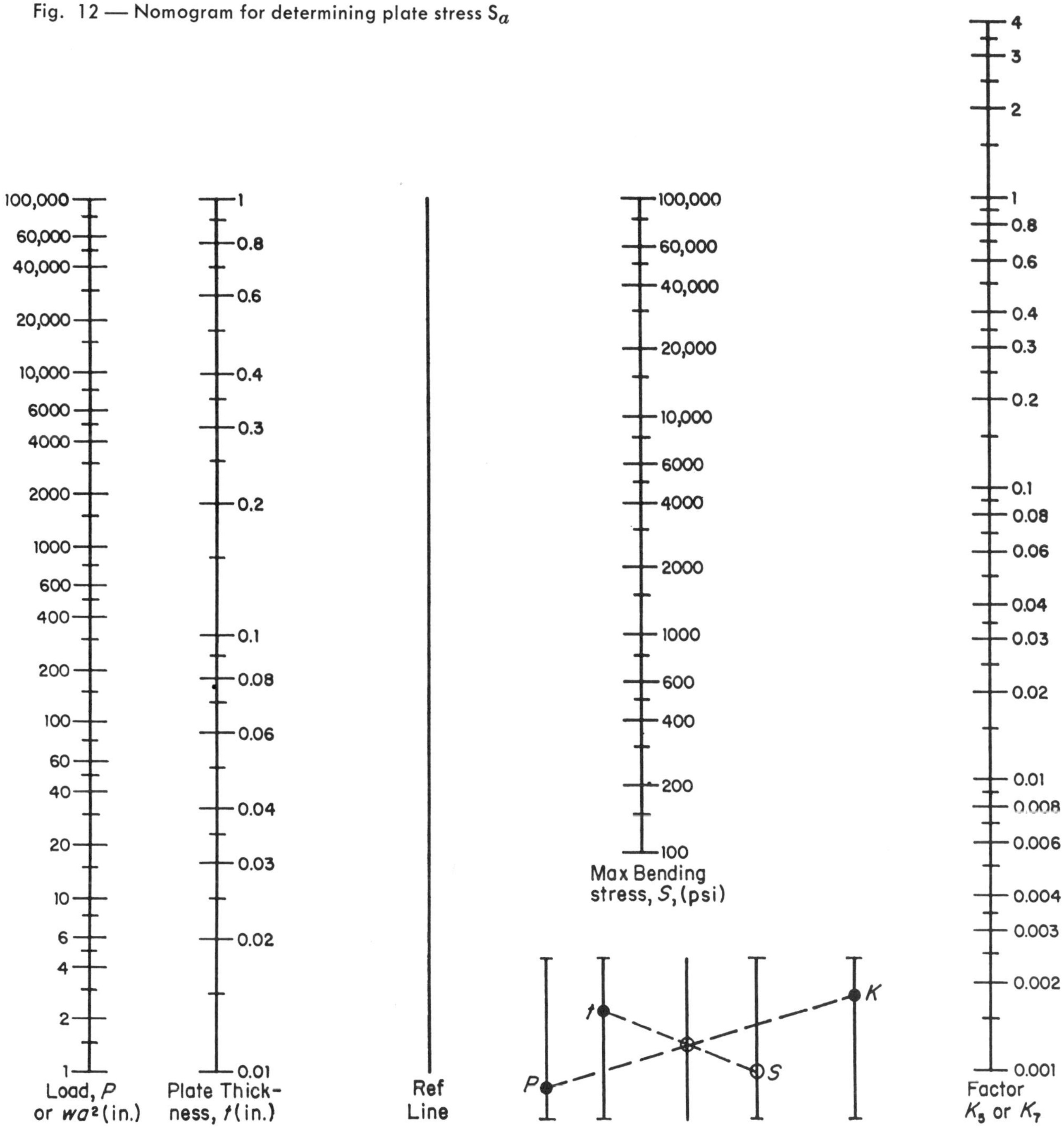

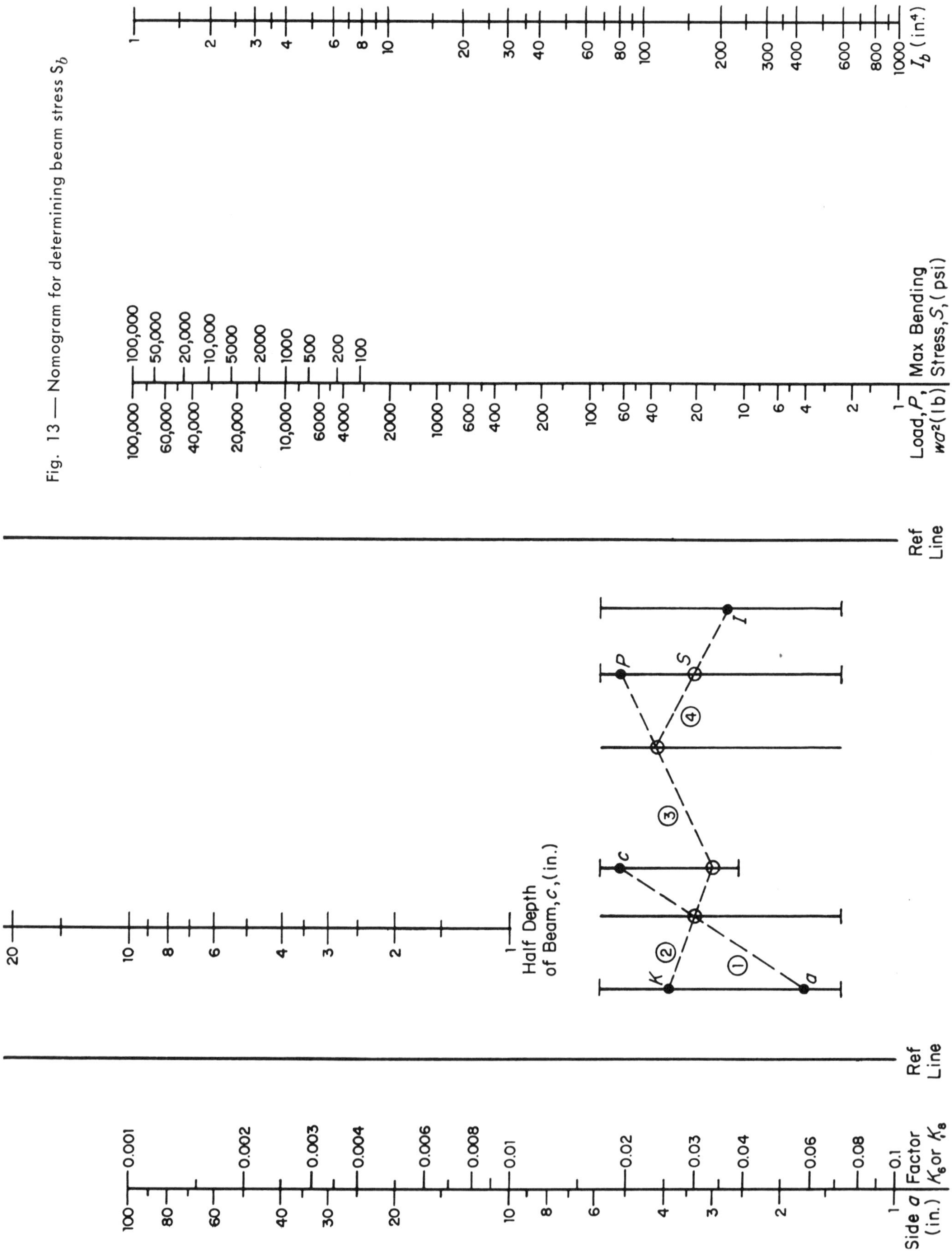

Fig. 13 — Nomogram for determining beam stress S_b

54. CIRCULAR PLATES WITH LINEAR VARYING THICKNESS

(*Ref. 1, 2*)

Equations for stress and deflection for this type of plate are tabulated together with their corresponding coefficients in Table 14.

TABLE 14

CIRCULAR SOLID PLATES WITH LINEARLY VARYING THICKNESS ($\nu = 0.25$)

MANNER OF LOADING AND SUPPORT	COEFF. K ①	RADIAL DISTANCE r	b/a 0.2	0.4	0.6	0.8	1.0	STRESS S	MAX. DEFLECTION y
1. CLAMPED, UNIFORM LOADING 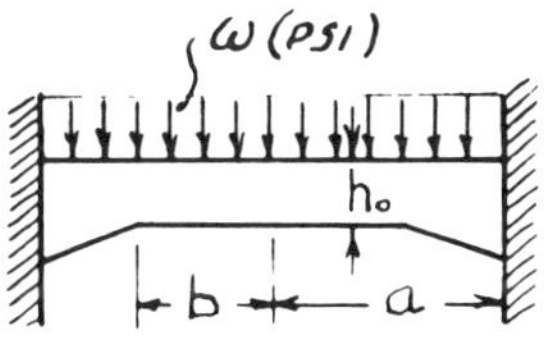	K_r	o	0.0122	0.0332	0.0543	0.0709	0.0781	$K\frac{6\omega a^2}{h_0^2}$	$K_1\frac{\omega a^4}{Eh_0^3}$
		b	0.0040	0.0007	-0.0188	-0.0591	-0.1250		
		a	-0.161	-0.156	-0.149	-0.140	-0.125		
	K_t	o	0.0122	0.0332	0.0543	0.0709	0.0781		
		b	0.0078	0.0157	0.0149	0.0009	-0.0310		
		a	-0.040	-0.039	-0.037	-0.035	-0.031		
	K_1		0.008	0.042	0.094	0.148	0.176		
2. CLAMPED, CENTER LOADING 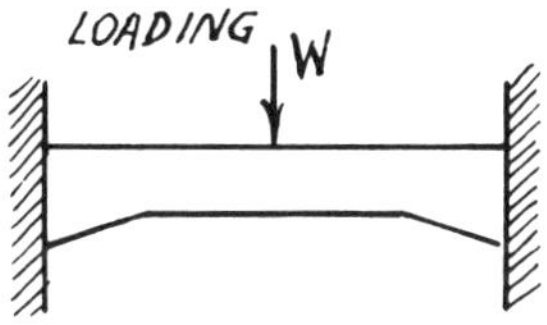	K_r	b	-0.034	-0.040	-0.050	-0.063	-0.080	$K\frac{6W}{h_0^2}$	$K_1\frac{Wa^2}{Eh_0^3}$
		a	-0.129	-0.112	-0.096	-0.084	-0.080		
	K_t	b	-0.028	-0.034	-0.044	-0.057	-0.020		
		a	-0.032	-0.028	-0.024	-0.021	-0.020		
	K_1		0.031	0.093	0.155	0.203	0.224		
			h_0/h_1 1.00	1.50	2.33				
3. PLATE SUPPORTED, UNIFORM LOADING 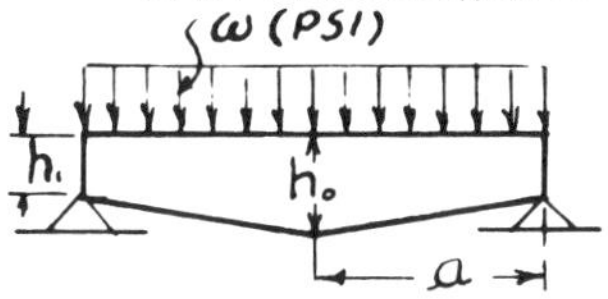	K_r	o	0.203	0.257	0.304			$K\frac{6\omega a^2}{h_0^2}$	$K_1\frac{\omega a^4}{Eh_0^3}$
		$a/2$	0.152	0.176	0.195				
	K_t	o	0.203	0.257	0.304				
		$a/2$	0.176	0.173	0.167				
		a	0.094	0.054	0.029				
	K_1		0.738	1.260	2.040				
4. SUPPORTED, CENTER LOADING 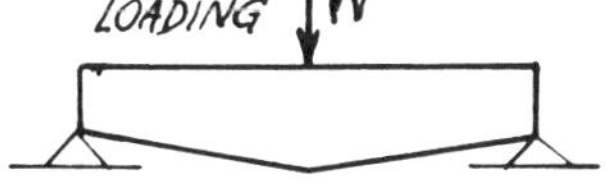	K_r	$a/2$	0.069	0.088	0.102			$K\frac{6W}{h_0^2}$	$K_1\frac{Wa^2}{Eh_0^3}$
	K_t	$a/2$	0.129	0.123	0.116				
		a	0.060	0.033	0.016				
	K_1		0.582	0.930	0.130				

① K_r – COEFFICIENT FOR RADIAL STRESS

K_t – COEFFICIENT FOR TANGENTIAL STRESS

55. ANNULAR PLATES WITH LINEAR VARYING THICKNESS

(*Ref. 7*)

In this section we present a solution to the problem of a symmetrically loaded circular plate with a central hole, the thickness of the plate at any section being proportional to the distance of the section from the center of the plate. Detailed investigation of six particular cases is reported in subject Ref. 7. The results of these investigations, in the form of tabulated coefficients for the maximum stress and deflection, calculated for various ratios of the external diameter of the plate to the diameter of the central hole, are presented here.

NOTATION

a = external radius of plate (in.)
b = radius of central hole (in.)
n = ratio a/b
P = concentrated load acting normally to middle surface and uniformly distributed around periphery of central hole (lb.)
q = intensity of uniformly distributed load (psi)
S = maximum (radial) stress (psi)
h = thickness of plate at external periphery (in.)
μ = Poisson's Ratio
y = maximum deflection (in.)
α = coefficient for the maximum stress (*See* Table 15)
β = coefficient for the maximum deflection (*See* Table 15)

ASSUMPTIONS

1. The effects of the stretching of the middle plane of the plate are negligible. This will be reasonable if the deflections are small everywhere compared with the thickness of the plate.

2. The maximum thickness of the plate is small compared with its other dimensions.

3. The effects on deflections of shearing stresses acting on normal sections of the plate are negligible.

4. Plane sections normal to middle surface of the plate before bending remain plane and normal after bending.

5. Stress is directly proportional to strain for the material of which the plate is made, and this material is assumed to be homogeneous and isotropic.

Cases 1, 2: $S = \alpha\, P/h^2$; $y = \beta\, Pa^2/Eh^3$.

Cases 3, 4, 5 and 6: $S = \alpha\, qa^2/h^2$; $y = \beta\, qa^4/Eh^3$.

Values of α and β are given in Table 15.

TABLE 15. VALUES OF α AND β FOR VARIOUS VALUES OF $n = a/b$
$\nu = 0.30$

	CASE	COEFFICIENT	$n = a/b$						MAXIMUM STRESS S	MAXIMUM DEFLECTION y
			1.25	1.5	2	3	4	5		
1.	EXTERNAL EDGE CLAMPED AND SUPPORTED INTERNAL EDGE CLAMPED	α	0.159	0.396	1.091	3.306	6.549	10.78	$\alpha \frac{P}{h^2}$	$\beta \frac{Pa^2}{Eh^3}$
		β	0.00174	0.0112	0.0606	0.261	0.546	0.876		
2.	EXTERNAL EDGE SIMPLY SUPPORTED INTERNAL EDGE CLAMPED	α	0.353	0.933	2.626	6.877	11.47	16.51	$\alpha \frac{P}{h^2}$	$\beta \frac{Pa^2}{Eh^3}$
		β	0.00816	0.0583	0.3448	1.358	2.387	3.268		
3.	EXTERNAL EDGE CLAMPED AND SUPPORTED INTERNAL EDGE CLAMPED	α	0.0785	0.208	0.52	1.27	1.94	2.515	$\alpha \frac{qa^2}{h^2}$	$\beta \frac{qa^4}{Eh^3}$
		β	0.00092	0.008	0.0495	0.193	0.346	0.482		
4.	EXTERNAL EDGE SIMPLY SUPPORTED INTERNAL EDGE CLAMPED	α	0.149	0.991	2.23	5.57	7.78	9.16	$\alpha \frac{qa^2}{h^2}$	$\beta \frac{qa^4}{Eh^3}$
		β	0.00551	0.0564	0.412	1.673	2.786	3.573		
5.	EXTERNAL EDGE FREE INTERNAL EDGE CLAMPED	α	0.249	0.638	3.96	13.64	26.00	40.63	$\alpha \frac{qa^2}{h^2}$	$\beta \frac{qa^4}{Eh^3}$
		β	0.00372	0.0453	0.401	2.119	4.245	6.283		
6.	BALANCED LOADING (PISTON) $P = \pi q(a^2 - b^2)$	α	0.1275	0.5145	2.051	7.965	17.35	30.00	$\alpha \frac{qa^2}{h^2}$	$\beta \frac{qa^4}{Eh^3}$
		β	0.00105	0.01145	0.0934	0.537	1.261	2.16		

56. LARGE DEFLECTION OF CIRCULAR PLATES WITHOUT HOLES*

(Ref. 8)

When the deflection becomes larger than about one-half the thickness, the stresses of the middle surface cannot be ignored. These stresses enable the plate to carry part of the load as a diaphragm in direct tension. Under such conditions, the plate is stiffer than indicated by ordinary theory. Stresses for a given load are less, and stresses for a given deflection are generally greater than the ordinary theory indicates.

The plate under consideration is a circular plate whose edge is clamped so that rotation and radial displacement are prevented at the edge. The plate has radius r and thickness h and is loaded by a uniformly distributed load ω per unit area that causes a maximum deflection which is large relative to the thickness of the plate.

It is assumed that the plate is so tightly clamped that no radial displacement at the edge is possible. The radial membrane stress at the edge is due to the tensile forces which must be applied radially to prevent edge displacement. As for the deflection, it is not a linear function of the load. Values of y/h for various values $\omega r^4/Eh^4$ are shown in Fig. 14.

Also, values of the stress multiplied by the quantity r^2/Eh^2 as ordinates and values of maximum deflection y divided by the thickness h as abscissas are shown in Fig. 15. It will be noted that the dimensionless ordinates and abscissas in Figs. 14 and 15 make it possible to use the curves for plates of many dimensions, provided that other conditions are the same.

It will be seen from Fig. 14 that variations in Poisson's Ratio have very little effect on the behavior of plate.

Example: A numerical example will illustrate the use of these curves. Consider a plate of thickness, $h = 0.02$ in., radius $r = 2$ in., and load $\omega = 3$ psi. Let $E = 30 \times 10^6$ psi and let $\nu = 0.3$

From Fig. 14 for $\omega r^4/Eh^4 = 10$ we obtain $y/h = 1.055$ so that $y = 0.0211$ in.

From Fig. 14 the stresses are:

1. Membrane stress at edge:

$$\frac{Sr^2}{Eh^2} = 0.56$$

$$S = 1{,}680 \text{ psi}$$

2. Bending stress at edge:

$$\frac{Sr^2}{Eh^2} = 5.85$$

$$S = 17{,}550 \text{ psi}$$

3. Membrane stress at center:

$$\frac{Sr^2}{Eh^2} = 1.07$$

$$S = 3{,}210 \text{ psi}$$

4. Bending stress at center:

$$\frac{Sr^2}{Eh^2} = 2.57$$

$$S = 7{,}710 \text{ psi}$$

*Abstracted by permission from Ref. 8.

57. STRESSES IN A THICK CIRCULAR PLATE WITH AN ECCENTRIC HOLE, WHEN SUBJECTED TO INSIDE OR OUTSIDE RADIAL PRESSURE

(Ref. 9)

(Results Applicable to Thick Cylinders)

A problem often encountered in engineering is the situation in which a thick circular plate with an eccentric hole is stressed by an inside or outside radial pressure. In order to determine the circumferential stresses for a similar plate with a concentric hole we use the standard Lame's equations for fixed cylinders. By transformation of Lame's equations it is possible to solve problems of circular plates with an eccentric hole as shown in Tables 16 and 17. These tabulated equations for circumferential stresses then also apply to thick cylinders.

NOTATION

R_2 = outside radius of plate (in.)
R_1 = inside radius of plate (in.)
S_o = circumferential stress along the contour of the outside circle of radius R_2 (psi)
S_i = circumferential stress along the contour of the inside circle of radius R_1 (psi)
d = location of hole from center of plate (in.)
θ = angle, as indicated in Tables 16 and 17 (deg.)
P_1 = inside pressure (psi)
P_2 = outside pressure (psi)

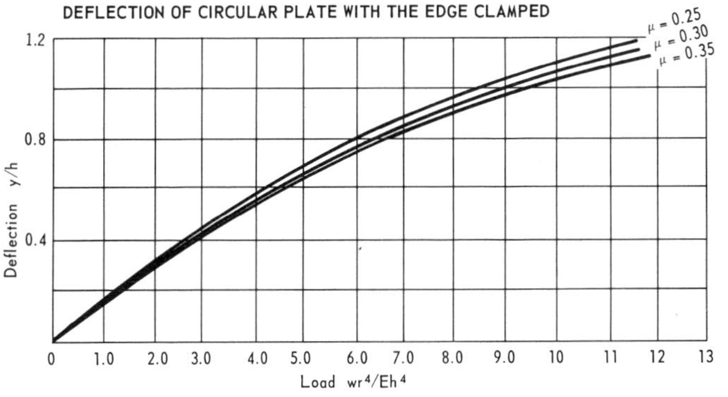

Fig. 14

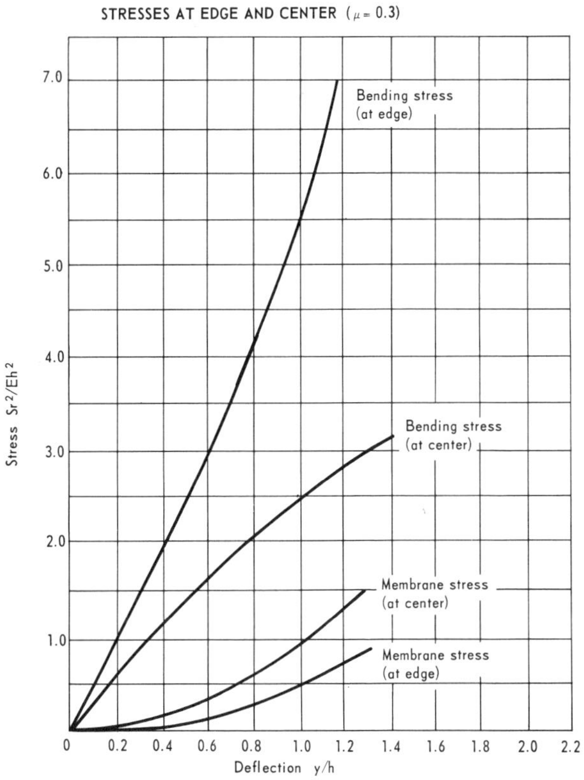

Fig. 15

Table 16. Cylinder or Plate with Hole Stressed by a Constant Inside Pressure P_1

Loading	*Location of Hole*	*Circumferential Stresses at Outer Surface, S_0, at inner S_i*
	d	$S_0 = 2P_1 \dfrac{R_1^2[R_2^2(R_2 - 2d\cos\theta)^2 - (R_1^2 - d^2)^2]}{(R_1^2 + R_2^2)[R_2^2 - (R_1 + d)^2][R_2^2 - (R_1 - d)^2]}$ (a)
		$\text{Max. } S_0 = 2P_1 \dfrac{R_1^2(R_2^2 + R_1^2 + 2R_2 d - d^2)}{(R_1^2 + R_2^2)[R_2^2 - R_1^2 - 2R_2 d + d^2]}$ (b)
	$d > \dfrac{R_2}{2}$	$\text{Max. } S_0 = 2P_1 \dfrac{R_1^2(R_2^2 + R_1^2 - 2R_2 d - d^2)}{(R_1^2 + R_2^2)(R_2^2 - R_1^2 + 2R_2 d + d^2)}$
	$d < \dfrac{R_2}{2}$	$\text{Max. } S_0 = 2P_1 \dfrac{R_1^2(R_2^2 + R_1^2 + 2R_2 d - d^2)}{(R_1^2 + R_2^2)[R_2^2 - R_1^2 - 2R_2 d + d^2]}$
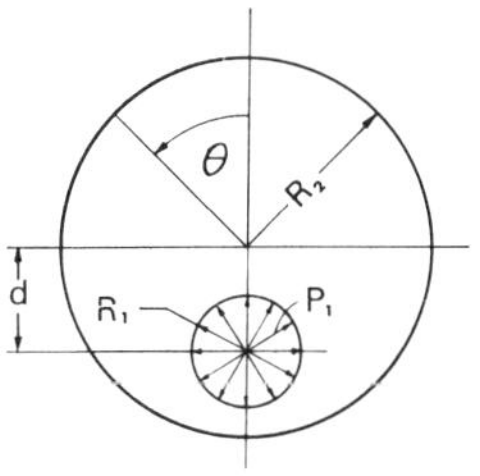	$d = 0$	$\text{Max. } S_0 = 2P_1 \dfrac{R_1^2}{R^2 - R_1^2}$
	d	$S_i = -P_1 + \dfrac{2P_1R_2^2[(R_2^2 - d^2)^2 - R_1^2(R_1^2 + 2d\cos\theta)^2]}{(R_1^2 + R_2^2)[R_2^2 - (R_1 - d)^2][R_2^2 - (R_1 + d)^2]}$
	$d < \dfrac{R_1}{2}$	$\text{Max. } S_i = -P_1\left[1 - \dfrac{2R_2^2(R_2^2 + R_1^2 + 2R_1 d - d^2)}{(R_1^2 + R_2^2)(R_2^2 - R_1^2 + 2R_1 d - d^2)}\right]$
	$d > \dfrac{R_1}{2}$	$\text{Max. } S_i = -P_1\left\{1 - \dfrac{2R_2^2(R_2^2 - d^2)^2}{(R_1^2 + R_2^2)[R_2^2 - (R_1 - d)^2][R_2^2 - (R_1 + d)^2]}\right\}$
	$d = 0$	$\text{Max. } S_i = P_1 \dfrac{R_2^2 + R_1^2}{R_2^2 - R_1^2}$

Table 17. Cylinder or Plate with Hole Stressed by a Constant Outside Pressure P_2

Loading	*Location of Hole*	S_0 — *Circumferential Stresses at Outer Surface* S_i — *Circumferential Stresses at Inner Surface*
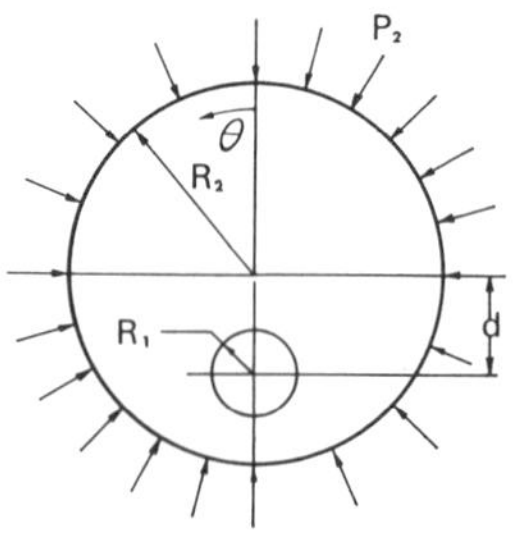	d	$S_0 = -P_2 \left\{1 + \frac{2R_1^2[R_2^2(R_2^2 - 2d\cos\theta)^2 - (R_1^2 - d^2)^2]}{(R_1^2 + R_2^2)[R_2^2 - (R_1 - d)^2][R_2^2 - (R_1 + d)^2]}\right\}$ $S_i = -2P_2R_2^2 \frac{(R_2^2 - d^2)^2 - R_1^2(R_1 + 2d\cos\theta)^2}{(R_1^2 + R_2^2)[R_2^2 - (R_1 - d)^2][R_2^2 - (R_1 + d)^2]}$
	$d < \frac{R_1}{2}$	Max. $S_i = -2P_2R_2^2 \frac{(R_2^2 - d^2)^2 - R_1^2(R_1 - 2d)^2}{(R_1^2 + R_2^2)[R_2^2 - (R_1 - d)^2][R_2^2 - (R_1 + d)^2]}$
	$d > \frac{R_1}{2}$	Max. $S_i = -2P_2R_2^2 \frac{(R_2^2 - d^2)^2}{(R_1^2 + R_2^2)[R_2^2 - (R_1 - d)^2][R_2^2 - (R_1 + d)^2]}$
	$d = 0$	Max. $S_0 = -P_2 \frac{R_2^2 + R_1^2}{R_2^2 - R_1^2}$ Max. $S_i = -\frac{2P_2R_2^2}{R_2^2 - R_1^2}$

Chap. V

REFERENCES

Section	*Ref. No.*	
50	1	ROARK, R. J.: *Formulas for Stress and Strain*, McGraw-Hill, N.Y., 1954
50	2	TIMOSHENKO, S.: *Theory of Plates and Shells*, McGraw-Hill, N.Y., 1959
50	3	GRIFFEL, W.: "Flat-Plate Design," *Product Engineering*, June 11, 1962
52	4	GRIFFEL, W.: "Slope for Circular Plates," *Design News*, Nov. 28, 1962
53	5	SEELY, F. B., and SMITH, J. O.: *Advanced Mechanics of Materials*, John Wiley, N.Y., 1952
53	6	GRIFFEL, W.: "Beam-Supported Plates," *Machine Design*, Feb. 1, 1962
55	7	CONWAY, H. D.: "The Bending of Symmetrically Loaded Circular Plates of Variable Thickness," A.S.M.E. Paper No. 47–A–35
56	8	WAY, S.: "Bending of Circular Plates With Large Deflection," *A.S.M.E. Transaction,* Aug. 1934, Vol. 56, #8, p. 627.
57	9	GRIFFEL, W.: "Stresses in a Thick Circular Plate With an Eccentric Hole, When Subjected to Inside or Outside Radial Pressure," *Product Design and Value Engineering*, Canada, March-April, 1965

Chapter VI

COLUMN DESIGN

58. SIMPLIFIED COLUMN DESIGN*

(*Ref. 1, 7*)

. . . A FAST, STRAIGHTFORWARD METHOD FOR FINDING THE BEST CROSS-SECTIONAL DIMENSIONS OF LONG, SLENDER MEMBERS UNDER COMPRESSION

Critical stress in straight, axially loaded column members is generally expressed as a function of slenderness ratio and material properties. In design, this relationship is often awkward to apply and usually leads to trial-and-error calculations because the slenderness ratio is not known at the beginning of the solution.

This section presents a simplified, direct approach to column design. Tables and a chart provided here eliminate guesswork in design calculations and, also, give a clearer picture of the effectiveness of specific column materials and cross-sectional shapes.

NOTATION

A = area of column cross section (in.^{2})
C = coefficient of constraint for column end connection (Table 2)
E = modulus of elasticity of column material (psi)
F = compressive load along column axis (lb.)
I = moment of inertia of column cross section about the neutral axis (in.^{4})
k = radius of gyration of column cross section (equation 3) (in.)
L = column length (in.)
N = factor of safety
Q = column design coefficient (equation 1)
R = slenderness ratio (equation 4)
s_b = compressive stress in critical column section due to bending moment (psi)
s_c = critical column stress (psi)
$= FN/A$
s_e = equivalent column stress for combined stress calculations (equation 6) (psi)
s_n = total compressive stress at critical column section (equation 7) (psi)
s_s = total shear stress at critical column section (psi)
s_y = compressive yield stress for column material (psi)
s_{ys} = torsional yield stress for column material (approximately 0.5 to 0.6 s_y) (psi)
Z = shape factor for column cross section
A^2/I
σ = maximum combined normal stress (psi)
τ = maximum combined shear stress (psi)

BASIC CONCEPTS

A mathematical study of the variables of column design shows that a relationship, similar to that for the slenderness ratio, can be established between critical column stress and certain known design requirements. These known requirements are: Column length, type of end connection, load, factor of safety, material properties and a so-called shape factor which depends on type and proportions of the column cross section. For simple cross sections—solid round, square, hexagon, etc.—the shape factor is a straightforward constant. For more complex shapes, it must be calculated. A procedure for calculating the shape factor for any cross-sectional shape is outlined in *Shape-Factor Calculations*. Shape factors for a number of commonly used cross sections are given in Table 1.

---

*Reprinted, with permission, from *Machine Design*, April 2, 1959.

Table 1—Shape Factors and Dimensional Relationships for Typical Column Cross Sections

Form	Shape Factor, Z	Dimensional Relationships	Form	Shape Factor, Z	Dimensional Relationships
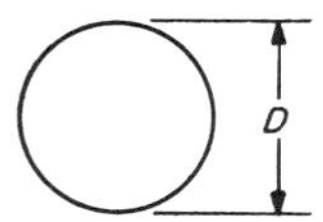	Solid Round 12.57	$D = 1.128A^{1/2}$	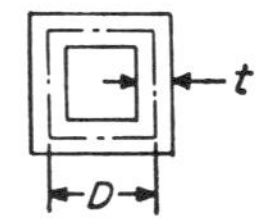	Square Tube $\dfrac{24}{n + 1/n}$	$t = \dfrac{1}{2}\left(\dfrac{A}{n}\right)^{1/2}$ $n = D/t$
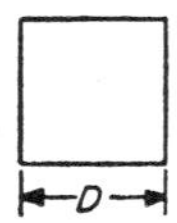	Solid Square 12	$D = A^{1/2}$	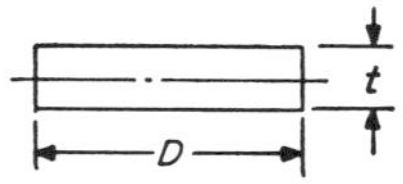	Flat Bar $12n$	$t = (A/n)^{1/2}$ $n = D/t$
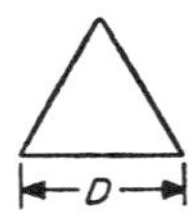	Solid Equilateral Triangle 10.39	$D = 1.520A^{1/2}$	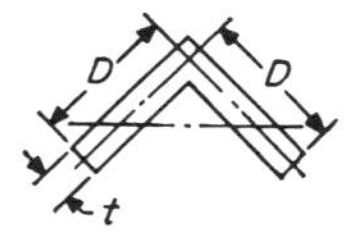	Equal-Leg Angle $\dfrac{48n}{n^2 + 2.5}$ (approx.) (Error is less than 0.5 per cent if $n \geqq 2.5$	$t = (A/2n)^{1/2}$ $n = D/t$
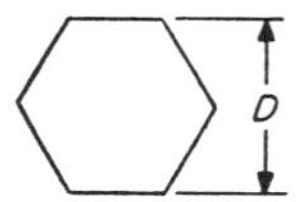	Solid Hexagon 12.47	$D = 1.075A^{1/2}$	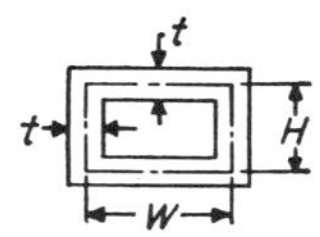	Rectangular Tubing—Horizontal Axis $\dfrac{24(N + n)^2}{N(3n^2 + 1) + n(n^2 + 3)}$	$t = \left[\dfrac{A}{2(N + n)}\right]^{1/2}$ $N = W/t,\ n = H/t$
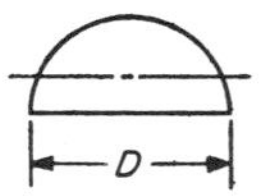	Solid Half-Round 22.48	$D = 1.596A^{1/2}$	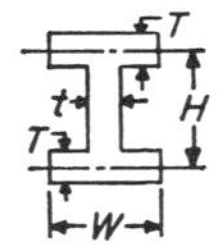	I-Beam—Major Axis $\dfrac{12[N(2m - n) + n]^2}{m(1 + N)^3 - (m - n)(1 - N)^3}$	$H = \left[\dfrac{A}{N(2m - n) + n}\right]^{1/2}$ $n = t/H,\ m = W/H,\ N = T/H$
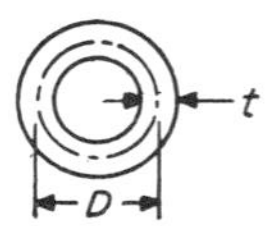	Round Tube $\dfrac{25.13}{n + 1/n}$	$t = [A/(\pi n)]^{1/2}$ $n = D/t$	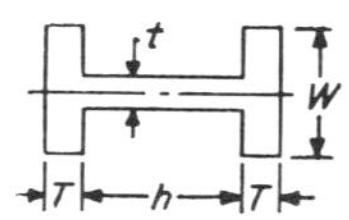	I-Beam—Minor Axis $\dfrac{12(2Nm + n)^2}{2Nm^3 + n^3}$	$h = \left[\dfrac{A}{2Nm + n}\right]^{1/2}$ $n = t/h,\ m = W/h,\ N = T/h$

Except for material properties, the known design requirements previously mentioned can be related in the form of a single expression for calculating a column-design coefficient. With this coefficient, the critical column stress can be found directly from the chart, Fig. 1. Area of the column cross section can then be determined and the dimensions calculated from relationships given in *Shape Factor Calculations* or Table 1.

The chart provided here, Fig. 1, is based on Johnson's parabolic column formula for the inelastic range and Euler's formula for the elastic range. It applies to steel. For other column formulas or materials a similar chart can be readily developed by a method which will be discussed later.

Since the approach presented here is merely a mathematical conversion of conventional techniques of column design, the results obtained will be the same as though the conventional methods were used.

SHAPE-FACTOR CALCULATIONS

The following procedure can be used to calculate shape factor Z for any cross section:

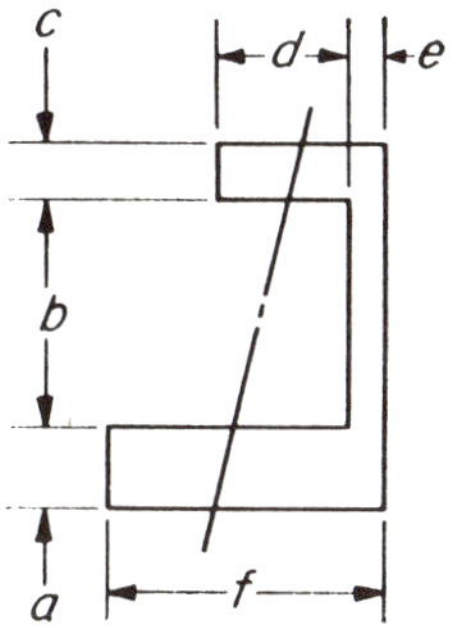

1. Choose any convenient set of dimensions, a_o, b_o, c_o, d_o, etc.
2. Calculate cross section area A_o from the dimensions selected in the preceding step.
3. Calculate moment of inertia I_o of the cross section about the buckling axis.
4. Calculate shape factor from

$$Z = \frac{A_o^2}{I_o}$$

General relationship for computing the dimensions of a cross section with the same shape factor Z but different area is:

$$x = x_o\sqrt{\frac{A}{A_o}} = x_oK$$

where x_o represents any dimension of a cross section with area A_o and shape factor Z, and x is the corresponding dimension for a cross section of area A. Thus, $a = a_oK$, $b = b_oK$, etc.

DESIGN PROCEDURE

Recommended sequence of steps for design of straight, axially loaded columns is:

1. Select a convenient form of column cross section. If thin walls or flanges are required, the possibility of local wrinkling should be considered. A review of codes, specifications, and the literature is usually advisable in such situations.

2. Determine shape factor Z. For simple cross-sectional shapes, read Z from Table 1. For complex shapes, compute Z from procedure in *Shape-Factor Calculations.*

3. Find coefficient of constraint C from Table 2.

4. Calculate column design coefficient Q from

$$Q = \frac{L^2Z}{NFC} \qquad (1)$$

5. Read critical column stress s_c from Fig. 1. In the chart, find line corresponding to value of Q computed in step 4. At the intersection of this line with the compressive yield-stress curve for the column material, read corresponding value of s_c from the vertical scale.

6. Calculate required cross-sectional area A from

$$A = \frac{NF}{s_c} \qquad (2)$$

If codes or specifications require or suggest a certain maximum permissible slenderness ratio, R_{max}, a slightly modified form of area calculation may be necessary. At the intersection point of the Q and s_y plots found on the chart in step 5, read the corresponding value of slenderness ratio R from the horizontal scale. Ideally, this value should not exceed the required value of R_{max}. However, if it does, compute cross-sectional area A from

$$A = \frac{L^2Z}{R_{max}^2\,C} \qquad (2a)$$

7. Calculate dimensions of column cross section from the area found in step 6, using the relationships given in Table 1 or *Shape-Factor Calculations.*

Permissible axial loads on straight columns that are subjected to column stresses only may be computed directly from the chart, Fig. 1, and Table 2. Procedure is:

Table 2—Coefficients of Constraint For Different End Connections

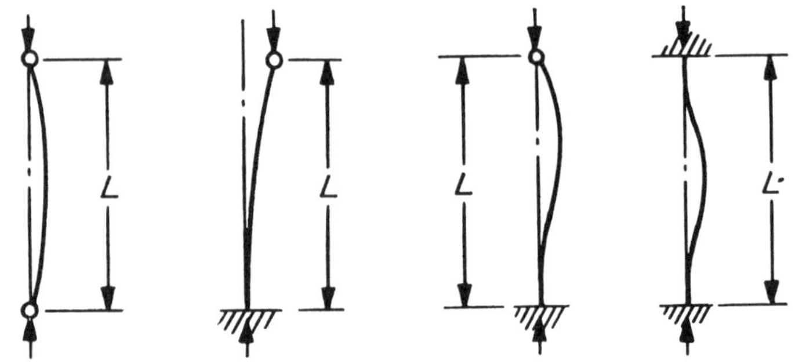

Recommended Values General design applications	1	1/4	1	1
Recommended Maximum Values Very long columns with rigid end connections	...	...	≦1.5	≦2.0*
Experimental Values				
$L/k = 100$	...	...	1.05	1.13
$L/k = 200$	...	...	1.78	3.0
Theoretical Values Very long columns	...	...	2	4

1. Read coefficient of constraint C from Table 2.

2. Determine radius of gyration k of column cross section from

$$k = \sqrt{\frac{I}{A}} \tag{3}$$

3. Calculate slenderness ratio R from

$$R = \frac{L}{k\sqrt{C}} \tag{4}$$

4. Read critical column stress s_c from Fig. 1. In the chart, locate point corresponding to slenderness ratio R on the compressive yield-stress curve for the column material. Read corresponding value of s_c from the vertical scale.

5. Calculate permissible load F from

$$F = \frac{As_c}{N} \tag{5}$$

Here again, local wrinkling may be a problem under certain conditions. See step 1 of the preceding design procedure for details.

The tables and the chart can also be used for design of columns subject to combined stresses. However, the approach is not as direct as that for columns with axial loads only. Recommended design procedure is:

1. Estimate approximate size and dimensions of column cross section. (Note: See local-wrinkling considerations in step 1 of design procedure for axially loaded columns.)

2. Read coefficient of constraint C from Table 2.

3. Determine radius of gyration k for assumed cross section from equation (3).

4. Calculate slenderness ratio R from equation (4).

5. Read critical column stress s_c corresponding to calculated slenderness ratio (step 4) and column-material yield stress from chart, Fig. 1.

6. Calculate equivalent column stress s_e from

$$s_e = \frac{Fs_y}{As_c} \tag{6}$$

where A is the estimated area and s_c is the critical column stress found in the foregoing step.

7. Calculate total normal stress s_n from

$$s_n = s_e + s_b \tag{7}$$

8. Calculate total shear stress s_n at the location where s_n occurs.

9. Calculate maximum shear stress τ from

$$\tau = \sqrt{s_S^2 + \left(\frac{s_n}{2}\right)^2} \tag{8}$$

and/or maximum normal stress σ from

$$\sigma = \frac{s_n}{2} + \tau. \tag{9}$$

10. Calculate factor of safety N for the assumed cross section from

$$N = \frac{s_{ys}}{\tau} \tag{10}$$

and/or

$$N = \frac{s_y}{\sigma} \tag{11}$$

depending on whether the shear or normal stress will be used as the criterion for design.

If the computed factor of safety, N, is not satisfactory, repeat steps 1 through 10 with another assumed cross section. Continue this process until a satisfactory solution is found.

PRACTICAL APPLICATIONS

Flexibility of the tables and the chart in column design calculations is apparent from the previous discussion. However, another important feature of this approach is the insight it provides on the effect of modification of column proportions and material. Full range of usefulness of the design procedures is best demonstrated by typical problem solutions.

Example 1: A piston rod 24 in. long is subjected to a compressive force of 15,000 lb. Type of application suggests a factor of safety of $N = 4$. Slenderness ratio should not exceed a value of $R_{max} = 200$. Find the required piston-rod diameter if material is SAE 1020 solid-round, cold-finished steel with $E = 30 \times 10^6$ psi and $s_y = 66{,}000$ psi.

From Table 1, shape factor $Z = 12.57$. From Table 2, coefficient of constraint $C = 1$.

From equation (1),

$$Q = \frac{(24)^2 12.57}{4(15{,}000)\,(1)} = 0.1207$$

From Fig. 1 at the intersection of $Q = 0.12$ and $s_y = 66{,}000$ psi, $s_c = 45{,}500$ psi and $R = 74$ which is well within the required limit, $R_{max} = 200$.

From equation (2), $A = 4(15{,}000)/45{,}500 = 1.32$ in.^{2} From Table 1 then, $D = 1.128(1.32)^{1/2} = 1.296$ in. Use $D = 1\frac{5}{16}$ in.

Fig. 1 shows that use of a higher strength steel would increase the critical column stress only about 4,000 psi at the most. This increase of approximately 9 percent also represents the potential reduction in weight that could be realized. If a tubular, rather than solid-round, cross section is used, shape factor Z and column design coefficient Q would both be reduced. This modification would lead to a greater increase in the value of critical column stress and, consequently, to a further reduction in area and weight, particularly for a steel with high yield strength.

Example 2: A feed hopper is to be supported by four angle-iron posts, each of which carries a compressive load of 30,000 lb. Determine the required size of the angles if the posts are 120 in. high and are structural steel with $s_y = 34{,}000$ psi. A factor of safety, $N = 2$, is considered adequate. Slenderness ratio must not exceed $R_{max} = 120$. For this application, a practical flange width-thickness ratio is 12:1. End connections should be considered as flexible.

From Table 1, D = flange width $- t/2 = 11.5t$, giving $n = 11.5$. Shape factor $Z \approx 48(11.5)/(11.5^2 + 2.5) \approx 4.10$.

From Table 2, $C = 1$, and from equation (1), $Q = 0.98$.

From Fig. 1, $s_c = 17{,}000$ psi and $R = 130$. This value of slenderness ratio exceeds the specified design limit. From equation (2a) then, $A = 4.10$ in.^{2}

From Table 1, $t = (4.10/23)^{1/2} = 0.422$ in. and flange width $= 5.06$ in.

If the design limit of $R_{max} = 120$ can be exceeded slightly, a standard 5 by 5 by $\frac{7}{16}$-in. angle might be used. For this angle, radius of gyration $k = 0.98$ in. From equation (4), slenderness ratio $R = 122.4$.

Example 3: A 30-in. connecting link is made from $\frac{1}{2}$ by 2 in. SAE 1020 hot-rolled flat strip steel. Find (1) the ultimate load capacity of the link and (2) the maximum permissible load for factor of safety $N = 2.5$. Yield stress $s_y = 43{,}000$ psi and coefficient of constraint $C = 1$.

From equation (3), $k = 0.1443$ in., and from equation (4), $R = 208$. From Fig. 1, $s_c = 6800$ psi.

Ultimate load $F_u = As_c = 2(\frac{1}{2})(6800) = 6800$ lb. and permissible load $F = F_u/N = 2{,}720$ lb.

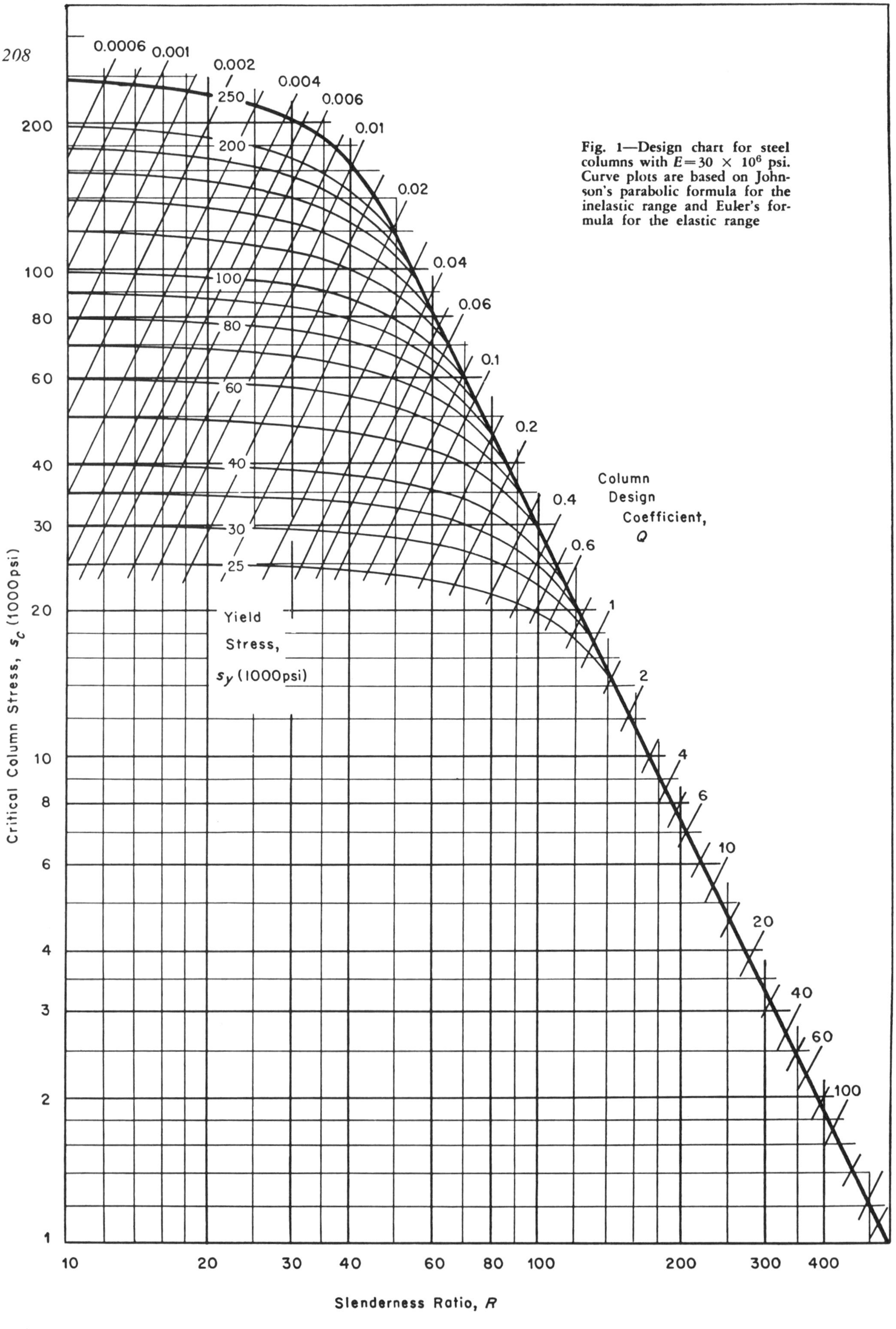

Fig. 1—Design chart for steel columns with $E = 30 \times 10^6$ psi. Curve plots are based on Johnson's parabolic formula for the inelastic range and Euler's formula for the elastic range

Example 4: A lever-operated screw jack with a 2-in. nominal-size Acme thread extends to a maximum lift height of 24 in. The load can move laterally. If a force of 100 lb. is acting on the 20-in. long lever arm, find the factor of safety under these conditions. The jack is rated for a load of 6,000 lb. and the screw material is C1045 steel, as rolled, with $s_y = 59{,}000$ psi and $s_{ys} = 35{,}400$ psi.

For a 2-in. Acme thread, minor or root diameter is 1.75 in. From Table 2, $C = \frac{1}{4}$.

Radius of gyration $k = 1.75/4 = 0.4375$ in. and, from equation (4), slenderness ratio $R \approx 110$. From Fig. 1, then, $s_c = 24{,}500$ psi. Root area of the Acme thread, $A = 2.41$ in.^{2}

From equation (6), equivalent column stress $s_e = 6{,}000$ psi. Bending moment $M = 100(24) = 2{,}400$ lb.-in. and section modulus of the thread root section = 0.526 in.^{3} Compressive stress produced by bending $s_b = 2{,}400/0.526 = 4{,}560$ psi.

From equation (7), $s_n = 10{,}560$ psi. For torque $T = 100(20) = 2{,}000$ lb.-in., shear stress is: $s_s = Tc/J = 2{,}000/1.052 = 1{,}901$ psi.

From equation (8), maximum shear stress $\tau = 5{,}610$ psi. If this stress value is used as the design criterion, then (equation 10) factor of safety $N = 6.31$.

From equation (9), maximum normal stress $\sigma = 10{,}890$ psi. If this stress value is the design criterion, then (equation 11) factor of safety $N = 5.42$.

MODIFIED CHART PLOTS

The approach presented here can be readily adapted to other column design formulas and materials by modifying the chart plot, Fig. 1.

1. Calculate critical column stress s_c for successive values of slenderness ratio R, using the appropriate column design formulas.
2. Plot s_c versus R on logarithmic graph paper in a manner similar to that for chart in Fig. 1.
3. Calculate maximum and minimum values of design coefficient Q for the new chart plot from $Q_{\min} = R_{\min}^2/s_{c\ \max}$ and $Q_{\max} = R_{\min}^2/s_{c\ \max}$.
4. Draw a new Q-scale by plotting the expression, $s_c = R^2/Q$, for successive convenient values of the parameter, Q, between the limit values determined in the preceding step, as shown in the present chart, Fig. 1.

59. ECCENTRICALLY LOADED COLUMNS*

A Quick, Simple Method for Calculating Maximum Stress

(*Ref. 8*)

Critical load of an axially loaded long column normally produces lateral instability in the column—even though the compressive stress produced by this load is less than the compressive yield stress of the material. For short columns the compressive stress produced by the Euler (critical) load may exceed the yield value, so Euler's column formula $P_{cr} = (K\pi^2EI)/l^2$, where K depends on end conditions—must be modified by the use of a reduced modulus for the material.

The unstable behavior characterized by the Euler load is based on the assumption that the member is initially straight and centrally loaded. In reality, production variations often introduce some eccentricity in loading and curvature in shape. It is sometimes even necessary to design these characteristics into a member to make it function properly.

An axial load applied eccentrically, or to a curved column, will generate bending stresses as well as uniform compressive stresses. This results in a gradually increasing deflection with increase in load, in contrast to a buckling behavior, where no deflection occurs until lateral instability takes place. Bending stresses, when added to the normal column compressive stresses, can produce stresses in excess of those developed by the Euler load.

BENDING ANALYSIS OF COLUMNS

To account for load eccentricity and initial curvature in a column, Table 3, it is necessary to determine the maximum moment generated by the end loads. This moment depends on the end-support conditions, eccentricity of loading, and initial curvature. A separate analysis, based on beam theory, is required for each set of conditions. From such an analysis, moment and deflection distribution can be plotted as functions of axial load.

*Reprinted, with permission, from *Machine Design*, April 2, 1959. Copyright 1959 by The Penton Publishing Co.

NOTATION

A = cross-sectional area of column (in.^{2})
c = distance from neutral axis to outermost fiber of column (in.)
D = mean diameter of column (in.)
E = modulus of elasticity (psi)
I = minimum moment of inertia of column cross section (in.^{4})
K = Euler constant depending on end supports of column
l = length of column (in.)
M = maximum bending moment caused by load eccentricity or initial curvature (lb.-in.)
P = axial column load (lb.)
P_{cr} = Euler (critical) column load (lb.)
ε = eccentricity of loading or amplitude of initial curvature (in.)
σ = stress (psi)
λl = load parameter $l(P/EI)^{1/2}$

Maximum moment can then be calculated, and maximum compressive stress found from

$$\sigma = \frac{P}{A} + \frac{Mc}{I} \tag{12}$$

The first term represents uniform compressive stress. The second term is the additional bending compressive stress produced by eccentricity or initial curvature.

USE OF CHARTS

Charts shown in Fig. 2 may be used only for columns where product EI is constant. In addition, eccentricities and curvatures must be in the plane of bending.

The following method may be used to determine total stress on an eccentrically loaded or initially curved column:

1. Determine load parameter $\lambda l = l(P/EI)^{1/2}$.
2. From the appropriate chart in Fig. 2 determine $(M/P\varepsilon)_{max}$. Multiply by $P\varepsilon$ to obtain maximum bending moment M.
3. From equation (12) calculate σ.

Example 1: Calculate the maximum compressive stress in a pinned-pinned aluminum tube with double eccentricity (Case 2), having the following dimensions and load conditions: Length, l = 8 ft.; mean diameter, D = 3 in.; wall thickness, $t = \frac{1}{8}$ in.; axial load, P = 6,500 lb.; eccentricity, ε = 1.5 in.; modulus, E = 10×10^6 psi.

TABLE 3 – Column Characteristics

	Case 1 -- Pinned-Pinned (single eccentricity) M at $y = \frac{\pi}{2\lambda}$
	Case 2 -- Pinned-Pinned (double eccentricity) M at $y = \frac{l}{2}$
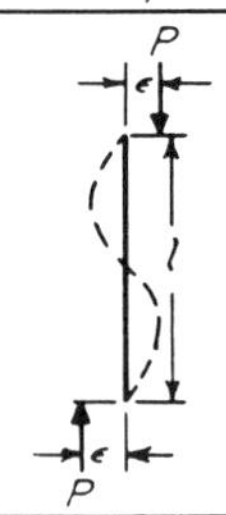	Case 3 -- Pinned-Pinned (compound eccentricity)
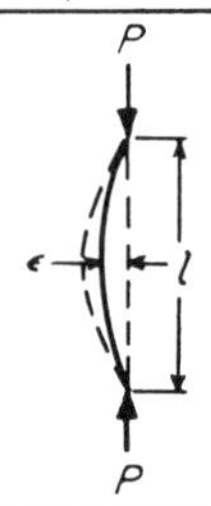	Case 4 -- Pinned-Pinned (initial curvature) M at $y = \frac{l}{2}$
	Case 5 -- Free-Fixed (single eccentricity) M at $y = 0$
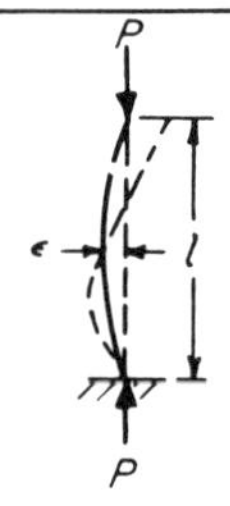	Case 6 -- Free-Fixed (initial curvature)
	Case 7 -- Fixed-Fixed (initial curvature) M at $y = 0, l$

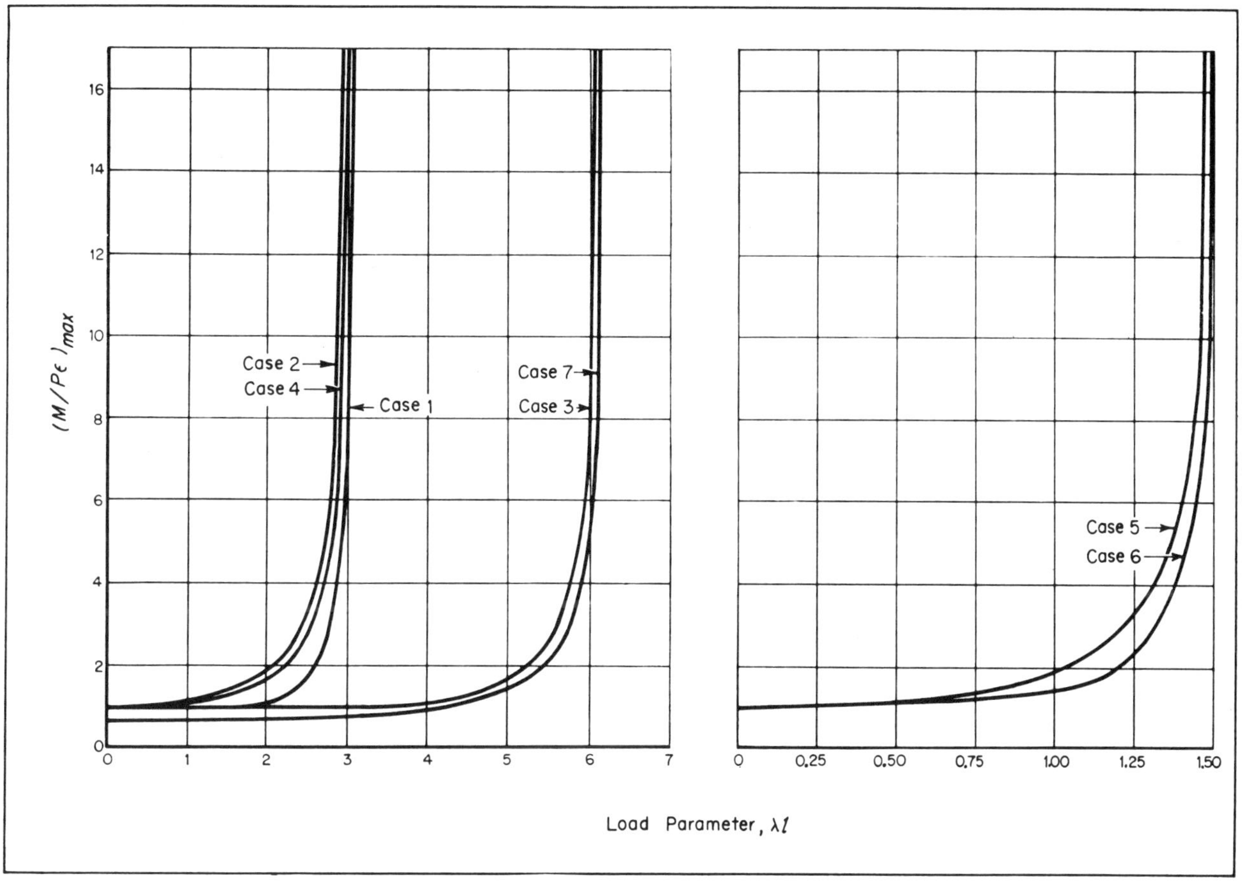

Fig. 2. Maximum moment of eccentrically loaded columns. As the axial load approaches the Euler, or critical load, the moment approaches infinity, indicating instability.

Section properties are:

$$A \cong \pi D t = 3.14(3)(1/8) = 1.175 \text{ in.}^2$$

$$I \cong \frac{\pi D^3 t}{8} = \frac{3.14(27)(1/8)}{8} = 1.325 \text{ in.}^4$$

Then,

$$\frac{P}{A} = \frac{6{,}500}{1.175} = 5{,}540 \text{ psi}$$

Load parameter is

$$\lambda l = l\sqrt{\frac{P}{EI}} = 96\sqrt{\frac{6{,}500}{1.325 \times 10^7}} = 2.25$$

From Fig. 2, case 2:

$$\frac{M}{P\varepsilon} = 2.35;\ M = 2.35(6{,}500)(1.5) = 22{,}900 \text{ lb.-in.}$$

Then,

$$\frac{Mc}{I} = \frac{22{,}900(1.5)}{1.325} = 25{,}900 \text{ psi}$$

and

$$\sigma = \frac{P}{A} + \frac{Mc}{I} = 5{,}540 + 25{,}900 = 31{,}440 \text{ psi}$$

The Euler load and stress for a centrally loaded pinned-pinned column, with $K = 1$, is:

$$P_{cr} = \frac{K\pi^2 EI}{l^2} = \frac{1(3.14)^2(10^7)(1.325)}{(96)^2} = 14{,}170 \text{ lb}$$

$$\sigma = \frac{P_{cr}}{A} = \frac{14{,}170}{1.175} = 12{,}050 \text{ psi}$$

Thus, even though the actual load is less than half the Euler load, the eccentricity increases maximum stress by 150 percent.

Example 2: Calculate permissible initial curvature ε (Case 7) for a fixed-fixed aluminum tube with the following dimensions: Length, $l = 45$ in.; nominal diameter, $D = 1$ in.; wall thickness, $t = \frac{1}{16}$ in.; modulus, $E = 10 \times 10^6$. Curvature ε is to be such that allowable load P is at least 80 percent of P_{cr} for a maximum combined compressive stress of 40,000 psi.

Section properties are:

$$A \cong \pi D t = \frac{3.14}{16} = 0.196 \text{ in.}^2$$

$$I \cong \frac{\pi D^3 t}{8} = \frac{3.14}{8(16)} = 0.0245 \text{ in.}^4$$

Euler load, with $K = 4$:

$$P_{cr} = \frac{K\pi^2 EI}{l^2} = \frac{4(3.14)^2(10^7)(0.0245)}{(45)^2} = 4{,}780 \text{ lb.}$$

and buckling stress is

$$\sigma = \frac{P_{cr}}{A} = \frac{4{,}780}{0.196} = 24{,}400 \text{ psi}$$

Since reduced load, $P = 0.80(P_{cr}) = 3{,}824$ lb., then $\lambda l = 45[3824/(0.0245 \times 10^7)]^{1/2} = 5.62$. From Fig. 2, case 7, $M/P\varepsilon = 2.35$.

Finally, based on equation (12)

$$\varepsilon = \frac{I}{c(m/P_\varepsilon)}\left\{\frac{\sigma}{P} - \frac{1}{A}\right\}$$

$$= \frac{0.0245}{0.5(2.35)}\left\{\frac{40{,}000}{3{,}824} - \frac{1}{0.196}\right\} = 0.111 \text{ in.}$$

Chap. VI

REFERENCES

Section	*Ref. No.*	
58	1	HOESCHEL, H. G.: "Simplified Column Design," *Machine Design*, April 2, 1959
58	2	SEELY, F. B. and SMITH, J. O.: *Advanced Mechanics of Materials*, John Wiley, N.Y., 1952
58	3	BRIDGET, F. J., JEROME, C. C., VOSSELLER, A. B., *A.S.M.E. Transaction*, Vol. 56, 1934
58	4	STOWELL, E. Z., HEIMERL, G. J., LIBOVE, C., LINDQUIST, E. E.: "Buckling Stresses for Flat Plates and Sections," *A.S.C.E. Proceedings*, Vol. 77, July, 1951
58	5	ROARK, R. J.: *Formulas for Stress and Strain*, McGraw-Hill, N.Y., 1954
58	6	FAIRES, V. M.: *Design of Machine Elements*, 3rd ed., The Macmillan Co., N.Y., 1955
58	7	KIRSCH: *Zeitschrift Verein Deutscher Ingeniere* (Germany), 1905
59	8	ZOROWSKI, C. F.: "Eccentrically Loaded Columns," *Machine Design*, May 24, 1962

Chapter VII

PRESSURE VESSELS

60. SHELL FORMULAS*

WEIGHT, GAS PRESSURE, AND LIQUID PRESSURE LOADING CONSIDERED.

(*Ref. 1 . . . 7*)

NOTATION

Forces and Moments:

p_x, p_θ, p_z, or $-p$ = applied loads in x, θ, z-directions (psi)

p_H, p_V = applied loads in horizontal and vertical directions, *see* equation (1) (psi)

p = internal pressure (psi)

V = vertical force (lb./in.)

W = total vertical force (lb.)

H = horizontal shear force (lb./in.)

$Q_x = H \sin \varphi$ = normal shear force (lb./in.) (not used)

N_x, N_θ = meridional and hoop tension force (lb./in.)

M_x, M_θ = meridional and hoop moment (lb.in./in.)

$\mathscr{M}, \mathscr{H}, \mathscr{V}$ = moment (lb.-in./in.), horizontal shear force (lb./in.) vertical force (lb./in.), applied at edge of shell

Coordinates and Geometric Parameters:

x, θ, z = coordinates in meridional, azimuthal, and inward normal directions

φ = latitude angle

R_x, R_θ, or R = radii or curvatures of shell middle surface in meridional and azimuthal directions

r = horizontal radius of shell middle surface (in.)

h = wall thickness (in.)

D = flexural rigidity of shell, *see* equation (12)

l = attenuation length, *see* equation (11)

l_v = axial component of l

$\mathscr{L}$ = meridional length of shell between apex (pole) and edge as in Fig. 1

n = distance from edge, in units of l; $n = (\bar{x} - x)/l$

ε, f = expansivity and flexibility coefficients, *see* equation (13)

μ = Poisson's Ratio

Bar over quantity = edge value

Subscripts i, o = inner fiber ($z = +h/2$), outer fiber ($z = -h/2$)

Deformations and Stresses:

u, v, w = displacements in x, θ, z-directions

δ = radial (outward) displacement, positive in same sense as $\mathscr{H}$

∇ = vertical (upward) displacement

χ = increase in latitude angle (angle of rotation), positive in same sense as $\mathscr{M}$

κ_x, κ_θ = increase in curvatures, $d\chi/dx$, $\chi \cot \varphi/R$ (not used)

$\varepsilon_x, \varepsilon_\theta$ = strains in x, θ-directions

σ_x, σ_θ = meridional and hoop stress

F_1, F_2, F_3, F_4 = functions defined in equation (10d)

---

*Reprinted, with permission, from the A.S.M.E. Paper No. 54-A-8.

Formulas for membrane and bending analyses of axisymmetrically loaded axisymmetrical conical and spherical shells are presented in Tables 1 and 2. Table 1, item 1, lists general membrane formulas; items 2 to 9 specialize these general formulas to the frequently occurring cases of conical and spherical shells loaded by weight, gas pressure, and liquid pressure.†

Table 2, item 10, lists shell-bending formulas in the familiar, Geckeler, approximation. The formulas express the shell displacements δ, χ, ∇, and the shell moments and forces M_x, M_θ, H, N_x, N_θ (per unit length), in terms of the applied edge moment $\mathscr{M}$ and applied edge shear $\mathscr{H}$, in the form of decaying sinusoidals $F_1 \ldots, F_4$. The argument n in the formulas denotes distance from the edge in multiples of

$$l = (\overline{rh}/\sin \overline{\varphi})^{1/2}/[3(1 - \mu^2)]^{1/4} \qquad (11)$$

Length l thus denotes the distance from the edge in which the exponential factor of F_k reduces from 1 to $1/e$; l may be called the "attenuation length" of the shell. The functions F_1, F_2, F_3, F_4 are tabulated in reference (3) as functions φ, ζ, ψ, θ, and in reference (4) as functions A, B, C, D.

Shell rotation $\bar{\chi}$ developed at the edge in the absence of force $\mathscr{H}$, under action of unit applied moment $\mathscr{M}$, may be called "flexibility," and is assigned the special symbol f. It is related to l, and to the shell stiffness

$$D = Eh^3/12(1 - \mu^2) \qquad (12)$$

by the formula‡

$$f = l/D \qquad (13a)$$

Shell displacement $\bar{\delta}$ developed at the edge in the absence of moment $\mathscr{M}$, under action of unit applied force $\mathscr{H}$, may be called "expansivity," and is assigned the special symbol ε. It is related to f and l_v (the axial component of l) by

$$\varepsilon = \tfrac{1}{2} f l_v^2 \qquad (l_v = l \sin \overline{\varphi}) \qquad (13b)$$

†We refer to the equations in items 1 to 10 as equations (1) to (10).

‡Equations (13a, b) are obtained by substituting $n = 0$, $\mathscr{H} = 0$, $\mathscr{M} = 1$, and $n = 0$, $\mathscr{H} = 1$, $\mathscr{M} = 0$, respectively, into the χ and δ expressions of equations (10a).

TABLE 1 MEMBRANE ANALYSIS

SHELL DESCRIPTION	TYPE OF LOADING	DESIGN PARAMETERS	DESIGN FORMULAS
General Shell $p_V = p_x \sin\phi + p_z \cos\phi$ $p_H = p_x \cos\phi - p_z \sin\phi$ 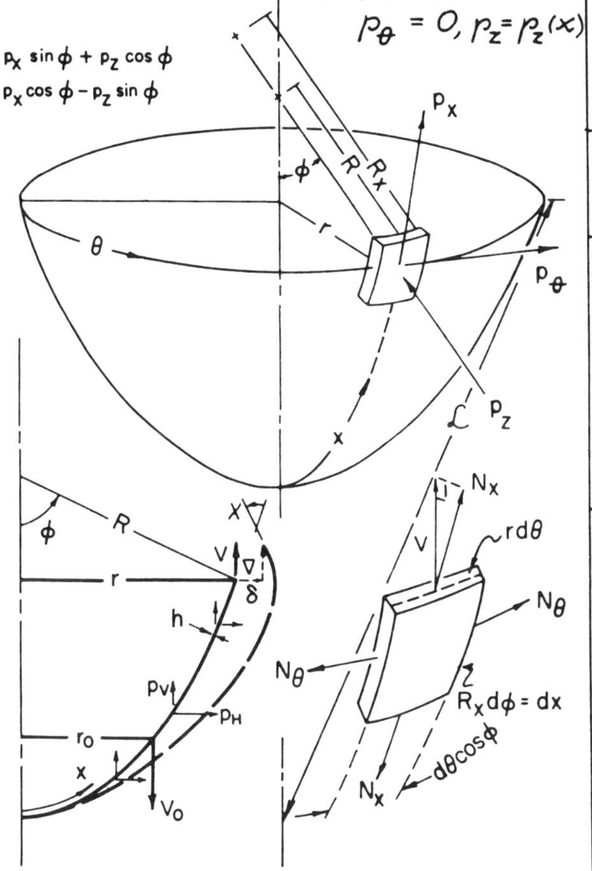	**1.** $p_x = p_x(x)$ $p_\theta = 0,\ p_z = p_z(x)$	(a) forces (lb./in.) $V,\ N_x,\ N_\theta$	$2\pi r V = 2\pi r_0 V_0 - 2\pi \int_{x_0}^{x} r p_V\, dx$ $N_x = \frac{V}{\sin\varphi}, \quad N_x(0) = -\frac{1}{2} R_x(0) p_z(0)$ $N_\theta = -R p_z - \frac{R}{R_x} N_x, \quad N_\theta(0) = -\frac{1}{2} R(0) p_z(0)$
		(b) strains (in./in.) $\varepsilon_x, \varepsilon_\theta$	$\varepsilon_x = \frac{N_x - \mu N_\theta}{Eh}, \quad \varepsilon_\theta = \frac{N_\theta - \mu N_x}{Eh}$
		(c) deformations (in., in/in, in.) $\delta,\ \chi,\ \nabla$	$\delta = r \varepsilon_\theta$ $\chi = (\varepsilon_x - \frac{R}{R_x}) \cot\varphi - \frac{d(R\varepsilon_\theta)}{dx}$ $\nabla \Big\|_{x_0}^{x} = \int_{x_0}^{x} (\varepsilon_x \sin\varphi + \chi \cos\varphi)\, dx$ $= -R\cos\varphi\, \varepsilon_\theta \Big\|_{x_0}^{x} + \int_{x_0}^{x} \frac{\varepsilon_x - R\varepsilon_\theta / R_x}{\sin\varphi}\, dx$
		(d) stresses (lb./sq.in) $\sigma_x,\ \sigma_\theta$	$\sigma_x = \frac{N_x}{h}, \quad \sigma_\theta = \frac{N_\theta}{h}$
Conical Shell 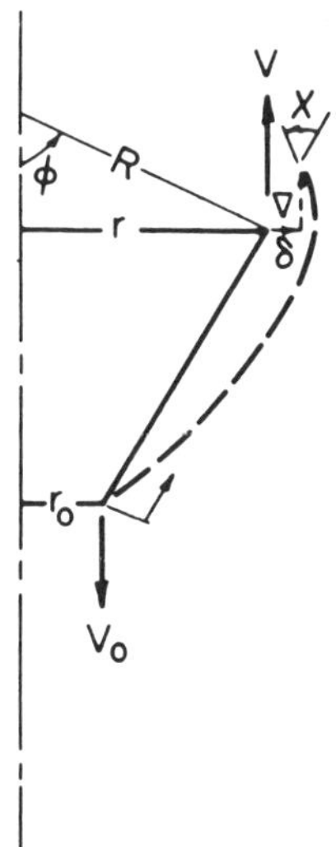	**2.** $p_x = p_x(x)$ $p_\theta = 0$ $p_z = p_z(x)$ $\{\ r = R\sin\varphi$ $R_x = \infty$ $dr = dx \cos\varphi$ $r - r_0 = x\cos\varphi\ \}$	(a) forces	$V = V_0 \frac{r_0}{r} - \frac{1}{r} \int_{r_0}^{r} r(p_x \tan\varphi + p_z)\, dr$ $= V_0 - \int_0^x p_x\, dx, \quad \text{for } \varphi = \pi/2$ $N_x = \frac{V}{\sin\varphi}, \quad N_\theta = -\frac{r p_z}{\sin\varphi}$
		(c) deformations	$\delta = r\varepsilon_\theta$ $\chi = \varepsilon_x \cot\varphi - \frac{d(R\varepsilon_\theta)}{dx} = (\varepsilon_x - \frac{d\delta}{dr}) \cot\varphi$ $\nabla = -R\varepsilon_\theta \cos\varphi \Big\|_0^x + \frac{1}{\sin\varphi} \int_0^x \varepsilon_x\, dx$ $= -\delta \cot\varphi \Big\|_{r_0}^{r} + \frac{1}{\sin\varphi \cos\varphi} \int_{r_0}^{r} \varepsilon_x\, dr$

TABLE 1 MEMBRANE ANALYSIS (*Continued*)

SHELL DESCRIPTION	TYPE OF LOADING	DESIGN PARAMETERS	DESIGN FORMULAS
Conical Shell 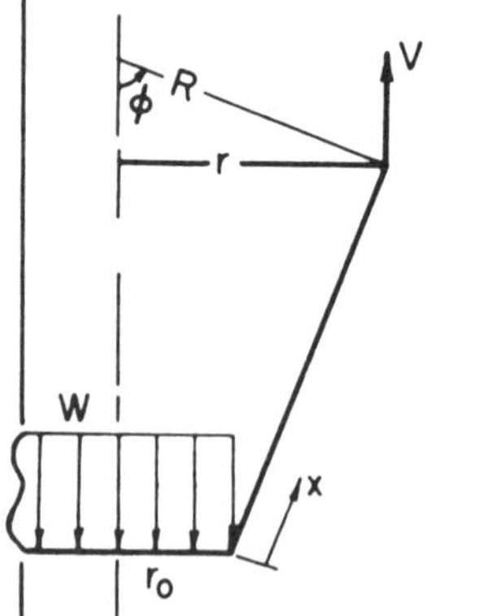	**3. Weight Loading** W (lb.) $p_x = p_\theta = p_z = 0$	(a) forces	$V = \frac{W}{2\pi r}$, $N_x = \frac{V}{\sin\varphi}$, $N_\theta = 0$
		(b) strains	$Eh\epsilon_x = \frac{V}{\sin\varphi}$, $Eh\epsilon_\theta = -\mu\frac{V}{\sin\varphi}$
		(c) deformations	$Eh\delta = -\frac{\mu V r}{\sin\varphi}$ $Eh\chi = \frac{V\cos\varphi}{\sin^2\varphi}$ $Eh\nabla = \frac{W \ln r/r_0}{2\pi \sin^2\varphi\cos\varphi}$ $= \frac{Wx}{2\pi r}$, for $\varphi = \pi/2$
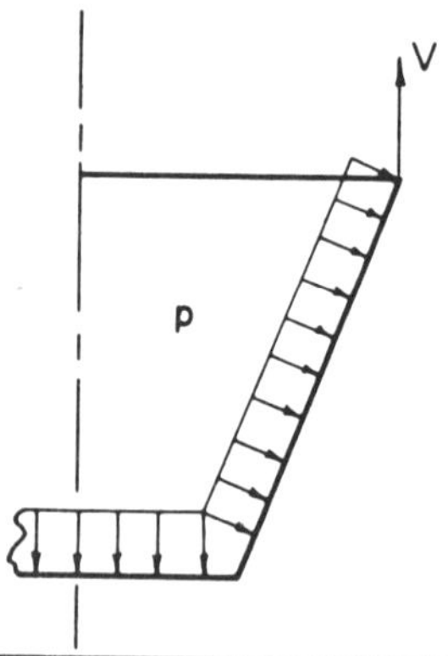	**4. Gas Pressure** p (psi) $p_x = p_\theta = 0$ $p_z = -p$	(a) forces	$V = \frac{pr}{2}$, $N_x = \frac{pr}{2\sin\varphi}$, $N_\theta = \frac{pr}{\sin\varphi}$
		(b) strains	$Eh\epsilon_x = (\frac{1}{2} - \mu)\frac{pr}{\sin\varphi}$, $Eh\epsilon_\theta = (1 - \frac{\mu}{2})\frac{pr}{\sin\varphi}$
		(c) deformations	$Eh\delta = (1 - \mu/2)\, pr^2/\sin\varphi$ $Eh\chi = -\frac{3}{2}\, pr\cot\varphi/\sin\varphi$ $Eh\nabla = p(r^2 - r_0^2)\frac{\cot\varphi}{\sin\varphi}\left[-1 + \frac{\mu}{2} + \frac{\frac{1}{4} - \mu/2}{\cos^2\varphi}\right]$ $= (\frac{1}{2} - \mu)prx$, for $\varphi = \pi/2$
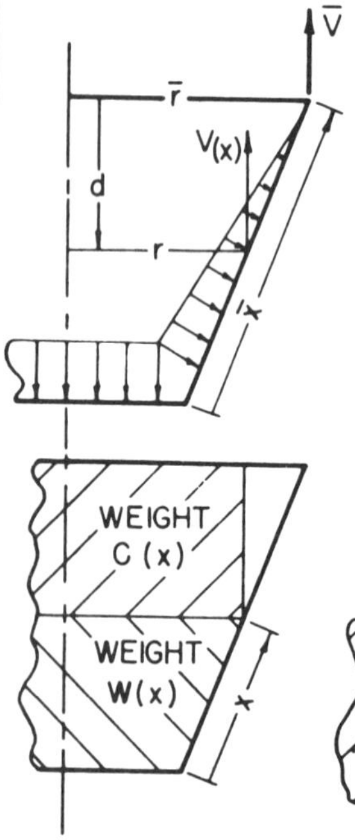	**5. Liquid Pressure** γd (lb/in²) $p_x = p_\theta = 0$ $p_z = -\gamma d$ $C(x) = \gamma r^2 \pi d(x)$, $\bar{C} = 0$. $W(x) = \frac{\gamma\pi}{3}(r^3 - r_0^3)\tan\varphi$ $= \frac{\gamma\pi}{3}(r^2 + r r_0 + r_0^2)\, x\sin\varphi$ $W^*(x) = \gamma\pi r^3$ 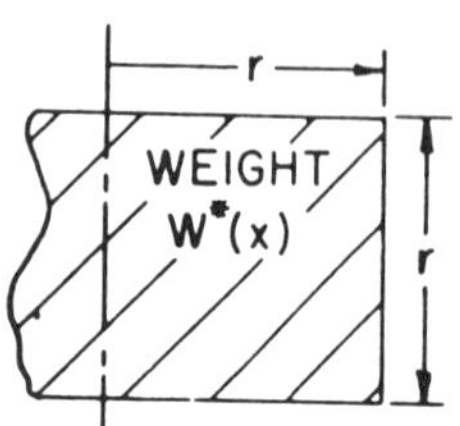	(a) forces	$V(x) = \frac{W(x) + C(x)}{2\pi r}$ $N_x = \frac{C + W}{2\pi r\sin\varphi}$, $N_\theta = \frac{2C}{2\pi r\sin\varphi}$
		(b) strains	$Eh\epsilon_x = \frac{(1-2\mu)C + W}{2\pi r\sin\varphi}$, $Eh\epsilon_\theta = \frac{(2-\mu)C - \mu W}{2\pi r\sin\varphi}$ For $\varphi = \pi/2$: $Eh\epsilon_x = \gamma r(\frac{\bar{x}}{2} - \mu d)$, $Eh\epsilon_\theta = \gamma r(d - \frac{\mu}{2}\bar{x})$
		(c) deformations	$Eh\delta = \frac{(2-\mu)C - \mu W}{2\pi\sin\varphi}$ $Eh\chi = \frac{2W^* + (W - 3C)\cot\varphi}{2\pi r\sin\varphi}$ For $\varphi = \pi/2$: $Eh\delta = \gamma r^2(d - \frac{\mu}{2}\bar{x})$ $Eh\chi = \gamma r^2$ $Eh\nabla = \frac{\bar{W}x}{2\pi r}(1 - 2\mu + \mu x/\bar{x})$

TABLE 1 MEMBRANE ANALYSIS (*Continued*)

SHELL DESCRIPTION	TYPE OF LOADING	DESIGN PARAMETERS	DESIGN FORMULAS
Spherical Shell 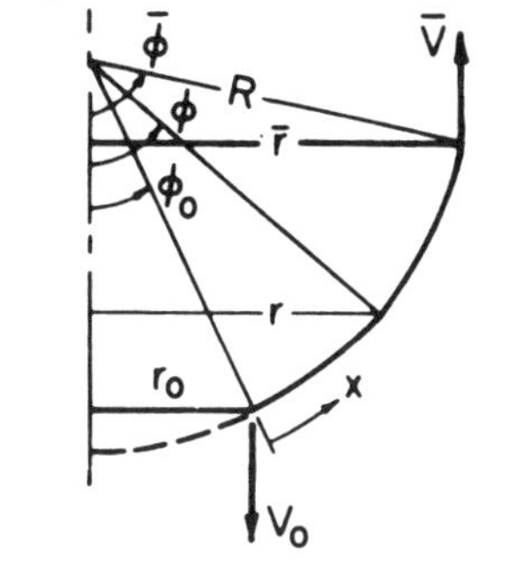	**6.** $p_x = p_x(\varphi)$ $p_\theta = 0$ $p_z = p_z(\varphi)$	(a) forces	$V(\varphi) = V_0 \frac{r_0}{r} - \frac{R}{\sin\varphi} \int_{\varphi_0}^{\varphi} (p_x \sin^2\varphi + p_z \sin\varphi\cos\varphi)\, d\varphi$ $N_x = \frac{V}{\sin\varphi}, \quad N_\theta = -Rp_z - N_x$
		(c) deformations	$Eh\delta = r\varepsilon_\theta$ $Eh\chi = (1+\mu)(N_x - N_\theta)\cot\varphi - \frac{d}{d\varphi}(\mu N_x - N_\theta)$ $\nabla\Big\vert_{\varphi_0}^{\varphi} = R\int_{\varphi_0}^{\varphi} (\varepsilon_x \sin\varphi + \chi\cos\varphi)\, d\varphi$ $= -R\varepsilon_\theta \cos\varphi\Big\vert_{\varphi_0}^{\varphi} + R\int_{\varphi_0}^{\varphi} \frac{\varepsilon_x - \varepsilon_\theta}{\sin\varphi}\, d\varphi$
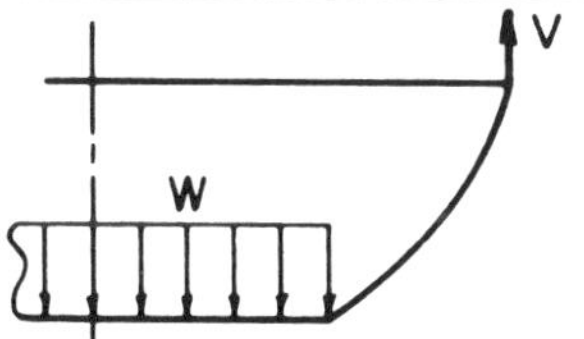	**7. Weight Loading** W (lb.) $p_x = p_\theta = p_z = 0$	(a) forces	$V = \frac{W}{2\pi r}, \quad N_x = -N_\theta = \frac{V}{\sin\varphi}$
		(b) strains	$Eh\varepsilon_x = -Eh\varepsilon_\theta = \frac{(1+\mu)V}{\sin\varphi}$
		(c) deformations	$Eh\delta = \frac{-(1+\mu)rV}{\sin\varphi}, \quad \chi = 0$
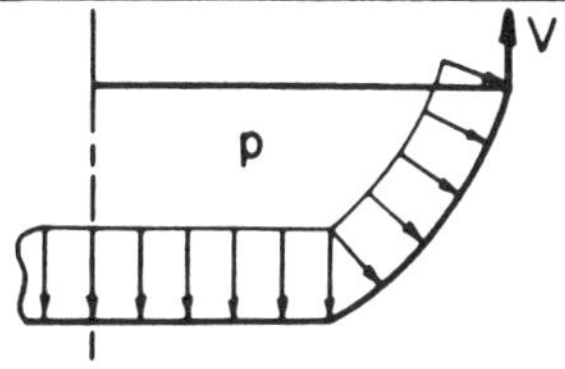	**8. Gas Pressure** p (psi) $p_x = p_\theta = 0, \quad p_z = -p$	(a) forces	$V = \frac{pr}{2}, \quad N_x = N_\theta = \frac{pR}{2}$
		(c) deformations	$Eh\delta = \frac{(1-\mu)prR}{2}, \quad \chi = 0$
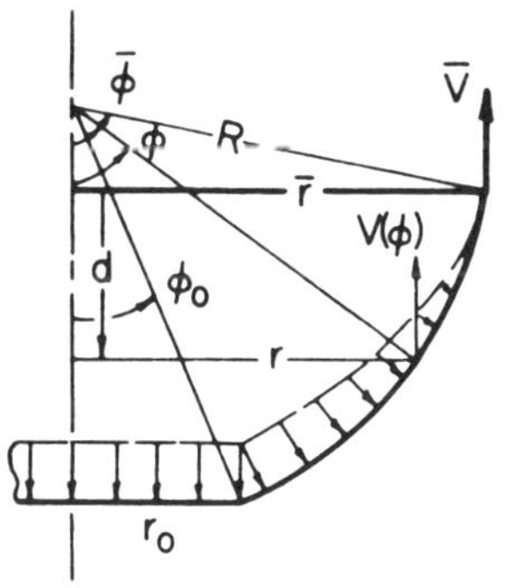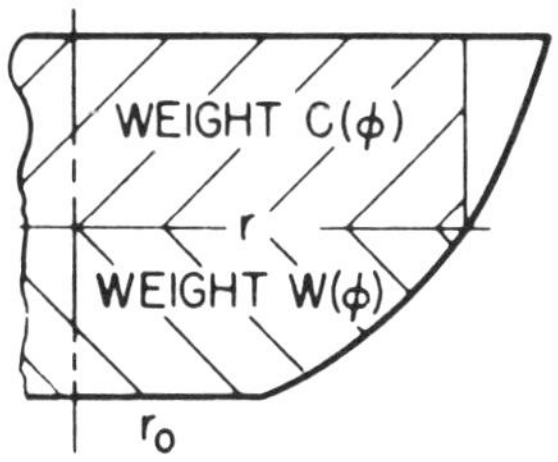	**9. Liquid Pressure** γd (psi) $p_x = p_\theta = 0$ $p_z = -\gamma d = -\gamma r(\cos\varphi - \cos\bar{\varphi})$ $C(\varphi) = \gamma R^3 \pi \sin^2\varphi \times (\cos\varphi - \cos\bar{\varphi})$ $W(\varphi) = \frac{1}{3}\gamma R^3 \pi \left[(1-\cos\varphi)^2 \times (2+\cos\varphi) - (1-\cos\varphi_0)(2+\cos\varphi_0)\right]$ $W^*(\varphi) = \gamma r^3 \pi$ 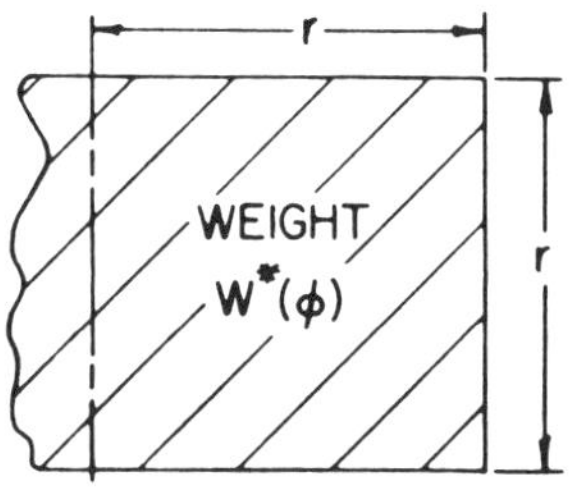	(a) forces	$V = \frac{C+W}{2\pi r}$ $N_x = \frac{C+W}{2\pi r\sin\varphi}, \quad N_\theta = \frac{C-W}{2\pi r\sin\varphi}$
		(b) strains	$Eh\varepsilon_x = \frac{C(1-\mu)+W(1+\mu)}{2\pi r\sin\varphi}$ $Eh\varepsilon_\theta = \frac{C(1-\mu)-W(1+\mu)}{2\pi r\sin\varphi}$
		(c) deformations	$Eh\delta = \frac{C(1-\mu)-W(1+\mu)}{2\pi\sin\varphi}$ $Eh\chi = \frac{2W^* - 4\mu W\cot\varphi}{2\pi r\sin\varphi}$

TABLE 2 EDGE LOAD-ANALYSIS

BENDING

SHELL DESCRIPTION	TYPE OF LOADING	DESIGN PARAMETERS	DESIGN FORMULAS

General Shell — **10. $\mathcal{H}, \mathcal{M}$**

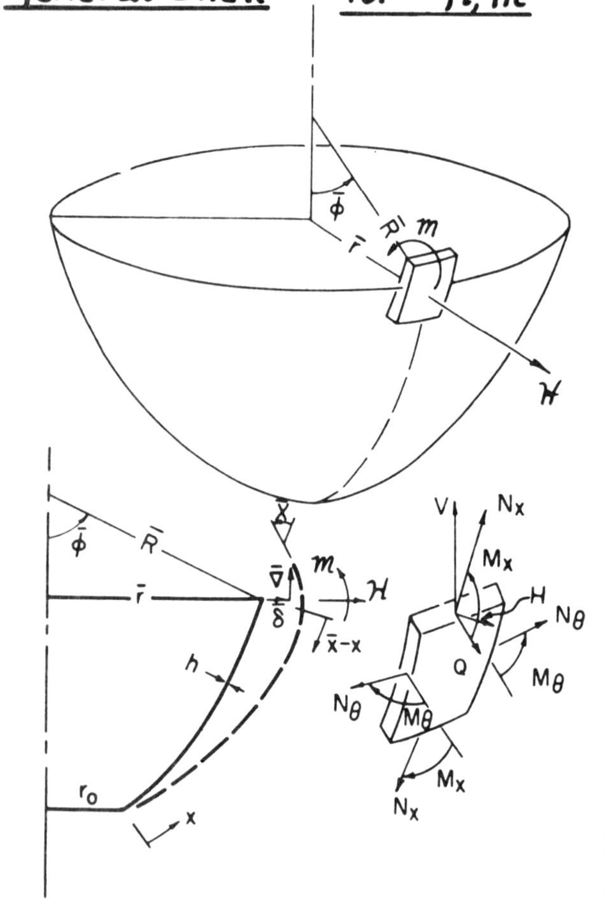

(a) deformations

$$\delta/f\ell_v = -\tfrac{1}{2}\mathcal{M}F_3(n) + \tfrac{1}{2}\mathcal{H}\ell_v F_4(n)$$

$$\chi/f = \mathcal{M}F_4(n) - \tfrac{1}{2}\mathcal{H}\ell_v F_1(n)$$

$$\nabla = -\delta\cot\varphi \qquad f = \ell/D$$

(b) moments & forces

$$M_x = \mathcal{M}F_1(n) - \mathcal{H}\ell_v F_2(n)$$

$$H\ell_v = 2\mathcal{M}F_2(n) + \mathcal{H}\ell_v F_3(n)$$

$$M_\theta = \mu M_x$$

$$N_x = H\cos\varphi$$

$$N_\theta = Eh\delta/r = \frac{2\sqrt{3(1-\mu^2)}}{h}\left[-\mathcal{M}F_3(n) + \mathcal{H}\ell_v F_4(n)\right]$$

$2\sqrt{3(1-\mu^2)} = 3.30$ for $\mu = 0.3$

(c) stresses

$$\sigma_{xo} = \frac{N_x}{h} + \frac{6M_x}{h^2}, \qquad \sigma_{xi} = \frac{N_x}{h} - \frac{6M_x}{h^2}$$

$$\sigma_{\theta o} = \frac{N_\theta}{h} + \frac{6M_\theta}{h^2}, \qquad \sigma_{\theta i} = \frac{N_\theta}{h} - \frac{6M_\theta}{h^2}$$

(d) The functions $F_1 = \sqrt{2}e^{-n}\cos(n-\frac{\pi}{4})$, $F_2 = e^{-n}\sin n$

$F_3 = -\sqrt{2}e^{-n}\sin(n-\frac{\pi}{4})$, $F_4 = e^{-n}\cos n$

n	F_1	F_2	F_3	F_4
0	1	0	1	1
.25	.947	.193	.562	.755
.50	.823	.291	.241	.532
.75	.668	.322	.024	.346
1.0	.508	.310	-.111	.199
1.5	.238	.223	-.207	.016
2.0	.067	.123	-.179	-.056
2.5	-.017	.049	-.115	-.066

Shell properties ($\mu = 0.3$)

ℓ = attenuation length $= 0.778\sqrt{\bar{r}h/\sin\bar{\varphi}}$

ℓ_v = vertical component of $\ell = \ell\sin\bar{\varphi}$

D = bending stiffness $= 0.0916\,Eh^3$

f = flexibility $= \ell/D$

ϵ = expansivity $= \frac{1}{2}f\ell_v^{\,2}$

n = distance from edge in units of ℓ $= \dfrac{\bar{x}-x}{\ell}$

61. DISCONTINUITY STRESSES IN PRESSURE VESSELS*

(Ref. 8, 9)

The pressure vessel considered in this section has the form of a surface of revolution with a wall thickness that is small compared to the radius of curvature of the wall. If the wall were very thin and offered no resistance to bending, it would be subjected only to direct stresses uniformly distributed over the thickness (membrane stresses).

If in a thin vessel there is an abrupt change in thickness or in meridional slope or curvature at any circumference, bending stresses occur in addition to the membrane stresses. These "discontinuity stresses" comprise: (1) a meridional bending stress S_1, which varies linearly throughout the thickness of the wall, (2) a circumferential or hoop bending stress S'_2, which varies linearly throughout the thickness of the wall, (3) an additional hoop stress S_2 uniform throughout the thickness of the wall and (4) a meridional shear stress S_s, assumed uniform throughout the thickness of the wall.

In order to determine the state of stress of a pressure vessel, it is necessary to find (1) membrane stresses and (2) discontinuity stresses. From the superposition of these two states of stress, the total stresses are obtained.

NOTATION

t = wall thickness (in.)
R = radius of discontinuity circle (in.)
R_1 = radius of curvature of meridian (in.)
R_2 = radius of curvature normal to meridian (in.)
ϕ_o = angle between axis of revolution and normal to wall
w = radial displacement (in.)
θ = angle of rotation of edges of vessel
x = distance measured along meridian from discontinuity circle
E = modulus of elasticity (psi)
D = flexural rigidity = $Et^3/12(1 - \mu^2)$
μ = Poisson's Ratio
λ = wall characteristic = $\sqrt[4]{3(1 - \mu^2)/R^2t^2}$
S_{1M} = meridional membrance stress (psi)
S_{2M} = hoop membrane stress (psi)

DETERMINATION OF STRESSES

The procedure to be followed in determining the stresses from data given in Tables 3, 4 and 5 is explained by considering the pressure vessel shown in Fig. 14. It consists of a cylinder closed by a semielliptical head and subjected to a uniform internal pressure, p.

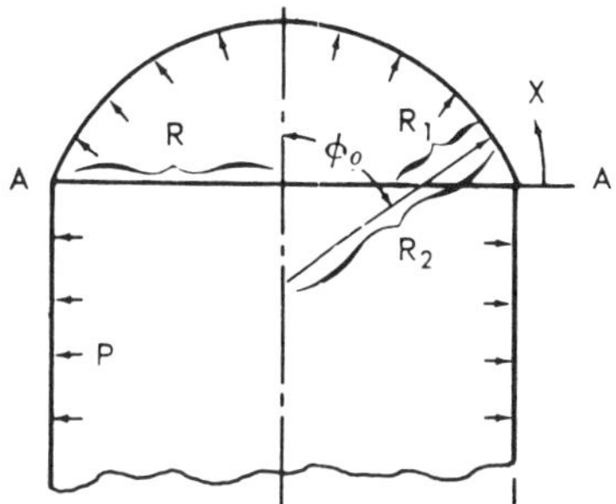

Fig. 14

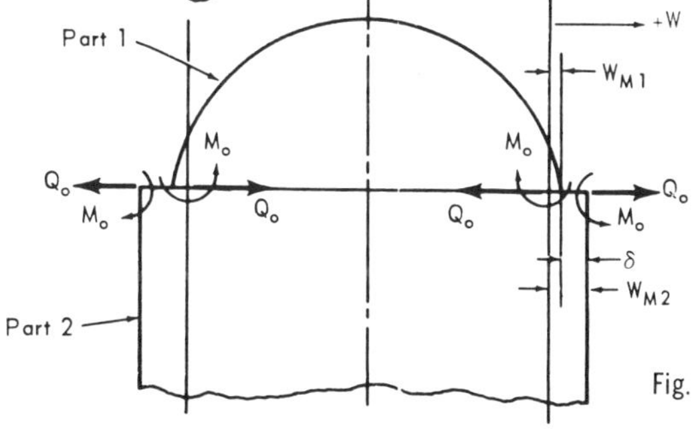

Fig. 15

The pressure vessel is assumed to be split into parts 1 and 2 along the circle where there is a discontinuity, line AA, Fig. 14 and 15. Because of action of the membrane stresses, each part expands radially by a different amount. Knowing the membrane stresses, which are found directly from the conditions of equilibrium, Table 3, these radial displacements may be computed.

In particular for the vessel of Fig. 14, the membrane stresses along circumference AA are, from Table 3:

In semielliptical head:

Meridional stress: $S_{1M} = pR/2t$ (14a)

Hoop stress: $S_{2M} = (pR/2t)[2 - (R/R_1)]$. (14b)

In cylindrical part:

Meridional stress: $S_{1M} = pR/2t$ (14c)

Hoop stress: $S_{2M} = pR/t$. (14d)

The radial displacements along AA are:

Taking $\mu = 0.3$

Semielliptical head:

$$w_{M1} = (pR^2/2Et)[1.7 - (R/R_1)] \quad (15a)$$

Cylindrical part:

$$w_{M2} = (pR^2/Et)(0.85) \quad (15b)$$

*Abstracted from Ref. 9.

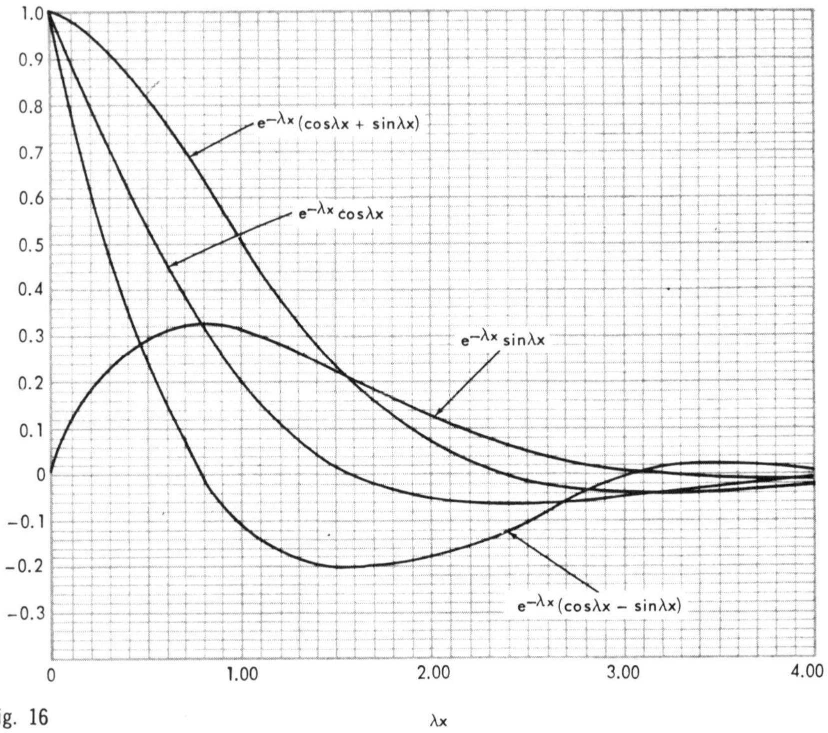

Fig. 16

Note that at the discontinuity circle AA: $R = R_2$ and $\phi_o = 90$ deg.

The difference in radial displacement δ between head and circular portions is then:

$$\delta = w_{M1} - w_{M2}. \tag{16}$$

In order to reassemble the two parts, it is necessary to apply uniformly distributed shearing forces Q_o and bending moments M_o along the edges. Q_o and M_o must fulfill the conditions:

1. The radial displacements of head w_1 and cylindrical part w_2 due to Q_o and M_o must cancel the membrane displacement δ found from equation (16):

$$w_1 - w_2 = -\delta \tag{17}$$

2. The angles of rotation θ_1 and θ_2 of the two edges must be the same

$$\theta_1 = \theta_2 \tag{18}$$

The equations for w_1, w_2, θ_1, θ_2 are given in Table 4.

Applying these formulas to the pressure vessel of Fig. 14

$$-Q_o/D\lambda^3 = -(pR^2/2Et)(R/R_1) \tag{19}$$

$$-\frac{M_0}{\lambda D} = +\frac{M_0}{\lambda D},$$

from which is found

$$Q_o = -(p/8\lambda)(R/R_1)$$

and

$$M_o = 0$$

Note that if the head and the cylindrical part have the same wall thickness and the same slope at the junction line, the bending moment M_o is zero.

Knowing Q_o and M_o, the discontinuity stresses at any point of the wall may be calculated from the equations given in Table 5 and from the curves of Fig. 16. The total stresses are obtained by adding the discontinuity stresses to the membrane stresses.

TABLE 3

Membrane Stresses and Radial Displacement Due to Uniform Internal Pressure p

Form of Vessel	Meridional Stress S_1	Hoop Stress S_2	Radial Displacement w_M
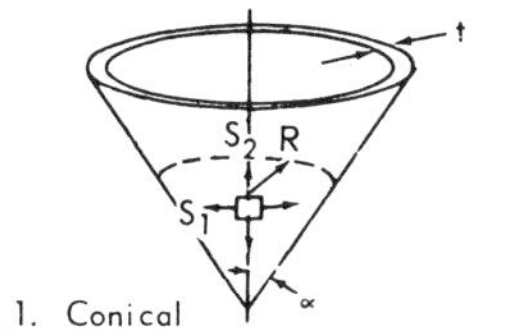 1. Conical	$S_{1M} = \frac{pR}{2t\cos\alpha}$	$S_{2M} = \frac{pR}{t\cos\alpha}$	$w_{M1} = \frac{pR^2}{Et\cos\alpha}\left[1 - \frac{\mu}{2}\right]$ $w_{M1} = \frac{R}{E}\left[S_2 - \mu S_1\right]$
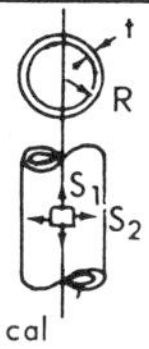 2. Cylindrical	$S_{1M} = \frac{pR}{2t}$	$S_{2M} = \frac{pR}{t}$	$w_{M2} = \frac{pR^2}{Et}\left[1 - \frac{\mu}{2}\right]$ $w_{M2} = \frac{R}{E}\left[S_2 - \mu S_1\right]$
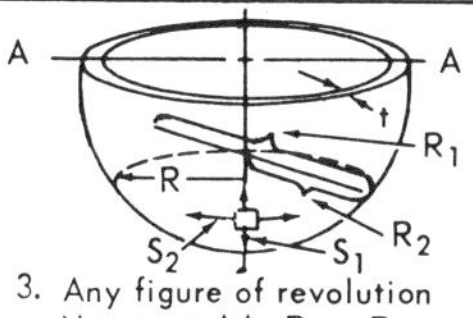 3. Any figure of revolution Note: at AA $R_2 = R$	$S_{1M} = \frac{pR_2}{2t}$	$S_{2M} = \frac{pR_2}{2t}\left[2 - \frac{R_2}{R_1}\right]$	$w_{M1} = \frac{R}{E}(S_2 - \mu S_1)$ At AA: $w_{M1} = \frac{pR^2}{Et}\left[2 - \mu - \frac{R}{R_1}\right]$
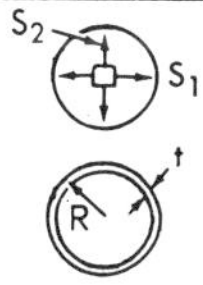 4. Spherical	$S_{1M} = \frac{pR}{2t}$	$S_{2M} = \frac{pR}{2t}$	$w_{M1} = \frac{pR^2}{2Et}\left[1 - \mu\right]$

TABLE 4

Radial Displacement and Angle of Rotation of Edge Due to Q_o and M_o

Form of vessel	Radial displacement	Angle of rotation
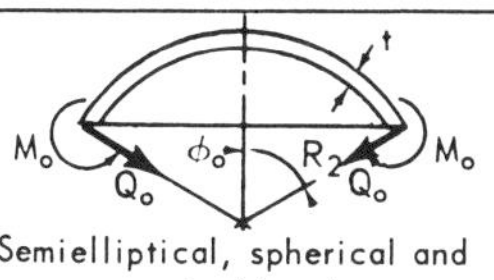 Semielliptical, spherical and conical head	$w_1 = -\frac{Q_o}{2D\lambda^3}\sin^2\phi_o - \frac{M_o}{2D\lambda^2}\sin\phi_o$	$\theta_1 = -\frac{Q_o}{2D\lambda^2}\sin\phi_o - \frac{M_o}{D\lambda}$
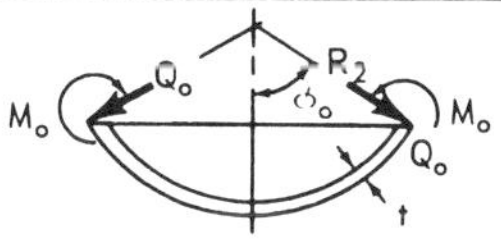 Cylindrical shell	$w_2 = +\frac{Q_o}{2D\lambda^3}\sin^2\phi_o - \frac{M_o}{2D\lambda^2}\sin\phi_o$	$\theta_2 = -\frac{Q_o}{2D\lambda^2}\sin\phi_o + \frac{M_o}{D\lambda}$

TABLE 5

Expressions for Discontinuity Stresses at Distance x Measured Along Meridian from Discontinuity Circle AA (see Fig. 1)

(1) Hoop Normal Stress

$$S_2 = \frac{2\lambda R}{t}\left(\frac{R}{R_2}\right)e^{-\lambda x}\left[Q_o\cos\lambda x - \lambda M_o(\cos\lambda x - \sin\lambda x)\right]$$

(2) Meridional Shear Stress:

$$S_s = \frac{R}{tR_2}\sqrt{\frac{R}{R_2}}\,e^{-\lambda x}\left[Q_o(\cos\lambda x - \sin\lambda x) + 2\lambda M_o\sin\lambda x\right]$$

(3) Maximum Meridional Bending Stress

$$S_1 = \frac{6R}{t^2R_2}\left(\frac{1}{\lambda}\right)e^{-\lambda x}\left[-Q_o\sin\lambda x + \lambda M_o(\cos\lambda x + \sin\lambda x)\right]$$

(4) Maximum Hoop Bending Stress

$$S'_2 = \mu S_1 + \left(\frac{6}{t^2}\right)D(1-\mu^2)\left(\frac{\cot\phi_o}{R_2}\right)\left(\frac{1}{2\lambda^2 D}\right)e^{-\lambda x}\left[Q_o(\cos\lambda x + \sin\lambda x) - 2\lambda M_o\cos\lambda x\right]$$

62. PRESSURE VESSELS WITH SPHERICAL CAPS

(*Ref. 10*)

A rapid and accurate formulation of the compatibility equations at the junction of cylinder and spherical cap is possible when using dimensionless coefficients. This assumes that the shell is subjected to uniform pressure and edge bending loading.

Pressure vessel analysis that takes into consideration the bending stresses due to the discontinuity at the junction of the cylinder and head is tedious and time consuming. As a rule, various Codes attempt to predict only the direct or membrane stress due to pressure ignoring the distribution of bending stresses throughout the shell. Therefore, if optimum use of material is desired, it is necessary to calculate the bending stresses in the vessel. The task of such calculations is greatly simplified when using dimensionless coefficients, known as edge influence coefficients, which determine the edge moment and edge shear.

CYLINDER—SPHERICAL CAP JUNCTION

Assume the pressure vessel shown in Fig. 17 as cut up into its two main structural parts, the cylindrical body and spherical head. Also, assume that each of these parts can freely deform under the action of the inside pressure on vessel walls. As a result of these deformations, we obtain along *A-A* a discontinuity in displacement and rotation between the neighborhood ends of the cylindrical body and the head. In order to maintain the continuity of the material in this region, we shall have to apply shearing force *H* and bending moment *M* uniformly distributed along the circle of juncture *A-A*. The determination of *H* and *M* is very easy if the edge influence coefficients are available.

For illustration of how *H* and *M* are calculated, the following notations will be used:

y_C^H, y_S^H, y_C^M, y_S^M = horizontal displacement of the edge *A* of the cylinder and spherical cap due to unit values of *H* and *M*

y_S^F = horizontal displacement of the edge *A* of the spherical cap due to horizontal component of the edge force

y_C^P, y_S^P = horizontal displacement of the edge *A* of the cylinder and spherical cap due to unit value of pressure *p*

θ_C^H, θ_S^H, θ_C^M, θ_S^M = rotation of edge *A* due to unit values of *H* and *M*

θ_C^P, θ_S^P = rotation of edge *A* due to unit value of pressure *p*

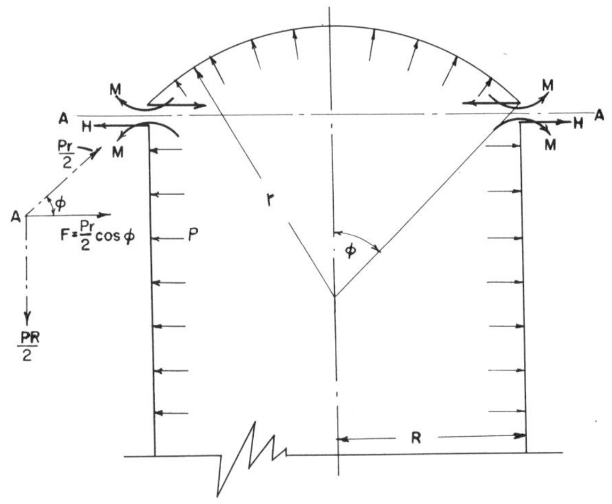

FIG. 17 DISCONTINUITY MOMENT AND SHEAR ACTING AT THE JUNCTION A-A

θ_S^F = rotation of edge *A* due to horizontal component of the edge force

Subscripts *C* and *S* refer to the cylinder and spherical cap.

The horizontal displacement and edge rotation of the two shells can then be written as follows:

1. Spherical cap

$$\left.\begin{aligned} y_S &= y_S^M M + y_S^H H + y_{SP}^P p + y_{SP}^F p \\ \theta_S &= \theta_S^M M + \theta_S^H H + \theta_{SP}^P p + \theta_{SP}^F p \end{aligned}\right\} \qquad (20)$$

2. Cylinder

$$\left.\begin{aligned} y_C &= y_C^M M + y_C^H H + y_{CP}^P p \\ \theta_C &= \theta_C^M M + \theta_C^H H + \theta_{CP}^P p \end{aligned}\right\} \qquad (21)$$

In order to insure that all parts of the pressure vessel fit together, the following compatibility equations at the junction *A-A* apply:

$$\left.\begin{aligned} y_C &= y_S \\ \theta_C &= \theta_S \end{aligned}\right\} \qquad (22)$$

It should be noted that the cylindrical shells considered here are sufficiently long so that edge loadings on one end produce no significant deformation at the other.

Substituting equations (20) and (21) into (22), we obtain two linear algebraic equations from which the unknown edge moment *M* and shear *H* can be determined. Once these have been calculated, the stress distribution throughout the two shells can be obtained easily by computing influence coefficients based on textbook equations.

EDGE INFLUENCE COEFFICIENTS

The equations for edge influence coefficient are based on the following expressions:

SPHERICAL CAP

Edge loading

$$H = 1 \quad y_S^H = -\frac{2\lambda r \sin^2 \varphi}{Eh_S}; \quad \theta_S^H = \frac{2\lambda^2 \sin \varphi}{Eh_S}$$

$$M = 1 \quad y_S^M = \frac{2\lambda^2 \sin \varphi}{Eh_S}; \quad \theta_S^M = -\frac{4\lambda^3}{Erh_S}$$

$$P = 1 \quad y_S^P = \frac{r^2(1-\nu)\sin \varphi}{2Eh_S}; \theta_S^P = 0 \tag{23}$$

$$P = 1 \quad y_S^F = \frac{-r^2\lambda \sin^2 \varphi \cos \varphi}{Eh_S};$$

$$\theta_S^F = \frac{r\lambda^2 \sin \varphi \cos \varphi}{Eh_S}$$

CYLINDER

Edge loading

$$H = 1 \quad y_C^H = \frac{1}{2D\beta^3} \quad ; \quad \theta_C^H = \frac{1}{2D\beta^2}$$

$$M = 1 \quad y_C^M = \frac{1}{2D\beta^2} \quad ; \quad \theta_C^M = \frac{1}{D\beta} \tag{24}$$

$$P = 1 \quad y_C^P = \frac{R^2(2-\nu)}{2Eh_C} \quad ; \quad \theta_C^P = 0$$

Where 2φ = total angle subtended by spherical cap

$$\lambda^4 = 3(1-\nu^2)\left(\frac{r}{h_S}\right)^2 \tag{25}$$

$$\beta^4 = 3(1-\nu^2)/R^2h_C^2$$

$$D = Eh_C^3/12(1-\nu^2) \tag{26}$$

$$\nu = 0.3 \text{ for steel}$$

By substitution of equations (25), (26) into (23), (24) and assuming the Poisson's Ratio $\nu = 0.3$ we obtain the following relationship for the edge influence coefficients:

Sphere

$$Ey_S^H = -2.570 \sin^2 \varphi \left(\frac{r}{h_S}\right)^{1.5}$$

$$Ey_S^M \times h_S = 3.305 \sin \varphi \left(\frac{r}{h_S}\right)$$

$$Ey_S^P/h_S = 0.350 \sin \varphi \left(\frac{r}{h_S}\right)^2$$

$$Ey_S^F/h_S = -1.286 \sin^2 \varphi \cos \varphi \left(\frac{r}{h_S}\right)^{2.5} \tag{27}$$

$$E\theta_S^H \times h_S = 3.305 \sin \varphi \left(\frac{r}{h_S}\right)$$

$$E\theta_S^M \times h_S^2 = -8.500 \left(\frac{r}{h_S}\right)^{1/2}$$

$$E\theta_S^F = 1.651 \sin \varphi \cos \varphi \left(\frac{r}{h_S}\right)^2$$

Cylinder

$$Ey_C^H = 2.570 \left(\frac{R}{h_C}\right)^{1.5}$$

$$Ey_C^M \times h_C = 3.305 \left(\frac{R}{h_C}\right)$$

$$Ey_C^P/h_C = 0.850 \left(\frac{R}{h_C}\right)^2 \tag{28}$$

$$E\theta_C^H \times h_C = 3.305 \left(\frac{R}{h_C}\right)$$

$$E\theta_C^M \times h_C^2 = 8.500 \left(\frac{R}{h_C}\right)^{1/2}$$

The use of these equations will be best demonstrated on the following example.

Example: The geometric characteristics of the shell shown in Fig. 17 are as follows: Spherical cap: r = 300 in., h_S = 0.625 in., r/h_S = 480, φ = 50 deg.; cylinder: R = 230 in., h_C = 0.5 in., R_C/h_C = 460. The internal pressure p = 150 psi; r and R are measured to the middle surface of the shell. Find the edge moment M and edge shear H.

Solution: From equations (27), (28) we may tabulate edge coefficients as shown in Table 6.

From equation (20):

$$Ey_S = \frac{1{,}223}{h_S} M - 15{,}723\, H + 61{,}773 h_S P - 2{,}426{,}530 h_S P$$

$$E\theta_S = -\frac{187}{h_S^2} M + \frac{1{,}223}{h_S} H + 188{,}442\, P$$

From equation (21):

$$Ey_C = \frac{1{,}520}{h_C} M + 25{,}365\, H + 179{,}860 h_C P$$

$$E\theta_C = \frac{182}{h_C^2} M + \frac{1{,}520}{h_C} H$$

TABLE 6 — EDGE COEFFICIENTS FOR CYLINDER AND SPHERICAL CAP (Example)

Edge Loading	Cylinder		Spherical Cap	
	Ey_C	$E\theta_C$	Ey_S	$E\theta_S$
$H = 1$	25,365	$1{,}520/h_C$	– 15,723	$1{,}223/h_S$
$M = 1$	$1{,}520/h_C$	$182/h_C^2$	$1{,}223/h_S$	$-\ 187/h_S^2$
$P = 1$	$179{,}800h_C$	0	$61{,}773h_S$	0
F	0	0	$2{,}426{,}534h_S$	188,442

Finally, from equation (22) for $h_S = 0.625$ in., $h_C = 0.5$ in., $P = 150$ psi, we obtain:

$$\left.\begin{aligned} M + 38\,H + 216{,}872 &= 0 \\ M + 0.90\,H - 23{,}391 &= 0 \end{aligned}\right\} \quad \begin{aligned} H &= -\ 6{,}495 \text{ lb./in.} \\ M &= 29{,}237 \text{ in.-lb./in.} \end{aligned}$$

63. CRITICAL STRESSES IN A PRESSURE VESSEL WITH A FLAT CLOSURE

(*Ref. 11*)

NOTATION

p = internal pressure (psi)
d = mean diameter of cylinder (in.)
t = thickness of cylinder (in.)
T = thickness of flat head closure (in.)
I_a = $S_{max}/(pd/2t)$, axial stress ratio
S_{max} = axial stress in the cylinder at the juncture (psi)

Means are outlined for calculating internal pressure stresses at the juncture of semi-infinite cylindrical vessel shells with flat head closure.

The calculation of critical stresses in a pressure vessel is a lengthy procedure that requires at times the solution of differential equations. As a rule, the procedure to follow is to calculate the shear force and axial bending moment at the head-cylinder junction from the continuity of the radial displacement and rotation at the junction. When these are known we may determine the shear, circumferential, and axial stresses at the junction in the cylinder and in the head.

In order to afford the designer a simplified method for calculating stresses at the juncture of the head and cylinder (also called discontinuity stresses) the critical stresses are expressed as multiples of hoop stresses, thus forming a stress index $I = S_{max}/pd/2t$).

The stress ratio, I, has distinct advantages. It reduces the magnitude of the results and enables one to tell immediately whether a stress is greater or less than the hoop stress in a thin cylinder.

The next step in the simplification procedure is to derive an equation for "I" in nondimensional form by taking into account the location of the joint between the head and shell, "nT" being the distance from the mid-plane of the head thickness to the joint. This was done by D. B. Westrom with the following results:

$$I_a = \frac{1}{2} + \frac{C_1(T/t)^3 - 2C_2n(T/t) + C_3(d/t)^{3/2}}{C_6(T/t)^3 + 2C_7n(T/t) + C_8/(T/t)} \\ \frac{+\ 2C_4n(T/t)(d/t)}{+\ C_9(d/t)^{1/2} + 4C_{10}n^2(T/t)^2/(d/t)^{1/2}} \\ \frac{+\ C_5(d/t)/(T/t)}{+\ C_{11}(T/t)^2/(d/t)^{1/2}} \qquad (29)$$

where:

$C_1 = 2.94317$ $C_6 = 1.90702$
$C_2 = 3.74071$ $C_7 = 4.84761$
$C_3 = 1.0000$ $C_8 = 1.026862$
$C_4 = 0.908912$ $C_9 = 2.66667$
$C_5 = 0.385077$ $C_{10} = 4.40610$
$C_{11} = 1.46869$

Of all the stresses of practical significance is the axial stress in the cylinder at the juncture; therefore only the equation for the axial stress ratio "I_a" is given.

Equation (29) is presented graphically in Figs. 18, 19, for $n = 0.5$ (junction at inner surface of head) and in Figs. 20, 21 for $n = 0$ (head fitted inside the shell). The theory of plates and shells fully verifies the exactness of equation (29).

In the derivation of the equation (29), the following was assumed:

1. The load is assumed to be entirely due to internal pressure.

2. The head is attached to a cylinder sufficiently long that whatever cylinder closure is used at the far end, it is too distant to affect the stresses in the flat closure.

3. The cylinder and the head are assumed to have uniform thickness.

Numerical Example: Assume $d/t = 50$ and $T/t = 3$ From Fig. 18 we find $I_a = 6.5$ or the maximum axial stress in the cylinder at the junction is found to be 6.5 times the hoop stress. From Fig. 19, we find that $S_{\max}/p = 165$ or the maximum critical stress $S_{\max} = 165p$, a very advantageous piece of information for design purpose.

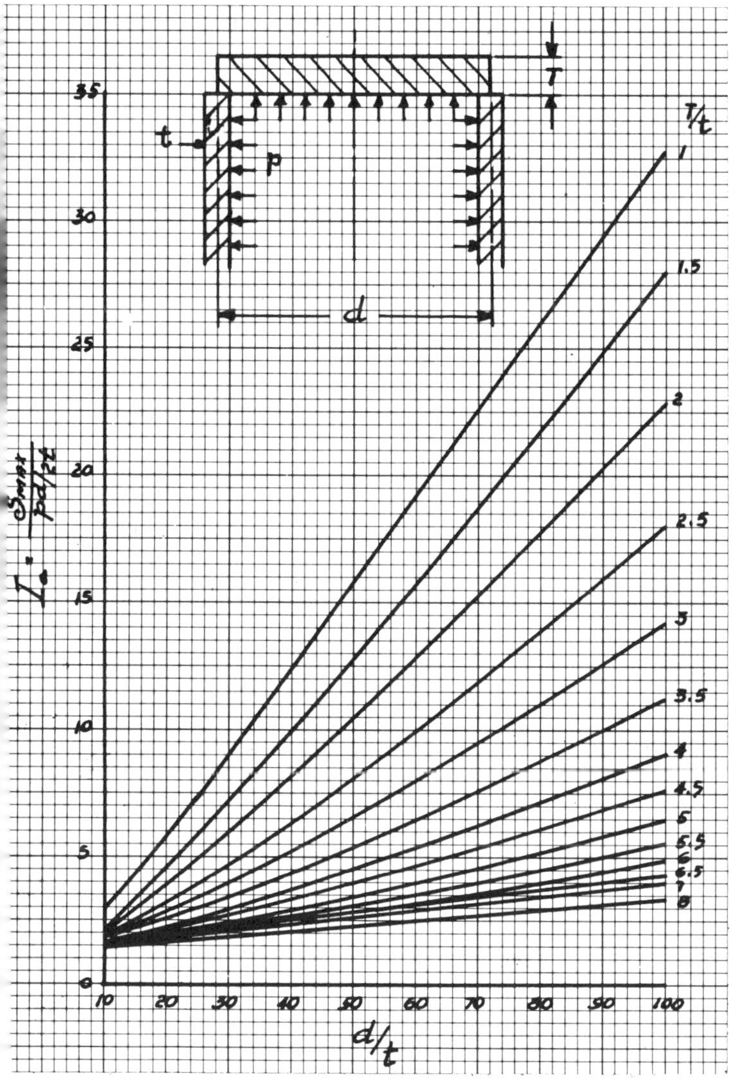

Fig. 18

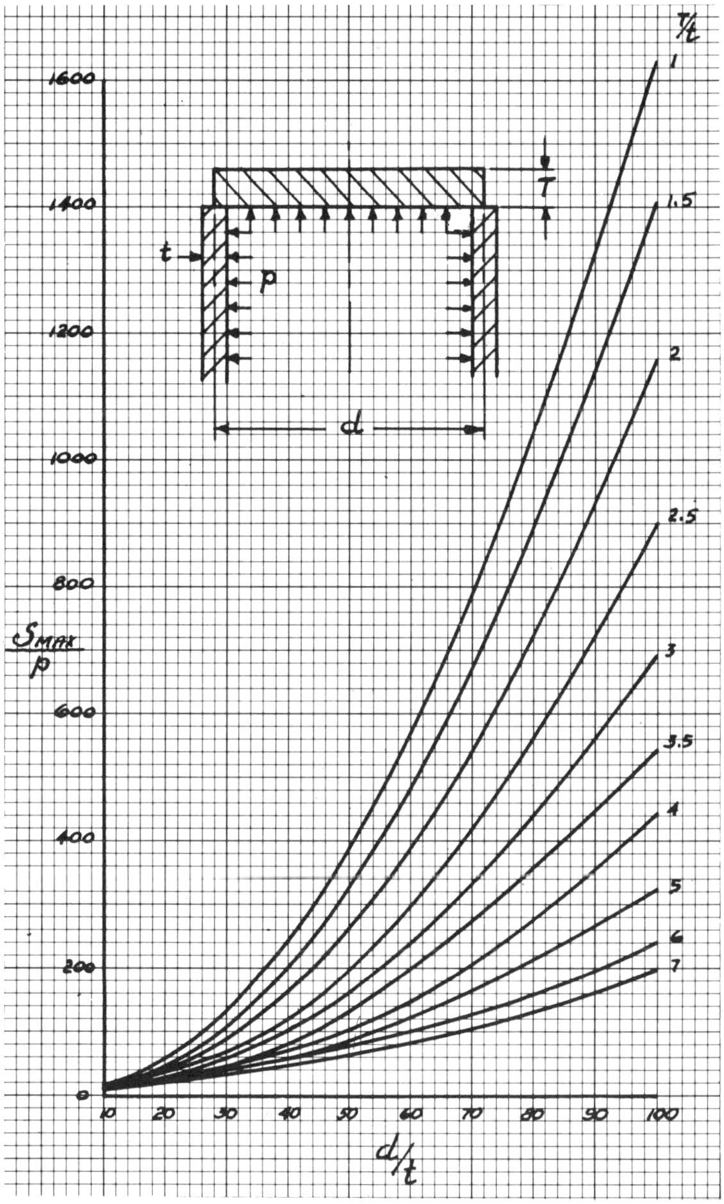

Fig. 19

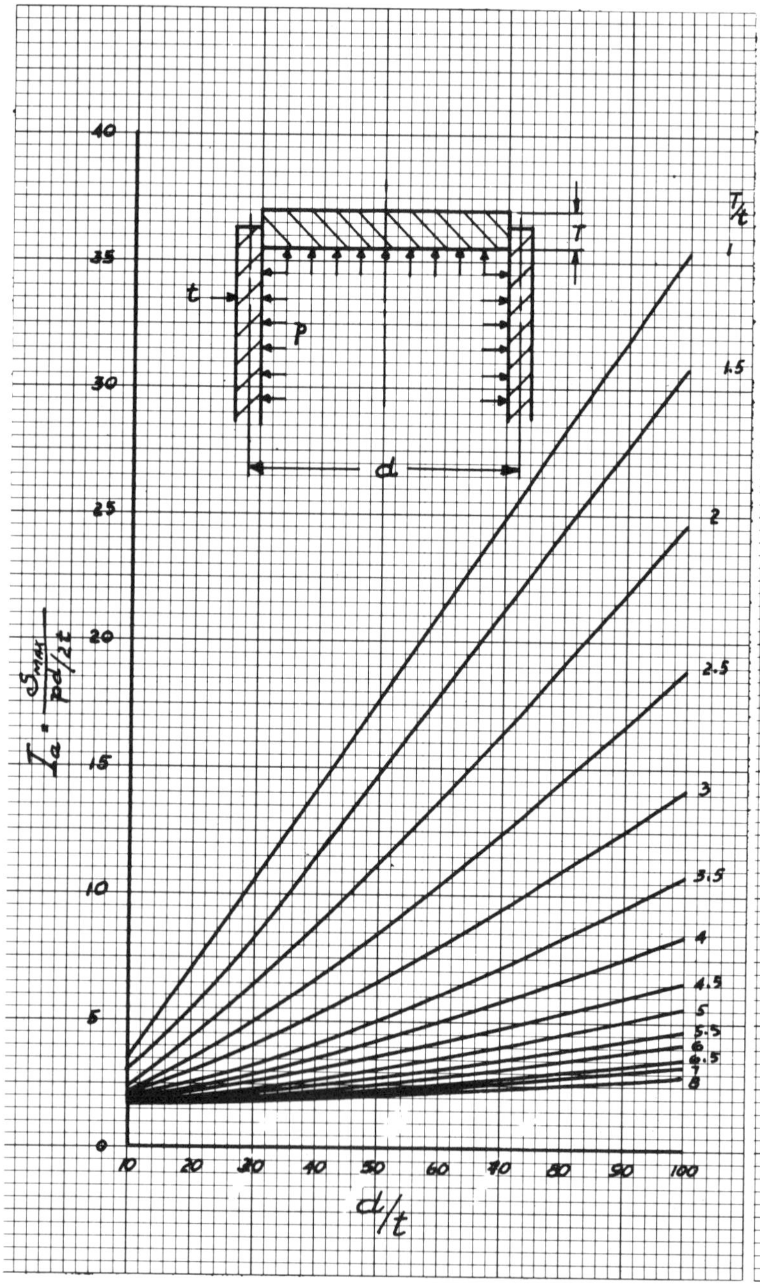
T/t
1
1.5
2
2.5
3
3.5
4
4.5
5
5.5
6
6.5
7
8
t
p
d
T
$I_a = \frac{S_{MAX}}{pd/2t}$
d/t
0
5
10
15
20
25
30
35
40
10
20
30
40
50
60
70
80
90
100

Fig. 20

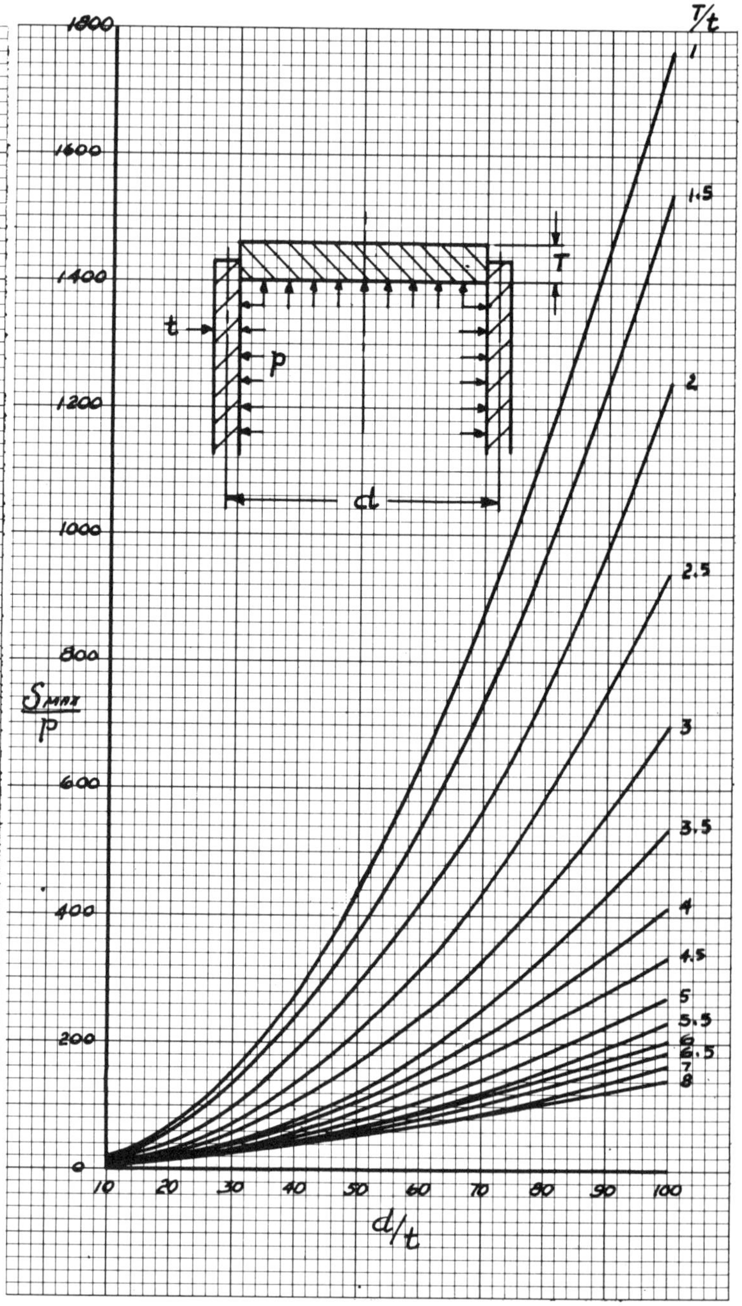
T/t
1
1.5
2
2.5
3
3.5
4
4.5
5
5.5
6
6.5
7
8
t
p
d
T
$\frac{S_{MAX}}{p}$
d/t
0
200
400
600
800
1000
1200
1400
1600
1800
10
20
30
40
50
60
70
80
90
100

Fig. 21

64. PRESSURE VESSEL WITH CONICAL HEAD CLOSURE

(*Ref. 12*)

NOTATION

p = internal pressure (psi)
T = thickness of cone (in.)
t = thickness of cylinder (in.)
α = half vertex angle of cone
d = diameter of cone base, diameter of cylinder (in.)
$I = S/(pd/2t)$ = stress index (nondimensional)
X = location of maximum circumferential stress in cylinder (in.)
M = discontinuity moment (in.-lb.)
Q = discontinuity shear (lb.)

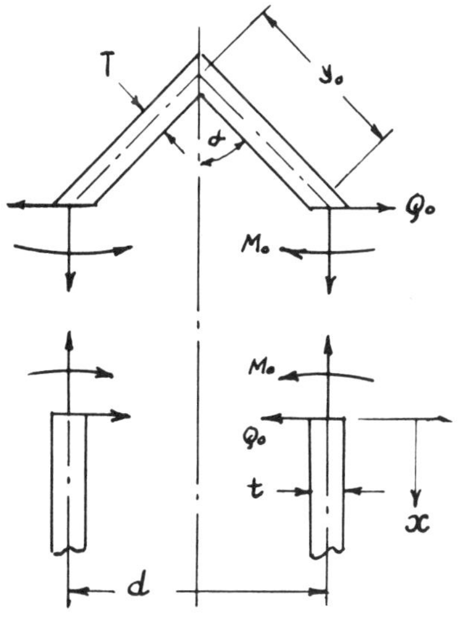

Fig. 22

This section presents a method of analysis for pressure vessels with conical heads that would not require a specialized knowledge of shell theory. One such method, used here, is expressing all the resulting stresses as multiples of $pd/2t$, the circumferential stress in a thin unrestrained cylinder under uniform pressure. This stress ratio, denoted by $I = S/(pd/2t)$, has two distinct advantages. It reduces the magnitude of results and enables one to tell immediately whether a stress is greater or less than the hoop stress in a thin cylinder.

The computation procedure of stress ratios, based on accurate bending theory of shells, is covered in detail in Ref. (*See* Ref. List). This article presents results of interest to designers, based on these references.

CRITICAL STRESSES

When internal pressure is applied to head-cylinder structure shown in Fig. 22, membrane stresses are set up in both the head and in the cylinder. These are well-known stresses that can be computed by available formulas, and their distribution over the wall thickness of the structures is assumed to be uniform.

Differences between the natural flexibility of the heads and that of the cylinders cause these two parts to attempt to deform radially and angularly at different rates under the effect of internal pressure. This is prevented at their junctures by the rigid attachment of the heads to cylinders. Such a disturbance of the natural deformations, arising from internal pressure, produces the following stresses: (1) circumferential; (2) axial; (3) axial shear. The first two are important because it is from them that the critical stress is selected. The third represents valuable information for designers. These juncture stresses exist in addition to membrane stresses just described.

Basically, the problem of calculating the juncture stresses consists of determining the: (1) axial bending moment M_o; (2) the shear force Q_o, both existing at the cone-cylinder junction (Fig. 22). They must be determined from the continuity of the radial displacement and rotation at the junction. Each of the three stresses, mentioned above, is divided by $pd/2t$ to form stress ratios I_a, for axial stress, I_{CM} for circumferential stress and I_S for shear stress.

For simplicity of analysis, eight curves (Figs. 23–29) given here will enable the designer to determine the magnitude of the discontinuity stresses in a pressure vessel with a conical head.

Figs. 23, 24, 25 and 26 give the maximum axial stress and maximum shear stress at the junction as multiples of $pd/2t$. These stresses are shown for various values of cone apex angle, ratio of conical head thickness to cylinder thickness (T/t), and ratio of cylinder diameter to cone thickness (d/T), which will include most of the vessels encountered in practice. Figures 27, 28, 29 and 30 show maximum circumferential stresses in the cylinder as multiples of $pd/2t$ and the location of these stresses.

It is important to the designer to note that when an internal pressure is applied, the cylinder always deflects outward and the conical head inward. Therefore, it is impossible to design a conical head to eliminate moment and shear at the junction. The graphs in Figs. 23, 24, 25

26, 27 and 28 show that the greatest stress is the axial stress at the junction. It may be located in the cylinder (indicated by broken lines) or in the cone (solid lines). The graphs also show that in order to minimize the stress, it is desirable to make the cone and cylinder equally thick if the angle α does not exceed 45 deg. Where this is not possible, the cone should have the greater thickness. Angles greater than 45 deg. require a thicker head. Reducing the cone vertex angle or the diameter-thickness ratio also reduces the stresses.

Example: Determine the discontinuity stresses, assuming a 5-ft. diameter vessel with a wall thickness in the cylinder of 0.3 in. The internal pressure is 50 psi and the conical head has an apex angle of 120 deg. Thus $d/t = 60/0.3 = 200$ and $\alpha = 60$ deg., $pd/2t = 5{,}000$ psi.

Using Fig. 26, the following table can be constructed:

$\frac{T}{t}$	0.8	1.0	1.2	1.6	2.0
$\frac{d}{T}$	250	200	167	125	100
$I_a = \frac{S}{pd/2t}$	12.6	8.6	9.0	8.6	8.2
S_a (psi)	63,000	43,000	45,000	43,000	41,000
$I_S = \frac{S}{pd/2t}$	0.478	0.395	—	0.488	—
S_S (psi)	2,400	1,970	—	2,440	—

From Fig. 30: For $d/t = 200$

$\frac{T}{t}$	0.8	1.0	1.2	1.6	2.0
$I_{CM} = \frac{S}{pd/2t}$	1.4	1.33	1.3	1.26	1.235
S_{CM} (psi)	7,000	6,650	6,500	6,300	6,200

$$\frac{X}{d} = 0.1 \quad \text{or } X = 6.0 \text{ in.}$$

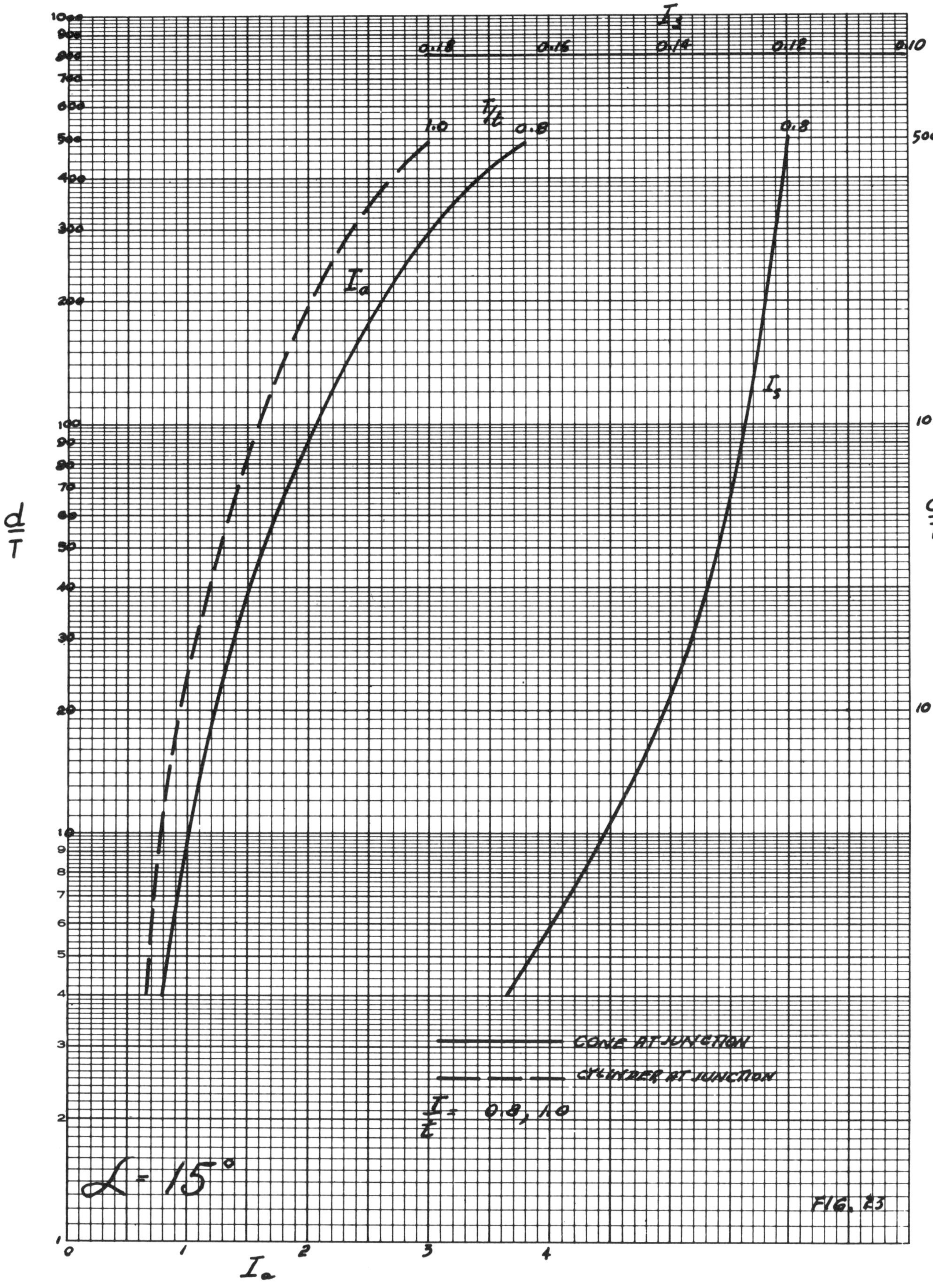
CONE AT JUNCTION
CYLINDER AT JUNCTION
T/t = 0.8, 1.0
α = 15°
FIG. 23
d/T
I_a
I_s
0.18
0.16
0.14
0.12
0.10
1.0
0.8
500
100
10

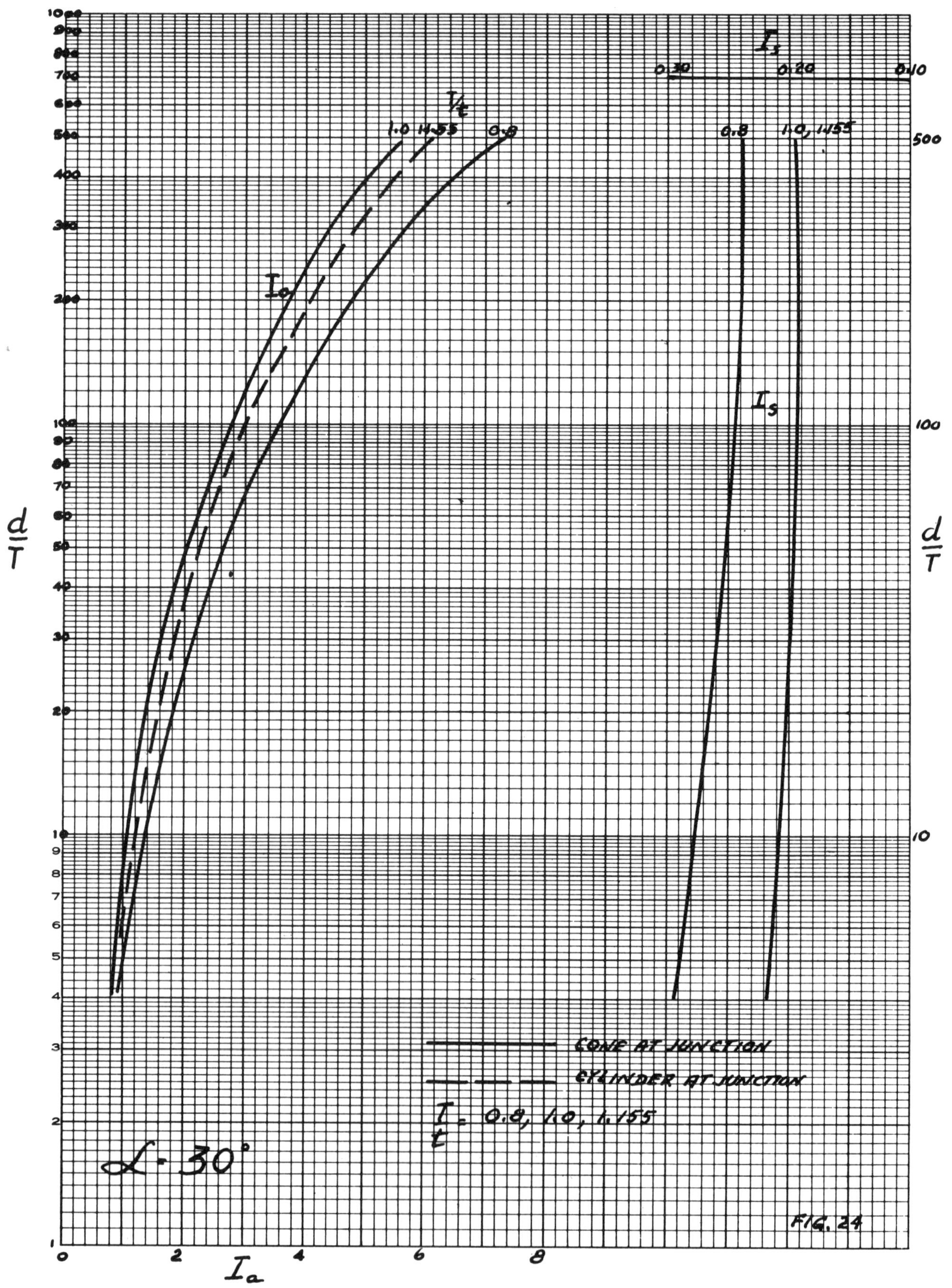

I_s
0.30
0.20
0.10
T/t
1.0 1.155
0.8
0.8
1.0, 1.155
I_a
I_s
d/T
d/T
CONE AT JUNCTION
CYLINDER AT JUNCTION
T/t = 0.8, 1.0, 1.155
α = 30°
FIG. 24

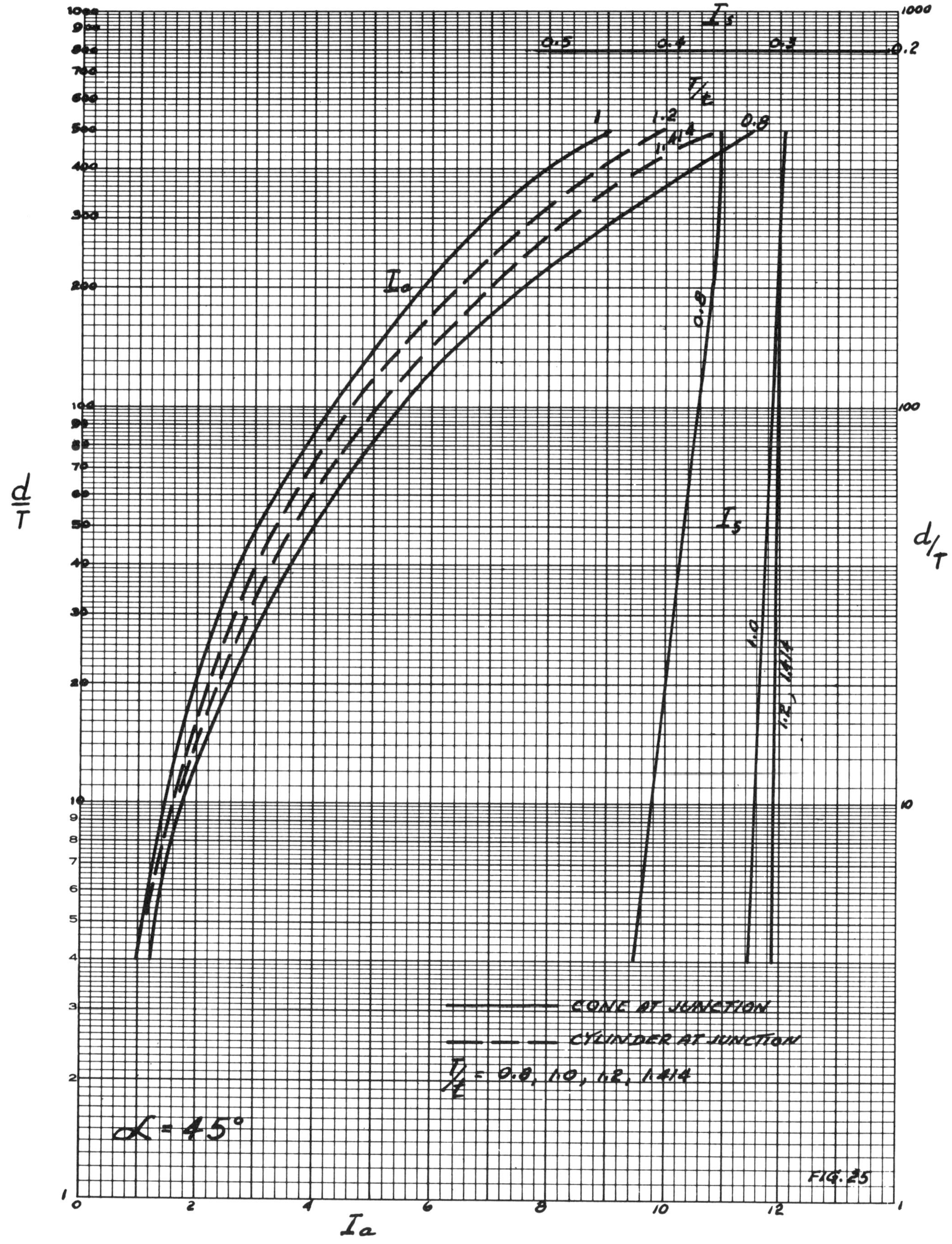

I_s
0.5
0.4
0.3
0.2
T/t
1
1.2
1.414
0.8
I_a
0.8
I_s
1.0
1.2, 1.414
CONE AT JUNCTION
CYLINDER AT JUNCTION
T/t = 0.8, 1.0, 1.2, 1.414
α = 45°
d/T
I_a

FIG. 25

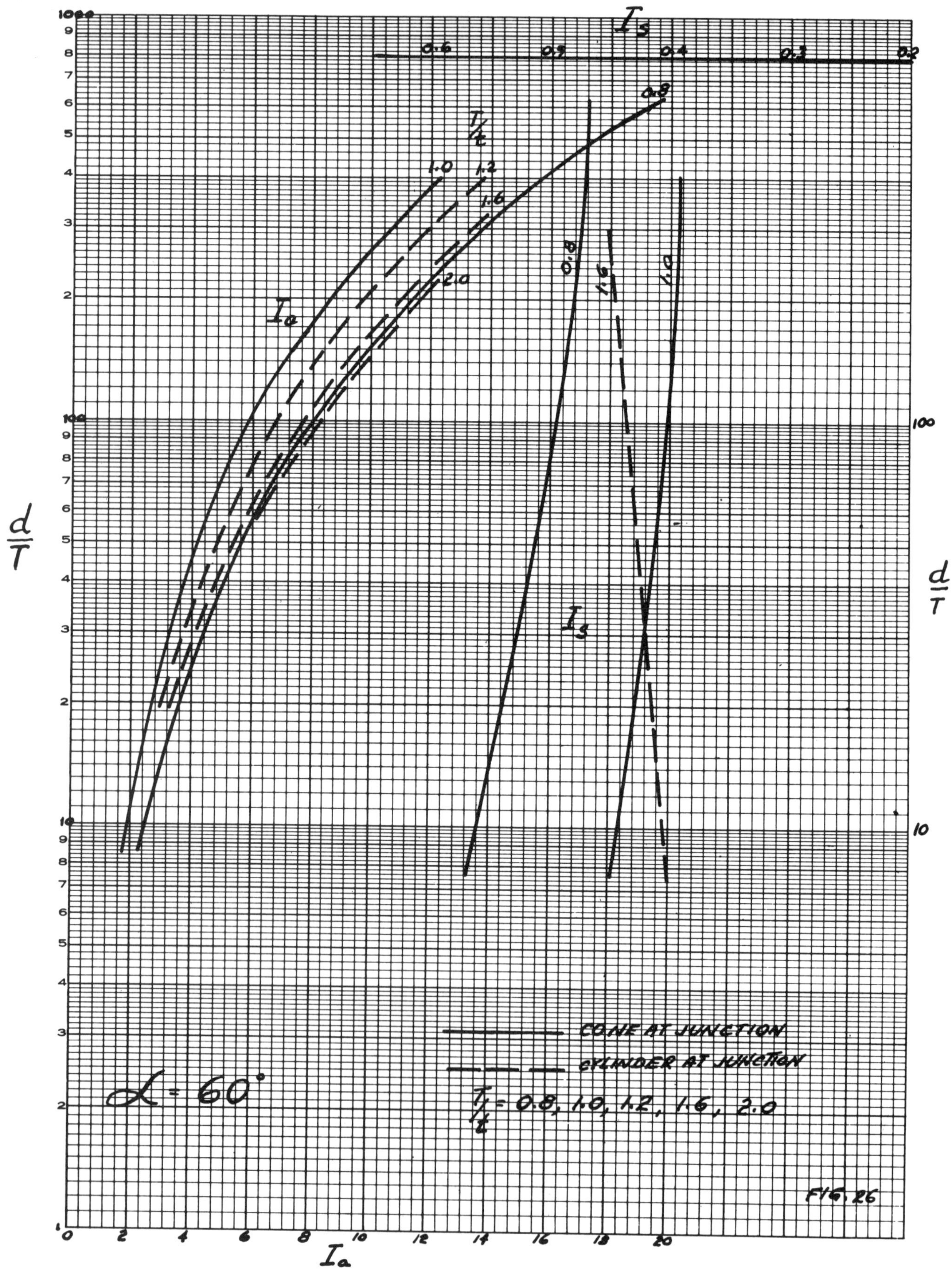

I_s
0.6
0.5
0.4
0.3
0.2
$\frac{T}{t}$
1.0
1.2
1.6
2.0
0.8
I_a
I_s
$\frac{d}{T}$
I_a
CONE AT JUNCTION
CYLINDER AT JUNCTION
$\frac{T}{t}$ = 0.8, 1.0, 1.2, 1.6, 2.0
α = 60°

FIG. 26

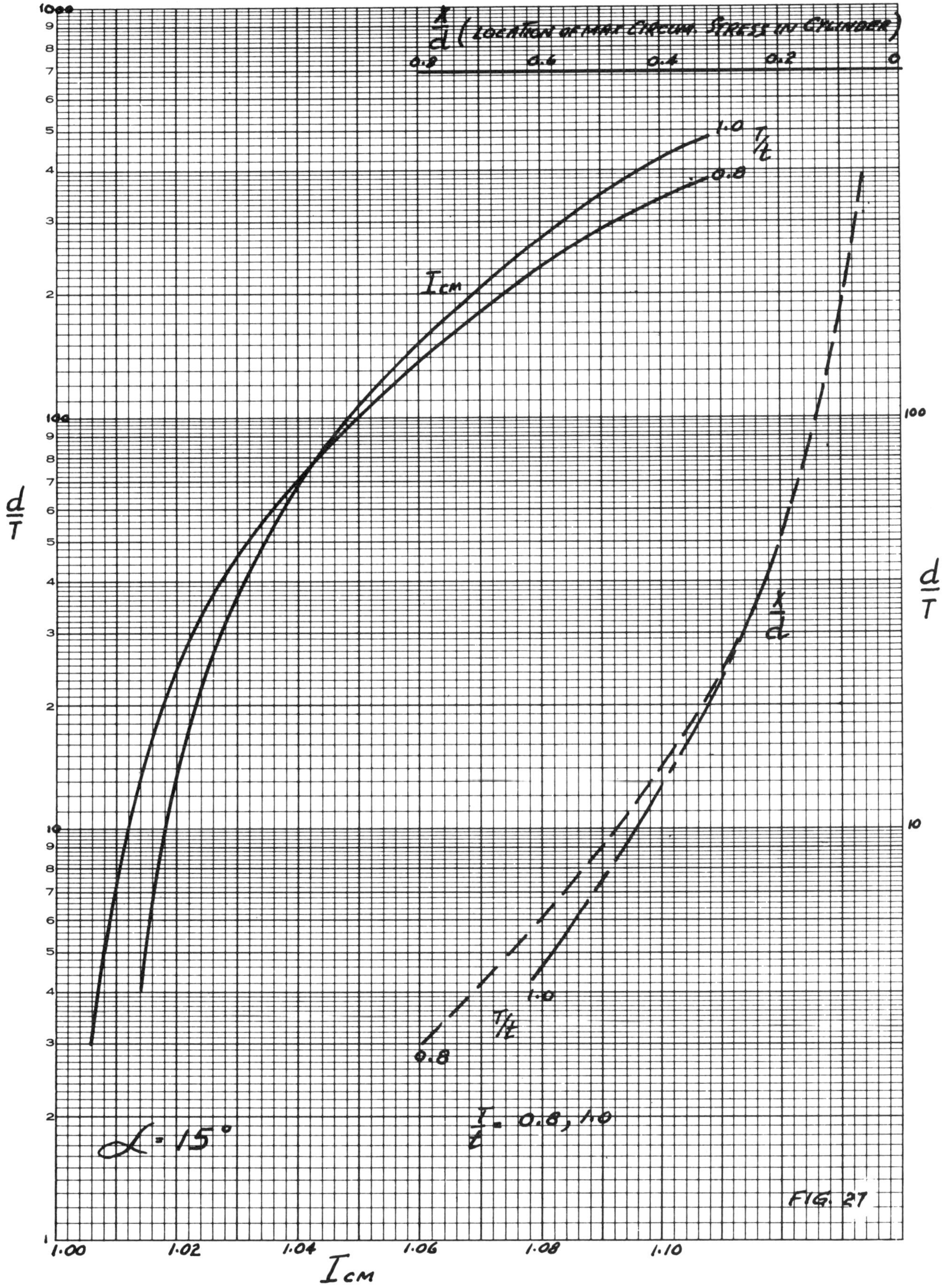
x/d (LOCATION OF MAX CIRCUM. STRESS IN CYLINDER)
0.8
0.6
0.4
0.2
0
d/T
d/T
I_{CM}
I_{CM}
T/t
1.0
0.8
x/d
$\alpha = 15°$
T/t = 0.8, 1.0
1.00
1.02
1.04
1.06
1.08
1.10
1
10
100
1000

FIG. 27

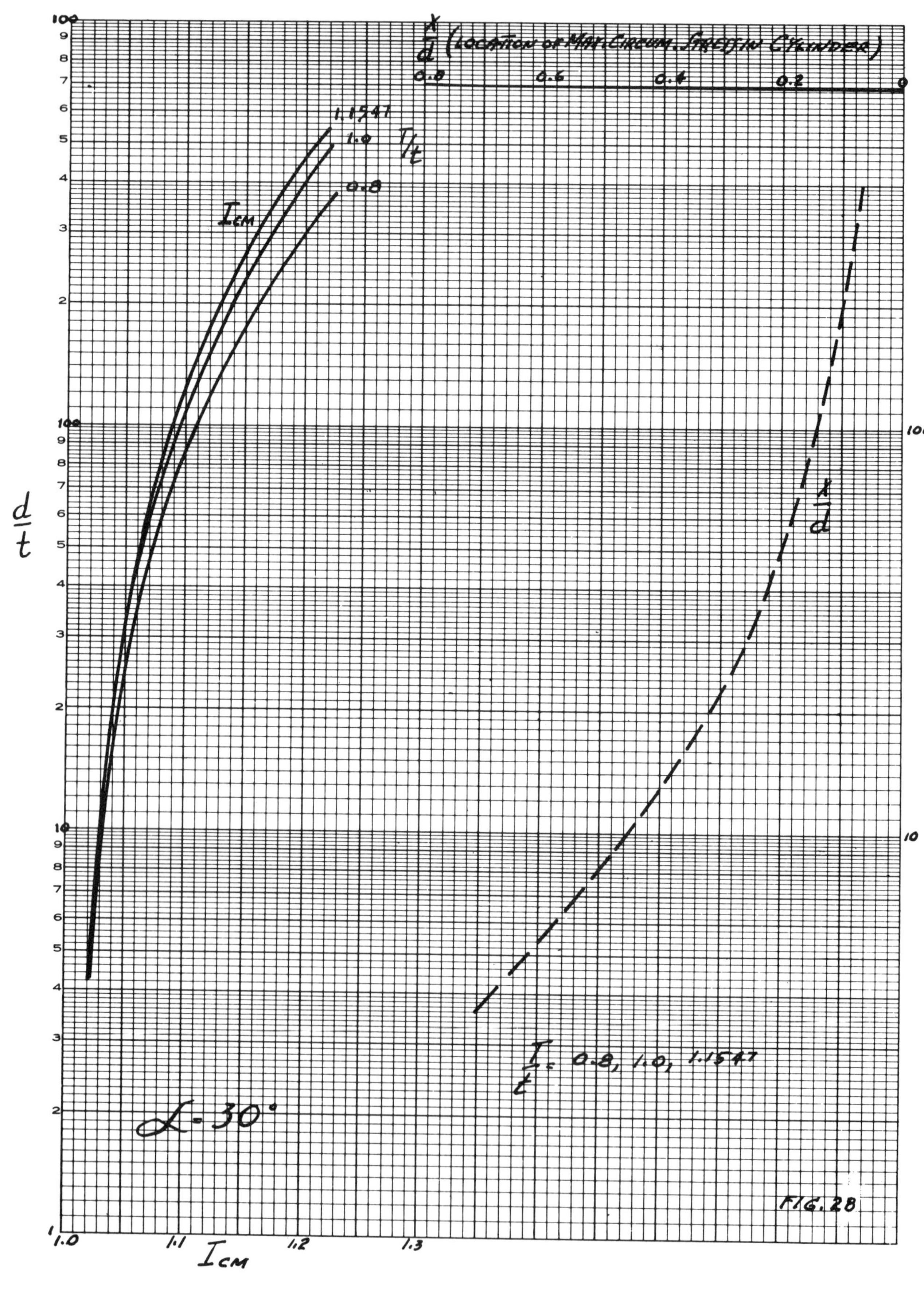

x/d (LOCATION OF MAX. CIRCUM. STRESS IN CYLINDER)
0.8 0.6 0.4 0.2 0
T/t
1.1547
1.0
0.8
I_{CM}
d/t
x/d
T/t = 0.8, 1.0, 1.1547
α = 30°
1.0 1.1 1.2 1.3
I_{CM}

FIG. 28

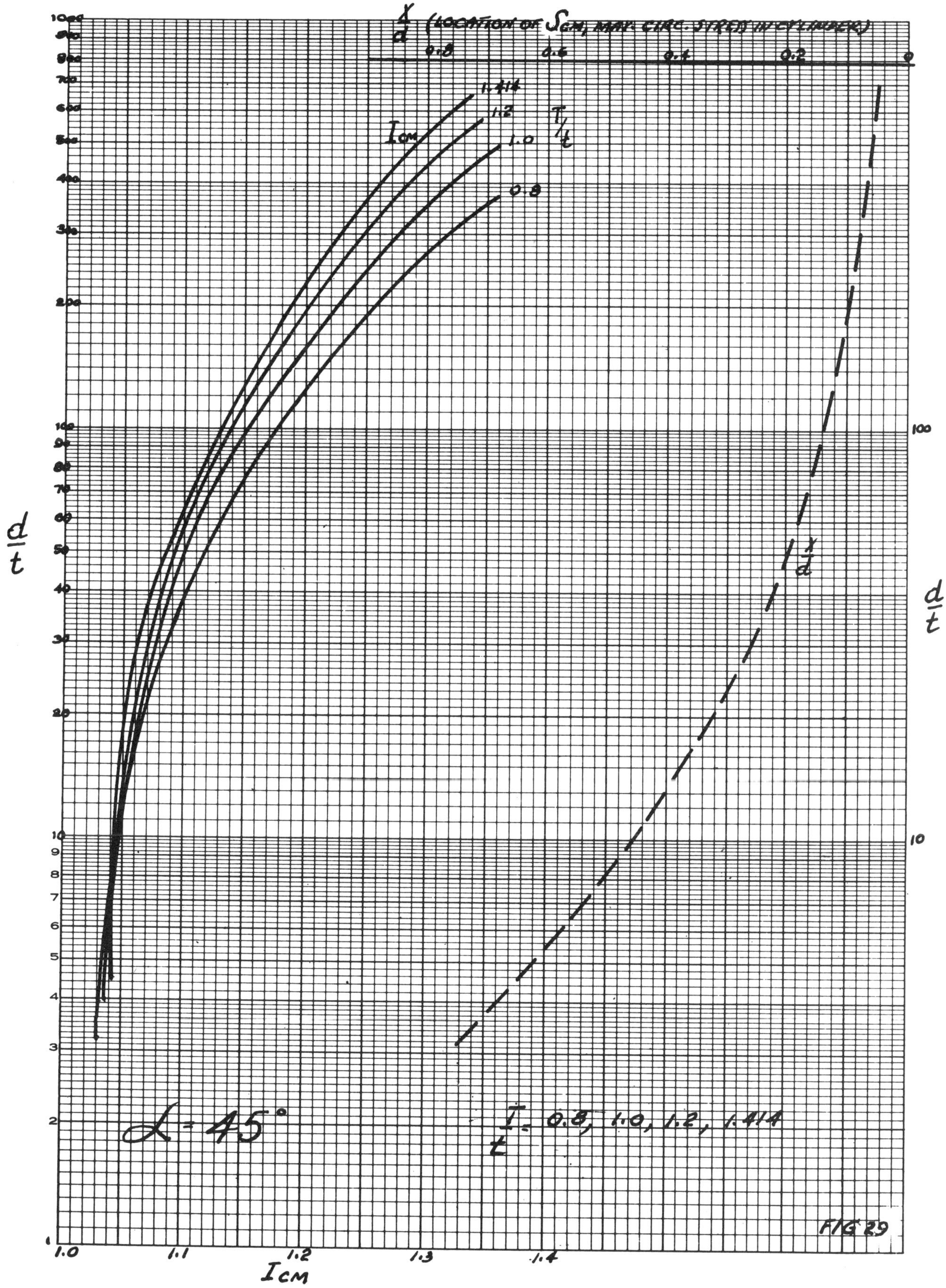
X/d (LOCATION OF S_{CM}, MAX. CIRC. STRESS IN CYLINDER)
0.8
0.6
0.4
0.2
0
1.414
1.2
1.0
0.8
I_{CM}
T/t
X/d
d/t
d/t
1.0
1.1
1.2
1.3
1.4
I_{CM}
$\alpha = 45°$
T/t = 0.8, 1.0, 1.2, 1.414

FIG 29

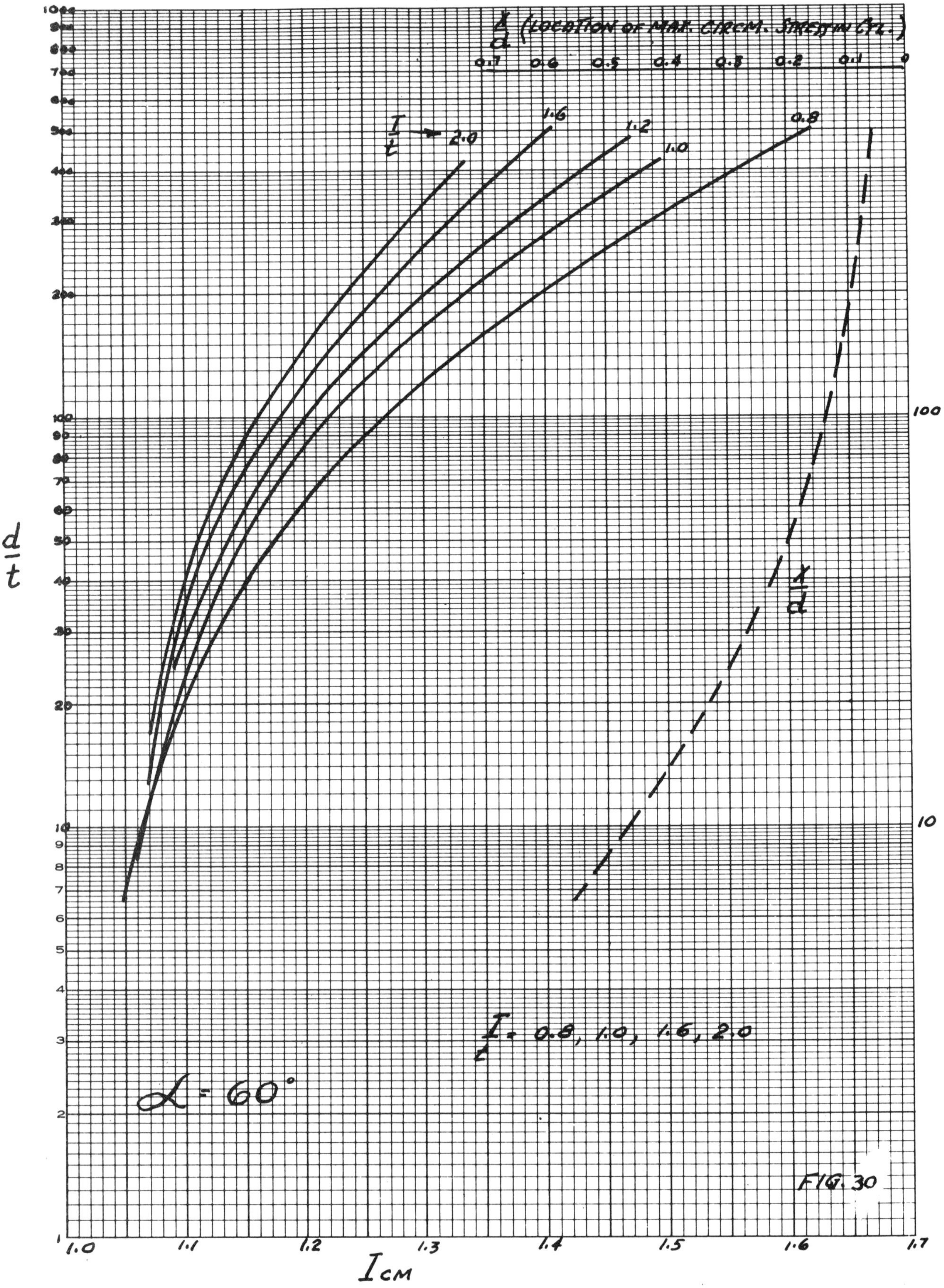

x/d (LOCATION OF MAX. CIRCM. STRESS IN CYL.)
0.7 0.6 0.5 0.4 0.3 0.2 0.1 0
I/t → 2.0
1.6
1.2
1.0
0.8
x/d
d/t
I/t = 0.8, 1.0, 1.6, 2.0
α = 60°
1.0 1.1 1.2 1.3 1.4 1.5 1.6 1.7
I_{CM}
100
10

FIG. 30

Chap. VII

REFERENCES

Section	*Ref. No.*	
60	1	HORVAY, G., and CLAUSEN, I. M.: "Membrane and Bending Analysis of Axisymmetrically Loaded Axisymmetrical Shells," A.S.M.E., Paper No. 54–A–8
60	2	ROARK, R. J.: *Formulas for Stress and Strain*, McGraw-Hill, N.Y., 1941
60	3	TIMOSHENKO, S.: *Strength of Materials*, Part II, Van Nostrand, Princeton, N.J., 1941
60	4	HETÉNYI, M.: *Beams on Elastic Foundation*, University of Michigan Press, Ann Arbor, Mich., 1946
60	5	HILDEBRAND, F. B.: "On Assymptotic Integration in Shell Theory," in *Proc. Symposia in Applied Mathematics*, Vol. III, McGraw-Hill, N.Y., 1950
60	6	REISSNER, E.: "On the Theory of Thin Elastic Shells," in *Reissner Anniversary Volume*, J. N. Edwards, Ann Arbor, Mich., 1949, p. 231
60	7	HORVAY, G., and CLAUSEN, I. M.: "Stresses and Deformations of Flanged Shells," *Journal of Applied Mech. Trans. A.S.M.E.*, Vol 76, 1954, p. 109–116
61	8	GRIFFEL, W.: "Discontinuity Stresses in Pressure Vessels," *Design News*, March 20, 1963
61	9	MAULBETSCH, S. H., and HETÉNYI, M.: "Stresses in Pressure Vessels," *Journal of Applied Mechanics,* September, 1936.
62	10	GALLETLY, G. D.: "Influence Coefficients and Pressure Vessel Analysis," A.S.M.E. Paper No. 59–A–163
63	11	WATTS, G. W., and LANG, H. A.: "The Stresses in a Pressure Vessel with a Flathead Closure," A.S.M.E. Paper No. 51–A–146
64	12	WATTS, G. W. and LANG, H. A.: "Stresses in a Pressure Vessel with a Conical Head," A.S.M.E., 1951

Chapter VIII

STRESS DUE TO PRESSURE BETWEEN ELASTIC BODIES

65. CONTACT STRESS OF TWO CYLINDERS*

(*Ref. 1*)

The mathematical theory for the surface stresses and deformations produced by pressure between curved bodies was developed by Hertz. Formulas based on his theories with the results supported by experiment give the maximum compressive stress, which occurs at the center of the surface of contact.

Formulas for compressive stress for three typical cases are given: (1) two cylinders; (2) cylinder in a circular groove; (3) cylinder on a flat plate. Simplified charts for each of the three cases are given assuming a contact between two elastic bodies of same material.

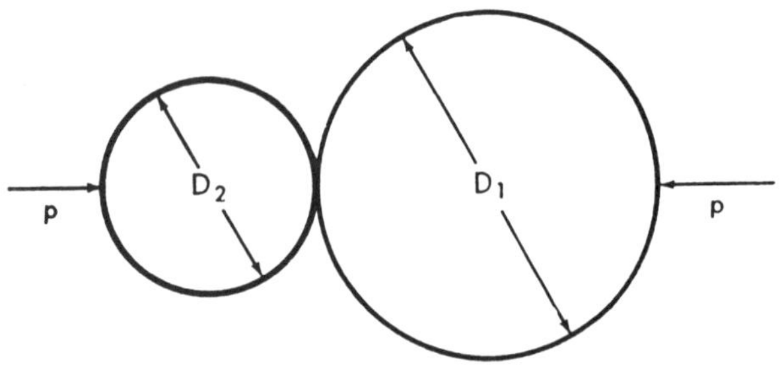

Fig. 1

The equation for the maximum contact stress which occurs between two cylinders with parallel axes, Fig. 1, is given by

$$\text{Max } S_c = 0.798 \sqrt{\frac{P(D_1 + D_2)/D_1 \times D_2}{(1-\gamma_1^2)/E_1 + (1-\gamma_2^2)/E_2}}$$

where S_c = compressive stress (psi)
p = load per linear inch
D_1 = diameter of cylinder 1 (in.)
D_2 = diameter of cylinder 2 (in.)
γ_1 = Poisson's Ratio for cylinder 1
γ_2 = Poisson's Ratio for cylinder 2
E_1 = modulus of elasticity for cylinder 1 (psi)
E_2 = modulus of elasticity for cylinder 2 (psi)

The chart, Fig. 2, has been drawn for $E_1 = E_2 = (30)(10^6)$ psi and $\gamma_1 = \gamma_2 = 0.25$. For any three given values of S_c, D_1, D_2 and p, the fourth can be found directly from the chart. The use of the chart is illustrated by the following example:

Example :

Given $D_1 = 4$ in.
$D_2 = 2$ in.
$S_{c\max} = 60{,}000$ psi

Determine maximum load per inch of cylinder.

Solution: For $D_1 = 4$ in., $D_2 = 2$ in., locate point A. Move horizontally to B for $S_c = 60{,}000$ psi. Read $p = 475$ lb./in.

*Reprinted with permission from *Design News*, Sept. 1, 1957. Copyright 1957 by the Rogers Publishing Co.

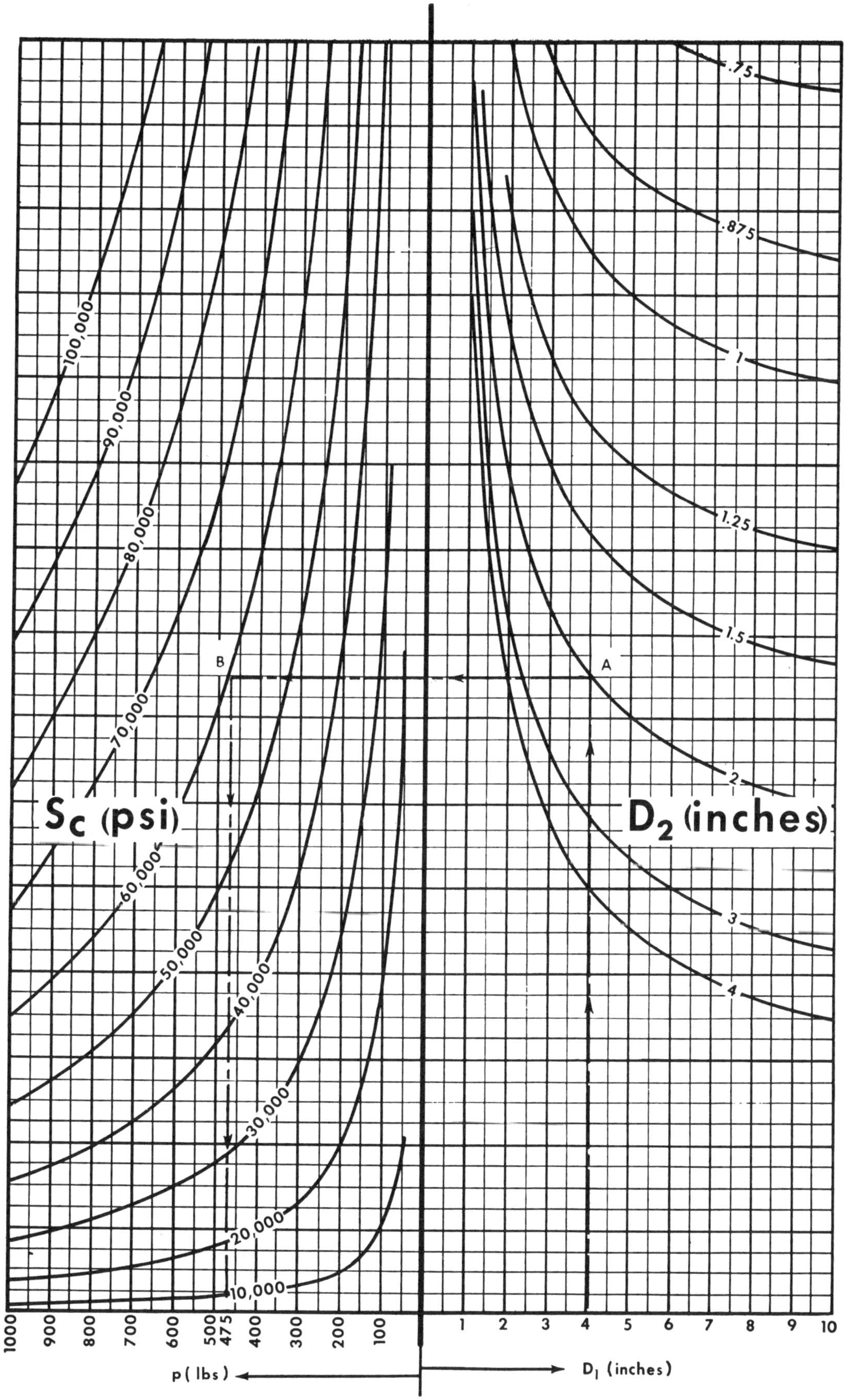
Sc (psi)
D2 (inches)
100,000
90,000
80,000
70,000
60,000
50,000
40,000
30,000
20,000
10,000
.75
.875
1
1.25
1.5
2
3
4
A
B
1000 900 800 700 600 500 475 400 300 200 100
1 2 3 4 5 6 7 8 9 10
p (lbs)
D1 (inches)

Fig. 2

66. CONTACT STRESS OF A CYLINDER IN A CIRCULAR GROOVE*

(*Ref. 2*)

The equation for the maximum contact stress which occurs between a cylinder in a circular groove, Fig. 3 is given by:

$$\text{Max } S_c = 0.798 \sqrt{\frac{P(D_1 - D_2)/D_1 \times D_2}{(1-\gamma_1^2)/E_1 + (1-\gamma_2^2)/E_2}}$$

where S_c = compressive stress (psi)
p = load per linear inch
D_1 = diameter of circular groove (in.)
D_2 = diameter of cylinder (in.)
γ_1 = Poisson's Ratio for member 1
γ_2 = Poisson's Ratio for cylinder 2
E_1 = modulus of elasticity for member 1
E_2 = modulus of elasticity for cylinder 2

The chart, Fig. 4, has been drawn for $E_1 = E_2 = 30 \times 10^6$ psi and $\gamma_1 = \gamma_2 = 0.25$. For any three given values of S_c, D_1, D_2, and p, the fourth can be found directly from the chart. The use of the chart is illustrated by the following example.

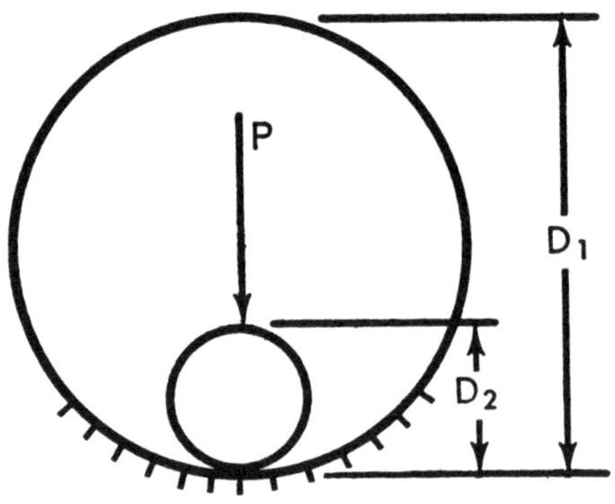

Fig. 4

Example:

Given $D_1 = 6$ in.
$D_2 = 1$ in.
S_c allowable = 80,000 psi.

Find the allowable load per inch of length of the cylinder.

Solution: For $D_1 = 6$ in. and $D_2 = 1$ in., locate point A. Move horizontally to locate B on the S_c = 80,000 psi line. Read $p = 760$ lb./in.

*Reprinted with permission from *Design News*, Sept. 1, 1957. Copyright 1957 by the Rogers Publishing Co.

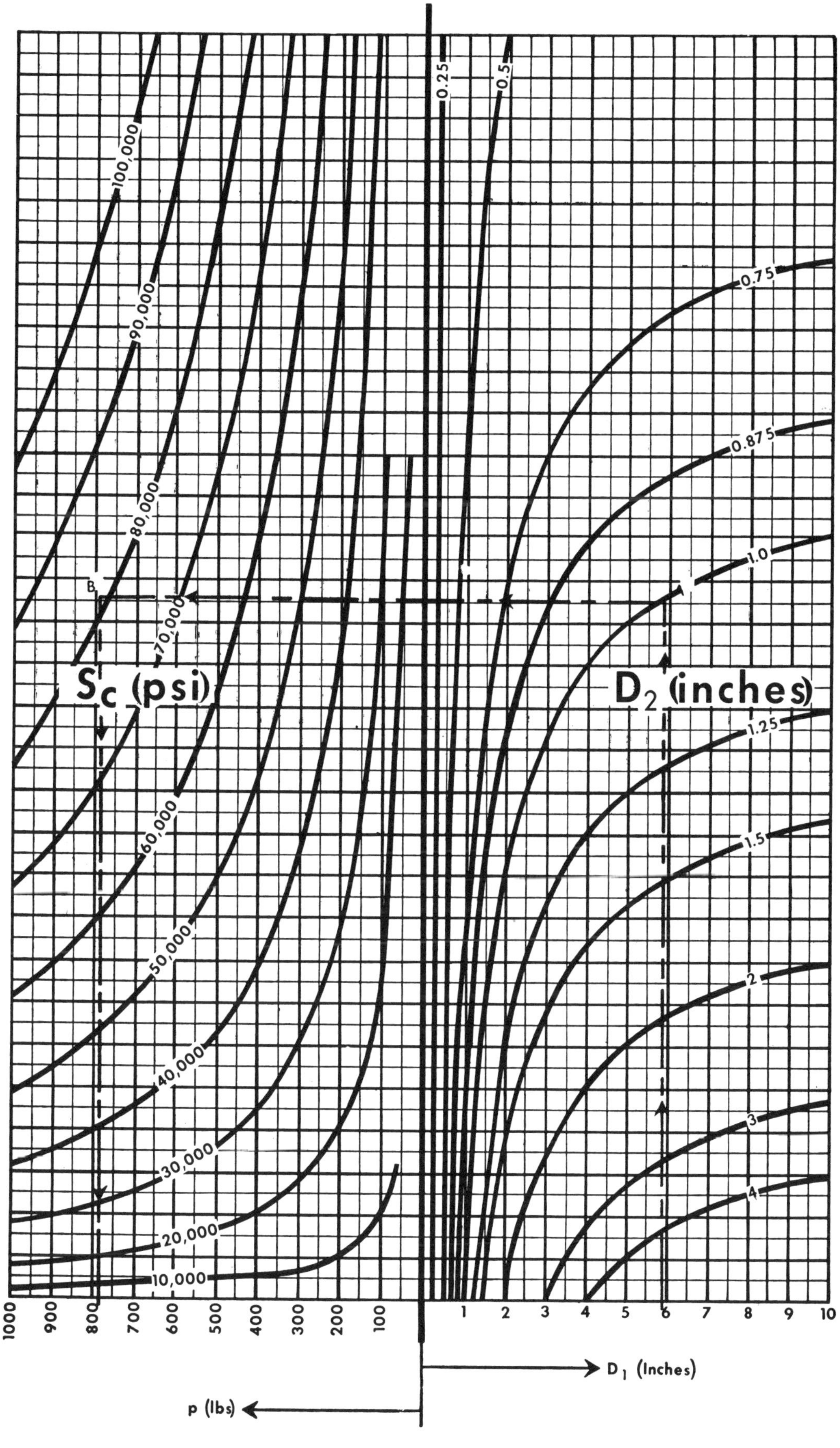

100,000
90,000
80,000
70,000
60,000
50,000
40,000
30,000
20,000
10,000
0.25
0.5
0.75
0.875
1.0
1.25
1.5
2
3
4
B
S_c (psi)
D_2 (inches)
1000 900 800 700 600 500 400 300 200 100
1 2 3 4 5 6 7 8 9 10
D_1 (Inches)
p (lbs)

Fig. 3

67. CONTACT STRESS OF A CYLINDER ON A FLAT PLATE*

(*Ref. 3*)

It is often necessary to check contact stresses where the design involves curved surfaces in contact with each other, as represented by a plate cam in contact with a flat faced follower. In addition to the necessity of checking for excessively high stresses because of plastic deformation, where the load is applied repetitively, fatigue failures must be guarded against by being certain that local contact stress is less than the allowable endurance stress.

The maximum compressive contact stress which occurs between a cylindrical surface and flat plate, as shown in Fig. 5, according to the Hertz stress equation, is given by

$$\text{Max } S_c = 0.798 \sqrt{\frac{P}{D[(1-\gamma_1^2)/E_1 + (1-\gamma_2^2)/E_2]}}$$

Calling either member, member 1, and the other member, member 2, one may define the symbols by:

S_c = compressive stress (psi)
p = load/linear in.
D = diameter of cylinder (in.)
γ_1 = Poisson's Ratio of member 1
γ_2 = Poisson's Ratio of member 2
E_1 = modulus of elasticity of member 1 (psi)
E_2 = modulus of elasticity of member 2 (psi)

The charts are drawn such that one may find quickly any one of the three, knowing the other two: stress, load/inch, and diameter. Two curves are drawn: one for cast iron and steel, with $E = 14\,(10^6)$ psi for cast iron, $E = 30\,(10^6)$ psi for steel, and Poisson's Ratio $= 0.25$ for both cast iron and steel; the other for $E_1 = E_2 = 30\,(10^6)$ psi and $\gamma_1 = \gamma_2 = 0.3$.

Two examples are given to illustrate the use of the chart.

Example 1: Given a cylinder 2 in. long and $\frac{1}{2}$-in. diameter supported by a flat plate. The load on the cylinder is 2,000 lb. $E_1 = 30\,(10^6)$ psi for the cylinder, $E_2 = 14\,(10^6)$ psi for the flat plate, Poisson's Ratio for both cast iron and steel is 0.25. What is the maximum compressive contact stress?

Solution: The load/linear inch = 2,000/2 = 1,000 lb. Enter the chart at $p = 1{,}000$ lb./in.; follow the value of $p = 1{,}000$ lb./in. horizontally until the line for $D = \frac{1}{2}$ in. is intersected at A. Move vertically to the left to intersect the proper curve at B. Read the compressive contact stress = 114,000 psi.

Example 2: Given a cylinder on a flat plate. If the maximum contact stress is limited to 160,000 psi, what size of cylinder should be used for a loading of 900 lb./linear inch? $E_1 = E_2 = 30\,(10^6)$ psi. $\gamma_1 = \gamma_2 = 0.3$.

Solution: Enter the chart at $S_c = 160{,}000$ psi. Follow the stress value until proper curve is intersected at point C. Draw horizontal line from point C to right until a line drawn vertically from $p = 900$ lb./in. is intersected at point D. Diameter of roller necessary is read just slightly less than $D = \frac{3}{8}$ in.

*Reprinted with permission from *Design News*, Sept. 1, 1957. Copyright 1957 by the Rogers Publishing Co.

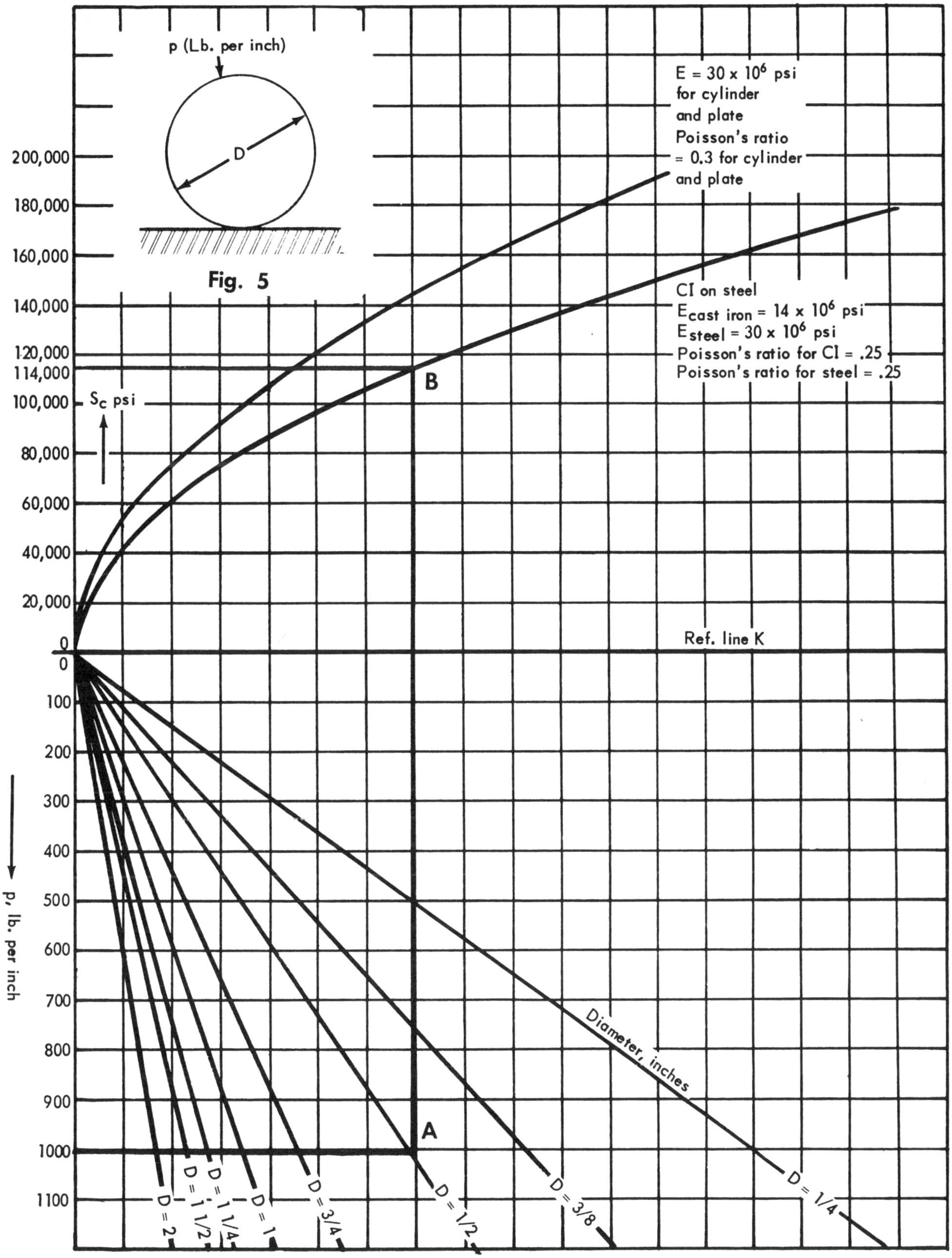
p (Lb. per inch)
D
Fig. 5
E = 30 x 10⁶ psi for cylinder and plate
Poisson's ratio = 0.3 for cylinder and plate
CI on steel
E cast iron = 14 x 10⁶ psi
E steel = 30 x 10⁶ psi
Poisson's ratio for CI = .25
Poisson's ratio for steel = .25
200,000
180,000
160,000
140,000
120,000
114,000
100,000
80,000
60,000
40,000
20,000
0
B
Sc psi
Ref. line K
0
100
200
300
400
500
600
700
800
900
1000
1100
p, lb. per inch
A
Diameter, inches
D = 2
D = 1 1/2
D = 1 1/4
D = 1
D = 3/4
D = 1/2
D = 3/8
D = 1/4

Fig. 6

Chap. VIII

REFERENCES

Section	*Ref. No.*	
65	1	LAUGHLIN, H. G., and HOLOWENKO, A. R.: "Contact Stress of Two Cylinders," *Design Data Manual*, 1958 (*Design News*)
66	2	HOLOWENKO, A. R., and LAUGHLIN, H. G.: "Contact Stress of a Cylinder in a Circular Groove," *Design Data Manual*, 1958 (*Design News*)
67	3	LAUGHLIN, H. G., and HOLOWENKO, A. R.: "Contact Stress of a Cylinder on a Flat Plate," *Design Data Manual*, 1958 (*Design News*)

Chapter IX

STRESS CONCENTRATION FACTORS

68. STRESS CONCENTRATION FACTORS FOR PLATES WITH HOLES

(*Ref. 1, 2*)

Abrupt geometric changes in plates, such as holes, distort the uniform stress distribution into complex patterns which localize the stresses in the regions of the holes. This localization, or *stress concentration*, causes a higher maximum stress than predicted with basic stress equations. Failure of a plate normally occurs in these regions.

Stress concentrations are highly deleterious when the parts are subjected to shock and vibration, to repeated or alternating stresses, and to extreme temperature conditions.

Much of the data presented here is based on recent work by the Russian scientist, academician G. N. Savin. An English translation of his book *Stress Concentration Around Holes* is available from Pergamon Press.

Savin derived mathematical equations and curves and then checked the results by employing experimental techniques such as photoelasticity.

This section presents stress distributions and stress-concentration factors for plates with elliptical, circular, square, and diamond holes. The plates are under tension. Both single and multiple hole patterns are considered. Subsequent sections cover additional types of loading, including plates subjected to bending, torsion, and buckling—as well as the bending of triangular and circular plates with holes—and also tension, shear, and bending of rectangular plates with holes strengthened by elastic ring.

NOTATION

a, b = dimensions of elliptical or rectangular holes (in.)

c = distance from neutral plane to outer fiber (in.)

C = perpendicular distance from base to center of gravity in equilateral triangle (in.)

d = distance from center of hole to cantilever support (in.)

h = plate thickness (in.)

H = torsional (twisting) couple

I = moment of inertia (in.^{4})

K = stress concentration factor; $K = M_\theta/M$ for moment loadings and S_θ/S_o for tensile loads

L = side of square or diamond (in.); length of beam (in.)

$m = (a - b)/(a + b)$

M = moment (lb.-in.)

M_θ = resulting moment due to the stress concentration around the hole

P = load (lb.)

q = ratio of major to minor axis of ellipse; $q = a/b$

Q = factor for designating distances between holes

r, R = radius of hole (in.)

r = radial distance to point of stress (in.) (also ρ)

r_c = corner radius (in.)

R = size constant. For an elliptical hole, $R = \frac{1}{2}(a + b)$; for a circle, R = radius; for other types of holes, see dimensions of the holes on the applicable charts

R_1 = internal diameter of reinforcing ring (in.)

S_c = compressive stress (psi)

S_o = nominal stress (the stress that would occur if the stress concentration around the hole is ignored) (psi)

S_{max} = actual maximum stress (psi)

S_r = radial stress (psi)

S_s = shear stress (psi)

S_θ = tangential stress along the contour of a hole at a point located by angle θ (psi)
θ = angle locating point of tangential stress (deg.)
ν = Poisson's Ratio

STRESS CONCENTRATION FACTOR

The stress concentration factor, K, is the ratio of the actual maximum stress, S_{max}, to the nominal stress, S_o. The nominal stress is the stress calculated from the elementary stress equations. Thus

$$S_{max} = K S_o$$

For tension and compression

$$S_{max} = K \frac{P}{A}$$

For bending

$$S_{max} = K \frac{M c}{I}$$

HOW INFINITE AN INFINITE PLATE?

The data in this section are highly accurate for cases where the plate is large in comparison to the holes. This condition is commonly referred to as an "infinite plate." Tests have shown, however, that a hole does not appreciably affect the stresses at a distance of a few diameters from the edge of the hole. Specifically, a distance of five diameters from the hole center can be regarded as an infinite distance—and the data can certainly be applied with assurance to cases where the hole diameter is three-fifths the total width of the plate and, in bent plates with triangular holes, to cases where the hole width is one-third the total width.

The following cases are for plates under uniform tension load along a single axis.

ELLIPTICAL HOLES

The holes are defined by the dimensions: $2a$ = length of axis across the plate (perpendicular to the direction of tension), $2b$ = length of axis parallel to the direction of tension. Two cases are covered in Fig. 1.

The stress that concerns us most is S_θ which is the circumferential (tangential) stress at the boundary of the hole. Angle θ is measured counter-clockwise from the y-axis, and the stress ratio S_θ/S_o varies with θ as shown by the chart. At $\theta = 90$ deg. (and at $\theta = 270$ deg.) the ratio

1 . . STRESS CONCENTRATION FACTORS

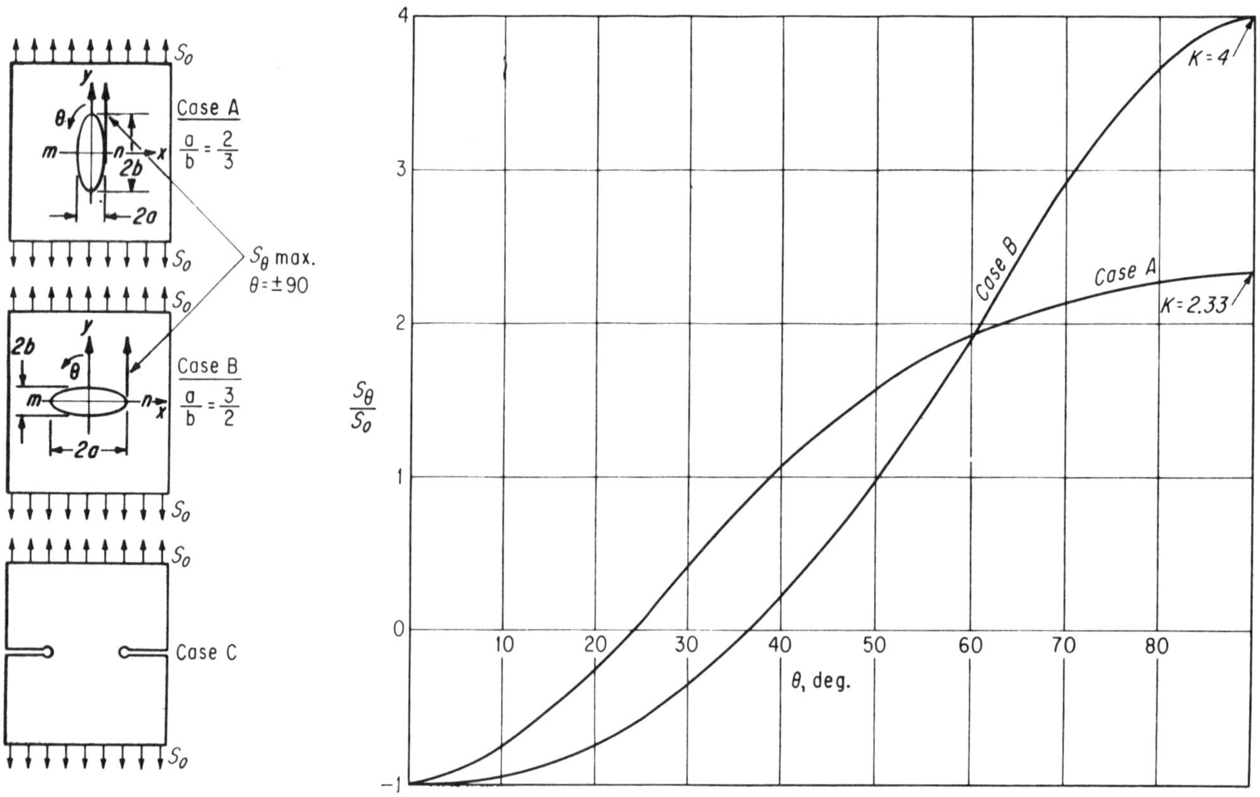

2 . . FOR ELLIPTICAL HOLES

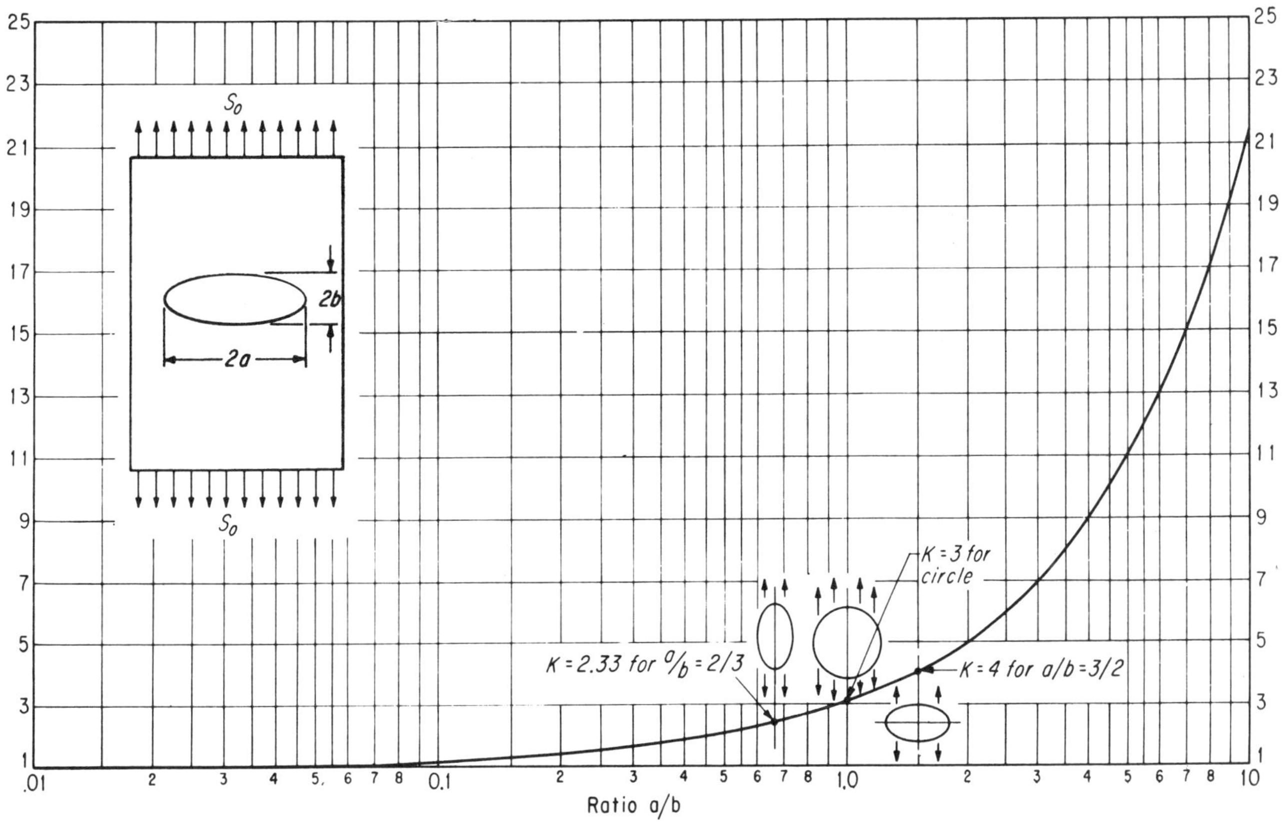

is a maximum. This maximum value is the stress concentration factor, K:

$$\frac{S_{\theta,\max}}{S_o} = \frac{S_{\max}}{S_o} = K$$

The maximum stress is also given by the equation

$$\frac{S_{\max}}{S_o} = 1 + 2\left(\frac{a}{b}\right) \tag{1}$$

Hence for cases $a/b = 3/2$ and $a/b = 2/3$, which are plotted in Fig. 1, the values of K are

$$K = 1 + 2\left(\frac{3}{2}\right) = 4$$

and

$$K = 1 + 2\left(\frac{2}{3}\right) = 2.33$$

The equation for the stress along the contour of any elliptical hole is

$$S_\theta = S_o\left(\frac{\sin^2\theta + 2q\sin^2\theta - q^2\cos^2\theta}{\sin^2\theta + q^2\cos^2\theta}\right), \tag{2}$$

where $q = a/b$.

Stress concentration factors for any ratio of a/b are given in Fig. 2. For a circle, $a = b$; hence the maximum stress from equation (1) is

$$S_{\max} = 3\,S_o \tag{3}$$

Figure 2 shows the location of this point on the curve. Note also that as ratio a/b increases, factor K increases and becomes, in the case of a narrow elliptical hole, very high. This explains why cracks running perpendicular to the direction of stress are likely to spread very quickly and why this spreading can be stopped by drilling holes at the ends of the crack, as shown in plate (C) of Fig. 1. Factor K is independent of the material, provided that the material is isotropic.

To clear up any possible misconception as to the use of K, check this typical problem. Given in Case (B), Fig. 1:

plate width = 4 in.
plate thickness = 0.125 in.
a = 2 in.
b = 1 in.
P = 8,000 lb.

Hence, the nominal stress is:

$$S_o = \frac{8{,}000}{4(0.125) - 2(0.125)} = 32{,}000 \text{ psi}$$

From Fig. 2, $K = 5$. Therefore, the actual maximum stress will be

$$S_{\max} = 5(32{,}000) = 160{,}000 \text{ psi}$$

Circular holes

Stress distribution around a circular hole in a plate under uniform tension, Fig. 3, can be determined from the equations:

Tangential stress:

$$S_\theta = \frac{S_o}{2}\left[\left(1 + \frac{r^2}{\rho^2}\right) - \left(1 + \frac{3r^4}{\rho^4}\right)\cos 2\theta\right], \quad (4)$$

Radial stress:

$$S_r = \frac{S_o}{2}\left[\left(1 - \frac{r^2}{\rho^2}\right) + \left(1 - \frac{4r^2}{\rho^2} + \frac{3r^4}{\rho^4}\right)\cos 2\theta\right], \quad (5)$$

Shear stress:

$$S_s = \frac{S_o}{2}\left(\sin 2\theta\right)\left(1 + \frac{2r^2}{\rho^2} - \frac{3r^4}{\rho^4}\right), \quad (6)$$

where r = radius of hole, ρ = radial distance to point of stress.

The highest stresses will be at the contour of the hole; hence the tangential stress is the stress to be concerned about: At the hole edge, $\rho = r$; hence

$$S_\theta = S_o(1 - 2\cos 2\theta) \quad (7)$$

The maximum value in equation (7) will occur when $\cos 2\theta = -1$ (when $\theta = \pm 90$ deg.) and will be equal to

$$S_{\theta,\max} = 3\,S_o \quad (8)$$

This equation is the same as equation (3). A plot of the stress distribution in Fig. 3 shows that $K = 3$ at points A and C. The stress then decreases, at first rapidly, then asymptotically, until at points B and D a compressive stress S_o is induced with an intensity equal to the nominal stress:

$$S_c = -S_o \quad (9)$$

Square holes

Two configurations are covered in Fig. 4: square holes in which the sides are parallel or perpendicular to the direction of loading (curves are shown as solid lines), and diamond holes in which the sides are at 45 deg. to the direction of loading (curves are dashed).

The corner radius is also varied as follows: $r_c = 0.06L$, $r_c = 0.025L$, $r_o = 0.014L$; where r_c = corner radius, and L = side of square or diamond.

For a square hole with $r_c = 0.06L$

$$S_\theta = S_o\left[\frac{4(AC + BD)}{C^2 + D^2}\right] \quad (10)$$

where

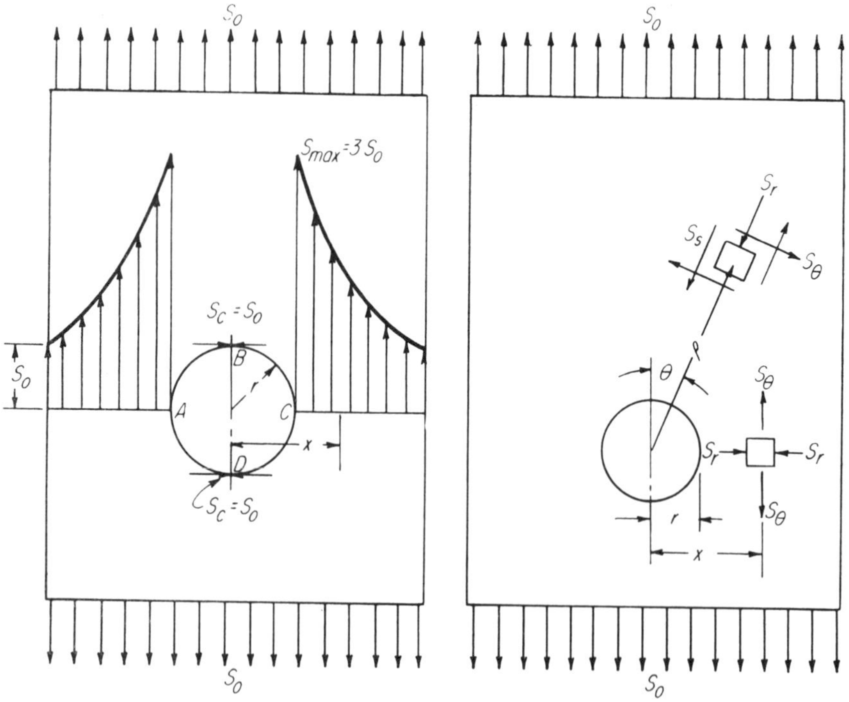

3 . . FOR CIRCULAR HOLES

$$A = 14 - 24\cos 2\theta - 7\cos 4\theta$$
$$B = -24\sin 4\theta - 7\sin 4\theta \qquad (11)$$
$$C = 56 + 28\cos 4\theta$$
$$D = 28\sin 4\theta$$

This is plotted as curve *I*.

For a diamond hole with $r_c = 0.06L$, equation (10) still applies, but with the following parameters:

$$A = 10 + 24\sin 2\theta - 5\cos 4\theta$$
$$B = -24\cos 2\theta - 5\sin 4\theta \qquad (12)$$
$$C = 40 + 20\cos 4\theta$$
$$D = 20\sin 4\theta$$

A plot of this group results in curve *IV*. From the curves it can be deduced that:

1. For the square and diamond patterns (curves *I* to *IV*) the maximum stress concentrations are near the corners.

4 . . FOR SQUARE AND DIAMOND HOLES

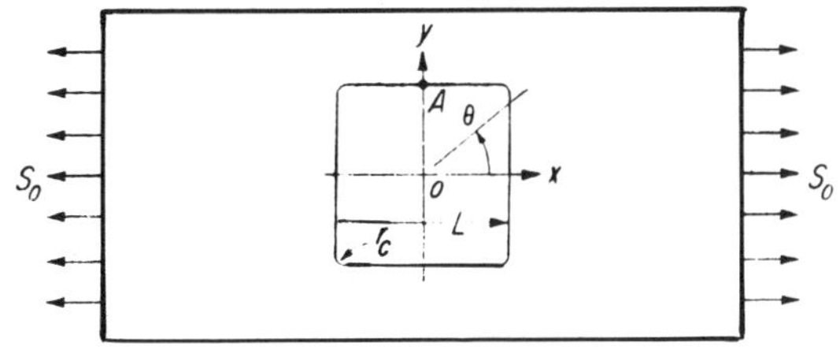

Square hole

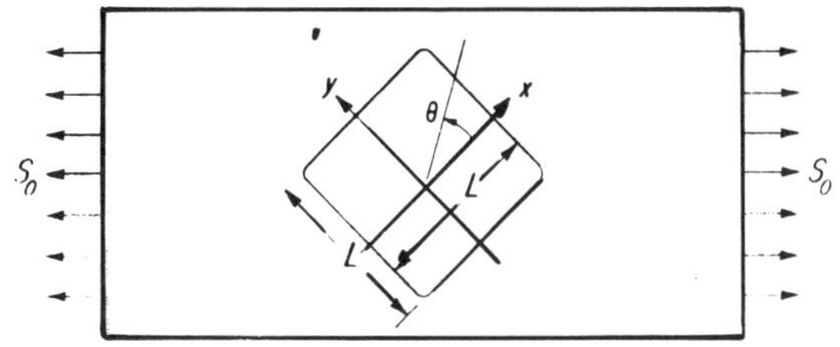

Diamond hole

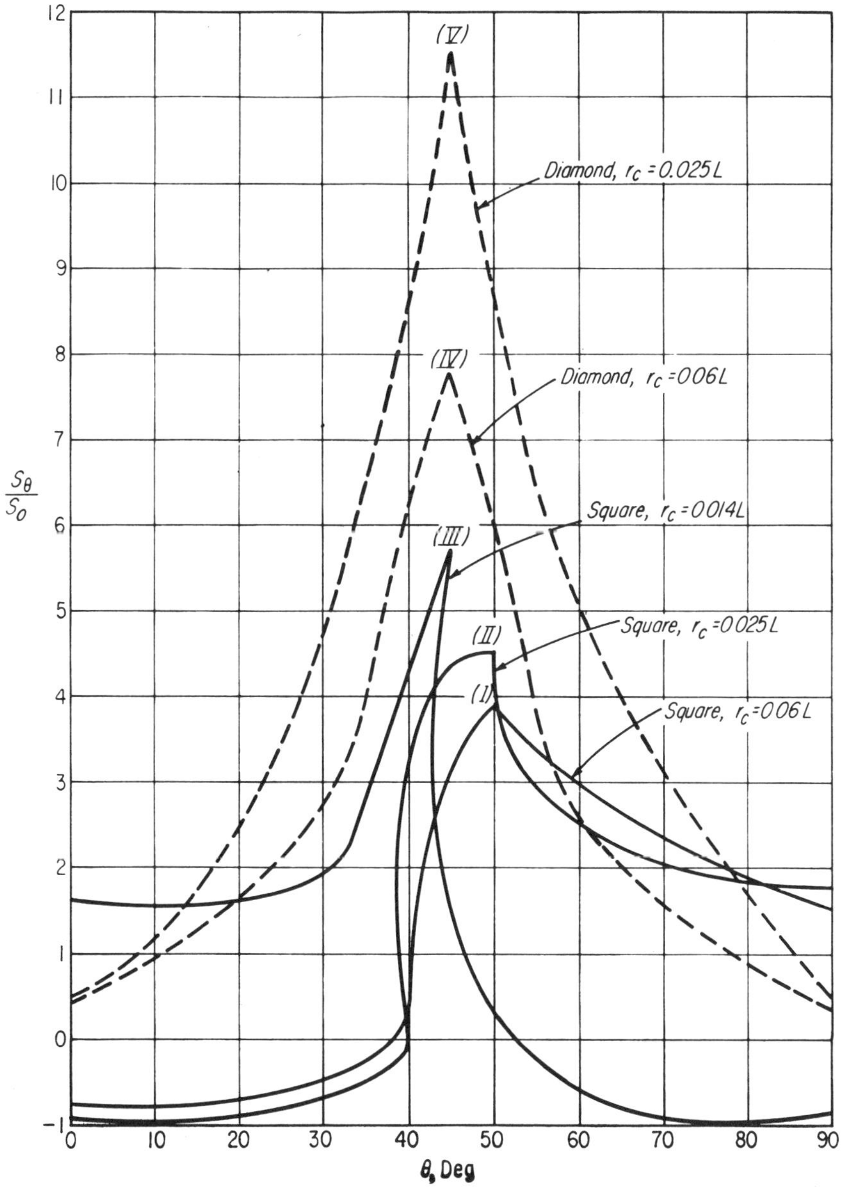

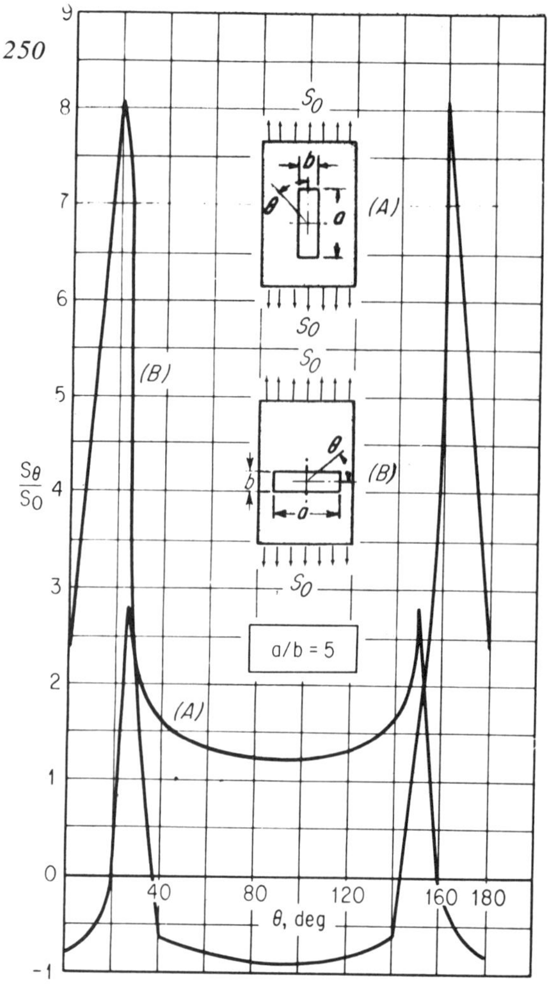

5 . . FOR RECTANGULAR HOLES

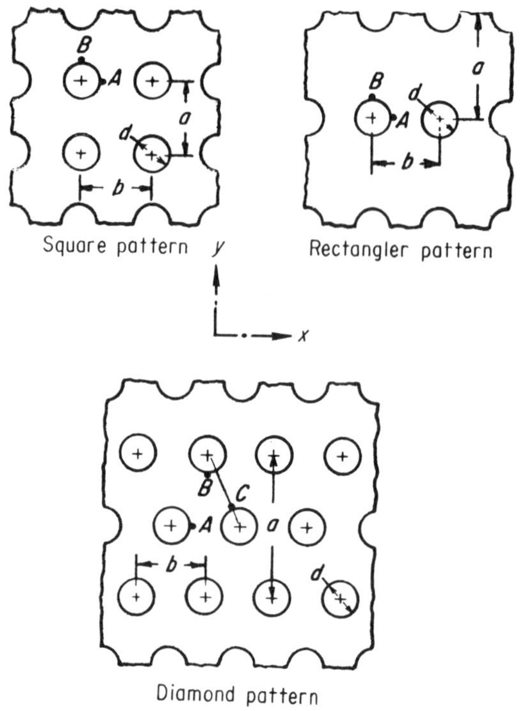

7 . . MULTI-HOLE PATTERNS

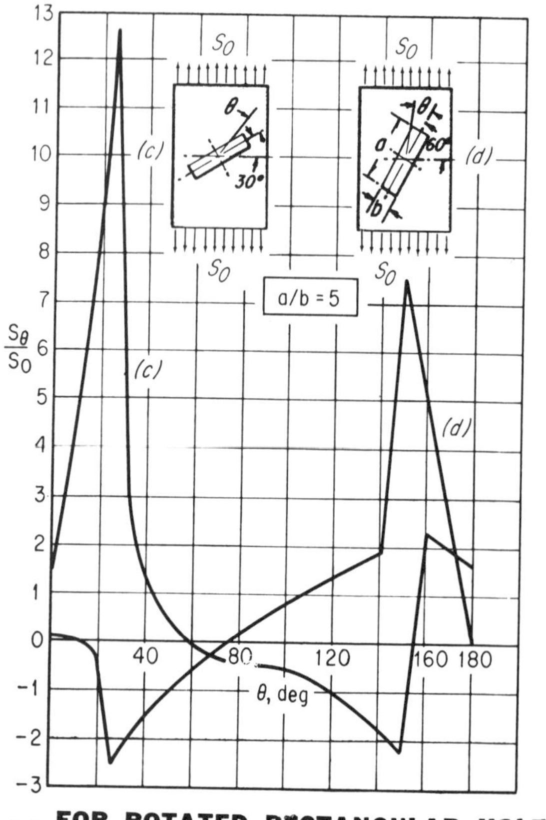

6 . . FOR ROTATED RECTANGULAR HOLES

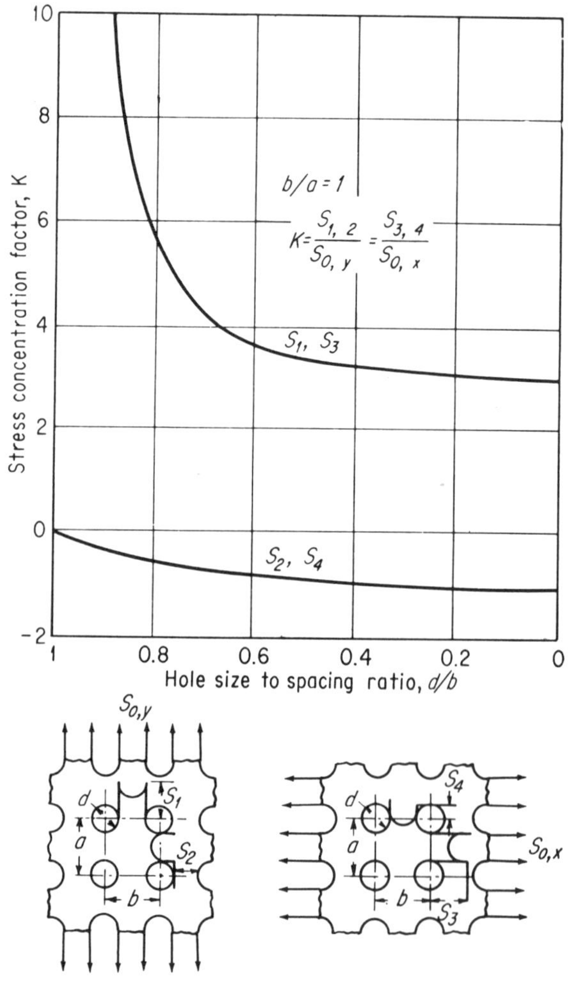

8 . . FOR SQUARE PATTERNS

2. Decreasing the corner radii increases the maximum stress.

Photoelastic studies have shown that the sides of a square which are parallel to the tension loading have a stress that is 1.4 times the nominal stress. For the "infinite" plate of curve *I*, the stress at these points ($\theta = 90$ deg.) is 1.5 S_o which is close to the experimental results.

Rectangular holes

Stress distributions around rectangular holes are shown in Figs. 5 and 6 for four different orientations of the hole relative to the tensile force: 0, 30, 60, and 90 deg. The ratio of the long side to the short side is $a/b = 5$. Here again there is a very high stress concentration at the corners; for example, $K = 8$ for case B.

For rectangular holes of various a/b ratios, see values below. Coefficient *K* gives the stress concentration at the corners; coefficient K' at the center of a side which is parallel to the direction of the applied tensile load. Ratios a/b when less than unity are for case *B*.

a/b	1:5	1:3	1:1	3:1	5:1
K'	2.5	2.2	1.5	1.35	1.2
K	8.0	6.2	3.0	3.2	3.0

TABLE I . . INTERPOLATION DATA

STRESS CONCENTRATION FACTOR	b/a = 1 (FIG. 8)	b/a = 0.63 (FIG. 9A)	b/a = 0.75 (INTERPOLATION)
K_1	3.32	3.40	3.37
K_2	−0.86	−0.90	−0.89
K_3	3.32	2.82	2.99
K_4	−0.86	−0.64	−0.71

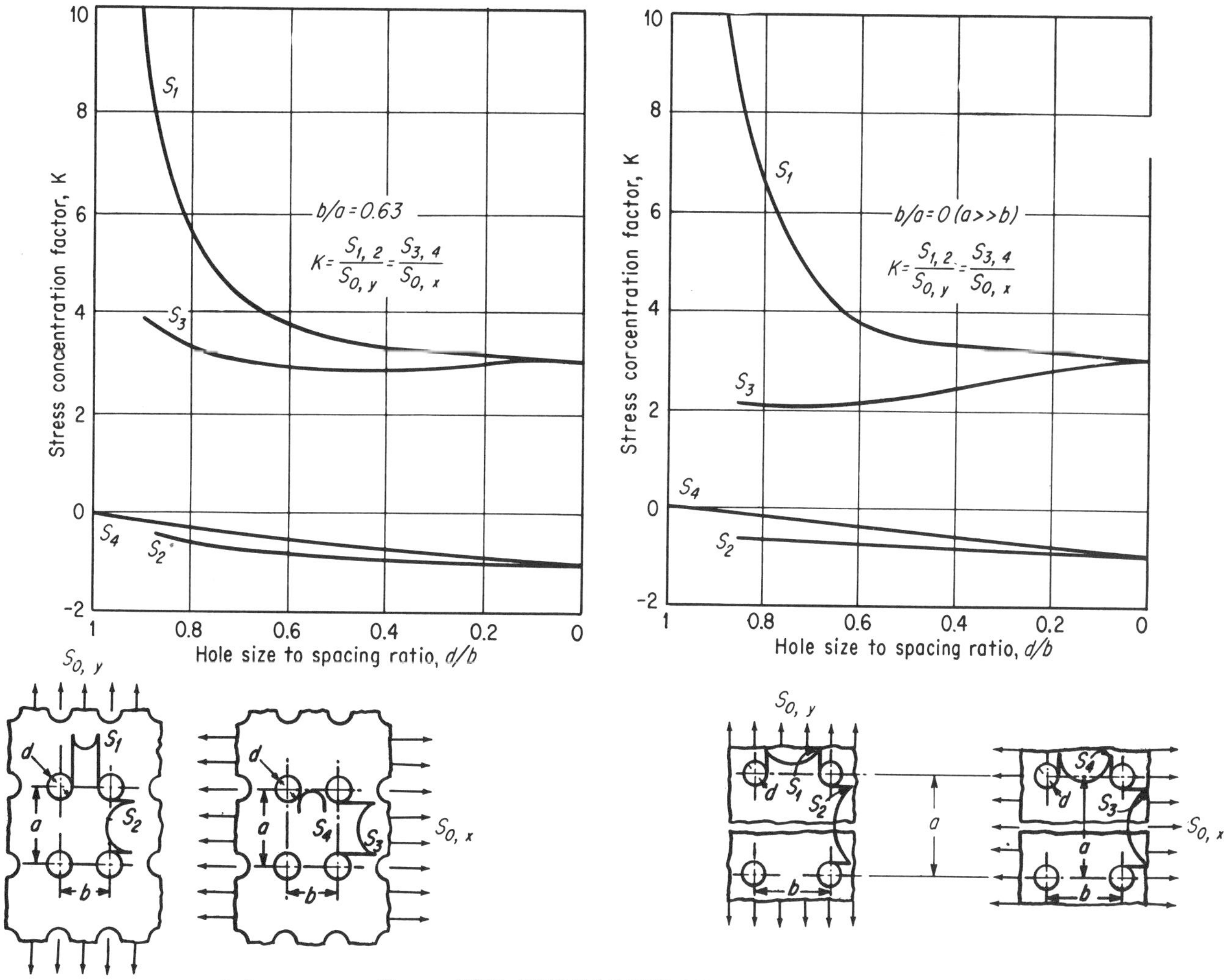

(A) 9 . . FOR RECTANGULAR PATTERNS (B)

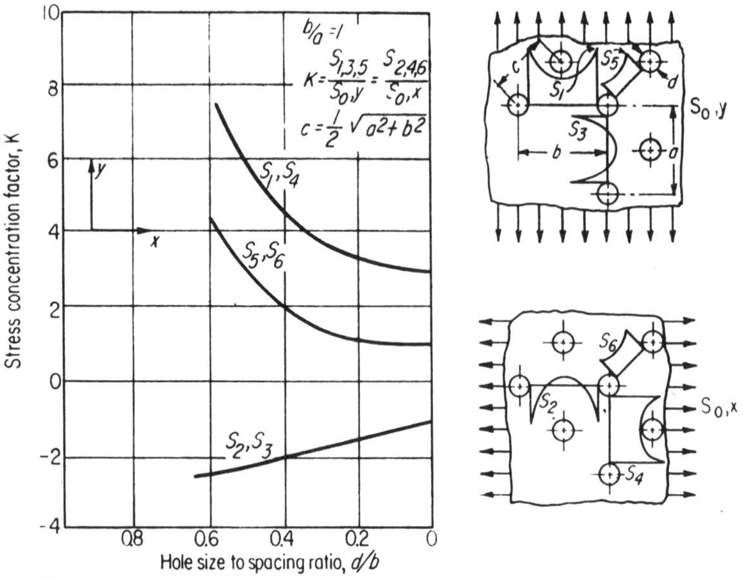

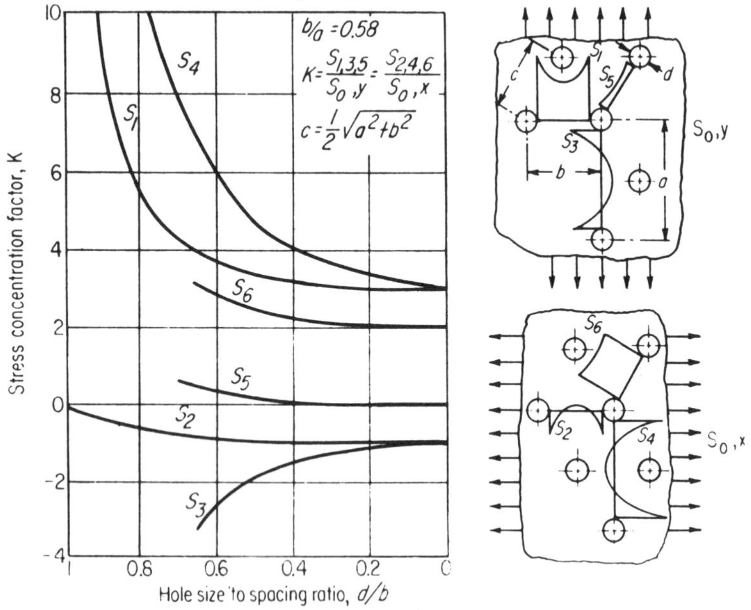

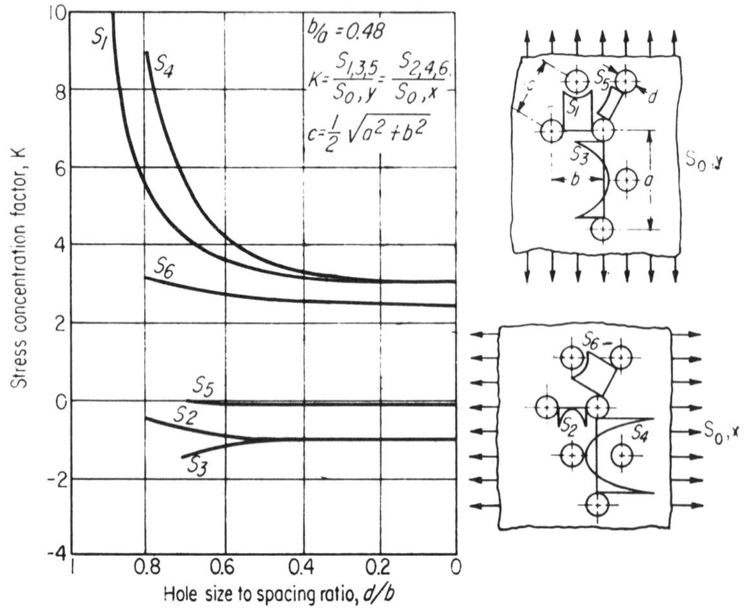

10 . . FOR DIAMOND PATTERNS

MULTI-HOLE PLATES

The best data available on holes arranged in square, rectangular or diamond patterns, Fig. 7, are based on work conducted by two German researchers in 1956. No complete English translation of their German paper is available, but some of their data were reported by A. J. Durelli and J. W. Dally, in *Product Engineering* (March 1956, p. 188). The work is quite important to design engineers and the key design curves, Figs. 7–10, are repeated here.

The stresses at points A, B, and C (S_A, S_B, S_C) can be computed as follows:

For square and rectangular patterns

$$\begin{aligned} S_A &= S_1 + S_4 = K_1 S_{o,y} + K_4 S_{o,x} \\ S_B &= S_2 + S_3 = K_2 S_{o,y} + K_3 S_{o,x} \end{aligned} \qquad (13)$$

For diamond patterns

$$\begin{aligned} S_A &= S_1 + S_2 = K_1 S_{o,y} + K_2 S_{o,x} \\ S_B &= S_3 + S_4 = K_3 S_{o,y} + K_4 S_{o,x} \\ S_C &= S_5 + S_6 = K_5 S_{o,y} + K_6 S_{o,x} \end{aligned} \qquad (14)$$

NOTE: The ordinates K_1, K_2, K_3, K_4, K_5, K_6, in Figs. 7, 8, 9, 10, should be read off the curves marked S_1, S_2, S_3, S_4, S_5, S_6.

Example 1: For example, consider a plate $\frac{1}{8}$ in. thick by 24 in. square with rectangularly arranged holes:

$$\begin{aligned} a &= 2 \text{ in.} \\ b &= 1.5 \text{ in.} \\ d &= 0.75 \text{ in.} \\ d/b &= 0.5 \\ b/a &= 0.75 \end{aligned}$$

The plate is subjected to a uniformly distributed load of 500 lb./in. in the y direction and 200 lb./in. in the x direction, the nominal stresses are

$$S_{o,y} = \frac{P}{A} = \frac{500(24)}{(1/8)(24)} = 4{,}000 \text{ psi}$$

$$S_{o,x} = \frac{P}{A} = \frac{200(24)}{(1/8)(24)} = 1{,}600 \text{ psi}$$

As b/a values of 0.75 are not given, interpolate between $b/a = 1.0$ and $b/a = 0.63$.

This interpolation is shown in Table 1. This gives

$$\begin{aligned} K_1 &= 3.37 \\ K_2 &= -0.89 \\ K_3 &= 2.99 \\ K_4 &= -0.71 \end{aligned}$$

Thus from equation (13), the stresses are:

$$S_A = (3.37)(4{,}000) + (-0.71)(1{,}600) = 13{,}245 \text{ psi}$$

$$S_B = (-0.89)(4{,}000) + (2.99)(1{,}600) = 1{,}220 \text{ psi}$$

The effect of the loading in the x direction is to lower the stress occurring at point A resulting from the loading in the y direction. Similarly the compressive stress at B, resulting from the load in the y direction, is changed to a relatively low tensile stress by the influence of the stresses resulting from the load in the x direction.

Example 2: A thin-walled perforated tube is subjected to a differential pressure between the inside and outside wall. The mean diameter is 3.19 in. and wall thickness is 0.250 in.; the tube is perforated with a diamond pattern of holes 0.50 in. in diameter. If it is assumed that the hoop and longitudinal directions correspond to the x and y directions respectively, then the spacing of the pattern can be expressed as $a = b = 1$ (10 holes about the tube spaced at $\frac{1}{2}$-in. levels). Further, if the pressure differential is 1,000 psi, the nominal stresses are:

$$S_{o,x} = k\frac{pR}{t}$$

$$S_{o,y} = \frac{pR}{2t}$$

where

$$k = \frac{2\pi Ra - n\pi d^2/2}{2\pi Ra}$$

$$= \frac{4Ra - nd^2}{4Ra}$$

$$S_{o\,x} = \left[\frac{4(1.595)(1) - 10(0.5)^2}{4(1.595)(1)}\right]\left[\frac{1{,}000(1.595)}{0.25}\right] = 3{,}860 \text{ psi}$$

$$S_{o,y} = \frac{1{,}000(1.595)}{2(0.25)} = 3{,}190 \text{ psi}$$

To determine the stress concentration factors K_{1-6} select the appropriate factors for the hole to size ratio of $d/b = 0.5$ in Fig. 10, for $C/a = 1$:

$$K_1 = K_4 = 5.62$$
$$K_2 = K_3 = -2.21$$
$$K_5 = K_6 = 2.82$$

The stress at points A, B and C can be computed from equation (14).

$$S_A = (5.62)(3{,}190) + (-2.21)(3{,}860) = 9{,}400 \text{ psi}$$
$$S_B = (-2.21)(3{,}190) + (5.62)(3{,}860) = 14{,}700 \text{ psi}$$
$$S_C = (2.82)(3{,}190) + (2.82)(3{,}860) = 19{,}900 \text{ psi}$$

Maximum stress occurs at point C in this example; however, it may also be at A or B depending on the spacing ratio, pattern spacing, and the relative values of the nominal stresses.

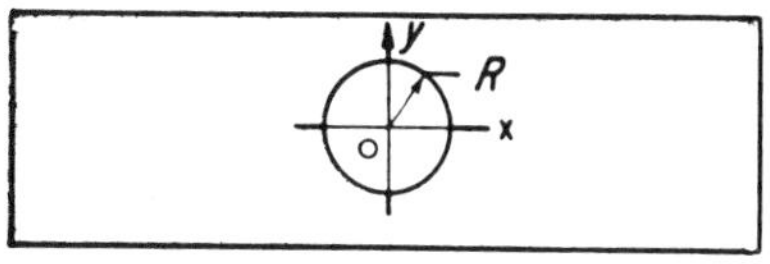

Circular hole

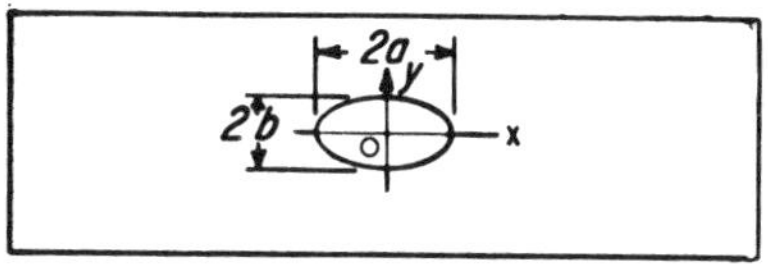

Elliptical hole a/b = 3

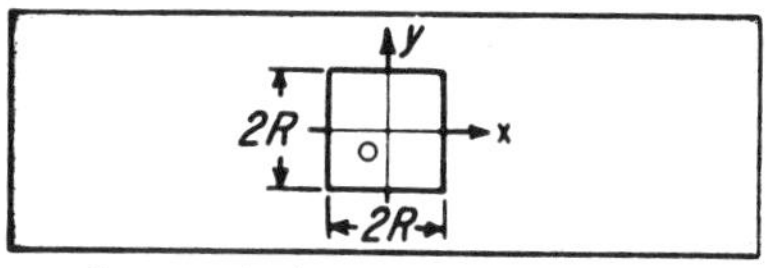

Square hole

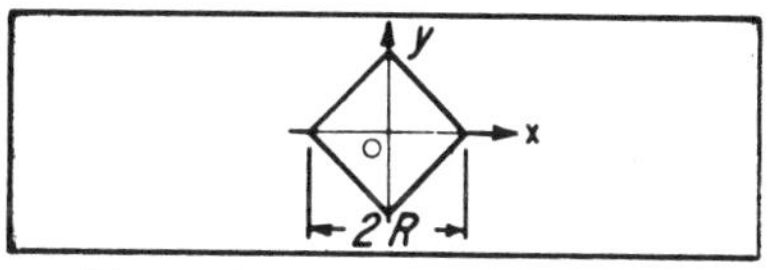

Diamond hole

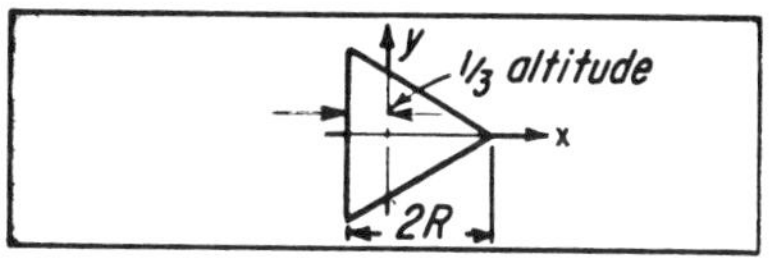

Triangular hole

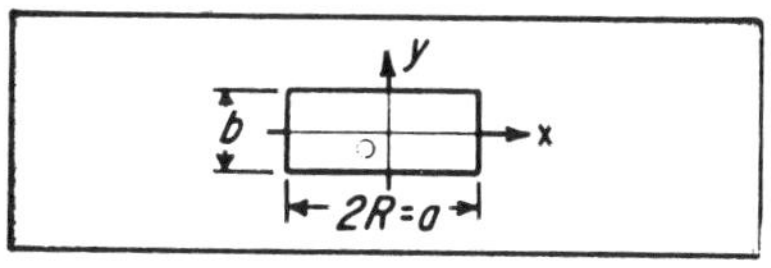

Rectangular hole a/b = 3.2

SIMPLE-BEAM LOADING

Here we have a case of pure bending resulting from applied moments at the ends of the plate or from the plate being loaded as a simple beam (Fig. 11). The data, theoretically, are for an "infinite" plate, but this should not restrict applications because tests have shown that a hole does not appreciably affect the stresses at a distance of more than a few diameters from its edge. Specifically, a distance of five diameters from the hole center can be regarded as an infinite distance, and the data will provide a good approximation to cases where the hole diameter is three-fifths or less of the total width of the plate. Note that the x axis coincides with the neutral axis of the plate.

For a plate without a hole and subjected to bending, the stress can be calculated from the well known basic formula

$$S = \frac{Mc}{I} \qquad (15)$$

11. BEAMS SUBJECTED TO BENDING

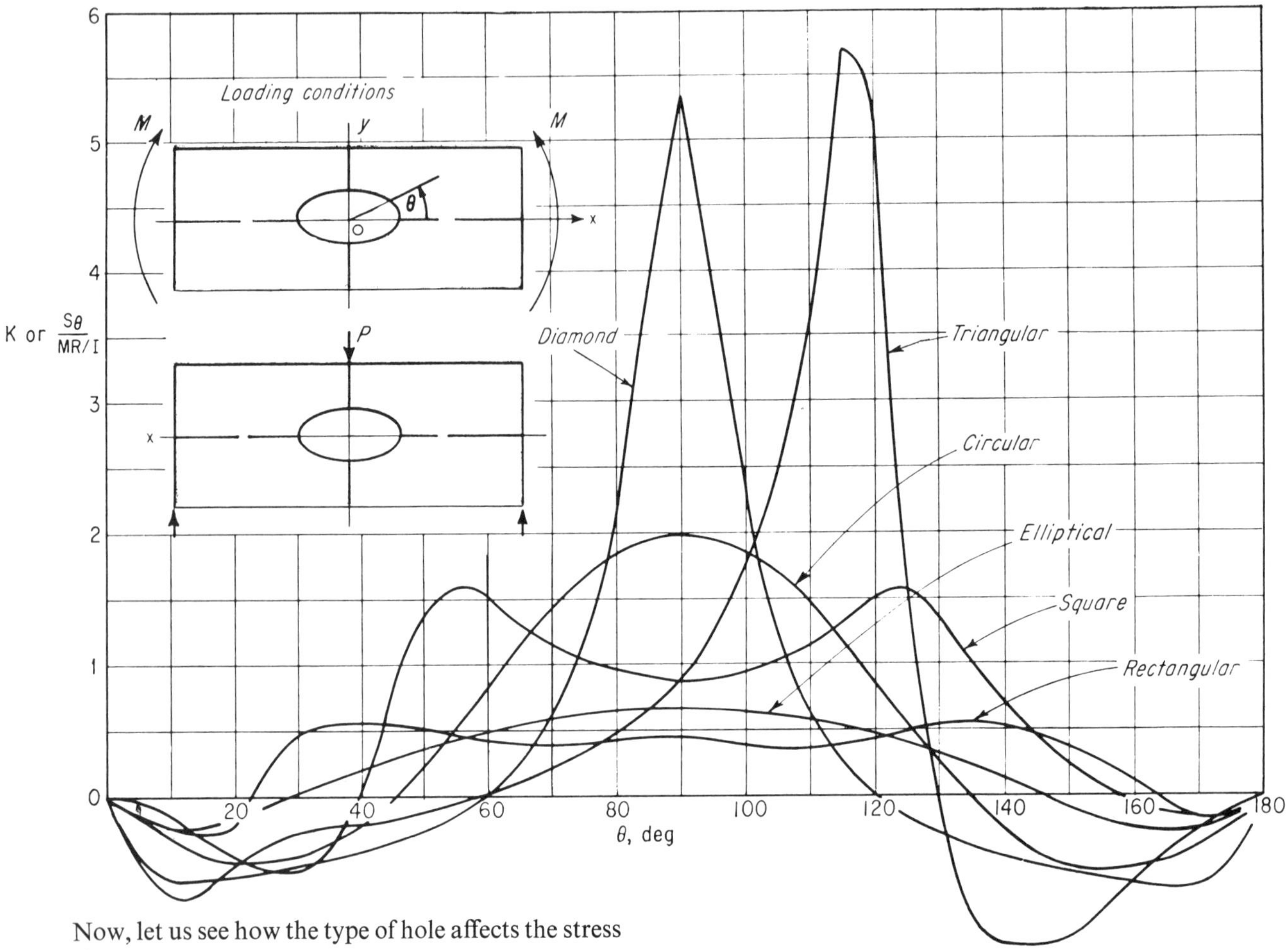

Now, let us see how the type of hole affects the stress equations.

Elliptical hole (Fig. 11)

The equation for the tangential stress along the contour of an elliptical hole is

$$S_\theta = \frac{MR}{I}\left[\frac{(1 + m - 2m^2)\sin\theta + (m - 1)\sin 3\theta}{1 + m^2 - 2m\cos 2\theta}\right] \quad (16)$$

where

a = semi axis along the x axis (in.)
b = semi axis along the y axis (in.)
$m = (a - b)/(a + b)$
R = constant characterizing the size of the hole (in.) For an ellipse, $R = \frac{1}{2}(a + b)$
S_θ = tangential stress along the contour of the hole at a point located by θ where θ is measured counter-clockwise from the x axis.

The curve in Fig. 11 is for an ellipse with 3:1 axis proportions. These proportions are most commonly employed. Factor m then equals $\frac{1}{2}$, and equation (16) reduces to

$$S_\theta = \frac{MR}{I}\left[\frac{4\sin\theta - 2\sin 3\theta}{5 - 4\cos 2\theta}\right] \quad (17)$$

The maximum value of S_θ occurs at 90 deg., in which case equation (16) reduces to

$$S_{\max} = \frac{Mb}{I}\left(1 + \frac{b}{a}\right) \quad (18)$$

If $b = 0$ but $a \neq 0$, as in the case of a slit along the o-x axis of length $2a$, $S_\theta = 0$. This means that a slit of length $2a$ parallel to the neutral axis of the strip (or beam) will not cause any stress concentration in the investigated plate.

If $a = 0$ and $b \neq 0$, as in the case of a slit of length $2b$ along the y axis, S_θ will be very large and the material will plastically deform or crack.

12. CANTILEVER BEAMS WITH HOLES

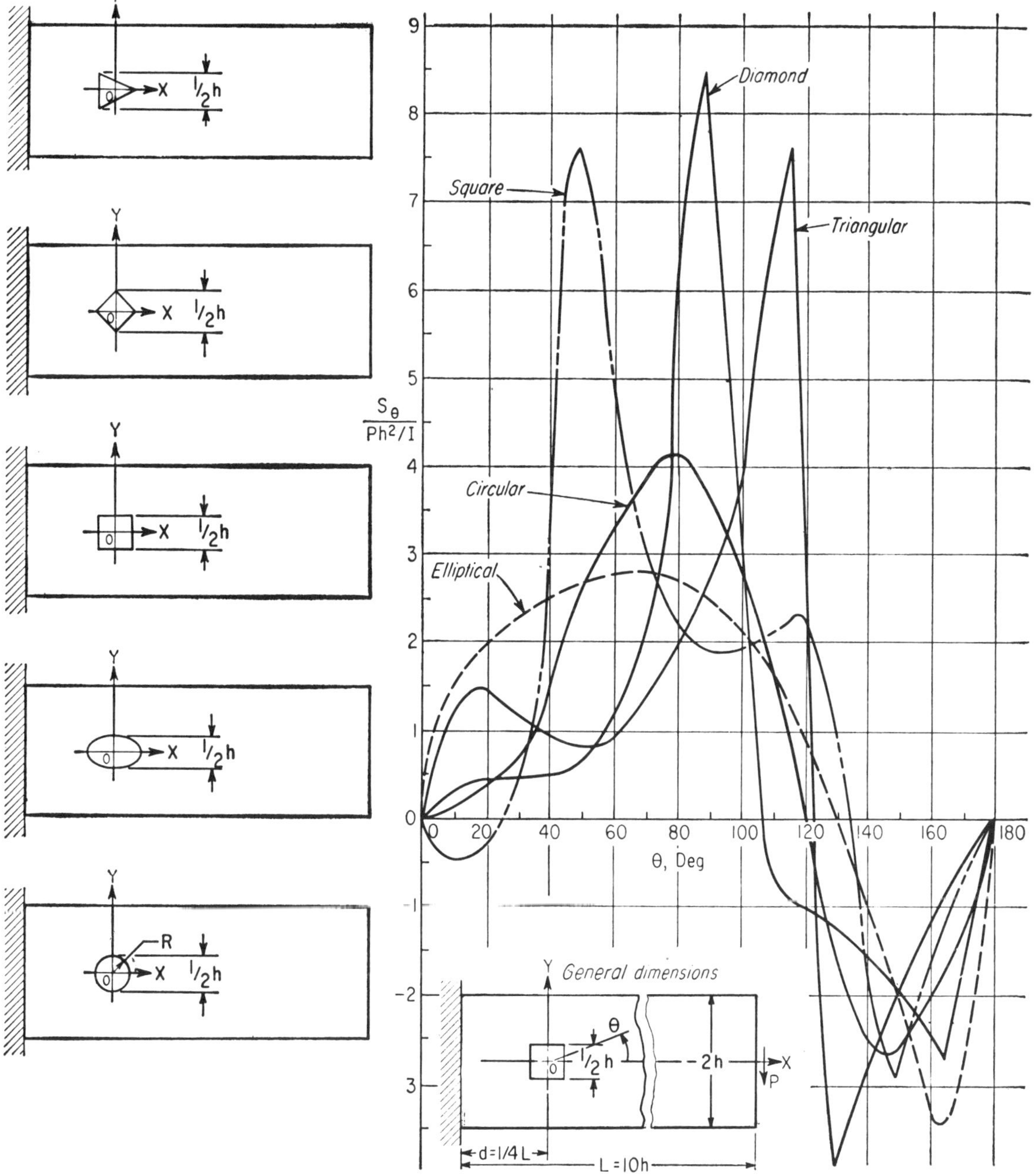

Circular hole (Fig. 11)

In this case $a = b$ and $m = 0$ in equation (16). Thus

$$S_\theta = \frac{MR}{I} [\sin \theta - \sin 3\theta] \tag{19}$$

where R = radius of the circular hole.

Square hole

The equation for S_θ is

$$S_\theta = \frac{MR}{I} \left[\frac{14 \sin \theta - 12 \sin 3\theta - 2 \sin 5\theta}{15 + 12 \cos 4\theta} \right] \tag{20}$$

where $2R$ = length of a side of the square.

13. PLATES SUBJECTED TO COMBINED LOADING (in x and y directions)

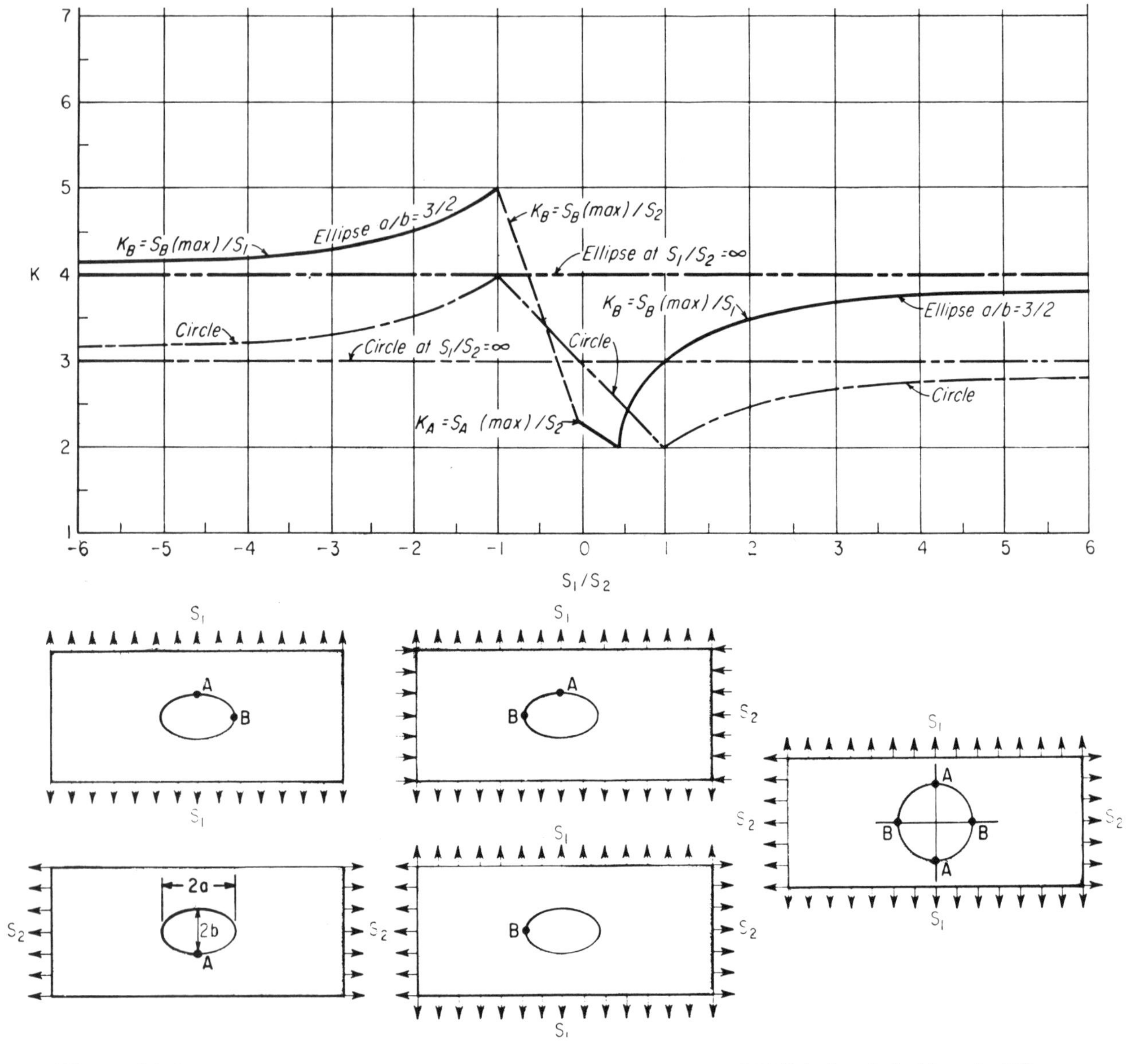

Diamond hole

This case is actually that of a square rotated 45 deg. The stress equation is as follows:

$$S_\theta = 2\frac{MR}{I}\left[\frac{\sin\theta - 6\sin 3\theta + \sin 5\theta}{15 - 12\cos 4\theta}\right] \quad (21)$$

where $2R$ = length of the diagonal of the diamond.

Triangular hole (Fig. 11)

This is the case of an equilateral triangle with the y axis through its center of gravity (at one-third the altitude)

$$S_\theta = 3\frac{MR}{I}\left[\frac{\sin\theta - 3\sin 3\theta + \sin 4\theta}{13 - 12\cos 3\theta}\right] \quad (22)$$

where $2R$ = height of the triangle.

Rectangular hole (Fig. 11)

This is the special case where $a/b = 3.2/1$

$$S_\theta = \frac{MR}{I} \times$$

$$\left[\frac{1{,}228\sin\theta - 490\sin 3\theta - 112\sin 5\theta - 48\sin 7\theta}{1{,}825 - 1{,}580\cos 2\theta + 720\cos 4\theta + 480\cos 6\theta}\right] \quad (23)$$

where $2R = a$.

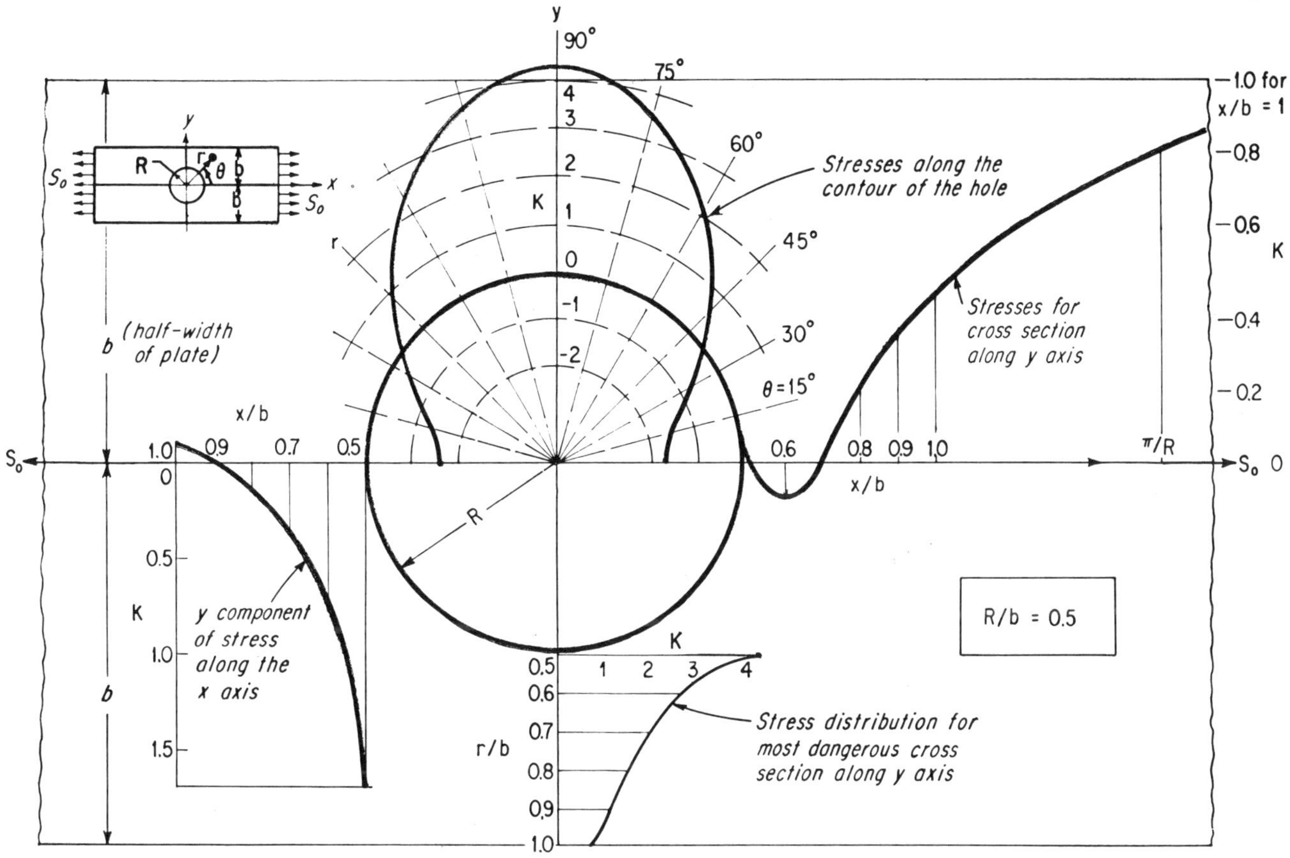

14. STRESS DISTRIBUTION IN PLATES WITH LARGE HOLES

There is a good agreement of the theoretical data in Fig. 11 with photoelastic experimental findings. In the case of circular holes, for example, Savin has found that the curve gives satisfactory results even when the diameter of the circular hole amounts to as much as three-fifths the width of the plate.

SAMPLE PROBLEM—DIAMOND HOLE

Consider a plate 14 in. high × 0.25 in. thick, with a diamond hole at its center. Diagonal $2R = 4$ in. Applied moment $M = 25{,}000$ in.-lb.:

$$I = 0.25\,(14 - 4)^3/12 = 20.8 \text{ in.}^4$$

From the curve in Fig. 11, the maximum value of K is 5.35. Hence

$$S_{\max} = K\frac{MR}{I} = \frac{5.35\,(25{,}000)\,(2)}{20.8} = 12{,}900 \text{ psi}$$

OFF-CENTERED AXES

If the center of the hole is not situated on the neutral axis of the beam (or plate) but at a distance d from it, as shown here, the problem will be reduced to: (1) calcula-

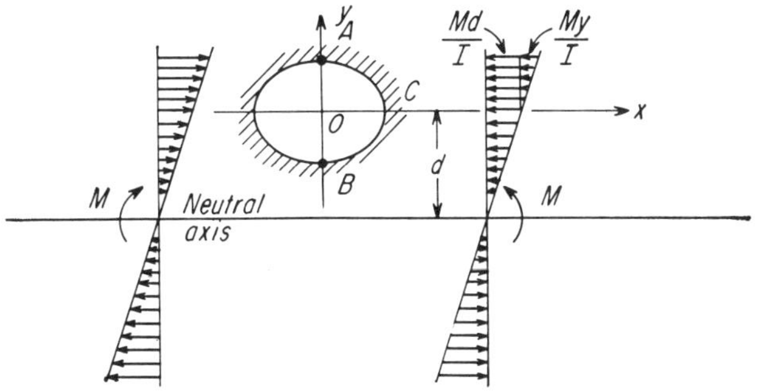

tion of tension in the direction of the neutral axis by the stresses $S = Md/I$, and (2) pure bending of the plate with a hole the center of which is in the neutral axis. $S_x =$

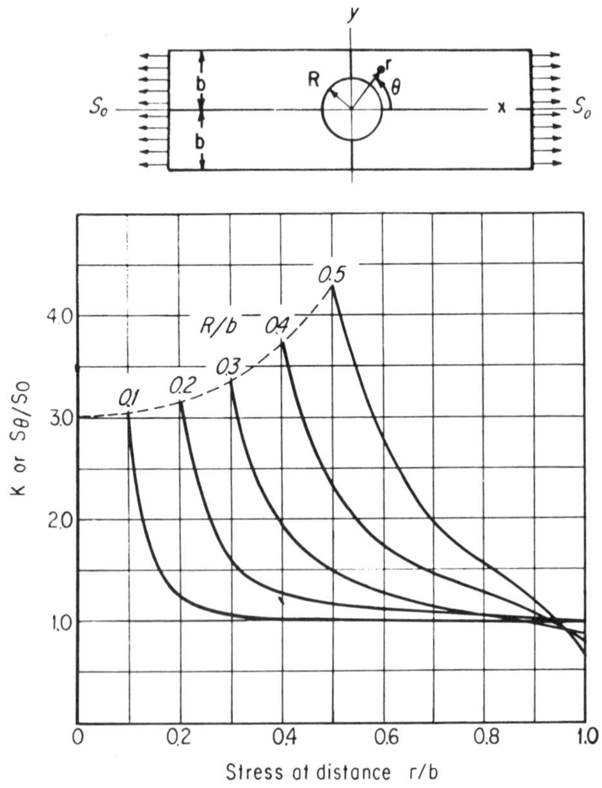

15. STRESS DISTRIBUTION IN PLATES WITH VARIOUS HOLE SIZES

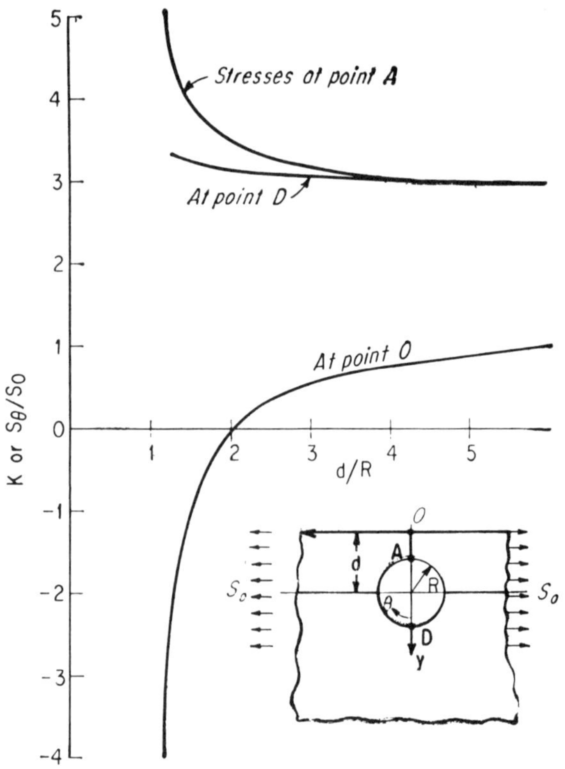

16. EFFECT OF EDGE HOLES

My/I. Thus for an elliptical hole with 3:1 axis proportions the stress at point A is

$$S_\theta = \frac{M}{I}\left[(b + d) + \frac{b}{a}(b + 2d)\right]$$

If $d = O$

$$S_\theta = \frac{Mb}{I}\left[1 + \frac{b}{a}\right],$$

which is the same as equation (18).

The stress at point B:

$$S_\theta = \frac{M}{I}\left[(d - b) + \frac{b}{a}(2d - b)\right]$$

The stress at point C:

$$S_\theta = \frac{Md}{I}$$

CANTILEVER BENDING

This is the case of a cantilever rigidly supported at one end (Fig. 12) with length L, height $2h$, and transverse load P applied at the free end. The hole center is on the neutral axis (x axis) and is at a distance d from the fixed end. The effect of the hole shape now follows:

Elliptical Hole

The stress distribution around an elliptical hole of semiaxis a and b (Fig. 12) can be determined by the following equation

$$S_\theta = -\frac{P(L - d)}{I} \times \left[\frac{(1 + m - 2m^2)\sin\theta + (m - 1)\sin 3\theta}{1 - 2m\cos 2\theta + m^2}\right] - \frac{P}{I}\left[\frac{[2h^2 + R^2(m^2 - 1)]\sin 2\theta + R^2(1 - m)\sin 4\theta}{1 - 2m\cos 2\theta + m^2}\right] \tag{24}$$

where

$$m = \frac{a - b}{a + b}$$

$$R = \frac{a + b}{2}$$

Comparing equation (24) with equation (18), it can be seen that the first part of equation (24) shows the stress due to bending moment $M = -P(L\text{-}d)$, while the second part characterizes the influence of the transverse force P.

Circular hole (Fig. 12)

The equation for the stress is obtained from equation (24) by letting $a = b$; therefore, $m = 0$ and R = radius of the circular hole:

$$S_\theta = -\frac{PR(L-d)}{I}(\sin\theta - \sin 3\theta) - \frac{P}{I}[(2h - R^2)\sin 2\theta + R^2 \sin 4\theta] \quad (25)$$

Photoelastic investigations by Savin of the stress distribution in a cantilever containing circular holes showed that:

1. Equation (25) is applicable to fairly large diameters of circular holes—as large as half of the plate height.
2. The stress S_θ along the contour of the hole is highest at the top point of the hole contour (in the zone of tension).
3. The stress on the outside contour of the beam is higher than the corresponding stress on the contour of the hole.
4. The most dangerous cross sections appear to be the parts of the plate that are weakened by the hole and at the point where the beam (plate) is built in.
5. The principal normal stresses occur in the cross section where the beam is supported.
6. If the cantilever has two holes, the influence of each hole on the stress along the beam axis extends to a distance not much larger than one hole diameter from the hole center. This state appears to be the result of the redistribution of stresses due to presence of second hole.

Square hole (Fig. 12)

$$S_\theta = -\frac{PR(L-d)}{I}\left[\frac{14\sin\theta - 12\sin 3\theta - 2\sin 5\theta}{3(5 + 4\cos 4\theta)}\right]$$

$$-\frac{P}{I}\left[\frac{(14.4h^2 - 7.4R^2)\sin 2\theta + 12R^2\sin 4\theta + 2R^2\sin 6\theta}{3(5 + 4\cos 4\theta)}\right]$$

where $2R$ = length of one side of the square.

17. EFFECT OF TWIN HOLES

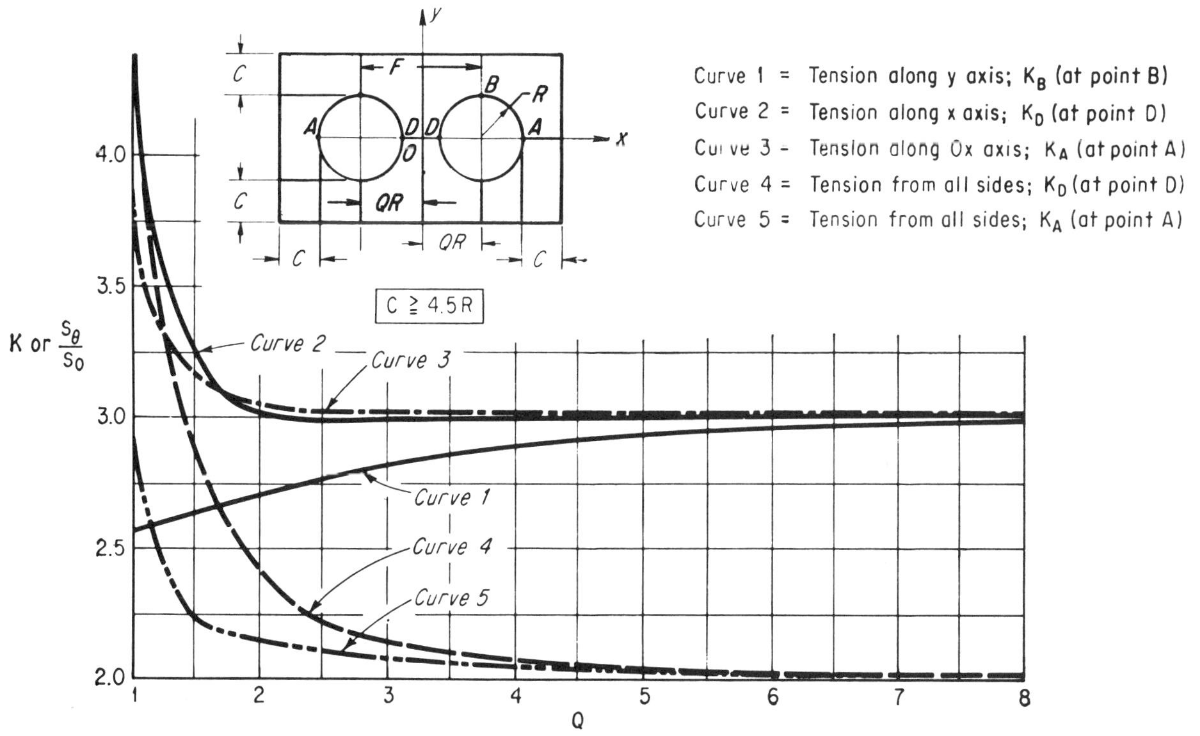

Diamond hole (Fig. 12)

$$S_\theta = -\frac{2PR(L-d)}{I}\left[\frac{\sin\theta - 6\sin 3\theta + \sin 5\theta}{3(5-4\cos 4\theta)}\right] \qquad (26)$$

$$-\frac{P}{I}\left[\frac{(216h^2 - 78.5R^2)\sin 2\theta + 84R^2\sin 4\theta - 14R^2\sin 6\theta}{21(5-4\cos 4\theta)}\right]$$

where $2R$ = length of the diagonal of the diamond.

Triangular hole (Fig. 12)

$$S_\theta = -\frac{3PR(L-d)}{I}\left[\frac{\sin\theta - 3\sin 3\theta + \sin 4\theta}{13 - 12\cos 3\theta}\right]$$

$$-\frac{P}{I}\left[\frac{(12h^2 - 29/3\,R^2)\sin\theta + 6(3h^2 - R^2)\sin 2\theta}{13 - 12\cos 3\theta}\right]$$

$$-\frac{P}{I}\left[\frac{9R^2\sin 4\theta - 3R^2\sin 5\theta}{13 - 12\cos 3\theta}\right] \qquad (27)$$

where $2R$ = height of the triangle.

SPECIAL CASES

To permit comparison of the influence of various types of holes on the stress distribution, in a cantilever beam, the stresses along the contour of the holes have been calculated for the cantilever bar with the following dimensions: height of beam = $2h$, $d = \frac{1}{4}L$, $L = 10h$, and maximum dimension of the hole in y axis = one quarter of beam height (or $\frac{1}{2}h$). Constant R was determined from these given conditions, and the results plotted in Fig. 12.

COMBINED LOADING

This group covers combined tension or compression in two perpendicular directions (Fig. 13). For ease of calculation, it is more convenient to solve separately the stress states for the individual uniaxial stresses and to obtain the sought solution of the biaxial stress state by superposition of separate solutions.

If equal tensile or compressive stresses are acting in two perpendicular directions, the stress concentration at the edge of a circular hole can be found by using Lame's formula for thick cylinders. On the basis of this theory the tensile stress S_θ in a tangential direction at the edge of a small hole is

$$S_{\theta(\max)} = 2\,S_o \qquad (28)$$

where S_o = nominal stress (the stress that will occur if the stress concentration around the hole is ignored), psi. This means a small circular hole causes the stress in the case under consideration to be doubled.

It was found previously (equation 3) that for simple tension

$$S_{\theta(\max)} = 3\,S_o \qquad (29)$$

18. TWIN HOLES SPACED AT VARIOUS DISTANCES

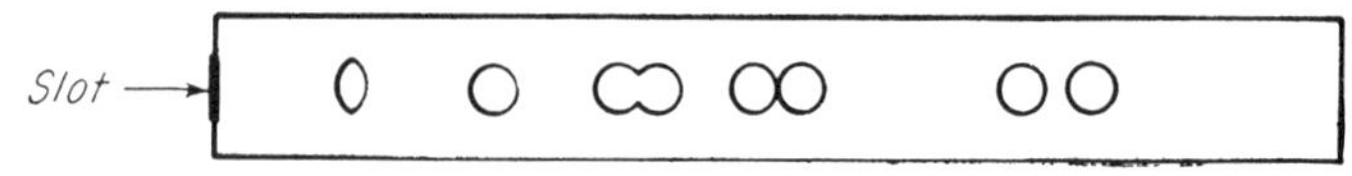

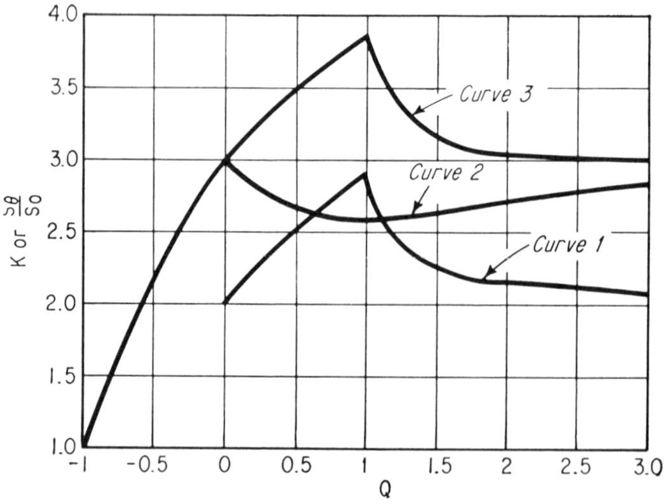

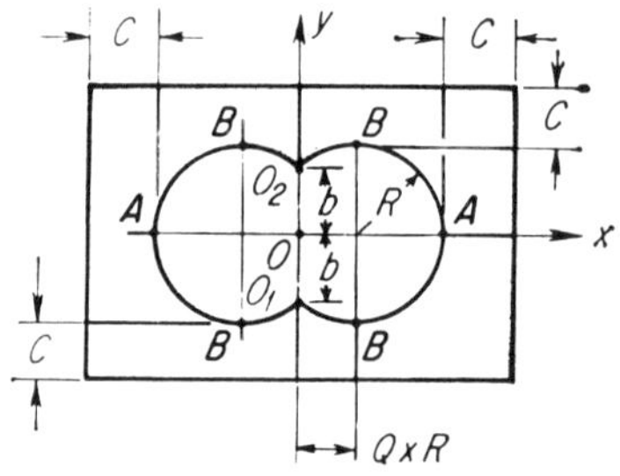

C ≥ 4.5R

Curve 1 = Biaxial tension in the plate; K_A (at point A)

Curve 2 = Tension along x axis; K_B (at point B)

Curve 3 = Tension along y axis; K_A (at point A)

Comparing equations (28) and (29), it can be concluded that for a case of simple tension a compressive stress S_o exists in which

$$S_c = -S_o \tag{30}$$

This stress acts in a tangential direction at points B in Fig. 13. Then, by the principle of superposition, equation (28) is obtained from equations (29) and (30). This approach also gives the solution to the case of the circular hole (Fig. 13). The tangential stress at points B at the edge of the hole is equal to $3\ S_1 - S_2$, and at points A is equal to $3\ S_2 - S_1$. When $S_2 = -S_1$, which is in the case of pure shear, the maximum stress at point B is

$$S_{A\ (\max)} = 3\ S_1 - (-S_1) = 4\ S_o$$

For points A

$$S_{B\ (\min)} = 3\ S_2 - S_1 = -4\ S_1$$

This means that in the case of twisting of a thin circular tube (which can be treated as a plate), a small circular hole produces a high stress concentration in which the maximum tensile stress at the edge of the hole becomes four times as large as the shearing stress uniformly distributed over the ends of the tube. This also happens in a solid shaft under torsion having holes in a radial direction, such as oil ducts in crankshafts.

The variation in K versus S_1/S_2 is shown in Fig. 13 for an elliptical hole of dimension $a/b = 3/2$, and also for a circular hole.

EFFECT OF LARGE HOLES IN PURE TENSION

If the hole is large in comparison to the width of the plate, specifically if the hole diameter is larger than one-fifth the plate width, then the plate width does influence the stress distribution to a significant amount.

19. INFINITE NUMBER OF HOLES—LOADING IN x DIRECTION

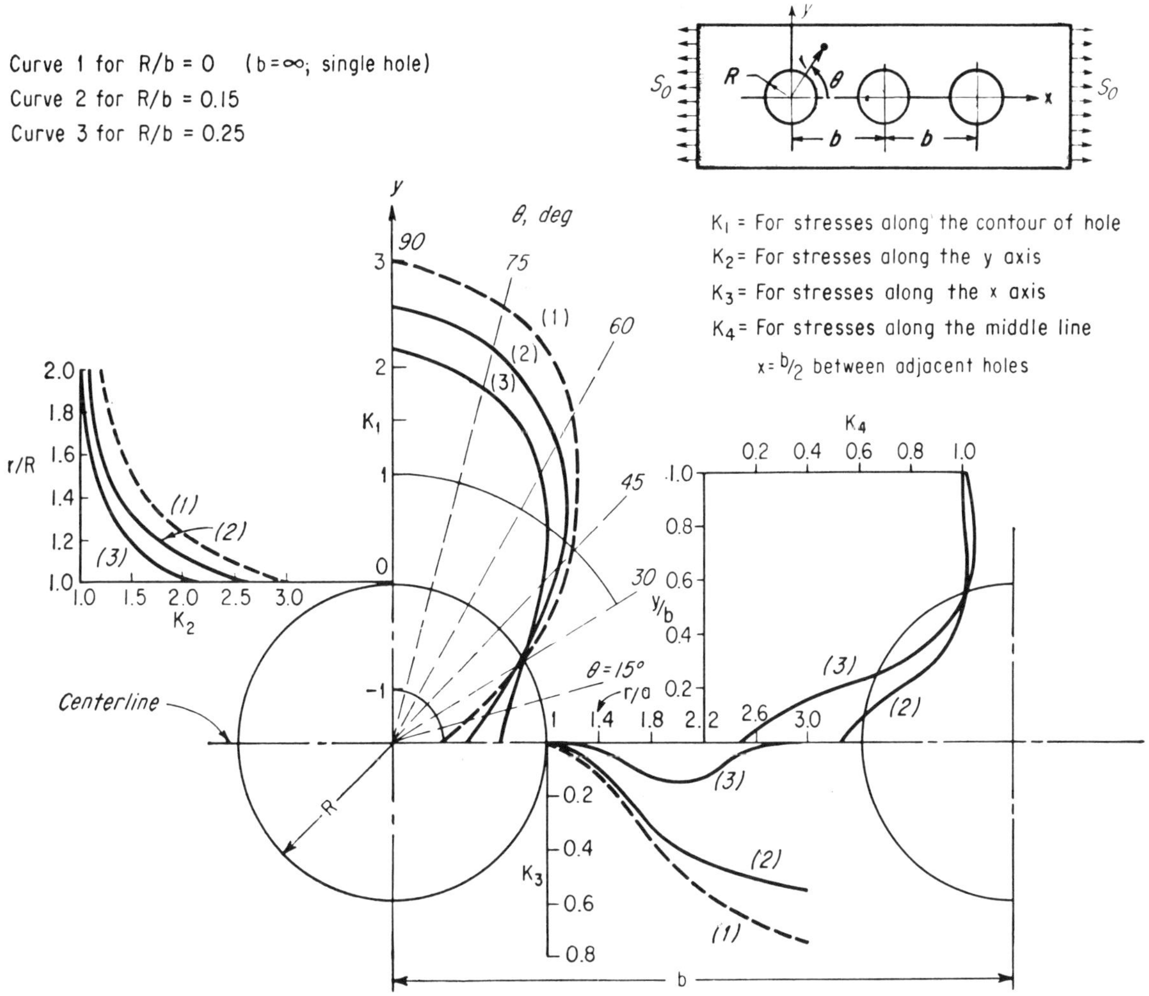

The plate under consideration (Fig. 14) has a width $2b$, with b its center on the axis of symmetry of the strip. The hole radius $= R$, and the plate is loaded with a uniform tensile stress S_o.

The equations for stresses in finite-width plates are very involved, so only the plotted results are summarized in Figs. 14 and 15. Figure 14 shows the stress distribution with a circular hole of size $R/b = 0.5$; i.e., the hole diameter is half the plate width. Figure 15 gives the stress concentration factor $K = S_\theta/S_o$ for various hole sizes at various distances r/b from the edge of the hole where r is the radial distance. In all cases, $\theta = \pi/2$, which is the most critical case. The dashed line gives the increase in stress concentration with increase in relative hole size R/b. Note that the stress concentration coefficient increases rapidly with the increasing radius of hole; thus, for $R/b = 0.5$ the stress coefficient equals $K = 4.32$, while for $R/b = 0$, $K = 3.0$. For values of R/b which differ from the values given in Fig. 15, the stress concentration factor can be determined directly by means of the dashed-lines curve in this figure.

It is usual to accept accuracies of up to 6 percent in stress analysis. On this basis, the data for infinite plates can be applied to plates of finite dimensions, provided that the ratio of the diameter to plate width is not less than 0.2; in other words, the data holds for widths at least five times the hole diameter. This conclusion also applies to cases of pure bending stresses.

EDGE HOLES

This is the case where the plate is large in comparison to the hole diameter, and the hole is at a close proximity to one edge (Fig. 16). Specifically the hole is at a distance d from the edge; angle θ in this case is measured counterclockwise from point D at the hole edge. The plate is under tension loading.

20. INFINITE NUMBER OF HOLES—LOADING IN y DIRECTION

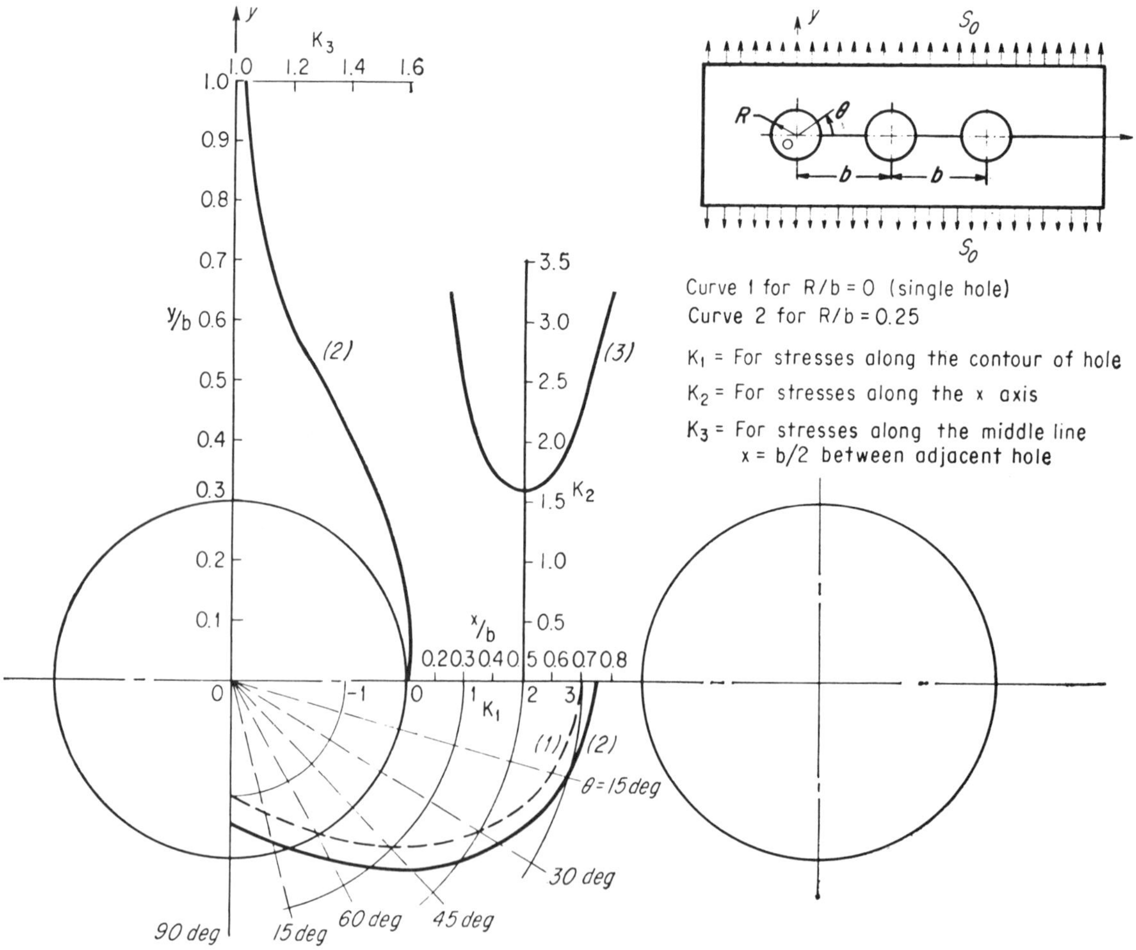

The stresses along the hole boundary are

$$S_\theta = S_o(1 + 2\cos 2\theta)$$

Values of K at points O, A, and D are given in Fig. 16 for various values of d/R.

SINGLE-ROW HOLES

Double holes: First of these cases involves two equal holes of radius R (Fig. 17). The stress state around these holes is determined for the cases where there is:

1. Tension S_o in the y direction only (curve 1).
2. Tension S_o in the x direction only (curves 2, 3).
3. Tension S_o in both direction (curves 4, 5).

The distance of the holes from the y axis is given by means of constant, Q.

Thus a hole is designated as being at a distance $Q \times R$ from the y-axis. Specifically, if F = distance between hole center lines, then $Q = F/2R$.

Here again the solution for the biaxial uniform stress state was obtained by superposition of the solutions for single-axis stress states. The maximum stress S_θ along the contour of the holes will be at points A and D (curves 4, 5). This holds true whether the tensile stress, S_o, is applied along the two axes, or it is applied in the direction of the x axis only (curves 3 and 2).

If the plate is under tension S_o in the direction of the y axis, the maximum stress will be at the points in which θ differs little from 90 deg., i.e., at points situated very near to point B (curve 1). The values of S_θ in Fig. 17 permit numerical evaluation of the influence of the holes on each other.

Within the accuracy of up to 6 percent the following can be said for almost all cases of tensile stresses along the x and y axes, and also for biaxial tensile force:

1. For $Q = 3$, where the center distance between the holes is equal to three diameters, the stress S_o along the contour of the holes will be equal to that prevailing when there is only one hole.
2. The results obtained for infinite plates can be applied to calculations in the case of plates of finite sizes perforated by two holes, provided that points A and B are at a distance of at least $4.5R$ from the edge.

Overlapping holes: The plate under consideration (Fig. 18) contains two overlapping holes. The final contour therefore is actually one hole consisting of two equal circular areas. It is assumed that the contour of the hole is free of external forces and that the stresses are either biaxial, or uniaxial along the x or y axis.

Factor Q, which when multiplied by the radius R locates the arc centers, is plotted against S_θ/S_o. The resulting shapes of the holes are illustrated. Note that the holes do not intersect until $Q < 1$. For $Q = 1$, the holes just touch; for $Q = 0$, the holes blend into one circular hole; for $Q = -1$ the hole has become a slot of length $2b$ along the y axis.

Since the equations are quite involved, only the results are summarized in the form of the three curves in Fig. 18. Curve 1 gives the stress concentration factor at points A for biaxial tension in the plate ($S_\theta/S_o = K_A$). Curve 2 is for uniaxial stress when tension is applied along the x axis. Here the points B are the points of interest. Curve 3 is for when tension is applied along the y axis. Points A are then the points of interest.

Infinite number of holes: The charts in Figs. 19 and 20 are for cases where there are more than two holes in a row. Figure 19 is when the tensile loading is along the x axis, and Fig. 20 when it is along the y axis. Distance b in these figures denotes the distance between hole centers; R is the hole radius and r locates the point of interest.

In Fig. 19, values of $S_\theta/S_o = K_1$ give the stresses along the contour of the hole; $S_y/S_o = K_2$ for stresses along the y axis; $S_r/S_o = K_3$ for stresses along the x axis, and $S_x/S_o = K_4$ for stresses along the centerline of the holes.

The size of the hole relative to spacing between holes is R/b. Two curves, $R/b = 0.15$ and $R/b = 0.25$, are plotted for each stress concentration factor K and compared to the case of a single circular hole where $R/b = 0$.

CYLINDRICAL BENDING

When a rectangular plate is subjected to moments along two opposing ends, its surface deflects to an almost cylindrical shape. Although the illustration shows the plate as being longer along the y-axis than along the

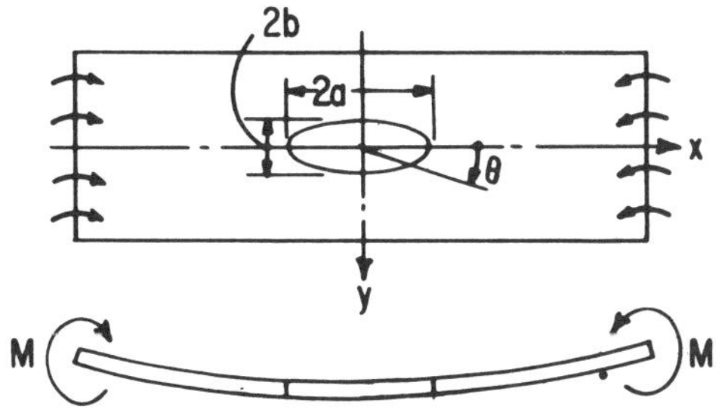

x-axis, the reverse may be true—provided that the moments are parallel to the y-axis. It is assumed that the cross section of the plate remains constant during bending, and normal to the plate surface, i.e., that there is pure bending.

The equations that follow are highly accurate for cases where the plate width is large in comparison to the holes, and the data can be applied with assurance to cases where the hole diameter is three-fifths the total width of the plate and, in bent plates with triangular holes, to cases where the hole width is one-third the total width. The equations give the resulting bending moment M_θ along the contour of the hole from ($\theta = 0$ to 360 deg.) as a function of the applied uniform moment M, and are applicable to all materials. The design charts, Figs. 21–36, are for plates where Poisson's Ratio $\nu = 0.3$ (thus applicable to steel).

To convert the resulting moment, M_θ, to a bending stress, S_θ use: $S = 6M_\theta/h^2$, where h = plate thickness.

Elliptical hole

21

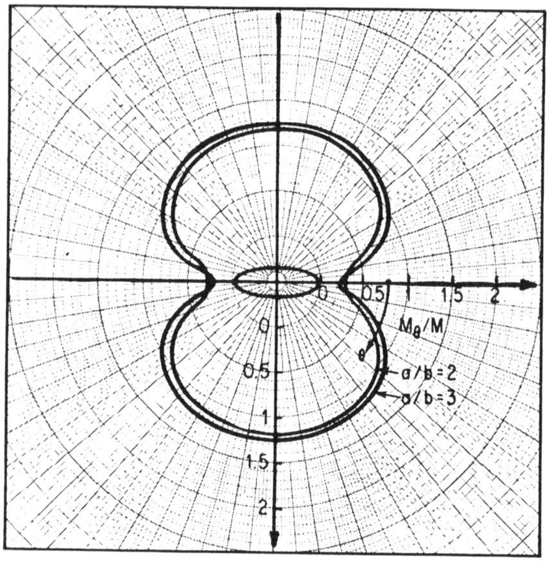

Elliptical hole: The ratio of the resulting moment, M_θ, to the applied moment, M, is given by

$$M_\theta = M\left[1 + \left(\frac{2(1+\nu)(1-m)}{3+\nu}\right) \times \left(\frac{m - \cos 2\theta}{m^2 - 2m\cos 2\theta + 1}\right)\right]$$

where $m = (a - b)/(a + b)$, and where a/b = semimajor axes ratio. Although the diagram shows a larger than b, a may be less than b. Ratio m then would have a negative value.

The distribution of the bending moment M_θ along the contour of the elliptical hole is shown in Fig. 21 for semimajor axes ratios of $a/b = 3$ and $a/b = 2$; i.e., $m = \frac{1}{2}$ and $\frac{1}{3}$, respectively. The maximum values, $M_\theta = 1.263\ M$ for $a/b = 3$ and $M_\theta = 1.22\ M$ for $a/b = 2$, occur at points $\theta = 90$ and 270 deg. (at $\theta = \pm 90$ deg.).

Note that the values for M_θ/M were plotted in polar coordinates. Thus, for any desired value of θ, which locates the point of interest on the hole edge, the respective value of M_o/M can be obtained directly from the chart.

Circular hole

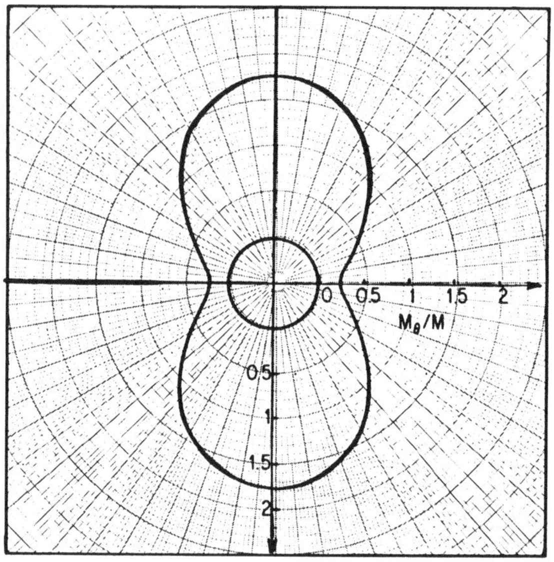

22

Circular hole: Here $a = b$, hence

$$M_\theta = M\left[1 - \frac{2(1+\nu)}{3+\nu}\cos 2\theta\right]$$

Maximum values occur at $\theta = \pm 90$ deg. (Fig. 22) and are equal to $M_\theta/M = (5 + 3\nu)/(3 + \nu)$.

The maximum value (M_{max}/M) is usually referred to as the *stress concentration factor*, *K*. For a circular hole, K values range from 1.667 to 1.857 (for values of $\nu = 0$ to $\nu = 0.5$, respectively).

Square hole:

$$\frac{M_\theta}{M} = \frac{2}{3+\nu} + \frac{2(1+\nu)}{5 + 4\cos 4\theta}\left[\frac{3}{2(3+\nu)} - \frac{36\cos 2\theta}{17 + 7\nu}\right]$$

At $\theta = \pm 90$ deg., $K = 1.28$ (Fig. 23). Maximum stress concentrations occur at the corners where $K = 2.244$.

Diamond hole: Rotating the square hole 45 deg. (Fig. 24) results in

$$M_\theta = M\left\{\frac{2}{3+\nu} + \frac{2(1+\nu)}{5 - 4\cos 4\theta} \times \left[\frac{3}{2(3+\nu)} - \frac{12}{19 + 5\nu}\cos 2\theta\right]\right\}$$

$K = 3.31$ at $\theta = \pm 90$ deg.

Square hole

23

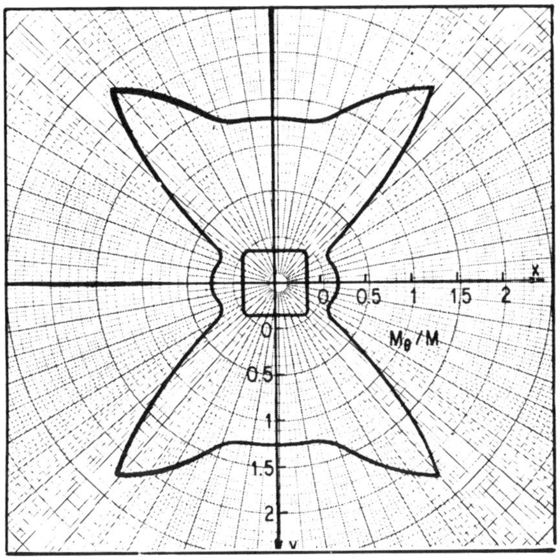

Diamond hole

24

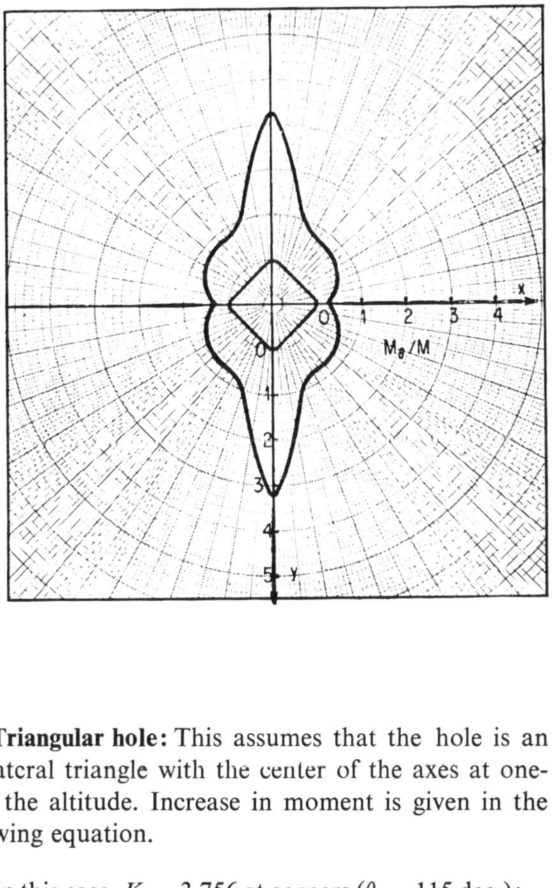

Triangular hole: This assumes that the hole is an equilateral triangle with the center of the axes at one-third the altitude. Increase in moment is given in the following equation.

In this case, $K = 3.756$ at corners ($\theta = 115$ deg.):

$$\frac{M_\theta}{M} = \frac{2}{3 + \nu} + \left(\frac{1 + \nu}{3 + \nu}\right)\left(\frac{5 + 12\cos\theta - 18\cos 2\theta}{13 - 12\cos 3\theta}\right)$$

Triangular hole

25

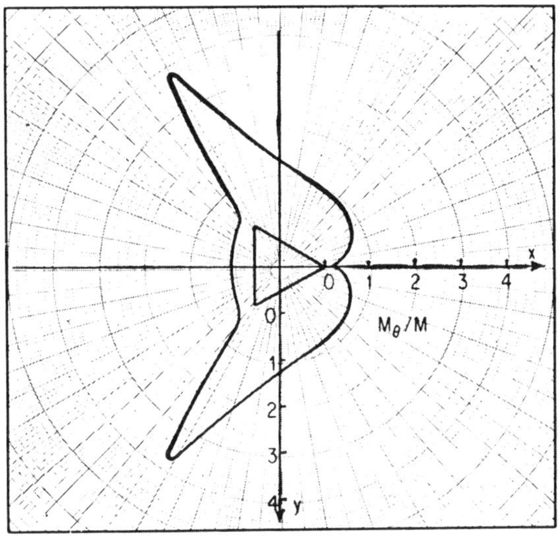

SPHERICAL BENDING

In the following cases the plates are loaded along both axes with equal moments M. Again, the constant loading results in a constant applied moment, $M = M_x = M_y$. Proceeding in the same manner:

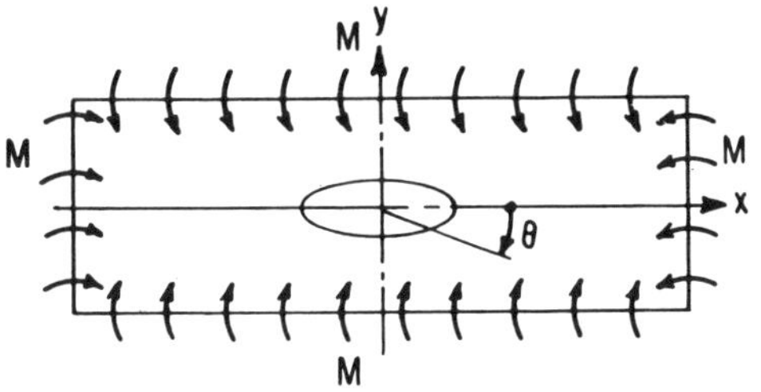

Elliptical hole

26

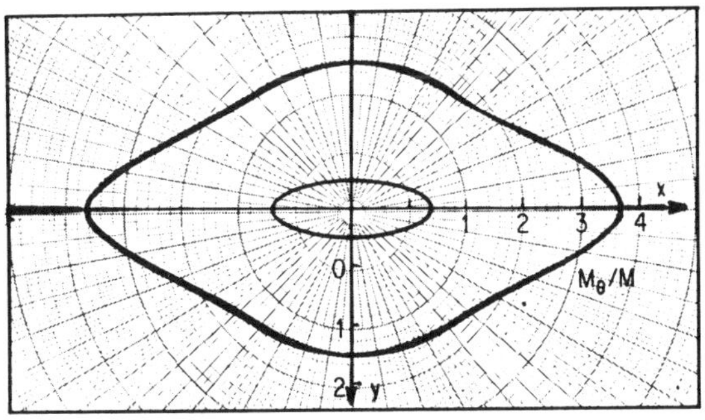

Elliptical hole:

$$\frac{M_\theta}{M} = 2 - \left(\frac{4m(1 + \nu)}{3 + \nu}\right)\left(\frac{m - \cos 2\theta}{m^2 - 2m\cos 2\theta + 1}\right)$$

Figure 26 shows the moment distribution for the case where $a/b = 3$, i.e., where $m = \frac{1}{2}$. Maximum values of $K = 3.58$ occur at points of intersection of the contour with the x axis, while at the intersection with the y axis (at $\theta = \pm 90$ deg.) the stress concentration is equal to $K = 1.475$.

Circular hole: Setting $m = 0$ in the previous equation shows that stress concentration around a circular hole is constant:

$$M_\theta/M = K = 2$$

Square hole

27

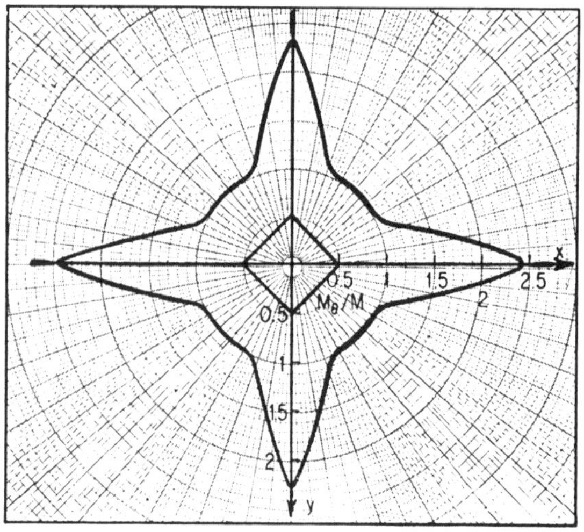

Square hole:

$$M_\theta = M\left[2 - \left(\frac{4(1+\nu)}{3+\nu}\right)\left(\frac{1 + 2\cos 4\theta}{5 + 4\cos 4\theta}\right)\right]$$

Maximum stress concentrations occur at the corners of the square hole and are equal to $K = 3.58$. At the center of the sides, $K = 1.47$.

Trianglar hole

28

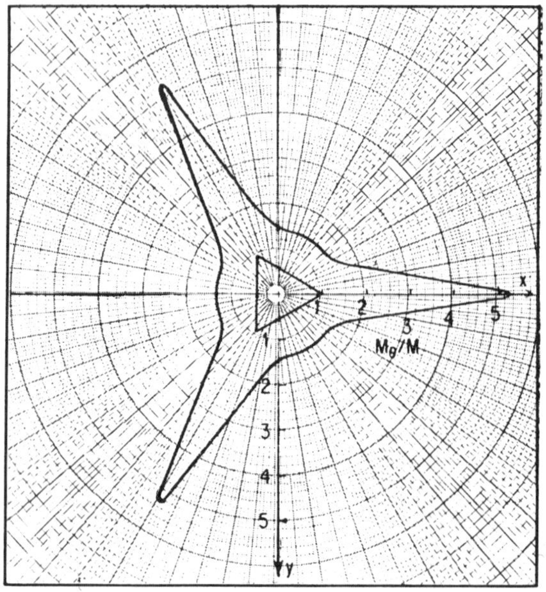

Triangular hole:

$$M_\theta = M\left[2 - \left(\frac{8(1+\nu)}{3+\nu}\right)\left(\frac{2 - 3\cos 3\theta}{13 - 12\cos 3\theta}\right)\right]$$

Maximum stress concentrations for the equilateral triangular hole occur at corners and are equal to $K = 5.15$. At the center of the sides, $K = 1.37$ (Fig. 28).

NOTE: Fig. 27 is not to scale.

TWISTING OF PLATES

The following equations apply to the case where there is a uniformly distributed torsional couple of moment H along all the four edges of a rectangular plate.

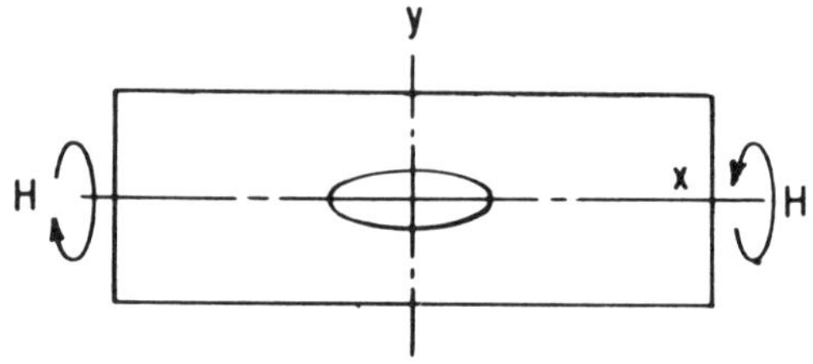

Elliptical hole: The distribution of resulting increase in moment M_o, Fig. 29, is for an elliptical hole with semi axes ratio $a/b = 3$:

$$M_\theta = H\left[\frac{4(1+\nu)\sin 2\theta}{(3+\nu)m^2 - 2m\cos 2\theta + 1}\right]$$

Elliptical hole

29

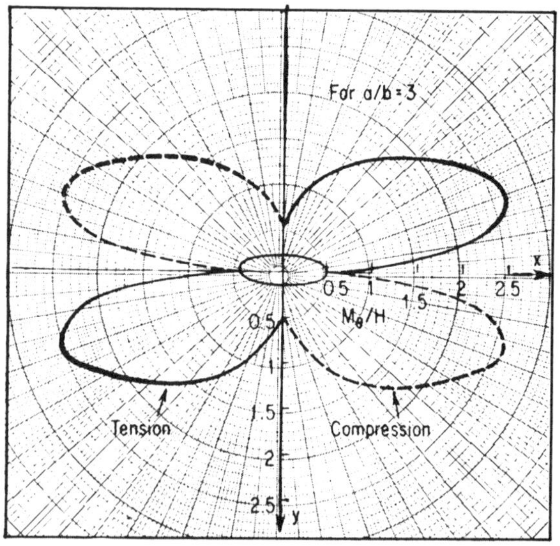

It is interesting to note that the maximum stress concentrations of $K = M_\theta/H = 2.11$ occur at approximately $\theta = 20$ and 160 deg., and that the stresses are zero at points A. The first and third quadrants of the contour of the hole are in compression (dashed lines), whereas the second and fourth quadrants are in tension.

(NOTE: Fig. 29 is not to scale.)

Circular hole

30

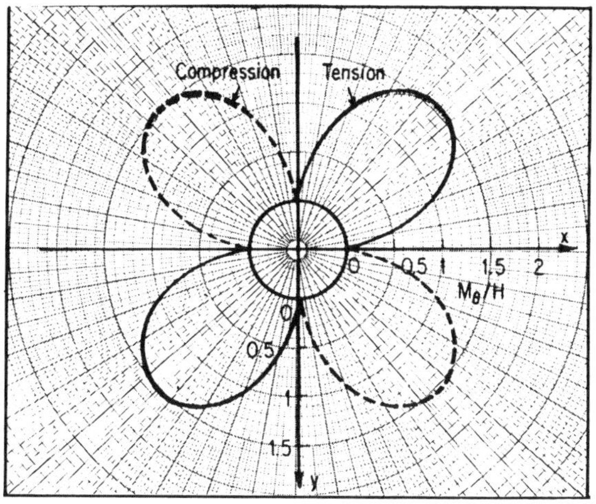

Circular hole:

$$M_\theta = H\left[\frac{1(+ \nu)4\sin 2\theta}{3 + \nu}\right]$$

Maximum stress concentrations occur at $\theta = \pm 45$ and ± 135 deg., with a value of $K = 1.58$ (Fig. 30).

Square hole

31

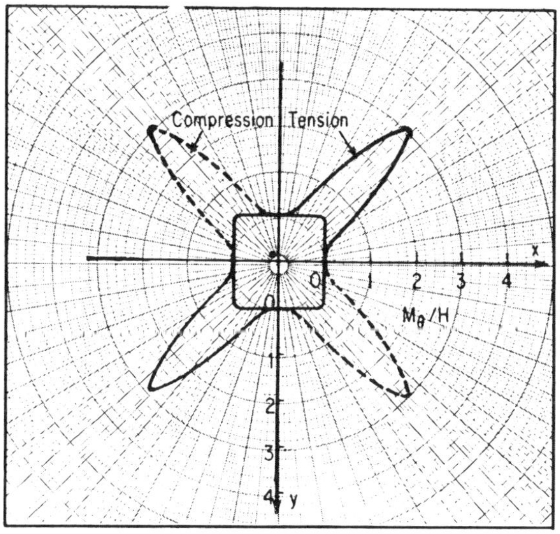

Square hole:

$$M_\theta = H\left[\frac{48\sin 2\theta(1 + \nu)}{(19 + 5\nu)(5 + 4\cos 4\theta)}\right]$$

Maximum values of $K = 3.04$ are at the corners (Fig. 31).

Diamond hole:

$$M_\theta = H\left[\frac{(1 + \nu) \times 144\sin 2\theta}{(17 + 7\nu)(5 - 4\cos 4\theta)}\right]$$

Diamond hole

32

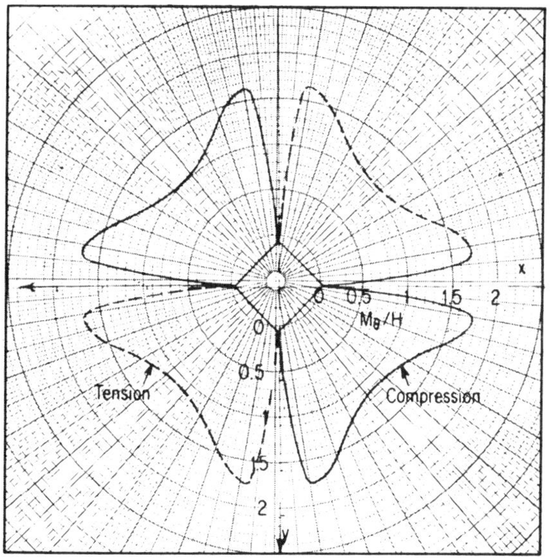

There are eight points of maximum stress concentrations: at $\theta = \pm 10$, 80, 100 and 170 deg., with values of $K = 1.72$. At the center of the sides of the hole, $K = 1.09$ (Fig. 32).

Trianglar hole

33

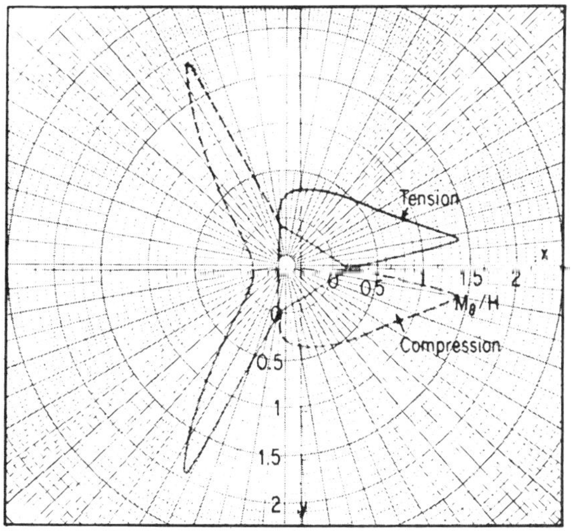

Triangular hole:

$$M_\theta = H\left[\frac{12(1 + \nu)(2\sin\theta + 3\sin 2\theta)}{(3 + \nu)(13 - 12\cos 3\theta)}\right]$$

The highest values occur at $\theta = \pm 120$ deg., where $K = 4.09$, and $\theta = \pm 10$ deg. where $K = 2.49$ (Fig. 33).

TRIANGULAR PLATES

These equations apply to the spherical bending of plates in the shape of equilateral triangles. The plates are subjected to uniformly distributed bending moments M, along the plate edges. Curves are for $\nu = 0.3$ (for steel).

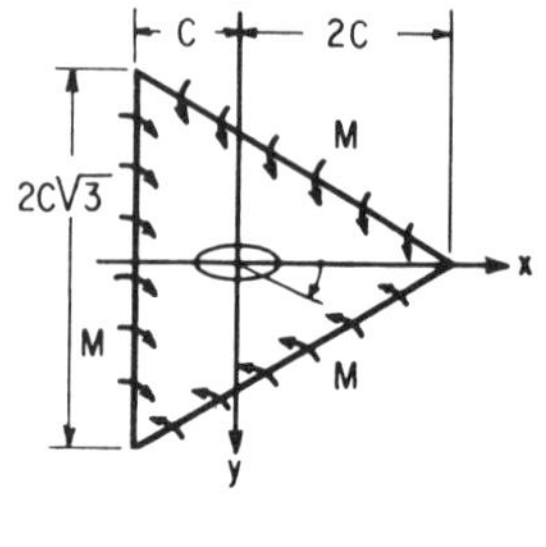

Elliptical hole

34

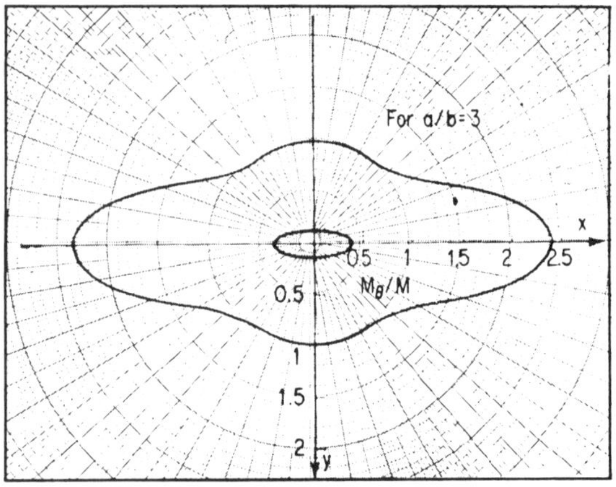

Elliptical hole:

$$M_\theta = M(1+\nu)\left[1 - \frac{2(1-\nu)R/C(m\cos\theta - \cos 3\theta) + 2m(1+\nu)(m - \cos 2\theta)}{(3+\nu)(m^2 - 2m\cos 2\theta + 1)}\right]$$

where $R = \frac{1}{2}(a+b)$. Factor C is distance to geometric center ($C = \frac{1}{3}$ altitude). The chart is for values of $a = 3$, $b = 1$, $C = 30$. Maximum $K = M_\theta/M = 2.4$ at $\theta =$ 0 deg. (Fig. 34).

Circular hole:

$$M_\theta = M\left[(1+\nu) + \left(\frac{2(1-\nu^2)\cos 3\theta}{3+\nu}\right)\left(\frac{R}{C}\right)\right]$$

where R = radius of hole. The chart is for values of $R/C = \frac{1}{10}$. Maximum value of $K = 1.35$ occurs at $\theta = 0$ (Fig. 35).

Circular hole

35

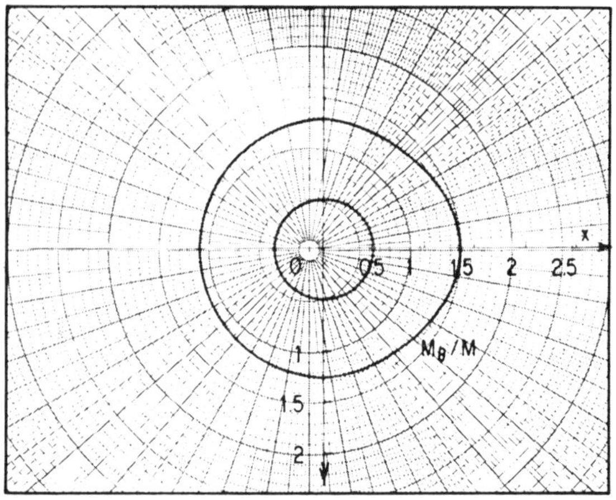

Square hole

36

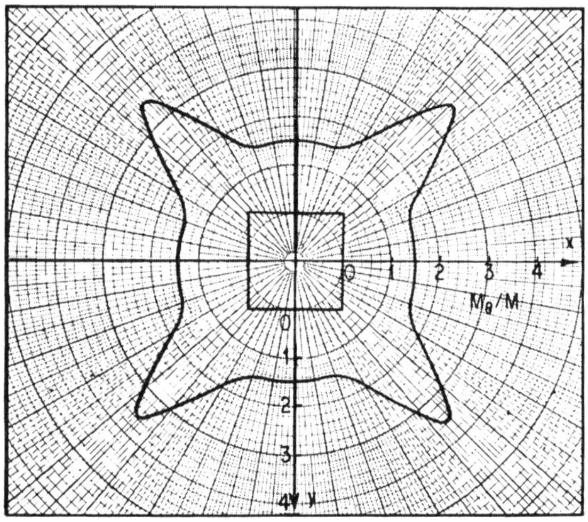

Square hole:

$$M_\theta = M\left[(1+\nu) - \frac{4(1-\nu^2)(R/C)(2\cos 3\theta + \cos\theta) - (1-\nu)^2(4\cos 4\theta + 2)}{(3+\nu)(5 + 4\cos 4\theta)}\right]$$

where R is the abscissa of the hole contour. Assuming $R/C = \frac{1}{10}$, the highest value $K = 2.42$ will occur at $\theta = \pm 135°$. At $\theta = 0°$, $K = 1.00$ and at $\theta = \pm 45°$, $K = 2.231$.

(Fig. 36 is not to scale).

REINFORCED HOLES

A very effective method for reducing the stress concentrations around holes is to insert a ring inside the hole. In other words, the hole can be made slightly larger than desired, say $\frac{5}{8}$ in. instead of $\frac{1}{2}$ in., and a ring can be inserted with a $\frac{1}{16}$ in. thick wall.

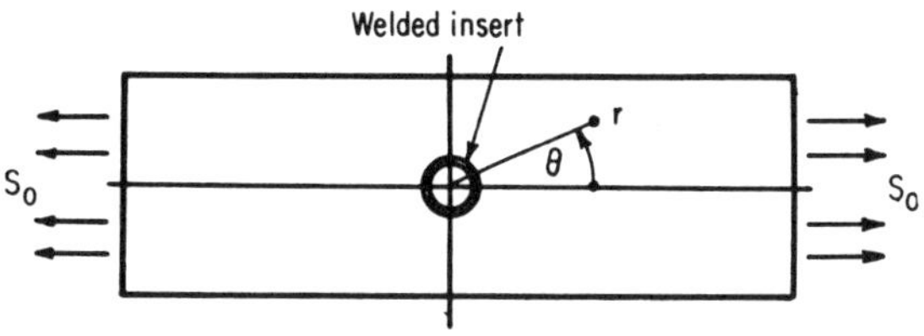

Contrary to some opinions, however, the simple insertion of a ring does not reduce the maximum stress around the hole. The ring must be retained—*and by some means other than with a press fit.* The method chiefly discussed by Savin is that of welding the ring into the hole. This generally requires that the ring protrudes somewhat on one or both sides of the plate to permit welding. The equations, therefore, also predict the stress concentrations around bushings or long rods which are inserted into the hole and welded in place.

The equations for the resulting stress concentrations look quite formidable, but they require merely straight substitution of values. Typical ratios of a ring-to-hole diameter are plotted for copper and steel rings.

Figure 37 shows a ring welded into a circular hole. The plate is stressed by a uniform tensile stress, S_o. Given below are the equations for resulting radial stress, S_R and tangential stress, S_θ, at any radial distance, r, from the center:

$$S_\theta = \frac{S_o}{2}\left[\left(1 + \frac{AR^2}{2r^2}\right) - \left(1 - \frac{3}{2}C\frac{R^4}{r^4}\right)\cos 2\theta\right]$$

37

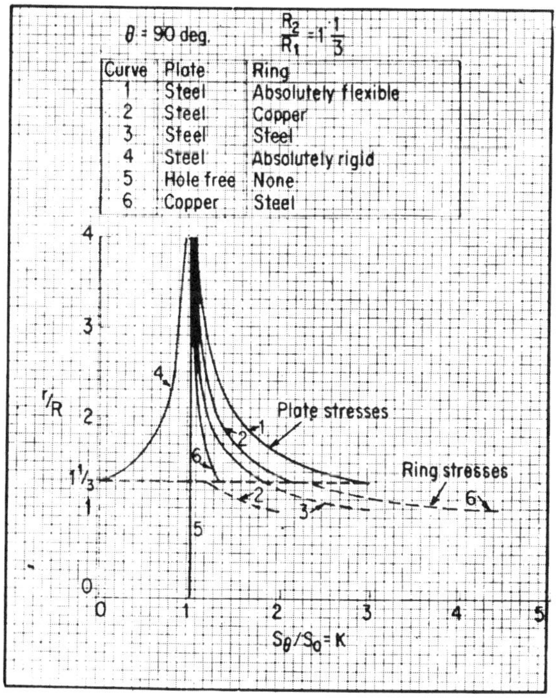

$$S_P = \frac{S_o}{2}\left[\left(1 - 2A\frac{R^2}{r^2}\right) + \left(1 - 2B\frac{R^2}{r^2} - \frac{3}{2}C\frac{R^4}{r^4}\right)\cos 2\theta\right]$$

where R is the outer radius of the ring (the actual size of the hole in the plate), and A, B, and C are constants given by the equations.

$$A = 2 - \frac{2(N^2 - 1)(1 + F)}{2(G - 1) - N^2(G - 2 - F_1G)}$$

$$B = 2 - \frac{2(1 + F)}{D}[(G - 1)(3N^6 - 6N^4 + 4N^2 - 1) + N^6(N^2 - 1)(1 + F_1G)]$$

$$C = -2 + \frac{2(1 + F)}{D}[(G - 1)(4N^6 - 7N^4 + 4N^2 - 1) + N^4(N^4 - 1)(1 + F_1G)]$$

where G = ratio of the modulus of rigidity of plate, G_P, to ring, G_R

$$G = G_P/G_R$$

$$F = \frac{3 - \nu}{1 + \nu}, \text{ for plate material}$$

$$F_1 = \frac{3 - \nu_1}{1 + \nu_1}, \text{ for ring material}$$

$N = R/R_1$, the ratio of outer to inner radius of ring and

$$D = (F + G)N^2[(G - 1)(3N^4 - 6N^2 + 4) + N^6(1 + F_1G)] + (F_1G - F)[(G - 1) + N^6(1 + GF_1)]$$

Absolutely flexible ring: This is as if the ring was omitted. Then $G_R = 0$, and $1/G = 0$.

The constants become: $A = 2$, $B = 2$, $C = -2$.

Absolutely rigid ring: Here $G_R = \infty$; hence $G = 0$. Thus

$$A = 1 - F$$
$$B = -2/F$$
$$C = 2/F$$

Plugged hole of elastic material: Assume that the hole is completely plugged with a dissimilar material that is completely elastic. Then $R_1 = 0$ and $1/N = 0$. Thus

$$A = 2 - \frac{2(1 + F)}{G - 2 - F_1G}$$

$$B = 2 - \frac{2(1 + F)}{F + G}$$

38

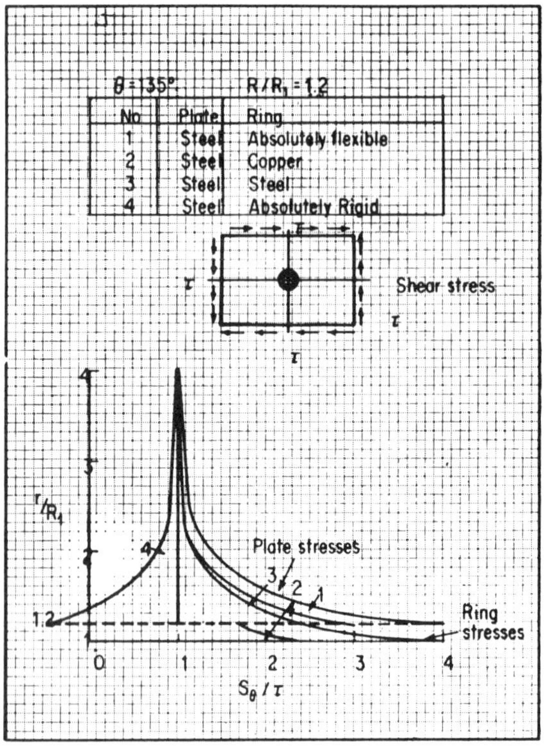

39

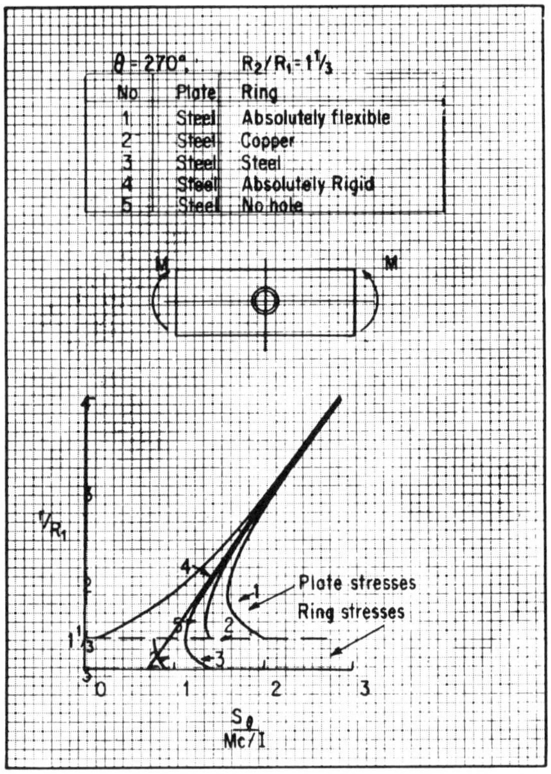

Plugged hole of rigid material: Here $G_P = \infty$ and $R_P = 0$. Thus

$$A = 1 - F \qquad B = -2/F \qquad C = 2/F,$$

which is the same as for the absolutely rigid ring.

Ring and plate of same material: Here $G = 1$, and $F = F_1$. Hence

$$A = B = 2/N^2 \qquad C = -2/N^4$$

Plugged hole of same material: Here $R_1 = 0$, $1/N = 0$, $G = 1$ and $F = F_1$, hence $A = 0$, $B = 0$, $C = 0$. This is as if the hole did not exist.

Comparison of data: Based on the above equations, curves are plotted of the stress concentrations around holes reinforced by rings of various materials (Fig. 37).

A ratio of $R/R_1 = 4/3 = 1\frac{1}{3}$ was taken as a typical size to employ. The stresses are tangential, as these are usually higher than the radial stresses. They are taken at $\theta = 90$ deg. where the maximum values are to be expected.

The ratio S_θ/S_o is the stress concentration factor, K. This is the value to be multiplied to the nominal stress (from $S_o = P/A$) to obtain the actual stress. For a plate without a hole (curve 5) this factor will be unity for all values of r/R_1. For a steel ring in a steel plate (curve 3), $K = 1.2$ at a distance of r/R_1 from the hole center, and 1.85 at the hole edge (at $r/R_1 = 1\frac{1}{3}$). These values are considerably lower than for round unreinforced holes. In one of the previous sections for plates under tension, it was shown that at the edge of the hole at $\theta = 90$ deg., the stress concentration $K = 3$.

The curves below the $r/R_1 = 1\frac{1}{3}$ line (the dashed-lines curves) represent stresses in the ring due to the deformation of the plate.

Figure 38 gives stress curves for plates under pure shear (in torsion); Fig. 39 is for plates under bending (as would occur with a simple-beam loading).

Chap. IX

REFERENCES

Section	*Ref. No.*	
68	1	Savin, G. N.: *Stress Concentration Around Holes*, Pergamon Press, N.Y., 1961
68	2	Griffel, W.: "Plates With Holes," *Product Engineering*, Sept., 1963, Nov., 1963, March, 1964, June, 1962

Chapter X

RIGID FRAMES

69. FORMULAS FOR SYMMETRICAL FRAMES

A SIMPLE METHOD FOR SOLVING CERTAIN TYPES, AND A ROUNDUP OF MOMENT EQUATIONS FOR 31 CASES OF THE MORE COMMON FRAME CONFIGURATIONS

(*Ref. 1, 2, 3*)

Frames are common structural components used in small machine tools and huge structural buildings alike. But the analysis of frames is generally omitted from standard strength-of-materials books. The main reason for this seems to be the complexity of most methods of finding the frame formulas.

What is offered here, therefore, is a compromise. It is a simple technique based on the use of a "dummy" load, which provides the moment equations for rectangular frames under vertical loading (such frames are by far the most common in use). In addition, to round out the designer's file on frames, a compilation is offered of 31 cases of rectangular, trapezoidal, triangular, and single-leg frames under horizontal and vertical loading, based on data from two books on frames (*see* ref.).

Also, with the technique of superposition (a numerical problem is included to show you how) these cases can be combined to solve frames subjected to still more complicated loading.

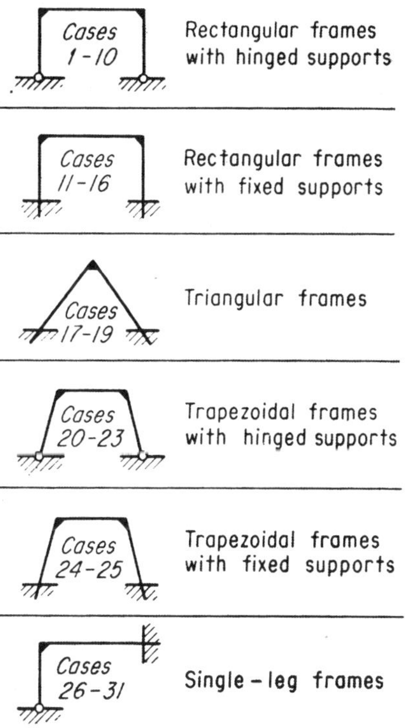

DERIVING THE EQUATIONS

Take two vertical columns, rigidly join them together by a cross tie (the horizontal member), and you have what is known as a rectangular rigid frame. The frame is either hinged or fixed to the support.

Now look at the hinged frame, shown in diagram *A*. Here it is assumed that the columns have a constant moment of inertia, I_1, and that the moment of inertia of the cross tie, I_2, is also constant throughout its length. Width of the frame, L, and height, h, determine the overall dimensions. The frame in this case is loaded by means of a point load, P. The forces that prevent the joints from separating are the horizontal reactions H_1 and H_2.

In employing the elastic energy method for analyzing a statically indeterminate structure such as a frame, one of the supports, joint 4, is assumed able to move freely in a horizontal direction, as shown in (*B*). Hence $H_2 = 0 = H_1$ and the structure becomes statically deter-

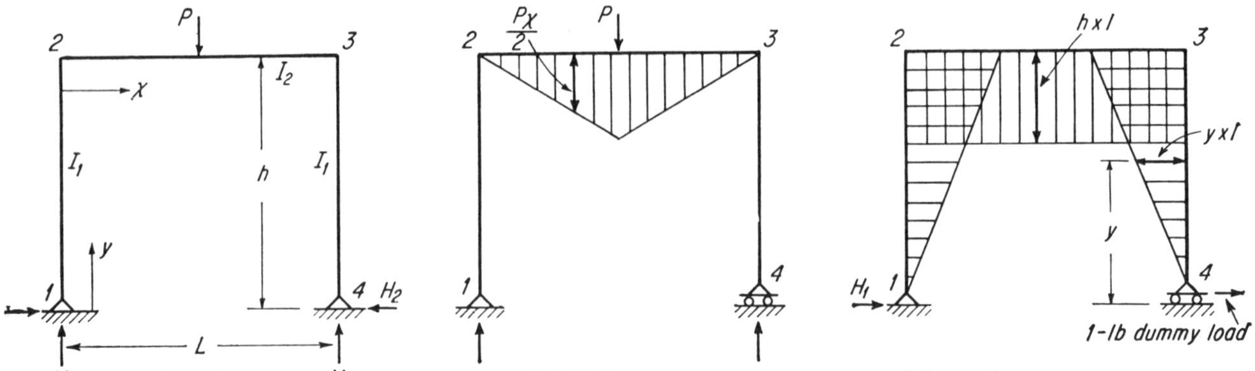

(A) Loading diagram (B) M-diagram (C) m-diagram

minate. The ends of the cross tie react as if they are part of a simple beam and you obtain the moment diagram shown under the cross tie. Here M is the moment at any point (located by distance x from joint 2) with the redundant force H removed (Diagrams A to C and subsequent diagrams show positive directions of forces and moment.)

Now, with the actual load (load P) acting, apply a hypothetical 1-lb. load (a dummy load) at the point where, and in line where, the deflection would occur. The actual direction of the load, right or left, does not matter because this is taken care of in the signs of the resulting moments, which are shown in Diagram C. Moment m, therefore, is the moment at any point, located by x, resulting from the dummy load's action at point where deflection occurs.

The theory of elastic energy then says that the deflection at joint 4, due to the applied load P, and with force H removed is:

$$y_1 = \int \frac{Mm}{EI}\,dx$$

and that the deflection of joint 4 produced by H is:

$$y_2 = -H \int \frac{m^2}{EI}\,dx$$

Because joint 4 does not move, $y_1 + y_2 = 0$, and hence

$$\frac{1}{EI}\int Mm\,dx - \frac{H}{EI}\int m^2\,dx = 0$$

or

$$H = \frac{1/EI \int Mm\,dx}{1/EI \int m^2\,dx}$$

What now remains is to evaluate M and m. From Diagram C, the value of m for a section of the column is

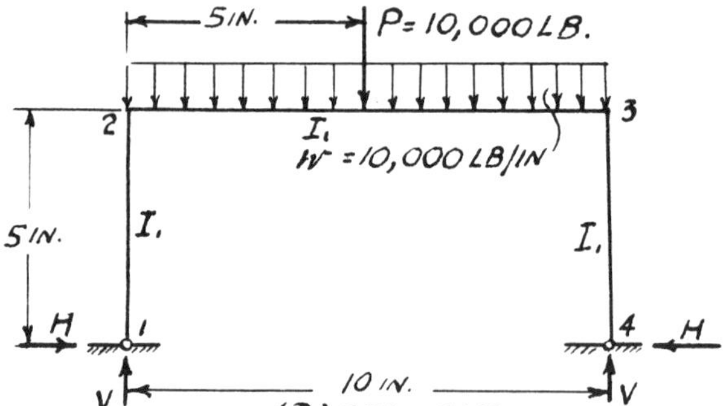

(D) EXAMPLE

equal to the distance of the section from the base, or $m = y$. The value for m for the cross tie is constant and equal to h.

Referring now to the values for M, these are zero for the frame columns and for the cross tie equal to the moment diagram for a simply supported beam. Hence

$$H = \frac{2\int_0^{L/2} \frac{Phx}{2EI_2\,dx}}{2\int_0^h y^2 \frac{dx}{EI_1} + \int_0^L h^2 \frac{dx}{EI_2}} = \frac{\frac{PL^2}{8}\,h/I_2}{\frac{2h^3}{3I_1} + \frac{Lh^2}{I_2}}$$

In the above equation, we recognize that $PL^2/8$ is the area under bending moment diagram of a simply supported beam. Thus, by letting

$$A = \frac{PL^2}{8} \text{ and } C = \frac{I_2}{I_1}\left(\frac{h}{L}\right)$$

we have

$$H = \frac{3A}{hL(2C+3)} \tag{1}$$

All other reactions, moments, and forces may be found by the usual equations of equilibrium.

Similarly for a built-in frame:

$$H = \frac{3A}{hL(C+2)} \tag{2}$$

Note that, from equations (1) and (2), the analysis of a symmetrical frame for a given loading condition depends only on the quantity A, the equation for which may be obtained from any standard strength-of-materials book.

In the derivation of the above equations, the distortion due to axial and shear forces is neglected.

Example

A frame with hinged supports (Diagram D) is made of 2-in. standard pipe and has concentrated and distributed loads. Find the maximum stresses. Thus $P = 10{,}000$ lb., $w = 1{,}000$ lb./in., $h/L = 5/10 = 0.5$.

This involves Cases 1 and 3.

FOR THE CROSS TIE, CASE 1

$$Mp = \frac{PL}{4} + M_2 = \frac{PL}{4}\left[1 - \frac{3}{2[3 + 2(h/L)]}\right]$$
$$= \frac{100{,}000}{4}\left[1 - \frac{3}{8}\right] = 15{,}600 \text{ lb.-in.}$$

FOR THE CROSS TIE, CASE 3

$$M_w = \frac{wL^2}{8} + M_2 = wL^2\left[\frac{1}{8} - \frac{1}{4[3 + 2(h/L)]}\right]$$
$$= 100{,}000(0.125 - 0.0625) = 6{,}300 \text{ lb.-in.}$$

Total moment due to P and w is 21,900 lb.-in. In addition, the moments H_1h and M_1 should be taken into consideration. Thus, the total moment is equal to

$$M_{\max} = 21{,}900 - Hh$$

Case 1:

$$H = \frac{3PL}{8h[3 + 2(h/L)]} = 1{,}875 \text{ lb./in.}$$

Case 3:

$$H = \frac{wL^2}{4h[3 + 2(h/L)]} = 1{,}250 \text{ lb./in.}$$

Total $H = 1{,}875 + 1{,}250 = 3{,}125$ lb./in.; hence

$$M_{\max} = 21{,}900 - 3{,}125(5)$$
$$= 6{,}300 \text{ lb.-in.}$$

A section of 2-in. standard pipe has an area of 1.075 in.^{2} and a section modulus of 0.561 in.^{3}

Thus, the bending stress at mid-span

$$S_B = \frac{M}{Z} = \frac{6{,}300}{0.561} = 11{,}230 \text{ psi}$$

For the frame columns, the maximum stress in the column is a combined axial and bending stress. The axial compressive stress is

$$S_A = \frac{10{,}000 + 10{,}000}{2(1.075)} = 9{,}300 \text{ psi}$$

The stress due to bending is again due to

$$S_B = \frac{H_1h}{(0.561)} = 27{,}850 \text{ psi}$$

Hence the maximum compressive stress in one column is

$$S_{\max} = 9{,}300 + 27{,}850$$
$$= 37{,}150 \text{ psi}$$

A tabulation of formulas for 31 cases of the most common frame configurations follows.

Frame constants and variations

$$F = \frac{I_2}{I_1}\left(\frac{s}{b}\right) \qquad C = \frac{I_2}{I_1}\left(\frac{h}{L}\right)$$

$$G = 3 + 2F \qquad D = 2 + C$$

$$J = 4C + 3 \qquad E = 1 + 6C$$

$$K = \frac{1}{3 + 2C} \qquad E_2 = F\left(1 - \frac{u^2}{h^2}\right)$$

h = frame height, in.
H = horizontal component of frame reaction, lb
I_1 = moment of inertia of the cross section of the frame, in.[4]; I_2 = for cross tie
L = span of frame
M_2, M_3 = bending moment at joints 2 and 3, lb-in.
M_m = maximum bending moment, lb-in.
M_x, M_y = bending moment at any section of frame, located by distance x or y, lb-in.
P = applied point load, lb
Q = shear load, lb
V = vertical component of frame reaction, lb
w = applied distributed load, lb/in.
x = distance from joint to point of interest

Rectangular frames with hinges supports—horizontal loading

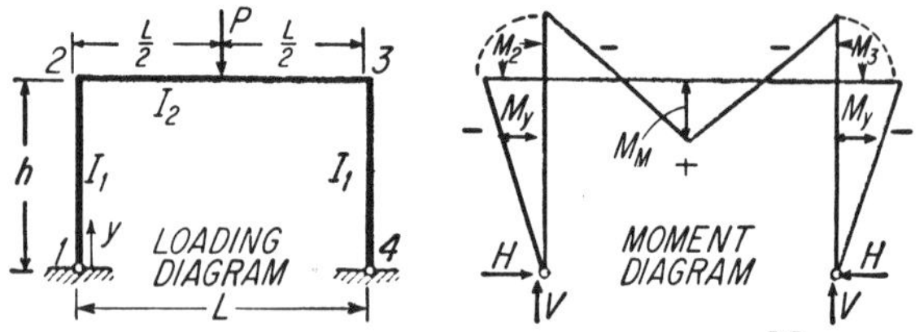

1. $M_2 = M_3 = -\frac{3}{8}PLK \qquad H = -\frac{M_2}{h}$

$$V = \frac{P}{2} \qquad M_{max} = M_m = \frac{PL}{4} + M_2$$

At any section of column: $M_y = M_2y/h$

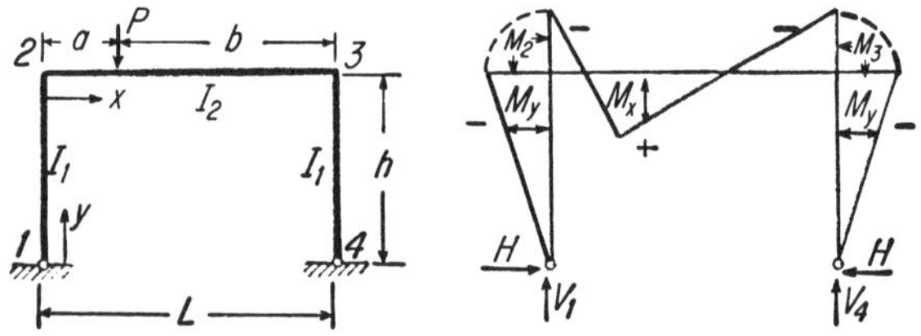

2. $M_2 = M_3 = -\frac{3PKab}{2L} \qquad M_y = M_2\frac{y}{h}$

$$V_1 = P\left(1 - \frac{a}{L}\right) \qquad V_4 = \frac{Pa}{L} \qquad H = -\frac{M_2}{h}$$

When $x \leqq a$, $M_x = \frac{Pbx}{L} + M_2$

When $x > a$, $M_x = Pa\left(1 - \frac{x}{L}\right) + M_2$

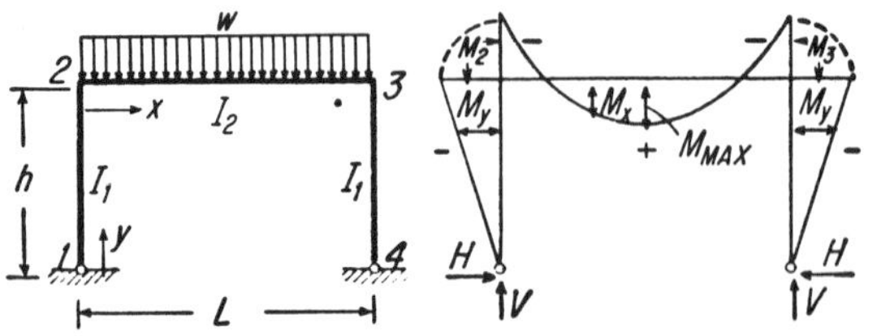

3. $M_2 = M_3 = -\frac{1}{4}wL^2K \qquad M_y = M_2\frac{y}{h}$

$$H = \frac{wL^2K}{4h} \qquad V = \frac{1}{2}wL$$

$$M_x = \frac{wLx}{2}\left(1 - \frac{x}{L}\right) + M_2$$

$$M_m = \frac{wL^2}{8} + M_2$$

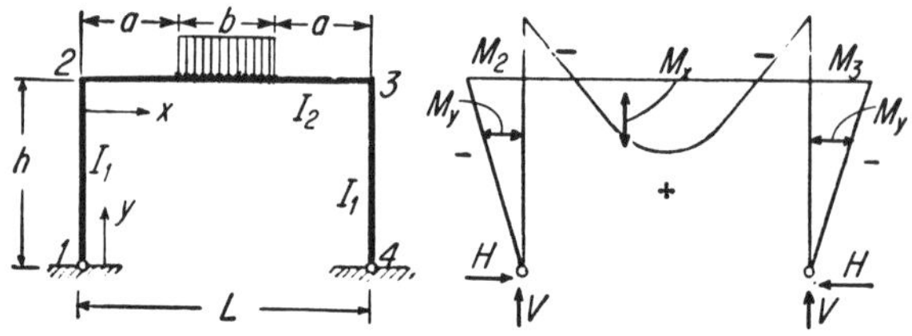

4. $M_2 = M_3 = -\frac{wbK(3L^2 - b^2)}{8L} \qquad M_y = \frac{M_2y}{h}$

$$H = -\frac{M_2}{h} \qquad V = \frac{wb}{2}$$

When $x \leqq a$, $M_x = \frac{wbx}{2} + M_2$

When $a < x \leqq \frac{L}{2}$,

$$M_x = \frac{wb}{2}\left[x - \frac{(x - a)^2}{b}\right] + M_2$$

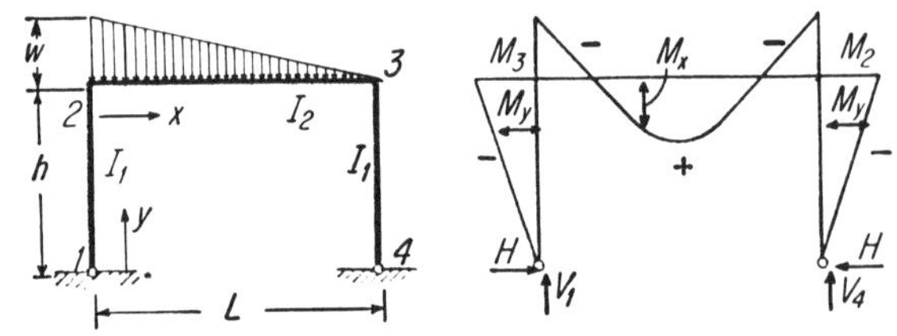

5. $M_2 = M_3 = -\frac{1}{8}wL^2K \qquad M_y = M\frac{y}{h}$

$$H = -\frac{M_2}{h} \qquad V_1 = \frac{1}{3}wL \qquad V_4 = \frac{1}{6}wL$$

$$M_x = \frac{wx}{6L}\left[(L - x)(2L - x)\right] + M_2$$

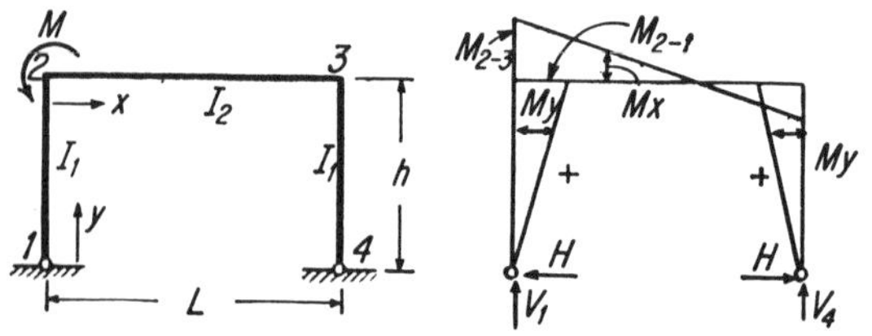

6. $M_{2\text{-}1} = M_3 = \frac{3}{2}MK \qquad M_{2\text{-}3} = M_{2\text{-}1} - M$

$$H = -\frac{M_2}{h} \qquad V_1 = \frac{M}{L} \qquad V_4 = -V_1$$

$$M_x = M_{2\text{-}3}\left(1 - \frac{x}{L}\right) + M_3\frac{x}{L} \qquad M_y = M_{2\text{-}1}\frac{h}{L}$$

Rectangular frames with hinges supports—vertical loading.

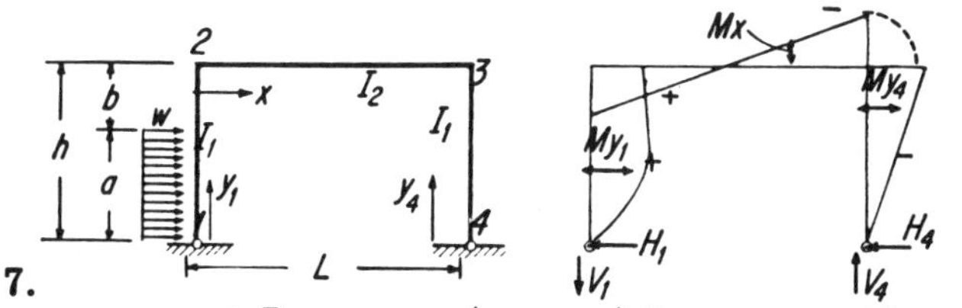

7.

$$M_2,\ M_3 = \frac{wa^2}{4}\left[\pm 1 - CK\left(1 - \frac{a^2}{2h^2}\right)\right] \qquad H_4 = -\frac{M_3}{h}$$

Use plus sign for M_2, minus sign for M_3

$$H_1 = H_4 - wa \qquad V_4 = \frac{wa^2}{2L} = -V_1$$

When $y_1 \leqq a$, $M_{y1} = \dfrac{way_1}{2}\left(\dfrac{b}{h} + \dfrac{a - y_1}{a}\right) + M_2\dfrac{y_1}{h}$

When $y_1 > a$, $M_{y1} = \dfrac{wa^2}{2}\left(1 - \dfrac{y_1}{h}\right) + M_2\dfrac{y_1}{h}$

$$M_{y4} = M_3\frac{y_4}{h} \qquad M_x = M_2\left(1 - \frac{x}{L}\right) + M_3\frac{x}{L}$$

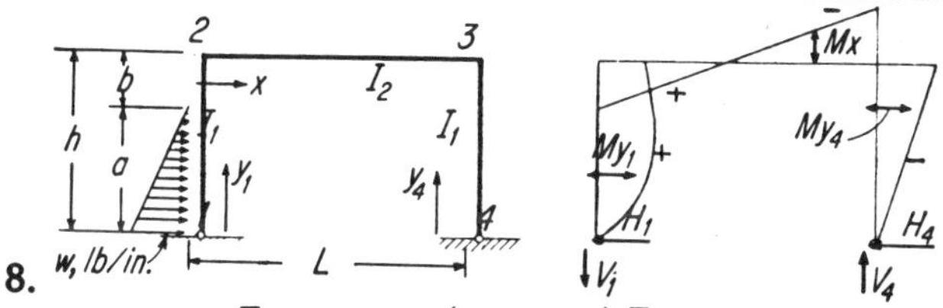

8.

$$M_2,\ M_3 = \frac{wa^2}{12}\left[\pm 1 - \frac{CK\left(10 - 3\dfrac{a^2}{h^2}\right)}{10}\right] \qquad H_4 = -\frac{M_3}{h}$$

Use plus sign for M_2, minus sign for M_3

$$H_1 = H_4 - \frac{wa}{2} \qquad V_4 = \frac{wa^2}{6L} \qquad V_1 = -V_4$$

When $y_1 \leqq a$,

$$M_{y1} = \frac{way_1}{6}\left[\frac{b}{h} + \frac{(a - y_1)(2a - y_1)}{a^2}\right] + M_2\frac{y_1}{h}$$

When $y_1 > a$, $M_{y1} = \dfrac{wa^2}{6}\left(1 - \dfrac{y_1}{h}\right) + M_2\dfrac{y_1}{h}$

$$M_{y4} = M_3\frac{y_4}{h} \qquad M_x = M_2\left(1 - \frac{x}{L}\right) + M_3\frac{x}{L}$$

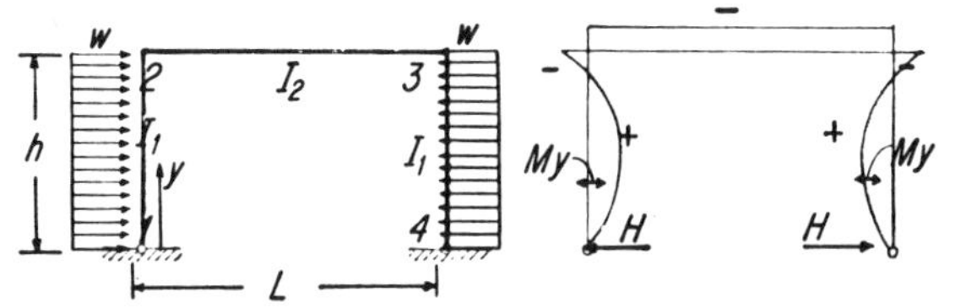

9.

$$M_2 = M_3 = -\tfrac{1}{4}wh^2CK \qquad V_1 = V_4 = 0$$

$$M_y = \frac{why}{2} + M_2\frac{y}{h} \qquad H = -\frac{wh}{2} - \frac{M_2}{h}$$

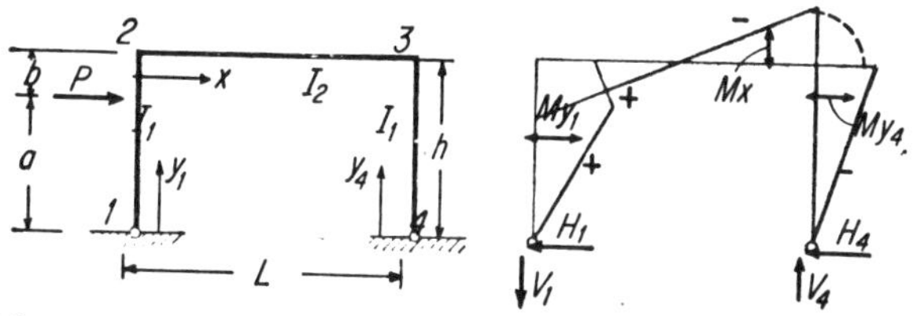

10.

$$M_2,\ M_3 = \frac{Pa}{2}\left[\pm 1 - CK\left(1 - \frac{a^2}{h^2}\right)\right]$$

Use plus sign for M_2, minus sign for M_3

$$H_4 = -\frac{M_3}{h} \qquad H_1 = H_4 - P$$

$$V_4 = \frac{Pa}{L} \qquad V_1 = -V_4$$

When $y_1 \leqq a$ $\quad M_{y1} = \dfrac{y_1}{h}(M_2 + Pb)$

When $y_1 > a$ $\quad M_{y1} = Pa\left(1 - \dfrac{y_1}{h}\right) + M_2\dfrac{y_1}{h}$

$$M_{y4} = M_3\frac{y_4}{h} \qquad M_x = M_2\left(1 - \frac{x}{L}\right) + M_3\frac{x}{L}$$

Rectangular frames with fixed supports

Rectangular frames with fixed supports—simple loading.

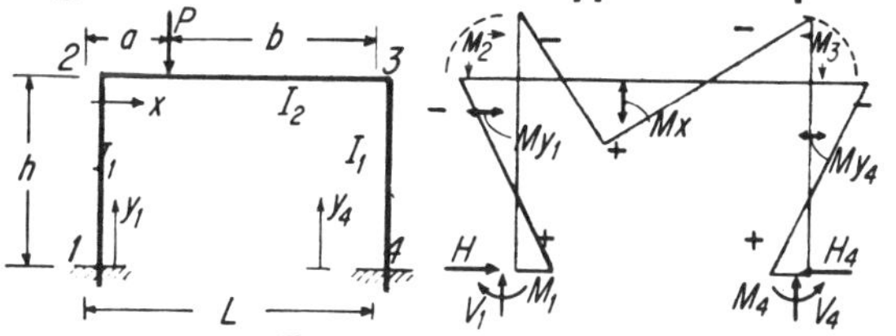

11.

$$M_1,\ M_4 = \frac{Pab}{2L}\left[\frac{1}{D} \pm \frac{L - 2a}{LE}\right] \qquad H = \frac{3Pab}{2hD}$$

For $\pm$ signs: $M_1 = -$; $\ M_4 = +$

$$M_2,\ M_3 = -\frac{Pab}{L}\left[\frac{1}{D} \pm \frac{L - 2a}{2LE}\right]$$

Where $M_2 = +$; $\ M_3 = -$.

$$V_4 = \frac{Pa}{L}\left[1 - \frac{b - (2ab/L)}{LE}\right] \qquad V_1 = P - V_4$$

When $x \leqq a$, $M_x = (Pb + M_3)\dfrac{x}{L} + M_2\left(1 - \dfrac{x}{L}\right)$

When $x > a$, $M_x = (Pa + M_2)\left(1 - \dfrac{x}{L}\right) + M_3\dfrac{x}{L}$

$$M_{y1} = M_2\frac{y_1}{h} + M_1\left(1 - \frac{y_1}{h}\right)$$

$$M_{y4} = M_4\left(1 - \frac{y_4}{h}\right) + M_3\frac{y_4}{h}$$

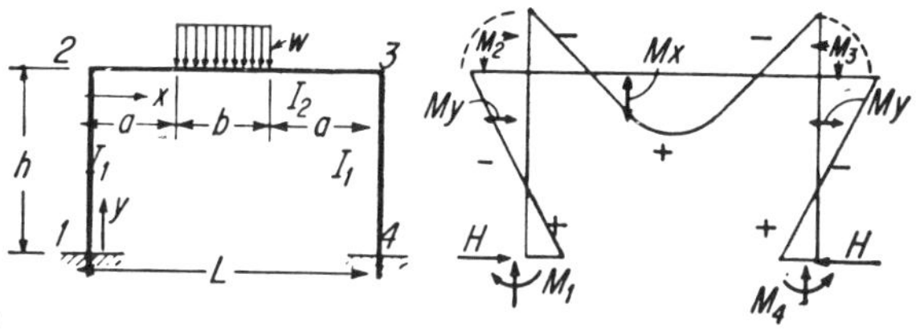

12.

$$M_1 = M_4 = \frac{wb\,(3L^2 - b^2)}{24LD} \qquad H = \frac{3M_1}{h}$$

$$M_2 = M_3 = -2M_1 \qquad V = \frac{wb}{2}$$

When: $x \leqq a$, $\quad M_x = \dfrac{wbx}{2} + M_2$

$a < x \leqq a + b$, $\quad M_x = \dfrac{wb}{2}\left[x - \dfrac{(x - a)^2}{b}\right] + M_2$

$x > a + b$, $\quad M_x = \dfrac{wb}{2}(L - x) + M_2$

$$M_y = M_2\frac{y}{h} + M_1\left(1 - \frac{y}{h}\right)$$

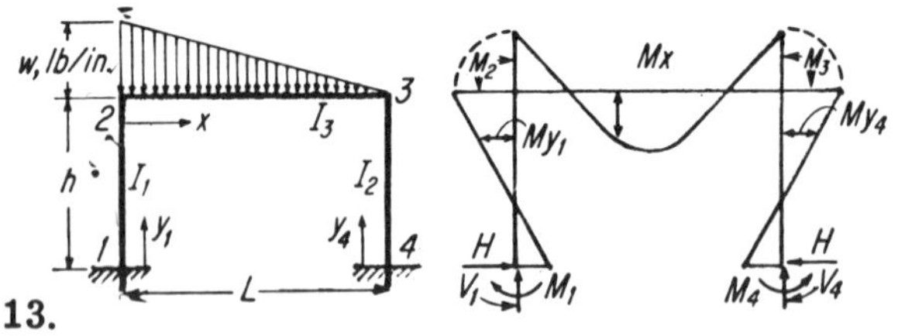

13.

$$M_1,\ M_4 = \frac{wL^2}{24}\left[\frac{1}{D} \pm \frac{1}{5E}\right]$$

$$M_2,\ M_3 = -\frac{wL^2}{12}\left[\frac{1}{D} \pm \frac{1}{10E}\right]$$

For $\pm$ signs: $M_1 = -$; $M_2 = +$; $M_3 = -$; $M_4 = +$

$$H = \frac{wL^2}{8hD} \qquad V_4 = \frac{wL}{6}\left[1 - \frac{1}{10E}\right]$$

$$V_1 = \frac{wL}{2} - V_4$$

$$M_x = \frac{wx}{6L}(L - x)(2L - x) + M_2\left(1 - \frac{x}{L}\right) + M_3\frac{x}{L}$$

$$M_{y1} = M_2\frac{y_1}{h} + M_1\left(1 - \frac{y_1}{h}\right)$$

$$M_{y4} = M_4\left(1 - \frac{y_4}{h}\right) + M_3\frac{y_4}{h}$$

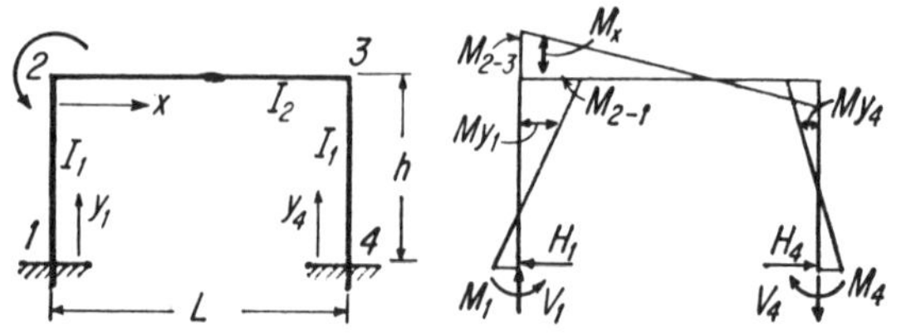

14.

$$M_1,\ M_4 = -\frac{M}{2}\left(\frac{1}{D} \pm \frac{1}{E}\right)$$

$$M_{2\text{-}1},\ M_3 = M\left[\frac{1}{D} \pm \frac{1}{2E}\right]$$

Signs: $M_1 = -$; $M_4 = +$ Signs: $M_{2\text{-}1} = +$; $M_3 = -$

$$M_{2\text{-}3} = M_{2\text{-}1} - M \qquad H = -\frac{3M}{2hD}$$

$$V_1 = -V_4 = \frac{6MC}{LE}$$

$$M_x = M_{2\text{-}3}\left(1 - \frac{x}{L}\right) + M_3\frac{x}{L}$$

$$M_{y1} = M_1\left(1 - \frac{y_1}{h}\right) + M_{2\text{-}1}\frac{y_1}{h};$$

$$M_{y4} = M_4\left(1 - \frac{y_4}{h}\right) + M_3\frac{y_4}{h}$$

Note: $M_{2\text{-}3}$ = moment acting on left end of member 2-3.

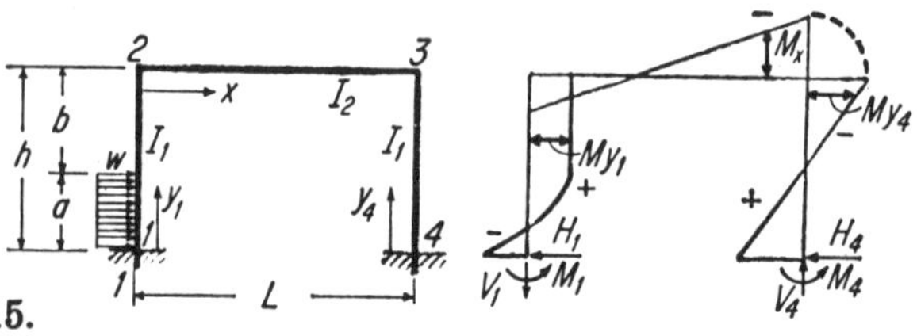

15.

$$M_1,\ M_4 = \frac{wa^3}{h}\left\{\frac{C}{24D}\left[\left(8 + \frac{6}{C}\right) - \frac{3a}{h}\left(1 + \frac{3}{C}\right) - \frac{6h}{a}\right] \pm \left[\frac{h}{4a} - \frac{C}{2E}\right]\right\}$$

For $\pm$ sign: $M_1 = -$; $M_4 = +$

$$M_2,\ M_3 = -\frac{wa^3}{h}\left[\frac{C}{24D}\left(4 - 3\frac{a}{h}\right) \pm \frac{1}{2E}\right]$$

For $\pm$ sign: $M_2 = -$; $M_3 = +$

$$H_4 = \frac{wa^2}{h}\left\{\frac{1}{4} + \frac{C}{24D}\left[6 - 3\frac{a^2}{h^2} - \left(2 - \frac{a}{h}\right)^2\left(3 + \frac{3}{C}\right)\right]\right\}$$

$$H_1 = H_4 - wa \qquad V_4 = -V_1 = \frac{wa^3C}{hLE}$$

When $y \leqq a$, $$M_{y1} = \frac{way_1}{2}\left(\frac{b}{h} + \frac{a - y_1}{a}\right) + M_1\left(1 - \frac{y_1}{h}\right) + M_2\frac{y_1}{h}$$

$y > a$, $$M_{y1} = \left(\frac{wa^2}{2} + M_1\right)\left(1 - \frac{y_1}{h}\right) + M_2\frac{y_1}{h}$$

$$M_{y4} = M_4\left(1 - \frac{y_4}{h}\right) + M_3\frac{y_4}{h}$$

$$M_x = M_2\left(1 - \frac{x}{L}\right) + M_3\frac{x}{L}$$

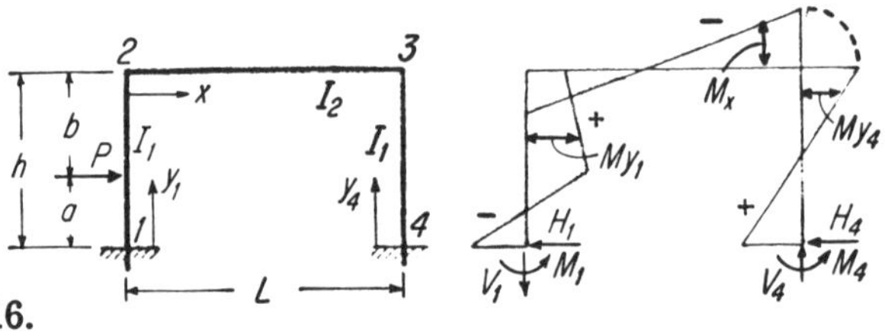

16.

$$M_1,\ M_4 = -\frac{Pa}{2}\left[\frac{bC}{h^2D}\left(b + \frac{h + b}{C}\right) \pm \frac{3C\left(1 - \frac{b}{h}\right)}{E} \pm 1\right]$$

Signs: $M_1 = + -$; $M_4 = - +$

$$M_2,\ M_3 = -\frac{Pa}{2}\left[\frac{abC}{h^2D} \pm \frac{3C\left(1 - \frac{b}{h}\right)}{E}\right]$$

Signs: $M_2 = -$; $M_3 = +$

$$H_4 = \frac{Pa}{2h}\left[\frac{bC}{hD}\left(a - b + \frac{h + b}{C}\right) + 1\right]$$

$$V_4 = -V_1 = \frac{3aCP\left(1 - \frac{b}{h}\right)}{LE} \qquad H_1 = H_4 - P$$

When $y \leqq a$;

$$M_{y1} = (M_2 + Pb)\frac{y_1}{h} + M_1\left(1 - \frac{y_1}{h}\right)$$

$y > a$;

$$M_{y1} = (M_1 + Pa)\left(1 - \frac{y_1}{h}\right) + M_2\frac{y_1}{h}$$

$$M_{y4} = M_4\left(1 - \frac{y_4}{h}\right) + M_3\frac{y_4}{h}$$

$$M_x = M_2\left(1 - \frac{x}{L}\right) + M_3\frac{x}{L}$$

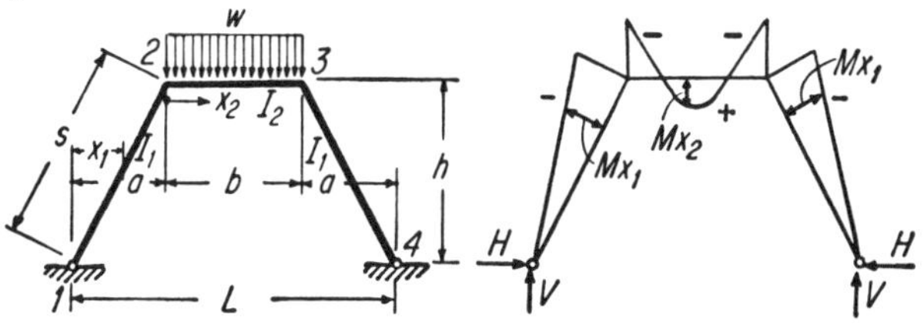

20.

$$M_2 = M_3 = -\frac{wb^2}{4G} \qquad H = \frac{wb}{2h}\left[a + \frac{b}{4G}\right]$$

$$M_{x1} = M_2\frac{x_1}{a} \qquad M_{x2} = \frac{wx_2}{2}[b - x_2] + M_2$$

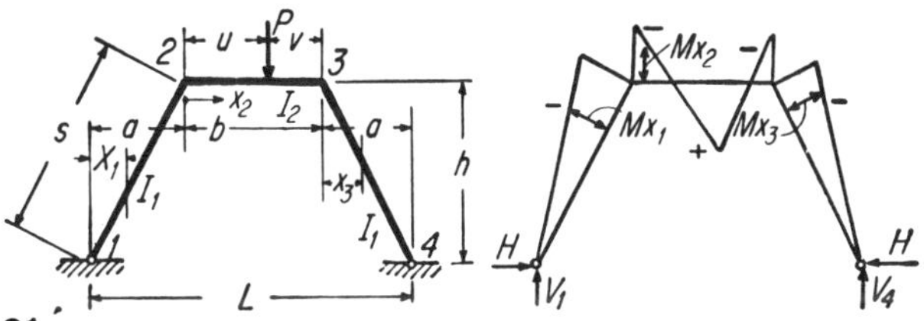

21.

$$M_2, M_3 = -\frac{P}{2}\left[\frac{3uv}{bG} \pm \frac{a(b-2u)}{L}\right]$$

For $\pm$ sign: $M_2 = -$; $M_3 = +$

$$H = \frac{P}{2h}\left[a + \frac{3uv}{bG}\right] \qquad V_1 = \frac{P}{2}\left(1 + \frac{b-2u}{L}\right)$$

$$V_4 = P - V_1 \quad M_{x1} = M_2\frac{x_1}{a} \quad M_{x2} = M_3\left(1 - \frac{x_3}{a}\right)$$

When $x_2 \leqq u$,

$$M_{x2} = (Pv + M_3)\frac{x_2}{b} + M_2\left(1 - \frac{x_2}{b}\right)$$

$x_2 > u$,

$$M_{x2} = (Pu + M_2)\left(1 - \frac{x_2}{b}\right) + M_3\frac{x_2}{b}$$

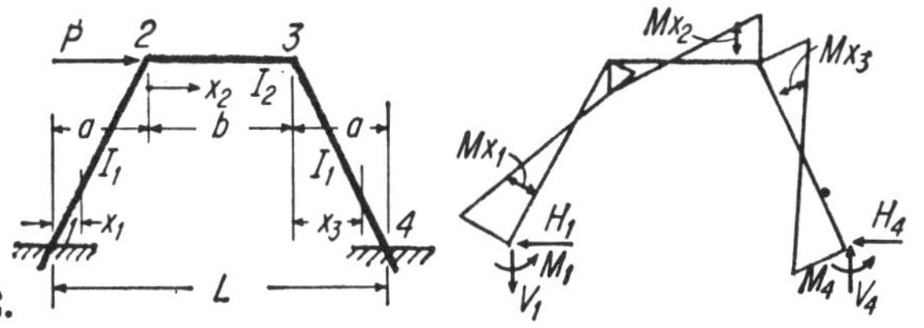

22.

$$\text{Let } z = \frac{\frac{1}{F} + 2 + \frac{L}{b}}{\frac{1}{F} + 2\left(1 + \frac{L}{b} + \frac{L^2}{b^2}\right)} \qquad M_3 = -\frac{Phb}{2L}[1 - z]$$

$$M_1 = -\frac{Phz}{2} \quad M_4 = \frac{Phz}{2} \quad M_2 = \frac{Phb}{2L}[1 - z]$$

$$H_1 = -\frac{P}{2} \quad H_4 = \frac{P}{2} \quad V_4 = -V_1 = \frac{Ph}{L}[1 - z]$$

$$M_{x1} = M_1\left(1 - \frac{x_1}{a}\right) + M_2\frac{x_1}{a}$$

$$M_{x2} = M_2\left(1 - \frac{x_2}{b}\right) + M_3\frac{x_2}{b}$$

$$M_{x3} = M_3\left(1 - \frac{x_3}{a}\right) + M_4\frac{x_3}{a}$$

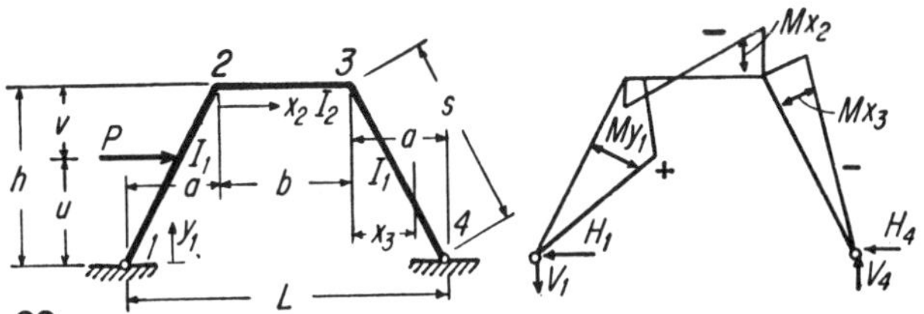

23.

$$M_2, M_3 = -\frac{Pu}{2}\left[\frac{E_2}{G} \pm \frac{b}{L}\right]$$

$$V_4 = -V_1 = \frac{Pu}{L}$$

For $\pm$ sign: $M_2 = -$; $M_3 = +$

$$H_4 = \frac{Pu}{2h}\left[1 + \frac{E_2}{G}\right] \qquad H_1 = H_4 - P$$

When $y_1 \leqq u$, $M_{y1} = (Pv + M_2)\frac{y_1}{h}$

When $y_1 > u$, $M_{y1} = Pu\left(1 - \frac{y_1}{h}\right) + M_2\frac{y_1}{h}$

$$M_{x2} = M_2\left(1 - \frac{x_2}{b}\right) + M_3\frac{x_2}{b}$$

$$M_{x3} = M_3\left(1 - \frac{x_3}{a}\right)$$

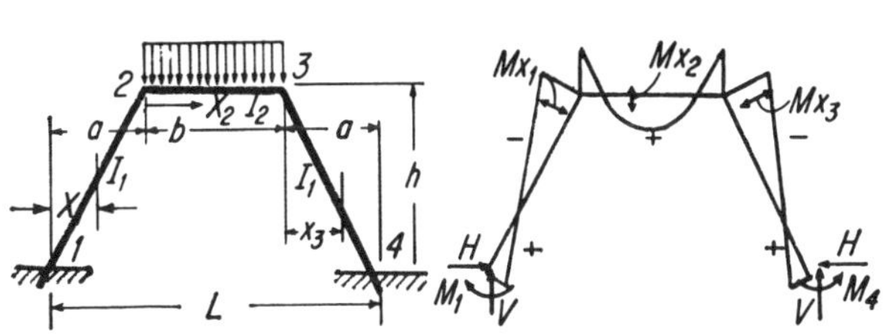

24.

$$M_1 = M_4 = \frac{wb^2}{12(2 + F)} \qquad M_2 = M_3 = -\frac{wb^2}{6(2 + F)}$$

$$H = \frac{wb}{2h}\left[a + \frac{b}{4 + 2F}\right] \qquad V = \frac{wb}{2}$$

$$M_{x1} = M_{x3} = M_1\left(1 - \frac{x_1}{a}\right) + M_2\frac{x_1}{a}$$

$$M_{x2} = \frac{wbx_2}{2}\left(1 - \frac{x_2}{b}\right) + M_2$$

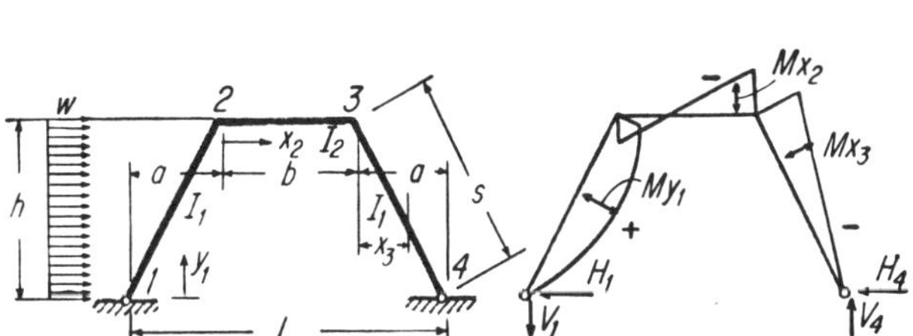

25.

$$M_2, M_3 = -\frac{wh^2}{4}\left[\frac{F}{2G} \pm \frac{b}{L}\right] \qquad V_4 = \frac{wh^2}{2L} = -V_1$$

For $\pm$ sign: $M_2 = -$; $M_3 = +$

$$H_4 = \frac{wh}{4}\left[1 + \frac{F}{2G}\right] \qquad H_1 = H_4 - wh$$

$$M_{y1} = \frac{why_1}{2}\left(1 - \frac{y_1}{h}\right) + M_2\frac{y_1}{h}$$

$$M_{x2} = M_2\left(1 - \frac{x_2}{b}\right) + M_3\frac{x_2}{b}$$

$$M_{x3} = M_3\left(1 - \frac{x_3}{a}\right)$$

Triangular frames

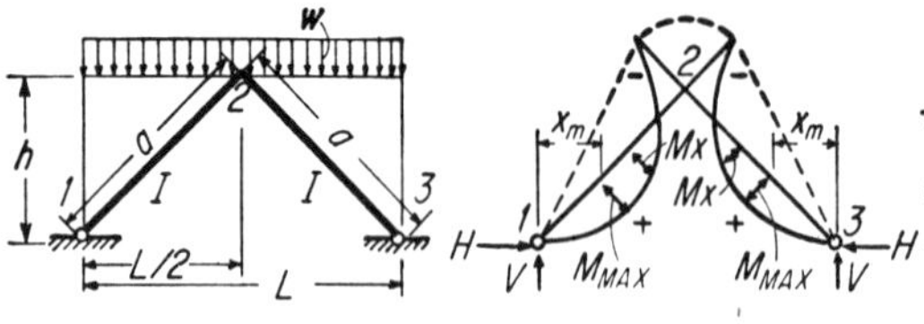

17.

$$M_2 = -\frac{wL^2}{32} \qquad M_x = \frac{wLx}{2}\left(\frac{3}{8} - \frac{x}{L}\right) \qquad x_m = \frac{3L}{16}$$

$$M_{max} = \frac{9wL^2}{512} \qquad V = \frac{wL}{2} \qquad H = \frac{5wL^2}{32h}$$

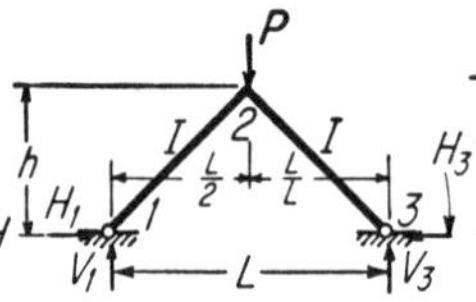

18.

$$V_1 = V_3 = \frac{P}{2}$$

$$H_1 = H_3 = \frac{PL}{4h}$$

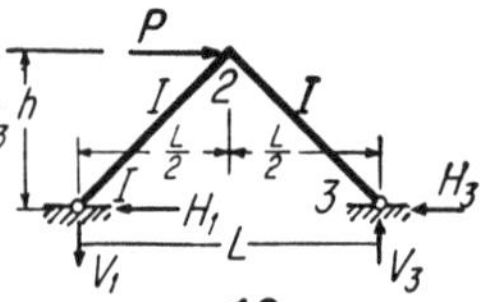

19.

$$H_1 = -H_3 = -\frac{P}{2}$$

$$V_1 = -V_3 = -\frac{Ph}{L}$$

Single-leg frames

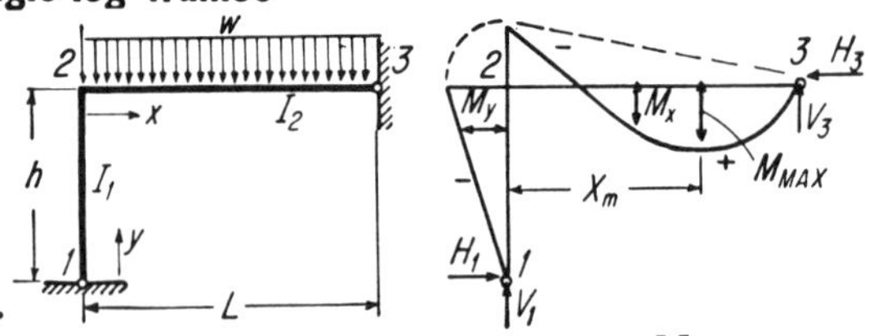

26.

$$M_2 = \frac{wL^2}{8(C+1)} \qquad H_1 = H_3 = \frac{M_2}{h}$$

$$V_1 = \frac{wL}{2} - \frac{M_2}{L} \qquad V_3 = \frac{wL}{2} + \frac{M_2}{L}$$

$$M_x = (L-x)\left[\frac{wx}{2} + \frac{M_2}{L}\right] \qquad x_m = L - \frac{V_3}{w}$$

$$M_{max} = \frac{V_3}{2}(L - x_m) \qquad M_y = \frac{y}{h} M_2$$

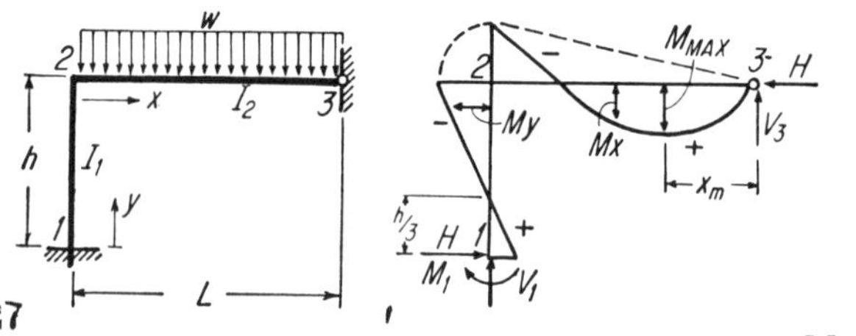

27

$$M_2 = -\frac{wL^2}{2J} \qquad M_1 = -\frac{M_2}{2} \qquad H = \frac{3M_1}{h}$$

$$V_1 = \frac{wL}{2} - \frac{M_2}{L} \qquad V_3 = wL - V_1$$

$$M_{max} = \frac{V_3 x_m}{2} \qquad \text{where } x_m = \frac{V_3}{w}$$

$$M_y = M_1 \frac{(h-y)}{h} + M_2 \frac{y}{h}$$

$$M_x = (L-x)\left(\frac{wx}{2} + \frac{M_2}{L}\right)$$

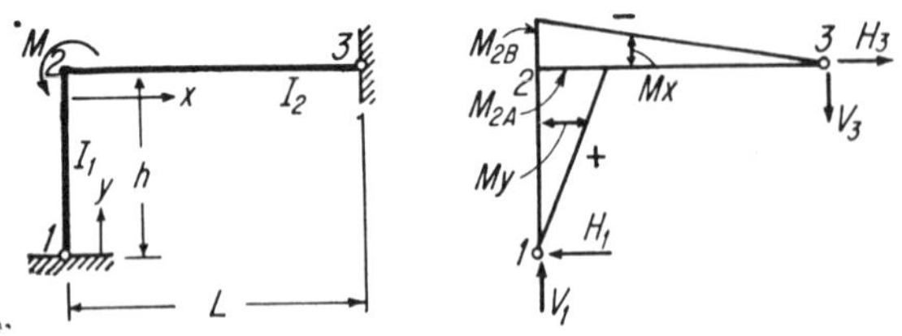

28.

$$M_{2A} = \frac{M}{C+1} \qquad M_{2B} = -\frac{MC}{C+1} \qquad M = M_{2A} - M_{2B}$$

$$M_3 = \frac{M_{2B}}{2}, \quad M_1 = -\frac{M_{2A}}{2} \qquad H_1 = H_3 = \frac{3\,M_1}{h}$$

$$V_1 = -V_3 = \frac{3\,M_3}{L}$$

$$M_x = M_{2B}\frac{L-x}{L} \qquad M_y = M_{2A}\frac{y}{h}$$

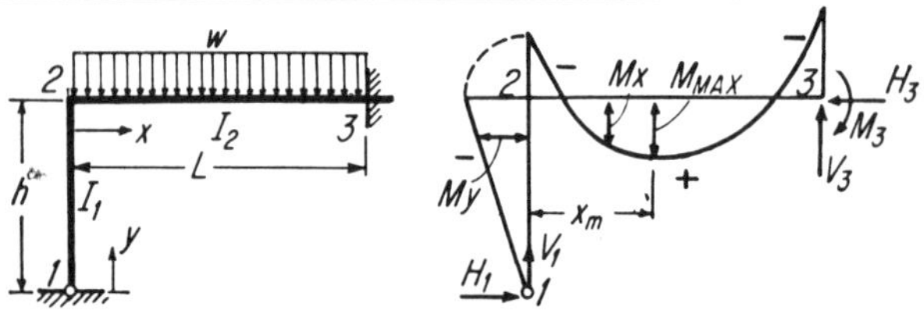

29.

$$M_2 = -\frac{wL^2}{4J} \qquad M_3 = -\frac{wL^2(2C+1)}{4J}$$

$$H_1 = H_3 = -\frac{M_2}{h} \qquad M_y = \frac{y}{h} M_2$$

$$V_1 = \frac{wL}{2} - \frac{M_2 - M_3}{L} \qquad V_3 = wL - V_1$$

$$M_{max} = \frac{V_1 x_m}{2} + M_2 \qquad \text{where } x_m = \frac{V_1}{w}$$

$$M_x = \frac{wx(L-x)}{2} + \frac{L-x}{L} M_2 + \frac{x}{L} M_3$$

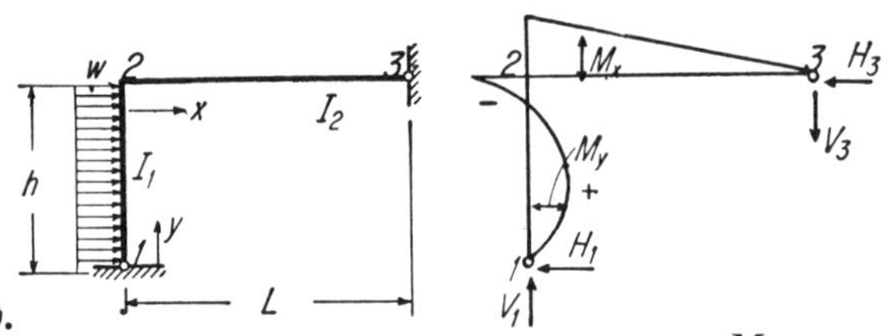

30.

$$M_2 = -\frac{wh^2C}{8(C+1)} \qquad V_1 = -V_3 = -\frac{M_2}{L}$$

$$H_1 = -\frac{wh}{2} - \frac{M_2}{h} \qquad H_3 = \frac{wh}{2} - \frac{M_2}{h}$$

$$M_x = M_2\left(1 - \frac{x}{L}\right) \qquad M_y = M_2\frac{y}{h} + \frac{wy(h-y)}{2}$$

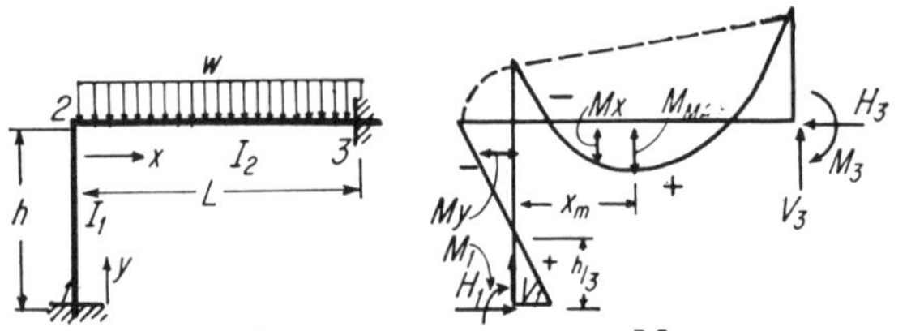

31.

$$M_2 = \frac{wL^2}{12(C+1)} \qquad M_1 = -\frac{M_2}{2}$$

$$M_3 = -\frac{wL^2(3C+2)}{24(C+1)} \qquad H_1 = H_3 = \frac{3M_1}{h}$$

$$V_1 = \frac{wL}{2} - \frac{M_2 - M_3}{L} \qquad V_3 = wL - V_1$$

$$M_{max} = \frac{V_1 x_m}{2} + M_2 \qquad \text{where } x_m = \frac{V_1}{w}$$

$$M_x = \frac{wx(L-x)}{2} + \frac{L-x}{L} M_2 + \frac{x}{L} M_3$$

$$M_y = \frac{h-y}{h} M_1 + \frac{y}{h} M_2$$

70. UNSYMMETRICAL FRAME

(*Ref.* 4, 5)

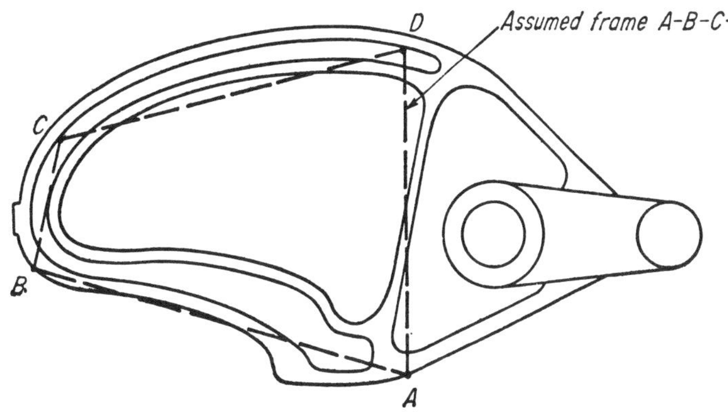

Fig 1

Elastic-center stress analysis is a method of finding stresses within a structure by first finding the moment distribution in an unsymmetrical frame, which has been *assumed*, as in Fig. 1.

This is a relatively new method. Once understood, it can become a powerful tool in the hands of a design engineer. It should be clearly understood that the term "frame" does not imply that the structure has the shape of the frame shown in the various analysis diagrams. More often than not, we have to use our good engineering judgment and assume a frame.

Distribution of internal resisting forces in a frame, such as assumed in Fig. 1, depends on section and physical properties. Stress analysis of the frame is based on certain redundant forces M_o, X_o, Y_o (Fig. 2), for which the solution can be obtained by the principle of continuity. There are many methods of applying the principle of continuity to obtain the solution for redundant forces in frames. The one that will be discussed here is known as elastic center method, which assumes that the redundant forces are acting at a special point called the elastic center of the frame. This center coincides with the centroid of the ds/EI value for the structure. (*See* Notation, for list of symbols.)

$$M_o = -\frac{\Sigma \phi_s}{\Sigma ds/I} \quad (3)$$

$$X_o = \frac{\Sigma \phi_s(y) - \Sigma \phi_s(x)\,[I_{xy}/I_y]}{I_x(1 - I^2_{xy}/I_xI_y)} \quad (4)$$

$$Y_o = \frac{-[\Sigma \phi_s(x) - \Sigma \phi_s(y)(I_{xy}/Iy)]}{I_y(1 - I^2_{xy}/I_xI_y)} \quad (5)$$

Derivation of the above forces is given in *Analysis and Design of Aircraft Structures*, by E. F. Bruhn and A. F. Schmitt.

ASSUMPTIONS

In the derivation of equations (3), (4) and (5), the distortion due to axial and shear forces is neglected. Furthermore it is assumed that the stress is proportional to strain.

Figure 2 shows a typical unsymmetrical frame with fixed supports at points A and B and carrying two loads P_1 and P_2. I, I_1, I_2 designate moments of inertia. It is assumed that the redundant forces M_o, X_o, Y_o are acting at the elastic center 0 of the frame. The problem is to determine the bending moment diagram under this loading.

REDUNDANT FORCES

In order to have a clear conception of the redundant forces as applied to frame analysis, consider an unsymmetrical curved beam fixed at ends A and B (Fig. 3) and carrying an external load P. This structure is statically indeterminate to the third degree.

In Fig. 4, the reaction at A has been replaced by its three components, namely, the forces x_A, y_A, and the moment M_A. Now treat the structure as a cantilever beam fixed at end B and carrying the redundant loads at A and known external load, P. Because joint A is fixed, it does not rotate when the structure is loaded. Therefore, three equations can be written stating that the horizontal, vertical, and angular deflection of point A must equal zero (principle of continuity), thus affording the solution to the problem.

The purpose here is to show how equations (3), (4) and (5), for the redundant forces, can be applied to the solution of an unsymmetrical frame.

NOTATION

To begin with, let us analyze these equations by taking each term separately:

1. M_o = moment, in.-lb., which is assumed positive when tension is applied on inside fiber of the frame.

2. $\Sigma\phi_s$ = total area of static M/I diagram; the angle change between the end faces of the ds element (Fig. 4) when acted upon by a constant static moment M_s is equal in value to the area of the M_s/I diagram on the element ds, that is $\phi_s = M_s/I$ (I is the moment of inertia).

3. $\Sigma ds/I = \Sigma\omega$, the total elastic weight of structure; the term ds/I of a beam element of length ds represents the angle change between its two end faces when the element is acted upon by a unit moment. This term is referred to as the elastic weight of the element and will be designated as ω. Physically, the elastic weight is the ability of the element to cause rotation when acted upon by a unit moment.

4. X_o, Y_o = forces (lb.)

5. x, y = distances from reference axes x^1-x^1, y^1-y^1 to the center of gravity of each member of a frame or, when barred, to elastic center.

6. $\Sigma\phi_s\,(y)$ = total moment of static M_s/I diagram about x axis.

7. $\Sigma\phi_s\,(x)$ = total moment of static M_s/I diagram about y axis.

8. I_x, I_y = elastic moments of inertia about the centroidal axes x-x, y-y of the frame. The terms i_x, i_y (*see* table) are the elastic moments of inertia of each member of the frame about its centroidal axes; $i_x = ds\,(ds_y)^2/12I = \omega(ds_y)^2/12$; $i_y = \omega(ds_x)^2/12$; ds_x, ds_y are horizontal and vertical projections of a member of the frame. Thus the moment of inertia about the reference axes x^1-x^1, y^1-y^1 (Fig. 2) is given by the following equations:

$$I_x^1 = \Sigma\,\omega y^2 + \Sigma\,i_x$$

$$I_y^1 = \Sigma\,\omega x^2 + \Sigma\,i_y$$

Having the moment of inertia about axes x^1-x^1, y^1-y^1, you can now find its value about the centroidal axes x-x, y-y of the frame by the use of the parallel axis theorem:

$$I_x = \Sigma\,I_x^1 - \Sigma\,\omega\bar{x}^2$$

$$I_y = \Sigma\,I_y^1 - \Sigma\,\omega\bar{y}^2$$

where $\bar{x}$, $\bar{y}$ are the distances from the two reference axes to the elastic center:

$$\bar{x} = \frac{\Sigma\,\omega(x)}{\Sigma\,\omega}\,; \qquad \bar{y} = \frac{\Sigma\,\omega(y)}{\Sigma\,\omega}$$

9. Product of inertia I_{xy}; if the structure is such that either the x or y axis through the elastic center is an axis of symmetry, then the product of inertia is zero.

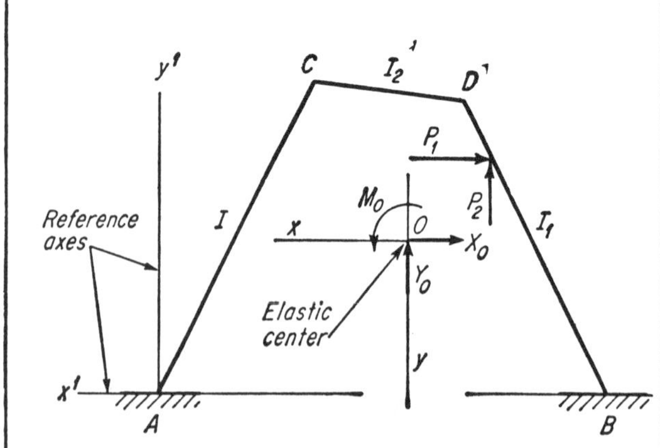

Fig 2

Elastic moments of inertia of

Member	I, in.^{4}	ds, in.
AB	0.00718	5.6
BC	0.0042	1.88
CD	0.00514	5.17
Σ	....	....

Location of elastic center: x =

Moment of inertia to axes through

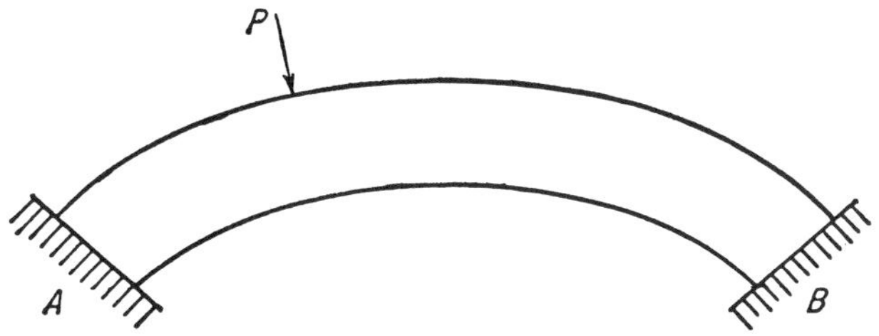

Fig 3

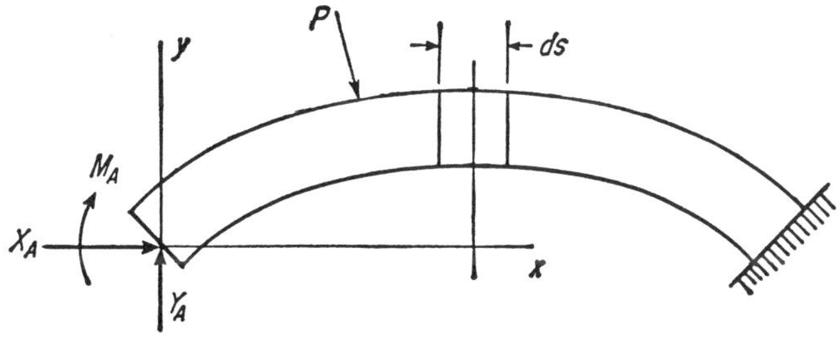

Fig 4

Finding stresses by the elastic-center method assumes that redundant forces in a "frame" (considered to be within the part under analysis) are acting at a special point called the elastic center of the frame. Here are the equations again:

$$\text{Moment (in.-lb.)}, M_o = \frac{-\Sigma \phi_s}{\Sigma ds/I} \quad (3)$$

$$\text{Force (lb.)}, X_o = \frac{\Sigma \phi_s(y) - \Sigma \phi_s(x)[I_{xy}/I_y]}{I_x[1 - (I^2_{xy}/I_x I_y)]} \quad (4)$$

$$\text{Force (lb.)}, Y_o = \frac{-\Sigma \phi_s(x) - \Sigma \phi_s(y)(I_{xy}/I_y)}{I_y[1 - (I^2_{xy}/I_x I_y)]} \quad (5)$$

frame members (see Fig 5)

ds_x, in.	ds_y, in.	$w = ds/I$	y in.	x, in.	wy	wy^2	$i_x = \frac{w(ds_y^2)}{12}$	wx	wx^2	$i_y = \frac{w(ds_x^2)}{12}$
1.48	5.4	780	2.7	0.74	2110	5700	1900	578	428	142
1.83	0.40	448	5.2	2.4	2330	12,116	6	1075	2580	122
1.30	5.0	1006	2.5	3.96	2515	6287	2098	3984	15,777	142
....	...	2234	...	....	6955	24,103	4004	5637	18,785	406

$\Sigma wx/\Sigma w = 2.51$ in.; $y = \Sigma wy/\Sigma w = 3.11$ in.

elastic center: $I_x = \Sigma i_x + \Sigma wy^2 - \bar{y}\,(\Sigma wy) = 6477$ in.^{4}; $I_y = \Sigma i_y + \Sigma wx^2 - \bar{x}\,(\Sigma wx) = 5042$ in.^{4}

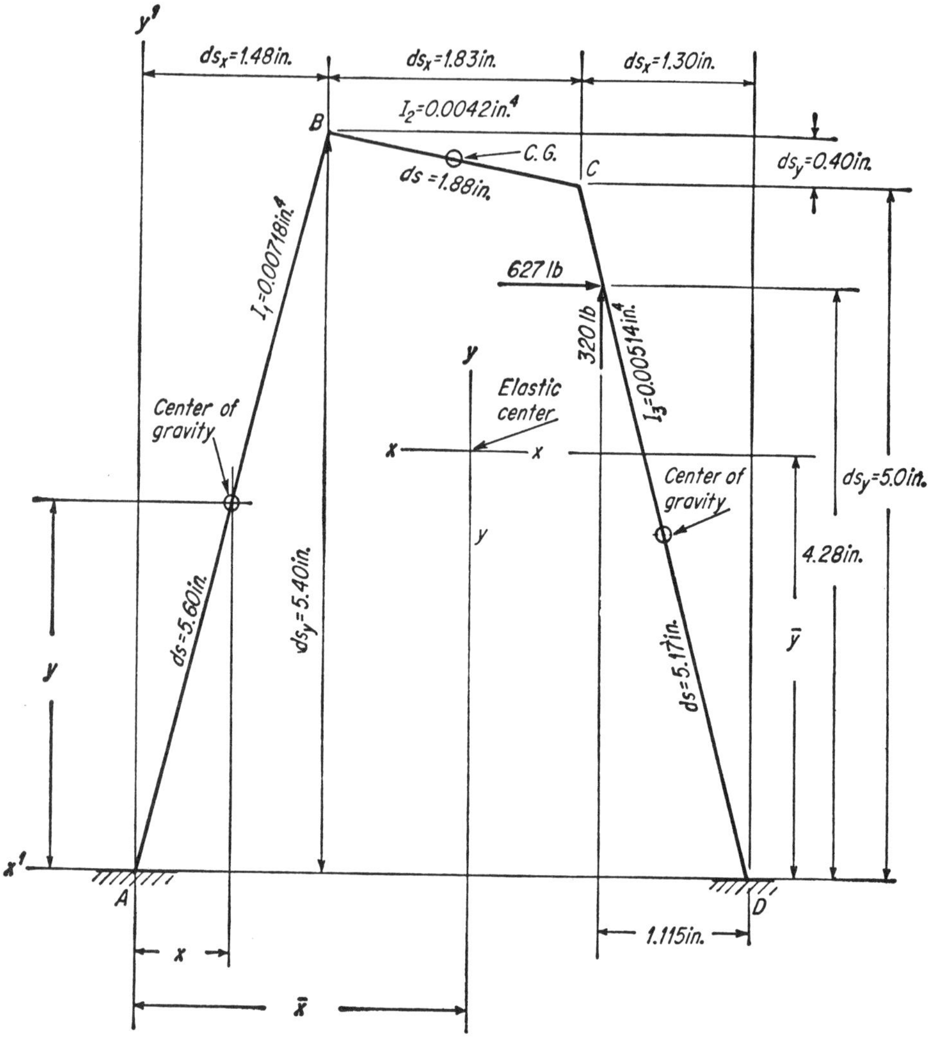

Fig 5
Geometry of the frame

where ϕ_s = total area of the static M/I diagram, I = moment of inertia, and x,y = distances from reference axes.

Example: Assume an asymmetrical frame, fixed at A, D, and loaded as shown in Fig. 5. Start by determining the bending-moment diagram.

Next step: Calculate the elastic product of inertia of the frame, I_{xy}. Figure 6 shows distances x_1y_1 of the CG, of each of the members of the frame, from the elastic center 0. Figure 7 shows a typical member.

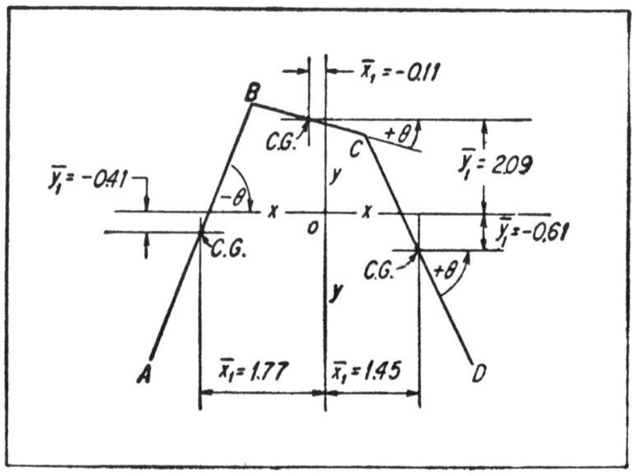

Fig. 6

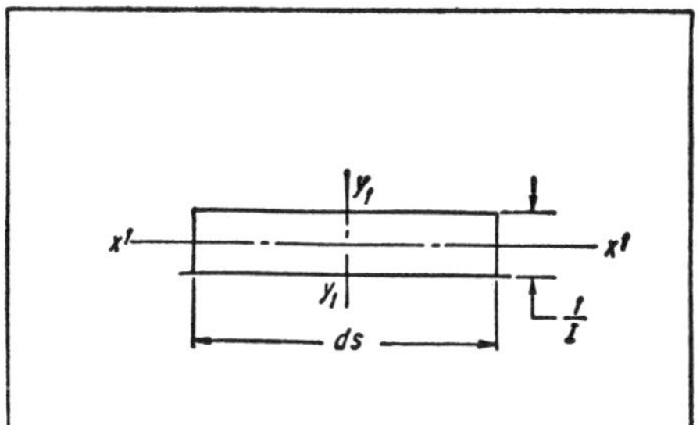

Typical member

Fig. 7

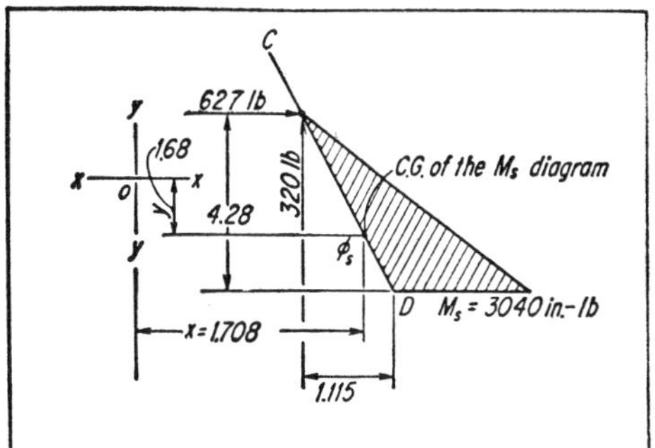

Moment diagram

Fig. 8

Rotating each member by the angle θ (counterclockwise direction is positive) parallel to axis x-x, then transferring to x-x we obtain:

$$I_{xy} = I^1_{xy}\cos 2\theta + \tfrac{1}{2}(I^1_x - I^1_y)\sin 2\theta + Ax_1y_1;$$
$$I_x = ds/12I^3 \text{ (negligible)};$$
$$I^1_y = ds^3/^1_y\ 12I.$$

The area $A = ds/I$. Because of the symmetry of the member, $I_{xy} = 0$; neglecting I^1_x we obtain: $I_{xy} = -\frac{1}{2} I^1_y \sin 2\theta + Ax_1y_1$. By substitution and remembering that for the member AB the angle θ is negative we obtain:

Member AB: I_{xy}	$= +1{,}045$ in.^{4}
Member BC: I_{xy}	$= -130$ in.^{4}
Member CD: I_{xy}	$= -1{,}419$ in.^{4}
I_{xy}	$= -504$ in.^{4}

Next step: Assume a static frame condition, draw moment M_s diagram, and compute the value ϕ_s which is the area of the M_s/I diagram (Fig. 8). $M_s = M^1_D = 627\ (4.28) + 320\ (1.115) = 3{,}040$ in.-lb.; $\phi_s = (3{,}040)(4.28)/(2)\ (0.00514) = 1{,}266{,}000$. This value is concentrated at the CG of the M_s diagram. From Fig. 8: $\phi_s(x) = 1{,}266{,}000(1.708) = 2{,}162{,}000$; $\phi_s(y) = 1{,}266{,}000(-1.68) = -2{,}127{,}000$. Thus, all the information that affords the calculation of the redundant forces is now available:

$$M_o = -\Sigma\phi_s/\Sigma\omega = -566 \text{ in.-lb.};$$
$$X_o = [\Sigma\phi_s y - \Sigma\phi_s(x)\ (I_{xy}/I_y)]/I_x(1 - I^2_{xy}/I_xI_y) = -297 \text{ lb.};$$
$$Y_o = [-\Sigma\phi_s(x) - \Sigma\phi_s(y)\ (I_{xy}/I_x)]/I_y(1 - I^2_{xy}/I_xI_y) = -408 \text{ lb.}$$

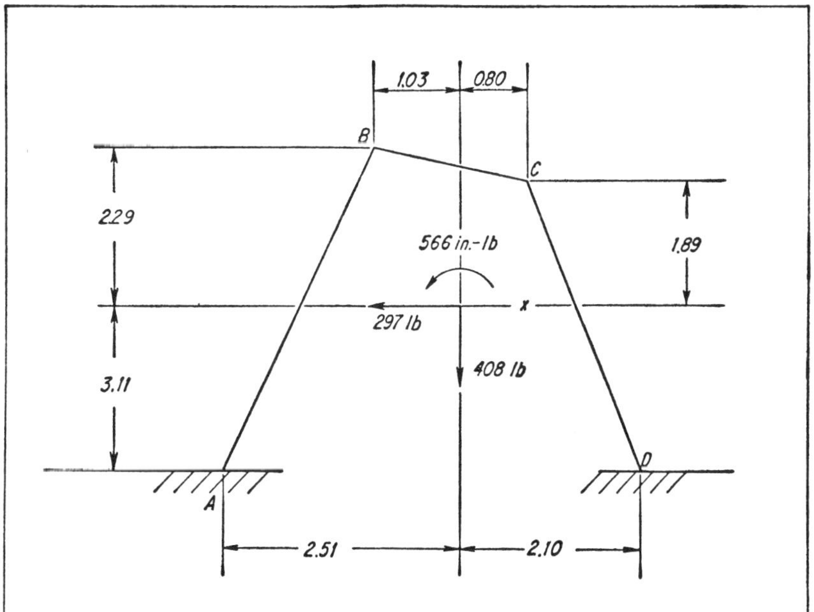

Load at elastic center

Fig. 9

The bending moment at any point of the frame is equal to original static moment M_s plus moments due to the redundant forces as shown in Fig. 9.

Point *A*: $M_s = 0$; $M_A = -566 - 297\,(3.11) + 408\,(2.51) = -466$ in.-lb.

Point *B*: $M_s = 0$; $M_B = -566 + 297\,(2.21) + 408\,(1.03) = +534$ in.-lb.

Point *C*: $M_s = 0$; $M_C = -566 + 297\,(1.89) - 408\,(0.80) = -331$ in.-lb.

Point *D*: $M_s = 3{,}040$ in.-lb.; $M_D = 2{,}474 - 297\,(3.11) - 408\,(2.08) = +700$ in.-lb.

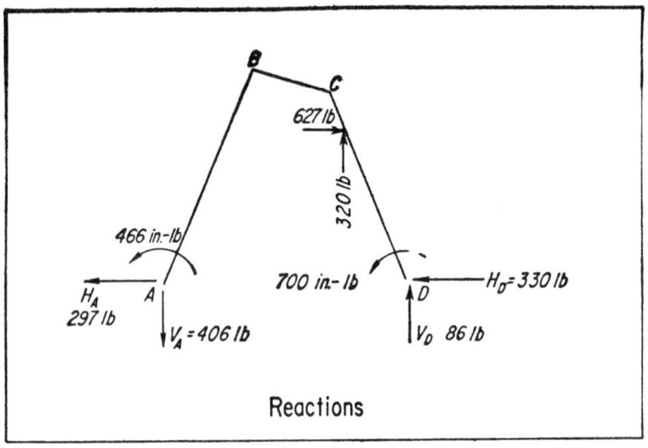

Fig. 10

Vertical and horizontal reactions at *A*, *D*, (Fig. 10): $\Sigma M_D = 0 = 627\,(4.28) + 320\,(1.115) - 700 - 466 - V_A(4.61)$. Therefore, $V_A = 406$ lb.; $V_D = 406 - 320 = 86$ lb.; $\Sigma M_B = -466 - 534 - 406\,(1.48) + H_A(5.4)$. Therefore, $H_A = 297$ lb., and $H_D = 627 - 297 = 330$ lb. Figure 11 illustrates the final results.

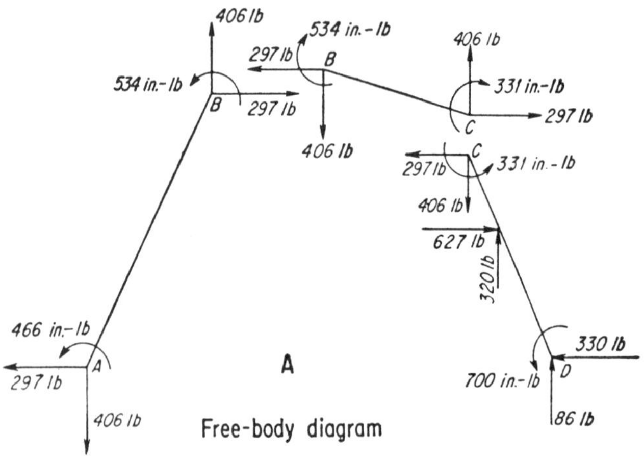

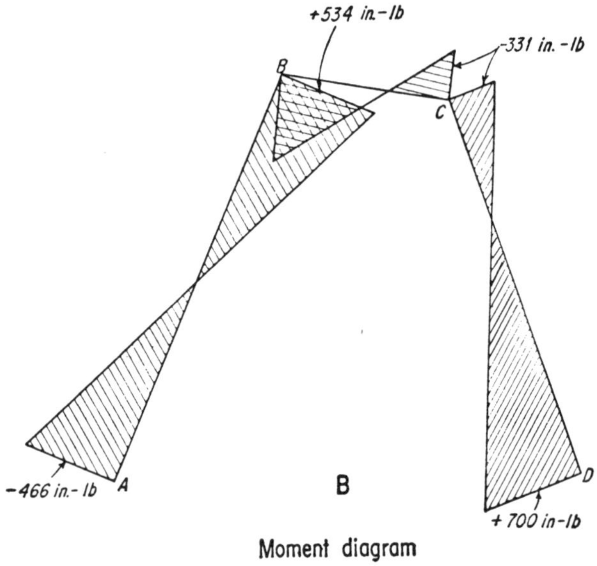

Fig. 11

Chap. X

REFERENCES

Section	*Ref. No.*	
69	1	GRIFFEL, W.: "Rigid Frame Formulas," *Product Engineering*, April 14, 1964
69	2	KLEINLOGEL, A., D.ENG.: *Rigid Frame Formulas*, Frederick Ungar Publishing Co., N.Y.
69	3	LEONTOVICH: *Frames and Arches*, McGraw-Hill, N.Y.
70	4	GRIFFEL, W.: "Stresses in Unsymmetrical Frames," *Product Engineering*, Nov. 11, 1963, Nov. 25, 1963
70	5	BRUHN, E. F., and SCHMITT, A. F.: *Analysis and Design of Aircraft Structures*, Vol. I, Tri-State Offset Co., 1958

Chapter XI

VIBRATION

71. VIBRATION FREQUENCY CHARTS*

(Ref. 1, 2, 3, 4, 5, 6, 7)

1. Rectangular plate
2. Square plate
3. Cantilever rectangular plate
4. Cantilever skew plate
5. Circular plate
6. Ring
7. Uniform beam
8. Continuous beam
9. Variable section beam
10. Combined beam

Vibration at or near resonance on such common structural elements as plates and beams creates high stresses and hastens the time to eventual failure. Standard practice is to design these elements so that their natural frequencies will not excite resonance.

The accompanying table of frequency constants and nomogram allow solutions for frequencies and modes of vibration for beams and plates with various boundary conditions. Effects of shear distortion and rotary inertia are neglected. Also, while applying the ordinary theory of thin plates, it is assumed that the plates are homogeneous and of uniform thickness.

The following general relationships of natural frequency apply here:

$$f = \frac{Kt}{2\pi a^2}\sqrt{\frac{gE}{12(1-\mu^2)\rho}} \quad \text{square plate} \quad (1)$$

$$f = \frac{Kt}{2\pi r^2}\sqrt{\frac{gE}{12(1-\mu^2)\rho}} \quad \text{circular plate} \quad (2)$$

$$f = \frac{K}{2\pi L^2}\sqrt{\frac{gEI}{w}} \quad \textit{beam} \quad (3)$$

Mode factor K, obtained from the Rayleigh-Ritz energy method, is available in reference literature.

Equations (1), (2) and (3) may be simplified by assuming the ratio $E/\rho = 105 \times 10^6$, a value used for steel, aluminum and magnesium, and $\mu = 0.3$. Also, by introducing the radius of gyration $k = \sqrt{I/A}$ the following equations are obtained:

$$f = (Ct/a^2) \times 10^5 \text{ rectangular plates} \quad (4)$$

$$f = (Ct/r^2) \times 10^5 \text{ circular plates} \quad (5)$$

$$f = (Ck/L^2) \times 10^5 \text{ beams} \quad (6)$$

$$f = (C_1h/L^2) \times 10^5 \text{ rectangular beams} \quad (7)$$

NOTATION

C = frequency constant
C_1 = frequency constant for rectangular beam
f = natural frequency (cps)
t = plate and ring thickness (in.)
a,b = plate dimensions (in.)
g = acceleration due to gravity (in./sec^2)
ρ = density of material (lb./in.^{3})
E = modulus of elasticity (psi)
μ = Poisson's Ratio
L = beam span (in.)
k = radius of gyration (in.) = $\sqrt{I/A}$
M = vibration mode number
n = number of spans
θ = skew angle (deg.)
r = plate radius, and mean ring radius (in.)
s = number of nodal circles
m = number of nodal diameters
I = moment of inertia (in.^{4})

*Abstracted from Ref. 6.

A = cross-sectional area of beam (in.^{2})
v = number of full waves around periphery
h = beam depth
b = beam width
w = weight of unit length beam (lb./in.)

These vibration natural frequencies, as expressed by equations (5), (6), (7) and (8), may be obtained directly from the nomogram. Thus, for a given value of the frequency constant C or C_1 (*see* Tables 1–8) corresponding to a mechanical structure under design, it is possible to assess the effect of changing the dimensions, type of supports and material.

The advantage of the natural frequency constants, as shown in Tables 1–8, results from the fact that they cover three most common materials: steel, aluminum and magnesium. For other materials, multiply the natural frequency obtained from the nomogram by the material correction factor:

$$(0.097 \times 10^{-3}) \sqrt{E/\rho}$$

Example: Determine the minimum thickness of a rectangular, steel, cantilever spring with a mounted free length of 2 in., which will exhibit no resonance within the frequency range from zero to 1,500 cps.

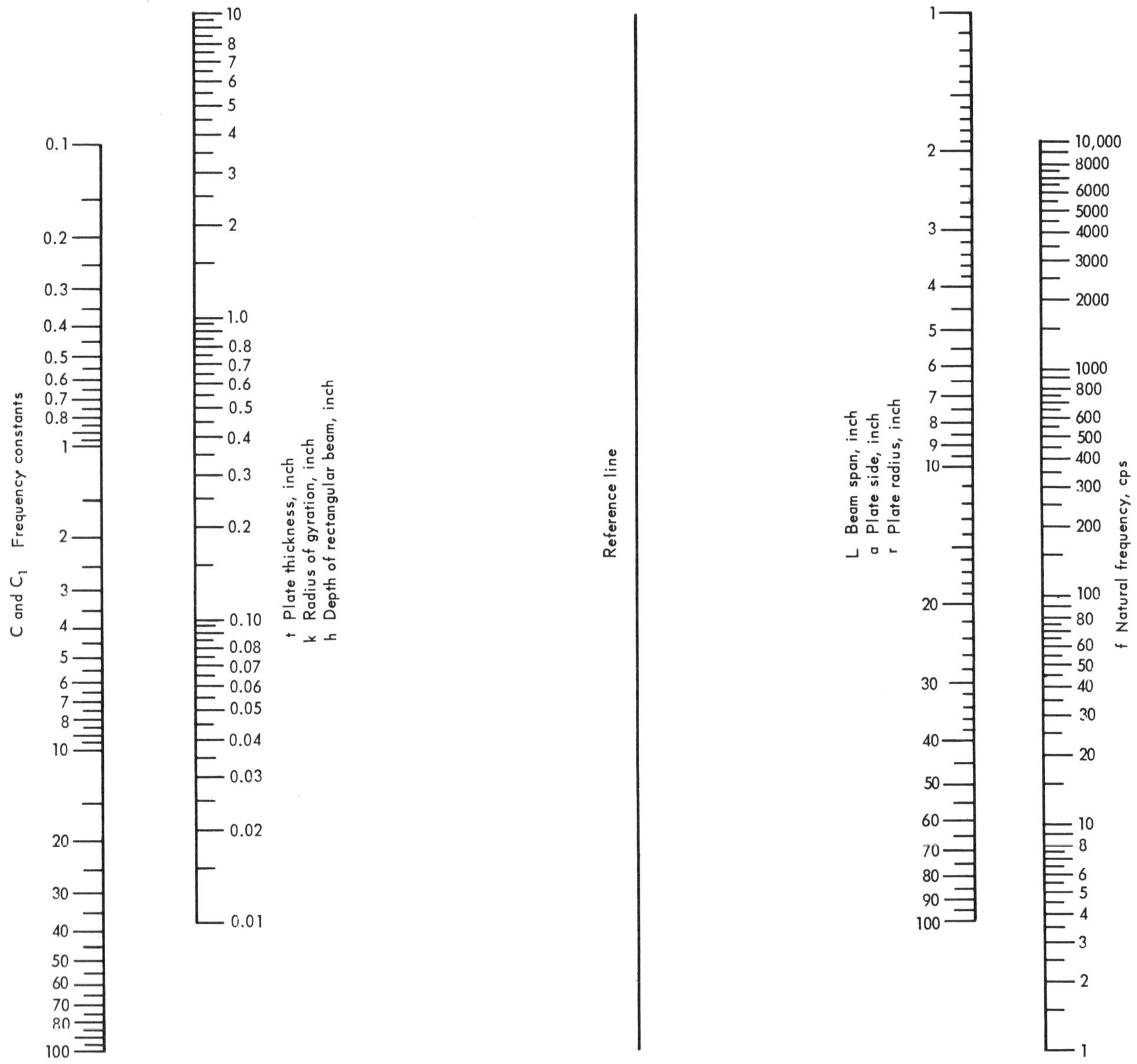

Solution: From Table 6, $C_1 = 0.34$ for fundamental mode. Using the nomogram, align $f = 1{,}500$ with $C_1 = 0.34$, intersecting the Reference line. Align this intersection with $L = 2$, intersecting h at 0.18 in., which is the thickness of the spring.

TABLE 1

Rectangular Plate, Frequency Constants - C

Plate support	a/b 1.0	1.5	2.0	2.5	3.0	∞	b/a 1.0	1.5	2.0	2.5	3.0	∞
S (top), S (left), S (right), S (bottom)	—	—	—	—	—	—	1.92	1.39	1.20	1.11	1.07	0.96
S (top), C (left), S (right), S (bottom)	2.31	1.51	1.26	1.14	1.08	0.96	2.31	1.84	1.69	1.62	1.58	1.50
S (top), C (left), C (right), S (bottom)	2.83	1.70	1.33	1.18	1.10	0.96	2.83	2.44	2.32	2.26	2.24	2.18
C (top), C (left), C (right), C (bottom)	—	—	—	—	—	—	3.49	2.63	2.39	2.31	2.26	2.18

S–supported
C–clamped
Note: use "t" and "a" with C

TABLE 2

Square Plate, Frequency Constants - C

Plate support	M=1	M=2	M=3	M=4	M=5
F (top), F (left), F (right), C (bottom)	0.34	0.83	2.08	2.66	3.02
C, C, C, C	3.49	7.12	10.50	12.77	12.83
F (top), C (left), F (right), C (bottom)	0.67	2.33	2.60	4.66	6.12
F, F, F, F	1.37	1.99	2.32	3.47	6.00
S, S, S, S	1.92	4.80	7.65	9.63	12.51
S (top), S (left), S (right), C (bottom)	2.31	5.03	5.71	8.38	9.76
S (top), C (left), C (right), S (bottom)	2.83	5.33	6.75	9.21	9.95

S–supported
F–free
C–clamped
Note: use "t" and "a" with C

TABLE 3

Cantilever Plate, Frequency Constants - C

Rectangular	a/b	M=1	M=2	M=3	M=4	M=5
	0.5	0.34	0.52	2.10	1.00	2.42
	1	0.34	0.83	2.10	2.67	3.03
	2	0.34	1.45	2.10	9.20	4.74
	5	0.34	3.38	2.09	54.86	10.30
Skewed	θ	M=1	M=2			
	15°	0.35	0.86			
	30°	0.38	0.99			
	45°	0.47	1.33			

Note: use "t" and "a" with C

TABLE 4

Circular Plates, Frequency Constant - C (2r)

Type of support	m	S=0	S=1	S=2	S=3
Clamped at boundary	0	0.99	3.86	8.62	—
	1	2.06	—	—	—
	2	3.38	—	—	—
Free boundary	0	—	0.88	3.74	—
	1	—	1.99	5.81	—
	2	0.51	3.42	—	—
	3	1.19	5.13	—	—
Clamped at center	—	0.36	2.03	5.88	11.61

m–number of nodal diameters
S–number of nodal circles
Note: use "t" and r with C

TABLE 5

Ring Vibrating in Its Own Plane, Frequency Constant - C

V=2	V=3	V=4	V=5	V=6
0.25	0.70	1.40	2.20	3.20

V–number of full waves around periphery

Note: use "t" and r with C

TABLE 6

Uniform Beam Frequency Constants:
C–Any Cross-Section C_1–Rectangular Cross-Section

	M=1 C	M=1 C_1	M=2 C	M=2 C_1	M=3 C	M=3 C_1	M=4 C	M=4 C_1	M=5 C	M=5 C_1
Cantilever	1.13	0.34	7.10	2.13	19.84	5.95	38.78	11.63	64.24	19.27
Simply supported ends	3.17	0.95	12.68	3.80	28.53	8.56	50.72	15.22	79.25	23.77
Fixed ends	7.19	2.16	19.84	5.95	38.78	11.63	64.24	19.27	95.89	28.77
Free ends	7.19	2.16	19.84	5.95	38.78	11.63	64.24	19.27	95.89	23.77
Fixed-hinged	4.94	1.48	16.04	4.81	33.47	10.04	57.25	17.17	87.36	26.21
Hinged-free	4.94	1.48	16.04	4.81	33.47	10.04	57.25	17.17	87.36	26.21

Note: (1) use "k" and "L" with C (2) use "h" and "L" with C_1

TABLE 7

Continuous Uniform Beam, Frequency Constants:
C—For Any Cross-Section C_1—For Rectangular Cross-Section

		M=1		M=2		M=3		M=4		M=5	
	n	C	C_1	C	C_1	C	C_1	C	C_1	C	C_1
Extreme ends simply supported n = 3, shown 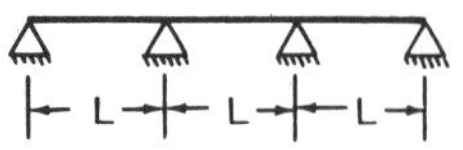	1	3.17	0.95	12.68	3.80	28.53	8.56	50.72	15.22	79.25	23.77
	2	3.17	0.95	4.94	1.48	12.68	3.80	16.04	4.81	28.53	8.56
	3	3.17	0.95	4.06	1.22	5.96	1.79	12.68	3.80	14.39	4.14
	4	3.17	0.95	3.71	1.11	4.94	1.48	6.40	1.92	12.68	3.80
	5	3.17	0.95	3.49	1.05	4.41	1.32	5.51	1.65	6.66	4.32
	6	3.17	0.95	3.42	1.03	4.06	1.22	4.94	1.48	5.96	1.79
	7	3.17	0.95	3.36	1.01	3.83	1.15	4.56	1.37	5.36	1.61
	8	3.17	0.95	3.30	0.99	3.71	1.11	4.25	1.27	4.94	1.48
	9	3.17	0.95	3.30	0.99	3.55	1.06	4.06	1.22	4.63	1.39
	10	3.17	0.95	3.30	0.99	3.49	1.05	3.90	1.17	4.06	1.22
Extreme ends clamped n = 3 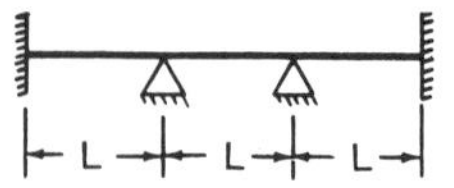	1	7.23	2.17	19.81	5.94	38.83	11.65	64.19	19.26	95.89	28.77
	2	4.94	1.48	7.23	2.17	16.04	4.81	19.81	5.94	33.47	10.04
	3	4.06	1.22	5.96	1.79	7.23	2.17	14.36	4.31	17.81	5.34
	4	3.71	1.11	4.94	1.48	6.40	1.92	7.23	2.17	13.73	4.12
	5	3.49	1.05	4.41	1.32	5.51	1.65	6.66	2.00	7.23	2.17
	6	3.42	1.03	4.06	1.22	4.94	1.48	5.96	1.79	6.75	2.02
	7	3.36	1.01	3.86	1.15	4.56	1.37	5.36	1.61	6.21	1.86
	8	3.30	0.99	3.71	1.11	4.25	1.27	4.94	1.48	5.67	1.70
	9	3.30	0.99	3.55	1.06	4.06	1.22	4.63	1.39	5.26	1.58
	10	3.30	0.99	3.49	1.05	3.90	1.17	4.06	1.22	4.94	1.48
Extreme ends clamped-supported n = 3 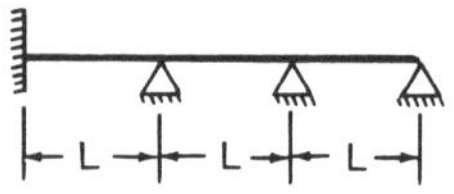	1	4.94	1.48	16.04	4.81	33.47	10.04	57.25	17.17	87.36	26.21
	2	3.71	1.11	6.40	1.92	13.73	4.12	18.58	5.57	30.08	9.02
	3	3.42	1.03	4.94	1.48	6.75	2.02	13.19	3.96	16.04	4.81
	4	3.30	0.99	4.25	1.27	5.67	1.70	6.94	2.08	12.93	3.88
	5	3.30	0.99	3.90	1.17	4.94	1.48	6.12	1.84	7.04	2.11
	6	3.23	0.97	3.71	1.11	4.50	1.35	5.45	1.63	6.40	1.92
	7	3.23	0.97	3.55	1.06	4.18	1.25	4.94	1.48	5.77	1.73
	8	3.23	0.97	3.49	1.05	3.96	1.19	4.56	1.37	5.36	1.61
	9	3.17	0.95	3.42	1.03	3.83	1.15	4.34	1.30	4.94	1.48
	10	3.17	0.95	3.36	1.01	3.71	1.11	4.12	1.24	4.63	1.39

Note: (1) use "k" and L with C (2) use "h" and L with C_1

TABLE 8

Variable Section Beams, Frequency Constant - C

Beam shape	b/b_o	h/h_o	M=1	M=2	M=3
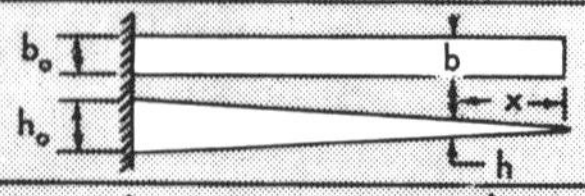	1	x/L	1.71	4.89	9.65
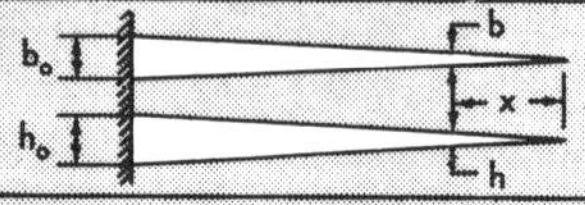	x/L	x/L	2.81	6.81	12.36
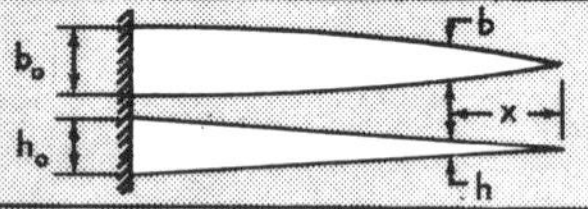	$(x/L)^{1/2}$	x/L	2.23	5.82	11.00
	$e^{x/L}$	1	1.52	7.78	20.60
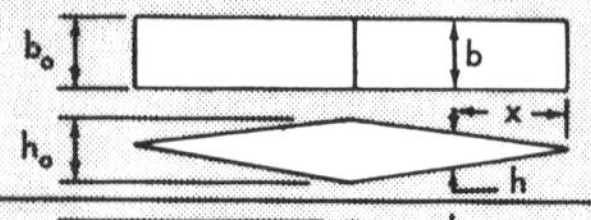	Symmetric mode of vibration of compound beam (double wedge)				
	—	—	3.27	7.65	—
	Antisymmetric mode of vibration of compound beam (double wedge)				
	1	x/L	4.95	—	—
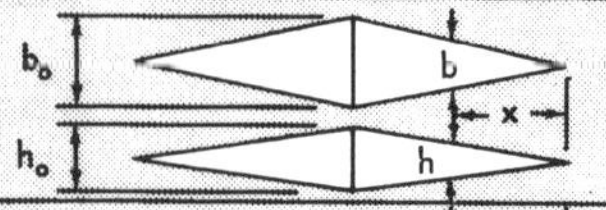	Symmetric mode of vibration of compound beam (double cone)				
	—	—	3.27	7.65	—
	Antisymmetric mode of vibration of compound beam (double cone)				
	x/L	x/L	4.95	—	—
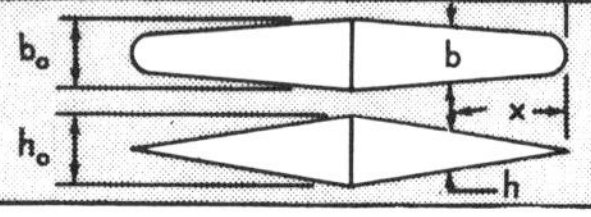	Symmetric mode of vibration of compound beam				
	—	—	2.56	6.60	—
	Antisymmetric mode of vibration of compound beam				
	$(x/L)^{1/2}$	x/L	4.20	—	—

Note: use "k" and "L" with C

72. CRITICAL SPEED OF SHAFTS

(Ref. 8, 9, 10)

Critical speed of a shaft depends on the magnitude and location of the applied load, on shaft diameter, and on the type of supporting bearings used. Although operating speed of a shaft should preferably be below its critical speed, this may not always be possible. In any case, a margin of 25 percent is usually recommended for safe operation.

The critical speed, N_{cr}, of a shaft can be found from

$$N_{cr} = \frac{K\sqrt{E\,d^2}}{\sqrt{WL^3}}$$

The nomogram shown in Fig. 1 solves this equation by means of a load-support factor K. This nomogram can be used to find the critical speed of shafts under seven different load and support conditions, Table 9.

In this method, if the shaft is rotating on anti-friction bearings, it is assumed to be simply supported. If the shaft is rotating on sleeve bearings, it is assumed to be fixed-ended. The method applies to horizontal as well as vertical shafts.

NOTATION

- d = diameter of shaft (in.)
- E = modulus of elasticity (psi)
- K = dimensionless load-support factor
- L = length of shaft (in.)
- N_{cr} = critical speed of shaft (rpm)
- W = concentrated load on shaft (lb.)
- w = load per unit length (lb./in.)
- ΔW = percentage of shaft weight added to concentrated load (lb.)

Data required for determining the critical speed of a shaft are the same as those required to calculate shaft deflection. Since a shaft is often of variable diameter, and since its stiffness may be increased by a long hub, it is necessary to assume a shaft of uniform diameter and stiffness. Under certain support and load conditions, a percentage of the shaft weight, ΔW, can be added to the applied load for greater accuracy in computing critical speed. Values of ΔW are given in Table 9.

While a shaft carrying a distributed load may have an infinite number of critical speeds, the first critical speed is usually most important to a designer. The nomogram, Fig. 1, used with load-support factor K, yields this value.

Example 1: A steel shaft has diameter $d = 1.5$ in., and length $L = 8$ in. It carries a uniformly distributed load $W = 60$ lb. Anti-friction-bearing supports are used.

This situation corresponds to Table 9, case 5. Thus, $K = 412$. Modulus of elasticity, E, of steel is 30×10^6 psi. Then, from the nomogram:

1. Align $K = 412$ with $E = 30 \times 10^6$, to intersect Index 1.
2. Line up the intersection on Index 1 with $d = 1.5$ to intersect Index 2.
3. Connect the point on Index 2 with $W = 60$ to intersect Index 3.
4. Line up the point on Index 3 with $L = 8$. The line intersects the N_{cr} line at $N_{cr} = 30{,}000$ rpm, which is the required critical speed.

Example 2: An aluminum shaft is cantilevered from a fixed journal bearing. It has an unsupported length $L = 10$ in., and diameter $d = 2$ in. If the critical speed $N_{cr} = 4{,}000$ rpm, what is the maximum concentrated load that the shaft can carry?

From Table 9, case 3, $K = 72$. Modulus of elasticity, $E = 10 \times 10^6$ psi. Then, from the nomogram:

1. Line up $K = 72$ with $E = 10 \times 10^6$ to intersect Index 1.
2. Connect the point on Index 1 to $d = 2$; mark the intersection with Index 2.
3. Align $N = 4{,}000$ with $L = 10$, intersecting Index 3.
4. Connect the points of intersection on Index 2 and 3 to give $W = 50$ lb.

TABLE 9 —Values of *K* for Various Load Conditions

Load-Support Condition*		Load-Support Factor, *K*	ΔW* (per cent)
Bearings simply supported, concentrated load at center	Case 1 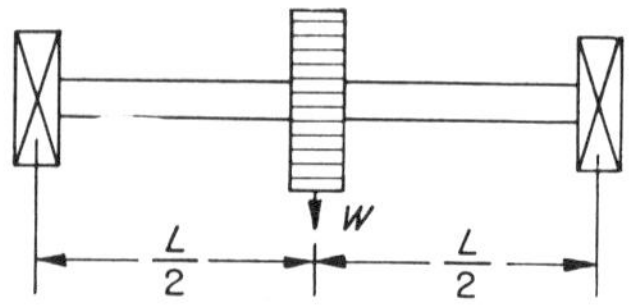	288	49
Bearings fixed, single concentrated load at center	Case 2 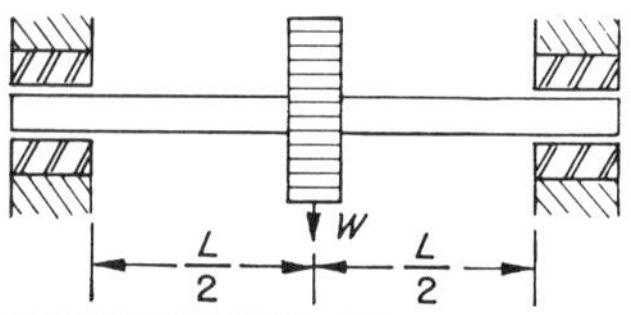	576	39
One end in fixed bearing, one end free, end load	Case 3 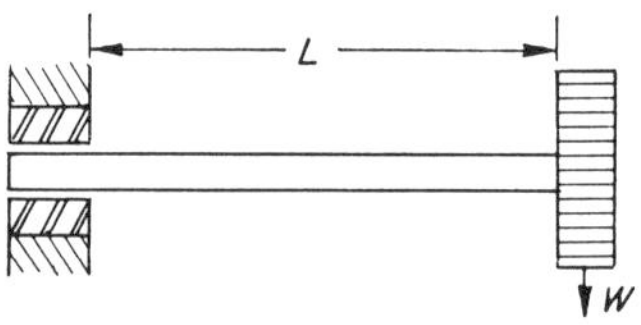	72	30
One end in fixed bearing, one end simply supported, concentrated load at center	Case 4 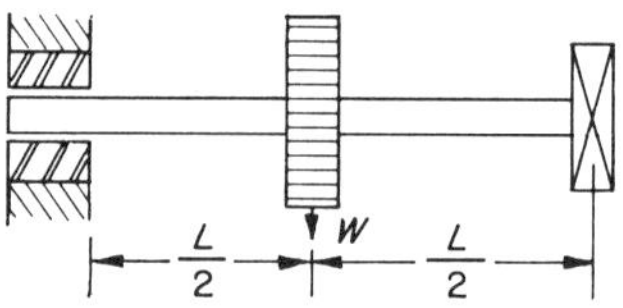	436	0
Bearings simply supported, uniform load	Case 5 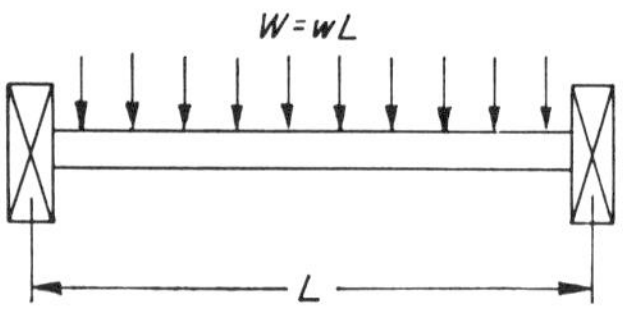	412	100
Bearings fixed, uniform load	Case 6 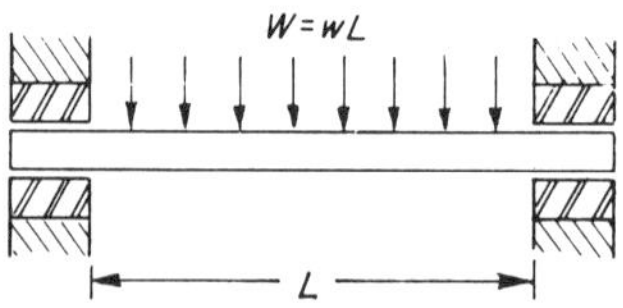	927	100
One end in fixed bearing, one end free, uniform load	Case 7 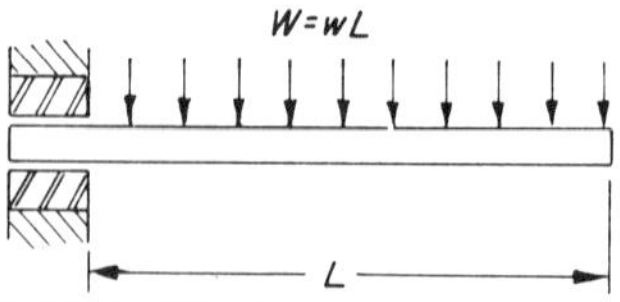	134	100

*Percentage of shaft weight to be added to concentrated load.

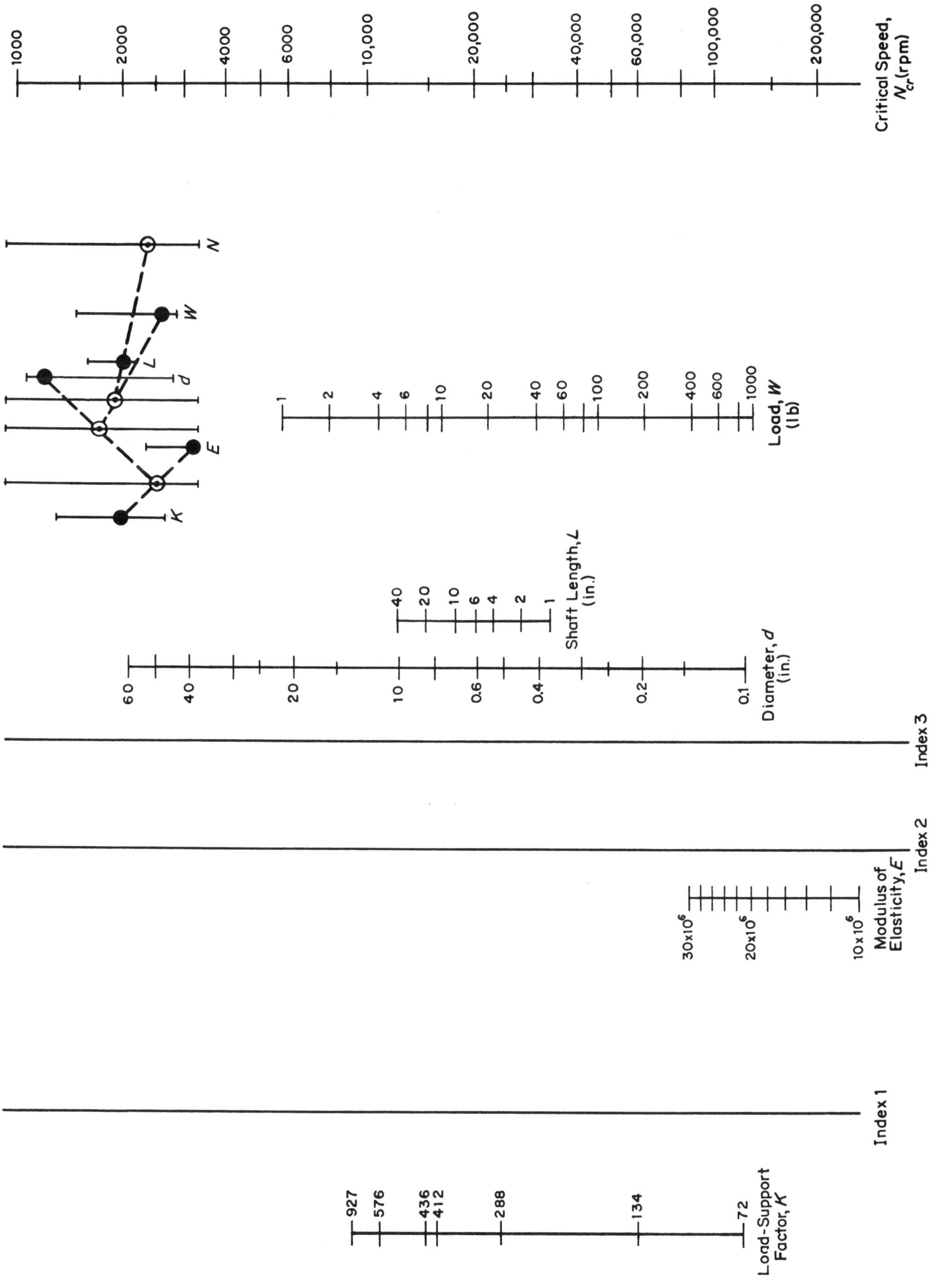

Fig. 1—Nomogram for determining critical speed of a shaft under various load conditions. Note that for Cases 5, 6, and 7, concentrated load $W = wL$.

73. CRITICAL SPEED OF MULTIPLE-LOADED SHAFTS

(Ref. 10, 11)

Operating speed of a shaft should be either above or below the critical speed. If the shaft operates at a uniform speed above critical speed, the shaft is described as flexible. If the operating speed is below critical speed, or varies through a wide range below critical speed, the shaft is rigid.

The critical speed of a shaft with multiple loads can be determined from Dunkerley's formula:

$$\frac{1}{N^2} = \frac{1}{n_0^2} + \frac{1}{n_1^2} + \frac{1}{n_2^2} + \frac{1}{n_3^2} + \ldots \qquad (8)$$

This equation can be solved by the use of a critical-speed constant k. Since k is a function of the static deflection, equation (8) can be rearranged to

$$N = \frac{18{,}770}{\sqrt{k_0 + k_1 + k_2 + k_3 + \ldots}} \qquad (9)$$

Constant k_0 is determined from Fig. 2; constants k_1, k_2, k_3, . . . from Fig. 3. Critical speed N is determined from Fig. 4 using $k = k_0 + k_1 + k_2 + k_3 + \ldots$ as a combined constant.

NOTATION

ρ = material density (lb./in.^{3})
d = diameter of shaft (in.)
d_1 = first value of shaft diameter (in.)
d_2 = corrected value of shaft diameter (in.)
K = dimensionless load-support factor
$k_0, k_1, k_2 \ldots$ = critical-speed constants corresponding to speeds $n_0, n_1, n_2, \ldots$
L = distance between supports (in.)
l = load-locating dimension (in.)
N = critical speed of shaft (rpm)
N_1, N_2 = critical speed of shaft corresponding to diameters d_1 and d_2 (rpm)
W = concentrated load on shaft (lb.)

Load-support factor K is determined from Fig. 2 for the appropriate load and support condition.

Allowance should be made for unavoidable inaccuracies in the assumed conditions. Factors which tend to cause uncertainty include: Difficulty of fixing the point of application of the load; the conditions at the supports; whether shaft is supported or partly fixed; inaccuracy of alignment; and obliquity of the rotating parts. An allowance of 25 percent should be included for safe operation.

Generally, if a shaft is rotating on sleeve bearings, it is considered to be fixed-ended. If a shaft is rotating on antifriction bearings, it is considered to be simply supported. The method applies to horizontal as well as vertical shafts.

A shaft of uniform constant diameter is assumed. The shaft size should first be calculated for strength and its diameter checked against the critical speed for safety. If the critical speed is unsuitable, a corrected diameter corresponding to a more suitable critical speed, can be found from

$$d_2 = d_1 \sqrt{\frac{N_2}{N_1}}$$

The nomogram in Fig. 2, for unloaded shafts, applies to steel, aluminum, magnesium, and any other material for which $\rho/E = 0.94 \times 10^{-8}$.

Example: A steel shaft has diameter $d = 10$ in. and length $L = 100$ in. Antifriction bearing supports are used. Loads are shown in Figs. 5, where $W_1 = 3{,}000$, $W_2 = 2{,}000$, $W_3 = 2{,}500$, $W_4 = 4{,}000$, and $W_5 = 1{,}500$.

Solution: This situation corresponds to case 2, Fig. 2. Thus, $K = 48$. Modulus of elasticity E of steel is 30×10^6 psi. Then, from nomogram, Fig. 2:

1. Align $K = 48$ with $L = 100$ to intersect Index 1.
2. Align intersection on Index 1 with $d = 10$.

Solution of Example Problem

Load (lb.)	Ratio l/L	Critical-Speed Constant
unloaded	—	$k_0 = 30$
$W_1 = 3{,}000$	0.20	$k_1 = 17$
$W_2 = 2{,}000$	0.35	$k_2 = 23$
$W_3 = 2{,}500$	0.50	$k_3 = 35$
$W_4 = 4{,}000$	0.44	$k_4 = 55$
$W_5 = 1{,}500$	0.28	$k_5 = 14$

This line intersects the k_0 line at 30.

To find $k_1 \ldots k_5$, first calculate values of the ratio l/L for each load. Then determine each value of k_n from

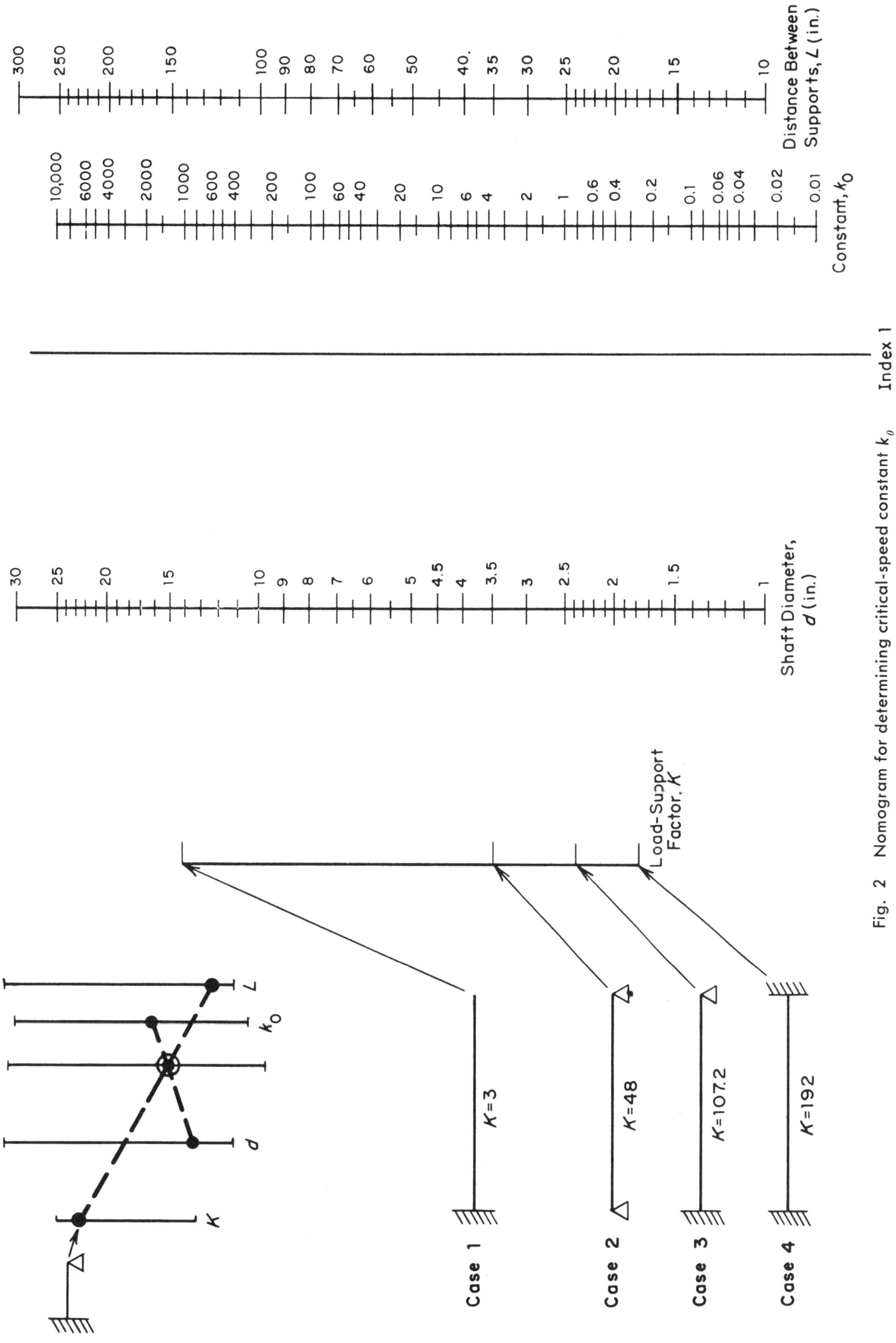

Fig. 2 Nomogram for determining critical-speed constant k_0 (for unloaded shaft)

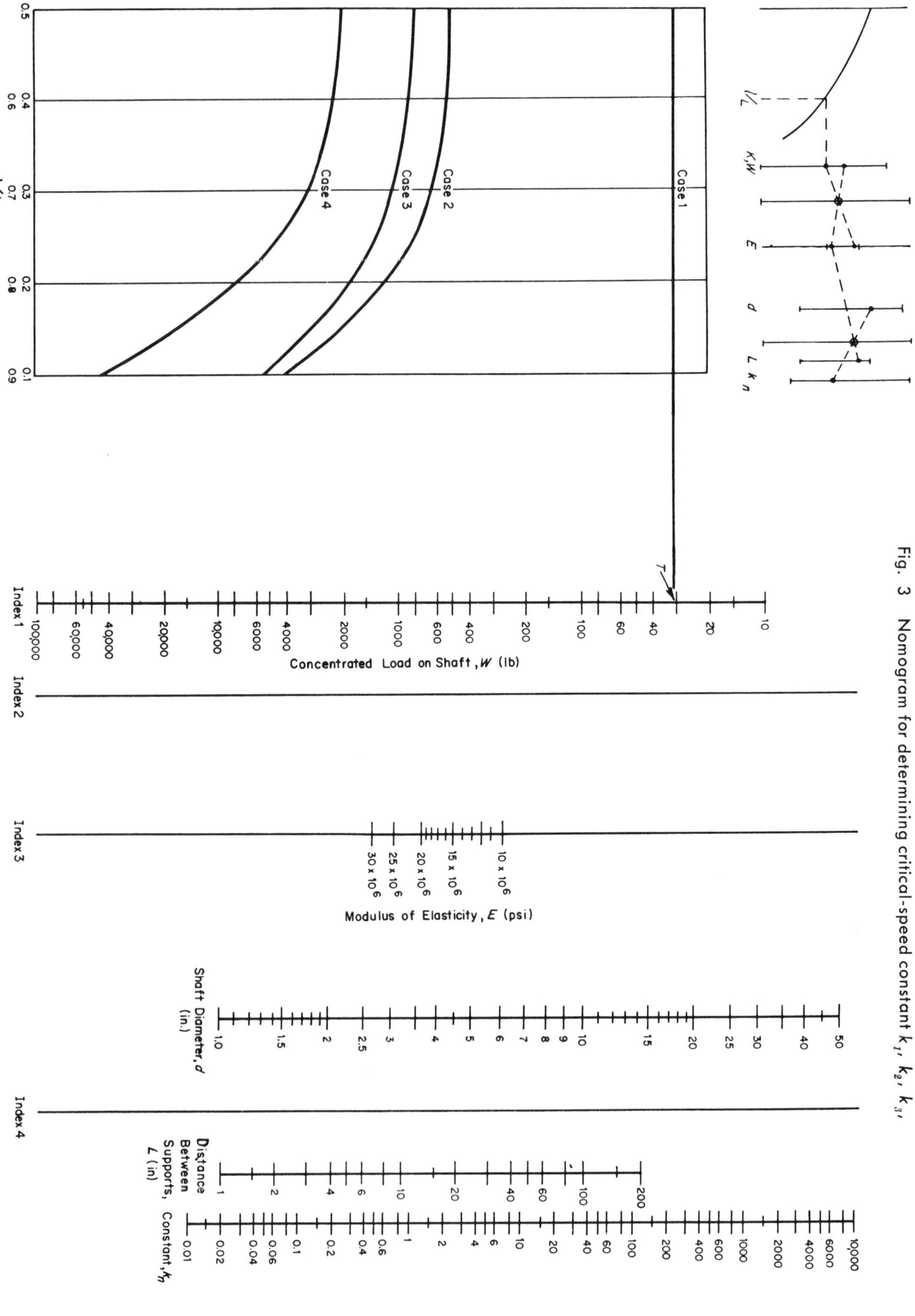

Fig. 3 Nomogram for determining critical-speed constant k_1, k_2, k_3,

nomogram, Fig. 3. Determination of the value of k_1 is given in detail. For example, for k_1:

1. Locate intersection of $l/L = 0.2$ and curve for case 2.

2. Move right to Index 1. For case 1, start at Point T, regardless of the values of l/L.

3. Line up intersection on Index 1 with $E = 30 \times 10^6$ psi, to intersect Index 2.

4. Align intersection on Index 2 with $W = 3{,}000$, to intersect Index 3.

5. Line up intersection on Index 3 with $L = 100$, to intersect Index 4.

6. Align intersection on Index 4 with $d = 10$, to intersect k_n at 17.

This is the value of k_1. The values of k_2, k_3, k_4, and k_5, Table 1, are found in a similar manner for the corresponding values W_2, W_3, W_4, and W_5.

By addition, $k = 174$. From Fig. 4, $N = 1{,}420$ rpm.

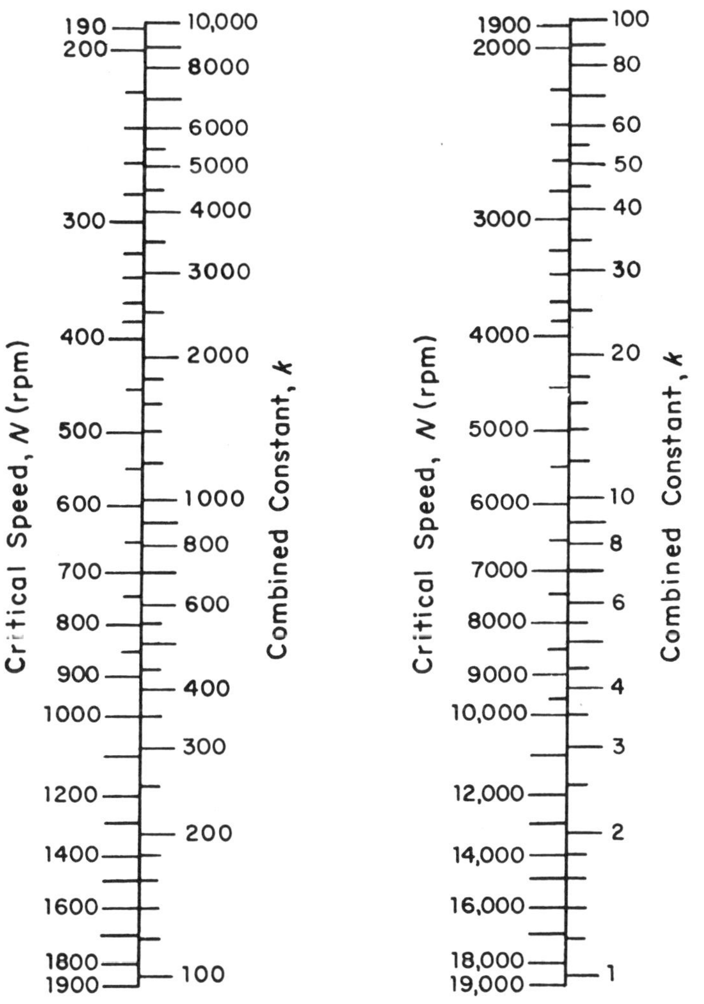

Fig. 4 Nomogram for determining critical-speed constant k_n for single-loaded shafts

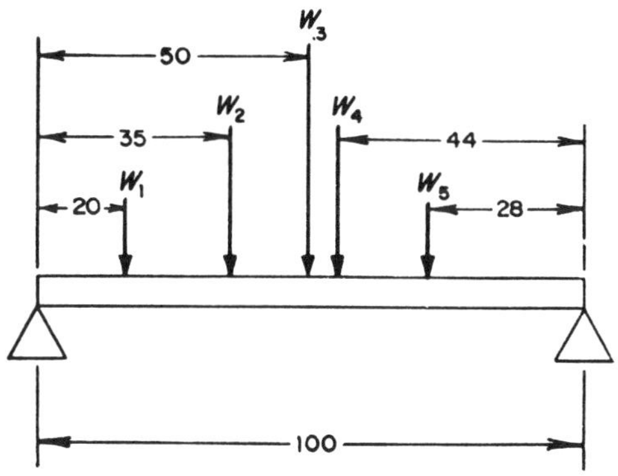

Fig. 5 Shaft loads used in Example problem

74. VIBRATION IN FLEXURE

(*Ref. 7, 13*)

Of Uniform, Unloaded Beams of Rectangular Cross Section

To avoid premature failure of equipment resulting from mechanical vibration, a structure should be designed so that resonance will not occur under the highest actual vibration frequency that may be encountered. Also, natural frequency must be low with respect to the lowest actual vibration frequency.

Many bodies or systems are able to vibrate in more than one manner, having, therefore, more than one natural frequency. Each is accompanied by its own mode or shape of vibration curve. The case considered is a beam having a uniform cross section and, therefore, a uniform weight distribution and elasticity.

The basic frequency equation for lateral beam vibration is:

$$f_n = C\sqrt{(gEI)/(wL^4)}, \qquad (10)$$

where:

f_n = natural frequency (cps)
C = dimensionless factor (*see* Table 8)
E = modulus of elasticity (psi)
g = gravitational acceleration = 386 in./sec^2
I = sectional moment of inertia (in.^{4})
w = weight of unit length beam (lb./in.)
L = beam length (in.)
ρ = density (lb./in.^{3})
b = beam width (in.)
h = beam height (in.)

Table 10 gives the values of mode factor C, and the deflection curve shape for six common beams.

For a uniform, rectangular cross-section beam:

$$I = bh^3/12 \text{ and } d = w\,bh$$

By substitution, equation (10) becomes:

$$f_n = 5.7\frac{Ch}{L^2}\sqrt{\frac{E}{\rho}} \qquad (11)$$

The ratio E/ρ, in equation (11), known as specific stiffness, is displayed in Fig. 6 for various classes of material. The bar graph shown has a plateau at $E/\rho = 10^8$ in., which contains most of the common structural materials. It is apparent that there is no difference in the "stiffness efficiency" of such materials as steel, aluminum, titanium, or magnesium; their specific stiffness $E/\rho = 105 \times 10^6$ in.

TABLE 10

	Fundamental mode	Second mode	Third mode	Fourth mode
CANTILEVER	C = 0.560	C = 3.51	C = 9.82	C = 19.2
SIMPLY SUPPORTED ENDS	C = 1.57	C = 6.28	C = 14.1	C = 25.2
FIXED ENDS	C = 3.56	C = 9.82	C = 19.2	C = 31.8
FREE ENDS	C = 3.56	C = 9.82	C = 19.2	C = 31.8
FIXED-HINGED	C = 2.45	C = 7.95	C = 16.6	C = 28.4
HINGED-FREE	C = 2.45	C = 7.95	C = 16.6	C = 28.4

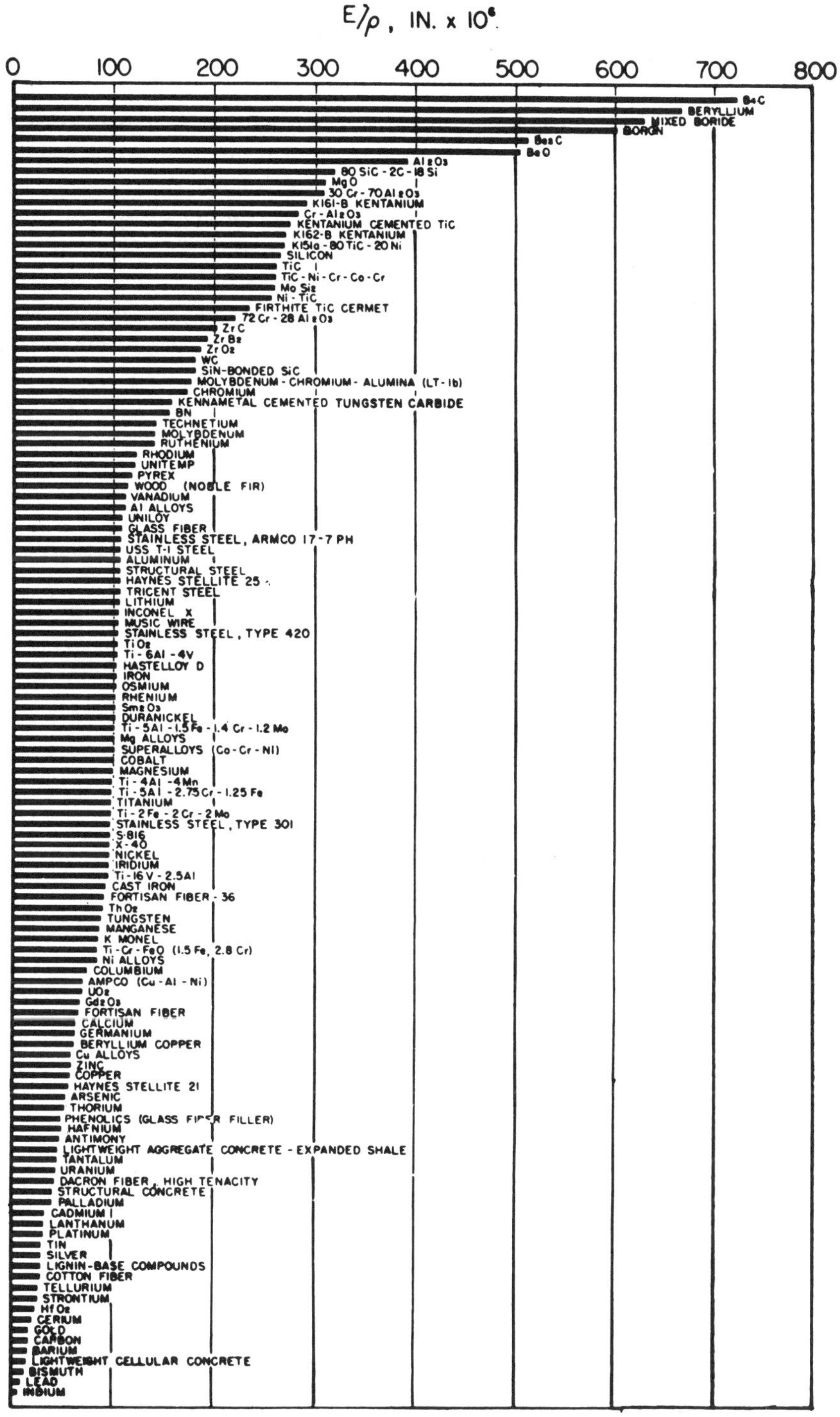

Fig. 6 · Specific stiffness at room temperature This chart is from "Selection of Material in Minimum Weight Design," Ralph L. Barnett, ASME, 63MD-25.

75. UNDAMPED FREE VIBRATION

(*Ref. 14*)

NATURAL FREQUENCIES OF WEIGHT "*W*" FOR MOST COMMON CASES

NOTATION

f = natural frequency (cps)
k = spring constant (Table 9)
W = total load (lb.)
g = acceleration due to gravity = 386 in./sec.^{2}
m = $W/386$ = mass
E = modulus of elasticity (psi)
I = sectional moment of inertia (in.^{4})
L = distance between supports (in.)
K = spring factor (*see* Table 9)
w = beam weight per unit length (lb./in.)
h = beam height (in.)
d = beam density (lb./in.^{3})

NATURAL FREQUENCE OF WEIGHT "*W*"

The basic equation for resonant frequency of any undamped system is:

$$f = \frac{1}{2\pi}\sqrt{\frac{K}{m}} \tag{12}$$

This equation can be applied to any structural element. It is only necessary to substitute the value of K (from Table 11) and m that apply to the particular structural configuration.

It is the purpose of this section to show how, by the use of a factor K, one single nomograph can provide solutions to vibration problems for 17 different cases as specified in Table 11.

By rearranging and substitution, equation (12) can be written thus:

$$f = 3.125\sqrt{\frac{KEI}{WL^3}} \tag{13}$$

The nomograph was based on equation (13). This nomograph also provides solutions to the natural period of vibration "T" in seconds during which a body or system, set in vibration, will complete one cycle. The natural period of vibration is the reciprocal of the natural frequency.

Example: A small steam turbine operates at speeds from 1,800 to 4,800 rpm. The turbine weights 250 lb. and is fastened to two 5-in., 9 lb.-per-ft. channels laid flat side up. Distance between the channel supports, functioning as simple beams, is 26 in. Determine the possibility of resonance and critical speed.

Solution: If the channels are considered to be simply supported at the ends and loaded uniformly over the full length, we find from Table 11, case 6, that $K = 76.8$. The total load on each beam is $W = \frac{1}{2}(250) + 1/12(26)(9) = 144.5$ lb. $I = 0.632$ in.^{4}, $E = 30 \times 10^6$ psi. By following the key diagram on the chart, find $f = 75.3$ rps. The number of vibrations per minute is $75.3 \times 60 = 4{,}518$. Therefore the critical turbine speed would be about 4,518 rpm. The channels are too flexible. Turning them 90 deg. would increase the moment of inertia to $I = 8.8$ in.^{4} For this new value of I the nomograph yields a safe resonant speed of 16,850 rpm.

THE INFLUENCE OF THE WEIGHT OF BEAM

What is the increase in load, ΔW, due to the weight of beam.

The formula for the undamped natural frequency in radians per second is:

$$\omega = 2\pi/\tau = \sqrt{g/y_1}$$

where:

τ = period of free vibration
y_1 = statical deflection of beam, under a load W_1

From Dunkerley's formula:

$$\frac{1}{\omega_A^2} = \frac{1}{\omega_1^2} + \frac{1}{\omega_2^2}$$

where:

ω_A = natural frequency of assembly
ω_1 = natural frequency of beam carrying only load W_1
ω_2 = natural frequency of beam only (weight of beam W_2).

By substitution we obtain:

$$\frac{1}{\omega_A^2} = \frac{y_1}{g} + \frac{y_2}{g} = \frac{1}{g}(y_1 + y_2)$$

The general equations of the maximum deflections are:

$$y_1 = W_1 L^3/K_1 EI, \quad y_2 = W_2 L^3/K_2 EI$$

or

$$y_1 + y_2 = \frac{L^3}{K_1 EI}\left[W_1 + \frac{K_1}{K_2} \times W_2\right]$$

From above equation we deduce that the increase in load $\Delta W = K_1/K_2\, W_2$, where K_1 and K_2 are factors equivalent to factor K in Table 11.

Example: Assume a cantilever triangular load. From Table 9, case 3, $K_2 = K = 15$. We assume that the weight of the beam is uniformly distributed along the length of the beam, therefore from Table 9, case 2, $K_1 = K = 8$. Finally $\Delta W = 15/8\, W_2 = 1.87\, W_2 = 187$ percent of W_2. Values of ΔW are tabulated in Table 11.

BEAM RESONANCE CHART

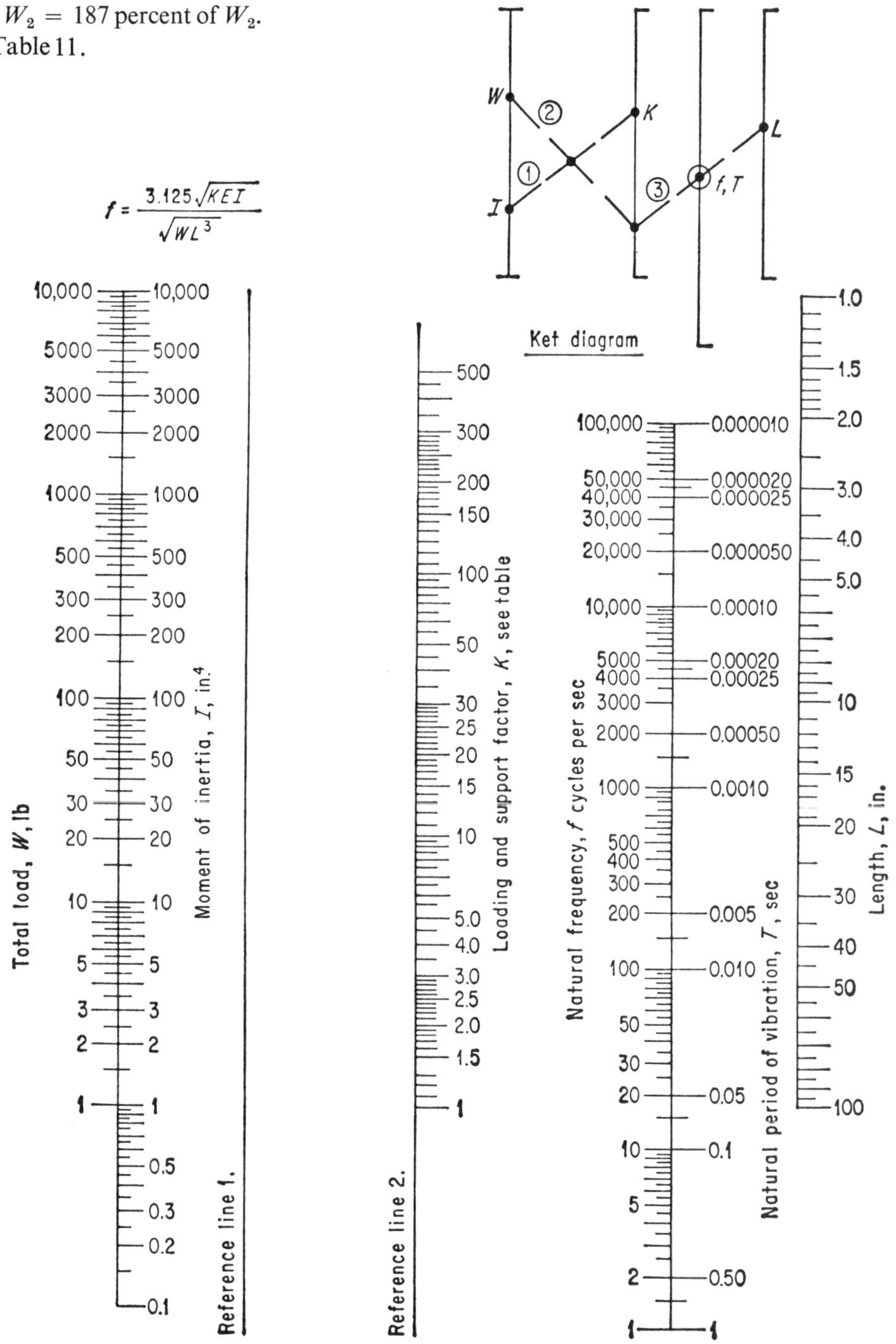

TABLE 11

LOADING & SUPPORTS	GEOMETRICAL CONFIGURATION	K	TOTAL LOAD, W	SPRING CONSTANTS, k	ΔW %
Cantilever, end load 1.	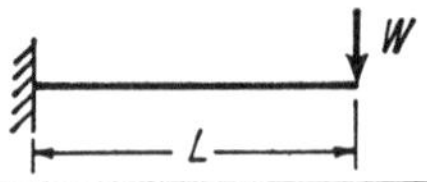	3	W	$\frac{3EI}{L^3}$	40
Cantilever, uniform load 2.	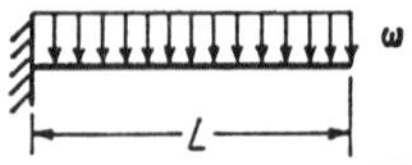	8	ωL	$\frac{8EI}{L^3}$	100
Cantilever, triangular load 3.	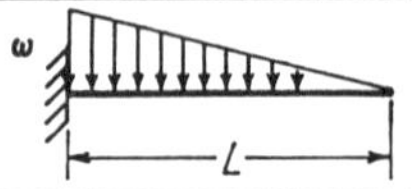	15	$\frac{1}{2}\omega L$	$\frac{15EI}{L^3}$	200
Cantilever, triangular load 4.	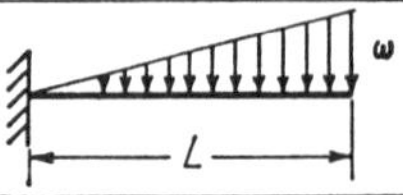	5.45	$\frac{1}{2}\omega L$	$\frac{5.45EI}{L^3}$	70
End supports, center load 5.	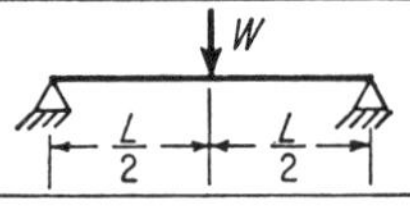	48	W	$\frac{48EI}{L^3}$	60
End supports, uniform load 6.	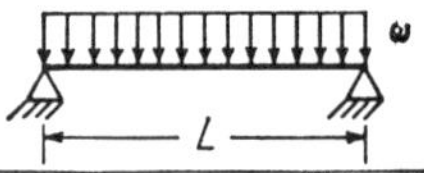	76.8	ωL	$\frac{76.8EI}{L^3}$	100
End supports, as shown 7.	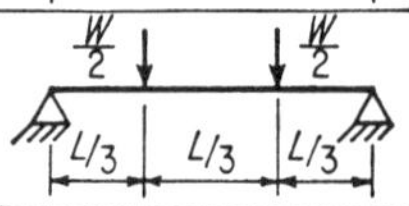	56.4	W	$\frac{56.4EI}{L^3}$	70
End supports, triangular load 8.	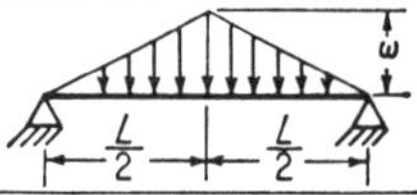	60	$\frac{1}{2}\omega L$	$\frac{60EI}{L^3}$	80
End supports, triangular load 9.	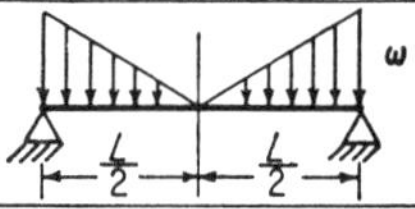	106.7	$\frac{1}{2}\omega L$	$\frac{106.7EI}{L^3}$	140
One end fixed, one end supported, center load 10.	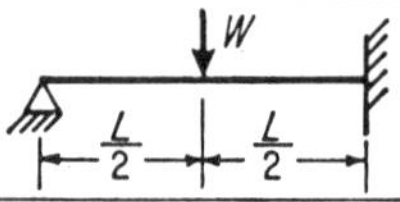	107.3	W	$\frac{107.3EI}{L^3}$	60
One end fixed, one end supported, uniform load 11.	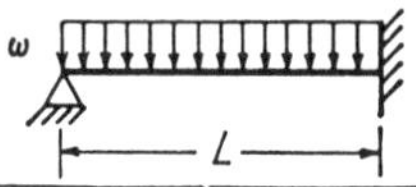	185	ωL	$\frac{185EI}{L^3}$	100
One end fixed, one end supported, as shown 12.	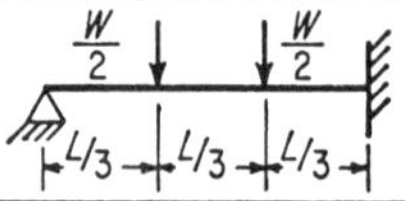	132	W	$\frac{132EI}{L^3}$	70
One end fixed, one end supported, triangular load 13.	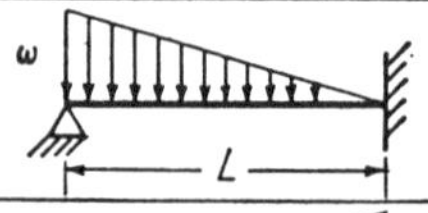	164	$\frac{1}{2}\omega L$	$\frac{164EI}{L^3}$	90
One end fixed, one end supported, triangular load 14.	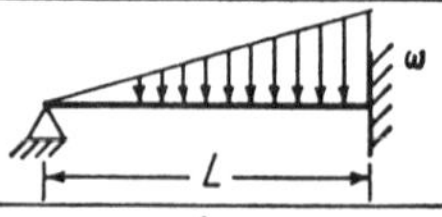	210	$\frac{1}{2}\omega L$	$\frac{210EI}{L^3}$	110
Both ends fixed, center load 15.	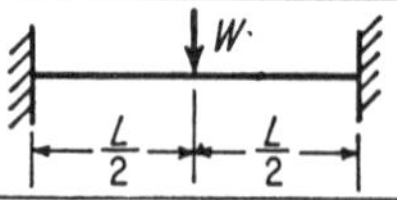	192	W	$\frac{192EI}{L^3}$	50
Both ends fixed, uniform load 16.	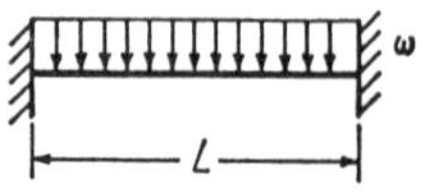	384	ωL	$\frac{384EI}{L^3}$	100
Both ends fixed, triangular load 17.	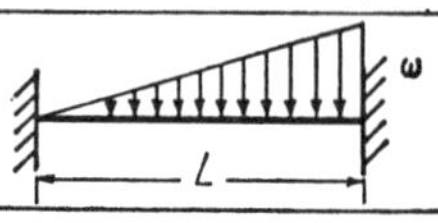	382	$\frac{1}{2}\omega L$	$\frac{382EI}{L^3}$	100

Chap. XI

REFERENCES

Section	*Ref. No.*	
71	1	GRIFFEL, W.: "Vibration Frequency Charts," *Design News*, April 3, 1963
71	2	YOUNG, D.: "Vibration of Rectangular Plates by the Ritz Method," A.S.M.E. Paper No. 50–APM–18, June 1950
71	3	BARTON, M. V.: "Vibration of Rectangular and Skew Cantilever Plates," A.S.M.E. Paper No. 50–A–12, Dec. 1950
71	4	AYRE, R. S., and JACOBSON, L. S.: "Natural Frequencies of Continuous Beams of Uniform Span Length," A.S.M.E. Paper No. 50–APM–21, June 1950
71	5	ADLER, A. A., and CRANCH, E. T.: "Vibration of Variable Section Beams," A.S.M.E. Paper No. 55–A–19
71	6	MACDUFF, J. N., and FELGAR, R. P.: "Vibration Design Charts," *A.S.M.E. Transaction*, Nov. 1958
71	7	CHURCH, A. H.: *Mechanical Vibrations*, John Wiley, N.Y., 1957
72	8	GRIFFEL, W.: "Critical Speed of Shafts," *Machine Design*, Oct. 12, 1961
72	9	VALANCE, A., and DOUGHTIE, V.: *Design of Machine Members*, McGraw-Hill, N.Y., 1951
72	10	*Mark's Mechanical Engineer's Handbook*, McGraw-Hill, N.Y., 6th ed., p. 105–106.
73	11	GRIFFEL, W.: "Critical Speed of Multiple-loaded Shafts," *Machine Design*, May 10, 1962
74	12	GRIFFEL, W.: "Vibration in Flexure," *Design News*, July 11, 1962
74	13	BARNETT, R. L.: "Selection of Material in Minimum Weight Design," A.S.M.E. Paper 63MP–25
75	14	GRIFFEL, W.: "Resonance of an Undamped Beam," *Product Engineering*, Aug. 20, 1962
72	15	HALSEY, A. F.: *Handbook for Machine Designers*, McGraw-Hill, N.Y., 1916.

Chapter XII

EQUATIONS FOR MASS AND AREA PROPERTIES

76. EQUATIONS FOR MASS AND AREA PROPERTIES OF VARIOUS GEOMETRICAL SHAPES*

(*Ref. 1*)

CONTENTS

*Reprinted, with permission, from NAVWEPS Report 7827, NOTS TP2838, April, 1962.

LIST OF DIAGRAMS

The following is a list of the geometrical shapes for which diagrams and equations are given.

NOMENCLATURE

An effort has been made to typify symbols for thickness, height, and base width for all shapes and plane areas considered. However, in some instances, additional nomenclature has been introduced for descriptive purposes. It is advisable to refer to the diagram associated with each item.

A Area, in^2

I Moment of inertia. The unit is generally expressed as follows: for a solid, $lb\text{-}ft^2$, $slug\text{-}ft^2$, etc; for a plane area, in^4, ft^4, etc.

I_p Polar moment of inertia

I_x Moment of inertia about the x-axis

I_y Moment of inertia about the y-axis

I_z Moment of inertia about the z-axis

I_{xy} Product of inertia in the x-y plane

I_{xz} Product of inertia in the x-z plane

I_{yz} Product of inertia in the y-z plane

k_x Radius of gyration about the x-axis

k_y Radius of gyration about the y-axis

k_z Radius of gyration about the z-axis

L Length, in. Where t (thickness) or d (diameter) of a thin rod is constant, V, W, and m are proportional. Statical moments and moments of inertia of the area or body may be functions of L

m Mass, lb or slugs

V Volume, in^3

W Weight, lb

$\bar{x}$ Centroidal distance along the x-axis, in.

$\bar{y}$ Centroidal distance along the y-axis, in.

$\bar{z}$ Centroidal distance along the z-axis, in.

ρ Mass density factor, lb/in^3; m/V

INTRODUCTION

In certain sections of this chapter, equations for moments of inertia and centroidal distances are developed through the use of the calculus, which, it is believed, will serve as a supplementary method for finding information not included herein. Other, simplified, forms for calculating properties are included that can be used in lieu of integral calculus.

As an aid in finding a particular item, the diagrams and accompanying equations for the various shapes are grouped under the following headings shown at the tops of the respective pages: solids, thin shells, thin rods, plane areas, and ogival shapes.

Certain structural shapes such as channels, I-beams, angles, and T-sections are omitted because these are normally found in a construction manual such as the AISC steel construction manual. However, variations of these sections, without fillets and bulbs, are included.

A separate section is devoted to the properties of ogival shapes, which are commonly used in the design of missile nose cones. The solid and thin-shelled tangent ogives are included because of their frequent use.

Special Notes

1. A thin-shelled body is one in which $t < (L/30)$, where t is the gage thickness of the material and L is the length, or radius, perpendicular to the axis of rotation measured at the maximum diameter.

2. A thin rod is one in which $L \geqq 30d$, where L is the length of the rod and d is the diameter of the rod.

3. Elliptic-area formulas may be used for circular complements such as half circles and quarter circles by substituting $a = b = R$.

4. Weight moments of inertia for plane areas can be obtained by multiplying the area moment of inertia by the area mass, M, and then dividing by the section area.

5. Linear dimensions are in inches in the sections that follow.

6. In most cases involving integration, cartesian, or rectangular, coordinates are used. Should the need arise to use polar coordinates for ease of integration of special integrals, it is advised that a review of a calculus text or similar reference be made. Also, double integration methods, if used, will in many instances reduce calculation time.

SOLIDS

SUMMARY OF EQUATIONS FOR MASS AND VOLUME PROPERTIES

Centroid by Integration (Homogeneous Mass)

$$\bar{x} = \frac{\int x\,dV}{\int dV}, \qquad \bar{y} = \frac{\int y\,dV}{\int dV}, \qquad \bar{z} = \frac{\int z\,dV}{\int dV}$$

Center of Gravity of a Coplanar System of Particles

$$\bar{x} = \frac{M_y}{m} = \frac{\Sigma Mx}{M}, \qquad \bar{y} = \frac{M_x}{m} = \frac{\Sigma My}{M}$$

Center of Gravity of a Non-coplanar System of Particles

$$\bar{x} = \frac{M_{yz}}{m} = \frac{\Sigma Mx}{M}, \qquad \bar{y} = \frac{M_{xz}}{m} = \frac{\Sigma My}{M}, \qquad \bar{z} = \frac{M_{xy}}{m} = \frac{\Sigma Mz}{M}$$

Moment of Inertia

$$I = \int r^2\,dm$$

Product of Inertia

$$I_{xy} = \int xy\,dm, \qquad I_{yz} = \int yz\,dm, \qquad I_{xz} = \int xz\,dm$$

Transfer of Axes

$$I = \bar{I} + md^2$$

$$k^2 = \bar{k}^2 + d^2$$

Radius of Gyration

$$k = \sqrt{I/m}$$

Centroid of a Composite Solid Body

$$\bar{x} = \frac{\Sigma W\bar{x}}{W} = \frac{\Sigma L\bar{x}}{L} = \frac{\Sigma V\bar{x}}{V}$$

$$\bar{y} = \frac{\Sigma W\bar{y}}{W} = \frac{\Sigma L\bar{y}}{L} = \frac{\Sigma V\bar{y}}{V}$$

$$\bar{z} = \frac{\Sigma W\bar{z}}{W} = \frac{\Sigma L\bar{z}}{L} = \frac{\Sigma V\bar{z}}{V}$$

CENTER OF GRAVITY OF A SYSTEM OF PARTICLES

Two systems of particles exist that will be defined as coplanar and non-coplanar.

Coplanar Particles

The first system, coplanar particles, can be resolved into a common mass located at such a position that the moment of its mass with respect to the x-axis would be equal to the moment-sum about the x-axis, and that the moment of its mass with respect to the y-axis would be equal to the moment-sum about the y-axis. The two coordinates presented would then locate the point that represents the center of gravity of the system in the x-y plane.

Therefore,

$$\bar{x}m = My \qquad \text{and} \qquad \bar{y}m = Mx$$

or

$$\bar{x} = \frac{My}{m} = \frac{m_1x_1 + m_2x_2 + \cdots + m_nx_n}{m_1 + m_2 + \cdots + m_n} = \frac{\Sigma mx}{\Sigma m}$$

and

$$\bar{y} = \frac{Mx}{m} = \frac{m_1y_1 + m_2y_2 + \cdots + m_ny_n}{m_1 + m_2 + \cdots + m_n} = \frac{\Sigma my}{\Sigma m}$$

Example.

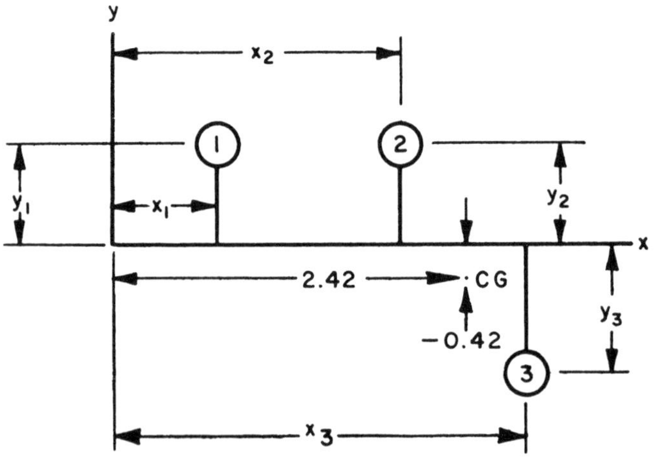

$x_1 = 1$ in.,	$y_1 = 1$ in.,	$m_1 = 1$ lb
$x_2 = 2$ in.,	$y_2 = 2$ in.,	$m_2 = 2$ lb
$x_3 = 3$ in.,	$y_3 = -2$ in.,	$m_3 = 4$ lb

$$\bar{x} = \frac{(1)(1) + (2)(2) + (4)(3)}{1 + 2 + 4} = 2.428 \text{ in.}$$

$$\bar{y} = \frac{(1)(1) + (2)(2) + (4)(-2)}{1 + 2 + 4} = -0.429 \text{ in.}$$

Note. Care must be taken to account for the proper signs when calculating the moment-sums.

Non-coplanar Particles

The second system, non-coplanar particles, can be resolved into a common mass located at a point in space represented by the coordinates $\bar{x}$, $\bar{y}$, and $\bar{z}$, which are readily calculated by the procedure used in the coplanar system and with the inclusion of the third reference-plane coordinate.

Thus,

$$\bar{x} = \frac{Myz}{M} = \frac{m_1x_1 + m_2x_2 + \cdots + m_nx_n}{m_1 + m_2 + \cdots + m_n} = \frac{\Sigma mx}{\Sigma m}$$

$$\bar{y} = \frac{Mxz}{M} = \frac{m_1y_1 + m_2y_2 + \cdots + m_ny_n}{m_1 + m_2 + \cdots + m_n} = \frac{\Sigma my}{\Sigma m}$$

$$\bar{z} = \frac{Mxy}{M} = \frac{m_1z_1 + m_2z_2 + \cdots + m_nz_n}{m_1 + m_2 + \cdots + m_n} = \frac{\Sigma mz}{\Sigma m}$$

Example.

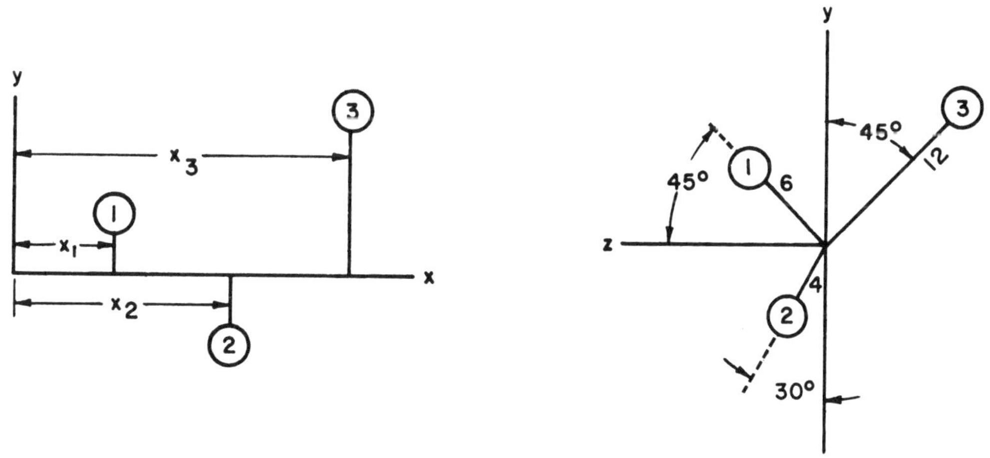

$x_1 = 2$ in., $y_1 = 6 \cos 45° = 4.24$, $z_1 = 6 \cos 45° = 4.24$

$x_2 = 4$ in., $y_2 = -4 \cos 30° = -3.46$, $z_2 = 4 \cos 60° = 2$

$x_3 = 6$ in., $y_3 = 12 \cos 45° = 8.48$, $z_3 = -12 \cos 45° = -8.48$

$m_1 = 1$ lb, $m_2 = 2$ lb, $m_3 = 3$ lb

$$\bar{x} = \frac{(1)(2) + (2)(4) + (3)(6)}{6} = 4.66 \text{ in. (ans.)}$$

$$\bar{y} = \frac{(1)(4.24) + (2)(-3.46) + (3)(8.48)}{6} = 3.79 \text{ in. (ans.)}$$

$$\bar{z} = \frac{(1)(4.24) + (2)(2) + (3)(-8.48)}{6} = -2.87 \text{ in. (ans.)}$$

The resulting coordinates of 4.66, 3.79, and -2.87 inches fix the location of the system's center of gravity.

TRANSFER OF AXES ON A SOLID BODY

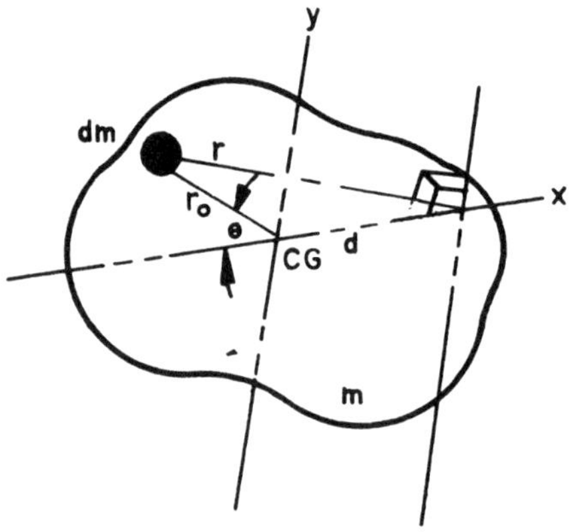

Let the radial distances from the two axes to any element of mass dm equal r and r_0, with the separation of the axes being d; applying the law of cosines, $r^2 = r_0^2 + d^2 + 2r_0 d \cos\theta$. The definition of the mass moment of inertia gives

$$I = \int r^2 dm = \int (r_0^2 + d^2 + 2r_0 d \cos\theta) dm$$
$$= \int r_0^2 dm + d^2 \int dm + 2d \int r_0 \cos\theta dm$$

Since the y-coordinate of the center of gravity with respect to an origin at θ is zero, the third integral drops out, leaving

$$I = \bar{I} + md^2$$

RADIUS OF GYRATION

The radius of gyration, k, of a body with respect to any axis is defined as the distance from the axis at which the mass may be conceived to be concentrated and to have the same moment of inertia with respect to the axis as does the actual whole, or distributed, mass.

Mathematically, $k = \sqrt{I/m}$ by definition. Substitution into $I = \bar{I} + md^2$ results in $k^2 = \bar{k}^2 + d^2$, which provides a method for transferring the centroidal radius of gyration to a parallel axis on the same body.

PRODUCT OF INERTIA

Generally, a three-dimensional body has three moments of inertia about the three mutually perpendicular coordinate axes and three products of inertia about the three coordinate planes.

The product of inertia of the body with respect to a pair of coordinate planes is the algebraic sum of the products obtained by multiplying the mass of each element of the body by its coordinates with reference to these planes. The value of the product of inertia can be positive, negative, or zero.

Mathematically, the products of inertia about the three planes are expressed as

$$I_{xy} = \int xydm, \qquad I_{yz} = \int yzdm, \qquad I_{xz} = \int xzdm$$

where dm is an element of mass. Or, the product of inertia may be calculated for an area where the two rectangular coordinate axes provide the system on which the computation is based, in the form

$$I_{xy} = \int xydA$$

where dA is an element of area, and x and y are the respective distances from the axes to the elements of area.

An application of the product of inertia may be seen on page 53 covering the properties of an angle.

CENTROIDS OF COMPOSITE VOLUMES

The determination of the centroid of a composite solid shape can be calculated by the application of the moment principle, in which the basic relationship takes the form

$$(w_1 + w_2 + w_3 + \cdot \cdot \cdot)\bar{X} = w_1\bar{x}_1 + w_2\bar{x}_2 + w_3\bar{x}_3 + \cdot \cdot \cdot$$

where w represents the weight of each part, $\bar{X}$ represents the x-coordinate of the center of gravity of the total body, and $\bar{x}$ represents the center of gravity of the individual parts. The resulting basic relationships are, therefore,

$$\bar{X} = \frac{\Sigma w\bar{x}}{W}, \qquad \bar{Y} = \frac{\Sigma w\bar{y}}{W}, \qquad \bar{Z} = \frac{\Sigma w\bar{z}}{W}$$

In each of the above relationships, weight values may be replaced with values for length, area, or volume, depending on the shape of the object.

It is likely that interest will be found in the determination of the centroidal distance of a hollow geometrical shape such as the frustum of a cone; the applicable equations will then be

$$\bar{X} = \frac{\Sigma v\bar{x}}{V}, \qquad \bar{Y} = \frac{\Sigma v\bar{y}}{V}, \qquad \bar{Z} = \frac{\Sigma v\bar{z}}{V}$$

where it is obvious that only one of these relationships is valid for a symmetrical body.

For hollow objects, or objects with holes in them, it should be realized that the void volume must be subtracted, as in the case of a hollow cone frustum.

$$\bar{X} = \frac{\Sigma v\bar{x}}{V} = \frac{(V\bar{x})_T - (V\bar{x})_H}{V_T - V_H}$$

where $(V\bar{x})_T$ represents the total solid frustum and $(V\bar{x})_H$ represents the inner, or hollow, frustum.

CENTROID OF A VOLUME

By expansion of the concepts used in the resolution of a system of particles, it is apparent that a summation, by integration, of a differential element of a body, leads to the determination of the centroid of the body.

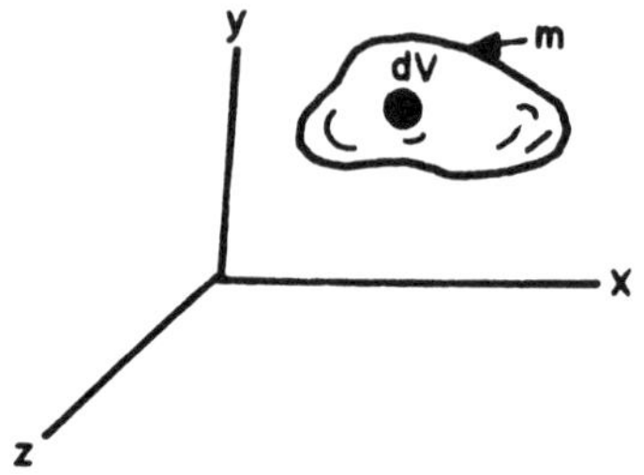

If the body is homogeneous, the density of the body, ρ, will be considered constant. Therefore, the element of mass is

$$dm = \rho dV$$

and, for the entire body,

$$m = \int \rho dV$$

Using previous equations and substituting,

$$\bar{x} = \frac{\int x dm}{m} = \frac{\int \rho x dV}{\int \rho dV} = \frac{\int x dV}{\int dV}$$

$$\bar{y} = \frac{\int y dm}{m} = \frac{\int \rho y dV}{\int \rho dV} = \frac{\int y dV}{\int dV}$$

$$\bar{z} = \frac{\int z dm}{m} = \frac{\int \rho z dV}{\int \rho dV} = \frac{\int z dV}{\int dV}$$

Therefore, it can be seen that the first moments of each summation are $V\bar{x}$, $V\bar{y}$, and $V\bar{z}$ for a homogeneous body.

Example. The centroid of the solid generated by revolving the area of the half parabola $y = 4 - x^2$ about the y-axis may be determined as follows.

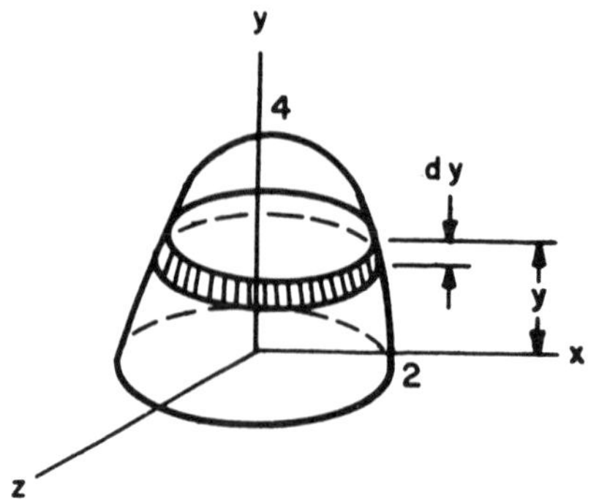

$$dV = \pi x^2 dy$$

$$\int dV = \int_0^4 \pi(4 - y)dy$$

$$\int dV = \pi\int_0^4 (4 - y)dy$$

$$V = \pi\Big[4y - (y^2/2)\Big]_0^4 = \pi(16 - 8) = 8\pi$$

$$V\bar{y} = \int y dV$$

$$= \int_0^4 \pi y x^2 dy = \pi\int_0^4 (4y - y^2)dy = \pi\Big[2y^2 - (y^3/3)\Big]_0^4$$

$$= \pi\Big[32 - (64/3)\Big] = 32\pi/3$$

Therefore,

$$\bar{y} = \frac{32\pi/3}{8\pi} = \frac{32}{24} = \frac{4}{3} \text{ (ans.)}$$

and $\bar{z} = \bar{x} = 0$, by symmetry.

MASS MOMENT OF INERTIA

The inertial resistance to rotational acceleration is that property of a body which is commonly known as its mass moment of inertia.

If a body of mass m is allowed to rotate about an axis at an angular acceleration α, an element of this mass, dm, will have a component of acceleration tangent to the circular path of $r\alpha$, with the tangential force on the element being $r\alpha dm$. Since the distance to the element is r, the resulting moment on the force equals $r^2\alpha dm$.

Integrating the elements of the body,

$$I = \int r^2 dm$$

an expression is obtained that is known as the mass moment of inertia of the body, where α is dropped out because it is constant for a given rigid body.

If the body is of constant mass density, the differential, dm, may be replaced with ρdV, since $dm = \rho dV$, and the following expression results

$$I = \rho\int r^2 dV$$

The units of mass moment of inertia are commonly expressed as lb-ft-sec^2 or slug-ft^2, or, dimensionally, ML^2.

Examples.

RIGHT CIRCULAR CYLINDER

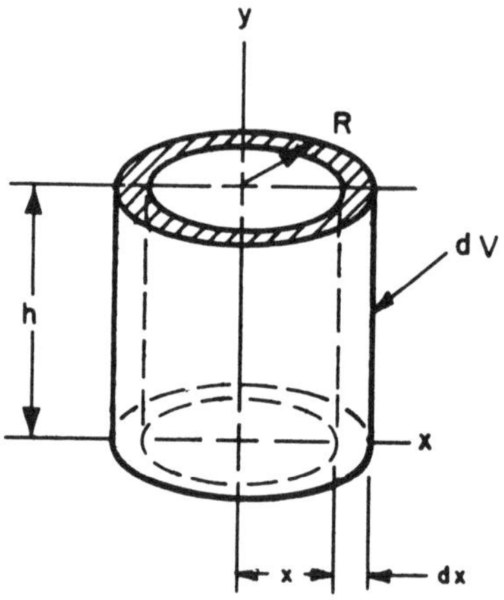

The mass moment of inertia for a right circular cylinder is found by the use of $I = \int r^2 dm$ and the equation for elemental cylinders, as follows

$$dm = \rho dV = \rho 2\pi x h dx$$

$$I_y = \rho \int_0^R (x^2)(2\pi x) h dx$$

$$= 2\pi\rho h \int_0^R x^3 dx$$

$$= \frac{2\pi\rho h}{4}\Big[x^4\Big]_0^R = \frac{\rho\pi h R^4}{2}$$

$$\rho = \frac{M}{V} = \frac{M}{\pi R^2 h}$$

$$I_y = \frac{M}{\pi R^2 h}\left(\frac{\pi h R^4}{2}\right) = \frac{MR^2}{2} \quad \text{(ans.)}$$

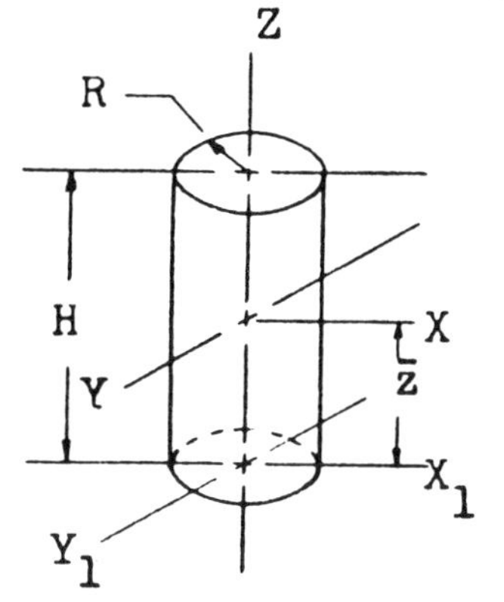

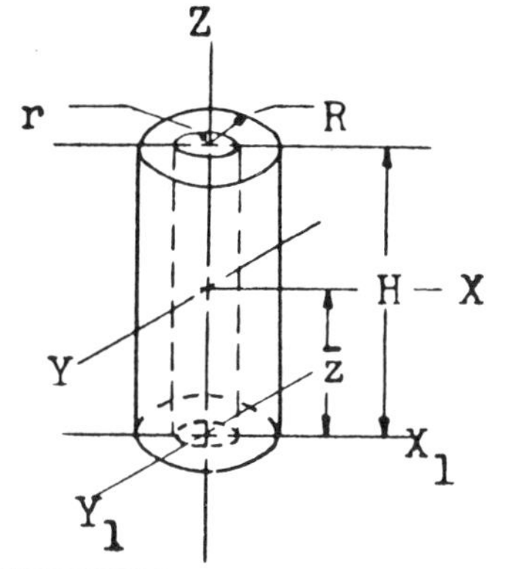

	RIGHT CIRCULAR CYLINDER	HOLLOW RIGHT CIRCULAR CYLINDER
VOLUME	$\pi R^2 H$	$\pi H(R^2 - r^2)$
CENTROID	$\bar{z} = \frac{H}{2}$	$\bar{z} = \frac{H}{2}$
WEIGHT MOMENT OF INERTIA	$I_x = I_y = \frac{W}{12}(3R^2 + H^2)$ $I_{x_1} = I_{y_1} = \frac{W}{12}(3R^2 + 4H^2)$ $I_z = \frac{WR^2}{2}$	$I_x = I_y = \frac{W}{12}\left[3(R^2 + r^2) + H^2\right]$ $I_{x_1} = I_{y_1} = W\left[\frac{R^2 + r^2}{4} + \frac{H^2}{3}\right]$ $I_z = \frac{W}{2}(R^2 + r^2)$
RADIUS OF GYRATION	$k_x = k_y = 0.289\sqrt{3R^2 + H^2}$ $k_{x_1} = k_{y_1} = 0.289\sqrt{3R^2 + 4H^2}$ $k_z = 0.707\ R$	$k_x = k_y = 0.289\sqrt{3(R^2 + r^2) + H^2}$ $k_z = 0.707\sqrt{R^2 + r^2}$ $k_{x_1} = k_{y_1} = \sqrt{\frac{R^2 + r^2}{4} + \frac{H^2}{3}}$

Adapted from Weight Handbook, Vol. 1, Society of Aeronautical Weight Engineers, Inc.

	RIGHT CIRCULAR CONE	FRUSTUM OF A CONE
	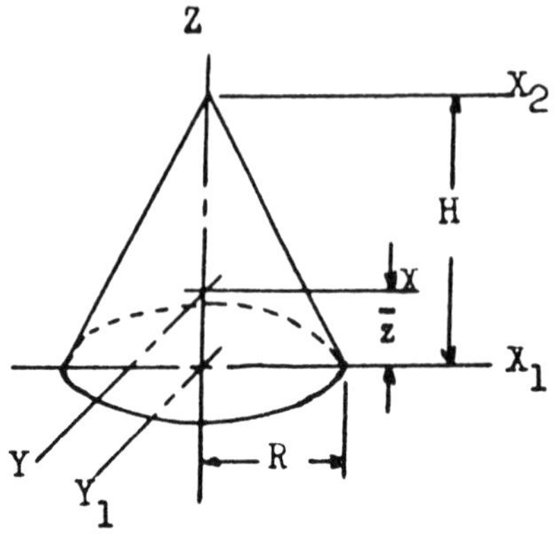	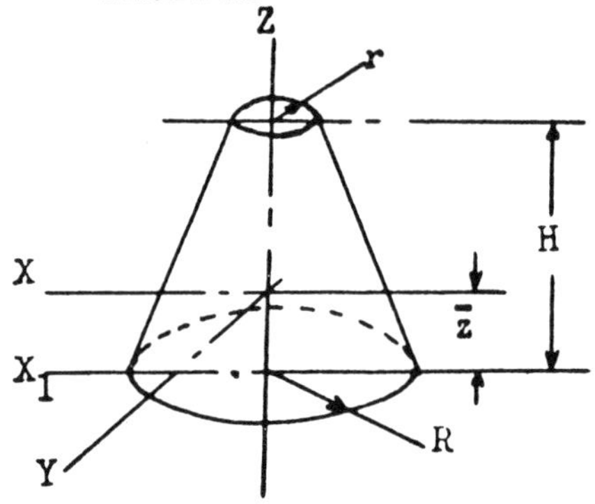
VOLUME	$\frac{\pi R^2 H}{3}$	$\frac{\pi H}{3}(R^2 + Rr + r^2)$
CENTROID	$\bar{z} = \frac{H}{4}$	$\bar{z} = \frac{H}{4}\left[\frac{R^2 + 2Rr + 3r^2}{R^2 + Rr + r^2}\right]$
WEIGHT MOMENT OF INERTIA	$I_x = I_y = \frac{3W}{20}(R^2 + \frac{H^2}{4})$ $I_{x_1} = I_{y_1} = \frac{W}{20}(3R^2 + 2H^2)$ $I_{x_2} = \frac{3W}{20}(R^2 + 4H^2)$ $I_z = \frac{3W}{10} R^2$	$I_z = \frac{3W}{10}\left[\frac{R^5 - r^5}{R^3 - r^3}\right]$
RADIUS OF GYRATION	$k_x = k_y = 0.387\sqrt{R^2 + \frac{H^2}{4}}$ $k_{x_1} = k_{y_1} = 0.224\sqrt{3R^2 + 2H^2}$ $k_{x_2} = 0.387\sqrt{R^2 + 4H^2}$ $k_z = 0.548\ R$	$k_z = 0.548\sqrt{\frac{R^5 - r^5}{R^3 - r^3}}$

Adapted from Weight Handbook, Vol. 1, Society of Aeronautical Weight Engineers, Inc.

CONE

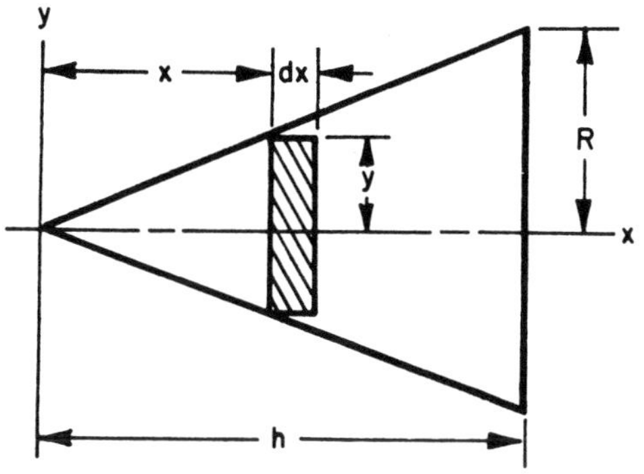

By similar triangles, $y = (Rx/h)$

$$dm = \rho dV$$

$$I_x \text{disk} = mR^2/2$$

$$r^2 = y^2$$

$$dI_x = dmr^2/2$$

$$= (\rho \pi y^2 dx)(y^2/2)$$

$$I_x = \int r^2 dm$$

$$I_x = \int_0^h \rho \pi y^4 dx/2$$

$$= \frac{\rho\pi}{2}\int_0^h \left(\frac{Rx}{h}\right)^4 dx = \frac{\rho\pi}{2}\left(\frac{R^4}{h^4}\right)\int_0^h x^4 dx$$

$$= \frac{\rho \pi R^4}{2h^4}\left[\frac{x^5}{5}\right]_0^h = \frac{\rho \pi R^4 h}{10}$$

$$\rho = \frac{m}{V} = \frac{m}{\pi R^2 h/3}$$

$$I_x = \frac{3m\pi R^4 h}{10\pi R^2 h} = \frac{3mR^2}{10} \text{ (ans.)}$$

FRUSTUM OF A CONE

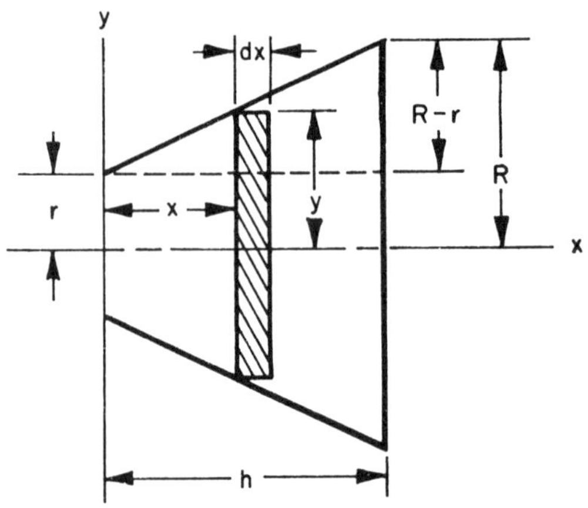

By similar triangles,

$$\frac{R - r}{h} = \frac{y - r}{x}$$

$$x = \frac{hy - hr}{R - r} = \frac{hy}{R - r} - \frac{hr}{R - r}$$

$$dx = \left(\frac{h}{R - r}\right)dy$$

$$I_x \text{disk} = mR^2/2$$

$$dI_x = y^2 dm/2$$

$$dm = \rho dV = \rho\pi y^2 dx$$

$$dI_x = \frac{y^2}{2}(\rho\pi y^2 dx) = \frac{\rho\pi y^4}{2}\left(\frac{h}{R - r}\right)dy$$

$$I_x = \frac{\rho\pi h}{2(R - r)} \int_r^R y^4 dy$$

$$= \frac{\rho\pi h}{2(R - r)}\left[\frac{y^5}{5}\right]_r^R = \frac{\rho\pi h}{10} \frac{(R^5 - r^5)}{(R - r)}$$

$$V = \frac{\pi h}{3} (R^2 + Rr + r^2)$$

$$I_x = \frac{m\pi h(R^5 - r^5)(3)}{10\pi h(R^2 + Rr + r^2)(R - r)}$$

$$= \frac{3m}{10} \frac{(R^5 - r^5)}{(R^3 - r^3)} \text{ (ans.)}$$

SPHERE

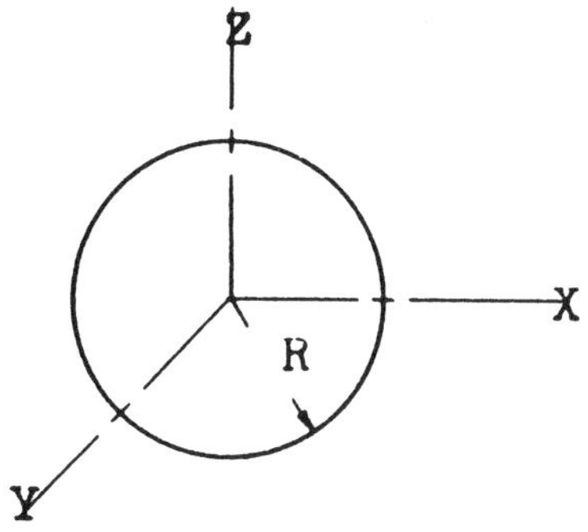

HOLLOW SPHERE

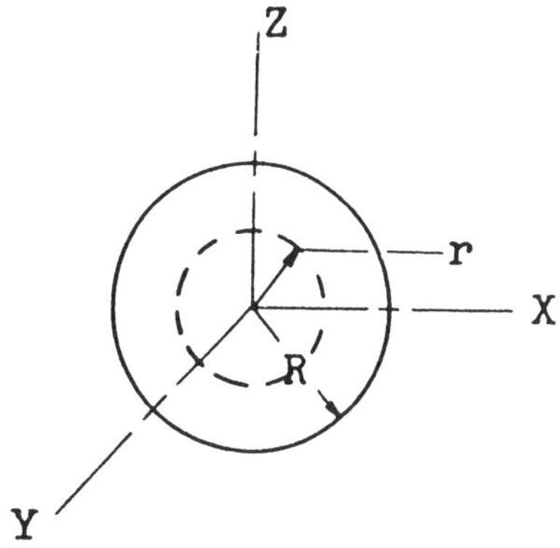

	SPHERE	HOLLOW SPHERE
VOLUME	$\frac{4}{3}\pi R^3$	$\frac{4}{3}\pi (R^3 - r^3)$
CENTROID	....	....
WEIGHT MOMENT OF INERTIA	$I_x = I_y = I_z = \frac{2}{5} WR^2$	$I_x = I_y = I_z = \frac{2}{5} W \left(\frac{R^5 - r^5}{R^3 - r^3}\right)$
RADIUS OF GYRATION	$k_x = k_y = k_z = 0.632\ R$	$k_x = k_y = k_z = 0.632\sqrt{\left(\frac{R^5 - r^5}{R^3 - r^3}\right)}$

Adapted from Weight Handbook, Vol. 1, Society of Aeronautical Weight Engineers, Inc.

SOLID HOMOGENEOUS SPHERE

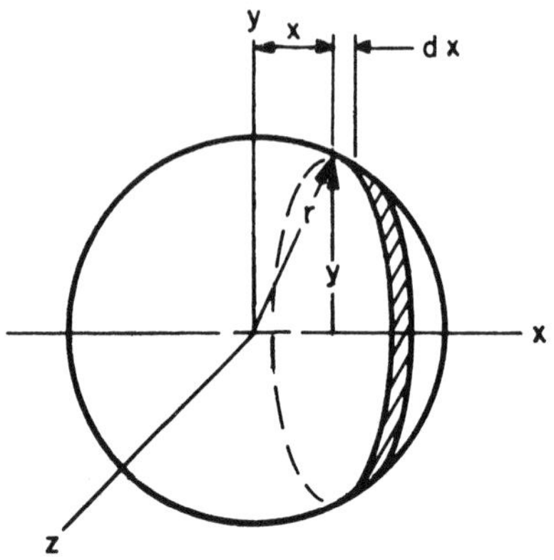

$$x = \sqrt{r^2 - y^2}$$

$$x^2 = r^2 - y^2$$

$$y^2 = r^2 - x^2$$

$$dm = \rho \pi y^2 dx$$

$$dI_x = \frac{dmy^2}{2} = \frac{(\rho \pi y^2 dx) y^2}{2} = \frac{\pi \rho (r^2 - x^2)^2 dx}{2}$$

$$I_x = \frac{\pi \rho}{2} \int_{-r}^{r} (r^2 - x^2)^2 dx = \frac{8 \pi \rho r^5}{15}$$

$$\rho = \frac{m}{V} = \frac{m}{(4/3) \pi r^3}$$

$$I_x = \frac{(8 \pi r^5 / 15) m}{(4/3) \pi r^3} = \frac{2}{5} mr^2 \text{ (ans.)}$$

	HEMISPHERE	ELLIPTICAL CYLINDER
	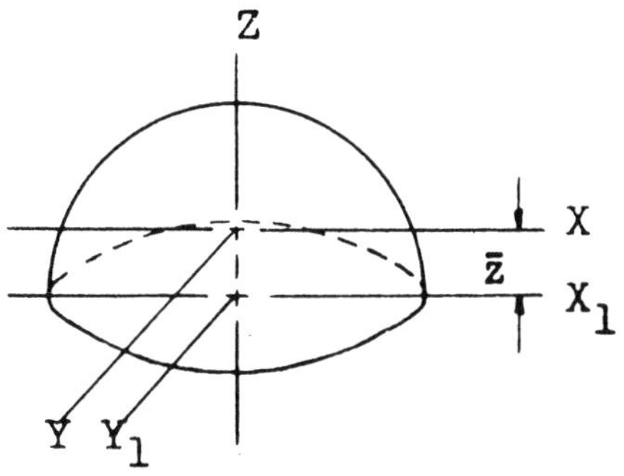	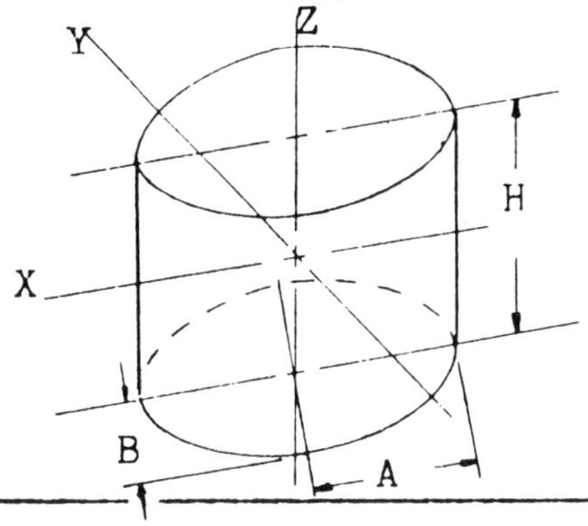
VOLUME	$\frac{2}{3} \pi R^3$	πABH
CENTROID	$\bar{z} = \frac{3}{8} R$	$\bar{z} = \frac{H}{2}$
WEIGHT MOMENT OF INERTIA	$I_x = I_y = 0.26 \ WR^2$ $I_{x_1} = I_{y_1} = I_z = 0.4WR^2$	$I_x = \frac{W}{12}(3B^2 + H^2)$ $I_y = \frac{W}{12}(3A^2 + H^2)$ $I_z = \frac{W}{4}(A^2 + B^2)$
RADIUS OF GYRATION	$k_x = k_y = 0.51R$ $k_{x_1} = k_{y_1} = k_z = 0.632R$	$k_x = 0.289\sqrt{3B^2 + H^2}$ $k_y = 0.289\sqrt{3A^2 + H^2}$ $k_z = \sqrt{\frac{A^2 + B^2}{2}}$

Adapted from Weight Handbook, Vol. 1, Society of Aeronautical Weight Engineers, Inc.

	ELLIPSOID	PARABOLOID OF REVOLUTION
	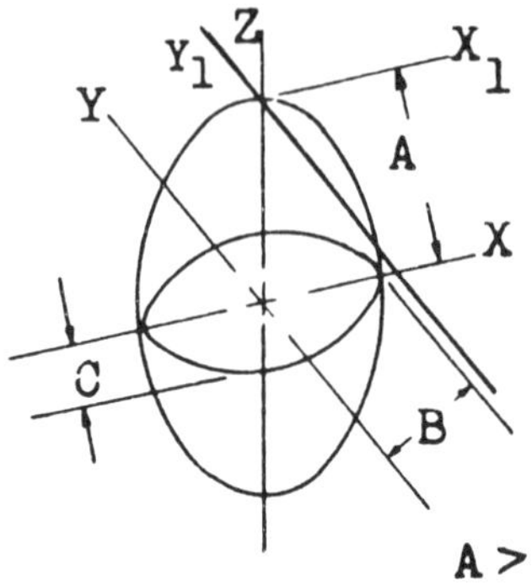A > B > C	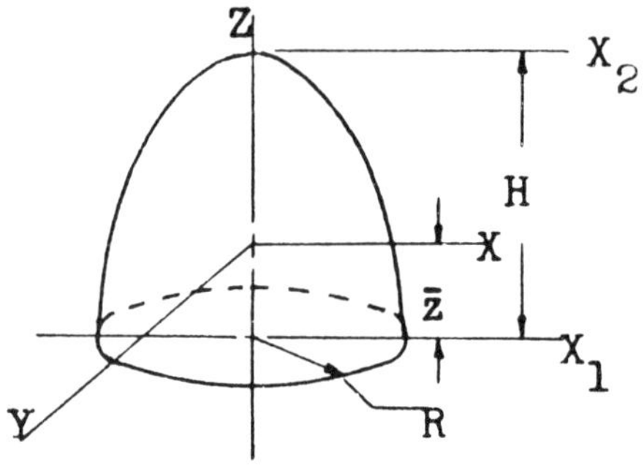
VOLUME	$\frac{4}{3}\pi ABC$	$\frac{\pi R^2 H}{2}$
CENTROID	....	$\bar{z} = \frac{H}{3}$
WEIGHT MOMENT OF INERTIA	$I_x = \frac{W}{5}(A^2 + C^2)$ $I_{x1} = \frac{W}{5}(6A^2 + C^2)$ $I_y = \frac{W}{5}(A^2 + B^2)$ $I_{y1} = \frac{W}{5}(6A^2 + B^2)$ $I_z = \frac{W}{5}(B^2 + C^2)$	$I_x = I_y = \frac{W}{18}(3R^2 + H^2)$ $I_{x1} = \frac{W}{6}(R^2 + H^2)$ $I_{x2} = \frac{W}{6}(R^2 + 3H^2)$ $I_z = \frac{WR^2}{3}$
RADIUS OF GYRATION	$k_x = 0.447\sqrt{A^2 + C^2}$ $k_{x_1} = 0.447\sqrt{6A^2 + C^2}$ $k_y = 0.447\sqrt{A^2 + B^2}$ $k_{y_1} = 0.447\sqrt{6A^2 + B^2}$ $k_z = 0.447\sqrt{B^2 + C^2}$	$k_x = k_y = 0.236\sqrt{3R^2 + H^2}$ $k_{x_1} = 0.408\sqrt{R^2 + H^2}$ $k_{x_2} = 0.408\sqrt{R^2 + 3H^2}$ $k_z = 0.577\ R$
SURFACE AREA	Ellipsoid of revolution or spheroid $B = C$ $A = \frac{4\pi B}{\sqrt{2}}\sqrt{A^2 + B^2}$	$A = \frac{2}{3}\pi R^2\left[m + \frac{1}{m+1}\right]$ $m = \sqrt{1 + \left(\frac{2H}{R}\right)^2}$
EXACT	$A = 2\pi A^2 + \pi\frac{B^2}{e}\log_e\frac{1+e}{1-e}$ e = eccentricity $= \frac{\sqrt{A^2 - B^2}}{A}$	

Adapted from Weight Handbook, Vol. 1, Society of Aeronautical Weight Engineers, Inc.

	ELLIPTIC PARABOLOID	THIN CIRCULAR LAMINA
	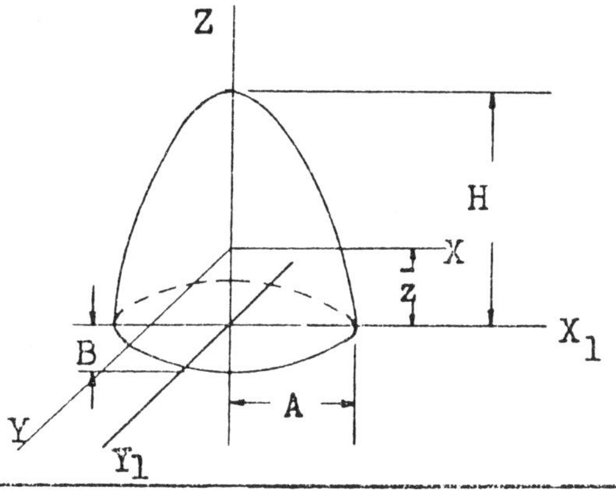	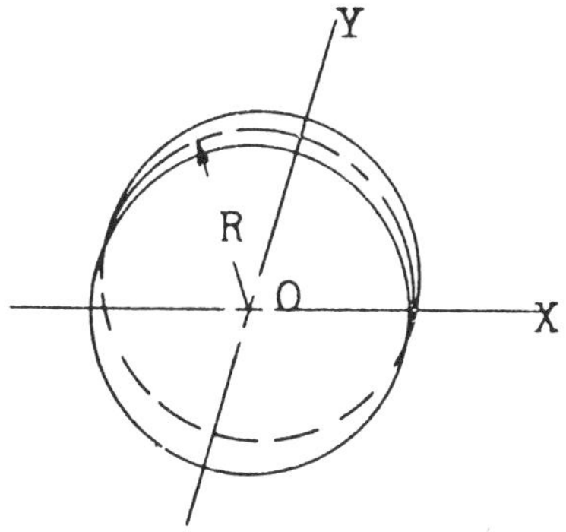
VOLUME	$\frac{\pi ABH}{2}$	$V = \pi R^2 H$
CENTROID	$\bar{z} = \frac{H}{3}$	AT O = GEOMETRICAL CENTER
WEIGHT MOMENT OF INERTIA	$I_x = \frac{W}{18}(3B^2 + H^2)$ $I_{x_1} = \frac{W}{6}(B^2 + H^2)$ $I_y = \frac{W}{18}(3A^2 + H^2)$ $I_{y_1} = \frac{W}{6}(A^2 + H^2)$ $I_z = \frac{W}{6}(A^2 + B^2)$	$I_x = \frac{WR^2}{4}$ $I_y = \frac{WR^2}{2}$
RADIUS OF GYRATION	$k_x = 0.236\sqrt{3B^2 + H^2}$ $k_{x_1} = 0.408\sqrt{B^2 + H^2}$ $k_y = 0.236\sqrt{3A^2 + H^2}$ $k_{y_1} = 0.408\sqrt{A^2 + H^2}$ $k_z = 0.408\sqrt{A^2 + B^2}$	$k_x = \frac{R}{2}$ $k_y = \frac{R}{\sqrt{2}}$
NOTE	....	Y AXIS IS PERPENDICULAR TO PLANE H = THICKNESS

Adapted from Weight Handbook, Vol. 1, Society of Aeronautical Weight Engineers, Inc.

	TORUS	SPHERICAL SECTOR
	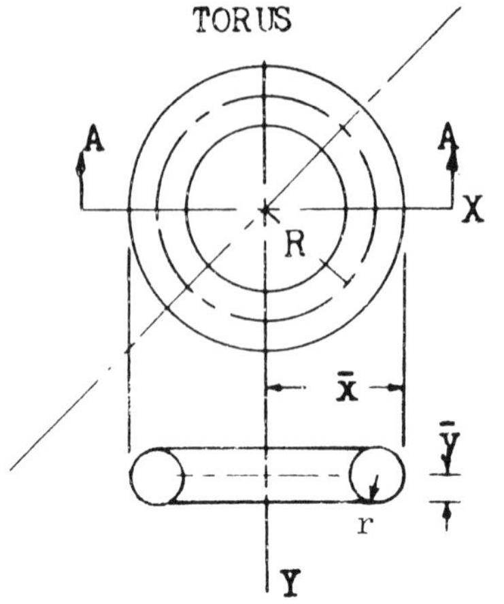	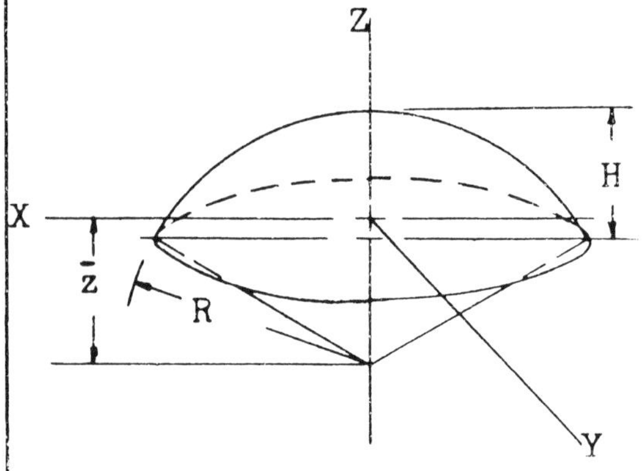
VOLUME	$2\pi^2 r^2 R$	$\frac{2}{3}\pi R^2 H$
CENTROID	$\bar{x} = \bar{z} = R + r$ $\bar{y} = r$	$\bar{z} = \frac{3}{8}(2R-H)$
WEIGHT MOMENT OF INERTIA	$I_x = I_z = \frac{W}{8}(4R^2 + 5r^2)$ $I_y = \frac{W}{4}(4R^2 + 3r^2)$	$I_z = \frac{WH}{5}(3R-H)$
RADIUS OF GYRATION	$k_x = k_z = 0.354\sqrt{4R^2 + 5r^2}$ $k_y = \sqrt{\frac{4R^2 + 3r^2}{2}}$	$k_z = 0.447\sqrt{(3R-H)(H)}$

Adapted from Weight Handbook, Vol. 1, Society of Aeronautical Weight Engineers, Inc.

SPHERICAL SEGMENT

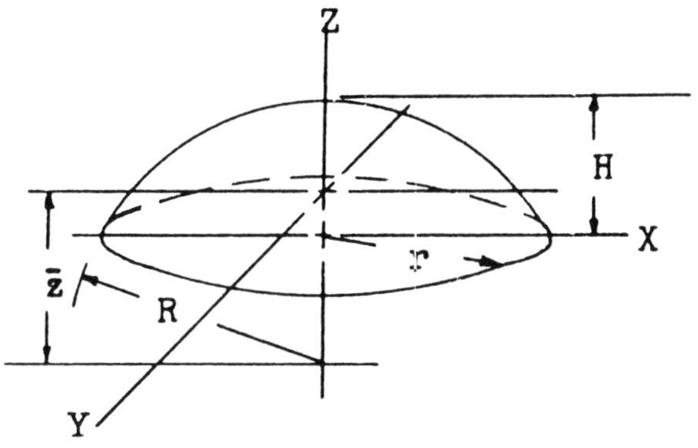

VOLUME	$\frac{\pi H}{6}(3r^2 + H^2) = \frac{\pi H^2}{3}(3R - H)$
CENTROID	$\bar{z} = \frac{3}{4}\frac{(2R - H)^2}{(3R - H)}$
WEIGHT MOMENT OF INERTIA	$I_z = \frac{2}{3}\frac{HW}{R-H}\left(R^2 - \frac{3}{4}RH + \frac{3}{20}H^2\right)$
RADIUS OF GYRATION	$k_z = \sqrt{\frac{2H}{3R-H}\left(R^2 - \frac{3}{4}RH + \frac{3}{20}H^2\right)}$

Adapted from Weight Handbook, Vol. 1, Society of Aeronautical Weight Engineers, Inc.

SEMICYLINDER

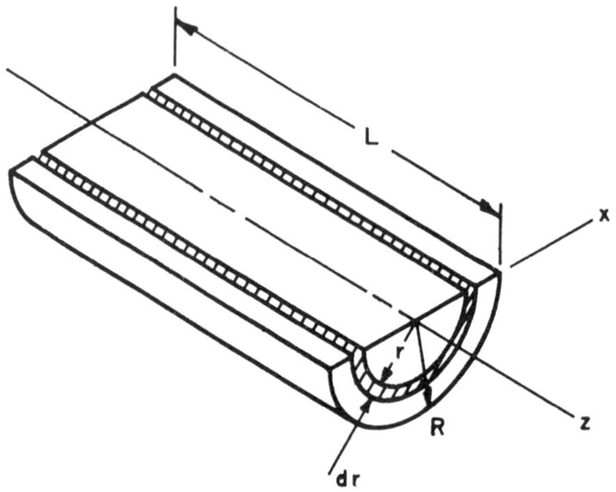

Elemental volume, half cylinder:

$$dm = \rho dV = \frac{\rho 2\pi r L dr}{2} = \rho\pi r L dr$$

$$I_z = \int r^2 dm = \int_0^R r^2(\rho\pi r L dr)$$

$$= \rho\pi L \int_0^R r^3 dr = \rho\pi L\left[\frac{r^4}{4}\right]_0^R = \frac{\rho\pi R^4 L}{4}$$

$$\rho = \frac{m}{V} = \frac{m}{\pi R^2 L/2}$$

$$I_z = \left(\frac{m}{\pi R^2 L/2}\right)\left(\frac{\pi R^4 L}{4}\right) = \frac{mR^2}{2} \text{ (ans.)}$$

SEMICYLINDER

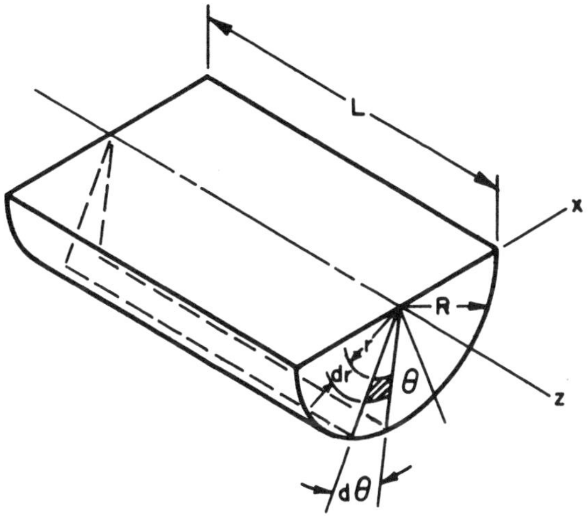

Elemental volume, wedge shape:

$$dm = \rho dV = \rho r d\theta dr L$$

$$I_z = \int r^2 dm = \int_0^R \int_0^\pi r^2 (\rho r L d\theta dr)$$

$$= \rho L \int_0^R \int_0^\pi r^3 d\theta dr = \rho L \pi \int_0^R r^3 dr$$

$$= \rho L \pi \left[\frac{r^4}{4} \right]_0^R = \frac{\rho L \pi R^4}{4}$$

$$\rho = \frac{m}{V} = \frac{m}{\pi R^2 L/2}$$

$$I_z = \left(\frac{\pi L R^4}{4} \right) \left(\frac{2m}{\pi R^2 L} \right) = \frac{mR^2}{2} \text{ (ans.)}$$

SEMICYLINDER

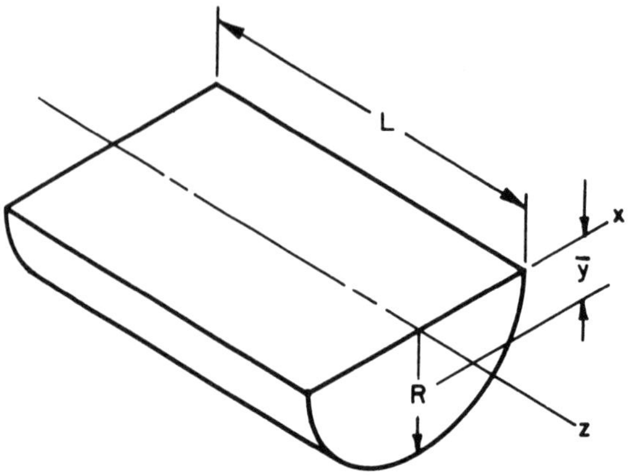

VOLUME	$\frac{\pi R^2 L}{2}$
CENTROID	$\bar{y} = \frac{4R}{3\pi}$
WEIGHT MOMENT OF INERTIA	$I_z = \frac{mR^2}{2}$
RADIUS OF GYRATION	$k_z = \frac{R}{\sqrt{2}} = 0.707R$

	RIGHT ANGLED WEDGE	ISOSCELES WEDGE
	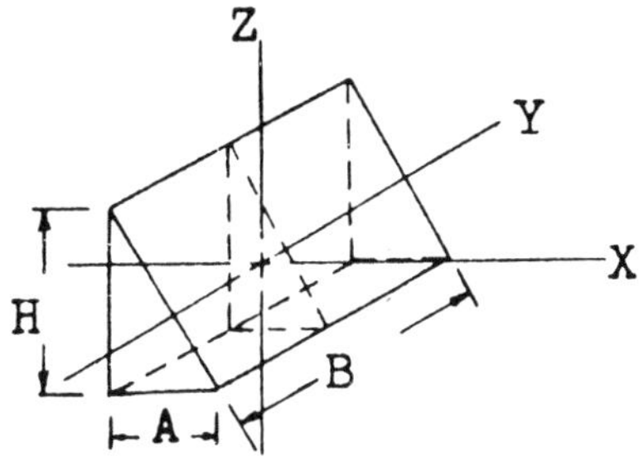	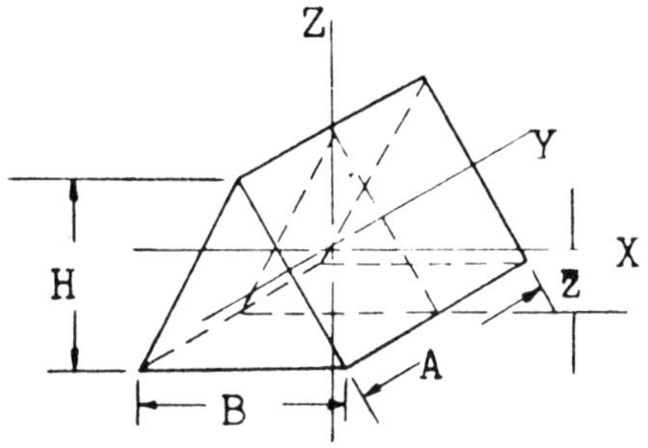
VOLUME	$\frac{ABH}{2}$	$\frac{ABH}{2}$
CENTROID	$\bar{x} = \frac{A}{3}$ $\bar{y} = \frac{B}{2}$ $\bar{z} = \frac{H}{3}$	$\bar{z} = \frac{H}{3}$
WEIGHT MOMENT OF INERTIA	$I_x = \frac{W}{36}(2H^2 + 3B^2)$ $I_y = \frac{W}{18}(A^2 + H^2)$ $I_z = \frac{W}{36}(2A^2 + 3B^2)$	$I_x = \frac{W}{36}(2H^2 + 3A^2)$ $I_y = \frac{W}{72}(4H^2 + 3B^2)$ $I_z = \frac{W}{24}(2A^2 + B^2)$
RADIUS OF GYRATION	$k_x = 0.167\sqrt{2H^2 + 3B^2}$ $k_y = 0.236\sqrt{A^2 + H^2}$ $k_z = 0.167\sqrt{2A^2 + 3B^2}$	$k_x = 0.167\sqrt{2H^2 + 3A^2}$ $k_y = 0.118\sqrt{4H^2 + 3B^2}$ $k_z = 0.204\sqrt{2A^2 + B^2}$

Adapted from Weight Handbook, Vol. 1, Society of Aeronautical Weight Engineers, Inc.

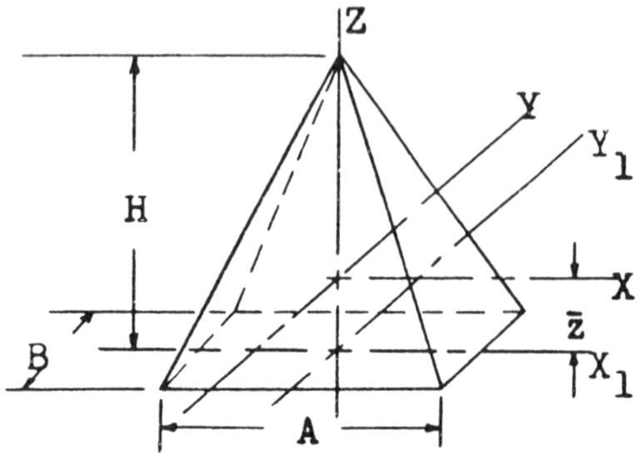

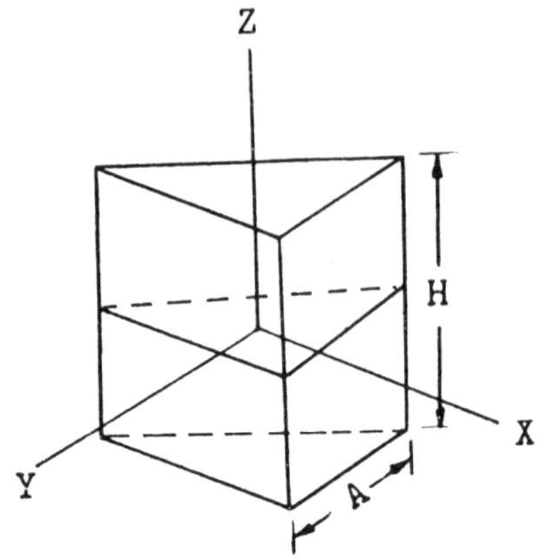

	RIGHT RECTANGULAR PYRAMID	REGULAR TRIANGULAR PRISM
VOLUME	$\frac{ABH}{3}$	$\frac{\sqrt{3}}{4} A^2H$
CENTROID	$\bar{z} = \frac{H}{4}$	$\bar{z} = \frac{H}{2}$
WEIGHT MOMENT OF INERTIA	$I_x = \frac{W}{20} (B^2 + \frac{3H^2}{4})$ $I_{x_1} = \frac{W}{20} (B^2 + 2H^2)$ $I_y = \frac{W}{20} (A^2 + \frac{3}{4} H^2)$ $I_{y_1} = \frac{W}{20} (A^2 + 2H^2)$ $I_z = \frac{W}{20} (A^2 + B^2)$	$I_x = I_y = \frac{W}{24} (A^2 + 2H^2)$ $I_z = \frac{WA^2}{12}$
RADIUS OF GYRATION	$k_x = 0.224 \sqrt{B^2 + \frac{3H^2}{4}}$ $k_{x_1} = 0.224 \sqrt{B^2 + 2H^2}$ $k_y = 0.224 \sqrt{A^2 + \frac{3H^2}{4}}$ $k_{y_1} = 0.224 \sqrt{A^2 + 2H^2}$ $k_z = 0.224 \sqrt{A^2 + B^2}$	$k_x = k_y = 0.204 \sqrt{A^2 + 2H^2}$ $k_z = 0.289A$

Adapted from Weight Handbook, Vol. 1, Society of Aeronautical Weight Engineers, Inc.

	CUBE	RECTANGULAR PRISM
	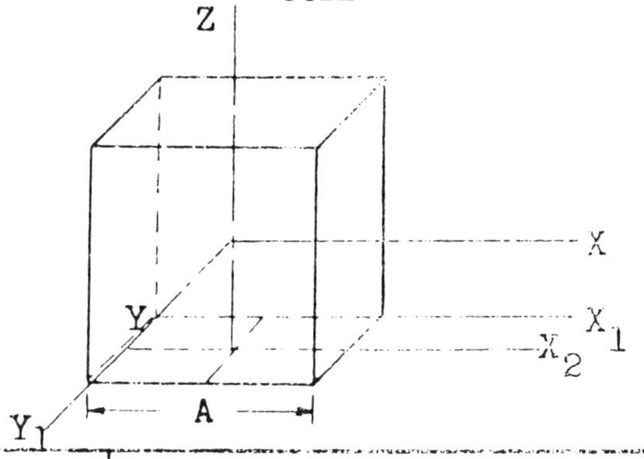	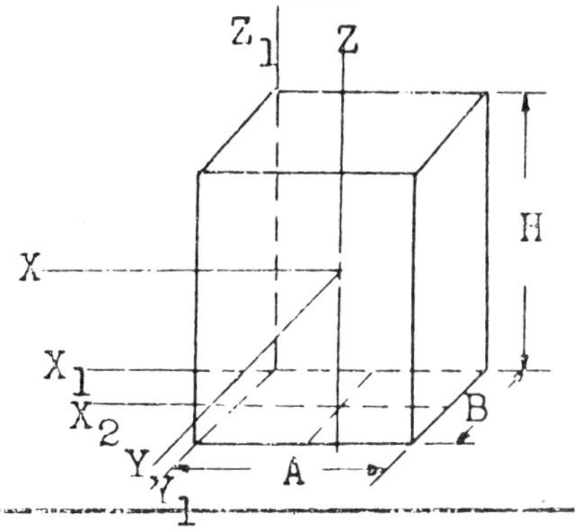
VOLUME	A^3	ABH
CENTROID	$\bar{x} = \bar{y} = \bar{z} = \frac{A}{2}$	$\bar{x} = \frac{A}{2}$ $\bar{y} = \frac{B}{2}$ $\bar{z} = \frac{H}{2}$
WEIGHT MOMENT OF INERTIA	$I_x = I_y = I_z = \frac{WA^2}{6}$ $I_{x_1} = I_{y_1} = I_{z_1} = \frac{2WA^2}{3}$ $I_{x_2} = \frac{5WA^2}{12}$	$I_x = \frac{W}{12}(B^2 + H^2)$ $I_{x_1} = \frac{W}{3}(B^2 + H^2)$ $I_{x_2} = \frac{W}{12}(B^2 + 4H^2)$ $I_y = \frac{W}{12}(A^2 + H^2)$ $I_{y_1} = \frac{W}{3}(A^2 + H^2)$ $I_z = \frac{W}{12}(A^2 + B^2)$ $I_{z_1} = \frac{W}{3}(A^2 + B^2)$
RADIUS OF GYRATION	$k_x = k_y = k_z = 0.408A$ $k_{x_1} = k_{y_1} = k_{z_1} = 0.816A$ $k_{x_2} = 0.646A$	$k_x = 0.289\sqrt{B^2 + H^2}$ $k_{x_1} = 0.577\sqrt{B^2 + H^2}$ $k_{x_2} = 0.289\sqrt{B^2 + 4H^2}$ $k_y = 0.289\sqrt{A^2 + H^2}$ $k_{y_1} = 0.577\sqrt{A^2 + H^2}$ $k_z = 0.289\sqrt{A^2 + B^2}$ $k_{z_1} = 0.577\sqrt{A^2 + B^2}$

Adapted from Weight Handbook, Vol. 1, Society of Aeronautical Weight Engineers, Inc.

THIN SHELLS

A thin shell can be developed by the subtraction of a smaller inner solid from a larger outer solid of similar shape, by the summation of elemental rings by integration, or by revolving an arc or a segment of an arc about a desired coordinate axis.

SUMMARY OF EQUATIONS

Surface Area Generated by Revolving Arc

If A(a,c) and B(b,d) are two points on a curve $F(x,y) = 0$. The area of the surface generated by revolving the arc AB about the x-axis is given by

$$S = 2\pi\int_{AB} y\,ds = 2\pi\int_a^b y\sqrt{1 + (dy/dx)^2}\,dx$$

or

$$2\pi\int_c^d y\sqrt{1 + (dy/dx)^2}\,dy$$

When revolved about the y-axis, the arc AB generates a surface area

$$S = 2\pi\int_{AB} x\,ds = 2\pi\int_a^b x\sqrt{1 + (dy/dx)^2}\,dx$$

or

$$2\pi\int_c^d x\sqrt{1 + (dx/dy)^2}\,dy$$

If A, given by $u = u_1$, and B, given by $u = u_2$, are two points on a curve that is defined by the parametric equations $x = f(u)$, $y = g(u)$, the surface area generated by revolving the arc AB about the x-axis is

$$S = 2\pi\int_{AB} y\,ds = 2\pi\int_{u_1}^{u_2} y\sqrt{(dx/du)^2 + (dy/du)^2}\,du$$

and the area generated about the y-axis bounded by AB is

$$S = 2\pi\int_{AB} x\,ds = 2\pi\int_{u_1}^{u_2} x\sqrt{(dx/du)^2 + (dy/du)^2}\,du$$

Centroid of a Surface of Revolution

The centroid of a surface of revolution generated by the arc of a curve $y = f(x)$ extending from A(a,c) to B(b,d) is defined by

$$S\bar{x} = 2\pi\int_a^b xy\,ds$$

where ds is an element of arc as described in the above equations as

$$ds = \sqrt{1 + (dy/dx)^2}\,dx$$

when taken about the x-axis, and

$$S\bar{y} = 2\pi\int_a^b xy\,ds$$

when taken about the y-axis.

SURFACE AREA AND CENTROIDAL DISTANCE

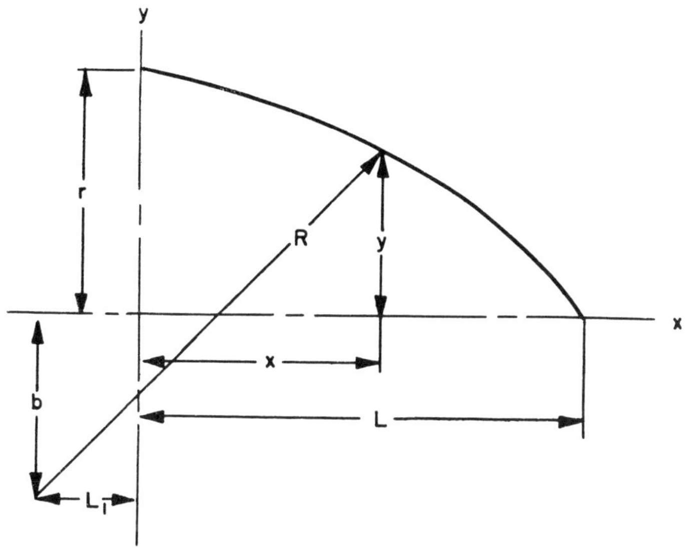

From the diagram above, the variable radius y is

$$(x + L_1)^2 + (y + b)^2 = R^2$$

$$y + b = \sqrt{R^2 - (x + L_1)^2}$$

$$y = \sqrt{R^2 - (x + L_1)^2} - b$$
$$= \left[(R^2 - L_1^2) - 2L_1x - x^2\right]^{1/2} - b$$

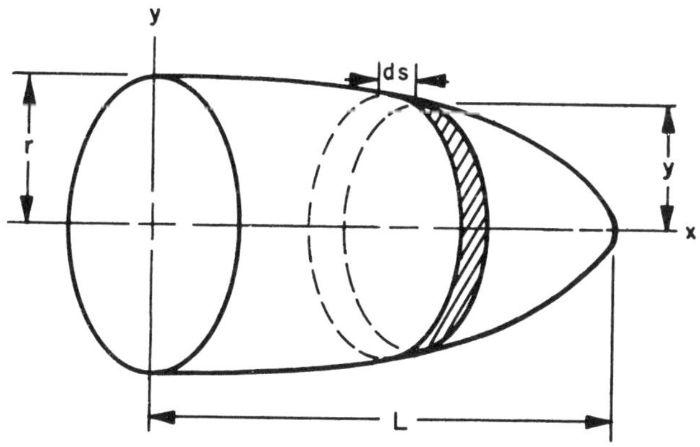

The surface area developed is

$$S = \int 2\pi y ds \qquad \text{where} \qquad ds = \sqrt{1 + (dy/dx)^2}dx$$

$$\frac{dy}{dx} = \left(\frac{1}{2}\right)\frac{-2x - 2L_1}{\left[(R^2 - L_1^2) - 2L_1x - x^2\right]^{1/2}} = -\frac{x + L_1}{(R^2 - L_1^2 - 2L_1x - x^2)^{1/2}}$$

$$1 + \left(\frac{dy}{dx}\right)^2 = \frac{(x + L_1)^2}{R^2 - L_1^2 - 2L_1x - x^2} + 1 = \frac{R^2dx}{R^2 - L_1^2 - 2L_1x - x^2}$$

$$S = 2\pi\int_0^L \left\{ \left[(R^2 - L_1^2) - 2L_1x - x^2\right]^{1/2} - b\right\}\left(\frac{R^2}{R^2 - L_1^2 - 2L_1x - x^2}\right)dx$$

$$S = 2\pi R\int_0^L \left[1 - \frac{b}{(R^2 - L_1^2 - 2L_1x - x^2)^{1/2}}\right]dx$$

$$= 2\pi R\left[x - b\sin^{-1}\frac{x + L_1}{R}\right]_0^L$$

$$= 2\pi R\left[L - b\left(\sin^{-1}\frac{L + L_1}{R} - \sin^{-1}\frac{L_1}{R}\right)\right] \quad \text{(ans.)}$$

The centroid of the shell can be determined by applying the basic mathematical statement

$$S\bar{x} = 2\pi\int_0^L xyds \qquad \text{where} \qquad ds = \sqrt{1 + (dy/dx)^2}dx$$

Then

$$S\bar{x} = 2\pi\int_0^L xy\sqrt{1 + (dy/dx)^2}dx$$

and substituting $y = \left[(R^2 - L_1^2) - 2L_1x - x^2\right]^{1/2} - b$,

$$S\bar{x} = 2\pi\int_0^L x\left\{\left[(R^2 - L_1^2) - 2L_1x - x^2\right]^{1/2} - b\right\}\left\{\frac{R}{\left[(R^2 - L_1^2) - 2L_1x - x^2\right]^{1/2}}\right\}dx$$

$$= 2\pi R\int_0^L \left\{x - \frac{xb}{\left[(R^2 - L_1^2) - 2L_1x - x^2\right]^{1/2}}\right\}dx$$

$$= 2\pi R\left[\frac{x^2}{2} + b\left\{\left[(R^2 - L_1^2) - 2L_1x - x^2\right]^{1/2} + L_1\sin^{-1}\frac{x + L_1}{R}\right\}\right]_0^L$$

$$= 2\pi R\left[\left(\frac{L^2}{2} + b\left\{\left[(R^2 - L_1^2) - 2L_1L - L^2\right]^{1/2} + L_1\sin^{-1}\frac{L + L_1}{R}\right\}\right) - \left\{b\left[(R^2 - L_1^2)^{1/2} + L_1\sin^{-1}\frac{L_1}{R}\right]\right\}\right]$$

$$= 2\pi R\left\{\left(\frac{L^2}{2} + b^2 + bL_1\sin^{-1}\frac{L + L_1}{R}\right) - b\left[(b + r) + L_1\sin^{-1}\frac{L_1}{R}\right]\right\}$$

$$= 2\pi R\left[\frac{L^2}{2} - br + bL_1\left(\sin^{-1}\frac{L + L_1}{R} - \sin^{-1}\frac{L_1}{R}\right)\right]$$

$$\bar{x} = \frac{2\pi R\left((L^2/2) - br + bL_1\left\{\sin^{-1}\left[(L + L_1)/R\right] - \sin^{-1}(L_1/R)\right\}\right)}{2\pi R\left(L - b\left\{\sin^{-1}\left[(L + L_1)/R\right] - \sin^{-1}(L_1/R)\right\}\right)}$$

$$= \frac{(L^2/2) - br + bL_1\left\{\sin^{-1}\left[(L + L_1)/R\right] - \sin^{-1}(L_1/R)\right\}}{L - b\left\{\sin^{-1}\left[(L + L_1)/R\right] - \sin^{-1}(L_1/R)\right\}} \quad \text{(ans.)}$$

Applying this result to a thin-shelled hemisphere, it can be seen that all terms containing b reduce to zero, which leaves

$$\bar{x} = L/2, \text{ but } L = R \text{ for the hemisphere;}$$

therefore,

$$\bar{x} = R/2 \quad \text{(ans.)}$$

An alternative method for determining the location of the centroid of a thin-shelled ogive is to perform volume subtractions: that is, to calculate the volume of the ogive that corresponds to the exterior dimensions desired and then to subtract a volume of the proportions that will ultimately leave the desired wall thickness. Combining the centroidal distances of the two volumes with their respective values of volume, the basic equation becomes

$$\bar{x} = \frac{\Sigma V\bar{x}}{V} = \frac{(V\bar{x})_O - (V\bar{x})_I}{V_O - V_I}$$

where $(V\bar{x})_O$ and $(V\bar{x})_I$ are the statical moments of the outer and inner volumes, respectively, with V_O and V_I representing the outer and inner volumes.

	LATERAL SURFACE OF A CIRCULAR CONE	LATERAL SURFACE OF FRUSTUM OF CIRCULAR CONE
	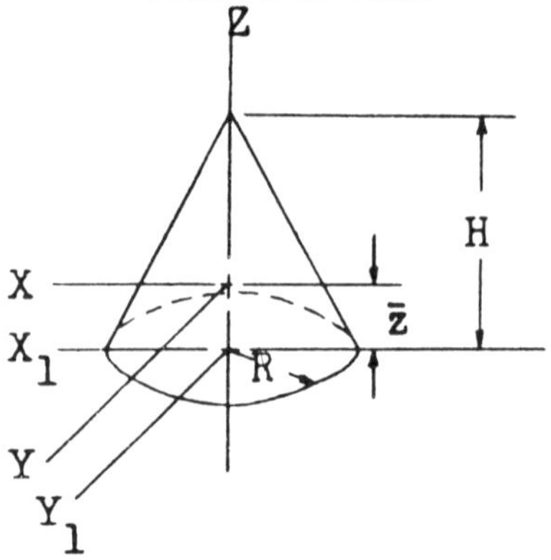 CONSIDER I_x AS THE SUM OF TWO MOMENTS OF INERTIA CIRCLE OF RADIUS R TRIANGLE OF ALTITUDE H WT. OF CIRCLE AND TRIANGLE EQUALS WT. OF SURFACE	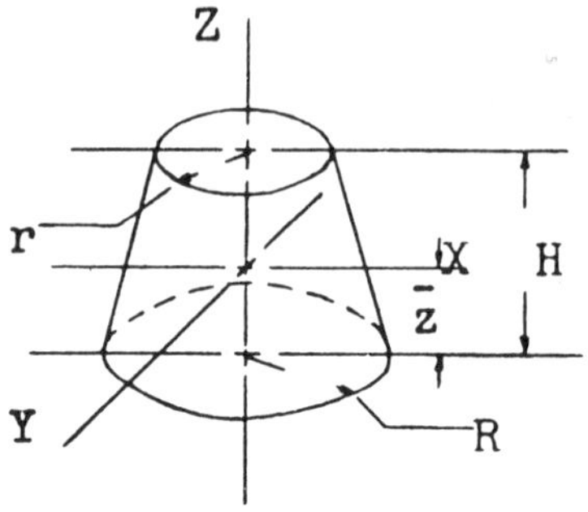 CONSIDER I_x AS THE SUM OF TWO MOMENTS OF INERTIA CIRCLE OF RADIUS R TRIANGLE OF ALTITUDE H WT. OF CIRCLE AND TRIANGLE EQUALS WT. OF SURFACE
SURFACE AREA	$\pi R \sqrt{R^2 + H^2}$	$\pi(R + r)\sqrt{H^2 + (R - r)^2}$
CENTROID	$\bar{z} = \frac{H}{3}$	$\bar{z} = \frac{H}{3}\left(\frac{2r + R}{r + R}\right)$
WEIGHT MOMENT OF INERTIA	$I_x = I_y = \frac{W}{4}\left(R^2 + \frac{2}{9}H^2\right)$ $I_z = \frac{WR^2}{2}$ $I_{x_1} = I_{y_1} = \frac{W}{12}\left(3R^2 + 2H^2\right)$	$I_x = I_y = \frac{W}{4}\left(R^2 + r^2\right) + \frac{WH^2}{18}\left\{1 + \frac{2Rr}{(R+r)^2}\right\}$ $I_z = \frac{W}{2}\left(R^2 + r^2\right)$
RADIUS OF GYRATION	$k_x = k_y = \sqrt{\frac{9R^2 + 2H^2}{6}}$ $k_{x_1} = k_{y_1} = 0.289\sqrt{3R^2 + 2H^2}$ $k_z = 0.707\ R$	$k_x = k_y = \sqrt{\frac{(R^2 + r^2)}{4} + \frac{H^2}{18}\left\{1 + \frac{2Rr}{(R+r)^2}\right\}}$ $k_z = 0.707\sqrt{R^2 + r^2}$

Adapted from Weight Handbook, Vol. 1, Society of Aeronautical Weight Engineers, Inc.

	LATERAL CYLINDRICAL SHELL	TOTAL CYLINDRICAL SHELL
	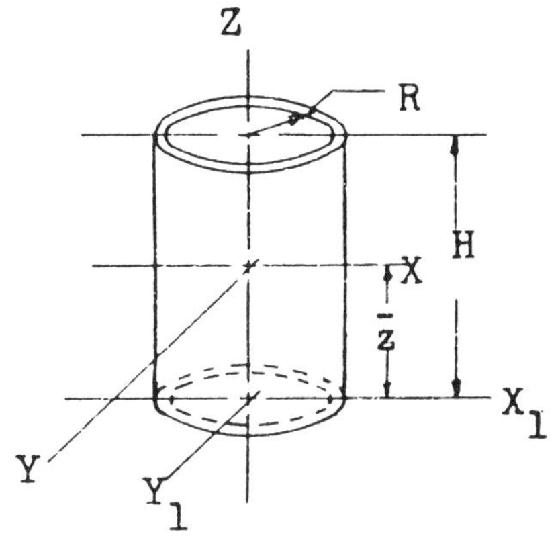	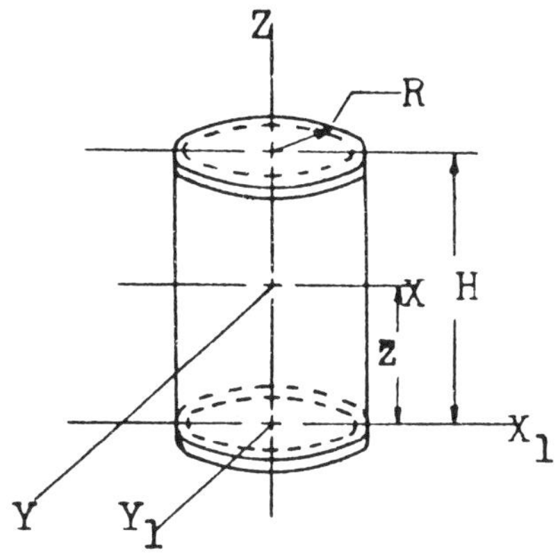
SURFACE AREA	$2\pi RH$	$2\pi R(R + H)$
CENTROID	$\bar{z} = \frac{H}{2}$	$\bar{z} = \frac{H}{2}$
WEIGHT MOMENT OF INERTIA	$I_x = I_y = \frac{W}{2}\left(R^2 + \frac{H^2}{6}\right)$ $I_z = WR^2$ $I_{x_1} = I_{y_1} = \frac{W}{6}\left(3R^2 + 2H^2\right)$	$I_x = I_y = \frac{W}{12(R+H)}\left[3R^2(R+2H)+H^2(3R+H)\right]$ $I_{x_1} = I_{y_1} = \frac{W}{12(R+H)}\left[3R^2(R+2H)+2H^2(3R+2H)\right]$ $I_z = \frac{WR^2}{2}\left[\frac{R+2H}{R+H}\right]$
RADIUS OF GYRATION	$k_x = k_y = 0.707\sqrt{R^2 + \frac{H^2}{6}}$ $k_{x_1} = k_{y_1} = 0.408\sqrt{3R^2 + 2H^2}$ $k_z = R$	$k_x = k_y = 0.289\sqrt{\frac{3R^2(R+2H)+H^2(3R+H)}{R+H}}$ $k_{x_1} = k_{y_1} = 0.289\sqrt{\frac{3R^2(R+2H)+2H^2(3R+2H)}{R+H}}$ $k_z = 0.707\,R\sqrt{\frac{R+2H}{R+H}}$

Adapted from Weight Handbook, Vol. 1, Society of Aeronautical Weight Engineers, Inc.

	SPHERICAL SHELL	HEMISPHERICAL SHELL
	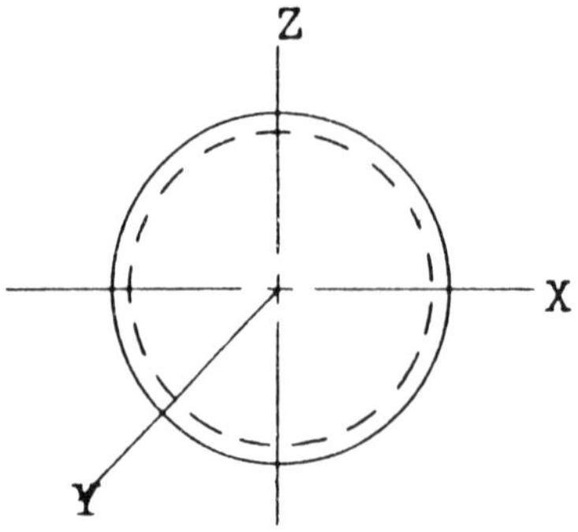	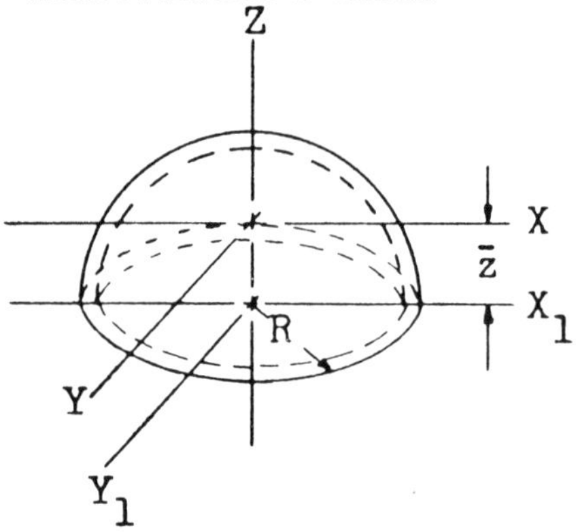
SURFACE AREA	$4\pi R^2$	$2\pi R^2$
CENTROID	$\bar{x} = \bar{y} = \bar{z} = 0$	$\bar{z} = \frac{R}{2}$
WEIGHT MOMENT OF INERTIA	$I_x = I_y = I_z = \frac{2}{3} WR^2$	$I_x = I_y = \frac{5}{12} WR^2$ $I_z = I_{x_1} = I_{y_1} = \frac{2}{3} WR^2$
RADIUS OF GYRATION	$k_x = k_y = k_z = 0.816\ R$	$k_x = k_y = 0.646\ R$ $k_z = k_{x_1} = k_{y_1} = 0.816\ R$

Adapted from Weight Handbook, Vol. 1, Society of Aeronautical Weight Engineers, Inc.

THIN RODS

A thin rod, or wire, is so designated when the length, L, is greater than 30 times the diameter, d. Lengths of curved arcs will be designated by s as noted.

SUMMARY OF EQUATIONS

Length of Arc

The length of arc of the curve F(x,y) = 0 limited by two selected points A(a,c) and B(b,d) is given by

$$s = \int_{AB} ds = \int_a^b \sqrt{1 + (dy/dx)^2}\,dx \qquad \text{or} \qquad \int_c^d \sqrt{1 + (dx/dy)^2}\,dy$$

Example. For the length of arc of the curve $y = x^{3/2}$ from x = 0 to x = 5,

$$\frac{dy}{dx} = \frac{3}{2}x^{1/2} \qquad \text{and} \qquad 1 + \left(\frac{dy}{dx}\right)^2 = 1 + \frac{9}{4}x$$

then

$$s = \int_a^b \sqrt{1 + \left(\frac{dy}{dx}\right)^2}\,dx = \int_0^5 \sqrt{1 + \frac{9}{4}x}\,dx = \frac{8}{27}\left(1 + \frac{9}{4}x\right)^{3/2}\Bigg]_0^5 = \frac{335}{27} \text{ (ans.)}$$

If A, given by $u = u_1$, and B, given by $u = u_2$, are points on a curve defined by the parametric equations x = f(u), y = g(u), the length of arc AB is given by

$$s = \int_{AB} ds = \int_{u_1}^{u_2} \sqrt{(dx/du)^2 + (dy/du)^2}\,du$$

Example. For the length of arc of the curve $x = t^2$, $y = t^3$ from t = 0 to t = 4,

$$\frac{dx}{dt} = 2t, \qquad \frac{dy}{dt} = 3t^2$$

and

$$\left(\frac{dx}{dt}\right)^2 + \left(\frac{dy}{dt}\right)^2 = 4t^2 + 9t^4 = 4t^2\left(1 + \frac{9}{4}t^2\right)$$

then

$$s = \int_0^4 \sqrt{1 + \frac{9}{4}t^2}\,(2t\,dt) = \frac{8}{27}\left(1 + \frac{9}{4}t^2\right)^{3/2}\Bigg]_0^4 = \frac{8}{27}(37\sqrt{37} - 1) \text{ (ans.)}$$

Centroid of an Arc

The centroidal coordinates $(\bar{x},\bar{y})$ of an arc of a plane curve of equation $F(x,y) = 0$ or $x = f(u)$, $y = g(u)$ can be determined by the relationships

$$\bar{x}s = \bar{x}\int ds = \int x ds \qquad \text{and} \qquad \bar{y}s = \bar{y}\int ds = \int y ds$$

where the limits of integration are determined from the extent of the desired integration.

Example. For the centroid of the first quadrant arc of the circle $x^2 + y^2 = 25$,

$$\frac{dy}{dx} = -\frac{x}{y} \qquad \text{and} \qquad 1 + \left(\frac{dy}{dx}\right)^2 = 1 + \frac{x^2}{y^2} = \frac{25}{y^2}$$

since $s = R\theta = 5\pi/2$

$$\frac{5\pi\bar{y}}{2} = \int_0^5 y\sqrt{1 + (dy/dx)^2}\,dx = \int_0^5 5dx = 25$$

and $\bar{y} = 10/\pi$ by symmetry, $\bar{x} = \bar{y}$, and the coordinates of the centroid are

$$\left(\frac{10}{\pi},\frac{10}{\pi}\right) \quad \text{(ans.)}$$

Moments of Inertia of an Arc

The moments of inertia of an arc, referred to the coordinate axes, are given by

$$I_x = \int_\alpha^\beta y^2 ds \qquad \text{and} \qquad I_y = \int_\alpha^\beta x^2 ds$$

Example. For the moment of inertia of the arc of a circle with respect to a fixed diameter,

$$\frac{dy}{dx} = -\frac{x}{y}, \qquad \sqrt{1 + (dy/dx)^2} = R/y, \qquad s = 2\pi R$$

The total moment of inertia is four times that of the first quadrant arc.

$$I_x = 4\int_0^R y^2 ds = 4\int_0^R y^2 \frac{R}{y} dx = 4R\int_0^R \sqrt{R^2 - x^2}\, dx = \pi R^3 = \frac{R^2 s}{2} \quad \text{(ans.)}$$

It should be noted that the value obtained contains only linear dimensions; therefore, to obtain the moment of inertia, I_x, in terms of mass, $m = \rho s$, the transformation should be made as $s = 2\pi R$. Then

$$I_x = \rho\pi R^3 \qquad \text{and} \qquad \rho = m/2\pi R$$

which results in

$$I_x = mR^2/2 \quad \text{(ans.)}$$

Centroid of an Area

The theorem of Pappus for the determination of the centroid of an area produced by the revolution of an arc is as follows.

If an arc of a curve is revolved about an axis in its plane and not crossing the arc, the area of the surface generated is equal to the product of the length of arc and the length of the path described by the centroid of the arc.

Example. For the centroid of the first quadrant arc of a circle of radius R,

$$s = 2\pi R^2 = (1/2\pi R)(2\pi\bar{x})$$

by symmetry, $\bar{x} = \bar{y}$, and the centroid has coordinates $(2R/\pi, 2R/\pi)$.

It follows that the appropriate surface area can be found from the same theorem if the centroidal distance is known.

Moment of Inertia of a Thin Rod

The moment of inertia of a long, slender rod can be calculated from the relationships

$$I_y = \int_0^L x^2\,dm \qquad \text{or} \qquad I_x = \int_0^L y^2\,dm$$

depending on the choice of axes.

Example. For the moment of inertia of a thin homogeneous rod about an end,

$$I_y = \int_0^L x^2\,dm = \rho\int_0^L x^2\,dx = \rho L^3/3$$

where $dm = \rho dx$

Recalling that $\rho = m/L$,

$$I_y = (m/L)(L^3/3) = mL^3/3 \quad \text{(ans.)}$$

SEGMENT OF A CIRCULAR ROD

α **in radians**

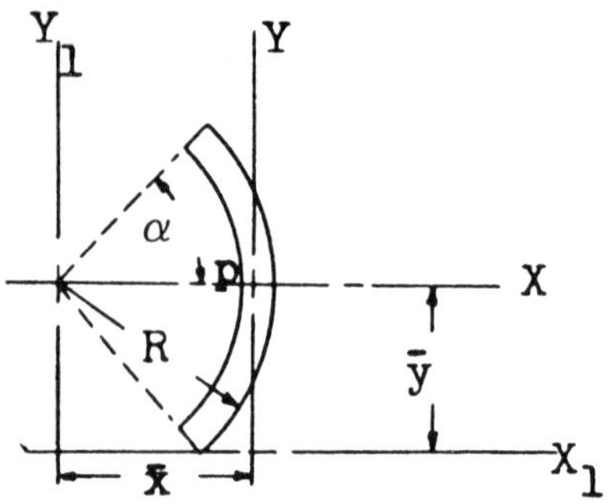

LENGTH	$2R\alpha$
CENTROID	$\bar{x} = \frac{R \sin\alpha}{\alpha}$ $\bar{y} = R\sin\alpha$
WEIGHT MOMENT OF INERTIA	$I_x = WR^2\left\{\frac{1}{2} - \frac{\sin\alpha\cos\alpha}{2\alpha}\right\}$ $I_{x1} = WR^2\left[\left\{\frac{1}{2} - \frac{\sin\alpha\cos\alpha}{2\alpha}\right\} + \sin^2\alpha\right]$ $I_y = WR^2\left[\left\{\frac{1}{2} + \frac{\sin\alpha\cos\alpha}{2\alpha}\right\} - \frac{\sin^2\alpha}{\alpha^2}\right]$ $I_{y1} = WR^2\left\{\frac{1}{2} + \frac{\sin\alpha\cos\alpha}{2\alpha}\right\}$ $I_p = WR^2\left\{1 - \frac{\sin^2\alpha}{\alpha^2}\right\}$
RADIUS OF GYRATION	$k_x = R\sqrt{\frac{1}{2} - \frac{\sin\alpha\cos\alpha}{2\alpha}}$ $k_{x_1} = R\sqrt{\frac{1}{2} - \frac{\sin\alpha\cos\alpha}{2\alpha} + \sin^2\alpha}$ $k_y = R\sqrt{\frac{1}{2} + \frac{\sin\alpha\cos\alpha}{2\alpha} - \frac{\sin^2\alpha}{\alpha^2}}$ $k_{y_1} = R\sqrt{\frac{1}{2} + \frac{\sin\alpha\cos\alpha}{2\alpha}}$ $k_p = R\sqrt{1 - \frac{\sin^2\alpha}{\alpha^2}}$

Adapted from Weight Handbook, Vol. 1, Society of Aeronautical Weight Engineers, Inc.

	CIRCULAR ROD	SEMICIRCULAR ROD
	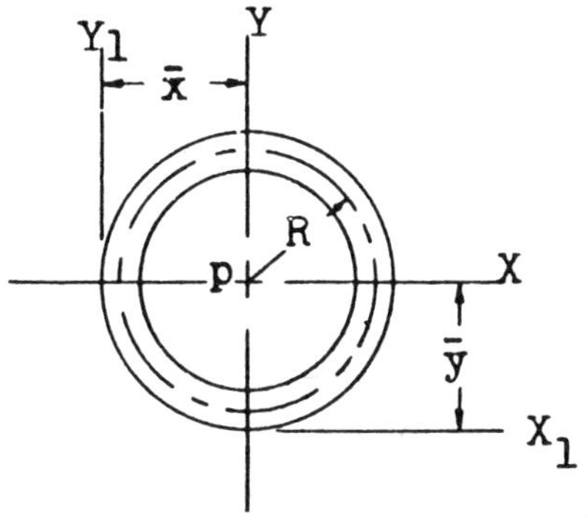	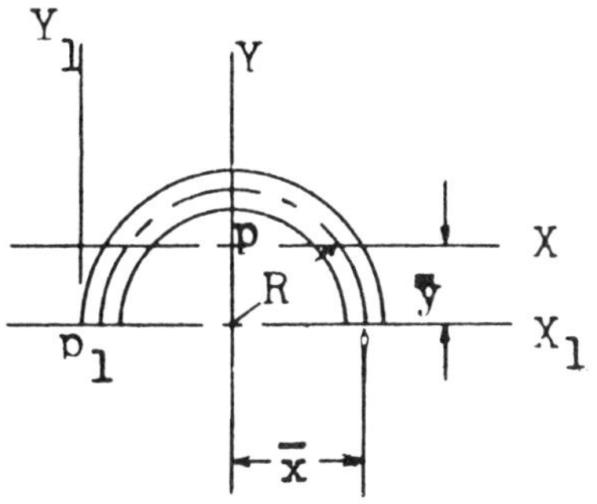
LENGTH	$2\pi R$	πR
CENTROID	$\bar{x} = \bar{y} = R$	$\bar{x} = R \quad \bar{y} = 0.6366R$
WEIGHT MOMENT OF INERTIA	$I_x = I_y = \frac{WR^2}{2}$ $I_z = WR^2$ $I_{x_1} = I_{y_1} = \frac{3WR^2}{2}$ $I_p = WR^2$	$I_x = 0.0947\ WR^2$ $I_{x_1} = 0.5\ WR^2$ $I_y = 0.5\ WR^2$ $I_{y_1} = 1.5\ WR^2$ $I_p = 0.5947\ WR^2$ $I_{p_1} = 2\ WR^2$
RADIUS OF GYRATION	$k_x = k_y = 0.707\ R$ $k_{x_1} = k_{y_1} = 1.225\ R$ $k_p = R$	$k_x = 0.308\ R$ $k_{x_1} = 0.707\ R$ $k_y = 0.707\ R$ $k_{y_1} = 1.225\ R$ $k_p = 0.771\ R$ $k_{p_1} = 1.414\ R$

Adapted from Weight Handbook, Vol. 1, Society of Aeronautical Weight Engineers, Inc.

	ELLIPTIC ROD	PARABOLIC ROD
	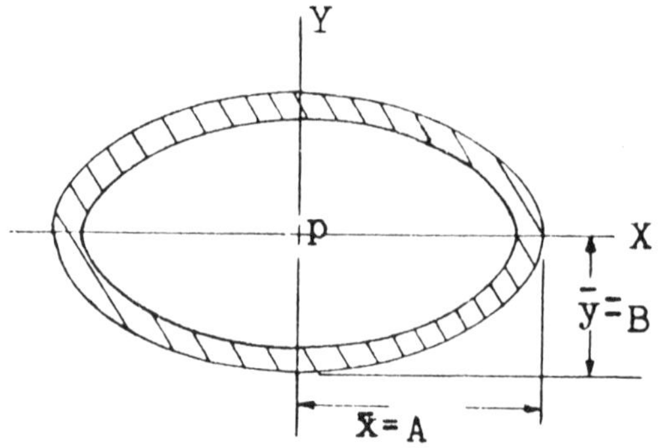	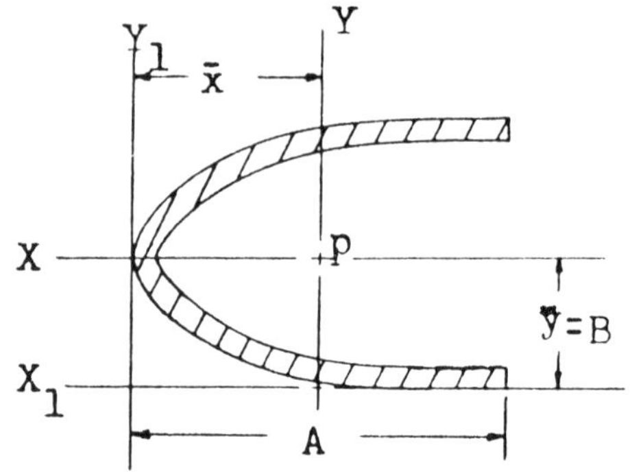
LENGTH	$\frac{\pi}{2A}(3A^2+B^2)$ (More Accurate) $\frac{\pi}{32A^3}\left[45A^4+22A^2B^2-3B^4\right]$	$L=\sqrt{4A^2+B^2}+\frac{B^2}{2A}\text{LOGe}\frac{2A+\sqrt{4A^2+B^2}}{B}$
CENTROID	$\bar{x}=A \qquad \bar{y}=B$	$\bar{x}=\frac{\sqrt{(4A^2+B^2)^3}}{8AL}-\frac{B^2}{16A} \qquad \bar{y}=B$
WEIGHT MOMENT OF INERTIA	$I_x=\frac{WB^2(55A^4+10A^2B^2-B^4)}{2(45A^4+22A^2B^2-3B^4)}$ $I_y=\frac{WA^2(35A^4+34A^2B^2-5B^4)}{2(45A^4+22A^2B^2-3B^4)}$ $I_p=I_x+I_y$	$I_x=\frac{WB^2}{8A^2}\left(\frac{\sqrt{(4A^2+B^2)^3}}{L}-\frac{B^2}{2}\right)$ $I_{x_1}=\frac{WB^2}{8A^2}\left(\frac{\sqrt{(4A^2+B^2)^3}}{L}-\frac{B^2}{2}\right)+WB^2$ $I_y=\frac{W\sqrt{(4A^2+B^2)^3}}{12L}-\frac{I_x}{8}-Wx^2$ $I_{y_1}=\frac{W\sqrt{(4A^2+B^2)^3}}{12L}-\frac{1}{8}I_x$ $I_p=I_x+I_y$
RADIUS OF GYRATION	$k_x=\sqrt{\frac{I_x}{W}}$ $k_y=\sqrt{\frac{I_y}{W}}$ $k_p=\sqrt{\frac{I_p}{W}}$	$k_x=\sqrt{\frac{I_x}{W}}$ $k_{x_1}=\sqrt{\frac{I_{x_1}}{W}}$ $k_y=\sqrt{\frac{I_y}{W}}$ $k_{y_1}=\sqrt{\frac{I_{y_1}}{W}}$ $k_p=\sqrt{\frac{I_p}{W}}$
NOTE	$A > B$	

Adapted from Weight Handbook, Vol. 1, Society of Aeronautical Weight Engineers, Inc.

	U-ROD	RECTANGULAR ROD
	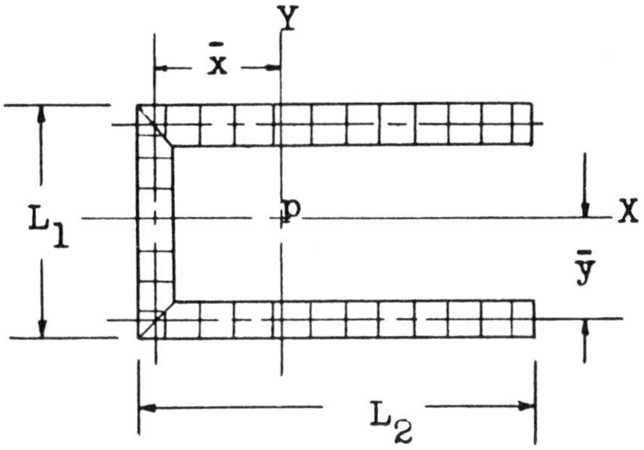	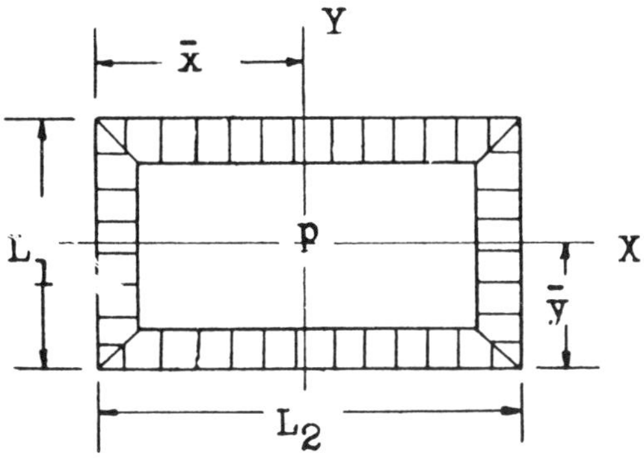
LENGTH	$L_1 + 2L_2$	$2(L_1 + L_2)$
CENTROID	$\bar{x} = \dfrac{L_2^2}{L_1+2L_2}$ $\bar{y} = \dfrac{L_1}{2}$	$\bar{x} = \dfrac{L_2}{2}$ $\bar{y} = \dfrac{L_1}{2}$
WEIGHT MOMENT OF INERTIA	$I_x = \dfrac{WL_1^2(L_1 + 6L_2)}{12(L_1 + 2L_2)}$ $I_y = \dfrac{WL_2^3\ (2L_1 + L_2)}{3\ (L_1+2L_2)^2}$ $I_p = I_x + I_y$	$I_x = \dfrac{WL_1^2(L_1+3L_2)}{12(L_1+L_2)}$ $I_y = \dfrac{WL_2^2(3L_1+L_2)}{12(L_1+L_2)}$ $I_p = I_x + I_y$
RADIUS OF GYRATION	$k_x = 0.289L_1 \sqrt{\dfrac{L_1+6L_2}{L_1+2L_2}}$ $k_y = \dfrac{0.577L_2}{L_1+2L_2} \sqrt{L_2(2L_1+L_2)}$ $k_p = \sqrt{\dfrac{I_p}{W}}$	$k_x = 0.289L_1 \sqrt{\dfrac{L_1+3L_2}{L_1+L_2}}$ $k_y = 0.289L_2 \sqrt{\dfrac{3L_1+L_2}{L_1 + L_2}}$ $k_p = \sqrt{\dfrac{I_p}{W}}$
NOTE	$W_1 = \dfrac{L_1}{L_2} W_2 \quad W = W_1 + 2W_2 = W_2\left(\dfrac{L_1}{L_2} + 2\right)$	$W_1 = \dfrac{L_1}{L_2} W_2 \quad W = 2W_1 + 2W_2 = 2W_2\left(\dfrac{L_1}{L_2} + 1\right)$

Adapted from Weight Handbook, Vol. 1, Society of Aeronautical Weight Engineers, Inc.

	V-ROD	L-ROD
	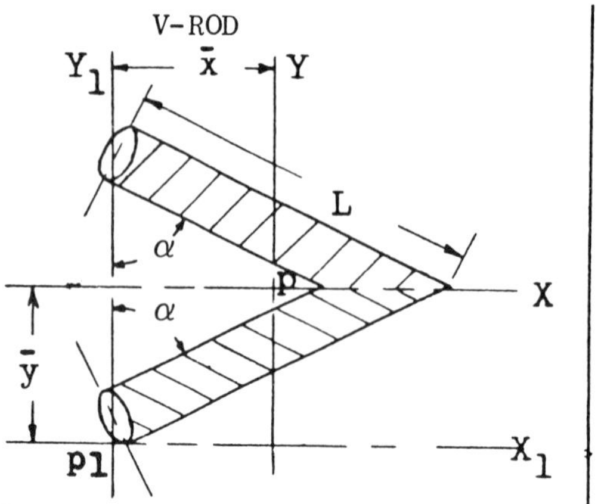	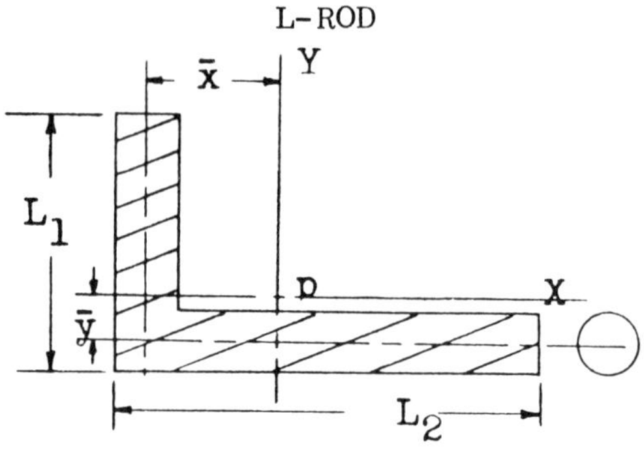
LENGTH	$2L$	$L_1 + L_2$
CENTROID	$\bar{x} = \frac{L}{2} \text{SIN } \alpha$ $\bar{y} = L \text{ COS } \alpha$	$\bar{x} = \frac{L_2^2}{2(L_1 + L_2)}$ $\bar{y} = \frac{L_1^2}{2(L_1 + L_2)}$
WEIGHT MOMENT OF INERTIA	$I_x = \frac{WL^2}{3} \text{COS}^2 \alpha$ $I_{x1} = \frac{4}{3} WL^2 \text{COS}^2 \alpha$ $I_y = \frac{WL^2}{12} \text{SIN}^2 \alpha$ $I_{y_1} = \frac{WL^2}{3} \text{SIN}^2 \alpha$ $I_p = I_x + I_y$ $I_{p_1} = I_{x1} + I_{y1}$	$I_x = \frac{W}{12} \frac{L_1^3 (L_1 + 4L_2)}{(L_1 + L_2)^2}$ $I_y = \frac{W}{12} \frac{L_2^3 (4L_1 + L_2)}{(L_1 + L_2)^2}$ $I_p = I_x + I_y$
RADIUS OF GYRATION	$k_x = 0.577 L \text{ COS } \alpha$ $k_{x1} = 1.15 L \text{ COS } \alpha$ $k_y = 0.289 L \text{ SIN } \alpha$ $k_{y_1} = 0.577 L \text{ SIN } \alpha$ $k_p = 0.289 L \sqrt{4 \text{ COS}^2 \alpha + \text{SIN}^2 \alpha}$ $k_{p_1} = 0.577 L \sqrt{4 \text{ COS}^2 \alpha + \text{SIN}^2 \alpha}$	$k_x = \frac{0.289 L_1}{L_1 + L_2} \sqrt{L_1 (L_1 + 4L_2)}$ $k_y = \frac{0.289 L_2}{L_1 + L_2} \sqrt{L_2 (4L_1 + L_2)}$ $k_p = \sqrt{\frac{I_p}{W}}$
NOTE	....	$W_1 + W_2 = W_2 \frac{L_1}{L_2} + 1 = W$

Adapted from Weight Handbook, Vol. 1, Society of Aeronautical Weight Engineers, Inc.

STRAIGHT ROD	INCLINED ROD NOT THROUGH C G AXIS

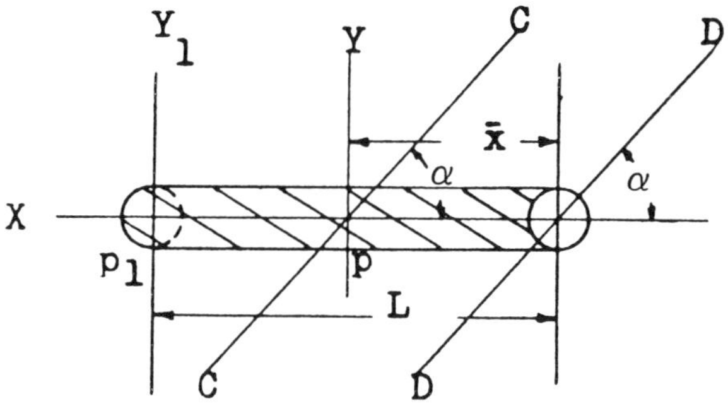

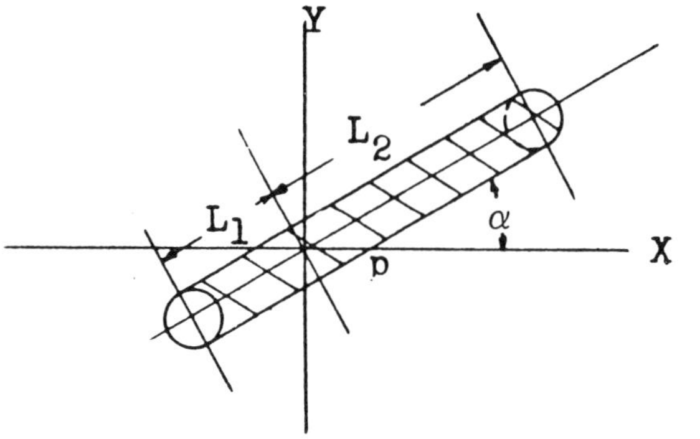

	STRAIGHT ROD	INCLINED ROD NOT THROUGH C G AXIS
LENGTH	L	$L_1 + L_2$
CENTROID	$\bar{x} = \frac{L}{2}$	
WEIGHT MOMENT OF INERTIA	I_x = NEGLIGIBLE $I_y = I_p = \frac{WL^2}{12}$ $I_{y_1} = I_{p_1} = \frac{WL^2}{3}$ $I_c = \frac{WL^2 \sin^2\alpha}{12}$ $I_D = \frac{WL^2 \sin^2\alpha}{3}$	$I_x = \frac{W}{3} \sin^2\alpha\,(L_1^2 - L_1 L_2 + L_2^2)$ $I_y = \frac{W}{3} \cos^2\alpha\,(L_1^2 - L_1 L_2 + L_2^2)$ $I_p = I_x + I_y$ NOTE: ASSUME: WT. OF $L_1 = W_1$ WT. OF $L_2 = W_2$ THEN: $W_1 = \frac{L1}{L2} W_2 \quad W = W_1 + W_2$ $W_1 + W_2 = W_2\left(\frac{L1}{L2} + 1\right)$
RADIUS OF GYRATION	k_x = NEGLIGIBLE $k_y = 0.289L$ $k_{y_1} = 0.577L$ $k_c = 0.289L \sin\alpha$ $k_D = 0.577L \sin\alpha$	$k_x = 0.577 \sin\alpha \sqrt{L_1^2 - L_1 L_2 + L_2^2}$ $k_y = 0.577 \cos\alpha \sqrt{L_1^2 - L_1 L_2 + L_2^2}$ $k_p = 0.577 \sqrt{L_1^2 - L_1 L_2 + L_2^2}$

Adapted from Weight Handbook, Vol. 1, Society of Aeronautical Weight Engineers, Inc.

PLANE AREAS

SUMMARY OF PLANE AREA PROPERTIES

Centroid by Integration

$$\bar{y} = \frac{\int y\,dA}{A} \qquad \bar{x} = \frac{\int x\,dA}{A}$$

Centroid by Area Moment Summation

$$\bar{y} = \frac{a_1 y_1 + a_2 y_2 + \cdot\cdot\cdot + a_n y_n}{a_1 + a_2 + \cdot\cdot\cdot + a_n}$$

$$\bar{x} = \frac{a_1 x_1 + a_2 x_2 + \cdot\cdot\cdot + a_n x_n}{a_1 + a_2 + \cdot\cdot\cdot + a_n}$$

Moments of Inertia

$$I_x = \int y^2 dA$$

$$I_y = \int x^2 dA$$

$$I_z = J_z = \int r^2 dA = I_x + I_y$$

Transfer of Axes

$$I_x = \bar{I}_x + Adx^2$$

$$I_y = \bar{I}_y + Ady^2$$

$$J_z = \bar{J}_z + Ad^2 = I_x + I_y$$

$$k^2 = \bar{k}^2 + d^2$$

Radius of Gyration

$$k = \sqrt{I/A}$$

CENTROID OF AN AREA

The centroid of an area is represented by a point whose distance from any axis times the total area is equal to the first moment of the area with respect to that axis.

The first moment of an area, often referred to as the statical moment, is the algebraic sum of the moments of the differential parts of the area, with the product of the differential area and the perpendicular distance from the differential area to the axis in question representing the moment of each segment.

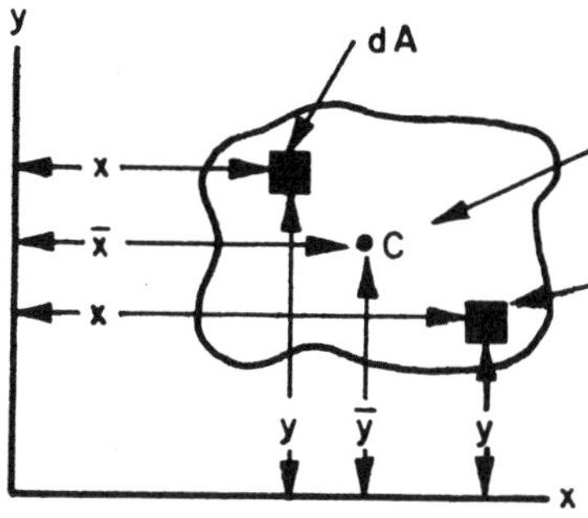

Letting Q represent the first moment of the area, the resulting mathematical expressions define the moment of area A

$$Q_x = \int y dA, \qquad Q_y = \int x dA$$

and, for the centroid,

$$A\bar{y} = \int y dA, \qquad A\bar{x} = \int x dA$$

or

$$\bar{y} = \frac{\int y dA}{A}, \qquad \bar{x} = \frac{\int x dA}{A}$$

Examples.

TRIANGLE

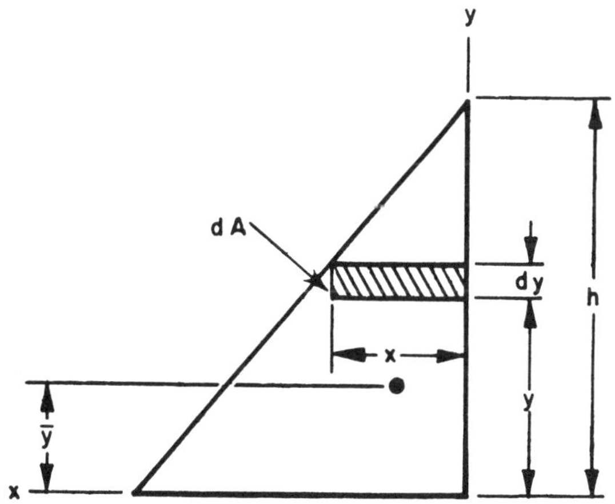

$$A = bh/2$$

$$A\bar{y} = \int y dA$$

$$dA = x dy$$

$$A\bar{y} = \int xy dy$$

From similar triangles,

$$\frac{x}{h - y} = \frac{b}{h}, \qquad \text{or} \qquad x = \frac{b(h - y)}{h}$$

$$A\bar{y} = \frac{b}{h}\int_0^h (h - y)y\,dy = \frac{bh^2}{6}$$

$$\bar{y} = \frac{bh^2/6}{bh/2} = \frac{h}{3} \quad \text{(ans.)}$$

CIRCULAR SECTOR

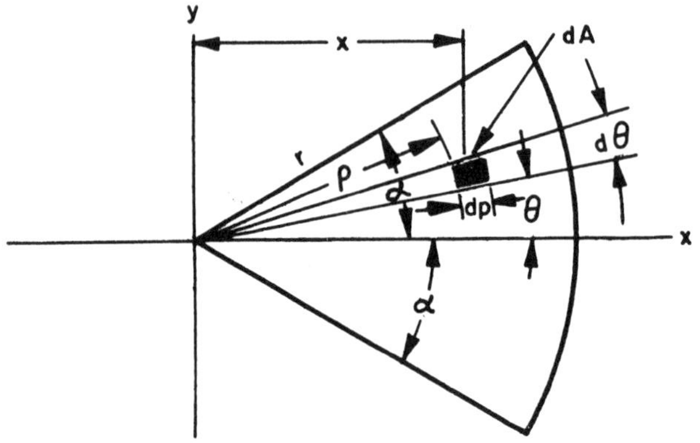

$$A\bar{x} = \int x\,dA$$

$$x = \rho \cos\theta$$

$$dA = \rho\,d\theta\,d\rho$$

$$A\bar{x} = \int_0^r\int_{-\alpha}^{\alpha} \rho \cos\theta\rho\,d\rho\,d\theta = \frac{2r^3 \sin\alpha}{3}$$

$$\bar{x} = \frac{(2/3)r^3 \sin\alpha}{A} = \frac{(2/3)r^3 \sin\alpha}{r^2\alpha} = \frac{2r \sin\alpha}{3\alpha} \quad \text{(ans.)}$$

HALF PARABOLA

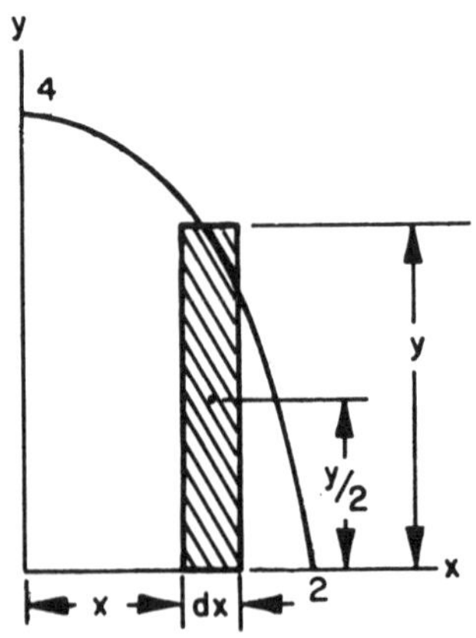

Parabola: $y = 4 - x^2$

$dA = y\,dx$

$$A = \int_0^2 y\,dx = \int_0^2 (4 - x^2)\,dx = \frac{16}{3}$$

$$Q_x = \int y\,dA = \int_0^2 \left(\frac{y}{2}\right)(y)\,dx = \int_0^2 \frac{y^2}{2}\,dx = \frac{1}{2}\int_0^2 (4 - x^2)^2\,dx = \frac{128}{15}$$

$$Q_y = \int x\,dA = \int_0^2 xy\,dx = \int_0^2 (4x - x^3)\,dx = 4$$

$$\bar{x} = \frac{Q_y}{A} = \frac{4}{16/3} = \frac{3}{4}, \qquad \bar{y} = \frac{Q_x}{A} = \frac{128/15}{16/3} = \frac{8}{5} \quad \text{(ans.)}$$

Note. When summing the elemental strips about the x-axis as shown in the diagram, the moment arm is equal to $y/2$. If the summation is made with respect to the y-axis, the moment arm is equal to x (not $x/2$).

CENTROID OF A COMPOSITE AREA

Composite areas have centroids, the coordinates of which may be determined by applying the basic definition to the total area as follows:

$$A\bar{y} = \Sigma ay \qquad \text{and} \qquad A\bar{x} = \Sigma ax$$

or

$$\bar{y} = \frac{\Sigma ay}{A} \qquad \text{and} \qquad \bar{x} = \frac{\Sigma ax}{A}$$

where x and y are the perpendicular distances from their respective axes to the areas in question, a is the area of each individual part, and A is the total area of the composite body.

Example.

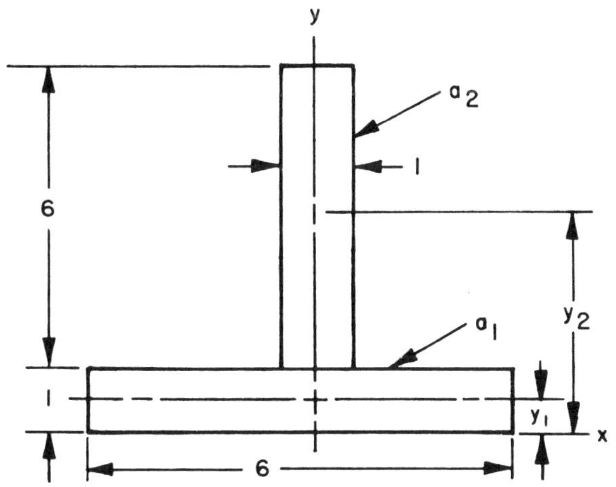

$$\bar{y} = \frac{a_1y_1 + a_2y_2}{A} = \frac{(1)(6)(1/2) + (6)(1)(4)}{6 + 6} = 2.25 \text{ in.} \quad \text{(ans.)}$$

$\bar{x} = 0$, by symmetry.

MOMENT OF INERTIA OF A COMPOSITE AREA

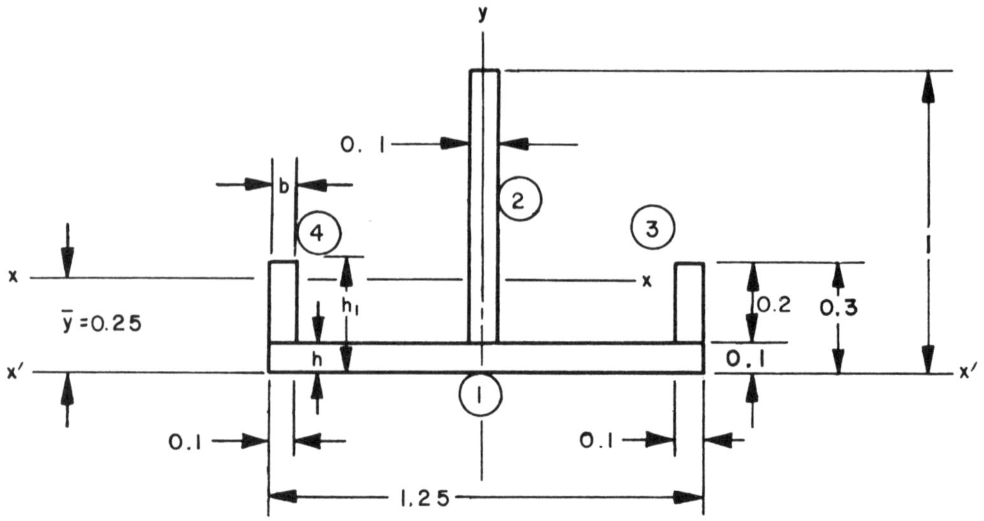

$$\bar{y} = \frac{\left[(1.25)(0.1)\right](0.05) + (0.1)(0.9)(0.55) + (2)(0.1)(0.2)(0.2)}{(1.25)(0.1) + (0.1)(0.9) + (2)(0.1)(0.2)}$$

$$= \frac{0.06375}{0.255} = 0.25 \text{ in.} \quad \text{(ans.)}$$

An alternative method for obtaining the moment of inertia of a composite area and the centroidal distance, d, as in the diagram above, is given in Table 1.

TABLE 1. COMPUTATION TABLE FOR THE MOMENT OF INERTIA AND CENTROIDAL DISTANCE OF A COMPOSITE AREA

Part	Area, in^2	b, in.	h_1^2, in^2	h^2, in^2	Moment, $\frac{b(h_1^2 - h^2)}{2}$, in^3	h_1^3, in^3	h^3, in^3	I_x' $b/3(h_1^3 - h^3)$, in^4
1	0.125	1.25	0.01	0	0.006	0.001	0	0.00003
2	0.09	0.1	1.0	0.01	0.049	1.000	0.001	0.03333
3	0.02	0.1	0.09	0.01	0.004	0.027	0.001	0.00087
4	0.02	0.1	0.09	0.01	0.004	0.027	0.001	0.00087
A = 0.255			M = 0.06375			$I_x' = 0.035$		

The centroidal distance may be found by either method. By definition,

$$d = \frac{M}{A} = \frac{0.06375}{0.255} = 0.25 = \bar{y}$$

$$I_x = I_{x'} - Ad^2$$

$$= 0.035 - 0.255(0.25)^2$$

$$= 0.0191 \text{ in}^4 \text{ (ans.)}$$

MOMENTS OF INERTIA OF A PLANE AREA

The moment of inertia of a plane area, mass, or volume is sometimes referred to as the second moment, since the first moment, Q, is multiplied by the differential area moment arm.

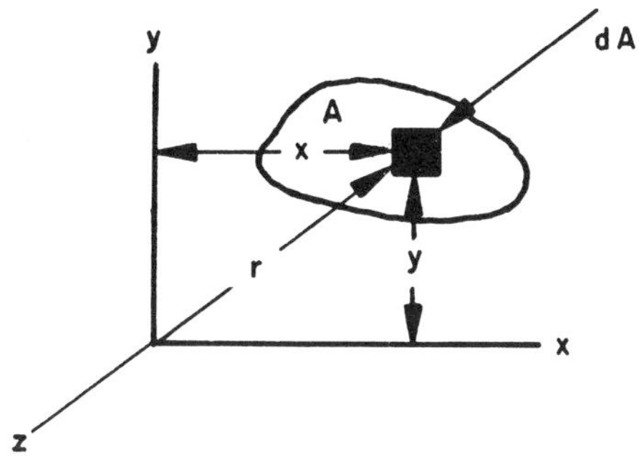

$$I_x = \int y^2 dA$$
$$I_y = \int x^2 dA$$

where the elements are integrated over the entire body.

The moment of inertia of the body about the z-axis is

$$I_z = J_z = \int r^2 dA$$

and, since $r^2 = x^2 + y^2$, $J_z = I_x + I_y$ where J_z is known as the polar moment of inertia of the body.

Example.

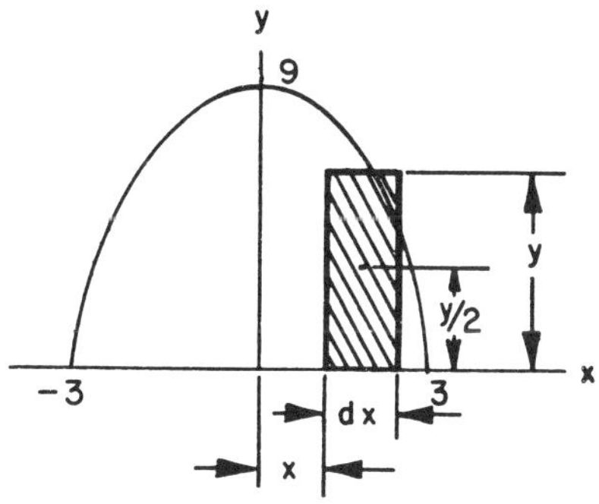

The moment of inertia of the parabola $y = 9 - x^2$ about the y-axis is calculated as follows

$$dA = ydx$$
$$x^2 = 9 - y$$
$$y = 9 - x^2$$
$$I_y = \int x^2 dA$$
$$= 2\int_0^3 (x^2)(9 - x^2)dx$$
$$= 2\int_0^3 (9x^2 - x^4)dx$$
$$= 2\left[3x^2 + \frac{x^5}{5}\right]_0^3 = \frac{324}{5} \text{ (ans.)}$$

TRANSFER OF AXES ON A PLANE AREA

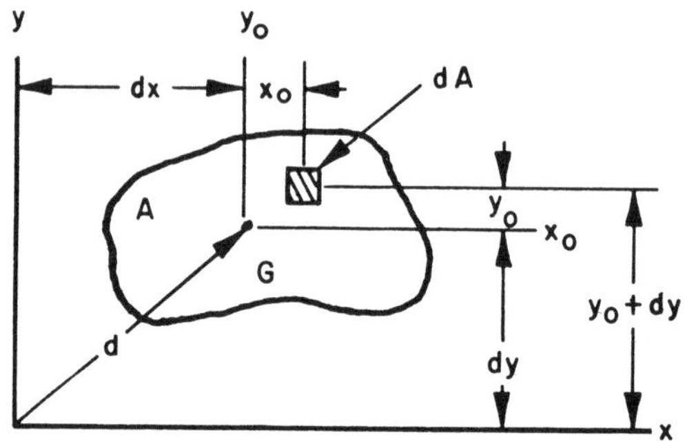

$$dI_x = (y_0 + dx)^2 dA$$

$$I_x = \int (y_0^2 + 2y_0 dx + dx^2) dA$$

$$= \int y_0^2 dA + 2dx \int y_0 dA + dx^2 \int dA$$

Since the second term in the equation above is zero, the resulting integrations leave

$$I_x = \bar{I}_x + Adx^2$$

and similarly,

$$I_y = \bar{I}_y + Ady^2$$

The sum of these two equations (from $J_z = I_x + I_y$) gives

$$J_z = \bar{J}_z + Ad^2$$

which is the polar moment of inertia of the body when transferred through the distance d.

RADIUS OF GYRATION

By definition, $k = \sqrt{I/A}$. Substituting into the moment of inertia equations yields

$$k^2 = \bar{k}^2 + d^2$$

where $\bar{k}$ is the radius of gyration about a centroidal axis parallel to the axis about which k applies, the axes being separated by d.

	SQUARE	HOLLOW SQUARE
	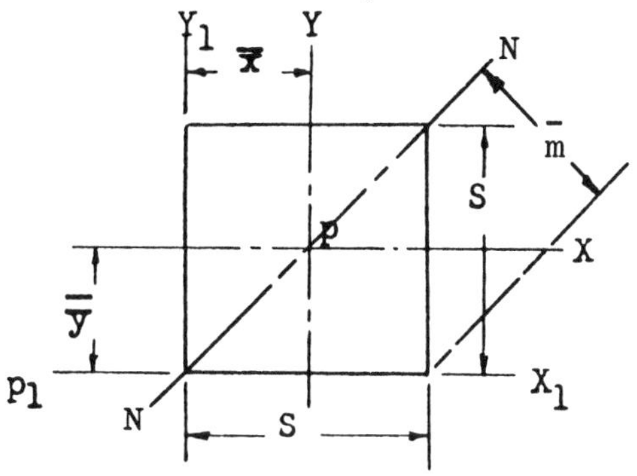	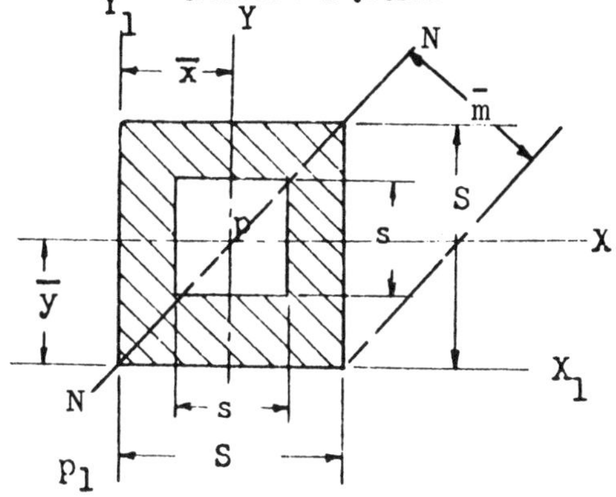
AREA	S^2	$S^2 - s^2$
CENTROID	$\bar{x} = \bar{y} = \frac{S}{2} \quad \bar{m} = \frac{S}{\sqrt{2}} = 0.7075$	$\bar{x} = \bar{y} = \frac{S}{2} \quad \bar{m} = \frac{S}{\sqrt{2}} = 0.7075$
WEIGHT MOMENT OF INERTIA	$I_x = I_y = \frac{WS^2}{12}$ $I_{x_1} = I_{y_1} = \frac{WS^2}{3}$ $I_p = I_x + I_y = \frac{WS^2}{6}$ $I_{p_1} = I_{x_1} + I_{y_1} = \frac{2WS^2}{3}$	$I_x = I_y = \frac{W(S^2 - s^2)}{12}$ $I_{x1} = I_{y_1} = \frac{W(4S^2 + s^2)}{12} = I_x + \frac{5}{12} WS^2$ $I_p = I_x + I_y = \frac{W(S^2 - s^2)}{6}$ $I_{p_1} = \frac{W(4S^2 + s^2)}{6} = 2 I_{x_1}$
AREA MOMENT OF INERTIA	$I_N = I_x = I_y = \frac{S^4}{12}$ $I_{x_1} = I_{y_1} = \frac{S^4}{3}$ $I_p = I_x + I_y = \frac{S^4}{6}$ $I_{p_1} = I_{x_1} + I_{y_1} = \frac{2S^4}{3}$	$I_N = I_x = I_y = \frac{S^4 - s^4}{12}$ $I_{x_1} = I_{y_1} = \frac{4S^4 - 3S^2s^2 - s^4}{12}$ $I_p = \frac{S^4 - s^4}{6}$ $I_{p_1} = \frac{4S^4 - 3S^2s^2 - s^4}{6}$
RADIUS OF GYRATION	$k_N = k_x = k_y = 0.289S$ $k_{x_1} = k_{y_1} = 0.577S$ $k_p = 0.408S$ $k_{p_1} = 0.816S$	$k_N = k_x = k_y = 0.289 \sqrt{S^2 + s^2}$ $k_{x_1} = \sqrt{\frac{I_{x1}}{S^2 - s^2}} = \frac{1}{6} \sqrt{3(4 S^2 + s^2)}$ $k_{y_1} = \sqrt{\frac{I_{y1}}{S^2 - s^2}} = \frac{1}{6} \sqrt{3(4 S^2 + s^2}$ $k_p = 0.408 \sqrt{S^2 + s^2}$ $k_{p_1} = \sqrt{\frac{I_{p1}}{S^2 - s^2}} = \frac{1}{6} \sqrt{6(4 S^2 + s^2}$

Adapted from Weight Handbook, Vol. 1, Society of Aeronautical Weight Engineers, Inc.

RECTANGLE

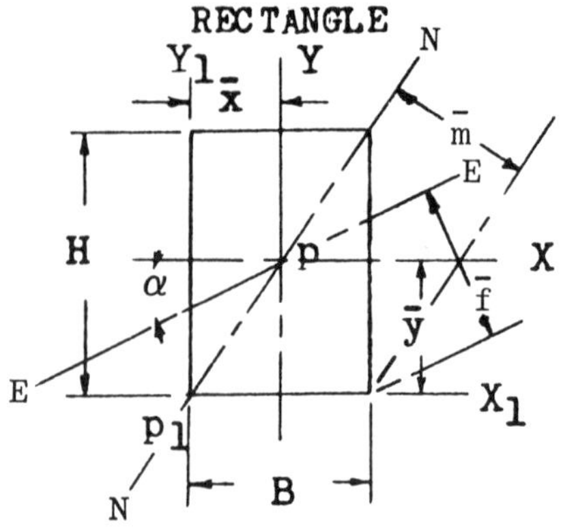

AREA	BH
CENTROID	$\bar{x} = \frac{B}{2}$ $\quad \bar{f} = \frac{B\sin\alpha + H\cos\alpha}{2}$ $\bar{y} = \frac{H}{2}$ $\quad \bar{m} = \frac{BH}{\sqrt{B^2 + H^2}}$
WEIGHT MOMENT OF INERTIA	$I_x = \frac{WH^2}{12}$ $\quad I_y = \frac{WB^2}{12}$ $\quad I_p = \frac{W}{12}(H^2 + B^2)$ $I_{x_1} = \frac{WH^2}{3}$ $\quad I_{y_1} = \frac{WB^2}{3}$ $\quad I_{p_1} = \frac{W}{3}(H^2 + B^2)$
AREA MOMENT OF INERTIA	$I_x = \frac{BH^3}{12}$ $\quad I_y = \frac{HB^3}{12}$ $\quad I_p = \frac{BH}{12}(H^2 + B^2)$ $\quad I_N = \frac{B^3H^3}{6(B^2 + H^2)}$ $I_{x_1} = \frac{BH^3}{3}$ $\quad I_{y_1} = \frac{HB^3}{3}$ $\quad I_{p_1} = \frac{BH}{3}(H^2 + B^2)$ $\quad I_E = \frac{BH(B^2\sin^2\alpha + H^2\cos^2\alpha)}{12}$
RADIUS OF GYRATION	$k_x = 0.289H$ $\quad k_y = 0.289B$ $\quad k_p = 0.289\sqrt{H^2 + B^2}$ $\quad k_N = \frac{BH}{\sqrt{6(B^2 + H^2)}}$ $k_{x_1} = 0.577H$ $\quad k_{y_1} = 0.577B$ $\quad k_{p_1} = 0.577\sqrt{H^2 + B^2}$ $\quad k_E = \sqrt{\frac{B^2\sin^2\alpha + H^2\cos^2\alpha}{12}}$

Adapted from Weight Handbook, Vol. 1, Society of Aeronautical Weight Engineers, Inc.

	HOLLOW RECTANGLE 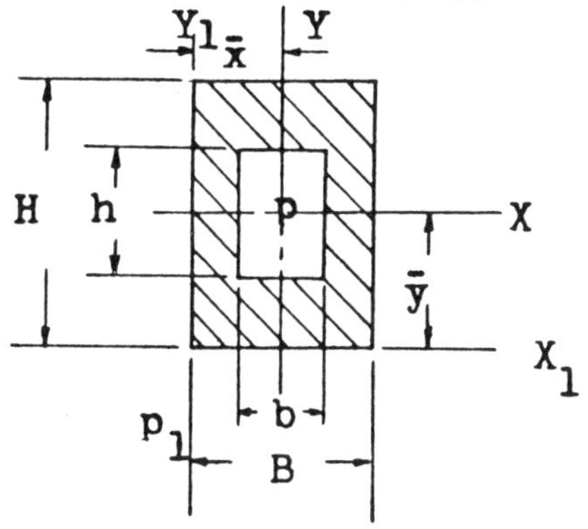
AREA	$BH - bh$
CENTROID	$\bar{x} = \frac{B}{2}$ $\quad \bar{y} = \frac{H}{2}$
WEIGHT MOMENT OF INERTIA	$I_x = \frac{W}{12}\left[\frac{BH^3 - bh^3}{BH - bh}\right]$ $\quad I_y = \frac{W}{12}\left[\frac{HB^3 - hb^3}{BH - bh}\right]$ $I_{x_1} = I_x + \frac{WH^2}{4}$ $\quad I_{y_1} = I_y + \frac{WB^2}{4}$ $I_p = I_x + I_y$ $\quad I_{p_1} = I_{x_1} + I_{y_1}$
AREA MOMENT OF INERTIA	$I_x = \frac{BH^3 - bh^3}{12}$ $\quad I_y = \frac{HB^3 - hb^3}{12}$ $I_{x_1} = \frac{BH^3}{3} - \frac{bh(3H^2 + h^2)}{12}$ $\quad I_{y_1} = I_y + \frac{(BH - bh)B^2}{4}$ $I_p = I_x + I_y$ $\quad I_{p_1} = I_{x_1} + I_{y_1}$
RADIUS OF GYRATION	$k_x = \sqrt{\frac{BH^3 - bh^3}{12(BH - bh)}}$ $\quad k_y = \sqrt{\frac{HB^3 - hb^3}{12(BH - bh)}}$ $\quad k_p = \sqrt{\frac{I_p}{BH - bh}}$ $k_{x_1} = \sqrt{\frac{I_{x_1}}{BH - bh}}$ $\quad k_{y_1} = \sqrt{\frac{I_{y_1}}{BH - bh}}$ $\quad k_{p_1} = \sqrt{\frac{I_{p_1}}{BH - bh}}$

Adapted from Weight Handbook, Vol. 1, Society of Aeronautical Weight Engineers, Inc.

ANGLE

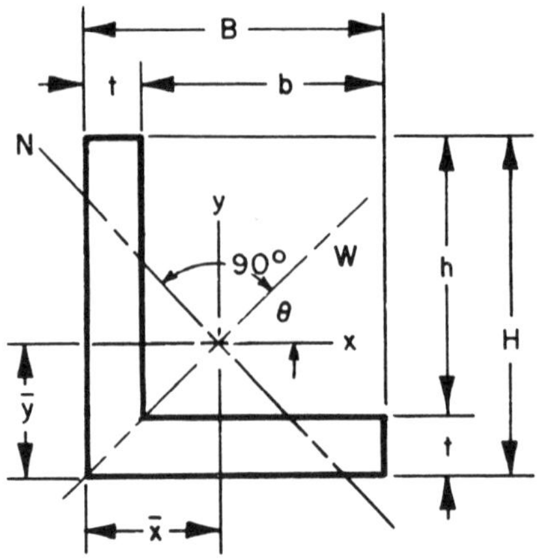

$$\text{Tan } 2\theta = \frac{2I_{xy}}{I_y - I_x}$$

I_{xy} = product of inertia about x-x and y-y

$$I_{xy} = \mp \frac{(bBhHt)}{4(B + h)}$$

I_{xy} is negative when the heel of the angle, with respect to the center of gravity, is in the first or third quadrant; positive when it is in the second or fourth quadrant

AREA	$t(B + h)$
CENTROID	$\bar{x} = \frac{B^2 + ht}{2(B + h)}$ $\bar{y} = \frac{H^2 + bt}{2(B + h)}$
WEIGHT MOMENT OF INERTIA	$I_x = \frac{m}{3}\left[\frac{t(H - \bar{y})^3 + B\bar{y}^3 - b(\bar{y} - t)^3}{t(H + B - t)}\right]$ $I_y = \frac{m}{3}\left[\frac{t(B - \bar{x})^3 + H\bar{x}^3 - h(\bar{x} - t)^3}{t(H + B - t)}\right]$ $I_N = \frac{m(I_x \sin^2\theta + I_y \cos^2\theta + I_{xy} \sin 2\theta)}{t(H + B - t)}$ $I_W = \frac{m(I_x \cos^2\theta + I_y \sin^2\theta - I_{xy} \sin 2\theta)}{t(H + B - t)}$
AREA MOMENT OF INERTIA	$I_x = 1/3\left[t(H - \bar{y})^3 + B\bar{y}^3 - b(\bar{y} - t)^3\right]$ $I_y = 1/3\left[t(B - \bar{x})^3 + H\bar{x}^3 - h(\bar{x} - t)^3\right]$ $I_N = I_x \sin^2\theta + I_y \cos^2\theta + I_{xy} \sin 2\theta$ $I_W = I_x \cos^2\theta + I_y \sin^2\theta - I_{xy} \sin 2\theta$
RADIUS OF GYRATION	$k = \sqrt{I/A}$

	EQUAL RECTANGLES	UNEQUAL RECTANGLES
	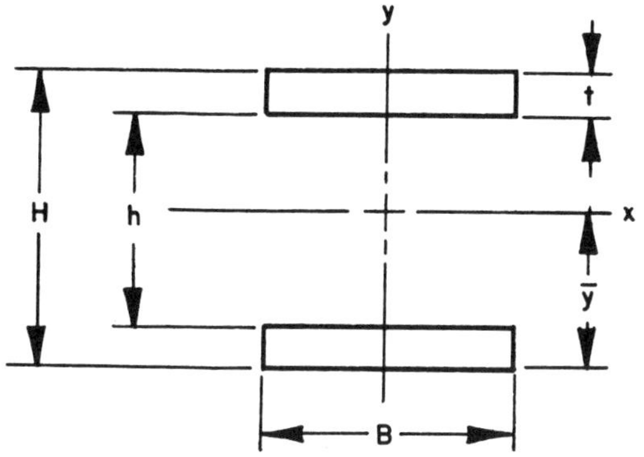	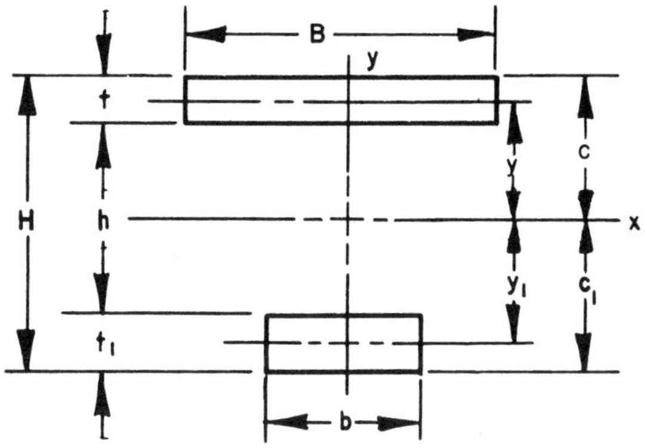
AREA	$B(H - h)$	$Bt + bt_1 \qquad c_1 = H - c$
CENTROID	$\bar{y} = H/2$	$c = \dfrac{(1/2)Bt^2 + bt_1\left[H - (1/2)t_1\right]}{Bt + bt_1}$
WEIGHT MOMENT OF INERTIA	$I_x = \dfrac{m(H^3 - h^3)}{12(H - h)}$ $I_y = \dfrac{mB^2}{12}$	$I_x = \dfrac{m(Bt^3 + 12Bty^2 + bt_1^3 + 12bt_1y_1^2)}{12(Bt + bt_1)}$ $I_y = \dfrac{m(tB^3 + t_1b^3)}{12(Bt + bt_1)}$
AREA MOMENT OF INERTIA	$I_x = \dfrac{B(H^3 - h^3)}{12}$ $I_y = \dfrac{B^3(H - h)}{12}$	$I_x = \dfrac{Bt^3}{12} + Bty^2 + \dfrac{bt_1^3}{12} + bt_1y_1^2$ $I_y = \dfrac{tB^3 + t_1b^3}{12}$
RADIUS OF GYRATION	$k_x = \sqrt{\dfrac{H^3 - h^3}{12B(H - h)}}$ $k_y = \sqrt{\dfrac{B^3(H - h)}{12B(H - h)}} = 0.289B$	$k_x = \sqrt{\dfrac{Bt^3 + 12Bty^2 + bt_1^3 + 12bt_1y_1^2}{12(B\ t + bt_1)}}$ $k_y = \sqrt{\dfrac{tB^3 + t_1b^3}{12(Bt + bt_1)}}$

H-SECTION

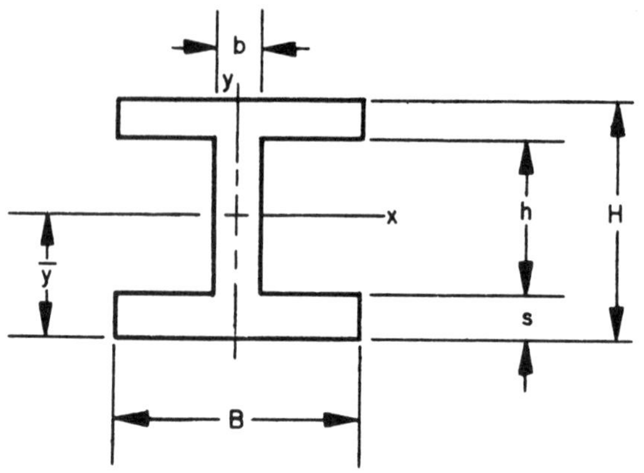

AREA	$BH - h(B - b)$
CENTROID	$\bar{y} = H/2 \qquad \bar{x} = 0$
WEIGHT MOMENT OF INERTIA	$I_x = \dfrac{m\left[BH^3 - h^3(B - b)\right]}{12\left[BH - h(B - b)\right]}$ $I_y = \dfrac{m(hb^3 + 2sB^3)}{12\left[BH - h(B - b)\right]}$
AREA MOMENT OF INERTIA	$I_x = \dfrac{BH^3 - h^3(B - b)}{12}$ $I_y = \dfrac{hb^3 + 2sB^3}{12}$
RADIUS OF GYRATION	$k_x = \sqrt{\dfrac{BH^3 - h^3(B - b)}{12\left[BH - h(B - b)\right]}}$ $k_y = \sqrt{\dfrac{hB^3 + 2sB^3}{12\left[BH - h(B - b)\right]}}$

	Z-SECTION	CROSSED RECTANGLES
	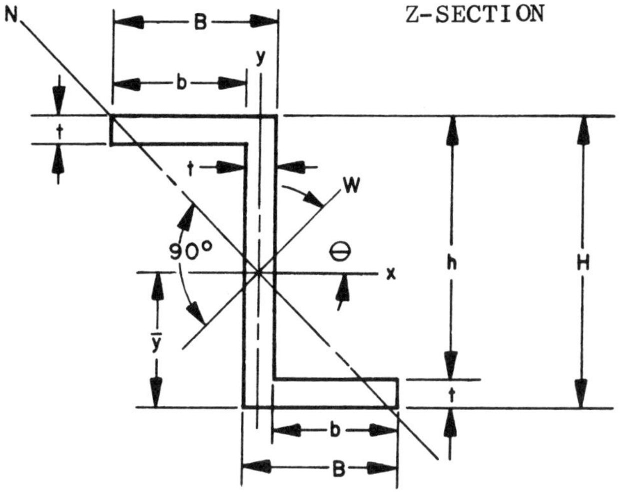 	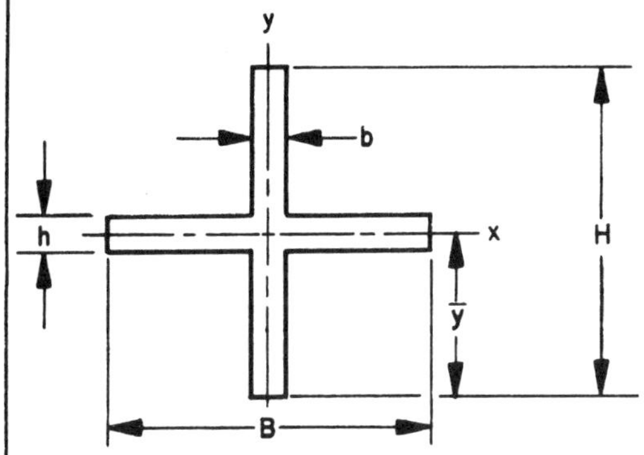
AREA	$t(H + 2b)$	$Hb + h(B - b)$
CENTROID	$\bar{y} = H/2 \qquad \bar{x} = 0$	$\bar{y} = H/2 \qquad \bar{x} = B/2$
WEIGHT MOMENT OF INERTIA	$\tan 2\theta = \dfrac{(Ht - t^2)(B^2 - Bt)}{I_x - I_y}$ $I_x = \dfrac{m\left[BH^3 - b(H - 2t)^3\right]}{12\left[t(H + 2b)\right]}$ $I_y = \dfrac{m\left[H(B + b)^3 - 2b^3h - 6bB^2h\right]}{12t(H + 2b)}$	$I_x = \dfrac{m\left[bH^3 + h^3(B - b)\right]}{12\left[Hb + h(B - b)\right]}$ $I_y = \dfrac{m\left[hB^3 + b^3(H - h)\right]}{12\left[Hb + h(B - b)\right]}$
AREA MOMENT OF INERTIA	$I_x = \dfrac{BH^3 - b(H - 2t)^3}{12}$ $I_y = \dfrac{H(B + b)^3 - 2b^3h - 6B^2bh}{12}$ $I_N = I_x \sin^2\theta + I_y \cos^2\theta + I_{xy} \sin 2\theta$ $I_W = I_x \cos^2\theta + I_y \sin^2\theta - I_{xy} \sin 2\theta$	$I_x = \dfrac{bH^3 + h^3(B - b)}{12}$ $I_y = \dfrac{hB^3 + b^3(H - h)}{12}$
RADIUS OF GYRATION	$k_x = \sqrt{I_x/A}$ $k_y = \sqrt{I_y/A}$	$k_x = \sqrt{\dfrac{bH^3 + h^3(B - b)}{12\left[Hb + h(B - b)\right]}}$ $k_y = \sqrt{\dfrac{hB^3 + b^3(H - h)}{12\left[Hb + h(B - b)\right]}}$

CHANNEL OR U-SECTION

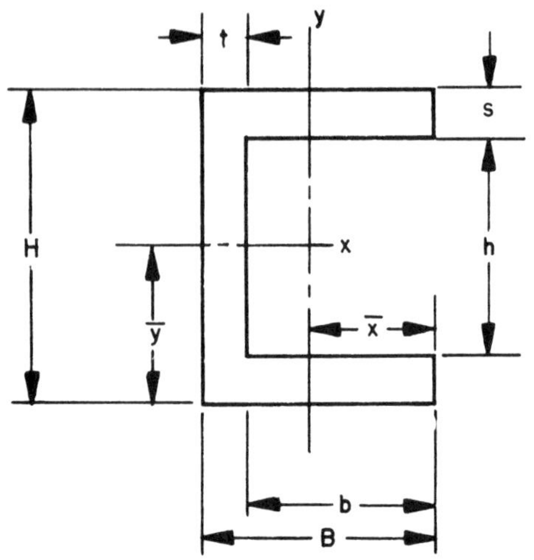

AREA	$BH - h(B - t) = A$
CENTROID	$\bar{y} = \frac{H}{2} \qquad \bar{x} = B - \frac{2B^2s + ht^2}{2BH - 2h(B - t)}$
WEIGHT MOMENT OF INERTIA	$I_x = \frac{m\left[BH^3 - h^3(B - t)\right]}{12\left[BH - h(B - t)\right]}$ $I_y = \frac{m}{3}\left[\frac{2sB^3 + ht^3 - 3A(B - \bar{x})^2}{BH - h(B - t)}\right]$
AREA MOMENT OF INERTIA	$I_x = \frac{BH^3 - h^3(B - t)}{12}$ $I_y = \frac{2sB^3 + ht^3}{3} - A(B - \bar{x})^2$
RADIUS OF GYRATION	$k_x = \sqrt{\frac{BH^3 - h^3(B - t)}{12\left[BH - h(B - t)\right]}}$ $k_y = \sqrt{\frac{2sB^3 + ht^3 - 3A(B - \bar{x})^2}{3\left[BH - h(B - t)\right]}}$

T-SECTION

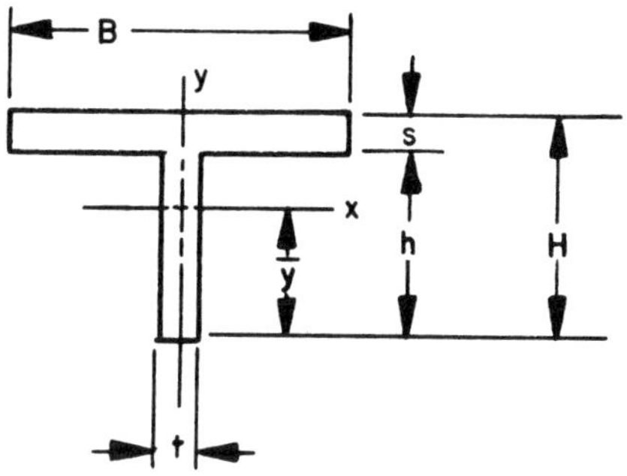

AREA	$Bs + ht$
CENTROID	$\bar{y} = H - \dfrac{H^2t + s^2(B - t)}{2(Bs + ht)}$
WEIGHT MOMENT OF INERTIA	$I_x = \dfrac{m}{3}\left[\dfrac{t\bar{y}^3 + B(H - \bar{y})^3 - (B - t)(H - \bar{y} - s)^3}{Bs + ht}\right]$ $I_y = \dfrac{m}{12}\left(\dfrac{sB^3 + ht^3}{Bs + ht}\right)$
AREA MOMENT OF INERTIA	$I_x = \dfrac{t\bar{y}^3 + B(H - \bar{y})^3 - (B - t)(H - \bar{y} - s)^3}{3}$ $I_y = \dfrac{sB^3 + ht^3}{12}$
RADIUS OF GYRATION	$k_x = \sqrt{I_x/A}$ $k_y = \sqrt{I_y/A}$

MODIFIED T-SECTION

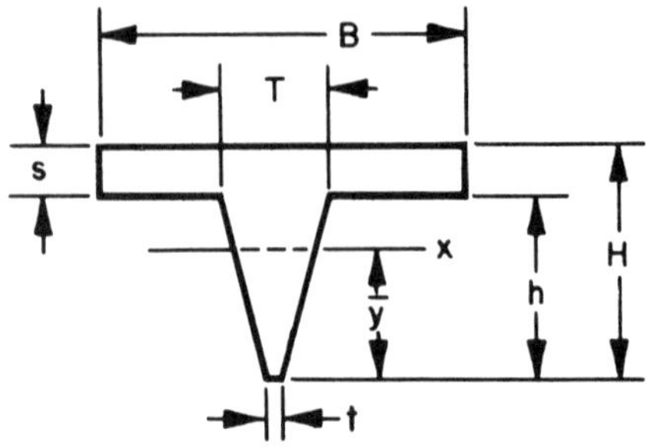

AREA	$Bs + \dfrac{h(T+t)}{2} = A$
CENTROID	$\bar{y} = H - \dfrac{\left[3Bs^2 + 3ht(H+s) + h(T+t)(h+3s)\right]}{6A}$
WEIGHT MOMENT OF INERTIA	$I_x = \dfrac{m\left\{\left[4Bs^3 + h^3(3t+T)\right] - 12A(H-\bar{y}-s)^2\right\}}{6\left[2Bs + h(T+t)\right]}$
AREA MOMENT OF INERTIA	$I_x = \dfrac{4Bs^3 + h^3(3t+T)}{12} - A(H-\bar{y}-s)^2$
RADIUS OF GYRATION	$k_x = \sqrt{I_x/A}$

REGULAR POLYGON

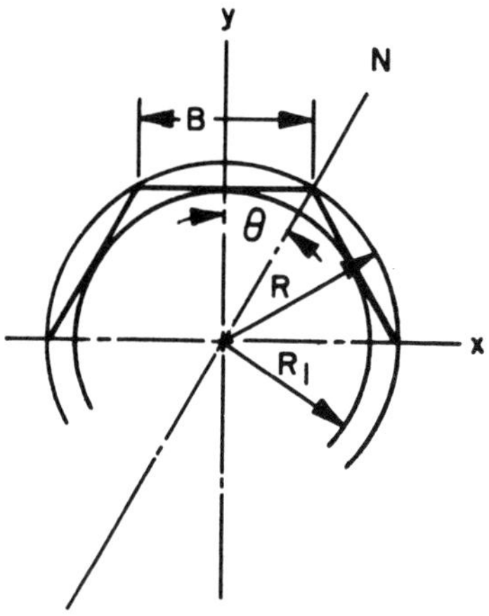

n = number of sides

$$\theta = \frac{180°}{n} \qquad B = 2\sqrt{R^2 - r_1^2}$$

AREA	$\frac{nB^2 \cot\theta}{4} = \frac{nR^2 \sin 2\theta}{2} = nR_1^2 \tan\theta$
CENTROID	$\bar{x} = \bar{y} = 0$
WEIGHT MOMENT OF INERTIA	$R = \frac{B}{2\sin\theta} \qquad R_1 = \frac{B}{2\tan\theta}$ $I_y = I_N = \frac{m(6R^2 - B^2)}{24} = \frac{m(12R_1^2 + B^2)}{48}$
AREA MOMENT OF INERTIA	$I_y = I_N = \frac{A(6R^2 - B^2)}{24} = \frac{A(12R_1^2 + B^2)}{48}$
RADIUS OF GYRATION	$k_y = k_N = \sqrt{\frac{6R^2 - B^2}{24}} = \sqrt{\frac{12R_1^2 + B^2}{48}}$

	REGULAR HEXAGON	REGULAR OCTAGON
	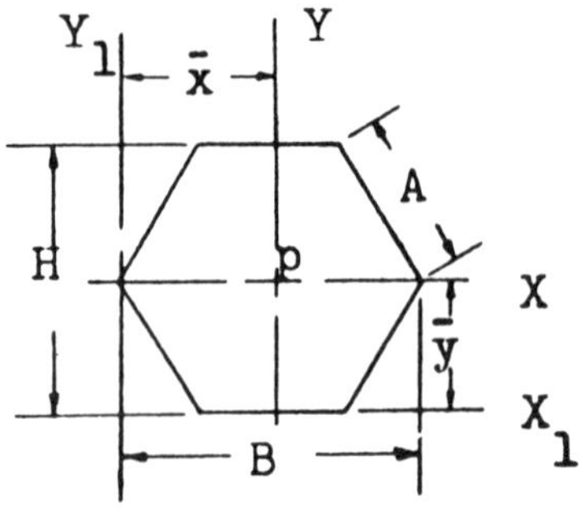	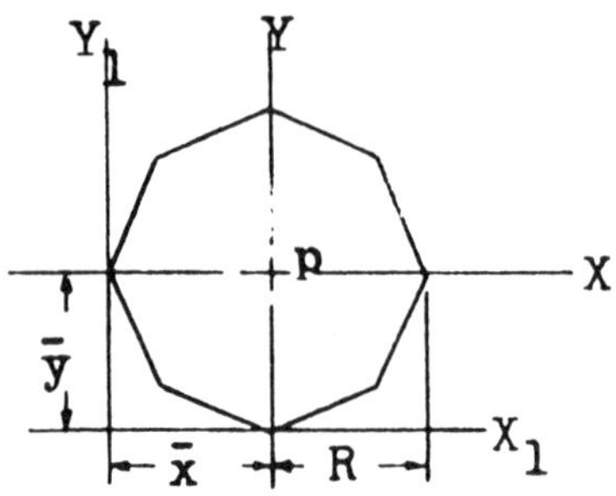
AREA	$0.866\ H^2$	$2.8284\ R^2$
CENTROID	$\bar{x} = \frac{B}{2} = A \qquad \bar{y} = \frac{H}{2}$	$\bar{x} = \bar{y} = R$
WEIGHT MOMENT OF INERTIA	$I_x = I_y = 0.0694WH^2 = 0.0521WB^2$ $I_{x_1} = 0.3194WH^2$ $I_p = 0.1389WH^2 = 0.1042WB^2$ $I_{y_1} = 0.4028WH^2 = 0.3021WB^2$	$I_x = I_y = 0.2256WR^2$ $I_{x_1} = I_{y_1} = 1.2256WR^2$ $I_p = 0.4512WR^2$
AREA MOMENT OF INERTIA	$I_x = I_y = 0.0601H^4$ $I_{x_1} = 0.2766H^4$ $I_{y_1} = 0.3488H^4$ $I_p = 0.1203H^4$	$I_x = I_y = 0.6381R^4$ $I_{x_1} = I_{y_1} = 3.4665R^4$ $I_p = 1.2761R^4$
RADIUS OF GYRATION	$k_x = k_y = 0.2635H = 0.2282B$ $k_{x_1} = 0.5652H$ $k_{y_1} = 0.6346H$ $k_p = 0.3727H$	$k_x = k_y = 0.4750R$ $k_{x_1} = k_{y_1} = 1.1071R$ $k_p = 0.672R$

Adapted from Weight Handbook, Vol. 1, Society of Aeronautical Weight Engineers, Inc.

	ISOSCELES TRAPEZOID	OBLIQUE TRAPEZOID
	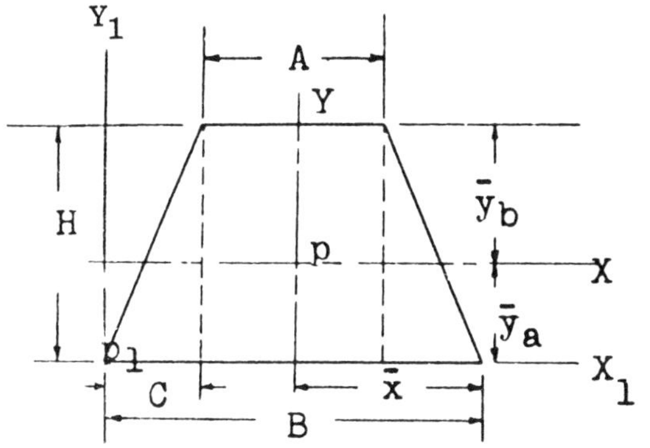	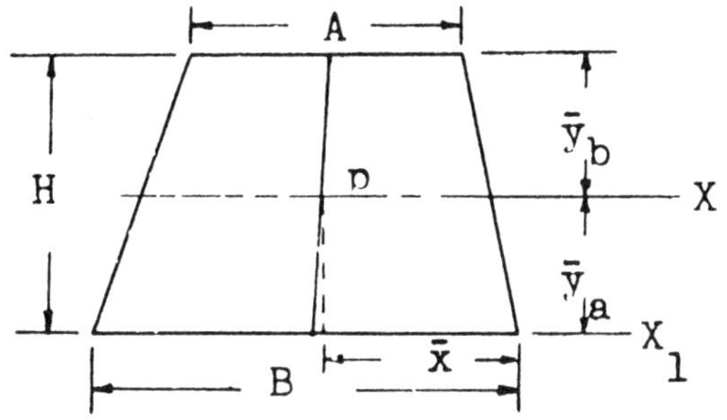
AREA	$\frac{H(A+B)}{2}$	$\frac{H(A+B)}{2}$
CENTROID	$\bar{x} = \frac{B}{2}$ $\bar{y}_a = \frac{H(B+2A)}{3(B+A)}$ $\bar{y}_b = \frac{H(A+2B)}{3(A+B)}$	$\bar{x}$ is on a line connecting mid-points of sides A & B. $\bar{y}_a = \frac{H(B+2A)}{3(A+B)}$ $\bar{y}_b = \frac{H(2B+A)}{3(A+B)}$
WEIGHT MOMENT OF INERTIA	$I_x = \frac{WH^2}{18}\left(1 + \frac{2AB}{(A+B)^2}\right)$ $I_{x_1} = \frac{WH^2(2A+C)}{6(A+C)}$ or $\frac{WH^2(3A+B)}{6(A+B)}$ $I_y = \frac{W}{24}(A^2+B^2)$ $I_{y_1} = \frac{W}{24}(A^2+7B^2)$ $I_p = I_x + I_y$ $I_{p_1} = I_{x_1} + I_{y_1}$	$I_x = \frac{WH^2}{18}\left(1 + \frac{2AB}{(A+B)^2}\right)$ $I_{x_1} = \frac{WH^2}{6}\left(\frac{3A+B}{A+B}\right)$
AREA MOMENT OF INERTIA	$I_x = \frac{H^3(A^2+4AB+B^2)}{36(A+B)}$ $I_{x_1} = \frac{H^3(A+B)(2A+C)}{12(A+C)}$ $I_y = \frac{H(A+B)(A^2+B^2)}{48}$ $I_{y_1} = \frac{H(A+B)(A^2+7B^2)}{48}$ $I_p = I_x + I_y$ $I_{p_1} = I_{x_1} + I_{y_1}$	$I_x = \frac{H^3(A^2+4AB+B^2)}{36(A+B)}$ $I_{x_1} = \frac{H^3(B+3A)}{12}$
RADIUS OF GYRATION	$k_x = H\sqrt{\frac{2(B^2+4AB+A^2)}{6(A+B)}}$ $k_{x_1} = \sqrt{\frac{H^2(2A+C)}{A+C}}$ $k_y = \sqrt{\frac{A^2+B^2}{24}}$ $k_{y_1} = \sqrt{\frac{A^2+7B^2}{24}}$ $k_p = \sqrt{\frac{2I_p}{H(A+B)}}$ $k_{p_1} = \sqrt{\frac{2I_{p_1}}{H(A+B)}}$	$k_x = H\sqrt{\frac{2(A^2+4AB+B^2)}{6(A+B)}}$ $k_{x_1} = H\sqrt{\frac{3A+B}{6(A+B)}}$

Adapted from Weight Handbook, Vol. 1, Society of Aeronautical Weight Engineers, Inc.

	PARALLELOGRAM	RIGHT-ANGLED TRAPEZOID
	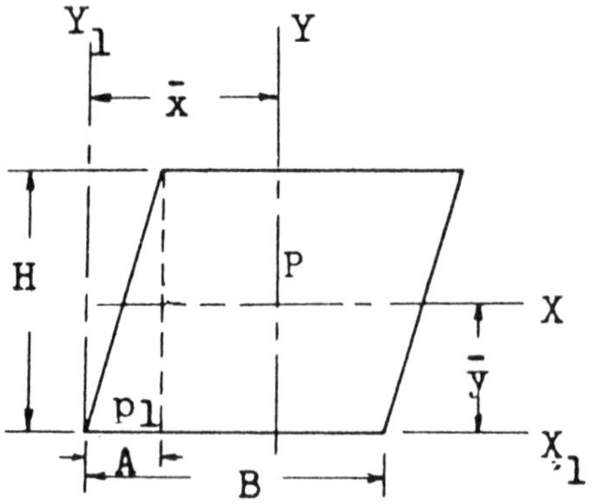	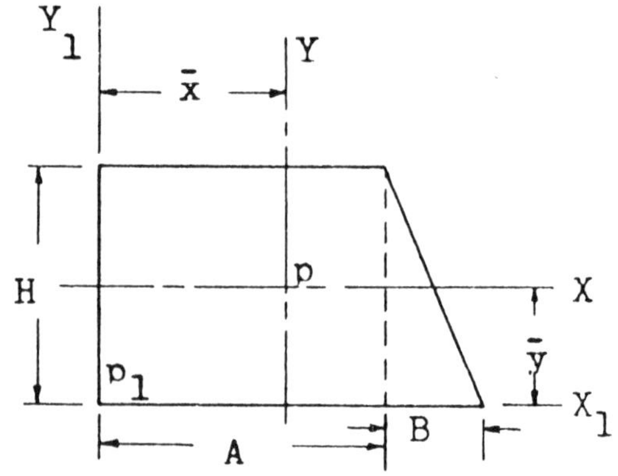
AREA	BH	$\frac{H}{2}(2A+B)$
CENTROID	$\bar{x} = \frac{A+B}{2}$ $\bar{y} = \frac{H}{2}$	$\bar{x} = \frac{3A^2+3AB+B^2}{3(2A+B)}$ $\bar{y} = \frac{H(3A+B)}{3(2A+B)}$
WEIGHT MOMENT OF INERTIA	$I_x = \frac{WH^2}{12}$ $I_{x_1} = \frac{WH^2}{3}$ $I_y = \frac{W}{12}(A^2+B^2)$ $I_{y_1} = \frac{W}{6}(2A^2+2B^2+3AB)$ $I_p = \frac{W}{12}(A^2+B^2+H^2)$ $I_{p_1} = \frac{W(2A^2+2B^2+3AB+2H^2)}{6}$	$I_x = \frac{WH^2(6A^2+6AB+B^2)}{18(2A+B)^2}$ $I_{x_1} = \frac{WH^2(4A+B)}{6(2A+B)}$ $I_y = I_{y_1} - W\bar{x}^2$ $I_{y_1} = W\left[\frac{6A^2(2A+3B)+B^2(12A+5B)}{18(2A+B)}\right]$ $I_p = I_x + I_y$ $I_{p_1} = I_{x_1} + I_{y_1}$
AREA MOMENT OF INERTIA	$I_x = \frac{BH^3}{12}$ $I_{x_1} = \frac{BH^3}{3}$ $I_y = \frac{BH(A^2+B^2)}{12}$ $I_{y_1} = \frac{BH}{6}(2A^2+2B^2+3AB)$ $I_p = \frac{BH}{12}(A^2+B^2+H^2)$ $I_{p_1} = \frac{BH}{6}(2A^2+2B^2+3AB+2H^2)$	$I_x = \frac{H^3(6A^2+6AB+B^2)}{36(2A+B)}$ $I_{x_1} = \frac{H^3(4A+B)}{12}$ $I_y = I_{y_1} - \frac{H(3A^2+3AB+B^2)^2}{18(2A+B)}$ $I_{y_1} = \frac{H[6A^2(2A+3B)+B^2(12A+5B)]}{36}$ $I_p = I_x + I_y$ $I_{p_1} = I_{x_1} + I_{y_1}$
RADIUS OF GYRATION	$k_x = 0.289H$ $k_{x_1} = 0.577H$ $k_y = 0.289\sqrt{A^2+B^2}$ $k_{y_1} = 0.408\sqrt{2A^2+2B^2+3AB}$ $k_p = 0.289\sqrt{A^2+B^2+H^2}$ $k_{p_1} = 0.408\sqrt{2A^2+2B^2+3AB+2H^2}$	$k_x = \frac{0.236H}{(2A+B)}\sqrt{6A^2+6AB+B^2}$ $k_{x_1} = 0.408H\sqrt{\frac{4A+B}{2A+B}}$ $k_y = \sqrt{\frac{2I_y}{H(2A+B)}}$ $k_{y_1} = \sqrt{\frac{6A^2(2A+3B)+B^2(12A+5B)}{18(2A+B)}}$ $k_p = \sqrt{\frac{2I_p}{H(2A+B)}}$ $k_{p_1} = \sqrt{\frac{2I_{p_1}}{H(2A+B)}}$

Adapted from Weight Handbook, Vol. 1, Society of Aeronautical Weight Engineers, Inc.

	OBTUSE-ANGLED TRIANGLE	RHOMBUS
	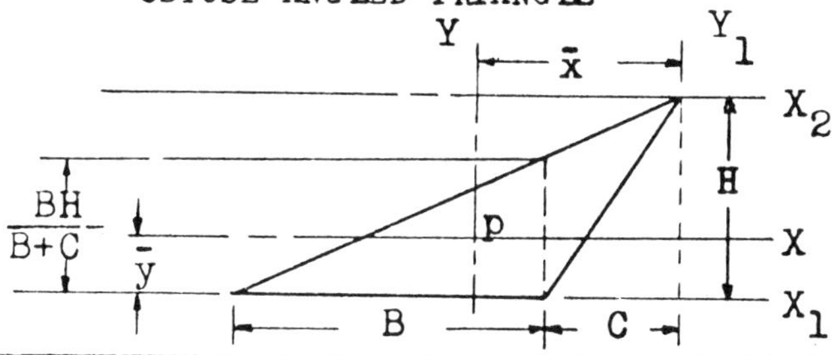	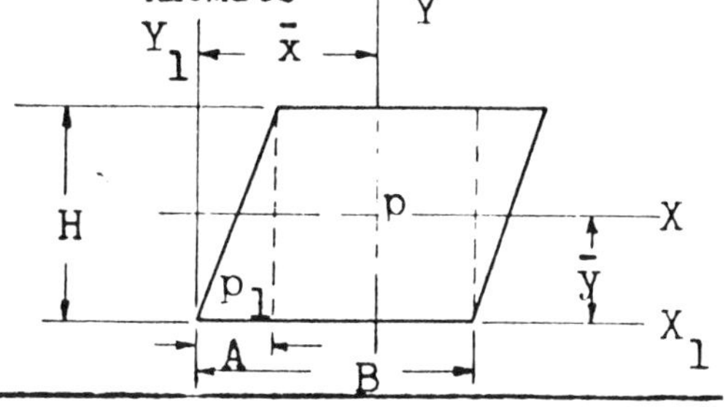
AREA	$\frac{BH}{2}$	BH
CENTROID	$\bar{x} = \frac{B + 2C}{3}$ $\bar{y} = \frac{H}{3}$	$\bar{x} = \frac{A+B}{2}$ $\bar{y} = \frac{H}{2}$
WEIGHT MOMENT OF INERTIA	$I_x = \frac{WH^2}{18}$ $I_{x_1} = \frac{WH^2}{6}$ $I_{x_2} = \frac{WH^2}{2}$ $I_y = \frac{W}{18}(B^2+BC+C^2)$ $I_{y_1} = \frac{W(B^2+3BC+3C^2)}{6}$ $I_p = \frac{W(H^2+B^2+BC+C^2)}{18}$	$I_x = \frac{WH^2}{12}$ $I_{x_1} = \frac{WH^2}{3}$ $I_y = \frac{W(A^2+B^2)}{12}$ $I_{y_1} = \frac{W(2A^2+2B^2+3AB)}{6}$ $I_p = \frac{WB^2}{6}$ $I_{p_1} = \frac{WB(3A + 4B)}{6}$
AREA MOMENT OF INERTIA	$I_x = \frac{BH^3}{36}$ $I_{x_1} = \frac{BH^3}{12}$ $I_{x_2} = \frac{BH^3}{4}$ $I_y = \frac{BH}{36}(B^2 + BC + C^2)$ $I_{y_1} = \frac{BH}{12}(B^2 + 3BC + 3C^2)$ $I_p = \frac{BH}{36}(H^2 + B^2 + BC + C^2)$	$I_x = \frac{BH^3}{12}$ $I_{x_1} = \frac{BH^3}{3}$ $I_y = \frac{BH(A^2 + B^2)}{12}$ $I_{y_1} = \frac{BH(2A^2 + 2B^2 + 3AB)}{6}$ $I_p = \frac{1}{6}B^3H$ $I_{p_1} = \frac{B^2H(3A + 4B)}{6}$
RADIUS OF GYRATION	$k_x = 0.236H$ $k_{x_1} = 0.408H$ $k_{x_2} = 0.707H$ $k_y = 0.236\sqrt{B^2 + BC + C^2}$ $k_{y_1} = 0.408\sqrt{B^2 + 3BC + C^2}$ $k_p = 0.236\sqrt{H^2 + B^2 + BC + C^2}$	$k_x = 0.289H$ $k_{x_1} = 0.577H$ $k_y = 0.289\sqrt{(A^2 + B^2)}$ $k_{y_1} = 0.408\sqrt{2A^2 + 2B^2 + 3AB}$ $k_p = 0.408B$ $k_{p_1} = 0.408\sqrt{B(3A + 4B)}$

Adapted from Weight Handbook, Vol. 1, Society of Aeronautical Weight Engineers, Inc.

	ISOSCELES TRIANGLE	OBLIQUE TRIANGLE
	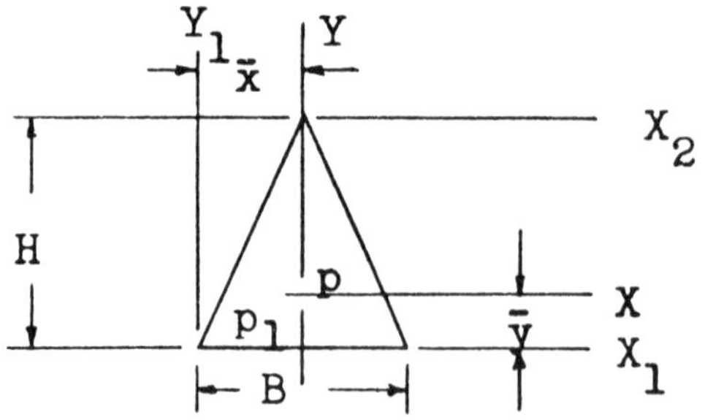	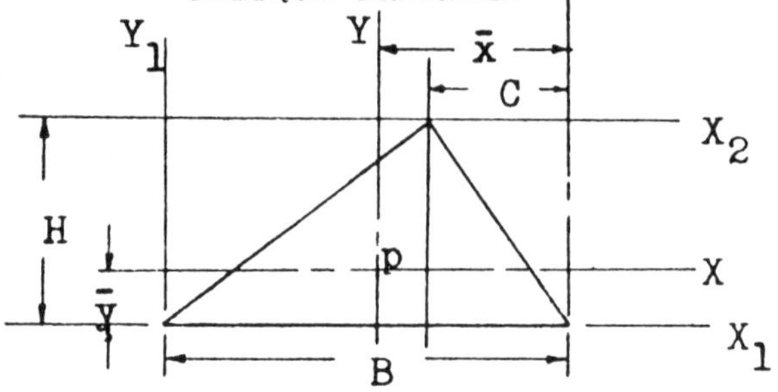
AREA	$\frac{BH}{2}$	$\frac{BH}{2}$
CENTROID	$\bar{x} = \frac{B}{2}$ $\bar{y} = \frac{H}{3}$	$\bar{y} = \frac{H}{3}$ $\bar{x} = \frac{B + C}{3}$
WEIGHT MOMENT OF INERTIA	$I_x = \frac{WH^2}{18}$ $I_{x_1} = \frac{WH^2}{6}$ $I_{x2} = \frac{WH^2}{2}$ $I_y = \frac{WB^2}{24}$ $I_{y1} = \frac{7WB^2}{24}$ $I_p = \frac{W(4H^2 + 3B^2)}{72}$ $I_{p_1} = \frac{W(7B^2 + 4H^2)}{24}$	$I_x = \frac{WH^2}{18}$ $I_{x_1} = \frac{WH^2}{6}$ $I_{x2} = \frac{WH^2}{2}$ $I_y = \frac{W}{18}(B^2 + C^2 - BC)$ $I_p = \frac{W}{18}(H^2 + B^2 + C^2 - BC)$
AREA MOMENT OF INERTIA	$I_x = \frac{BH^3}{36}$ $I_{x_1} = \frac{BH^3}{12}$ $I_{x2} = \frac{BH^3}{4}$ $I_y = \frac{B^3H}{48}$ $I_{y_1} = \frac{7B^3H}{48}$ $I_p = \frac{4BH^3 + 3B^3H}{144}$ $I_{p_1} = \frac{4BH^3 + 7B^3H}{48}$	$I_x = \frac{BH^3}{36}$ $I_{x1} = \frac{BH^3}{12}$ $I_{x2} = \frac{BH^3}{4}$ $I_y = \frac{BH}{36}(B^2 + C^2 - BC)$ $I_p = \frac{BH}{36}(H^2 + B^2 + C^2 - BC)$
RADIUS OF GYRATION	$k_x = 0.236H$ $k_{x_1} = 0.408H$ $k_{x_2} = 0.707H$ $k_y = 0.204B$ $k_{y_1} = 0.540B$ $k_p = 0.118\sqrt{4H^2 + 3B^2}$ $k_{p_1} = 0.204\sqrt{4H^2 + 7B^2}$	$k_x = 0.236H$ $k_{x_1} = 0.408H$ $k_{x_2} = 0.707H$ $k_y = 0.236\sqrt{B^2 + C^2 - BC}$ $k_p = 0.236\sqrt{H^2 + B^2 + C^2 - BC}$

Adapted from Weight Handbook, Vol. 1, Society of Aeronautical Weight Engineers, Inc.

	RIGHT TRIANGLE	EQUILATERAL TRIANGLE
	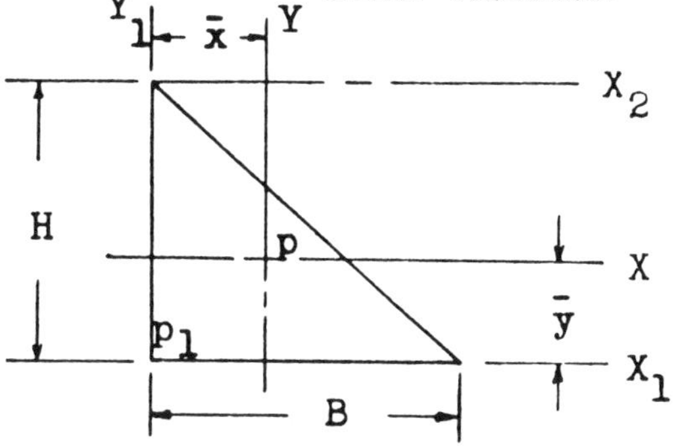	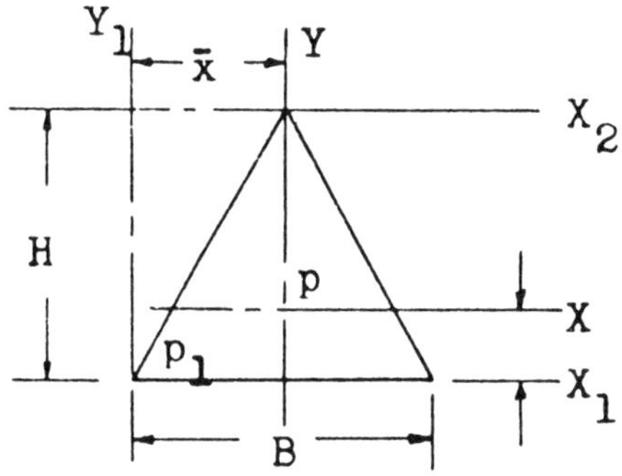
AREA	$\frac{BH}{2}$	$\frac{BH}{2}$
CENTROID	$\bar{x} = \frac{B}{3}$ $\bar{y} = \frac{H}{3}$	$\bar{x} = \frac{B}{2}$ $\bar{y} = \frac{H}{3}$
WEIGHT MOMENT OF INERTIA	$I_x = \frac{WH^2}{18}$ $I_{x_1} = \frac{WH^2}{6}$ $I_{x_2} = \frac{WH^2}{2}$ $I_y = \frac{WB^2}{18}$ $I_{y_1} = \frac{WB^2}{6}$ $I_p = \frac{W(H^2+B^2)}{18}$ $I_{p_1} = \frac{W(H^2+B^2)}{6}$	$I_x = I_y = \frac{WB^2}{24}$ $I_{x_1} = \frac{WB^2}{8}$ $I_{x2} = \frac{WH^2}{2}$ $I_{y_1} = \frac{7WB^2}{24}$ $I_p = \frac{WB^2}{12}$ $I_{p_1} = \frac{5WB^2}{12}$
AREA MOMENT OF INERTIA	$I_x = \frac{BH^3}{36}$ $I_{x_1} = \frac{BH^3}{12}$ $I_{x2} = \frac{BH^3}{4}$ $I_y = \frac{B^3H}{36}$ $I_{y1} = \frac{B^3H}{12}$ $I_p = \frac{BH(B^2 + H^2)}{36}$ $I_{p1} = \frac{BH(B^2+H^2)}{12}$	$I_x = I_y = \frac{B^3H}{48}$ $I_{x1} = \frac{B^3H}{16}$ $I_{x2} = \frac{BH^3}{4}$ $I_{y1} = \frac{7B^3H}{48}$ $I_p = \frac{B^3H}{24}$ $I_{p1} = \frac{5B^3H}{24}$
RADIUS OF GYRATION	$k_x = 0.236H$ $k_{x_1} = 0.408H$ $k_{x_2} = 0.707H$ $k_y = 0.236B$ $k_{y_1} = 0.408B$ $k_p = 0.236 \sqrt{B^2 + H^2}$ $k_{p_1} = 0.408 \sqrt{B^2 + H^2}$	$k_x = 0.204B$ $k_{x_1} = 0.354B$ $k_y = 0.204B$ $k_{x_2} = 0.707H$ $k_{y_1} = 0.540B$ $k_p = 0.289B$ $k_{p_1} = 0.456B$

Adapted from Weight Handbook, Vol. 1, Society of Aeronautical Weight Engineers, Inc.

CIRCLE

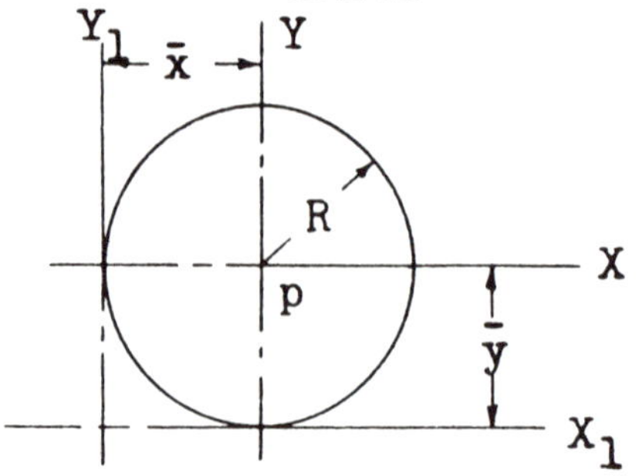

HOLLOW CIRCLE

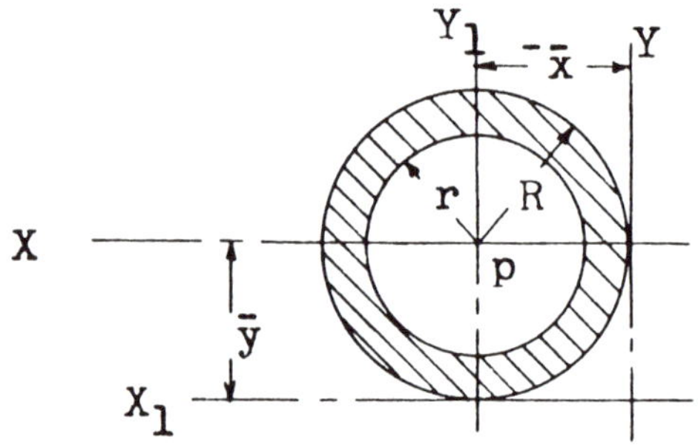

	CIRCLE	HOLLOW CIRCLE
AREA	$0.7854\ D^2$	$\pi\ (R^2 - r^2)$
CENTROID	$\bar{x} = \bar{y} = R$	$\bar{x} = \bar{y} = R$
WEIGHT MOMENT OF INERTIA	$I_x = I_y = \frac{WD^2}{16} = \frac{WR^2}{4}$ $I_{x_1} = I_{y_1} = 1.25\ WR^2$ $I_p = \frac{WD^2}{8} = \frac{WR^2}{2}$	$I_x = I_y = \frac{W(R^2 + r^2)}{4}$ $I_p = \frac{W(R^2 + r^2)}{2}$ $I_{x_1} = I_{y_1} = \frac{W(5R^2 + r^2)}{4}$
AREA MOMENT OF INERTIA	$I_x = I_y = 0.0491D^4$ $I_{x_1} = I_{y_1} = 0.2454D^4$ $I_p = 0.0982D^4$	$I_x = I_y = \frac{\pi(R^4 - r^4)}{4}$ $I_p = \frac{\pi(R^4 - r^4)}{2}$ $I_{x_1} = I_{y_1} = \frac{\pi(5R^4 - 4R^2r^2 - r^4)}{4}$
RADIUS OF GYRATION	$k_x = k_y = \frac{D}{4}$ $k_{x_1} = k_{y_1} = 0.5590D = 1.118R$ $k_p = 0.3536D$	$k_x = k_y = \frac{1}{2}\sqrt{R^2 + r^2}$ $k_p = \sqrt{\frac{R^2 + r^2}{2}}$ $k_{x_1} = k_{y_1} = \frac{1}{2}\sqrt{5R^2 + r^2}$

Adapted from Weight Handbook, Vol. 1, Society of Aeronautical Weight Engineers, Inc.

	SEMICIRCLE	HOLLOW SEMICIRCLE
	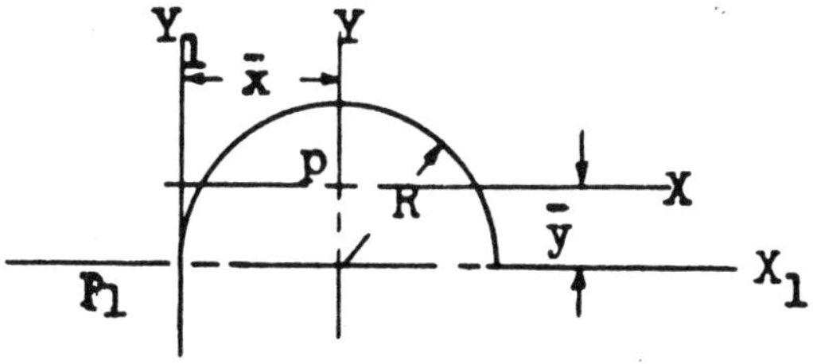	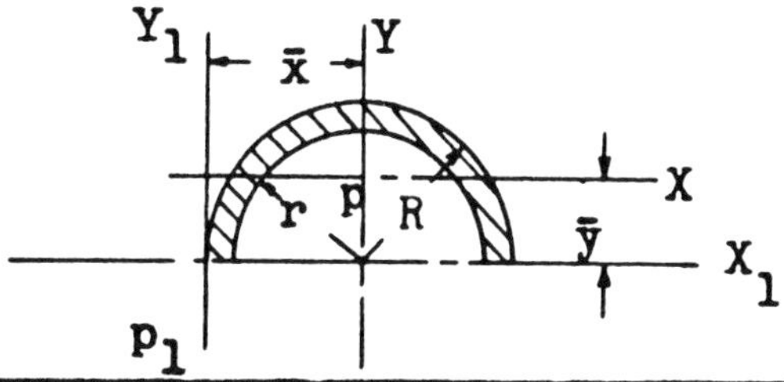
AREA	$0.3927\ D^2 = 1.571R^2$	$\frac{\pi(R^2 - r^2)}{2}$
CENTROID	$\bar{x} = R \quad \bar{y} = 0.2122D = 0.4244R$	$\bar{x} = R \quad \bar{y} = 0.4244\left(R + \frac{r^2}{R + r}\right)$
WEIGHT MOMENT OF INERTIA	$I_x = 0.06987\ WR^2 = 0.01747\ WD^2$ $I_{x_1} = 0.25\ WR^2 = 0.0625\ WD^2$ $I_y = 0.25\ WR^2 = 0.0625\ WD^2$ $I_{y_1} = 1.25\ WR^2 = 0.3125\ WD^2$ $I_p = 0.3199\ WR^2 = 0.0800\ WD^2$ $I_{p_1} = 1.50\ WR^2 = 0.3750\ WD^2$	$I_x = \frac{W(R^2 + r^2)}{4} - W\bar{y}^2 \quad I_y = \frac{W(R^2+r^2)}{4}$ $I_p = I_x + I_y$ $I_{x1} = \frac{W(R^2 + r^2)}{4} \quad I_{y1} = \frac{W(5R^2 + r^2)}{4}$ $I_{p1} = \frac{W(3R^2 + r^2)}{2}$
AREA MOMENT OF INERTIA	$I_x = 0.1098\ R^4 = 0.00686\ D^4$ $I_{x_1} = 0.3927\ R^4 = 0.02454\ D^4$ $I_y = 0.3927\ R^4 = 0.02454\ D^4$ $I_{y_1} = 1.9635\ R^4 = 0.12272\ D^4$ $I_p = 0.5025\ R^4 = 0.03140\ D^4$ $I_{p_1} = 2.3562\ R^4 = 0.14726\ D^4$	$I_x = \frac{\pi}{8}(R^4 - r^4) - \frac{\pi(R^2 - r^2)}{2}\bar{y}^2$ $I_{x1} = \frac{\pi}{8}(R^4 - r^4)$ $I_y = \frac{\pi}{8}(R^4 - r^4)$ $I_{y1} = \frac{\pi(R^2 - r^2)(5R^2 + r^2)}{8}$ $I_p = I_x + I_y$ $I_{p1} = I_{x_1} + I_{y_1}$
RADIUS OF GYRATION	$k_x = 0.264R = 0.132D$ $k_{x_1} = 0.5R = 0.25D$ $k_y = 0.5R = 0.25D$ $k_{y_1} = 1.118R = 0.559D$ $k_p = 0.566R = 0.2828D$ $k_{p_1} = 1.225R = 0.6124D$	$k_x = \sqrt{\frac{2I_x}{\pi(R^2 - r^2)}} \quad k_{y_1} = \sqrt{\frac{2I_{y1}}{\pi(R^2 - r^2)}}$ $k_{x_1} = \sqrt{\frac{R^2 + r^2}{2}} \quad k_p = \sqrt{\frac{2I_p}{\pi(R^2 - r^2)}}$ $k_y = \sqrt{\frac{R^2 + r^2}{2}} \quad k_{p_1} = \sqrt{\frac{2I_{p1}}{\pi(R^2 - r^2)}}$

Adapted from Weight Handbook, Vol. 1, Society of Aeronautical Weight Engineers, Inc.

	NOSE RIB	CIRCULAR SECTOR
	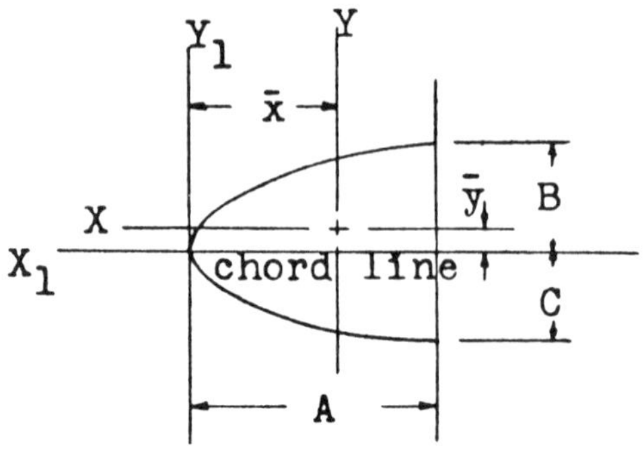 based on parabolic segment	α in radians 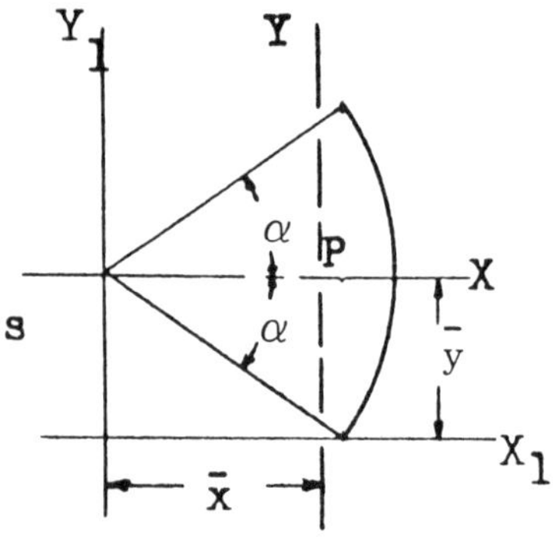
AREA	$\frac{2}{3} A (B + C)$	$R^2 \alpha$
CENTROID	$\bar{x} = 0.6A$ $\quad \bar{y} = 0.375 (B-C)$	$\bar{x} = \frac{2}{3}\left[\frac{R \sin \alpha}{\alpha}\right]$ $\quad \bar{y} = R \sin \alpha$
WEIGHT MOMENT OF INERTIA	$I_x = \frac{W}{320}(19B^2 + 26BC + 19C^2)$ $I_{x1} = \frac{W}{5}(B^2 - BC + C^2)$ $I_y = 0.0686WA^2$ $I_{y1} = 0.4286WA^2$ $I_p = I_x + I_y$ $I_{p1} = I_{x1} + I_{y1}$	$I_x = \frac{WR^2}{4\alpha}(\alpha - \sin \alpha \cos \alpha)$ $I_{x1} = I_x + WR^2 \sin^2 \alpha$ $I_y = \frac{WR^2}{4\alpha}\left(\alpha - \frac{16 \sin^2 \alpha}{9\alpha} + \frac{\sin 2\alpha}{2}\right)$ $I_{y1} = \frac{WR^2}{4\alpha}(\alpha + \sin \alpha \cos \alpha)$ $I_p = I_x + I_y$
AREA MOMENT OF INERTIA	$I_x = \frac{A(B + C)}{480}(19B^2 + 26BC + 19C^2)$ $I_{x1} = 0.1333(AB + AC)(B^2 - BC + C^2)$ $I_y = 0.0457 A^3 (B + C)$ $I_{y1} = 0.2857 A^3 (B + C)$ $I_p = I_x + I_y$ $I_{p1} = I_{x1} + I_{y1}$	$I_x = \frac{R^4}{4}(\alpha - \sin \alpha \cos \alpha)$ $I_{x1} = \frac{R^4}{4}(\alpha - \sin \alpha \cos \alpha) + R^4 \alpha \sin^2 \alpha$ $I_y = \frac{R^4}{4}(\alpha - \frac{16 \sin^2 \alpha}{9\alpha} + \frac{\sin 2\alpha}{2}$ $I_{y1} = \frac{R^4}{4}(\alpha + \sin \alpha \cos \alpha)$ $I_p = \frac{R^4}{4}\left(2\alpha - \frac{16 \sin^2 \alpha}{9\alpha}\right)$
RADIUS OF GYRATION	$k_x = \sqrt{\frac{3I_x}{2A(B+C)}}$ $k_y = \sqrt{\frac{3I_y}{2A(B+C)}}$	$k_x = \frac{R}{2}\sqrt{1 - \frac{\sin \alpha \cos \alpha}{\alpha}}$ $k_{x1} = \sqrt{\frac{I_{x1}}{R^2 \alpha}}$ $k_y = \frac{R}{2}\sqrt{1 + \frac{\sin \alpha \cos \alpha}{\alpha} - \frac{16 \sin^2 \alpha}{9\alpha^2}}$ $k_{y1} = \frac{R}{2}\sqrt{1 + \frac{\sin \alpha \cos \alpha}{\alpha}}$ $k_p = \frac{R}{2}\sqrt{2 - \frac{16 \sin^2 \alpha}{9\alpha^2}}$

Adapted from Weight Handbook, Vol. 1, Society of Aeronautical Weight Engineers, Inc.

HOLLOW CIRCULAR SECTOR

α in radians

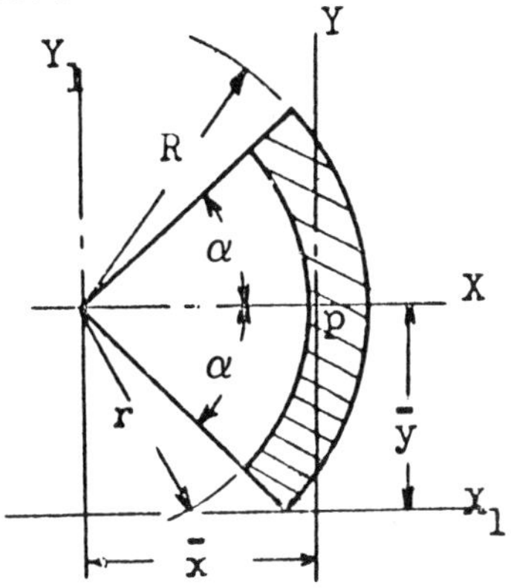

AREA	$(R^2 - r^2)\alpha$
CENTROID	$\bar{x} = \dfrac{2\text{SIN}\,\alpha\,(R^3-r^3)}{3\alpha\,(R^2-r^2)}$ $\quad \bar{y} = R\,\text{SIN}\,\alpha$
WEIGHT MOMENT OF INERTIA	$I_x = \dfrac{W}{4}(R^2+r^2)\left(1 - \dfrac{\text{SIN}\,\alpha\,\text{COS}\,\alpha}{\alpha}\right)$ $I_{x_1} = I_x + WR^2\,\text{SIN}^2\alpha$ $I_y = I_{y_1} - W\left[\dfrac{2\text{SIN}\alpha\,(R^3-r^3)}{3\alpha\,(R^2-r^2)}\right]^2$ $I_{y_1} = \dfrac{W}{4}(R^2+r^2)\left(1 + \dfrac{\text{SIN}\alpha\text{COS}\alpha}{\alpha}\right)$ $I_p = I_x + I_y$
AREA MOMENT OF INERTIA	$I_x = \dfrac{\alpha}{4}(R^4-r^4)\left(1 - \dfrac{\text{SIN}\alpha\text{COS}\alpha}{\alpha}\right)$ $I_{x_1} = I_x + \alpha\,(R^4-r^2\,R^2)\,\text{SIN}^2\,\alpha$ $I_y = I_{y_1} - \dfrac{1}{\alpha(R^2-r^2)}\left[\dfrac{2\text{SIN}\alpha(R^3-r^3)}{3}\right]^2$ $I_{y_1} = \dfrac{\alpha}{4}(R^4-r^4)\left(1 + \dfrac{\text{SIN}\alpha\,\text{COS}\alpha}{\alpha}\right)$ $I_p = I_x + I_y$
RADIUS OF GYRATION	$k_x = \sqrt{\dfrac{R^2+r^2}{4}\left[1 - \dfrac{\text{SIN}\,\alpha\,\text{COS}\,\alpha}{\alpha}\right]}$ $k_{x_1} = \sqrt{\dfrac{I_{x_1}}{(R^2-r^2)\,\alpha}}$ $k_y = \sqrt{\dfrac{I_y}{(R^2-r^2)\,\alpha}}$ $k_{y_1} = \sqrt{\left[\dfrac{R^2+r^2}{4}\right]\left[1 + \dfrac{\text{SIN}\,\alpha\,\text{COS}\,\alpha}{\alpha}\right]}$ $k_p = \sqrt{\dfrac{I_p}{(R^2 - r^2)\alpha}}$

Adapted from Weight Handbook, Vol. 1, Society of Aeronautical Weight Engineers, Inc.

CIRCULAR SEGMENT

α in radians
A = area

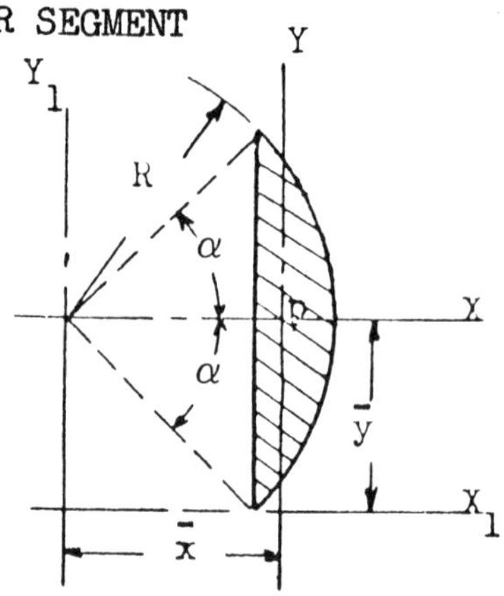

AREA	$\frac{R^2}{2}(2\alpha - \mathrm{SIN}\ 2\alpha)$
CENTROID	$\bar{x} = \frac{4R\ \mathrm{SIN}^3\alpha}{3(2\alpha - \mathrm{SIN}\ 2\alpha)}$ $\quad \bar{y} = R\,\mathrm{SIN}\ \alpha$
WEIGHT MOMENT OF INERTIA	$I_x = \frac{WR^2}{4}\left[1 - \frac{2\ \mathrm{SIN}^3\alpha\ \mathrm{COS}\ \alpha}{3(\alpha - \mathrm{SIN}\ \alpha\ \mathrm{COS}\ \alpha)}\right]$ $\quad I_{y_1} = \frac{WR^2}{4}\left[1 + \frac{2\ \mathrm{SIN}^3\alpha\ \mathrm{COS}\ \alpha}{\alpha - \mathrm{SIN}\ \alpha\ \mathrm{COS}\ \alpha}\right]$ $I_{x_1} = I_x + WR^2\mathrm{SIN}^2\alpha$ $\quad I_p = I_x + I_y$ $I_y = I_{y_1} - W\left[\frac{2R\ \mathrm{SIN}^3\alpha}{3(\alpha - \mathrm{SIN}\ \alpha\ \mathrm{COS}\ \alpha)}\right]^2$
AREA MOMENT OF INERTIA	$I_x = \frac{AR^2}{4}\left[1 - \frac{2}{3}\left[\frac{\mathrm{SIN}^3\alpha\ \mathrm{COS}\ \alpha}{\alpha - \mathrm{SIN}\ \alpha\ \mathrm{COS}\ \alpha}\right]\right]$ $\quad I_y = \frac{AR^2}{4}\left[1 + \frac{2\ \mathrm{SIN}^3\alpha\ \mathrm{COS}\ \alpha}{\alpha - \mathrm{SIN}\ \alpha\ \mathrm{COS}\ \alpha}\right]$ $I_{x_1} = I_x + \frac{R^4}{2}(2\alpha - \mathrm{SIN}\ 2\alpha)(\mathrm{SIN}^2\alpha)$ $\quad I_p = I_x + I_y$ $I_y = I_{y_1} - \frac{4R^6\ \mathrm{SIN}^6\alpha}{9A}$
RADIUS OF GYRATION	$k_x = \sqrt{\frac{R^2}{4}\left[1 - \frac{2\ \mathrm{SIN}^3\alpha\ \mathrm{COS}\ \alpha}{3(\alpha - \mathrm{SIN}\ \alpha\ \mathrm{COS}\ \alpha)}\right]}$ $k_y = \sqrt{\frac{2I_y}{R^2(2\alpha - \mathrm{SIN}\,2\alpha)}}$ $k_p = \sqrt{\frac{2I_p}{R^2(2\alpha - \mathrm{SIN}\ 2\alpha)}}$

Adapted from Weight Handbook, Vol. 1, Society of Aeronautical Weight Engineers, Inc.

CIRCULAR COMPLEMENT

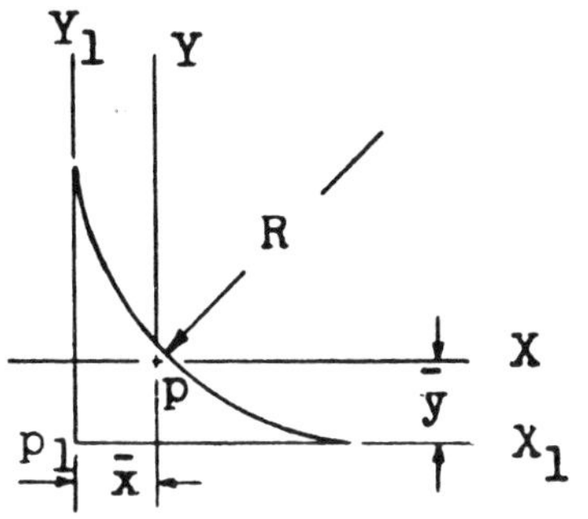

ELLIPTIC COMPLEMENT

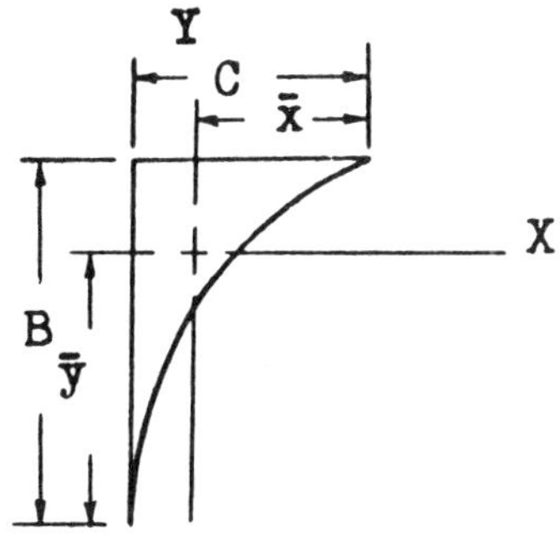

	CIRCULAR COMPLEMENT	ELLIPTIC COMPLEMENT
AREA	$0.2146\ R^2$	$0.2146\ BC$
CENTROID	$\bar{x} = \bar{y} = 0.2234\ R$	$\bar{x} = \frac{C}{1.288} = 0.7766C$ $\bar{y} = \frac{B}{1.288} = 0.7766B$
WEIGHT MOMENT OF INERTIA	$I_x = I_y = 0.035\ WR^2$ $I_{x_1} = I_{y_1} = 0.085\ WR^2$ $I_p = 0.070\ WR^2$ $I_{p_1} = 0.170\ WR^2$	$I_x = 0.035\ WB^2$ $I_y = 0.035\ WC^2$ $I_p = 0.035\ W(B^2 + C^2)$
AREA MOMENT OF INERTIA	$I_x = I_y = 0.0075\ R^4$ $I_{x_1} = I_{y_1} = 0.0183\ R^4$ $I_p = 0.0151\ R^4$ $I_{p_1} = 0.0365\ R^4$	$I_x = 0.0075B^3C$ $I_y = 0.0075BC^3$ $I_p = 0.0075\ BC(B^2 + C^2)$
RADIUS OF GYRATION	$k_x = k_y = 0.188\ R$ $k_{x_1} = k_{y_1} = 0.292\ R$ $k_p = 0.265\ R$ $k_{p_1} = 0.412\ R$	$k_x = 0.188B$ $k_y = 0.188C$ $k_p = \sqrt{0.035(B^2 + C^2)}$

Adapted from Weight Handbook, Vol. 1, Society of Aeronautical Weight Engineers, Inc.

	SEMIELLIPSE	HOLLOW SEMIELLIPSE
	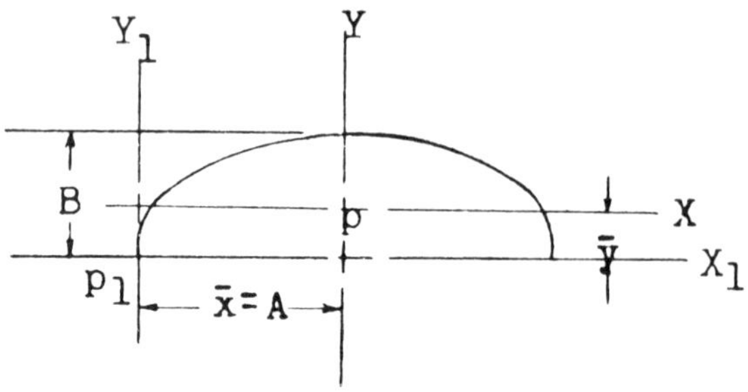	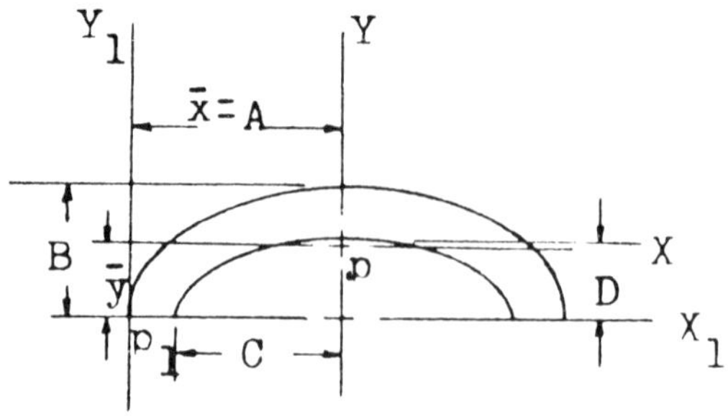
AREA	$\frac{\pi AB}{2}$	$\frac{\pi (AB-CD)}{2}$
CENTROID	$\bar{x} = A$ $\bar{y} = 0.424\ B$	$\bar{x} = A$ $\bar{y} = \frac{4}{3\pi}\left[\frac{AB^2-CD^2}{AB-CD}\right]$
WEIGHT MOMENT OF INERTIA	$I_x = 0.070\ WB^2$ $I_{x_1} = 0.25\ WB^2$ $I_y = 0.25\ WA^2$ $I_{y_1} = 1.25\ WA^2$ $I_p = \frac{W(A^2 + 0.280B^2)}{4}$ $I_{p_1} = \frac{W(5A^2 + B^2)}{4}$	$I_x = \frac{W}{4}\left(\frac{AB^3-CD^3}{AB-CD}\right) - W\left(\frac{4}{3\pi}\left[\frac{AB^2-CD^2}{AB-CD}\right]\right)^2$ $I_{x_1} = \frac{W}{4}\left(\frac{AB^3-CD^3}{AB-CD}\right)$ $I_y = \frac{W}{4}\left(\frac{A^3B-C^3D}{AB-CD}\right)$ $I_{y_1} = \frac{W}{4}\left(\frac{A^3B-C^3D}{AB-CD}\right) + WA^2$ $I_p = I_x + I_y$
AREA MOMENT OF INERTIA	$I_x = 0.1098\ AB^3$ $I_{x_1} = 0.3927\ AB^3$ $I_y = 0.3927\ A^3B$ $I_{y_1} = 1.9635\ A^3B$ $I_p = AB\ (0.3927\ A^2 + 0.1098\ B^2)$ $I_{p_1} = AB\ (1.9635\ A^2 + 0.3927\ B^2)$	$I_x = \frac{\pi}{8}(AB^3-CD^3) - \frac{\pi(AB-CD)}{2}\left[\frac{4(AB^2-CD^2)}{3\pi(AB-CD)}\right]^2$ $I_{x_1} = \frac{\pi}{8}(AB^3-CD^3)$ $I_y = \frac{\pi}{8}(A^3B-C^3D)$ $I_{y_1} = \frac{\pi}{8}(A^3B-C^3D) + \frac{\pi A^2(AB-CD)}{2}$ $I_p = I_x + I_y$ $I_{p_1} = I_{x_1} + I_{y_1}$
RADIUS OF GYRATION	$k_x = 0.2643\ B$ $k_{x_1} = \frac{B}{2}$ $k_y = \frac{A}{2}$ $k_{y_1} = 1.118\ A$ $k_p = \sqrt{\frac{A^2}{4} + \frac{B^2}{4} - \frac{16\ B^2}{9\pi^2}}$ $k_{p_1} = \sqrt{\frac{2I_{p1}}{\pi\ AB}}$	$k_x = \sqrt{\frac{2I_x}{\pi(AB-CD)}}$ $k_{x_1} = \sqrt{\frac{AB^3-CD^3}{4(AB-CD)}}$ $k_y = \sqrt{\frac{2I_y}{\pi(AB-CD)}}$ $k_{y_1} = \sqrt{\frac{2I_{y1}}{\pi(AB-CD)}}$ $k_p = \sqrt{\frac{2I_p}{\pi(AB-CD)}}$ $k_{p_1} = \sqrt{\frac{2I_{p1}}{\pi(AB-CD)}}$

Adapted from Weight Handbook, Vol. 1, Society of Aeronautical Weight Engineers, Inc.

	ELLIPSE	HOLLOW ELLIPSE
	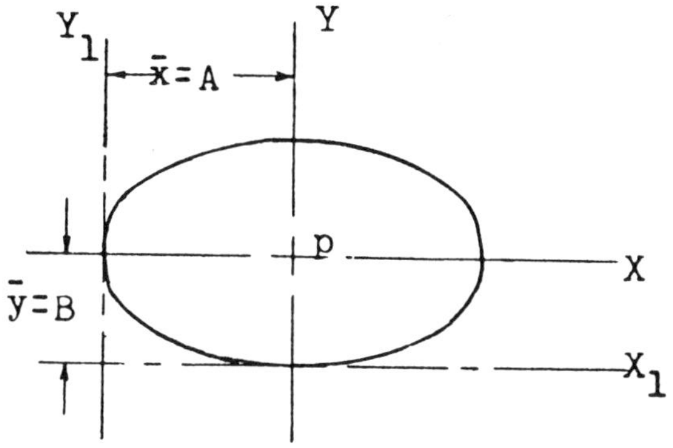	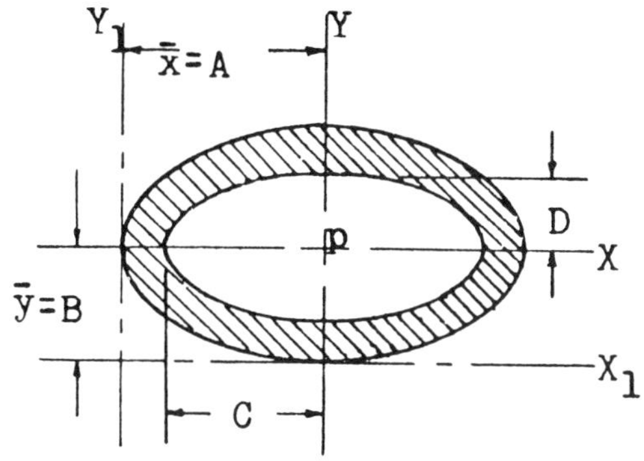
AREA	πAB	$\pi (AB - CD)$
CENTROID	$\bar{x} = A$ $\quad \bar{y} = B$	$\bar{x} = A$ $\quad \bar{y} = B$
WEIGHT MOMENT OF INERTIA	$I_x = \frac{WB^2}{4}$ $I_{x_1} = 1.25\, WB^2$ $I_y = \frac{WA^2}{4}$ $I_{y_1} = 1.25\, WA^2$ $I_p = \frac{W(A^2 + B^2)}{4}$	$I_x = \frac{W}{4}\left[\frac{AB^3 - CD^3}{AB - CD}\right]$ $I_{x_1} = \frac{W}{4}\left[\frac{AB^3 - CD^3}{AB - CD}\right] + WB^2$ $I_y = \frac{W}{4}\left[\frac{A^3B - C^3D}{AB - CD}\right]$ $I_{y_1} = \frac{W}{4}\left[\frac{A^3B - C^3D}{AB - CD}\right] + WA^2$ $I_p = I_x + I_y$
AREA MOMENT OF INERTIA	$I_x = \frac{\pi AB^3}{4} = 0.7854\, AB^3$ $I_{x_1} = 1.25 \quad \pi AB^3 = 3.927\, AB^3$ $I_y = \frac{\pi A^3B}{4} = 0.7854\, A^3B$ $I_{y_1} = 1.25 \quad \pi A^3B = 3.927\, A^3B$ $I_p = \frac{\pi AB(A^2 + B^2)}{4}$	$I_x = \frac{\pi}{4}(AB^3 - CD^3)$ $I_{x_1} = \frac{\pi}{4}(AB^3 - CD^3) + \pi (AB - CD) \quad (B^2)$ $I_y = \frac{\pi}{4}(A^3B - C^3D)$ $I_{y_1} = \frac{\pi}{4}(A^3B - C^3D) + \pi (AB - CD)\,(A^2)$ $I_p = I_x + I_y$
RADIUS OF GYRATION	$k_x = \frac{B}{2}$ $k_{x_1} = 1.118\, B$ $k_y = \frac{A}{2}$ $k_{y_1} = 1.118\, A$ $k_p = \frac{\sqrt{A^2 + B^2}}{2}$	$k_x = \sqrt{\frac{AB^3 - CD^3}{4(AB - CD)}}$ $k_{x_1} = \sqrt{\frac{I_{x_1}}{\pi (AB - CD)}}$ $k_y = \sqrt{\frac{A^3B - C^3D}{4(AB - CD)}}$ $k_{y_1} = \sqrt{\frac{I_{y_1}}{\pi (AB - CD)}}$ $\quad k_p = \sqrt{\frac{I_p}{\pi (AB - CD)}}$

Adapted from Weight Handbook, Vol. 1, Society of Aeronautical Weight Engineers, Inc.

	QUARTER ELLIPSE	HALF ELLIPSE
	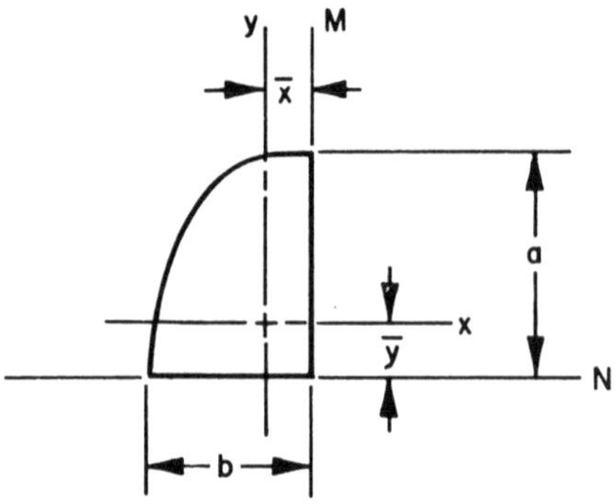	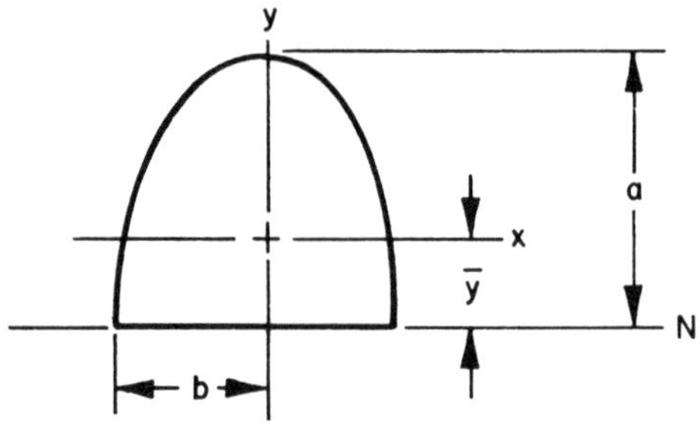
AREA	$\pi ab/4$	$\pi ab/2$
CENTROID	$\bar{y} = \frac{4a}{3\pi}$ $\quad \bar{x} = \frac{4b}{3\pi}$	$\bar{y} = \frac{4a}{3\pi}$ $\quad \bar{x} = 0$
WEIGHT MOMENT OF INERTIA	$I_x = \frac{4ma}{\pi}\left(\frac{\pi}{16} - \frac{4}{9\pi}\right)$ $I_y = \frac{4mb^2}{\pi}\left(\frac{\pi}{16} - \frac{4}{9\pi}\right)$ $I_N = \frac{ma^2}{4}$ $I_M = \frac{mb^2}{4}$	$I_x = \frac{2ma^2}{\pi}\left(\frac{\pi}{8} - \frac{8}{9\pi}\right)$ $I_y = \frac{mb^2}{4}$ $I_N = \frac{ma^2}{4}$
AREA MOMENT OF INERTIA	$I_x = a^3b\left(\frac{\pi}{16} - \frac{4}{9\pi}\right)$ $I_y = ab^3\left(\frac{\pi}{16} - \frac{4}{9\pi}\right)$ $I_N = \frac{\pi a^3 b}{16}$ $I_M = \frac{\pi a b^3}{16}$	$I_x = a^3b\left(\frac{\pi}{8} - \frac{8}{9\pi}\right)$ $I_y = \frac{\pi a b^3}{8}$ $I_N = \frac{\pi a^3 b}{8}$
RADIUS OF GYRATION	$k_x = \sqrt{I_x/A}$ $k_y = \sqrt{I_y/A}$ $k_N = \sqrt{I_N/A}$ $k_M = \sqrt{I_M/A}$	$k_x = \sqrt{I_x/A}$ $k_y = \sqrt{I_y/A}$ $k_N = \sqrt{I_N/A}$

	PARABOLIC SEGMENT	PARABOLIC HALF-SEGMENT
	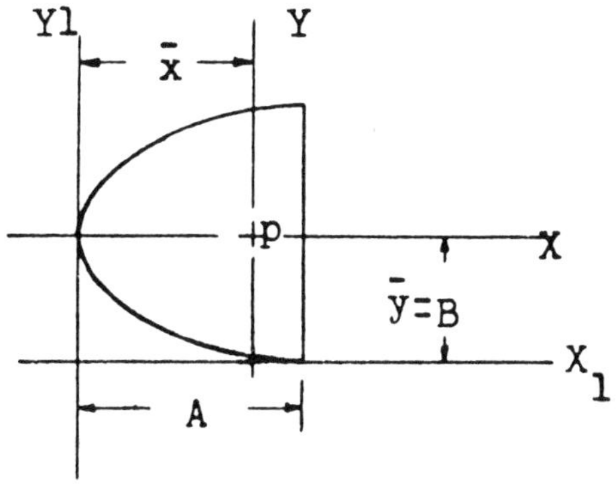	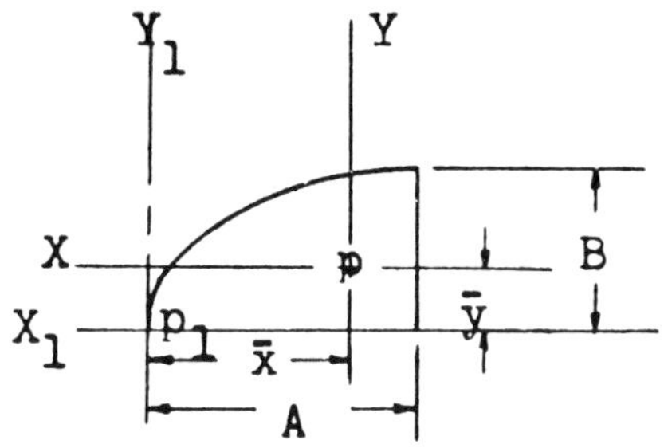
AREA	$\frac{4}{3}AB$	$\frac{2AB}{3}$
CENTROID	$\bar{x} = 0.6A$ $\bar{y} = B$	$\bar{x} = 0.6A$ $\bar{y} = 0.375B$
WEIGHT MOMENT OF INERTIA	$I_x = 0.2WB^2$ $I_{x_1} = 1.2\ WB^2$ $I_y = 0.0686WA^2$ $I_{y_1} = 0.4286WA^2$ $I_p = I_x + I_y$	$I_x = 0.0594WB^2$ $I_{x_1} = 0.2WB^2$ $I_y = 0.0686WA^2$ $I_{y_1} = 0.4286WA^2$ $I_p = I_x + I_y$ $I_{p_1} = I_{x_1} + I_{y_1}$
AREA MOMENT OF INERTIA	$I_x - 0.2667AB^3$ $I_{x_1} = 1.6AB^3$ $I_y = 0.0914A^3B$ $I_{y_1} = 0.5714A^3B$ $I_p = I_x + I_y$	$I_x = 0.0396AB^3$ $I_{x_1} = 0.1333AB^3$ $I_y = 0.0457A^3B$ $I_{y_1} = 0.2857A^3B$ $I_p = I_x + I_y$ $I_{p_1} = I_{x_1} + I_{y_1}$
RADIUS OF GYRATION	$k_x = 0.4472B$ $k_{x_1} = 1.095B$ $k_y = 0.2619A$ $k_{y_1} = 0.6547A$ $k_p = \sqrt{\frac{3I_p}{4AB}}$	$k_x = 0.2437B$ $k_{x_1} = 0.4472B$ $k_y = 0.2619A$ $k_{y_1} = 0.6547A$ $k_p = \sqrt{\frac{3I_p}{2AB}}$ $k_{p_1} = \sqrt{\frac{3I_{p1}}{2AB}}$

Adapted from Weight Handbook, Vol. 1, Society of Aeronautical Weight Engineers, Inc.

	COMPLEMENT OF HALF PARABOLA	PARABOLIC FILLET IN RIGHT ANGLE
	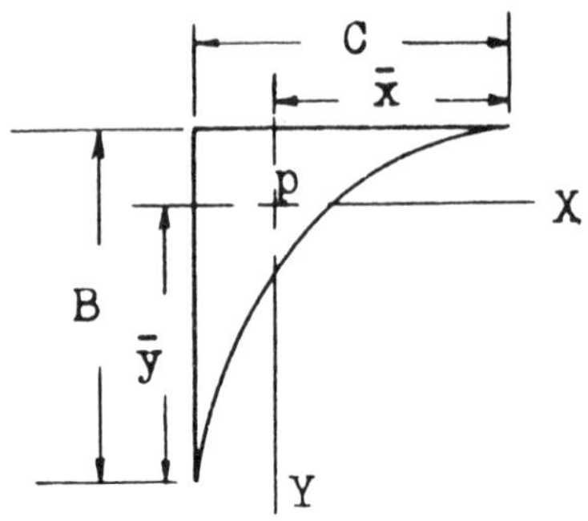	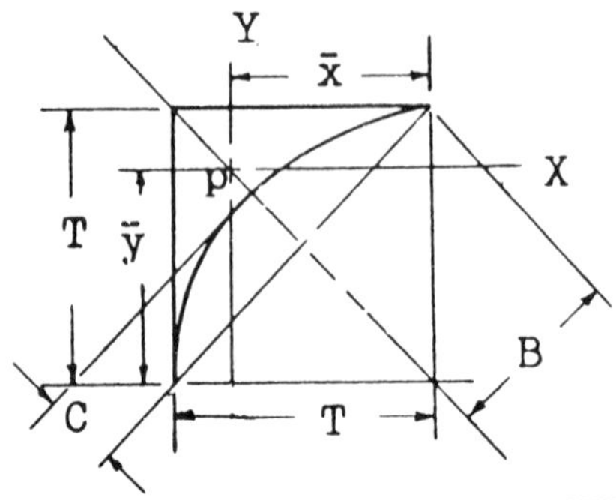
AREA	$\frac{1}{3} BC$	$\frac{T^2}{6}$
CENTROID	$\bar{x} = \frac{3}{4} C$ $\bar{y} = \frac{7}{10} B$	$\bar{x} = \bar{y} = \frac{4}{5} T$ $B = \frac{T}{\sqrt{2}}$ $C = \frac{T}{2\sqrt{2}}$
WEIGHT MOMENT OF INERTIA	$I_x = \frac{37}{700} WB^2$ $I_y = \frac{3}{80} WC^2$ $I_p = \frac{W}{20}\left(\frac{37}{35} B^2 + \frac{3C^2}{4}\right)$	$I_x = I_y = \frac{3WT^2}{100}$ $I_p = \frac{3WT^2}{50}$
AREA MOMENT OF INERTIA	$I_x = \frac{37}{2100} B^3C$ $I_y = \frac{1}{80} BC^3$ $I_p = \frac{BC}{20}\left(\frac{37}{105} B^2 + \frac{C^2}{4}\right)$	$I_x = I_y = \frac{11}{2100} T^4$ $I_p = \frac{T^4}{100}$
RADIUS OF GYRATION	$k_x = 0.230 B$ $k_y = 0.194 C$ $k_p = \sqrt{\frac{3I_p}{BC}}$	$k_x = k_y = 0.173 T$ $k_p = 0.2449 T$

Adapted from Weight Handbook, Vol. 1, Society of Aeronautical Weight Engineers, Inc.

OGIVAL SHAPES

An ogival shape is one that is developed as a convex solid of revolution.

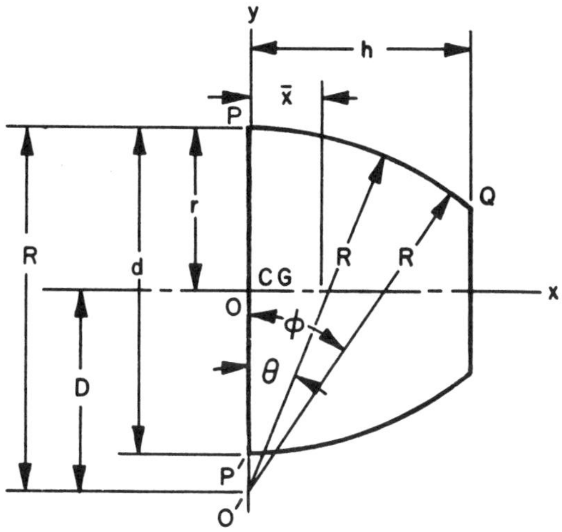

The solid of revolution developed in the diagram above is bounded by the arc PQ of a circle of radius R (radius of longitudinal curvature) whose center, o′, lies on the side of the axis of revolution, ox, opposite to the arc PQ and distant D = oo′ from the axis of revolution ox.

If the arc PQ cuts the axis ox for the desired length, h, the ogive is said to be pointed. The diameter of the base, d, is known as the diameter of swell, which is effectively the maximum transverse diameter of the body.

A tangent ogive is one that includes its base as illustrated in the above diagram.

A secant ogive is one that does not include its base; that is, the length h does not extend to the point where the maximum swell occurs.

If the radial center of the arc PQ, designated as o′, is moved to a location on the arc (PQ) side of the axis of revolution ox, a non-standard ogive will result as illustrated below.

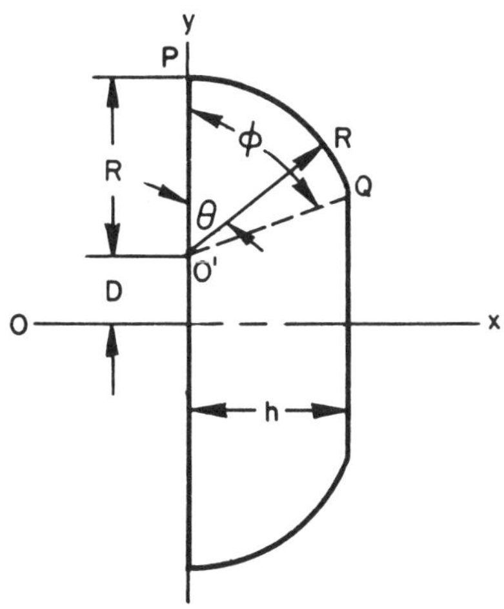

The properties of ogival shapes are included because of their common usage in the design of missiles, artillery shells, and other systems requiring this family of geometrical shapes.

Sample derivations are given so that the reader can quickly review methods of approach in the basic steps involved in the development of a general equation used to describe an ogival property.

The section entitled Properties of a Solid Ogive contains the equations describing the volume, moment, and moments of inertia about the central axis and the base plane for both the truncated and the complete ogive.

These equations have been developed in both the exact and approximate forms. Examination of the equations will show that the term $\theta/\sin\theta$ found in the exact equations has been replaced by a convergent series expansion, thereby leading to the simpler approximate forms.

The use of the approximate equations is left to the reader's discretion; however, tables of expected error are included as a guide in determining the number of terms to be used in the equations.

PROPERTIES OF A SOLID OGIVE

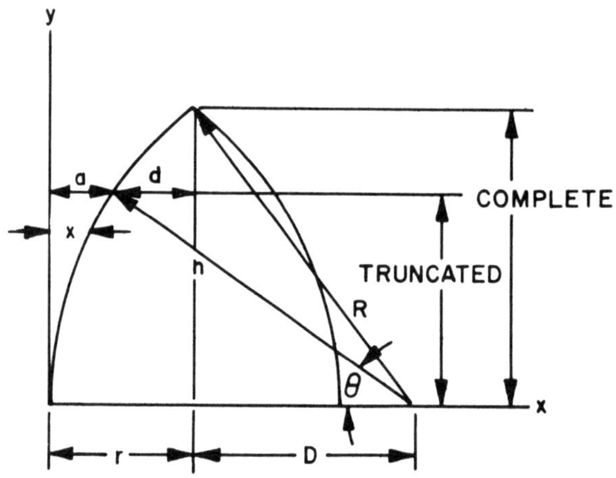

R = ogive radius

h = length of ogival head (truncated or complete)

r = radius of base

d = radius of truncated nose (d = 0 for complete ogive)

$D = R - r$

$a = r - d$

$\sin\theta = h/R$

SUMMARY OF EQUATIONS FOR A TRUNCATED OGIVE

Volume

$$V = \pi h\left[R^2 - \frac{h^2}{3} - D\left(d + R\frac{\theta}{\sin\theta}\right)\right]$$

$$V = \frac{\pi h}{9}\left\{(3r - a)^2 + 2a^2 - 6a^2\frac{R - r}{R}\left[\frac{1}{5} + \frac{3}{35}\frac{a}{R} + \frac{4}{105}\left(\frac{a}{R}\right)^2 + \frac{4}{231}\left(\frac{a}{R}\right)^3 + \frac{8}{1,001}\left(\frac{a}{R}\right)^4 + \frac{8}{2,145}\left(\frac{a}{R}\right)^5 + \frac{64}{36,465}\left(\frac{a}{R}\right)^6 + \frac{192}{230,945}\left(\frac{a}{R}\right)^7 + \cdots\right]\right\}$$

$$= \frac{\pi h a^2}{9}\left\{\left(\frac{3r}{a} - 1\right)^2 + 2 - 6\left(1 - \frac{r}{R}\right)\left[\frac{1}{5} + \frac{3}{35}\frac{a}{R} + \frac{4}{105}\left(\frac{a}{R}\right)^2 + \frac{4}{231}\left(\frac{a}{R}\right)^3 + \frac{8}{1,001}\left(\frac{a}{R}\right)^4 + \frac{8}{2,145}\left(\frac{a}{R}\right)^5 + \frac{64}{36,465}\left(\frac{a}{R}\right)^6 + \frac{192}{230,945}\left(\frac{a}{R}\right)^7 + \cdots\right]\right\}$$

First Moment

$$\bar{y}V = \pi\left[\frac{h^2}{2}\left(r^2 + \frac{2}{3}RD - \frac{h^2}{2}\right) - \frac{2}{3}Da(R - a)^2\right]$$

$$= \pi a\left[Rrd + \frac{1}{3}a^2(R + 2r) - \frac{1}{2}a\left(r^2 + \frac{1}{2}a^2\right)\right]$$

Moment of Inertia About the Central Axis

$$I_A = \frac{\pi}{2}h\left[(R^2 + D^2)^2 + 2D^2(2R^2 - h^2) - h^2\left(\frac{2}{3}R^2 - \frac{1}{5}h^2\right)\right] - \frac{\pi}{4}Dh\left[\left(R\frac{\theta}{\sin\theta} + R - a\right)(4D^2 + 3R^2) + 2(R - a)^3\right]$$

$$I_A = \frac{\pi}{2}h\left\{r^3\left(r - \frac{4}{3}a\right) + \frac{2}{15}r^2a^2\left(9 + 2\frac{r}{R}\right) + \frac{1}{5}a^3(R - 5r + a) - a^4\left[7 - 15\frac{r}{R} + 12\left(\frac{r}{R}\right)^2 - 4\left(\frac{r}{R}\right)^3\right]\left[\frac{1}{35}\frac{R}{a} + \frac{4}{315} + \frac{4}{693}\frac{a}{R} + \frac{8}{3,003}\left(\frac{a}{R}\right)^2 + \frac{8}{6,435}\left(\frac{a}{R}\right)^3 + \frac{64}{109,395}\left(\frac{a}{R}\right)^4 + \frac{64}{230,945}\left(\frac{a}{R}\right)^5 + \cdots\right]\right\}$$

$$= \frac{\pi}{2}ha^4\left\{\left(\frac{r}{a}\right)^3\left(\frac{r}{a} - \frac{4}{3}\right) + \frac{2}{15}\left(\frac{r}{a}\right)^2\left(9 + 2\frac{r}{R}\right) + \frac{1}{5}\left(\frac{R}{a} - \frac{5r}{a} + 1\right) - \left[7 - 15\frac{r}{R} + 12\left(\frac{r}{R}\right)^2 - 4\left(\frac{r}{R}\right)^3\right]\left[\frac{1}{35}\frac{R}{a} + \frac{4}{315} + \frac{4}{693}\frac{a}{R} + \frac{8}{3,003}\left(\frac{a}{R}\right)^2 + \frac{8}{6,435}\left(\frac{a}{R}\right)^3 + \frac{64}{109,395}\left(\frac{a}{R}\right)^4 + \frac{64}{230,945}\left(\frac{a}{R}\right)^5 + \cdots\right]\right\}$$

Note. Since a density of unity was used in the derivation of the equations, it would be well to recall the relationship $\rho = m/V$. The use of this relationship when coupled with the moment of inertia equations leads to the determination of the mass moment of inertia.

Moment of Inertia About the Base Plane

$$I_B = \frac{\pi}{15}h^3\left[5(R^2 + D^2) - 3h^2\right] - \frac{\pi}{4}Dh\left[R^3\frac{\theta}{\sin\theta} - (R - a)(R^2 - 2h^2)\right]$$

$$I_B = \pi ha\left\{\frac{2}{7}a^2\left(R + \frac{9}{5}r\right) + \frac{1}{3}r^2(2R - a) - \frac{1}{5}a(4Rr + a^2) - 2a^3\left(\frac{R - r}{R}\right)\left[\frac{1}{315} + \frac{1}{693}\frac{a}{R} + \frac{2}{3,003}\left(\frac{a}{R}\right)^2 + \frac{2}{6,435}\left(\frac{a}{R}\right)^3 + \frac{16}{109,395}\left(\frac{a}{R}\right)^4 + \frac{16}{230,945}\left(\frac{a}{R}\right)^5 + \cdots\right]\right\}$$

$$= \pi ha^4\left\{\frac{2}{7}\left(\frac{R}{a} + \frac{9}{5}\frac{r}{a}\right) + \frac{1}{3}\left(\frac{r}{a}\right)^2\left(\frac{2R}{a} - 1\right) - \frac{1}{5}\left(\frac{4Rr}{a^2} + 1\right) - 2\left(1 - \frac{r}{R}\right)\left[\frac{1}{315} + \frac{1}{693}\frac{a}{R} + \frac{2}{3,003}\left(\frac{a}{R}\right)^2 + \frac{2}{6,435}\left(\frac{a}{R}\right)^3 + \frac{16}{109,395}\left(\frac{a}{R}\right)^4 + \frac{16}{230,945}\left(\frac{a}{R}\right)^5 + \cdots\right]\right\}$$

SUMMARY OF EQUATIONS FOR A COMPLETE OGIVE

Volume

$$V = \pi h\left[R^2 - \frac{h^2}{3} - DR\frac{\theta}{\sin\theta}\right]$$

$$V = \frac{2}{3}\pi hr^2\left[\frac{4}{5} + \frac{4}{35}\frac{r}{R} + \frac{1}{21}\left(\frac{r}{R}\right)^2 + \frac{8}{385}\left(\frac{r}{R}\right)^3 + \frac{4}{429}\left(\frac{r}{R}\right)^4 + \frac{64}{15,015}\left(\frac{r}{R}\right)^5 + \frac{24}{12,155}\left(\frac{r}{R}\right)^6 + \frac{128}{138,567}\left(\frac{r}{R}\right)^7 + \frac{64}{146,965}\left(\frac{r}{R}\right)^8 + \cdots\right]$$

First Moment

$$\bar{y}V = \frac{\pi}{3}r^3\left(R - \frac{r}{4}\right)$$

Moment of Inertia About the Central Axis

$$I_A = \frac{\pi}{2}h\left[R^2\left(R^2 + \frac{5}{2}D^2\right) - h^2\left(\frac{2}{3}D^2 + \frac{7}{15}h^2\right) - \frac{RD}{2}\frac{\theta}{\sin\theta}(4D^2 + 3R^2)\right]$$

$$I_A = 4\pi hr^4\left[\frac{16}{315} + \frac{32}{3,465}\frac{r}{R} + \frac{8}{2,145}\left(\frac{r}{R}\right)^2 + \frac{8}{5,005}\left(\frac{r}{R}\right)^3 + \frac{181}{255,255}\left(\frac{r}{R}\right)^4 + \frac{8}{24,871}\left(\frac{r}{R}\right)^5 + \frac{28}{188,955}\left(\frac{r}{R}\right)^6 + \cdots\right]$$

Moment of Inertia About the Base Plane

$$I_B = \frac{\pi}{60}h^3(8R^2 + 17D^2) - \frac{\pi}{4}Dh\left(R^3\frac{\theta}{\sin\theta} - D^3\right)$$

$$I_B = 2\pi hr^4\left[\frac{8}{105}\frac{R}{r} - \frac{4}{315} + \frac{2}{1,155}\frac{r}{R} + \frac{1}{1,287}\left(\frac{r}{R}\right)^2 + \frac{16}{45,045}\left(\frac{r}{R}\right)^3 + \frac{2}{12,155}\left(\frac{r}{R}\right)^4 + \frac{32}{415,701}\left(\frac{r}{R}\right)^5 + \frac{16}{440,895}\left(\frac{r}{R}\right)^6 + \frac{128}{7,436,429}\left(\frac{r}{R}\right)^7 + \cdots\right]$$

TABLE 2. UPPER BOUND FOR ERROR WITH THE USE OF APPROXIMATE EQUATIONS FOR THE VOLUME OF A TRUNCATED OGIVE

Number of terms	Upper bound for error, %			
	a/r = 0.3	a/r = 0.5	a/r = 0.7	a/r = 0.9
		r/R = 0.1		
None	1.3	4.1	8.8	15.2
One	0.02	0.09	0.3	0.7
Two	......	......	0.009	0.03
		r/R = 0.3		
None	1.1	3.4	7.3	13.1
One	0.04	0.2	0.7	1.7
Two	......	0.02	0.07	0.2
Three	......	......	......	0.03
		r/R = 0.5		
None	0.8	2.5	5.7	10.4
One	0.05	0.3	0.9	2.2
Two	......	0.03	0.2	0.4
Three	......	......	0.02	0.09
		r/R = 0.7		
None	0.5	1.6	3.7	7.1
One	0.04	0.3	0.8	2.0
Two	......	0.04	0.2	0.6
Three	......	......	0.04	0.2
Four	......	......	......	0.05

EXPECTED ERROR WITH THE USE OF APPROXIMATE EQUATIONS

Tables 2–7 give the upper bounds for error incurred with the use of the approximate equations. Examination of the tables will show that including a sufficient number of terms in the convergent series expansion will reduce the error markedly.

TABLE 3. UPPER BOUND FOR ERROR WITH THE USE OF APPROXIMATE EQUATIONS FOR THE MOMENT OF INERTIA ABOUT THE CENTRAL AXIS FOR A TRUNCATED OGIVE

Number of terms	Upper bound for error, %			
	a/r = 0.3	a/r = 0.5	a/r = 0.7	a/r = 0.9
		r/R = 0.1		
None	6.0	26.5	53.9	74.1
One	0.08	0.8	3.5	10.1
Two	......	0.02	0.1	0.5
Three....	......	......	0.004	0.02
		r/R = 0.3		
None	1.32	7.1	20.2	38.4
One	0.05	0.5	2.3	6.9
Two	......	0.03	0.2	0.9
Three....	......	......	0.02	0.1
Four......	......	......	......	0.02
		r/R = 0.5		
None	0.5	2.7	8.5	18.7
One	0.03	0.3	1.4	4.4
Two	......	0.04	0.2	0.9
Three....	......	......	0.04	0.2
Four......	......	......	......	0.04
		r/R = 0.7		
None	0.2	1.0	3.4	8.2
One	0.02	0.2	0.8	2.4
Two	......	0.03	0.2	0.7
Three....	......	......	0.04	0.2
Four......	......	......	......	0.05

TABLE 4. Upper Bound for Error With the Use of Approximate Equations for the Moment of Inertia About the Base Plane for a Truncated Ogive

Number of terms	Upper bound for error, %			
	a/r = 0.3	a/r = 0.5	a/r = 0.7	a/r = 0.9
		r/R = 0.1		
None	0.004	0.02	0.08	0.3
One	0.00005	0.0005	0.003	0.01
		r/R = 0.3		
None	0.01	0.06	0.2	0.7
One	0.0004	0.004	0.02	0.08
		r/R = 0.5		
None	0.01	0.07	0.3	0.9
One	0.0007	0.008	0.05	0.2
Two	......	......	......	0.04
		r/R = 0.7		
None	0.01	0.07	0.3	0.9
One	0.001	0.01	0.06	0.3
Two	......	......	......	0.08

TABLE 5. Upper Bound for Error With the Use of Approximate Equations for the Volume of a Complete Ogive

Number of terms	Upper bound for error, %			
	r/R = 0.1	r/R = 0.3	r/R = 0.5	r/R = 0.7
One	1.5	5.1	9.5	15.4
Two	0.06	0.6	1.9	4.1
Three	0.003	0.08	0.4	1.2
Four	......	0.01	0.09	0.4
Five	......	......	0.002	0.1
Six	......	......	......	0.04

TABLE 6. Upper Bound for Error With the Use of Approximate Equations for the Moment of Inertia About the Central Axis for a Complete Ogive

Number of terms	Upper bound for error, %			
	r/R = 0.1	r/R = 0.3	r/R = 0.5	r/R = 0.7
One	1.9	6.4	12.1	19.6
Two	0.08	0.7	2.2	4.9
Three	0.003	0.1	0.5	1.4
Four	......	0.01	0.1	0.5
Five	......	......	0.03	0.2
Six	......	......	......	0.05

TABLE 7. Upper Bound for Error With the Use of Approximate Equations for the Moment of Inertia About the Base Plane for a Complete Ogive

Number of terms	Upper bound for error, %			
	r/R = 0.1	r/R = 0.3	r/R = 0.5	r/R = 0.7
One	1.7	5.0	8.3	11.7
Two	0.03	0.3	0.8	1.9
Three	0.001	0.03	0.2	0.6
Four	......	0.005	0.04	0.2
Five	......	......	0.01	0.06
Six	......	......	......	0.02

ALTERNATIVE EQUATIONS FOR VOLUME, MOMENT, AND MOMENT OF INERTIA OF AN OGIVE

Another and perhaps more concise set of methods for determining the volume, moment, and moment of inertia of an ogival shape is shown below. Again, complete derivations are included that can be used as guidelines in equation development. Special note should be made that angles are measured in radians.

Complete Ogive

Volume.

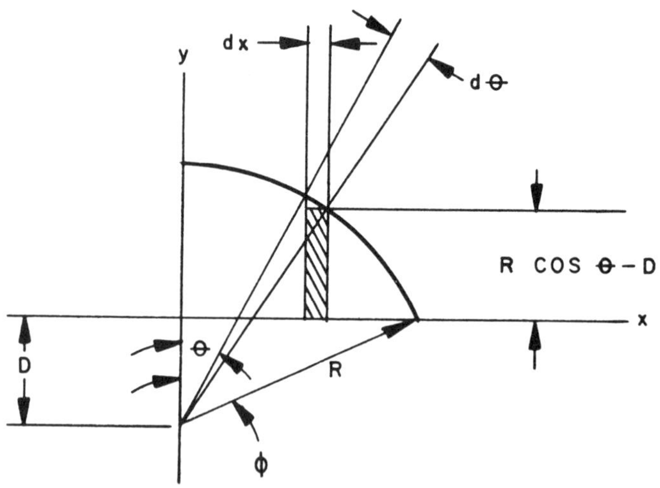

$$m = D/R$$

$$a = \sin\phi = \sqrt{1 - m^2}$$

$$V = \pi\int y^2 dx$$

$$V = \pi\int (R\cos\theta - D)^2 dx$$

$$= \pi\int (\cos\theta - D/R)^2 R^2 dx$$

$$= \pi\int R^2 (\cos\theta - m)^2 dx$$

$$dx = R(\cos\theta) d\theta$$

$$V = \pi\int R^2(\cos\theta - m)^2 R\cos\theta d\theta$$

$$= \pi R^3\int_0^{\phi}(\cos\theta - m)^2\cos\theta d\theta$$

$$= \pi R^3\int_0^{\phi}(\cos^2\theta - 2m\cos\theta + m^2)\cos\theta d\theta$$

$$= \pi R^3\int_0^{\phi}(\cos^3\theta - 2m\cos^2\theta + m^2\cos\theta)d\theta$$

$$= \pi R^3\left[\sin\theta - \frac{\sin^3\theta}{3} - 2m\left(\frac{1}{2}\theta + \frac{1}{2}\sin\theta\cos\theta\right) + m^2\sin\theta\right]_0^{\phi}$$

$$= \pi R^3\left(-\frac{\sin^3\phi}{3} + \sin\phi + m^2\sin\phi - m\sin\phi\cos\phi - m\phi\right)$$

$$= \pi R^3\left(-\frac{a^3}{3} + a - m\phi\right)$$

<u>Moment</u>.

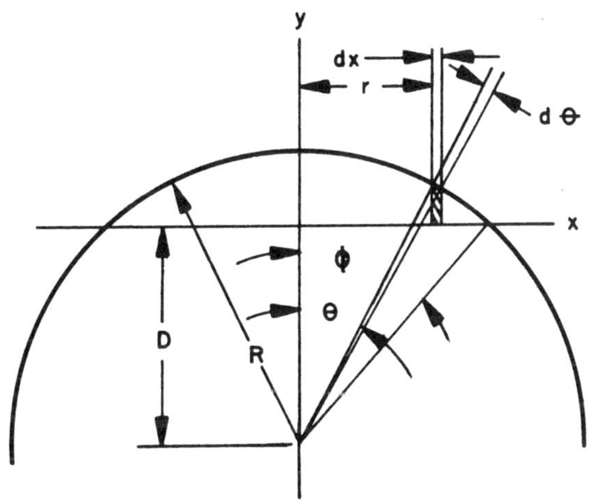

$$m = D/R$$

$$dV = \pi R^3(\cos\theta - m)^2(\cos\theta)d\theta$$

$$r = R\sin\theta$$

$$M = \int r dV$$

$$M = \int_0^{\phi}(R\sin\theta)\left[\pi R^3(\cos\theta - m)^2(\cos\theta)\right]d\theta$$

$$= \pi R^4\int_0^{\phi}(\cos^3\theta - 2m\cos^2\theta + m\cos\theta)\sin\theta d\theta$$

$$= \pi R^4\left[-\frac{\cos^4\theta}{4} + \frac{2m\cos^3\theta}{3} - \frac{m^2\cos^2\theta}{2} + \frac{1}{4} - \frac{2m}{3} + \frac{m^2}{2}\right]_0^{\phi}$$

$$m = D/R = \cos\theta$$

$$M = \pi R^4\left(-\frac{m^4}{4} + \frac{2m^4}{3} - \frac{m^4}{2} + \frac{1}{4} - \frac{2m}{3} + \frac{m^2}{2}\right)$$

$$= \pi R^4\left(\frac{-3m^4 + 8m^4 - 6m^4}{12} + \frac{m^2}{2} - \frac{2m}{3} + \frac{1}{4}\right)$$

$$= \pi R^4\left(-\frac{m^4}{12} + \frac{m^2}{2} - \frac{2m}{3} + \frac{1}{4}\right)$$

Moment of Inertia About the Central Axis.

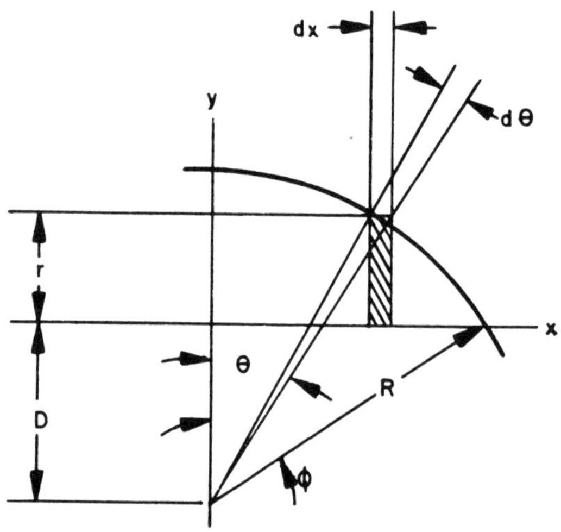

$$I_A = \frac{1}{2}\int r^2 dv$$

$$dv = \pi R^3(\cos\theta - m)^2 \cos\theta d\theta$$

$$r = R(\cos\theta - m)$$

$$m = D/R, \qquad \cos\phi = m, \qquad \sin\phi = a$$

$$I_A = \frac{\pi R^5}{2}\int_0^\phi (\cos\theta - m)^4 \cos\theta d\theta$$

$$= \frac{\pi R^5}{2}\int_0^\phi(\cos^4\theta - 4\cos^3\theta m + 6\cos\theta^2 m^2 - 4\cos\theta m^3 + m^4)\cos\theta d\theta$$

$$= \frac{\pi R^5}{2}\int_0^\phi(\cos^5\theta - 4m\cos^4\theta + 6m^2\cos^3\theta - 4m^3\cos^2\theta + m^4\cos\theta)$$

$$\int \cos^5\theta = \left[\frac{\cos^4\theta\sin\theta}{5} + \frac{4}{5}\left(\sin\theta - \frac{\sin^3\theta}{3}\right)\right]_0^\phi$$

$$\int \cos^4\theta = \left[\frac{\cos^3\theta\sin\theta}{4} + \frac{3}{4}\left(\frac{\theta}{2} + \frac{\sin\theta\cos\theta}{2}\right)\right]_0^\phi$$

$$\int \cos^3 \theta = \left[\sin \theta - \frac{\sin^3 \theta}{3}\right]_0^{\phi}$$

$$\int \cos^2 \theta = \left[\frac{\theta}{2} + \frac{\sin \theta \cos \theta}{2}\right]_0^{\phi}$$

$$I_A = \frac{\pi R^5}{2}\left[-\frac{9m^4 a}{5} + \frac{9m^2 a}{2} - 2m^2 a^3 + \frac{4a}{5} - \frac{4a^3}{15} - \left(2m^3 + \frac{3m}{2}\right)\phi\right]$$

<u>Moment of Inertia About the Base Plane.</u>

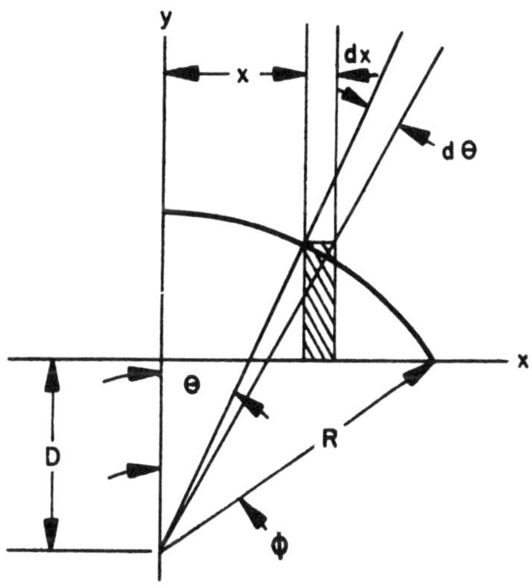

dI_C = differential of inertia about the centroid

$$I_B = \int dI_C + \int x^2 dv$$

$$dI_C = \frac{dv}{12}\left[3y^2 + (dx)^2\right]$$

$$dv = \pi y^2 dx$$

$$dI_C = (y^2 dv/4)$$

$$m = D/R = \cos \phi$$

$$a = L/R = \sin \phi$$

$$I_B = \int \frac{y^2 \pi y^2 dx}{4} + \int x^2 \pi y^2 dx$$

$$= \pi \int \left(\frac{y^4}{4} + y^2 x^2\right) dx$$

$$I_B = \pi R^5 \int \left[\frac{(\cos \theta - m)^4 \cos \theta d\theta}{4} + (\cos \theta - m)^2 \sin^2 \theta \cos \theta d\theta\right]$$

$$I_B = \pi R^5\left[-\frac{9m^4a}{20} + \frac{9m^2a}{8} - \frac{m^2a^3}{2} + \frac{a}{5} - \frac{a^3}{15} - \frac{1}{4}\left(2m^3 + \frac{3m}{2}\right)\phi + \frac{3m^4a}{10} - \frac{m^2a}{4} + \frac{m^2a^3}{3} + \frac{a}{5} - \frac{a^3}{15} - \frac{m}{4}\phi\right]$$

$$= \pi R^5\left[-\frac{3m^4a}{20} + \frac{7m^2a}{8} - \frac{m^2a^3}{6} + \frac{2a}{5} - \frac{2a^3}{15} - \left(\frac{m^3}{2} + \frac{5m}{8}\right)\phi\right]$$

Truncated Ogive

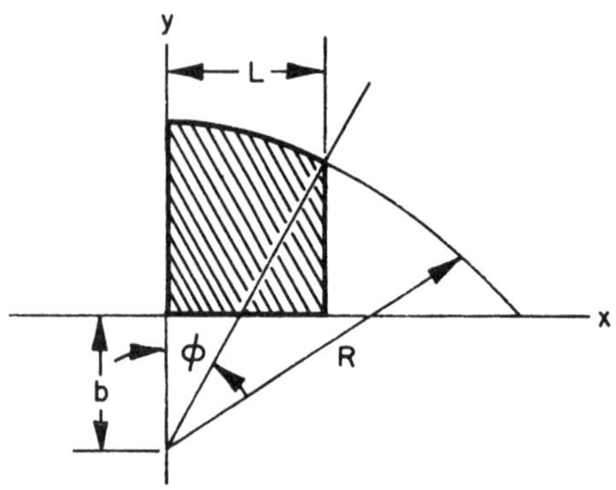

Moment.

$$M = \pi R^4\left(-\frac{b^4}{4} + \frac{2mb^3}{3} - \frac{m^2b^2}{2} + \frac{m^2}{2} - \frac{2m}{3} + \frac{1}{4}\right)$$

Volume.

$$V = \pi R^3\left(-\frac{a^3}{3} + a + m^2a - mab - m\phi\right)$$

Moment of Inertia About the Central Axis.

$$I_A = \frac{\pi R^5}{2}\left[\frac{b^4a}{5} - b^3ma - 2bm^3a - \frac{3bma}{2} - 2a^3m^2 - \frac{4a^3}{15} + \frac{4a}{5} + m^4a + 6am^2 - \left(2m^3 + \frac{3m}{2}\right)\phi\right]$$

Moment of Inertia About the Base Plane.

$$I_B = \pi R^5\left[-\frac{3b^4a}{20} + \frac{mb^3a}{4} - \frac{5mab}{8} - \frac{m^3ab}{2} - \frac{2a^3}{15} - \frac{m^2a^3}{6} + \frac{m^4a}{4} + \frac{3m^2a}{2} + \frac{2a}{5} - \left(\frac{m^3}{2} + \frac{5m}{8}\right)\phi\right]$$

where

$$a = L/R = \sin \phi$$

$$b = \sqrt{1 - a^2} = \cos \phi$$

$$m = D/R$$

THIN-SHELLED OGIVE

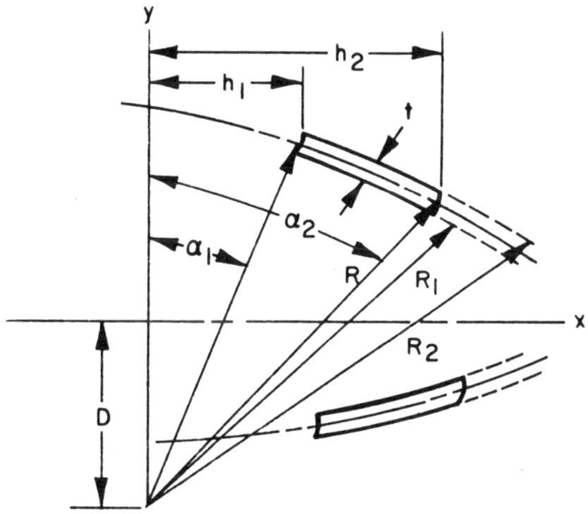

The thin ogival shell illustrated is one commonly encountered in the design of a missile nose cone. Simplified approximations have been developed that offer the designer a short-cut method of determining the properties of this type of section. These relationships are

$$V = 2\pi R^2 t\left[(a_2 - a_1) - m(\alpha_2 - \alpha_1)\right]$$

$$M = 2\pi R^3 t\left[\frac{(a_2^2 - a_1^2)}{2} + m\left(\sqrt{1 - a_2^2} - \sqrt{1 - a_1^2}\right)\right]$$

$$I_p = 2\pi R^4 t\left\{(a_2 - a_1)(1 + 3m^2) - \frac{(a_2^3 - a_1^3)}{3} - m^3(\alpha_2 - \alpha_1) - \frac{3m}{2}\left[(\alpha_2 - \alpha_1) + \left(a_2\sqrt{1 - a_2^2} - a_1\sqrt{1 - a_1^2}\right)\right]\right\}$$

$$I'_{TB} = 2\pi R^4 t\left\{\frac{(a_2^3 - a_1^3)}{3} - \frac{m}{2}\left[(\alpha_2 - \alpha_1) - \left(a_2\sqrt{1 - a_2^2} - a_1\sqrt{1 - a_1^2}\right)\right]\right\}$$

where

$$t = R_2 - R_1$$

$$R = \frac{R_1 + R_2}{2}$$

$$m = D/R$$

$$a = h/R$$

SOURCES

Material from the following sources was used by permission in compiling this handbook.

American Institute of Steel Construction. Steel Construction, 5th ed. New York, AISC, 1955.

Ayres, F. Theory and Problems of Differential and Integral Calculus. New York, Schaum Publishing Co., 1950.

Computation Laboratory of Harvard University. ''Tables for the Design of Missiles,'' in Annals of the Computation Laboratory of Harvard University, Vol. XVII. Cambridge, Mass., Harvard University Press, 1948.

Hudson, Ralph G. The Engineers' Manual. New York, Wiley, 1955.

Meriam, J. L. Mechanics. Part I. Statics. New York, Wiley, 1955.

————. Mechanics. Part II. Dynamics. New York, Wiley, 1957.

Oberg, Erik, and F. D. Jones. Machinery's Handbook, 15th ed., Holbrook L. Horton, ed. New York, The Industrial Press, 1954.

Society of Aeronautical Weight Engineers, Inc. Weight Handbook, Vol. 1, 3rd ed., Sydney S. Brown and Brian C. Wildsmith, eds. Los Angeles, SAWE, 1944.

U. S. Naval Ordnance Test Station. ''Formulas for Ogive Volume and Moments,'' by John W. Odle, Science Department, Mathematics Section. Paper No. 21. China Lake, Calif., NOTS, 7 January 1947.

77. Moments of Inertia and Centroidal Distances of T-Beams

(Ref. 2)

Graphs for usual and unusual ratios of $b_0 : b$ and $d : d_0$.

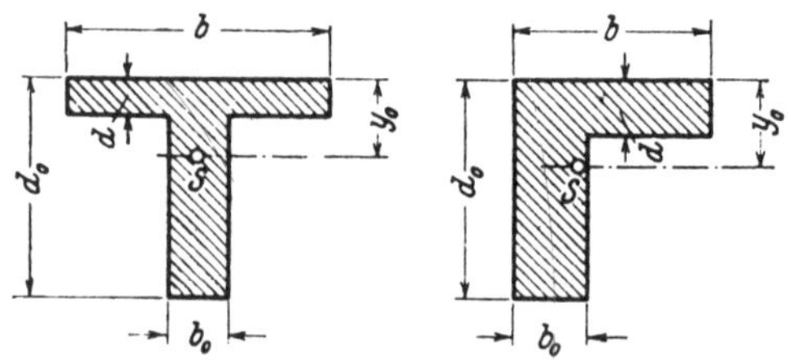

Fig. 1
T-Beam Cross Sections

b, b_0, d_0, d are the dimensions of the T-beam according to the above figure.
y_0 is the centroidal distance from the top fiber.
I is the moment of inertia about the centroidal axis parallel to b.

Using $$\beta = \frac{b_0}{b}, \quad \text{and} \quad \delta = \frac{d}{d_0},$$

we obtain $$y_0 = \eta . d_0, \quad I = \mu . \frac{bd_0^3}{12},$$

where η is obtained from Table I, μ from Table II.

NOTE: For T-beams having a flange on one side only, the centroidal distances and moments of inertia may also be obtained for the direction parallel to d_0, when b is replaced by d_0, b_0 by d, and vice versa.

Table I

Values of η. $$\left(\eta = \frac{1}{2} \cdot \frac{\beta + [1-\beta]\,\delta^2}{\beta + [1-\beta]\,\delta}\right)$$

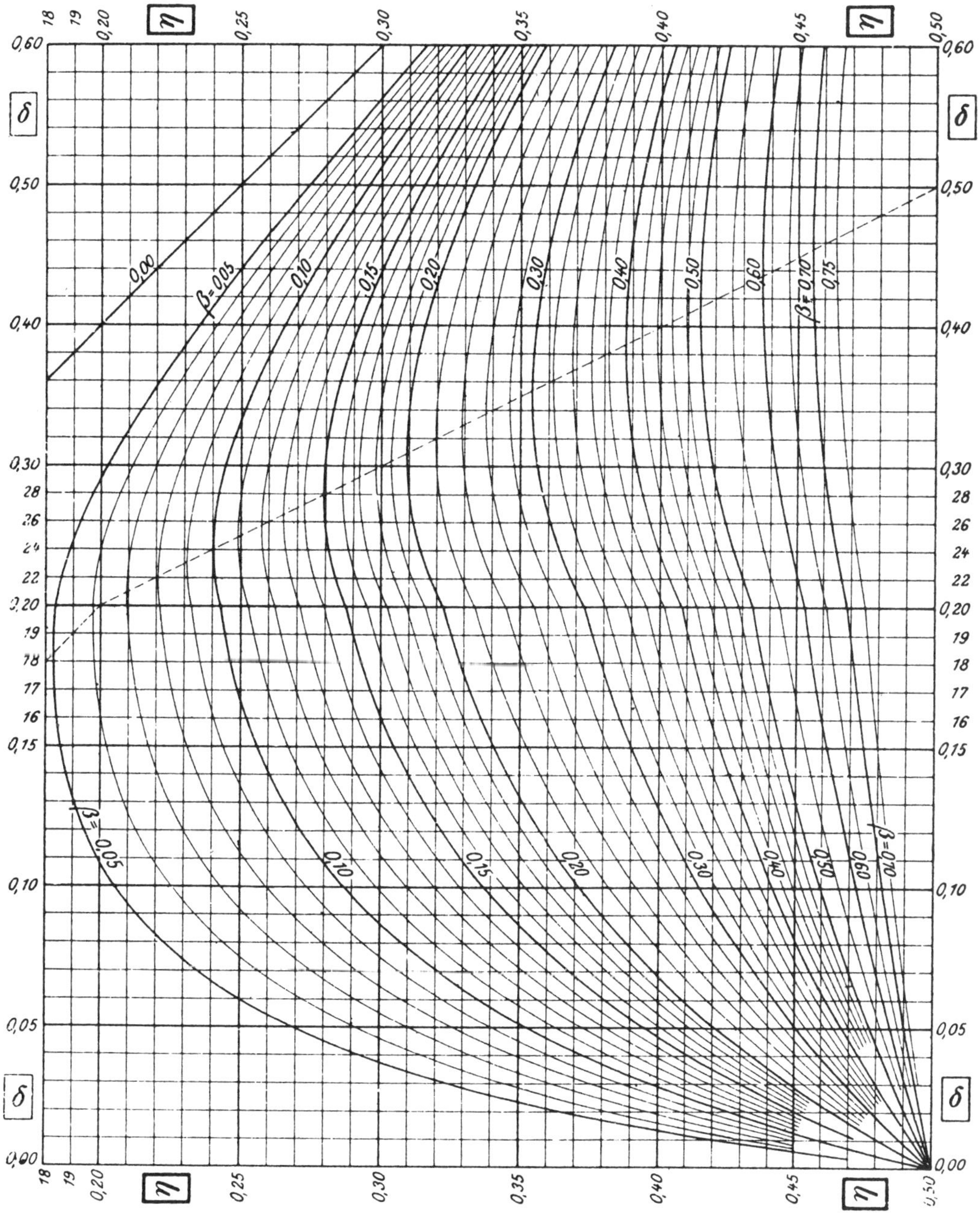

Table II (left half)

Values of μ.

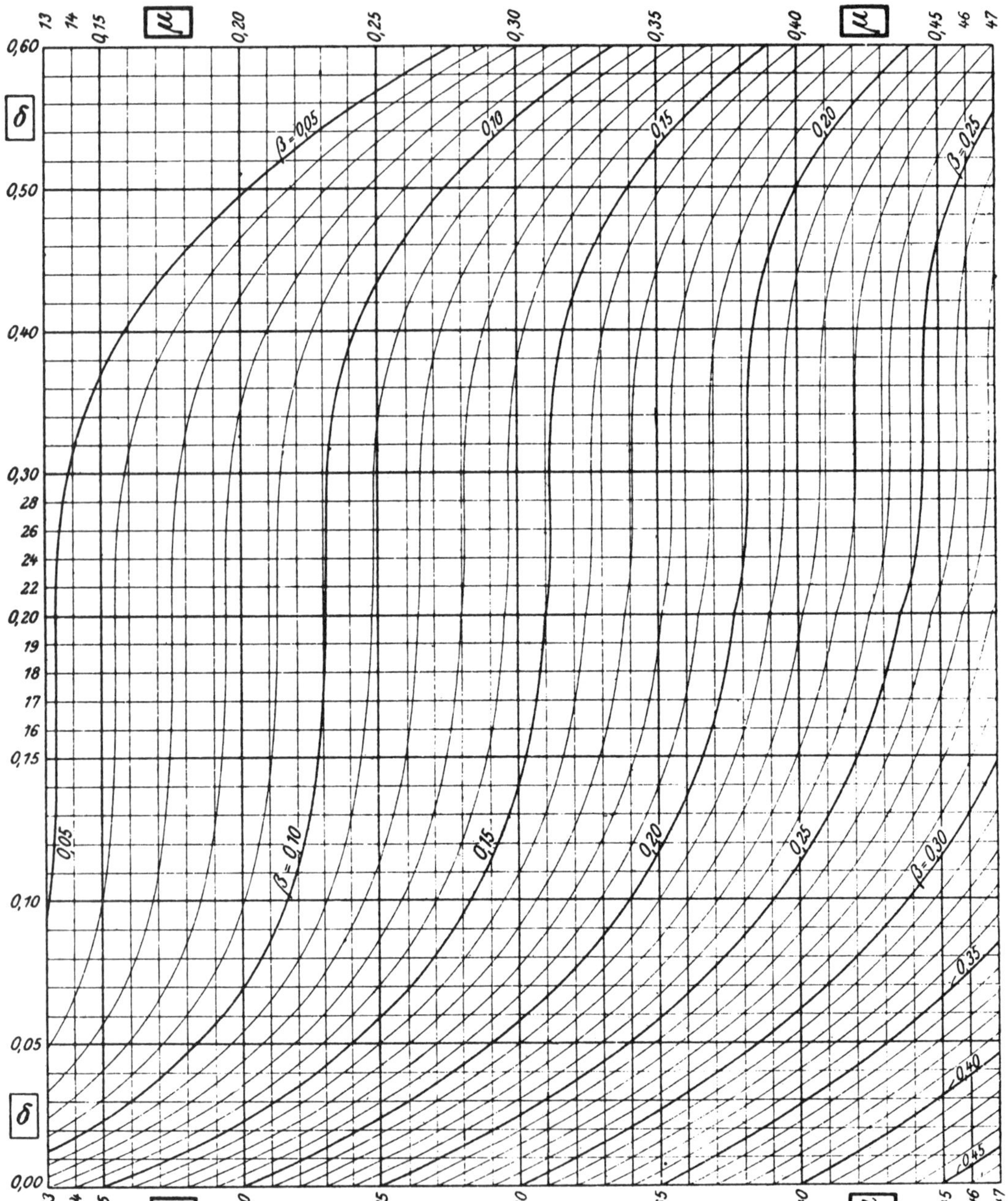

Table II (right half)

$(\mu = 4[\eta^3 - (1 - \beta)(\eta - \delta)^3 + \beta(1 - \eta)^3]$; for η see Table I.)

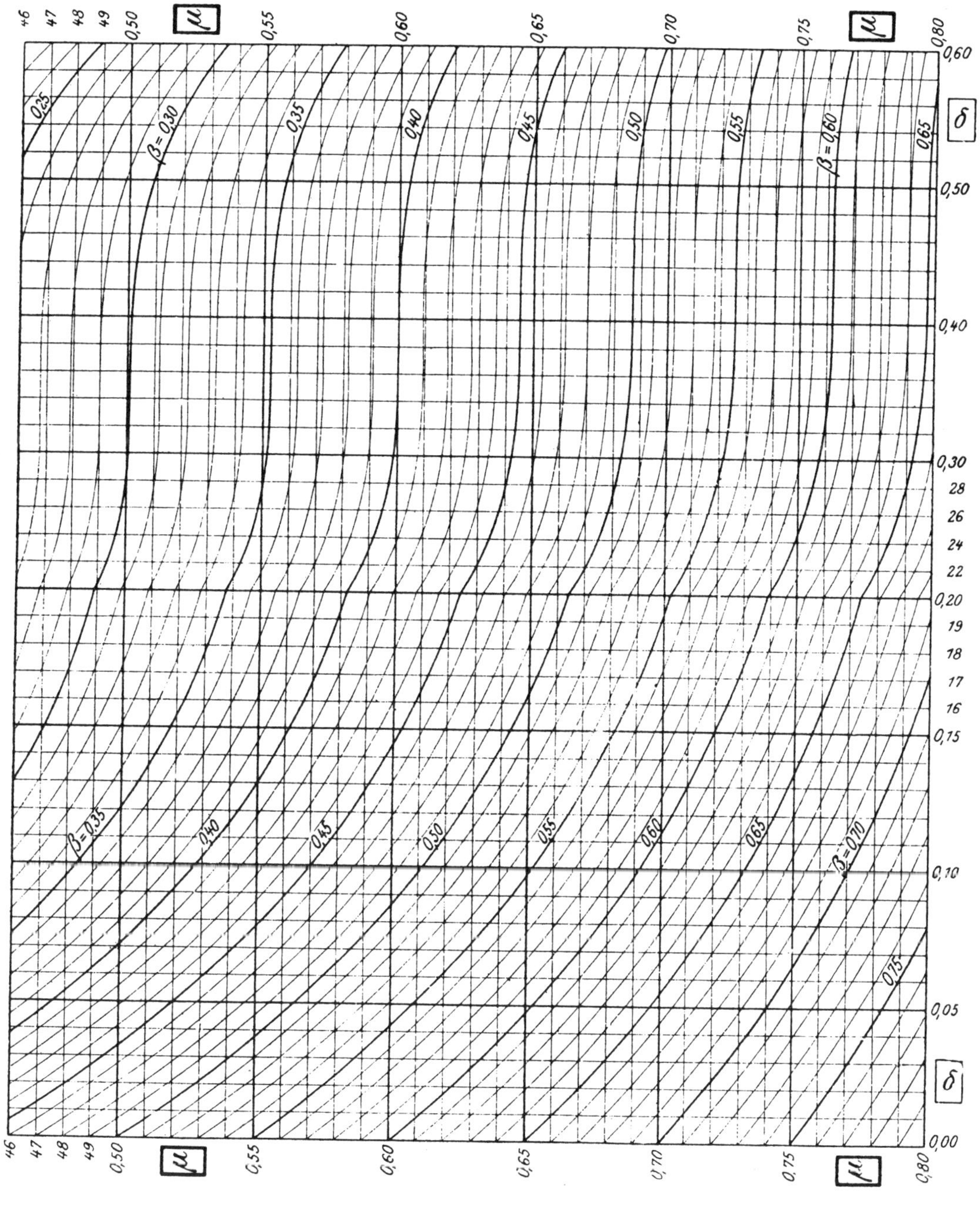

Chap. XII

REFERENCES

Section	*Ref. No.*	
76	1	MEYERS, J. A.: "Handbook of Equations for Mass and Area Properties of Various Geometrical Shapes," Navweps Report 7827, Nots Tp 2838, April, 1962
76	2	*Weight Handbook*, 3rd. edition. Sydney S. Brown and Brian C. Wildsman, editors, Los Angeles, S.A.W.E., 1944.
77	3	KLEINLOGEL, A. D.ENG.: *Beam Formulas*, Frederick Ungar Publishing Co., N.Y., 1953

Appendix A

THE EFFECT OF POISSON'S RATIO ON STRESSES IN FLAT PLATES

(*See Chapter V*)

(*Ref.* S. Kravitz: "Radial Stress and Deflection of Flat Plates," *Design News*, Nov. 25, 1964.)

In Chapter V, the maximum stresses, and maximum deflections, were conveniently expressed by simple formulas with numerical coefficients K that depend upon the ratio of plate dimensions and upon the chosen value of Poisson's Ratio $\mu = 0.3$, a value used for steel. The Poisson's Ratio of other materials are:

1. Brass 0.33
2. Copper 0.33
3. Cast Iron 0.27
4. Wrought Iron 0.28
5. Aluminum alloy 0.33
6. Brass 0.33
7. Copper 0.36
8. Malleable Iron 0.27

It was suggested that with the Poisson's Ratio of 0.3 and by tolerating a 5 percent discrepancy, the numerical coefficients tabulated in section 50 may be used for *any* structural material.

In what follows, we present four tables taken, by permission, from the reference article, that show the values of constants K for a range of Poisson's Ratio from 0.2 to 0.4.

NOTATION

W = total load acting on the plate (lb.)
w = unit load (psi)
a = radius of the plate (in.)
b = radius of the uniform concentric load (in.)
S_r = maximum radial stress (psi)
y = maximum deflection (in.)
t = thickness of the uniform plate (in.)
μ = Poisson's Ratio

CASE 1 EDGE ON FREE SUPPORT: UNIFORM LOAD ON CONCENTRIC CIRCLE

$W = w\pi b^2$

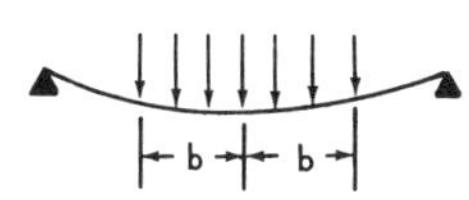

$S_r = K_1 \frac{W}{t^2}$ $\qquad$ S_r occurs at the center as a compressive stress

$y = L_1 \frac{Wa^2}{Et^3}$

b/a	K_1			L_1		
	$\mu=0.2$	$\mu=0.3$	$\mu=0.4$	$\mu=0.2$	$\mu=0.3$	$\mu=0.4$
0.05	2.194	2.337	2.480	0.609	0.549	0.485
0.10	1.796	1.906	2.016	0.602	0.543	0.479
0.15	1.562	1.653	1.744	0.593	0.535	0.472
0.20	1.396	1.473	1.550	0.582	0.524	0.462
0.25	1.266	1.333	1.400	0.569	0.512	0.451
0.30	1.159	1.217	1.276	0.554	0.498	0.439
0.35	1.067	1.119	1.170	0.537	0.483	0.425
0.40	0.987	1.033	1.079	0.519	0.467	0.411
0.45	0.916	0.956	0.997	0.501	0.450	0.395
0.50	0.851	0.887	0.923	0.481	0.431	0.379
0.55	0.791	0.823	0.855	0.460	0.412	0.362
0.60	0.736	0.764	0.793	0.438	0.393	0.344
0.65	0.684	0.710	0.735	0.416	0.372	0.326
0.70	0.635	0.658	0.681	0.393	0.352	0.308
0.75	0.589	0.609	0.629	0.370	0.331	0.289
0.80	0.544	0.563	0.581	0.346	0.309	0.270
0.85	0.502	0.518	0.534	0.322	0.288	0.251
0.90	0.460	0.475	0.490	0.298	0.266	0.232
0.95	0.421	0.434	0.447	0.273	0.244	0.213
1.00	0.382	0.394	0.406	0.248	0.221	0.193

CASE 2 EDGE ON FREE SUPPORT: UNIFORM LOAD ON CONCENTRIC RING

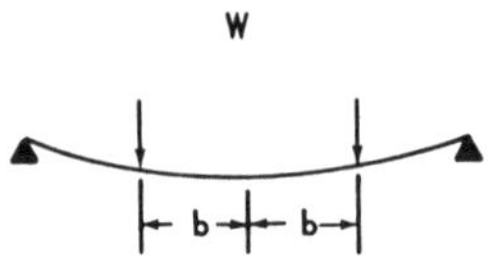

$S_r = K_2 \frac{W}{t^2}$ S_r occurs as a compressive stress

$y = L_2 \frac{Wa^2}{Et^3}$

b/a	K_2			L_2		
	$\mu=0.2$	$\mu=0.3$	$\mu=0.4$	$\mu=0.2$	$\mu=0.3$	$\mu=0.4$
0.05	1.907	2.026	2.145	0.606	0.547	0.483
0.10	1.508	1.595	1.681	0.594	0.536	0.473
0.15	1.274	1.341	1.408	0.578	0.521	0.459
0.20	1.105	1.159	1.213	0.557	0.501	0.442
0.25	0.973	1.017	1.061	0.533	0.479	0.422
0.30	0.864	0.899	0.935	0.506	0.455	0.400
0.35	0.769	0.798	0.827	0.477	0.428	0.376
0.40	0.685	0.709	0.733	0.446	0.400	0.350
0.45	0.610	0.629	0.648	0.413	0.370	0.324
0.50	0.540	0.556	0.571	0.379	0.338	0.296
0.55	0.476	0.488	0.500	0.343	0.306	0.267
0.60	0.415	0.424	0.433	0.307	0.273	0.238
0.65	0.357	0.364	0.371	0.270	0.239	0.208
0.70	0.302	0.307	0.311	0.232	0.205	0.178
0.75	0.248	0.252	0.255	0.193	0.171	0.148
0.80	0.197	0.199	0.201	0.155	0.136	0.118
0.85	0.146	0.147	0.148	0.116	0.102	0.088
0.90	0.097	0.097	0.098	0.077	0.068	0.058
0.95	0.048	0.048	0.048	0.038	0.034	0.029
1.00	0.000	0.000	0.000	0.000	0.000	0.000

CASE 3 EDGE ON FIXED SUPPORT: UNIFORM LOAD ON CONCENTRIC CIRCLE

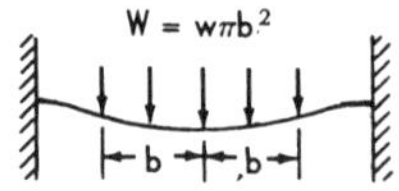

$S_r = K_3 \frac{W}{t^2}$ If b/a is less than 0.588, S_r occurs as a compressive stress at the center

$y = L_3 \frac{Wa^2}{Et^3}$ If b/a is greater than 0.588, S_r occurs as a tensile stress at the edge

b/a	K_3			L_3		
	$\mu=0.2$	$\mu=0.3$	$\mu=0.4$	$\mu=0.2$	$\mu=0.3$	$\mu=0.4$
0.05	1.717	1.860	2.003	0.227	0.215	0.199
0.10	1.321	1.431	1.541	0.222	0.211	0.194
0.15	1.090	1.181	1.272	0.216	0.204	0.189
0.20	0.928	1.005	1.083	0.208	0.197	0.182
0.25	0.803	0.870	0.937	0.199	0.188	0.174
0.30	0.703	0.761	0.820	0.189	0.179	0.165
0.35	0.619	0.671	0.722	0.179	0.169	0.156
0.40	0.548	0.594	0.639	0.168	0.159	0.147
0.45	0.487	0.527	0.568	0.157	0.149	0.138
0.50	0.433	0.469	0.505	0.146	0.139	0.128
0.55	0.386	0.418	0.450	0.136	0.129	0.119
0.60	0.392	0.392	0.392	0.125	0.119	0.110
0.65	0.377	0.377	0.377	0.115	0.109	0.100
0.70	0.360	0.360	0.360	0.105	0.099	0.092
0.75	0.343	0.343	0.343	0.095	0.090	0.083
0.80	0.325	0.325	0.325	0.086	0.082	0.076
0.85	0.305	0.305	0.305	0.078	0.074	0.068
0.90	0.284	0.284	0.284	0.070	0.067	0.062
0.95	0.262	0.262	0.262	0.063	0.060	0.056
1.00	0.239	0.239	0.239	0.057	0.054	0.050

CASE 4 EDGE ON FIXED SUPPORT: UNIFORM LOAD ON CONCENTRIC RING

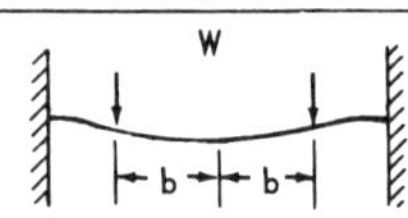

$S_r = K_4 \frac{W}{t^2}$ If b/a is less than 0.31, S_r occurs as a compressive stress

$y = L_4 \frac{Wa^2}{Et^3}$ If b/a is greater than 0.31, S_r occurs as a tensile stress at the edge

b/a	K_4			L_4		
	$\mu=0.2$	$\mu=0.3$	$\mu=0.4$	$\mu=0.2$	$\mu=0.3$	$\mu=0.4$
0.05	1.431	1.550	1.669	0.225	0.213	0.197
0.10	1.036	1.122	1.208	0.216	0.205	0.189
0.15	0.807	0.874	0.941	0.204	0.194	0.179
0.20	0.647	0.701	0.755	0.191	0.181	0.167
0.25	0.526	0.570	0.613	0.175	0.166	0.153
0.30	0.429	0.465	0.501	0.159	0.151	0.139
0.35	0.419	0.419	0.419	0.142	0.135	0.124
0.40	0.401	0.401	0.401	0.125	0.119	0.110
0.45	0.381	0.381	0.381	0.109	0.103	0.095
0.50	0.358	0.358	0.358	0.092	0.088	0.081
0.55	0.333	0.333	0.333	0.077	0.073	0.067
0.60	0.306	0.306	0.306	0.062	0.059	0.055
0.65	0.276	0.276	0.276	0.049	0.046	0.043
0.70	0.244	0.244	0.244	0.037	0.035	0.032
0.75	0.209	0.209	0.209	0.026	0.025	0.023
0.80	0.172	0.172	0.172	0.0170	0.0162	0.0149
0.85	0.132	0.132	0.132	0.0098	0.0093	0.0086
0.90	0.091	0.091	0.091	0.0044	0.0042	0.0039
0.95	0.047	0.047	0.047	0.0011	0.0011	0.0010
1.00	0.000	0.000	0.000	0.0000	0.0000	0.0000

Appendix B

BEAMS ON ELASTIC FOUNDATION

Tables of constants, for deflection (K_y), slope (K_θ), moment (K_M) and shear (K_Q), to be used together with equations tabulated in Tables 11, and 12, Chapter II. These constants were based on: Hetényi, M. "Beams on Elastic Foundation", University of Michigan Press, Ann Arbor, Michigan, 1958.

CASE 1

Hinged Ends, Center Load

λL	X/L	K_y	K_Θ	K_M
2.0	0.	0.8122	0.	1.3557
2.0	0.0625	0.7923	0.1544	1.1183
2.0	0.1250	0.7373	0.2806	0.9056
2.0	0.1875	0.6540	0.3817	0.7159
2.0	0.2500	0.5483	0.4604	0.5465
2.0	0.3125	0.4255	0.5191	0.3943
2.0	0.3750	0.2903	0.5596	0.2553
2.0	0.4375	0.1471	0.5833	0.1254
2.0	0.5000	0.	0.5911	0.
3.0	0.	1.0880	0.	1.1191
3.0	0.0625	1.0529	0.1771	0.7822
3.0	0.1250	0.9623	0.2979	0.5189
3.0	0.1875	0.8348	0.3759	0.3231
3.0	0.2500	0.6842	0.4227	0.1858
3.0	0.3125	0.5203	0.4486	0.0966
3.0	0.3750	0.3494	0.4613	0.0439
3.0	0.4375	0.1753	0.4666	0.0158
3.0	0.5000	0.0000	0.4679	0.0000
4.0	0.	1.0522	0.	0.9954
4.0	0.0625	0.9997	0.1918	0.5606
4.0	0.1250	0.8759	0.2907	0.2500
4.0	0.1875	0.7196	0.3259	0.0486
4.0	0.2500	0.5564	0.3225	-0.0629
4.0	0.3125	0.4003	0.3003	-0.1049
4.0	0.3750	0.2567	0.2742	-0.0966
4.0	0.4375	0.1249	0.2546	-0.0562
4.0	0.5000	0.0000	0.2475	-0.0000
5.0	0.	1.0090	0.	0.9832
5.0	0.0625	0.9317	0.2198	0.4554
5.0	0.1250	0.7625	0.3035	0.1081
5.0	0.1875	0.5699	0.3028	-0.0907
5.0	0.2500	0.3931	0.2584	-0.1779
5.0	0.3125	0.2498	0.1997	-0.1877
5.0	0.3750	0.1424	0.1462	-0.1482
5.0	0.4375	0.0634	0.1099	-0.0804
5.0	0.5000	0.0000	0.0972	-0.0000

CASE 2

Hinged Ends, Uniform Load

λL	X/L	K_y	K_Θ	K_M
2.0	0.	0.	-0.8122	0.
2.0	0.0625	0.1007	-0.7923	0.1544
2.0	0.1250	0.1966	-0.7373	0.2806
2.0	0.1875	0.2838	-0.6540	0.3817
2.0	0.2500	0.3592	-0.5483	0.4604
2.0	0.3125	0.4202	-0.4255	0.5191
2.0	0.3750	0.4650	-0.2903	0.5596
2.0	0.4375	0.4924	-0.1471	0.5833
2.0	0.5000	0.5017	0.	0.5911
2.0	0.5625	0.4924	0.1471	0.5833
2.0	0.6250	0.4650	0.2903	0.5596
2.0	0.6875	0.4202	0.4255	0.5191
2.0	0.7500	0.3592	0.5483	0.4604
2.0	0.8125	0.2838	0.6540	0.3817
2.0	0.8750	0.1966	0.7373	0.2806
2.0	0.9375	0.1007	0.7923	0.1544
2.0	1.0000	0.0000	0.8122	0.
3.0	0.	0.	-1.0880	0.
3.0	0.0625	0.2017	-1.0529	0.1771
3.0	0.1250	0.3914	-0.9623	0.2979
3.0	0.1875	0.5603	-0.8348	0.3759
3.0	0.2500	0.7030	-0.6842	0.4227
3.0	0.3125	0.8161	-0.5203	0.4486
3.0	0.3750	0.8977	-0.3494	0.4613
3.0	0.4375	0.9469	-0.1753	0.4666
3.0	0.5000	0.9633	-0.0000	0.4679
3.0	0.5625	0.9469	0.1753	0.4666
3.0	0.6250	0.8977	0.3494	0.4613
3.0	0.6875	0.8161	0.5203	0.4486
3.0	0.7500	0.7030	0.6842	0.4227
3.0	0.8125	0.5603	0.8348	0.3759
3.0	0.8750	0.3914	0.9623	0.2979
3.0	0.9375	0.2017	1.0529	0.1771
3.0	1.0000	0.0000	1.0880	0.
4.0	0.	0.	-1.0522	-0.
4.0	0.0625	0.2585	-0.9997	0.1918
4.0	0.1250	0.4940	-0.8759	0.2907
4.0	0.1875	0.6938	-0.7196	0.3259
4.0	0.2500	0.8532	-0.5564	0.3225
4.0	0.3125	0.9726	-0.4003	0.3003
4.0	0.3750	1.0544	-0.2567	0.2742
4.0	0.4375	1.1019	-0.1249	0.2546
4.0	0.5000	1.1175	0.	0.2475
4.0	0.5625	1.1019	0.1249	0.2546
4.0	0.6250	1.0544	0.2567	0.2742
4.0	0.6875	0.9726	0.4003	0.3003
4.0	0.7500	0.8532	0.5564	0.3225
4.0	0.8125	0.6938	0.7196	0.3259
4.0	0.8750	0.4940	0.8759	0.2907
4.0	0.9375	0.2585	0.9997	0.1918
4.0	1.0000	0.0000	1.0522	0.
5.0	0.	0.	-1.0090	-0.
5.0	0.0625	0.3068	-0.9317	0.2198
5.0	0.1250	0.5729	-0.7625	0.3035
5.0	0.1875	0.7810	-0.5699	0.3028
5.0	0.2500	0.9308	-0.3931	0.2584
5.0	0.3125	1.0302	-0.2498	0.1997
5.0	0.3750	1.0906	-0.1424	0.1462
5.0	0.4375	1.1222	-0.0634	0.1099
5.0	0.5000	1.1319	0.	0.0972
5.0	0.5625	1.1222	0.0634	0.1099
5.0	0.6250	1.0906	0.1424	0.1462
5.0	0.6875	1.0302	0.2498	0.1997
5.0	0.7500	0.9308	0.3931	0.2584
5.0	0.8125	0.7810	0.5699	0.3028
5.0	0.8750	0.5729	0.7625	0.3035
5.0	0.9375	0.3068	0.9317	0.2198
5.0	1.0000	0.0000	1.0090	0.

CASE 3

Hinged Ends, Equal Moment at Both Ends

λL	X/L	K_y	K_Θ	K_M
2.0	0.	0.	1.3557	1.0000
2.0	0.0625	0.1544	1.1183	0.8993
2.0	0.1250	0.2806	0.9056	0.8034
2.0	0.1875	0.3817	0.7159	0.7162
2.0	0.2500	0.4604	0.5465	0.6408
2.0	0.3125	0.5191	0.3943	0.5798
2.0	0.3750	0.5596	0.2553	0.5350
2.0	0.4375	0.5833	0.1254	0.5076
2.0	0.5000	0.5911	-0.	0.4983
2.0	0.5625	0.5833	-0.1254	0.5076
2.0	0.6250	0.5596	-0.2553	0.5350
2.0	0.6875	0.5191	-0.3943	0.5798
2.0	0.7500	0.4604	-0.5465	0.6408
2.0	0.8125	0.3817	-0.7159	0.7162
2.0	0.8750	0.2806	-0.9056	0.8034
2.0	0.9375	0.1544	-1.1183	0.8993
2.0	1.0000	0.	-1.3557	1.0000
3.0	0.	0.	1.1191	1.0000
3.0	0.0625	0.1771	0.7822	0.7983
3.0	0.1250	0.2979	0.5189	0.6086
3.0	0.1875	0.3759	0.3231	0.4397
3.0	0.2500	0.4227	0.1858	0.2970
3.0	0.3125	0.4486	0.0966	0.1839
3.0	0.3750	0.4613	0.0439	0.1023
3.0	0.4375	0.4666	0.0158	0.0531
3.0	0.5000	0.4679	-0.0000	0.0367
3.0	0.5625	0.4666	-0.0158	0.0531
3.0	0.6250	0.4613	-0.0439	0.1023
3.0	0.6875	0.4486	-0.0966	0.1839
3.0	0.7500	0.4227	-0.1858	0.2970
3.0	0.8125	0.3759	-0.3231	0.4397
3.0	0.8750	0.2979	-0.5189	0.6086
3.0	0.9375	0.1771	-0.7822	0.7983
3.0	1.0000	0.	-1.1191	1.0000
4.0	0.	0.	0.9954	1.0000
4.0	0.0625	0.1918	0.5606	0.7415
4.0	0.1250	0.2907	0.2500	0.5060
4.0	0.1875	0.3259	0.0486	0.3062
4.0	0.2500	0.3225	-0.0629	0.1468
4.0	0.3125	0.3003	-0.1049	0.0274
4.0	0.3750	0.2742	-0.0966	-0.0544
4.0	0.4375	0.2546	-0.0562	-0.1019
4.0	0.5000	0.2475	-0.0000	-0.1175
4.0	0.5625	0.2546	0.0562	-0.1019
4.0	0.6250	0.2742	0.0966	-0.0544
4.0	0.6875	0.3003	0.1049	0.0274
4.0	0.7500	0.3225	0.0629	0.1468
4.0	0.8125	0.3259	-0.0486	0.3062
4.0	0.8750	0.2907	-0.2500	0.5060
4.0	0.9375	0.1918	-0.5606	0.7415
4.0	1.0000	0.	-0.9954	1.0000
5.0	0.	0.	0.9832	1.0000
5.0	0.0625	0.2198	0.4554	0.6932
5.0	0.1250	0.3035	0.1081	0.4271
5.0	0.1875	0.3028	-0.0907	0.2190
5.0	0.2500	0.2584	-0.1779	0.0692
5.0	0.3125	0.1997	-0.1877	-0.0302
5.0	0.3750	0.1462	-0.1482	-0.0906
5.0	0.4375	0.1099	-0.0804	-0.1222
5.0	0.5000	0.0972	-0.0000	-0.1319
5.0	0.5625	0.1099	0.0804	-0.1222
5.0	0.6250	0.1462	0.1482	-0.0906
5.0	0.6875	0.1997	0.1877	-0.0302
5.0	0.7500	0.2584	0.1779	0.0692
5.0	0.8125	0.3028	0.0907	0.2190
5.0	0.8750	0.3035	-0.1081	0.4271
5.0	0.9375	0.2198	-0.4554	0.6932
5.0	1.0000	0.	-0.9832	1.0000

CASE 4

Hinged Ends, Concentrated Moment at One End

λL	X/L	K_y	K_Θ	K_M
2.0	0.	0.	1.0030	1.0000
2.0	0.0625	0.1104	0.7678	0.8820
2.0	0.1250	0.1932	0.5618	0.7674
2.0	0.1875	0.2520	0.3836	0.6587
2.0	0.2500	0.2902	0.2317	0.5579
2.0	0.3125	0.3109	0.1039	0.4661
2.0	0.3750	0.3171	-0.0021	0.3840
2.0	0.4375	0.3112	-0.0889	0.3117
2.0	0.5000	0.2955	-0.1588	0.2492
2.0	0.5625	0.2721	-0.2142	0.1958
2.0	0.6250	0.2425	-0.2574	0.1510
2.0	0.6875	0.2082	-0.2904	0.1137
2.0	0.7500	0.1703	-0.3148	0.0829
2.0	0.8125	0.1298	-0.3323	0.0574
2.0	0.8750	0.0874	-0.3439	0.0360
2.0	0.9375	0.0440	-0.3505	0.0173
2.0	1.0000	0.	-0.3526	-0.0000
3.0	0.	0.	1.0062	1.0000
3.0	0.0625	0.1557	0.6662	0.8139
3.0	0.1250	0.2541	0.3944	0.6384
3.0	0.1875	0.3075	0.1853	0.4805
3.0	0.2500	0.3270	0.0314	0.3440
3.0	0.3125	0.3222	-0.0756	0.2304
3.0	0.3750	0.3011	-0.1442	0.1393
3.0	0.4375	0.2700	-0.1827	0.0693
3.0	0.5000	0.2340	-0.1986	0.0183
3.0	0.5625	0.1965	-0.1985	-0.0162
3.0	0.6250	0.1602	-0.1881	-0.0370
3.0	0.6875	0.1263	-0.1722	-0.0464
3.0	0.7500	0.0957	-0.1545	-0.0470
3.0	0.8125	0.0683	-0.1378	-0.0408
3.0	0.8750	0.0438	-0.1245	-0.0297
3.0	0.9375	0.0214	-0.1159	-0.0156
3.0	1.0000	0.	-0.1130	-0.0000
4.0	0.	0.	0.9992	1.0000
4.0	0.0625	0.1925	0.5612	0.7544
4.0	0.1250	0.2904	0.2409	0.5320
4.0	0.1875	0.3215	0.0232	0.3451
4.0	0.2500	0.3090	-0.1109	0.1980
4.0	0.3125	0.2714	-0.1811	0.0892
4.0	0.3750	0.2222	-0.2057	0.0142
4.0	0.4375	0.1710	-0.2000	-0.0330
4.0	0.5000	0.1237	-0.1763	-0.0587
4.0	0.5625	0.0836	-0.1439	-0.0690
4.0	0.6250	0.0520	-0.1091	-0.0687
4.0	0.6875	0.0289	-0.0763	-0.0618
4.0	0.7500	0.0135	-0.0479	-0.0512
4.0	0.8125	0.0044	-0.0254	-0.0389
4.0	0.8750	0.0003	-0.0092	-0.0259
4.0	0.9375	-0.0007	0.0006	-0.0129
4.0	1.0000	-0.	0.0038	-0.
5.0	0.	0.	1.0000	1.0000
5.0	0.0625	0.2249	0.4712	0.6962
5.0	0.1250	0.3132	0.1208	0.4342
5.0	0.1875	0.3156	-0.0840	0.2319
5.0	0.2500	0.2718	-0.1818	0.0905
5.0	0.3125	0.2094	-0.2082	0.0019
5.0	0.3750	0.1460	-0.1926	-0.0459
5.0	0.4375	0.0911	-0.1568	-0.0649
5.0	0.5000	0.0486	-0.1152	-0.0659
5.0	0.5625	0.0188	-0.0764	-0.0573
5.0	0.6250	0.0002	-0.0444	-0.0448
5.0	0.6875	-0.0098	-0.0204	-0.0321
5.0	0.7500	-0.0134	-0.0039	-0.0212
5.0	0.8125	-0.0128	0.0067	-0.0129
5.0	0.8750	-0.0097	0.0128	-0.0070
5.0	0.9375	-0.0051	0.0158	-0.0030
5.0	1.0000	-0.	0.0167	0.0000

CASE 5

Fixed Ends, Center Load

λL	X/L	K_y	K_θ	K_M	K_Q
2.0	0.	0.2967	-0.	-0.9211	1.0000
2.0	0.0625	0.2836	-0.0640	-0.6757	0.9635
2.0	0.1250	0.2494	-0.1054	-0.4391	0.9300
2.0	0.1875	0.2014	-0.1265	-0.2103	0.9017
2.0	0.2500	0.1468	-0.1294	0.0123	0.8799
2.0	0.3125	0.0926	-0.1164	0.2302	0.8650
2.0	0.3750	0.0456	-0.0892	0.4453	0.8564
2.0	0.4375	0.0125	-0.0498	0.6589	0.8530
2.0	0.5000	-0.	0.	0.8720	0.8524
3.0	0.	0.6967	-0.	-1.0885	1.0000
3.0	0.0625	0.6627	-0.1346	-0.7377	0.8716
3.0	0.1250	0.5765	-0.2115	-0.4333	0.7547
3.0	0.1875	0.4597	-0.2417	-0.1692	0.6573
3.0	0.2500	0.3307	-0.2350	0.0626	0.5831
3.0	0.3125	0.2060	-0.2003	0.2711	0.5330
3.0	0.3750	0.1003	-0.1452	0.4651	0.5047
3.0	0.4375	0.0272	-0.0764	0.6518	0.4933
3.0	0.5000	0.	0.	0.8363	0.4916
4.0	0.	0.9292	-0.	-1.0538	1.0000
4.0	0.0625	0.8731	-0.1887	-0.6113	0.7726
4.0	0.1250	0.7396	-0.2791	-0.2771	0.5698
4.0	0.1875	0.5704	-0.2985	-0.0350	0.4057
4.0	0.2500	0.3960	-0.2703	0.1359	0.2851
4.0	0.3125	0.2382	-0.2133	0.2571	0.2063
4.0	0.3750	0.1122	-0.1421	0.3482	0.1633
4.0	0.4375	0.0296	-0.0680	0.4248	0.1466
4.0	0.5000	0.	0.	0.4972	0.1441
5.0	0.	0.9898	-0.	-1.0093	1.0000
5.0	0.0625	0.9100	-0.2235	-0.4796	0.6995
5.0	0.1250	0.7336	-0.3077	-0.1260	0.4412
5.0	0.1875	0.5304	-0.3037	0.0847	0.2437
5.0	0.2500	0.3420	-0.2511	0.1916	0.1081
5.0	0.3125	0.1899	-0.1786	0.2310	0.0261
5.0	0.3750	0.0823	-0.1051	0.2327	0.0152
5.0	0.4375	0.0200	-0.0428	0.2175	0.0301
5.0	0.5000	0.	0	0.1977	0.0321

CASE 6

Fixed Ends, Uniform Load

λL	X/L	K_y	K_θ	K_M
2.0	0.	1.0000	0.	0.5991
2.0	0.0625	0.9918	-0.0611	0.3844
2.0	0.1250	0.9715	-0.0974	0.2007
2.0	0.1875	0.9449	-0.1126	0.0473
2.0	0.2500	0.9167	-0.1105	-0.0765
2.0	0.3125	0.8908	-0.0946	-0.1717
2.0	0.3750	0.8702	-0.0687	-0.2391
2.0	0.4375	0.8570	-0.0360	-0.2792
2.0	0.5000	0.8524	0.	-0.2925
2.0	0.5625	0.8570	0.0360	-0.2792
2.0	0.6250	0.8702	0.0687	-0.2391
2.0	0.6875	0.8908	0.0946	-0.1717
2.0	0.7500	0.9167	0.1105	-0.0765
2.0	0.8125	0.9449	0.1126	0.0473
2.0	0.8750	0.9715	0.0974	0.2007
2.0	0.9375	0.9918	0.0611	0.3844
2.0	1.0000	1.0000	0.	0.5991
3.0	0.	1.0000	0.	0.9722
3.0	0.0625	0.9704	-0.1462	0.5990
3.0	0.1250	0.8983	-0.2289	0.2938
3.0	0.1875	0.8051	-0.2603	0.0516
3.0	0.2500	0.7080	-0.2518	-0.1340
3.0	0.3125	0.6200	-0.2132	-0.2697
3.0	0.3750	0.5508	-0.1534	-0.3618
3.0	0.4375	0.5067	-0.0800	-0.4149
3.0	0.5000	0.4916	0.	-0.4323
3.0	0.5625	0.5067	0.0800	-0.4149
3.0	0.6250	0.5508	0.1534	-0.3618
3.0	0.6875	0.6200	0.2132	-0.2697
3.0	0.7500	0.7080	0.2518	-0.1340
3.0	0.8125	0.8051	0.2603	0.0516
3.0	0.8750	0.8983	0.2289	0.2938
3.0	0.9375	0.9704	0.1462	0.5990
3.0	1.0000	1.0000	0.	0.9722
4.0	0.	1.0000	-0.	1.0570
4.0	0.0625	0.9443	-0.2036	0.5920
4.0	0.1250	0.8133	-0.3058	0.2442
4.0	0.1875	0.6508	-0.3341	-0.0022
4.0	0.2500	0.4876	-0.3115	-0.1673
4.0	0.3125	0.3448	-0.2556	-0.2713
4.0	0.3750	0.2354	-0.1794	-0.3318
4.0	0.4375	0.1672	-0.0921	-0.3624
4.0	0.5000	0.1441	0.	-0.3716
4.0	0.5625	0.1672	0.0921	-0.3624
4.0	0.6250	0.2354	0.1794	-0.3318
4.0	0.6875	0.3448	0.2556	-0.2713
4.0	0.7500	0.4876	0.3115	-0.1673
4.0	0.8125	0.6508	0.3341	-0.0022
4.0	0.8750	0.8133	0.3058	0.2442
4.0	0.9375	0.9443	0.2036	0.5920
4.0	1.0000	1.0000	0.	1.0570
5.0	0.	1.0000	-0.	1.0262
5.0	0.0625	0.9187	-0.2322	0.4916
5.0	0.1250	0.7386	-0.3258	0.1348
5.0	0.1875	0.5297	-0.3315	-0.0781
5.0	0.2500	0.3344	-0.2878	-0.1873
5.0	0.3125	0.1746	-0.2212	-0.2307
5.0	0.3750	0.0594	-0.1472	-0.2392
5.0	0.4375	-0.0094	-0.0730	-0.2353
5.0	0.5000	-0.0321	0.	-0.2326
5.0	0.5625	-0.0094	0.0730	-0.2353
5.0	0.6250	0.0594	0.1472	-0.2392
5.0	0.6875	0.1746	0.2212	-0.2307
5.0	0.7500	0.3344	0.2878	-0.1873
5.0	0.8125	0.5297	0.3315	-0.0781
5.0	0.8750	0.7386	0.3258	0.1348
5.0	0.9375	0.9187	0.2322	0.4916
5.0	1.0000	1.0000	0.	1.0262

CASE 7

Cantilever Beam, End Load

λL	X/L	K_y	K_Θ	K_M
2.0	0.	-0.	-0.0000	0.1334
2.0	0.0625	0.0022	0.0368	0.1607
2.0	0.1250	0.0095	0.0804	0.1880
2.0	0.1875	0.0226	0.1307	0.2149
2.0	0.2500	0.0424	0.1877	0.2411
2.0	0.3125	0.0698	0.2512	0.2659
2.0	0.3750	0.1055	0.3205	0.2886
2.0	0.4375	0.1502	0.3952	0.3079
2.0	0.5000	0.2044	0.4741	0.3225
2.0	0.5625	0.2688	0.5559	0.3307
2.0	0.6250	0.3435	0.6388	0.3305
2.0	0.6875	0.4284	0.7202	0.3195
2.0	0.7500	0.5234	0.7974	0.2951
2.0	0.8125	0.6274	0.8664	0.2543
2.0	0.8750	0.7394	0.9229	0.1939
2.0	0.9375	0.8574	0.9614	0.1103
2.0	1.0000	0.9788	0.9758	0.
3.0	0.	-0.	-0.0000	-0.0830
3.0	0.0625	-0.0025	-0.0243	-0.0465
3.0	0.1250	-0.0082	-0.0348	-0.0098
3.0	0.1875	-0.0147	-0.0315	0.0276
3.0	0.2500	-0.0192	-0.0140	0.0659
3.0	0.3125	-0.0190	0.0181	0.1056
3.0	0.3750	-0.0115	0.0653	0.1465
3.0	0.4375	0.0064	0.1280	0.1882
3.0	0.5000	0.0375	0.2064	0.2294
3.0	0.5625	0.0847	0.2997	0.2678
3.0	0.6250	0.1508	0.4065	0.3002
3.0	0.6875	0.2378	0.5236	0.3218
3.0	0.7500	0.3474	0.6458	0.3266
3.0	0.8125	0.4798	0.7655	0.3069
3.0	0.8750	0.6336	0.8717	0.2532
3.0	0.9375	0.8050	0.9498	0.1550
3.0	1.0000	0.9869	0.9808	0.
4.0	0.	-0.	-0.	-0.0516
4.0	0.0625	-0.0030	-0.0228	-0.0396
4.0	0.1250	-0.0109	-0.0395	-0.0272
4.0	0.1875	-0.0222	-0.0498	-0.0134
4.0	0.2500	-0.0352	-0.0525	0.0032
4.0	0.3125	-0.0477	-0.0458	0.0242
4.0	0.3750	-0.0571	-0.0272	0.0511
4.0	0.4375	-0.0600	0.0065	0.0851
4.0	0.5000	-0.0522	0.0591	0.1265
4.0	0.5625	-0.0286	0.1341	0.1743
4.0	0.6250	0.0169	0.2340	0.2254
4.0	0.6875	0.0905	0.3591	0.2740
4.0	0.7500	0.1983	0.5061	0.3111
4.0	0.8125	0.3447	0.6660	0.3229
4.0	0.8750	0.5310	0.8218	0.2912
4.0	0.9375	0.7530	0.9461	0.1928
4.0	1.0000	0.9981	0.9989	0.
5.0	0.	0.	0.0000	-0.0091
5.0	0.0625	-0.0010	-0.0064	-0.0115
5.0	0.1250	-0.0042	-0.0143	-0.0136
5.0	0.1875	-0.0100	-0.0233	-0.0149
5.0	0.2500	-0.0188	-0.0325	-0.0142
5.0	0.3125	-0.0302	-0.0403	-0.0098
5.0	0.3750	-0.0435	-0.0435	0.0006
5.0	0.4375	-0.0565	-0.0378	0.0194
5.0	0.5000	-0.0655	-0.0170	0.0492
5.0	0.5625	-0.0647	0.0264	0.0916
5.0	0.6250	-0.0458	0.1002	0.1464
5.0	0.6875	0.0018	0.2112	0.2097
5.0	0.7500	0.0903	0.3621	0.2720
5.0	0.8125	0.2317	0.5474	0.3157
5.0	0.8750	0.4340	0.7472	0.3132
5.0	0.9375	0.6961	0.9211	0.2249
5.0	1.0000	0.9999	1.0000	0.

CASE 8

Cantilever Beam, Uniform Load

λL	X/L	K_y	K_Θ	K_M
2.0	0.	0.0000	0.	-0.9758
2.0	0.0625	0.0141	-0.1082	-0.7599
2.0	0.1250	0.0520	-0.1913	-0.5748
2.0	0.1875	0.1079	-0.2531	-0.4192
2.0	0.2500	0.1771	-0.2972	-0.2915
2.0	0.3125	0.2554	-0.3270	-0.1894
2.0	0.3750	0.3396	-0.3455	-0.1106
2.0	0.4375	0.4274	-0.3555	-0.0525
2.0	0.5000	0.5169	-0.3594	-0.0122
2.0	0.5625	0.6068	-0.3592	0.0130
2.0	0.6250	0.6963	-0.3567	0.0259
2.0	0.6875	0.7850	-0.3531	0.0293
2.0	0.7500	0.8729	-0.3496	0.0259
2.0	0.8125	0.9599	-0.3468	0.0186
2.0	0.8750	1.0464	-0.3450	0.0101
2.0	0.9375	1.1325	-0.3442	0.0030
2.0	1.0000	1.2186	-0.3441	0.0000
3.0	0.	0.0000	0.	-0.9808
3.0	0.0625	0.0304	-0.1515	-0.6468
3.0	0.1250	0.1066	-0.2468	-0.3806
3.0	0.1875	0.2100	-0.2981	-0.1771
3.0	0.2500	0.3261	-0.3166	-0.0290
3.0	0.3125	0.4446	-0.3120	0.0717
3.0	0.3750	0.5582	-0.2922	0.1333
3.0	0.4375	0.6627	-0.2639	0.1637
3.0	0.5000	0.7557	-0.2323	0.1704
3.0	0.5625	0.8369	-0.2011	0.1599
3.0	0.6250	0.9070	-0.1731	0.1378
3.0	0.6875	0.9674	-0.1499	0.1091
3.0	0.7500	1.0201	-0.1323	0.0781
3.0	0.8125	1.0673	-0.1205	0.0484
3.0	0.8750	1.1111	-0.1139	0.0235
3.0	0.9375	1.1532	-0.1112	0.0064
3.0	1.0000	1.1948	-0.1108	0.0000
4.0	0.	0.0000	0.	-0.9989
4.0	0.0625	0.0527	-0.1924	-0.5611
4.0	0.1250	0.1767	-0.2904	-0.2409
4.0	0.1875	0.3319	-0.3215	-0.0234
4.0	0.2500	0.4909	-0.3090	0.1106
4.0	0.3125	0.6368	-0.2715	0.1809
4.0	0.3750	0.7605	-0.2224	0.2055
4.0	0.4375	0.8588	-0.1712	0.2000
4.0	0.5000	0.9323	-0.1239	0.1765
4.0	0.5625	0.9839	-0.0837	0.1443
4.0	0.6250	1.0174	-0.0519	0.1099
4.0	0.6875	1.0372	-0.0285	0.0776
4.0	0.7500	1.0472	-0.0127	0.0498
4.0	0.8125	1.0510	-0.0031	0.0279
4.0	0.8750	1.0511	0.0018	0.0123
4.0	0.9375	1.0497	0.0035	0.0030
4.0	1.0000	1.0478	0.0038	0.
5.0	0.	0.0000	0.	-1.0000
5.0	0.0625	0.0789	-0.2249	-0.4713
5.0	0.1250	0.2528	-0.3132	-0.1211
5.0	0.1875	0.4526	-0.3158	0.0837
5.0	0.2500	0.6379	-0.2721	0.1812
5.0	0.3125	0.7890	-0.2099	0.2075
5.0	0.3750	0.9002	-0.1468	0.1919
5.0	0.4375	0.9742	-0.0921	0.1562
5.0	0.5000	1.0178	-0.0497	0.1149
5.0	0.5625	1.0390	-0.0199	0.0768
5.0	0.6250	1.0450	-0.0009	0.0461
5.0	0.6875	1.0419	0.0098	0.0239
5.0	0.7500	1.0339	0.0149	0.0099
5.0	0.8125	1.0239	0.0167	0.0025
5.0	0.8750	1.0134	0.0169	-0.0003
5.0	0.9375	1.0028	0.0168	-0.0004
5.0	1.0000	0.9924	0.0167	-0.0000

CASE 9

Cantilever Beam, Triangular Load

λL	X/L	K_y	K_Θ	K_M	K_Q
2.0	0.	-4.0000	2.0000	1.9457	-2.7280
2.0	0.0625	-3.7230	2.4063	1.3248	-2.2448
2.0	0.1250	-3.4043	2.6721	0.8202	-1.7990
2.0	0.1875	-3.0597	2.8253	0.4219	-1.3948
2.0	0.2500	-2.7016	2.8910	0.1191	-1.0347
2.0	0.3125	-2.3396	2.8919	-0.0992	-0.7196
2.0	0.3750	-1.9805	2.8475	-0.2444	-0.4496
2.0	0.4375	-1.6289	2.7748	-0.3277	-0.2241
2.0	0.5000	-1.2874	2.6879	-0.3601	-0.0420
2.0	0.5625	-0.9571	2.5981	-0.3522	0.0982
2.0	0.6250	-0.6376	2.5142	-0.3144	0.1977
2.0	0.6875	-0.3280	2.4425	-0.2566	0.2580
2.0	0.7500	-0.0263	2.3868	-0.1886	0.2801
2.0	0.8125	0.2694	2.3483	-0.1197	0.2648
2.0	0.8750	0.5614	2.3262	-0.0593	0.2129
2.0	0.9375	0.8515	2.3172	-0.0163	0.1245
2.0	1.0000	1.1411	2.3158	-0.	-0.0000
3.0	0.	-6.0000	2.0000	3.9113	-4.9241
3.0	0.0625	-5.5079	3.1464	2.2706	-3.8419
3.0	0.1250	-4.8539	3.7513	1.0162	-2.8687
3.0	0.1875	-4.1265	3.9512	0.1026	-2.0262
3.0	0.2500	-3.3901	3.8645	-0.5208	-1.3218
3.0	0.3125	-2.6890	3.5904	-0.9057	-0.7527
3.0	0.3750	-2.0504	3.2089	-1.1011	-0.3095
3.0	0.4375	-1.4884	2.7826	-1.1519	0.0210
3.0	0.5000	-1.0068	2.3581	-1.0976	0.2537
3.0	0.5625	-0.6019	1.9683	-0.9720	0.4033
3.0	0.6250	-0.2651	1.6345	-0.8037	0.4836
3.0	0.6875	0.0153	1.3679	-0.6165	0.5063
3.0	0.7500	0.2523	1.1720	-0.4300	0.4806
3.0	0.8125	0.4590	1.0432	-0.2612	0.4136
3.0	0.8750	0.6471	0.9721	-0.1244	0.3097
3.0	0.9375	0.8262	0.9442	-0.0332	0.1715
3.0	1.0000	1.0027	0.9400	-0.	-0.0000
4.0	0.	-8.0000	2.0000	5.9947	-6.9965
4.0	0.0625	-7.1931	4.2044	2.9825	-5.0859
4.0	0.1250	-6.0040	5.1320	0.8656	-3.4315
4.0	0.1875	-4.6989	5.1950	-0.5020	-2.0933
4.0	0.2500	-3.4505	4.7279	-1.2816	-1.0771
4.0	0.3125	-2.3577	3.9855	-1.6282	-0.3550
4.0	0.3750	-1.4654	3.1490	-1.6780	0.1105
4.0	0.4375	-0.7809	2.3382	-1.5425	0.3951
4.0	0.5000	-0.2882	1.6230	-1.3073	0.5250
4.0	0.5625	0.0414	1.0370	-1.0345	0.5528
4.0	0.6250	0.2417	0.5878	-0.7654	0.5151
4.0	0.6875	0.3460	0.2667	-0.5256	0.4399
4.0	0.7500	0.3842	0.0551	-0.3283	0.3475
4.0	0.8125	0.3808	-0.0696	-0.1786	0.2513
4.0	0.8750	0.3546	-0.1314	-0.0764	0.1590
4.0	0.9375	0.3184	-0.1533	-0.0183	0.0748
4.0	1.0000	0.2795	-0.1564	-0.	-0.
5.0	0.	-10.0000	2.0000	7.9998	-8.9985
5.0	0.0625	-8.7612	5.4412	3.3210	-6.0392
5.0	0.1250	-6.8460	6.5064	0.3425	-3.5920
5.0	0.1875	-4.8422	6.1478	-1.3000	-1.7687
5.0	0.2500	-3.0768	5.0789	-1.9928	-0.5402
5.0	0.3125	-1.6909	3.7828	-2.0783	0.1942
5.0	0.3750	-0.7054	2.5506	-1.8271	0.5586
5.0	0.4375	-0.0745	1.5284	-1.4322	0.6721
5.0	0.5000	0.2769	0.7639	-1.0183	0.6343
5.0	0.5625	0.4286	0.2448	-0.6551	0.5198
5.0	0.6250	0.4510	-0.0722	-0.3736	0.3798
5.0	0.6875	0.3990	-0.2405	-0.1790	0.2457
5.0	0.7500	0.3107	-0.3121	-0.0618	0.1342
5.0	0.8125	0.2094	-0.3303	-0.0050	0.0528
5.0	0.8750	0.1065	-0.3269	0.0109	0.0035
5.0	0.9375	0.0053	-0.3210	0.0060	-0.0140
5.0	1.0000	0.0946	-0.3196	0.	0.0000

CASE 10

Free Ends, Equal Concentrated Loads at Both Ends

λL	X/L	K_y	K_Θ	K_M	K_Q
2.0	0.	0.7376	-0.5991	0.	-1.0000
2.0	0.0625	0.6634	-0.5844	0.1139	-0.8249
2.0	0.1250	0.5926	-0.5439	0.2070	-0.6680
2.0	0.1875	0.5283	-0.4825	0.2816	-0.5281
2.0	0.2500	0.4727	-0.4045	0.3396	-0.4032
2.0	0.3125	0.4277	-0.3139	0.3829	-0.2908
2.0	0.3750	0.3946	-0.2141	0.4127	-0.1883
2.0	0.4375	0.3744	-0.1085	0.4302	-0.0925
2.0	0.5000	0.3676	0.	0.4360	0.
2.0	0.5625	0.3744	0.1085	0.4302	0.0925
2.0	0.6250	0.3946	0.2141	0.4127	0.1883
2.0	0.6875	0.4277	0.3139	0.3829	0.2908
2.0	0.7500	0.4727	0.4045	0.3396	0.4032
2.0	0.8125	0.5283	0.4825	0.2816	0.5281
2.0	0.8750	0.5926	0.5439	0.2070	0.6680
2.0	0.9375	0.6634	0.5844	0.1139	0.8249
2.0	1.0000	0.7376	0.5991	0.	1.0000
3.0	0.	0.8936	-0.9722	0.	-1.0000
3.0	0.0625	0.7133	-0.9408	0.1582	-0.6989
3.0	0.1250	0.5438	-0.8598	0.2662	-0.4637
3.0	0.1875	0.3929	-0.7459	0.3359	-0.2887
3.0	0.2500	0.2654	-0.6114	0.3777	-0.1660
3.0	0.3125	0.1644	-0.4650	0.4008	-0.0863
3.0	0.3750	0.1914	-0.3122	0.4122	-0.0393
3.0	0.4375	0.0475	-0.1567	0.4169	-0.0141
3.0	0.5000	0.0328	-0.0000	0.4181	0.0000
3.0	0.5625	0.0475	0.1567	0.4169	0.0141
3.0	0.6250	0.0914	0.3122	0.4122	0.0393
3.0	0.6875	0.1644	0.4650	0.4008	0.0863
3.0	0.7500	0.2654	0.6114	0.3777	0.1660
3.0	0.8125	0.3929	0.7459	0.3359	0.2887
3.0	0.8750	0.5438	0.8598	0.2662	0.4637
3.0	0.9375	0.7133	0.9408	0.1582	0.6989
3.0	1.0000	0.8936	0.9722	0.	1.0000
4.0	0.	1.0046	-1.0570	-0.	-1.0000
4.0	0.0625	0.7449	-1.0043	0.1927	-0.5632
4.0	0.1250	0.5083	-0.8799	0.2920	-0.2512
4.0	0.1875	0.3076	-0.7229	0.3274	-0.0488
4.0	0.2500	0.1474	-0.5589	0.3240	0.0632
4.0	0.3125	0.0275	-0.4021	0.3016	0.1053
4.0	0.3750	-0.0547	-0.2579	0.2755	0.0971
4.0	0.4375	-0.1024	-0.1255	0.2558	0.0564
4.0	0.5000	-0.1180	0.	0.2486	0.0000
4.0	0.5625	-0.1024	0.1255	0.2558	-0.0564
4.0	0.6250	-0.0547	0.2579	0.2755	-0.0971
4.0	0.6875	0.0275	0.4021	0.3016	-0.1053
4.0	0.7500	0.1474	0.5589	0.3240	-0.0632
4.0	0.8125	0.3076	0.7229	0.3274	0.0488
4.0	0.8750	0.5083	0.8799	0.2920	0.2512
4.0	0.9375	0.7449	1.0043	0.1927	0.5632
4.0	1.0000	1.0046	1.0570	0.	1.0000
5.0	0.	1.0171	-1.0262	-0.	-1.0000
5.0	0.0625	0.7050	-0.9476	0.2235	-0.4632
5.0	0.1250	0.4344	-0.7756	0.3087	-0.1099
5.0	0.1875	0.2227	-0.5796	0.3080	0.0923
5.0	0.2500	0.0704	-0.3998	0.2628	0.1809
5.0	0.3125	-0.0308	-0.2540	0.2031	0.1909
5.0	0.3750	-0.0922	-0.1448	0.1487	0.1508
5.0	0.4375	-0.1243	-0.0645	0.1118	0.0818
5.0	0.5000	-0.1341	0.	0.0989	0.0000
5.0	0.5625	-0.1243	0.0645	0.1118	-0.0818
5.0	0.6250	-0.0922	0.1448	0.1487	-0.1508
5.0	0.6875	-0.0308	0.2540	0.2031	-0.1909
5.0	0.7500	0.0704	0.3998	0.2628	-0.1809
5.0	0.8125	0.2227	0.5796	0.3080	-0.0923
5.0	0.8750	0.4344	0.7756	0.3087	0.1099
5.0	0.9375	0.7050	0.9476	0.2235	0.4632
5.0	1.0000	1.0171	1.0262	0.	1.0000

CASE 11

Free Ends, Equal Moments at Both Ends

λL	X/L	K_y	K_Θ	K_M	K_Q
2.0	0.	0.5991	-0.9211	1.0000	0.
2.0	0.0625	0.3844	-0.7965	0.9918	-0.0611
2.0	0.1250	0.2007	-0.6737	0.9715	-0.0974
2.0	0.1875	0.0473	-0.5539	0.9449	-0.1126
2.0	0.2500	-0.0765	-0.4375	0.9167	-0.1105
2.0	0.3125	-0.1717	-0.3246	0.8908	-0.0946
2.0	0.3750	-0.2391	-0.2146	0.8702	-0.0687
2.0	0.4375	-0.2792	-0.1067	0.8570	-0.0360
2.0	0.5000	-0.2925	0.	0.8524	0.
2.0	0.5625	-0.2792	0.1067	0.8570	0.0360
2.0	0.6250	-0.2391	0.2146	0.8702	0.0687
2.0	0.6875	-0.1717	0.3246	0.8908	0.0946
2.0	0.7500	-0.0765	0.4375	0.9167	0.1105
2.0	0.8125	0.0473	0.5539	0.9449	0.1126
2.0	0.8750	0.2007	0.6737	0.9715	0.0974
2.0	0.9375	0.3844	0.7965	0.9918	0.0611
2.0	1.0000	0.5991	0.9211	1.0000	0.
3.0	0.	0.9792	-1.0885	1.0000	0.
3.0	0.0625	0.5990	-0.9029	0.9704	-0.1462
3.0	0.1250	0.2938	-0.7272	0.8983	-0.2289
3.0	0.1875	0.0516	-0.5673	0.8051	-0.2603
3.0	0.2500	-0.1340	-0.4255	0.7080	-0.2518
3.0	0.3125	-0.2697	-0.3012	0.6200	-0.2132
3.0	0.3750	-0.3618	-0.1918	0.5508	-0.1534
3.0	0.4375	-0.4149	-0.0931	0.5067	-0.0800
3.0	0.5000	-0.4323	0.	0.4916	0.
3.0	0.5625	-0.4149	0.0931	0.5067	0.0800
3.0	0.6250	-0.3618	0.1918	0.5508	0.1534
3.0	0.6875	-0.2697	0.3012	0.6200	0.2132
3.0	0.7500	-0.1340	0.4255	0.7080	0.2518
3.0	0.8125	0.0516	0.5673	0.8051	0.2603
3.0	0.8750	0.2938	0.7272	0.8983	0.2289
3.0	0.9375	0.5990	0.9029	0.9704	0.1462
3.0	1.0000	0.9722	1.0885	1.0000	0.
4.0	0.	1.0570	-1.0538	1.0000	-0.
4.0	0.0625	0.5920	-0.8087	0.9443	-0.2036
4.0	0.1250	0.2442	-0.5879	0.8133	-0.3058
4.0	0.1875	-0.0022	-0.4046	0.6508	-0.3341
4.0	0.2500	-0.1673	-0.2626	0.4876	-0.3115
4.0	0.3125	-0.2713	-0.1591	0.3448	-0.2556
4.0	0.3750	-0.3318	-0.0874	0.2354	-0.1794
4.0	0.4375	-0.3624	-0.0380	0.1672	-0.0921
4.0	0.5000	-0.3716	-0.	0.1441	0.
4.0	0.5625	-0.3624	0.0380	0.1672	0.0921
4.0	0.6250	-0.3318	0.0874	0.2354	0.1794
4.0	0.6875	-0.2713	0.1591	0.3448	0.2556
4.0	0.7500	-0.1673	0.2626	0.4876	0.3115
4.0	0.8125	-0.0022	0.4046	0.6508	0.3341
4.0	0.8750	0.2442	0.5879	0.8133	0.3058
4.0	0.9375	0.5920	0.8087	0.9443	0.2036
4.0	1.0000	1.0570	1.0538	1.0000	0.
5.0	0.	1.0262	-1.0093	1.0000	-0.
5.0	0.0625	0.4916	-0.7057	0.9187	-0.2322
5.0	0.1250	0.1348	-0.4453	0.7386	-0.3258
5.0	0.1875	-0.0781	-0.2470	0.5297	-0.3315
5.0	0.2500	-0.1873	-0.1127	0.3344	-0.2878
5.0	0.3125	-0.2307	-0.0343	0.1746	-0.2212
5.0	0.3750	-0.2392	0.0011	0.0594	-0.1472
5.0	0.4375	-0.2353	0.0077	-0.0094	-0.0730
5.0	0.5000	-0.2326	-0.	-0.0321	0.
5.0	0.5625	-0.2353	-0.0077	-0.0094	0.0730
5.0	0.6250	-0.2392	-0.0011	0.0594	0.1472
5.0	0.6875	-0.2307	0.0343	0.1746	0.2212
5.0	0.7500	-0.1873	0.1127	0.3344	0.2878
5.0	0.8125	-0.0781	0.2470	0.5297	0.3315
5.0	0.8750	0.1348	0.4453	0.7386	0.3258
5.0	0.9375	0.4916	0.7057	0.9187	0.2322
5.0	1.0000	1.0262	1.0093	1.0000	0.

CASE 12

Free Ends, Concentrated Load at One End

λL	X/L	K_y	K_Θ	K_M	K_Q
2.0	0.	1.1376	1.1341	0.	1.0000
2.0	0.0625	0.9964	1.1200	0.1080	0.7333
2.0	0.1250	0.8586	1.0828	0.1848	0.5015
2.0	0.1875	0.7264	1.0298	0.2347	0.3035
2.0	0.2500	0.6015	0.9673	0.2620	0.1377
2.0	0.3125	0.4848	0.9004	0.2704	0.0021
2.0	0.3750	0.3764	0.8333	0.2637	-0.1054
2.0	0.4375	0.2763	0.7695	0.2452	-0.1868
2.0	0.5000	0.1838	0.7115	0.2180	-0.2442
2.0	0.5625	0.0981	0.6610	0.1851	-0.2793
2.0	0.6250	0.0182	0.6192	0.1491	-0.2937
2.0	0.6875	-0.0571	0.5865	0.1125	-0.2887
2.0	0.7500	-0.1288	0.5628	0.0776	-0.2654
2.0	0.8125	-0.1981	0.5474	0.0468	-0.2245
2.0	0.8750	-0.2659	0.5389	0.0222	-0.1665
2.0	0.9375	-0.3330	0.5356	0.0059	-0.0916
2.0	1.0000	-0.3999	0.5351	0.	-0.0000
3.0	0.	1.0066	1.0004	0.	1.0000
3.0	0.0625	0.8210	0.9695	0.1543	0.6575
3.0	0.1250	0.6459	0.8919	0.2508	0.3829
3.0	0.1875	0.4882	0.7870	0.3018	0.1709
3.0	0.2500	0.3515	0.6698	0.3184	0.0141
3.0	0.3125	0.2371	0.5514	0.3101	-0.0956
3.0	0.3750	0.1444	0.4394	0.2850	-0.1664
3.0	0.4375	0.0716	0.3389	0.2496	-0.2063
3.0	0.5000	0.0164	0.2528	0.2091	-0.2223
3.0	0.5625	-0.0242	0.1823	0.1673	-0.2205
3.0	0.6250	-0.0529	0.1271	0.1272	-0.2057
3.0	0.6875	-0.0727	0.0864	0.0908	-0.1819
3.0	0.7500	-0.0861	0.0584	0.0594	-0.1519
3.0	0.8125	-0.0953	0.0411	0.0340	-0.1178
3.0	0.8750	-0.1021	0.0321	0.0154	-0.0807
3.0	0.9375	-0.1077	0.0287	0.0039	-0.0414
3.0	1.0000	-0.1130	0.0282	0.	-0.0000
4.0	0.	1.0008	1.0015	0.	1.0000
4.0	0.0625	0.7550	0.9488	0.1926	0.5616
4.0	0.1250	0.5323	0.8247	0.2907	0.2411
4.0	0.1875	0.3452	0.6693	0.3218	0.0234
4.0	0.2500	0.1979	0.5101	0.3093	-0.1108
4.0	0.3125	0.0890	0.3641	0.2717	-0.1810
4.0	0.3750	0.0140	0.2403	0.2226	-0.2055
4.0	0.4375	-0.0332	0.1419	0.1715	-0.1996
4.0	0.5000	-0.0590	0.0682	0.1243	-0.1758
4.0	0.5625	-0.0692	0.0163	0.0843	-0.1432
4.0	0.6250	-0.0687	-0.0176	0.0529	-0.1084
4.0	0.6875	-0.0615	-0.0379	0.0299	-0.0756
4.0	0.7500	-0.0505	-0.0488	0.0147	-0.0475
4.0	0.8125	-0.0376	-0.0536	0.0057	-0.0255
4.0	0.8750	-0.0239	-0.0552	0.0014	-0.0101
4.0	0.9375	-0.0101	-0.0555	0.0001	-0.0016
4.0	1.0000	0.0038	-0.0555	-0.	0.0000
5.0	0.	1.0003	1.0003	0.	1.0000
5.0	0.0625	0.6964	0.9214	0.2249	0.4711
5.0	0.1250	0.4342	0.7476	0.3131	0.1206
5.0	0.1875	0.2317	0.5479	0.3155	-0.0842
5.0	0.2500	0.0901	0.3628	0.2716	-0.1817
5.0	0.3125	0.0013	0.2121	0.2093	-0.2079
5.0	0.3750	-0.0466	0.1013	0.1461	-0.1919
5.0	0.4375	-0.0659	0.0277	0.0915	-0.1556
5.0	0.5000	-0.0671	-0.0157	0.0494	-0.1133
5.0	0.5625	-0.0584	-0.0368	0.0203	-0.0738
5.0	0.6250	-0.0456	-0.0435	0.0026	-0.0412
5.0	0.6875	-0.0321	-0.0419	-0.0063	-0.0169
5.0	0.7500	-0.0197	-0.0369	-0.0088	-0.0008
5.0	0.8125	-0.0090	-0.0317	-0.0075	0.0081
5.0	0.8750	0.0002	-0.0279	-0.0044	0.0108
5.0	0.9375	0.0086	-0.0261	-0.0014	0.0079
5.0	1.0000	0.0167	-0.0259	-0.	0.0000

CASE 13

Free Ends, Concentrated Moment at One End

λL	X/L	K_y	K_Θ	K_M	K_Q
2.0	0.	-1.1341	1.0762	1.0000	0.
2.0	0.0625	-0.8807	0.9519	0.9836	0.1256
2.0	0.1250	-0.6579	0.8314	0.9397	0.2214
2.0	0.1875	-0.4644	0.7178	0.8751	0.2913
2.0	0.2500	-0.2982	0.6133	0.7959	0.3387
2.0	0.3125	-0.1569	0.5192	0.7073	0.3669
2.0	0.3750	-0.0376	0.4366	0.6138	0.3788
2.0	0.4375	0.0624	0.3658	0.5191	0.3771
2.0	0.5000	0.1463	0.3068	0.4262	0.3639
2.0	0.5625	0.2168	0.2591	0.3379	0.3411
2.0	0.6250	0.2767	0.2220	0.2564	0.3101
2.0	0.6875	0.3286	0.1946	0.1834	0.2722
2.0	0.7500	0.3747	0.1757	0.1208	0.2282
2.0	0.8125	0.4171	0.1640	0.0698	0.1787
2.0	0.8750	0.4572	0.1578	0.0318	0.1241
2.0	0.9375	0.4963	0.1554	0.0082	0.0645
2.0	1.0000	0.5351	0.1551	0.0000	0.
3.0	0.	-1.0004	1.0038	1.0000	0.
3.0	0.0625	-0.6589	0.8183	0.9690	0.1545
3.0	0.1250	-0.3854	0.6433	0.8913	0.2514
3.0	0.1875	-0.1743	0.4857	0.7862	0.3029
3.0	0.2500	-0.0184	0.3493	0.6684	0.3202
3.0	0.3125	0.0905	0.2352	0.5491	0.3127
3.0	0.3750	0.1607	0.1429	0.4360	0.2887
3.0	0.4375	0.2002	0.0710	0.3339	0.2544
3.0	0.5000	0.2162	0.0168	0.2458	0.2150
3.0	0.5625	0.2147	-0.0222	0.1728	0.1744
3.0	0.6250	0.2011	-0.0489	0.1148	0.1353
3.0	0.6875	0.1793	-0.0661	0.0709	0.0995
3.0	0.7500	0.1524	-0.0763	0.0396	0.0684
3.0	0.8125	0.1227	-0.0816	0.0189	0.0426
3.0	0.8750	0.0916	-0.0839	0.0069	0.0225
3.0	0.9375	0.0599	-0.0846	0.0014	0.0083
3.0	1.0000	0.0282	-0.0847	0.0000	0.
4.0	0.	-1.0015	1.0021	1.0000	0.
4.0	0.0625	-0.5624	0.7567	0.9472	0.1929
4.0	0.1250	-0.2409	0.5344	0.8229	0.2910
4.0	0.1875	-0.0220	0.3478	0.6673	0.3219
4.0	0.2500	0.1136	0.2010	0.5082	0.3090
4.0	0.3125	0.1855	0.0926	0.3626	0.2705
4.0	0.3750	0.2118	0.0178	0.2397	0.2200
4.0	0.4375	0.2079	-0.0295	0.1430	0.1671
4.0	0.5000	0.1858	-0.0558	0.0720	0.1176
4.0	0.5625	0.1545	-0.0674	0.0242	0.0749
4.0	0.6250	0.1200	-0.0696	-0.0043	0.0406
4.0	0.6875	0.0858	-0.0665	-0.0178	0.0149
4.0	0.7500	0.0538	-0.0616	-0.0206	-0.0025
4.0	0.8125	0.0242	-0.0568	-0.0166	-0.0122
4.0	0.8750	-0.0033	-0.0535	-0.0096	-0.0148
4.0	0.9375	-0.0296	-0.0520	-0.0029	-0.0106
4.0	1.0000	-0.0555	-0.0517	-0.0000	-0.
5.0	0.	-1.0003	1.0002	1.0000	0.
5.0	0.0625	-0.4715	0.6964	0.9211	0.2250
5.0	0.1250	-0.1210	0.4343	0.7472	0.3133
5.0	0.1875	0.0840	0.2320	0.5472	0.3158
5.0	0.2500	0.1818	0.0907	0.3620	0.2720
5.0	0.3125	0.2084	0.0022	0.2111	0.2095
5.0	0.3750	0.1930	-0.0455	0.1002	0.1461
5.0	0.4375	0.1575	-0.0644	0.0267	0.0910
5.0	0.5000	0.1163	-0.0654	-0.0161	0.0482
5.0	0.5625	0.0778	-0.0567	-0.0361	0.0180
5.0	0.6250	0.0461	-0.0444	-0.0409	-0.0012
5.0	0.6875	0.0223	-0.0321	-0.0365	-0.0117
5.0	0.7500	0.0055	-0.0221	-0.0276	-0.0159
5.0	0.8125	-0.0059	-0.0150	-0.0175	-0.0157
5.0	0.8750	-0.0139	-0.0110	-0.0086	-0.0125
5.0	0.9375	-0.0201	-0.0094	-0.0023	-0.0072
5.0	1.0000	-0.0259	-0.0091	-0.0000	-0.

CASE 14

Free Ends, Center Load

λL	X/L	K_y	K_Θ	K_M	K_Q
2.0	0.	1.1785	0.	0.9211	-1.0000
2.0	0.0625	1.1654	-0.1003	0.6895	-0.8532
2.0	0.1250	1.1306	-0.1739	0.4942	-0.7095
2.0	0.1875	1.0803	-0.2253	0.3343	-0.5712
2.0	0.2500	1.0195	-0.2589	0.2080	-0.4399
2.0	0.3125	0.9520	-0.2786	0.1136	-0.3166
2.0	0.3750	0.8810	-0.2885	0.0490	-0.2021
2.0	0.4375	0.8083	-0.2920	0.0119	-0.0965
2.0	0.5000	0.7352	-0.2925	-0.	0.
3.0	0.	1.0904	0.	1.0885	-1.0000
3.0	0.0625	1.0563	-0.1713	0.7516	-0.7977
3.0	0.1250	0.9690	-0.2865	0.4887	-0.6072
3.0	0.1875	0.8468	-0.3588	0.2937	-0.4365
3.0	0.2500	0.7037	-0.4003	0.1581	-0.2909
3.0	0.3125	0.5491	-0.4212	0.0720	-0.1734
3.0	0.3750	0.3893	-0.4297	0.0244	-0.0853
3.0	0.4375	0.2276	-0.4321	0.0042	-0.0275
3.0	0.5000	0.0655	-0.4323	-0.	0.
4.0	0.	1.0800	0.	1.0538	-1.0000
4.0	0.0625	1.0238	-0.2066	0.6207	-0.7349
4.0	0.1250	0.8887	-0.3210	0.3148	-0.4946
4.0	0.1875	0.7132	-0.3732	0.1195	-0.2939
4.0	0.2500	0.5217	-0.3881	0.0132	-0.1393
4.0	0.3125	0.3280	-0.3852	-0.0279	-0.0332
4.0	0.3750	0.1373	-0.3776	-0.0280	0.0249
4.0	0.4375	-0.0501	-0.3726	-0.0109	0.0358
4.0	0.5000	-0.2360	-0.3716	0.	0.
5.0	0.	1.0444	0.	1.0093	-1.0000
5.0	0.0625	0.9645	-0.2283	0.4849	-0.6824
5.0	0.1250	0.7869	-0.3226	0.1473	-0.4072
5.0	0.1875	0.5780	-0.3363	-0.0371	-0.1938
5.0	0.2500	0.3745	-0.3111	-0.1086	-0.0454
5.0	0.3125	0.1910	-0.2761	-0.1065	0.0424
5.0	0.3750	0.0278	-0.2485	-0.0668	0.0761
5.0	0.4375	-0.1226	-0.2349	-0.0215	0.0611
5.0	0.5000	-0.2683	-0.2326	0.	0.

Index

D

E

ERRATA

Handbook of Formulas for Stress and Strain

William Griffel

Page 2 — Section 6, equation should read: $\beta = \gamma = -\alpha v = -\dfrac{S}{E}v$

Page 9 — Case 5 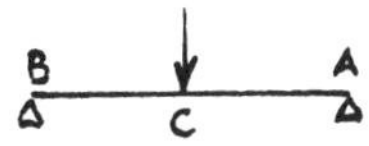 should be: —48 at C.

Case 8 should be: —60 at C.

Case 9 should be: —106.7 at C.

Page 11 — Section 17, second column, last equation should read: $y = \dfrac{2SL^2}{K_2 dE}$.

Page 12 — First column, line 9 should read: in Fig. 2.

Page 13 — First column, line 5 should read: in Fig. 2.

Page 14 — First column, line 18 should read: in Fig. 7.

Second column, lines 9 and 10 should read: point *C*, section *BC*.

Page 17 — 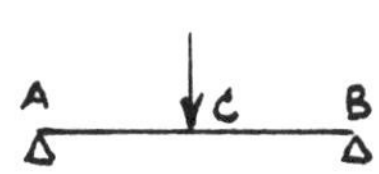 $\theta_2 = \ldots$ at B.

Page 18 — 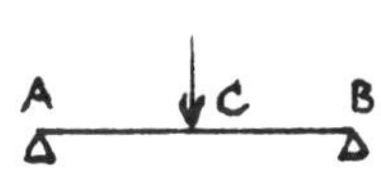

Page 19 — Moment equation should read: $M_x = \dfrac{1}{3}W\left(x - \dfrac{x^3}{L^2}\right)$

Page 21 — 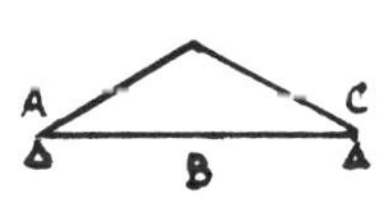

Page 24 — $\theta_2 = \ldots$ at D.

Page 25 — Second column, (*C* to *D*):

$$y_x = \frac{1}{6EI}\left\{R_1(x^3 - L^2x) - W\left[\left(x - \frac{1}{3}a - \frac{2}{3}b\right)^3 - d^3\frac{x}{L} - \frac{1}{6}be^2\left(1 - \frac{x}{L}\right) + \frac{17}{270}e^3\left(1 - \frac{x}{L}\right)\right]\right\}.$$

Page 28 — $y_x = -\dfrac{1}{24}\dfrac{W}{EIL}(x^4 - 4L^3 + 3L^4).$

Page 29 — Deflection, equation should read: $y_{max} = -\dfrac{WL^3}{3EI}$

Max. Resulting Stress . . . : $S_{max} = \dfrac{WLc}{I}$

Max. Deflection . . . : $y_{max} = -\dfrac{1}{3}\dfrac{SL^2}{Ec}$ at $x = 0$

Page 45 — First column: $R_1 = \dfrac{1}{20}\dfrac{W}{L^3}[(10ab + 15a^2 + 5b^2)L - 4a^3 - 2ab^2 - 3a^2b - b^3].$

Page 46 — 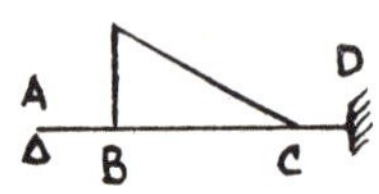

Page 47 — Deflection, equation should read: $y_x = \dfrac{1}{24}\dfrac{Wx^2}{EIL}(2Lx - L^2 - x^2)$

Page 52 — 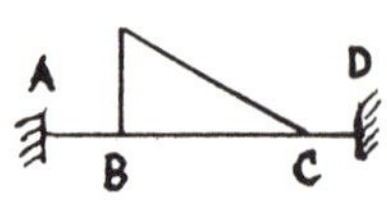

; (B to C): $y_x = \dfrac{1}{EI}\left(\dfrac{1}{6}R_1x^3 - \dfrac{1}{2}M_1x^2 - \dfrac{1}{60}W\dfrac{(x-a)^5}{e^2}\right).$

Page 54 — Second column, line 7: $C = (15\text{-}3)/15.$

Page 61 — 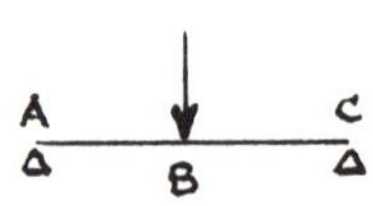

Page 65 — First column, last line: $U = 240(20{,}000/30 \times 10^6 \times 6)^{1/2} = 2.53.$

Page 78 — Second column, line 7 from bottom: $y_t = \dfrac{\ldots}{\ldots 10^6}\ldots.$

Page 81 — Equation 40: $d = 1.72\sqrt[3]{\dfrac{M_B + \sqrt{M_B^2 + M_T^2}}{S}}.$

Page 89 — Second column, line 6 from bottom: $K = 48EI/L^3.$

Page 91 — Second column, line 4, should read: case 15.
Second column, line 6, should read: Fig. 51.
Second column, line 15, should read: case 15.
Second column, line 16, should read: Fig. 51.

Page 117 — First column, line 9, should read: Table 12.
Second column, line 8, should read: Table 12.
Second column, line 16: $y = \dfrac{pR^2}{Et}.$

Page 118 — Second column, line 1: . . . values of M.

Page 131 — Second column, line 18: $\gamma = 6.0$ deg.

Page 137 — Second column, line 8: . . . $EI\alpha/T_oR$, $EI\phi/T_oR$.

Page 138 Second column, line 4 from bottom: $\alpha = -0.0004$ radian. $\phi = 0.03$ radian.

Second column, line 6 from bottom: $\frac{EI}{T_oR}\alpha = \cdots; \frac{EI}{T_oR}\phi = \cdots$.

Page 139 Table 19 should have the following headings: $EIy/T_oR^2, EI\alpha/T_oR, EI\phi/T_oR$.

Page 148 case 8

Θ Degrees	Θ Radians	K_S	K'_S	K_M	K'_M	K_T	K'_T
0	0.	0.	0.	0.	0.	0.	0.
5	0.0873	0.	0.	0.0146	0.0291	11.4693	11.4256
10	0.1746	0.	0.	0.0292	0.0583	5.7566	5.6691
15	0.2619	0.	0.001	0.0440	0.0877	3.8623	3.7305
20	0.3492	0.001	0.002	0.0590	0.1174	2.9227	2.7463
25	0.4365	0.002	0.003	0.0744	0.1474	2.3654	2.1436
30	0.5238	0.003	0.006	0.0902	0.1779	1.9993	1.7312
35	0.6111	0.005	0.010	0.1065	0.2090	1.7429	1.4274
40	0.6984	0.008	0.015	0.1234	0.2407	1.5552	1.1911
45	0.7857	0.012	0.022	0.1410	0.2734	1.4138	0.9994
50	0.8730	0.017	0.031	0.1596	0.3069	1.3050	0.8385
55	0.9603	0.024	0.042	0.1791	0.3417	1.2205	0.6997
60	1.0476	0.032	0.057	0.1999	0.3777	1.1544	0.5768
65	1.1349	0.042	0.076	0.2220	0.4154	1.1032	0.4658
70	1.2222	0.055	0.100	0.2458	0.4548	1.0640	0.3634
75	1.3095	0.071	0.129	0.2715	0.4962	1.0351	0.2674
80	1.3968	0.092	0.167	0.2994	0.5401	1.0153	0.1758
85	1.4841	0.117	0.214	0.3300	0.5869	1.0038	0.0869
90	1.5714	0.149	0.273	0.3636	0.6370	1.0000	−0.0006

Page 158 First column, line 1 should read: Equations (4), (5), (6) ⋯.

Page 162 Fig. 6 upside down.

Page 174 Add a note: "K_2 is tabulated as C and C_1 on p. 188."

Page 178 Table III, case 1: $K_1 = 6$.

Page 201 Table 16, for $d = 0$: Max. $S_0 = 2P_1 \frac{R_1^2}{R_2^2 - R_1^2}$.

Page 247 Second column, line 3 from bottom: $2b = 1$ in.
Second column, line 4 from bottom: $2a = 2$ in.

Page 269 Second column, line 1: $S_R = \cdots$.

Page 272 Figure D: $W = 1000$ LB/in.

Page 275 Example 9, second line should read: $M_y = \frac{why}{2} - \frac{wy^2}{2} + M_2\frac{y}{h}$

Page 296 First column: $C = \cdots$ (see Table 10).

Page 298 First column, under "Notation," Table 9 should read: Table 11.